D0745444

ARC
LIBRARY

## The New Physics for the Twenty-First Century

Recent scientific advances have led to a dramatic reappraisal of our understanding of the world around us, and made a significant impact on our lifestyle. Underpinning all the other branches of science, physics affects the way we live our lives.

This book investigates the key frontiers in modern-day physics, exploring our Universe – from the particles inside an atom to the stars that make up a galaxy, from brain research to the latest advances in high-speed electronic research networks.

Each of the nineteen self-contained chapters written by leading international experts in a lively and accessible style will fascinate scientists of all disciplines and anyone wanting to know more about the world of physics today.

Gordon Fraser worked in science publications at CERN for over 25 years, where he was editor of *Cern Courier*, the international monthly news magazine of high-energy physics. After gaining his Ph.D. in theoretical particle physics at Imperial College, London, he branched out into technical journalism and communications. He is author, co-author, and editor of several science books, including *Antimatter – The Ultimate Mirror*, published in 2000.

# The New PHYSICS for the Twenty-First Century

Edited by

GORDON FRASER

CAMBRIDGE UNIVERSITY PRESS
Cambridge, New York, Melbourne, Madrid, Cape Town, Singapore, São Paulo

Cambridge University Press
The Edinburgh Building, Cambridge CB2 2RU, UK

Published in the United States of America by Cambridge University Press, New York

www.cambridge.org
Information on this title: www.cambridge.org/9780521816009



First published 2006

Printed in Singapore

*A catalog record for this publication is available from the British Library*

*Library of Congress Cataloging in Publication data*

The new physics for the twenty-first century/edited by Gordon Fraser.
p. cm.
Includes bibliographical references and index.
ISBN-13: 978 0 521 81600 9
ISBN-10: 0 521 81600 9
1. Physics–History – 20th century. 2. Physics–History – 21st century. 3. Quantum theory.
I. Fraser, Gordon, 1943– II. Title.
QC7.5. N49 2005 530′.09′05–dc22 2005006466

ISBN-13 978-0-521-81600-9 hardback
ISBN-10 0-521-81600-9 hardback

# Contents

# Contributors

Wendy Freedman, Carnegie Observatories, Pasadena
Rocky Kolb, Chicago/Fermilab
Ronald Adler, Stanford
Arnon Dar, Technion Haifa
Chris Quigg, Fermilab
Michael Green, Cambridge
Claude Cohen-Tannoudji, ENS Paris
Jean Dalibard, ENS Paris
Christopher Foot, Oxford
William Phillips, NIST
Henry Hall, Manchester
Subir Sachdev, Harvard
Anton Zeilinger, Vienna
Artur Ekert, Cambridge and Singapore
Yoseph Imry, Weizmann Institute
Henry Abarbanel, UC San Diego
Antonio Politi, Florence
Tony Hey, Microsoft
Anne Trefethen, UK e-Science Core Programme
Cyrus Safinya, UC Santa Barbara
Nikolaj Pavel, Humboldt, Berlin
Robert Cahn, Cambridge
Ugo Amaldi, Milan-Bicocca and TERA Foundation

# Editor's acknowledgements

My sincere thanks go to Simon Mitton of Cambridge University Press for the original invitation to undertake this challenging work. Subsequent progress was supervised by Simon Capelin who first had the idea for such a physics anthology 20 years ago and who worked closely with Paul Davies on the 1989 edition of *New Physics*.

With material from so many sources, it was a difficult book to put together. However, the enthusiastic and diligent CUP production team transformed an immense pile of amorphous material into an attractive final product.

Many thanks also go to all the contributors. It was a privilege to work with so many distinguished scientists and to learn so much about new physics and its impact on the twenty-first century.

# Introduction
# The new physics for the Twenty-First Century

Physics is the science of matter – the stuff of the Universe around us, and of energy – the capacity of matter to act in different ways. Physics is the systematic study of how this matter and energy behave, the explanation of what this reveals, and the understanding it brings. A magnificent allegory of what a physicist does can be found in the Old Testament, the Book of Job, Chapter 28.

*For he looketh to the ends of the earth, and seeth under the whole heaven; To make the weight for the winds; and he weigheth the waters by measure. When he sought a decree for the rain, and a way for the lightning and the thunder. Then he did see it and declare it. . .*

If our surroundings are seen as being built up of matter, much of Nature is ultimately physics, so physics underpins many other branches of science. It is difficult to be more ambitious than that. But as though such boldness were not enough of a challenge, new physics has gone on to reveal that matter and energy can exist in forms and behave in ways very different from those we know in everyday life. The goal becomes even more ambitious. Nature, and therefore physics, has become much wider than what we normally see around us.

Progress in science comes from not looking at Nature at face value, but undertaking some voyage of discovery to reveal a different viewpoint. From this new vantage, the landscape takes on new aspects and dimensions, leading to fresh insights and new satisfaction. With this vision, the next step is perhaps even more fulfilling – predicting what can be seen from a higher standpoint.

Historically, there have been several major watersheds for physics, each of which could have warranted the publication of a book called *New Physics.* Many of these opened up fresh aspects of the Universe. One was the renaissance of learning in Europe in the middle of the second millennium. New insights from giant figures such as Kepler, Copernicus, and Galileo showed that the Earth is not the hub of the Universe. This was not easy to swallow, and the implications rumbled for many centuries, but for scientists the logical culmination came with Newton's masterpiece picture of mechanics and gravity. (The first edition of Newton's *Principia,* which embodied the new dynamics, was published by the Royal Society of London in 1687. An updated second edition, produced this time at Cambridge, appeared in 1713.) The flow of results from Newton's work encouraged the idea of determinism – that everything that happens follows from what went before, and that the future is merely history waiting to happen.

A subsequent watershed in physics strangely coincided with the hinge between the nineteenth and twentieth centuries. Until this point, physicists had tried to picture the world around us as an assembly of precision components that fit neatly together like a Swiss watch and moved according to immutable laws, giving a vivid mental image of how Nature worked on any scale in space and in time. However, the dawn of the twentieth century seemed to fire imagination across the entire spectrum of human culture. Science, as well as art and music, became imbued with new and unconventional directions. Within the space of a few years, the discovery of the electron, the introduction of the quantum principle, and the arrival of relativity theory had revealed unexpected aspects of Nature.

The first few decades of the twentieth century uncovered the implications of relativity and quantum mechanics. Still notoriously difficult for beginners, these major innovations dramatically showed how we have been conditioned by our limited everyday experience, and how Nature can appear enigmatic under unfamiliar conditions. In another physics direction, big telescopes revealed new depths of the Universe, showing that our Galaxy is only one of a multitude, and is hidden in an insignificant back row. This major reappraisal of astronomy had a similar impact to the revelations of Copernicus four hundred years earlier.

For many generations of twentieth-century physicists, relativity and quantum mechanics were the "New Physics." Cosy models with gearwheels and springs had to be jettisoned, and the early twentieth century showed that Nature instead works on a different plan. Simple models and familiar preconceptions of cause and effect may once have been useful, but taken at face value were a major obstacle to further understanding. The Universe is governed ultimately by probabilities. The implications of this uncertainty began to take root soon after the discovery of quantum mechanics, but it took several decades before other important concepts such as quantum entanglement were understood and demonstrated.

During this time, the central goal of physics remained as trying to explain as much as possible from some minimal subset of hypotheses. With the theory of general relativity and the Schrödinger equation, and later with the help of powerful computers, perhaps it would be possible to work out all the implications and probabilities.

## Simplicity and complexity

Because of its broad scope, physics has traditionally diverged along two mutually opposite frontiers – the study of the infinitely small, and that of the immensely large. Looking inwards into matter, elementary particle physicists attempt to identify the innermost constituents of Nature – quarks, electrons, . . . – while their field-theoretical colleagues try to make sense of it all and explain how these particles interact to form substance and matter. Looking outwards, astronomers and astrophysicists chart the large-scale map of a Universe composed of stars and galaxies clustered on a dizzying array of scales, and cosmologists attempt to understand how all this came about.

The first section of the book – "Matter and the Universe" – deals with the composition of what we see around us. It begins with a chapter that explains how the cosmos itself has come under systematic study, complementing what has been learned by looking at its various components at different scales. This wider view reveals what is perhaps the biggest enigma of all – the Universe has to contain much more than we can see, a new Copernican revolution. The familiar matter and energy that we know and understand seems only to be the tip of a vast cosmic iceberg, the rest – "dark matter" and "dark energy" – being hidden from view. The visible tip of the cosmic iceberg is underpinned by a vast invisible plinth, complemented by mysterious forces that blur the familiar attraction of gravity. Despite all our efforts so far, we do not even know what, or even where, most of the Universe is!

Whatever and wherever it is, the Universe is enormous, but is ultimately made up of very small things. On the large scale, the Universe is described by relativistic gravity, and on the small scale by quantum physics. The sheer difference in scale between these two descriptions and the resultant polarity of their approaches kept them apart for many years. But the latter years of the twentieth century brought the realization that the Universe had been created in a Big Bang – an enormous but highly concentrated burst of energy where quantum physics had ruled and where gravity had played a major role. The Universe is the ultimate physics experiment. This realization heralded a new synthesis. A better knowledge of the Universe's microstructure brings a deeper understanding of the Big Bang and therefore its subsequent evolution and large-scale structure. The first chapters in this book cover this ongoing synthesis of the large and the small. The stubborn reconciliation of gravity with quantum physics remains a key objective in this work.

With such a synthesis, some ambitious physicists began to seek a "Theory of everything," an iconic set of succinct equations that had governed the Big Bang and everything thereafter. Such a theory would have to reconcile the quantum world with gravity. The theory of "superstrings" is often described as a candidate for such a theory. Author Michael Green does not fully share this view; however, the long-sought unification of gravity with quantum mechanics does seem to be viable in this approach. But such an ambitious venture has its price, namely the inclusion of many invisible extra dimensions that complement the four dimensions of conventional spacetime.

Despite concessions to probability, to extra dimensions, and to sheer calculational difficulties, the advances of the twentieth century went on to show that a bold reductionist approach is not the only solution, and in some cases is not even possible.

Pioneer attempts to understand physics through constituent behavior used the kinetic theory of gases, viewed as unstructured assemblies of microscopic elastic billiard balls. Confined to its stated range of validity, this modest theory was highly successful. With the realization that constituent motion ultimately has to be quantized, the subsequent picture of quantum statistical mechanics brought new understanding.

Looking at the Universe as a gas of galaxies may give some useful results, but its limitations are uncomfortably obvious. Everywhere around us we see structure – planetary systems, crystals, atoms, . . . – each level of which has its own characteristic behavior. New physics has revealed that matter can be made to exist in structured forms that are less familiar but nevertheless remarkable – superfluids, semiconductors, plasmas, liquid crystals, . . ., new types of matter whose appearance is often marked by phase transitions – abrupt onsets of new behavior. Many of these forms are not easily visible or even present in the Universe around us and can behave in curious ways. Many are due to microscopic quantum effects, which can nevertheless exist only in bulk matter (there is no such thing as a lone superconducting electron, and nobody has yet found a magnetic monopole). This new synthetic material is described in the second section – "Quantum matter."

Such bulk behavior can become relatively easy to identify once it has been discovered, but its explanation from first principles can be much less apparent. Ferromagnetism, known since antiquity, is a classic example. On a different timescale, about half a century elapsed between the discovery of superconductivity and superfluidity and their initial explanation in quantum terms. Quantum mechanics, developed in the early years of the twentieth century to explain the interactions of a few electrons with each other, showed how our confident familiarity with everyday Nature can be deceptive. Despite its apparent unfamiliarity, quantum mechanics helps us to understand and explain more aspects of the behavior of matter.

The third section – "Quanta in action" – includes chapters on quantum entanglement; on quanta, ciphers and computers; and on small-scale structures and nanoscience, providing good examples of how quantum physics can no longer be confined to a remote, cosily inaccessible realm and is becoming increasingly interwoven with everyday phenomena, enabling more phenomena to be explained, new ones to be discovered, novel materials engineered, and fresh approaches explored. The quantum world is no longer just a blurry backdrop visible only to the front row of the audience, and the realization that Nature can be controlled and manipulated at the quantum level has become a new driving force in physics and in technology.

The subtle interplay of quantum physics and measurement was for a long time a difficult conceptual obstacle, and even objections by such great minds as Niels Bohr and Albert Einstein were dismissed in some quarters as philosophical or pedantic. In his contribution on quantum entanglement, Anton Zeilinger explains how these questions were resolved and how the resulting deeper understanding is being exploited in new investigations, and even in potential applications, such as teleportation, quantum cryptography, and quantum computation. Such new methods could handle information as quantum bits, "qubits," while the deeper exploration of quantum mechanics continually tells us more about the enigma of reality.

One particularly important underlying quantum effect is Bose–Einstein condensation, in which atoms cooled to very low temperatures descend the quantum ladder and pile up on the energy floor. Like the questions which ultimately led to the discovery of quantum entanglement and related phenomena, such behavior was first predicted three-quarters of a century ago but has only recently been achieved, the technical barrier in this case having been the development of subtle techniques needed to cool atoms to within a fraction of a degree of absolute zero.

The chapters by William Phillips and Christopher Foot and by Claude Cohen-Tannoudji and Jean Dalibard describe how ultra-cold atoms have become a focus of attention for new physics at the start of a new century/millennium. This physics requires special skills, but does not require major resources and is well within the reach of university research laboratories. It is not Big Science. The resonance of interest in Bose–Einstein condensation is redolent of how the fashion for studying cathode rays at the end of the nineteenth century paved the way for much of the new physics of the twentieth century. As explained in the chapter by Subir Sachdev, Bose–Einstein condensation is the underlying mechanism in many other phenomena.

Theoretical physics is inexorably linked with mathematics, the natural language of Nature. When words have not yet been invented to describe many unfamiliar phenomena that can occur in Nature, mathematics provides the script (although this book contains a minimum of formalism). Riemannian geometry, for example, provided the ideal canvas on which to paint Einstein's ideas on relativity and gravity. Encouraged by this success, and others, such as the application of group theory, physicists hoped that further mathematical tools would turn up.

However, some developments began to suggest otherwise. Nature is complex as well as enigmatic, and this frontier of complexity provides a further and very new dimension for physics, complementing the traditional polarity of the very small and the very large, and possibly opening up new avenues for understanding the phenomena of the early Universe. Some of these insights into complexity come from the study of systems that are very familiar but nevertheless remain enigmatic (turbulent flow, grain dynamics). In such systems, small changes do not produce small effects. A seemingly insignificant footprint can provoke a mighty avalanche, but tracing such a catastrophe from first principles is difficult. While some modern physicists rely on high technology, modest experiments with piles of sand or mustard seeds still provide valuable insights.

In their continual quest for understanding, physicists make simplifying assumptions, stripping down problems to their bare essentials. A two-body problem is the easiest to analyze, and breaking down a complicated picture into the sum of interacting pairs of constituents can and does produce incisive results. But there is a danger that, if this analysis goes too far, the resulting component parts lose some of their connectivity. The exact original system cannot be recreated – only an approximation to it. The world cannot be forced to become simpler than it really is.

In his chapter on nonlinearity, Henry Abarbanel points out that a hundred years ago, before the advent of quantum theory, Poincaré discovered that just three fully interacting bodies could produce highly complicated results. As if afraid of the consequences, many physicists shunned this awkward revelation and preferred to investigate systems that, even if they were less familiar, were more predictable. But predictability is the ultimate criterion of understanding. As more systems came under investigation, calculations became more difficult and exact predictability increasingly a luxury.

Describing these nonlinear and often chaotic systems meanwhile calls for a reappraisal of the underlying objectives and the exploitation of radically new techniques. General examples that could link the physics of microscopic objects with more empirical laws underlying highly complex systems have been identified, providing valuable alternative viewpoints. Powerful computers take over when analytical mathematical treatment becomes impossible, and simulations provide an alternative, or possibly new, level of understanding. These implications of nonlinearity and complexity are elegantly summarized in the contributions by Henry Abarbanel and Antonio Politi. These chapters, together with that on collaborative physics and e-science by Tony Hey and Anne Trefethen, make up the fourth section of the book – "Calculation and computation."

As well as spanning the many orders of magnitude which separate the apparent diameter of a quark from that of the Universe, physicists now have to try to reconcile their traditional goal of seeking simplicity with the innate natural complexity which appears to surround us at all levels, and the ability of bodies, whether they be in crystals, galaxies, ferromagnets or biological cells, to "organize" themselves at very different scales into self-perpetuating structures. The insights gained from the physics of complexity have important implications in other disciplines, and could help explain how life itself functions and even how the human brain perceives and understands its environment. The natural signals inside the human nervous system are very different from synthetic pulses transmitted through circuits or glass fibers. Nikolaj Pavel's contribution on medical physics outlines how brain function is beginning to be explored, while Antonio Politi's on complex systems tries to understand our understanding, where physics techniques provide useful analogies.

Cyrus Safinya's intriguing chapter on biophysics shows how the behavior of complex biomolecules and biological function is ultimately underpinned by physics. Supramolecular biophysics contributes to the design of DNA carriers for gene therapy and for studying chromosome structure and function. It can also elucidate the structure and dynamics of nerve cells, and the mechanisms controlling DNA. Here is surely a subject full of physics potential for the twenty-first century.

## Ingenuity

Experimental physics is firmly planted in ingenuity, and new discoveries can often be sparked by technological breakthroughs and innovative instrumentation. The advent of the telescope in the early seventeenth century revolutionized astronomy. Many of the developments at the hinge of the nineteenth and twentieth centuries were rooted in a recently acquired mastery of high-voltage and high-vacuum techniques. In the twentieth century, the liquefaction of helium opened the door to the cryogenic world of superconductivity and superfluidity.

There have been suggestions that Einstein's ideas of relativity could have been catalyzed by the increasing awareness in the late nineteenth and early twentieth centuries of the need to synchronize clocks. Local time was once very arbitrary, but mass transport needed accurate timetables valid over wide distances. The advent of radio communications substantially extended this requirement.

Later in the twentieth century, World War II brought major advances in nuclear physics, microwave techniques, and digital computers. This was initially science applied to the war effort, but with the war over, pure science reaped a huge reward from these advances. The scale of this applied science had also set a new scale for pure research. "Big Science" had arrived, requiring massive support and international collaboration. Space research is a good example, and the ability to carry out experiments beyond the stifling blanket of the Earth's atmosphere opens up a huge new window. Cosmology has now become an experimental science as well as a theoretical one.

Big Science has traditionally been involved with the physics frontiers of the very large, such as space research with its requirements for major instrumentation and satellites, and the very small, notably particle physics based on enormous high-energy accelerators. However, major efforts also attack the frontier of complexity. New international collaborations focus on the long-term goal of controlling thermonuclear fusion. On a smaller scale, applied physics laboratories such as Bell Labs and IBM Research have contributed an impressive array of pure-science breakthroughs. These developments have produced such fruits as semiconductors and transistors, with a major impact on industry and lifestyle.

The transistor was developed through an applied-physics effort by an industrial laboratory before the underlying science of semiconductors had fully been understood. However, pure research can also bring valuable spinoffs – a classic example being Röntgen's discovery of X-rays, which were used for medical imaging even before the atomic-physics origin of this radiation had become clear.

In another sphere, the advent of modern telecommunications and the ready availability of powerful computers has revolutionized research methods. No longer do scientists have to work alone in isolated laboratories, however large these might be. While the central laboratory remains the focus, some of the research effort can be "subcontracted" to teams dispersed across the globe. At CERN, the European particle-physics laboratory in Geneva, Switzerland, the World Wide Web was developed in the late 1980s to enable scientists to access research information remotely and share it with their colleagues via the Internet, without necessarily having to come to the laboratory. In the 1990s, the "Web" went on to take the world by storm. New "Grid" developments aim to do for raw data what the Web did for basic information.

In the chapter on collaborative physics and e-science, Tony Hey and Anne Trefethen outline how computers are increasingly being used in industry and commerce as well as science. The "virtual organizations" needed to handle the computing applications of tomorrow will one day become as commonplace as today's Internet. Such infrastructures will have to be very complex, but at the same time must be reliable and secure. Data-hungry physics applications are a driving force in the evolution of this e-science. The development of computation and computer techniques has been staggering. A few decades ago, few could have imagined the speed or the range of this impact. While this will continue, a new horizon could soon open up with the advent of true quantum computation, described by Artur Ekert in his contribution on quanta, ciphers and computers.

Physics and physicists play key roles in society. In his contribution on the physics of materials, Robert Cahn points out that transistors are now more widespread, and cheaper, than grains of rice. Invented only some sixty years ago, the transistor has become the quantum of modern electronics. Another example of a physics-based applications explosion is the laser, invented a decade after the transistor, but which now plays a key role in storing and retrieving digital data in increasingly compact forms. (The basic principles of this device are explained in the chapter by Phillips and Foot, which explores the deep analogy between Bose–Einstein condensates and lasers.) Lasers are not as ubiquitous as transistors, but nevertheless are widely used in home and office electronics equipment, and are extensively employed in medicine for microsurgery. In chemistry, lasers have opened up a new sector where they are used to "freeze-frame" the ultra-high-frequency kinetics of atomic or molecular mechanisms (see the contribution "Chemistry in a new light" by Jim Baggott in the companion volume *The New Chemistry*, edited by Nina Hall (Cambridge University Press, 2000)).

Lasers have also opened up new areas of physics and enable ultra-precision measurement to be made. The contribution by Claude Cohen-Tannoudji and Jean Dalibard describes the ingenious techniques used to cool atoms to very low energy levels and explore new quantum aspects of atomic behavior.

Less widespread than transistors and lasers, but also important, are powerful techniques such as neutron scattering, and X-ray analysis using beams of increasingly short wavelength. These tools were conceived inside physics laboratories, but quickly developed into valuable probes of matter for use in other areas of science and in engineering. Half a century ago, the unraveling of the molecular structure of DNA and of proteins by X-ray analysis was a milestone in biology and chemistry. Today large sources of neutrons and of X-rays (synchrotron radiation) have been built in many countries to serve large user communities. They provide vivid examples of the growing usefulness of physics.

This book cannot cover all such burgeoning applications fields, and concentrates instead on a few examples: biophysics, medical physics, and the physics of materials, which together make up the final section of the book – science in action. The earlier chapters which cover computational physics (Hey and Trefethen), nanoscience (Imry), and quantum cryptography (Ekert) provide other valuable examples of applications. Most of these areas are very new, but the history of medical physics can be traced back to Volta's eighteenth century demonstration that external electric current pulses applied to a frog's leg stimulate muscle contraction. Together, these examples illustrate how physics, even its extreme quantum description, has permeated modern life and will continue to do so.

The final chapter by Ugo Amaldi on physics and society was designed as an off-beat "after-dinner speech" for the book rather than a formal conclusion. It places the science in a contemporary setting, examines some current issues, and suggests what needs to be done to ensure the welfare of physics and physicists. Amaldi boldly predicts some unusual and interesting developments and spinoffs.

## Multidisciplinary science and society

The diversity of all these new frontiers means that physics has also become highly multidisciplinary. Not that long ago, geniuses of the stature of Kelvin or Fermi could patrol many experimental and theoretical frontiers, contributing to several areas of the subject and correlating developments in apparently different spheres. A more recent example was Richard Feynman (1918–88). As well as making pioneer contributions to field theory, Feynman also produced key insights in the study of superfluidity, in quantum computation, and in nanoscience, making new objectives emerge from muddled thinking. Although he was a theorist, his suggestions went on to make major impacts in experimental science.

The sheer weight of accumulated knowledge means that physics has become polarized along distinct directions. There are dichotomies between big and small science and between basic and applied physics. There are contrasts between theory and experiment and between emergent and reductive approaches. Propeled to the distant frontier of a discipline, physicists run the risk of becoming isolated from society and even their scientific colleagues. One goal of this book is to provide a broad overview of the subject so that those working in one field can appreciate better what is happening elsewhere, even in their own university or research center.

The book has cast its net wide. It spans a wide spectrum of research at the frontiers of the very large, the very small, the very complex, and the most ingenious. However, the coverage of applications areas presented here is necessarily only a selection of the fields in which physics makes important contributions. There are more, and there will be even more. In July 1945, Vannevar Bush, Director of the US Office of Scientific Research and Development, submitted a report to President Truman. The document – "Science – The Endless Frontier" – went on to become extremely influential. The words used in 1945 have not lost their implication. The science underpinned by physics today is still an endless frontier.

There is another reason for appreciating modern physics. Because it is the basis of Nature, understanding this science is a vital component of today's civilization. It is no accident that almost all physics Nobel prizewinners and all the contributors to this book have worked or work in developed nations. There is a correlation between physics capability and national prosperity. Training new physicists will not immediately make nations more prosperous, but an intellectual core community is needed to ensure high academic standards and catalyze scientific and technological development. The Pakistani physicist Abdus Salam (1926–96), who shared the 1979 Nobel Prize, worked tirelessly to promote the cause of physics in Third World nations.

In exploring its frontiers, the increasing awareness of scale and complexity in modern physics brings a profound sense of humility at our own puny role in the nature of things and the inadequacy of our understanding. The implications of physics are fundamental but bewildering, ranging from the origins of the Universe to the latest applications of semiconductor technology. This impresses the layman but could initially awe the student. However, a deeper appreciation of the role and power of physics provides fresh opportunities for imagination.

The success of the first edition of *New Physics*, edited by Paul Davies and published in 1989, showed the value of an authoritative anthology of frontier science, with contributions from internationally recognized communicators who have all made distinguished contributions to the topics they write about. This book continues that tradition. Editorial intervention has been minimized, so the chapters retain individual styles and conventions. Inevitably, there is some overlap between chapters, but this has been cross-referenced. All chapters should be accessible to non-specialists.

We hope that you will enjoy reading, and thinking, about the newest *New Physics*.

**Gordon Fraser, Divonne-les-Bains, July 2006**

## Gordon Fraser

Gordon Fraser studied physics and went on to a doctorate in the theory of elementary particles at Imperial College, London. After a brief research career, he moved to science writing. For 25 years he worked on publications at CERN, the European particle-physics laboratory in Geneva. He is author, co-author, and editor of several popular science books and has taught science communication as a visiting lecturer in several universities.

# I Matter and the Universe

# 1 Cosmology

Wendy L. Freedman and Edward W. Kolb

## 1.1 Introduction

Cosmology is the study of the origin, evolution, composition, and structure of the Universe. As a scientific discipline cosmology began only in the twentieth century. Among the fundamental theoretical and observational developments that established the Big-Bang model were the general theory of relativity proposed by Albert Einstein in 1915, the development of the theory of an expanding relativistic cosmology by Alexander Friedmann in 1922 and Georges Lemaître in 1927, the observation of the expansion of the Universe by Edwin Hubble in 1929, the development of the theory of Big-Bang nucleosynthesis by Ralph Alpher, George Gamow, and Robert Herman in the early 1950s, and the discovery of the cosmic background radiation by Arno Penzias and Robert Wilson in 1964.

Traditionally, cosmology has been a data-starved science, but cosmology today is experiencing a fertile interplay between observation and theory. Precision measurements of the expansion rate of the Universe, the large-scale homogeneity and isotropy of the distribution of galaxies, the existence and high degree of isotropy of the 3-K cosmic microwave background radiation, and the abundances of the light elements support the basic picture of an expanding hot-Big-Bang Universe.

Still, while credence for the basic hot-Big-Bang model is stronger than ever, the picture is far from complete.

- Our Universe is observed to consist of matter. Today there is no evidence of antimatter in the form of stars, galaxies, galaxy clusters or any form of cosmological antimatter. However, antimatter should have been plentiful when the temperature of the Universe was high enough to create matter–antimatter pairs. In order to prevent complete annihilation of matter and antimatter, the early Universe must have had a very slight asymmetry between the matter density and the antimatter density. Theories for the origin of the matter–antimatter asymmetry all involve charge-conservation–parity (CP) violation and violation of the law of conservation of baryon number. While CP violation has been observed in reactions involving kaons and B-mesons, unfortunately there is no experimental evidence for violation of baryon number (such as proton decay).
- Observations reveal that the predominant form of matter in the Universe is dark (it emits no light) and seemingly without strong or electromagnetic interactions.

*The New Physics for the Twenty-First Century*, ed. Gordon Fraser.

The most plausible theoretical candidate for this matter is a new species of particle that was produced in the early Universe and interacts only weakly with ordinary matter. This new species of matter must be "cold," i.e., it must have a typical velocity much less than the velocity of light. For this reason, it is referred to as "cold dark matter" or CDM. Although there are plausible theoretical candidates, relic dark-matter particles have not been detected, nor have they been produced and detected in accelerator experiments.

- Visible matter in the form of galaxies is distributed in large clusters and long filaments, with voids and bubbles on scales up to hundreds of millions of light years. As the Universe is sampled on increasingly larger scales, the distribution of matter becomes more and more uniform. To a very good approximation, the cosmic background radiation is isotropic about us. The standard hot-Big-Bang model assumes that the Universe is homogeneous and isotropic on large scales, but offers no explanation for why.
- In homogeneous, isotropic cosmological models, the spatial geometry of the Universe may be flat, spherical, or hyperbolic. The standard hot-Big-Bang model is consistent with the observed high degree of spatial flatness of the Universe, but cannot explain it.
- Observations of the large-scale distribution of galaxies (here, "large-scale" refers to scales larger than galaxies) confirm the idea that galaxies and other large-scale structure formed under the influence of gravity acting on primordial seed fluctuations in the temperature and density of the Universe. The standard cosmological model is silent on the origin of the seed perturbations.
- Observations of the cosmic microwave background from the Big Bang reveal a Universe that is very smooth, but not perfectly smooth. The temperature of the Universe is very nearly the same in all directions, but there are tiny fluctuations (at a level of only a few parts in 100 000) in the temperature of the background in different directions of the sky. These small temperature fluctuations cannot be explained by the standard cosmology.
- The early-Universe theory of inflation potentially provides an explanation for the observed homogeneity and isotropy of the Universe, the spatial flatness, the seeds of structure, and the temperature anisotropies. However, inflation involves particle physics at energies beyond the experimental reach of terrestrial accelerators, and hence it is untested.
- Various recent observations suggest that the present expansion velocity of the Universe is *larger* than it was in the recent cosmological past; the expansion of the Universe is accelerating. This increase with time of the expansion velocity of the Universe is believed to result from the mass–energy of the Universe dominated by a type of "lambda" term (or a cosmological constant) in the gravitational field equations as proposed by Einstein in 1917. More generally, the lambda term is referred to as "dark energy." At the current time, the physical explanation for the dark energy is unknown.
- Dark energy and dark matter seem to comprise 95% of the mass–energy density of the present Universe (70% dark energy and 25% dark matter). Only about 5% of the mass–energy of the Universe is made of ordinary neutrons and protons, i.e., familiar matter (see Figure 1.1). We do not yet understand the basic forms of matter and energy in the Universe.

The good news of modern cosmology is the success of a cosmological model, lambda-CDM, which seems capable of explaining how the primordial Universe evolved to its present state. The lambda-CDM model (Table 1.1) is an expanding hot-Big-Bang model that assumes a present Universe with the aforementioned composition

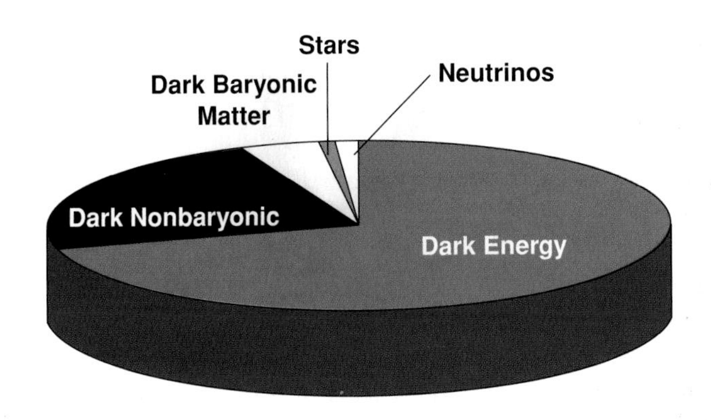


**Figure 1.1.** The Universe today in the standard lambda-CDM model, with one-third matter, two-thirds dark energy. Most of the matter is dark. About 5% of matter is in baryonic form, while only 1% is in stars. The contribution from neutrinos is comparable to that in stars.

**Table 1.1.** *Cosmological parameters for the lambda-CDM cosmological model*

| |
|---|
| Expansion rate: $H_0 = 72 \pm 8$ km s$^{-1}$ Mpc$^{-1}$ |
| Temperature: $T = 2.725 \pm 0.002$ K |
| Critical density: $\rho_C = 1.88h^2 \times 10^{-29}$ g cm$^{-3}$ |
| Total mass-energy density: $\Omega_{\text{TOTAL}} = 1.02 \pm 0.02$ |
| Dark-energy density: $\Omega_\Lambda = 0.73 \pm 0.04$ |
| Matter density: $\Omega_M = 0.27 \pm 0.04$ |
| Baryon density: $\Omega_B = 0.044 \pm 0.004$ |
| Massive neutrinos: $\Omega_\nu h^2 \leq 0.0076$ (95% CL) |
| Age of the Universe: $t_0 = 13.7 \pm 0.2$ Gyr |

The number for the expansion rate is taken from the Hubble Key Project (Freedman *et al.*, 2001, *Ap. J.* **553**, 47). Other parameters are from the WMAP collaboration (Spergel *et al.*, 2003, *Ap. J. Suppl.* **148**, 175).

and small primordial perturbations in the temperature and density produced by inflation. The lambda-CDM model has the potential to connect quantitatively the Universe on macroscopic scales to the fundamental laws of physics for elementary particles on microscopic scales. Not only do new directions in cosmology arise from ideas in elementary particle physics, but also the Universe is used as a laboratory to test new particle theories. The troubling aspect of the lambda-CDM model is that it requires the existence of mysterious dark matter and dark energy, and it assumes an early episode of inflation controlled by high-energy physics beyond our capability to test.

This chapter reviews the standard Big-Bang cosmology and discusses the unsolved problems and the modern developments that attempt to solve them. The final section looks forward to possible theoretical and observational developments that will be part of the quest for a deeper understanding of our Universe.

## 1.2 Hot-Big-Bang cosmology

The overall framework for understanding the evolution of the Universe is the hot-Big-Bang model. This model is based on Einstein's theory of general relativity, and assumes that on the largest scales the Universe is homogeneous and isotropic. A Universe that is homogeneous has the same properties (density, temperature, expansion rate, etc.) at every point in space, while an isotropic Universe would appear the same if viewed in any direction. The metric describing homogeneous/isotropic spacetimes is the Robertson–Walker metric, developed by H. P. Robertson and A. G. Walker in 1935.

**BOX 1.1 THE FRIEDMANN EQUATION FOR THE EXPANSION RATE OF THE UNIVERSE**

A Friedmann–Robertson–Walker (FRW) cosmology can be characterized by a number of parameters that describe the expansion, the global geometry, and the general composition of the Universe. These parameters are all related via the Friedmann equation, which comes from the Einstein equations:

$$H^2(t) = \frac{8\pi G}{3} \sum_i \rho_i(t) - \frac{k}{a^2(t)},$$

where $G$ is Newton's gravitational constant, $H(t)$ is the expansion rate of the Universe ($H_0$ is the expansion rate today), $\rho_i(t)$ is the mass–energy density in individual components, $a(t)$ is the scale factor describing the separation of galaxies during the expansion, and $k = 0$ or $\pm 1$ is the spatial curvature term. Dividing by $H^2$, and defining $\Omega_{\rm TOT} = \sum_i \Omega_i$, where $\Omega_i = \rho_i/[3H^2/(8\pi G)]$, we may rewrite the Friedmann equation as $\Omega_{\rm TOT} - 1 = \Omega_k = k/(a^2 H^2)$. For the case of a spatially flat Universe ($k = 0$), $\Omega_{\rm TOT} = 1$. The factor $\sum_i \Omega_i$ characterizes the composition of the Universe: the sum of the densities of ordinary matter (or baryons), cold dark matter, hot dark matter, radiation, and other sources of energy (for example dark energy). The values for these parameters must be determined empirically.

Observations of the temperature fluctuations in the cosmic microwave background radiation suggest a spatially flat (or very nearly spatially flat) Universe, i.e., the $k/a^2$ term in the Friedmann equation may be neglected, and $\Omega_{\rm TOT} = 1$.

The homogeneous/isotropic Big-Bang model is known as the Friedmann–Robertson–Walker (or FRW) cosmology (Box 1.1).

The dynamics of a homogeneous, isotropic Universe is described by the evolution with time of a quantity known as the *cosmic scale factor*, $a(t)$. The physical distance between objects in a homogeneous, isotropic Universe is proportional to the cosmic scale factor. For most of its history, the temperature of the Universe decreased as the inverse of the cosmic scale factor. The expansion rate of the Universe, $H(t)$, is defined in terms of the time derivative of the scale factor, $H(t) = \dot{a}/a$, where the overdot denotes the derivative with respect to time. The present expansion rate of the Universe, $H_0$, is Hubble's constant.

The evolution of the scale factor, and hence the evolution of the expansion rate, is found from Friedmann's equation. The expansion rate is governed by the matter and energy density of the Universe contributed by the individual components of matter and energy.

The matter density of the Universe is diluted by the expansion, as is the radiation energy density. The Universe of earlier epochs was hotter and denser than the Universe today. In the standard cosmological model, the early Universe was radiation-dominated, and the present Universe is matter-dominated (see Boxes 1.2 and 1.3).

Principal observational support for the basic Big-Bang picture is provided by (1) the expansion of the Universe, (2) the large-scale homogeneity and isotropy of galaxies and large-scale structure, (3) the cosmic microwave background radiation, and (4) the abundances of the light elements.

### 1.2.1 The expansion of the Universe

Shortly after discovering the universality of gravity, Newton discovered the phenomenon of *gravitational instability*. In an inhomogeneous Universe, regions with

### BOX 1.2 THE MASS-ENERGY DENSITY OF THE UNIVERSE

The mass–energy density of a FRW cosmology is homogeneous and isotropic. Its evolution may be modeled by assuming that the mass–energy density consists of several components: a density of nonrelativistic matter, a density of relativistic matter, and a density due to vacuum energy (or equivalently, a cosmological constant).

The change in the mass–energy density can be found from the conservation of the stress-energy tensor in the expanding background. The evolution is most easily expressed in terms of the scale factor $a(t)$. The mass–energy density of nonrelativistic matter decreases as the cube of the scale factor, the mass–energy density of a relativistic component of the stress-energy tensor decreases as the fourth power of the scale factor, and the contribution to the mass–energy density from a cosmological constant is unchanged by expansion of the Universe.

As a consequence of the different forms of evolution of the components, the stress-energy tensor was radiation-dominated at early times, and then matter-dominated, and finally dominated by the vacuum term.

The scale factor had different dependences on time during the various epochs. During the radiation-dominated era the scale factor increased as $a(t) \propto t^{1/2}$, during the matter-dominated era $a(t) \propto t^{2/3}$, and during a vacuum-dominated era $a(t) \propto \exp t$.

### BOX 1.3 COSMOLOGICAL EPOCHS

| Epoch | Age of the Universe | Significant events |
|---|---|---|
| Planck | $10^{-43}$ seconds | Initial spacetime singularity<br>Quantum gravity determines evolution<br>Strings/branes present in the Universe |
| Inflation | $10^{-35}$ seconds | Origin of density perturbations<br>Origin of temperature fluctuations<br>Origin of background gravitational waves |
| Radiation | Earlier than about 10 000 years | Production of relic dark matter<br>Production of neutrino background<br>Primordial nucleosynthesis |
| Matter | Later than about 10 000 years | Recombination of electrons and nuclei<br>Last scattering of background radiation<br>Growth of density inhomogeneities |
| Dark energy | About a billion years ago | Present acceleration of the Universe |

It is useful to divide the expansion history of the Universe into a series of "cosmological epochs." Very little is known of the Planck epoch, which is expected to have occurred when the age of the Universe was the Planck time, $10^{-43}$ seconds after the bang; possible events during the Planck time are very speculative. The time associated with the inflation epoch is also uncertain. Events during inflation are less uncertain because their imprint can be observed today. The radiation and matter epochs are part of the standard cosmological model. Finally, the most recent epoch started when the mass–energy density of the Universe became dominated by dark energy.

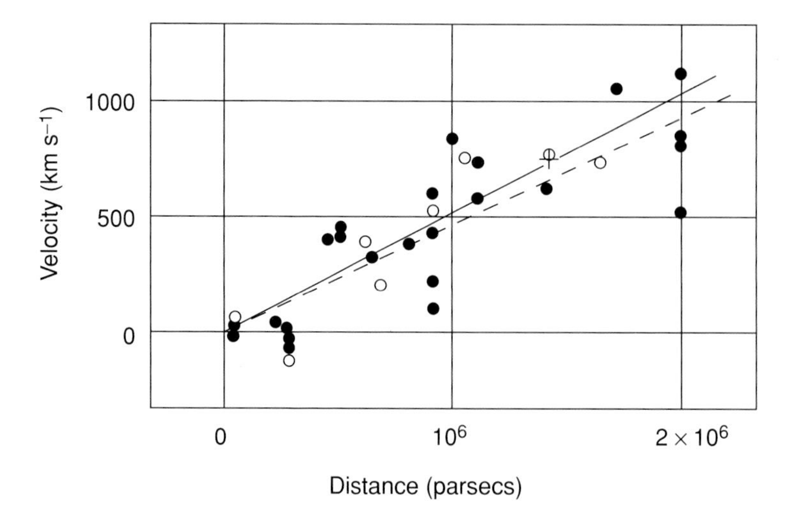


**Figure 1.2.** Hubble's original data providing evidence for the expansion of the Universe.

density larger than average will eventually accrete matter from under-dense regions, and the density inhomogeneities will grow. A Newtonian static Universe is unstable against even the slightest departures from homogeneity in the mass distribution.

In 1915 Einstein recognized that in a relativistic cosmology even an exactly homogeneous Universe would be static. This was his motivation for introducing the "cosmological constant," the famous parameter that prevented the Universe from contracting or expanding. Friedmann pointed out that Einstein's static cosmological model is unstable against small perturbations in the scale factor. Subsequently, Friedmann, and independently Lemaître, developed expanding cosmological models.

In 1929, Edwin Hubble empirically established the form of the law for the expansion of the Universe. Using the 100-inch telescope at Mount Wilson, Hubble measured the distances to a sample of nearby galaxies. These distances allowed him to establish that these objects were indeed remote, and not connected to our own Galaxy. Making use of published radial velocities, Hubble went on to discover a correlation between galaxy distance and recession velocity – evidence for the expanding Universe. The slope of the velocity versus distance relation yields the Hubble constant, the current expansion rate of the Universe. Hubble's original data are shown in Figure 1.2. For comparison, a modern version of the "Hubble diagram" is shown in Figure 1.3.

Hubble's constant, $H_0$, sets the scale for distances in the present Universe. Hubble's law relates the distance to an object and its redshift: $cz = H_0 d$, where $d$ is the distance to the object and $z$ is its redshift. Hubble's constant is usually expressed in units of kilometers per second per megaparsec (a megaparsec, abbreviated Mpc, is one million parsecs, where a parsec is approximately 3.26 light years). Equivalently, Hubble's constant may be expressed in terms of an inverse distance or an inverse age: $H_0 = 72$ km s$^{-1}$ Mpc$^{-1}$; $H_0^{-1} = 13.6 \times 10^9$ yr; $cH_0^{-1} = 4160$ Mpc, where the 10% observational uncertainty in the determination of $H_0$ should be recalled. Often the Hubble constant is expressed in dimensionless units, $h = H_0/100$ km s$^{-1}$ Mpc$^{-1}$. Hubble's law relating the distance and the redshift (see Box 1.4) holds in any FRW cosmology for redshifts less than unity. As the redshift becomes unity or greater, the distance–redshift relationship in a FRW model depends on the nature of the mass–energy content of the Universe.

The inverse of Hubble's constant also sets the scale for the expansion age of the Universe (see Box 1.5). Again, the exact relation between the expansion age and Hubble's constant depends on the nature of the mass–energy content of the Universe. However, the fact that $H_0^{-1}$ is approximately the age of the Universe supports the standard Big-Bang model.

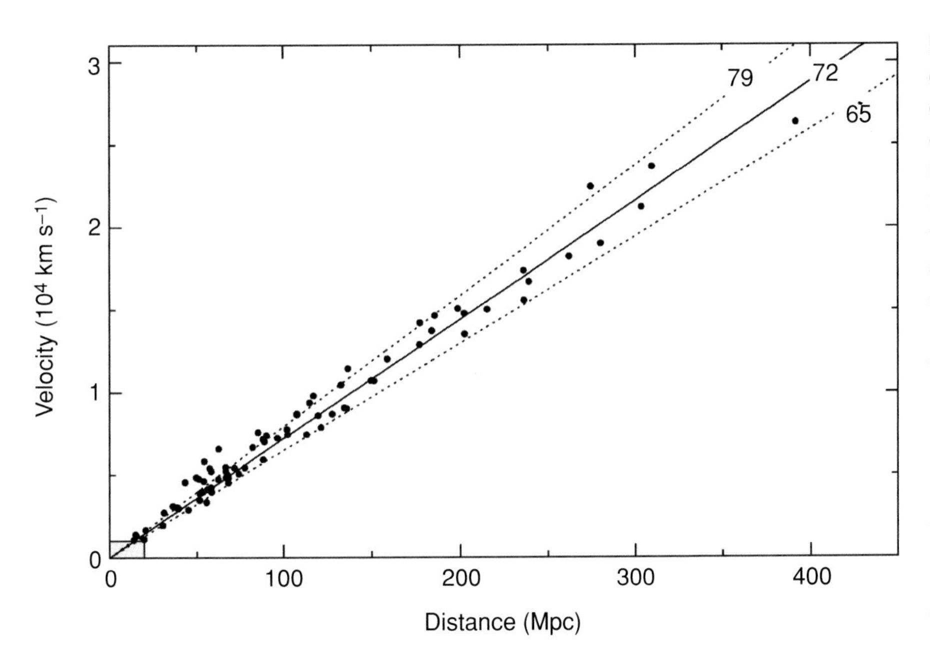


**Figure 1.3.** Accurate determination of the Hubble constant has always been a challenge, and only recently have improvements in instrumentation, the launch of the Hubble Space Telescope, and the development of several different measurement methods led to a convergence on its value. Distances to nearby galaxies obtained as part of a Hubble Key Project (Freedman *et al.*, 2001, *Ap. J.* 553, 47) have allowed calibration of five different methods for measuring the Hubble constant, all showing very good agreement to within their respective uncertainties. These yield a value of the Hubble constant of 72 km s$^{-1}$ Mpc$^{-1}$, with a total uncertainty of plus or minus 10%. Not shown are the errors associated with the individual data points (see Freedman *et al.* for a thorough discussion of the errors). The shaded box in the lower-left-hand corner indicates the extent of the data at the time of Hubble's original paper.

BOX 1.4 THE REDSHIFT

The wavelength of a photon traveling in an expanding Universe increases with time, or is "redshifted." The spectral lines of distant galaxies are systematically shifted to the red end of the spectrum. If the laboratory wavelength of a spectral line is denoted by $\lambda_0$, the measured wavelength of the spectral line of a distant galaxy, $\lambda$, will be larger than $\lambda_0$. The redshift of the spectral line is denoted by $z$, and is given by

$$\frac{\lambda}{\lambda_0} = 1 + z.$$

Traditionally, the redshift of galaxies is expressed in terms of an equivalent recessional velocity of the galaxy that would lead to a Doppler shift of the spectral lines. For velocities much less than the velocity of light, $c$, the recessional velocity is related to the redshift by $v_R = cz$.

The redshift of the Universe, $z$, is defined in terms of the ratio of the scale factor of the Universe to its present value. Denoting the scale factor by $a(t)$ and the present value of the scale factor by $a_0$, the redshift is defined by $1 + z = a_0/a(t)$. Therefore, "the Universe at a redshift of $z$" refers to the Universe when the scale factor was given by $a(t) = a_0/(1 + z)$.

For most of the history of the Universe the temperature decreased in proportion to the inverse of the scale factor: $T(t) = (1 + z)T_0$, where $T_0$ is the present temperature.

## 1.2.2 The homogeneity and isotropy of the Universe

The standard FRW cosmology describes a Universe in which the distribution of matter and energy is homogeneous and isotropic. The strongest evidence for homogeneity and isotropy is the remarkable isotropy of the cosmic background radiation (see the next section). Additional evidence for isotropy is the uniformity of the Hubble expansion in different directions on the sky, as well as the isotropy of distant radio sources, X-ray sources, and gamma-ray bursts.

BOX 1.5 THE AGE OF THE UNIVERSE

There are two ways of determining the age of the Universe, the time since the Big Bang. The first is based on the cosmological model and the expansion of the Universe, and the second is by using a chronometer based on models of stellar evolution applied to the oldest stars in the Universe. Both methods are completely independent of each other, and so offer an important consistency check. Obviously, there can be no objects in the Universe older than the Universe itself.

The expansion of the Universe is governed by the rate at which the Universe is expanding, and the extent to which gravity slows the expansion, and dark energy causes it to speed up. For a flat Universe with no dark energy ($\Omega_M \simeq 1, \Omega_\Lambda \simeq 0$), the age is simply two-thirds of the Hubble time, or only 9.3 billion years for $h = 0.7$. A Universe with one-third matter and two-thirds dark energy ($h = 0.7$, $\Omega_M \simeq 0.3$, $\Omega_\Lambda \simeq 0.7$) has an age comparable to the Hubble time, or 13.5 billion years.

The most well developed of the stellar chronometers employs the oldest stars in the Milky Way, which are found in globular clusters, systems of about a million or so stars bound together by gravity. Recent, detailed models computing the rate at which hydrogen is converted to helium within stellar cores, compared with observations of stars in globular clusters, yield ages of about 12.5 billion years, with an uncertainty of about 10%. An estimate for the age for the Universe requires allowing for the time taken to form these globular clusters: from theoretical considerations this is estimated to be about a billion years. This age estimate for the Universe agrees well with the expansion age. Three other cosmic chronometers – the cooling of the oldest white dwarf stars, the decay of several radioactive isotopes made in stellar explosions, and the cosmic microwave background – yield similar ages. At present, all of the independent ages calculated for the Universe are consistent with an age of about 13.5 billion years.

## Large-scale structure in the Universe

Of course the Universe is neither homogeneous nor isotropic on all scales. There is a rich structure to the distribution of matter in the Universe. Matter (or at least visible matter) is concentrated into gas, stars, galaxies, galaxy clusters, superclusters, and other large features. Galaxies seem to be distributed on an enormous web-like structure. For this reason, the homogeneous/isotropic model is assumed on very large scales, and inhomogeneities are treated as perturbations.

The spatial distribution of galaxies shows a high degree of clustering and inhomogeneity. Measuring the actual three-dimensional distribution of this structure has advanced considerably with the development of new spectrographs with multiple fiber feeds and slits that allow hundreds of galaxy redshifts (and hence the distance to the object via Hubble's law) to be measured simultaneously (see Figure 1.4). These observations have revealed not only galaxy clusters and superclusters, but also large voids, walls, and filaments extending over 300 million light years. It appears that the typical largest structures have now been identified, with the most recent surveys failing to turn up yet larger structures. Two recent large surveys, the Anglo-Australian Two-degree Field Galaxy Redshift Survey (2dFGRS), and the Sloan Digital Sky Survey (SDSS), have measured a quarter of a million redshifts each, with the SDSS still working toward the goal of a map of the Universe containing the three-dimensional locations of 660 000 galaxies. The statistical properties of the galaxy distribution in the Universe may be inferred from these maps of the distribution of galaxies, using

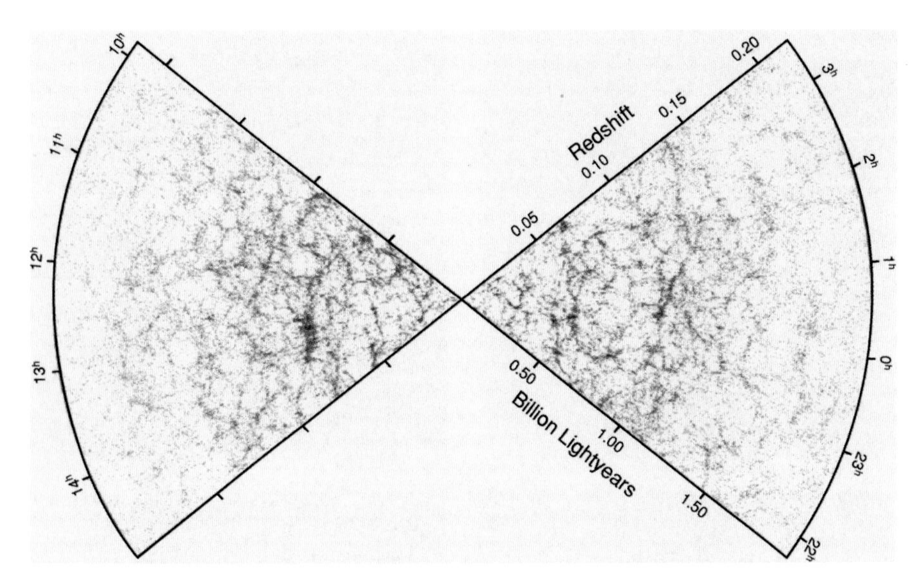


**Figure 1.4.** The redshift-cone diagram from the Two-degree Field Galaxy Redshift Survey. The fields are observed in two long strips in the northern and southern galactic-cap regions. Large voids and walls in the large-scale structure are evident with dimensions of order 100 $h^{-1}$ Mpc.

correlation functions, power spectra, and so on for quantitative comparison with models.

The study of large-scale structure also provides constraints on the nature of dark matter and inflation through computer simulations. The properties of simulated universes can now be calculated for up to a billion individual particles, with different cosmological parameters and initial conditions, and then compared with observations (see Figure 1.5). These simulations have found, for example, that flat (zero-curvature) lambda-CDM models compare extremely well with the cosmic structure observed today (as do open universes with cold dark matter). However, model universes for which the matter is comprised only of high-velocity (or hot) dark matter or baryonic matter, or spatially flat universes with only cold dark matter (with no dark energy) produce distributions of galaxies on large scales that are completely incompatible with the observations.

Cold-dark-matter particles interact only gravitationally, so computations are relatively straightforward. There are still limitations in the numerical simulations to date, however. Models have become more complicated, incorporating additional astrophysics governing ordinary matter: the dynamics of the gas, the formation of molecules, their destruction and cooling, radiative transfer, star formation, heating, and feedback. Observations over a range of redshifts and rest wavelengths will be critical in providing constraints on these models. Lambda-CDM models provide an excellent description for the distribution of galaxies on large scales, but seem to predict the formation of too many satellite companion galaxies by about an order of magnitude. The explanation for this difference may involve gas- and star-formation feedback, processes that are not yet well understood. The central densities of galaxies are another area where CDM model predictions are not yet precise, and observations are limited by resolution.

## 1.2.3 The cosmic microwave background radiation

Earlier than about 380 000 years after the Big Bang (a redshift of about 1100), the temperature of the Universe was sufficiently high to ionize completely all of the matter in the Universe. The photons in this hot ionized plasma continually scattered off the free electrons, and reached a state of thermal equilibrium. As the Universe expanded

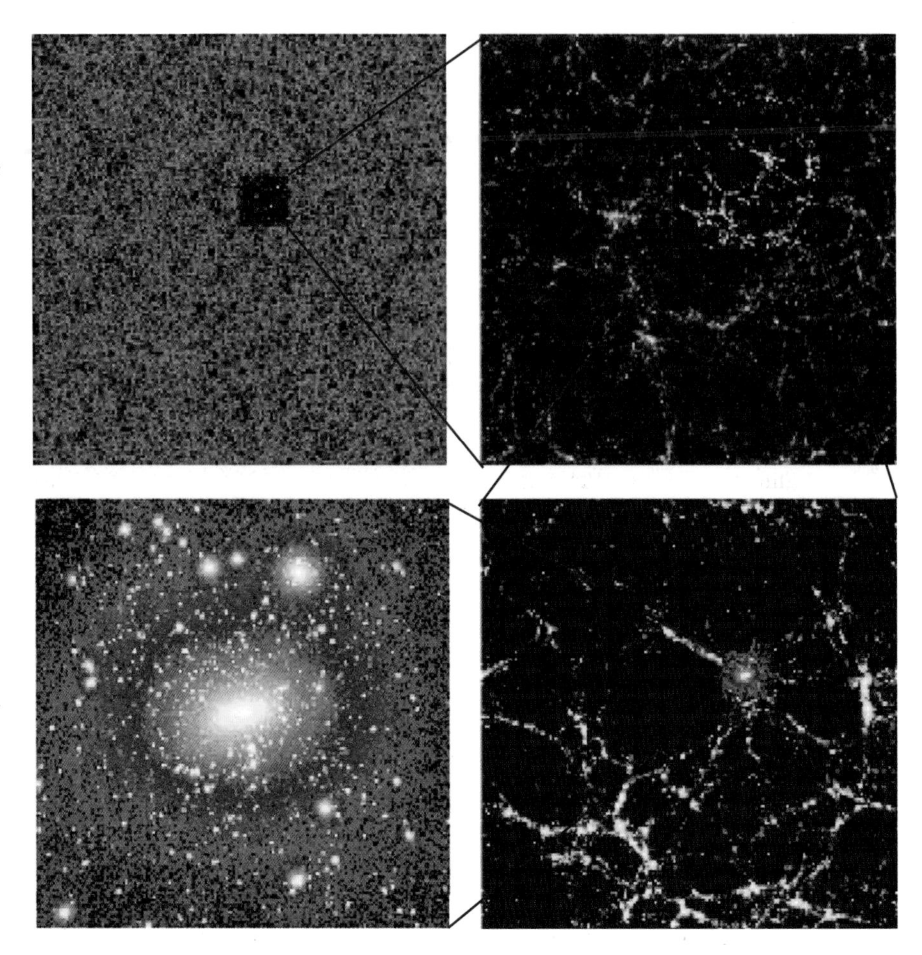

**Figure 1.5.** Four different simulations of cold dark matter in lambda-CDM, showing the emergence of large-scale structure and galaxies. Clockwise from the top left, the images are shown at decreasing scales and increasing resolution. The first simulation is for $10^9$ particles; the others are for roughly $10^7$ particles. The dimensions of the boxes decrease from 3000 to 250, 140, and 0.5 $h^{-1}$ Mpc. This image was produced by the Virgo consortium for cosmological simulations, an international collaboration of researchers in the UK, Germany, and Canada. The simulations were carried out at the Max Planck Rechenzentrum in Garching and at the Institute for Computational Cosmology, University of Durham. Courtesy of Carlos Frenk.

and cooled to about 3000 K, electrons began to combine with protons into atoms in a process known as *recombination* (of course, recombination is a misnomer since the electrons and nuclei had never previously been combined). As a result of recombination, the photon mean free path grew, and radiation and matter decoupled, allowing photons to journey unimpeded through the Universe. The radiation that last scattered off the electrons when the Universe was at 3000 K has now cooled to a present temperature of $2.725 \pm 0.002$ K, and has been redshifted by the expansion so that it is now detected at microwave wavelengths.

The serendipitous 1964 discovery of the cosmic microwave background radiation by two astronomers, Arno Penzias and Robert Wilson, provides one of the strongest pieces of evidence for the Big-Bang theory. This radiation has a characteristic blackbody spectrum as predicted for radiation that was once in thermal equilibrium (see Figure 1.6). It is also very isotropic, exhibiting a very high degree of uniformity across the sky.

We can summarize our knowledge of the cosmic background radiation as follows. (1) The spectrum of the background radiation is extremely accurately described by a blackbody. (2) The average temperature of the Universe is $2.725 \pm 0.002$ K. There are about 410 background photons in every cubic centimeter of the Universe. (3) There is a dipole moment in the radiation pattern with peak magnitude $\Delta T = 0.003\,365 \pm 0.000\,027$ K associated with the peculiar velocity of our local group of galaxies of $627 \pm 22$ km s$^{-1}$ with respect to the background radiation. (4) Upon removal of the dipole pattern there is a spectrum of fluctuations in the background radiation. The

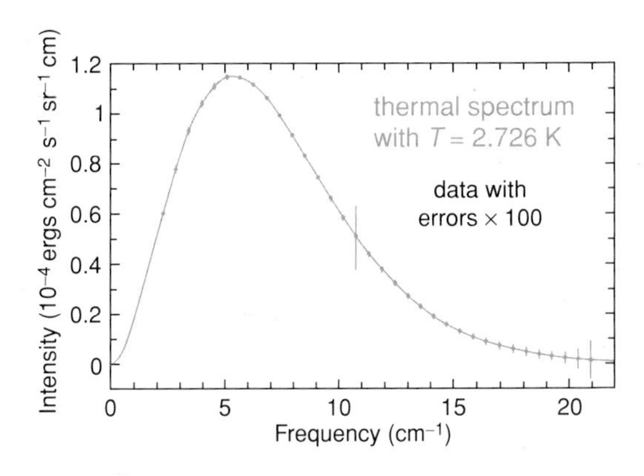


**Figure 1.6.** The solid curve is the theoretical intensity from a blackbody, as predicted by the Big-Bang theory. The data were taken by the FIRAS instrument on COBE. The FIRAS data match the curve very well; without expanding the experimental error bars it would be impossible to distinguish the data from the theoretical curve.

typical magnitudes of these temperature fluctuations are a few parts in 100 000. The amplitude of the fluctuations depends on the angular scale. The largest fluctuations (other than the dipole moment) occur around an angular scale of a degree. (5) The microwave background radiation is also polarized as a result of Thomson scattering of the radiation off electrons when the Universe was about 380 000 years old. This polarization has also recently been detected.

## Fluctuations in the cosmic microwave background (CMB) radiation

The Cosmic Background Explorer (COBE) satellite in 1992 first detected the residual fluctuations in the temperature of the background radiation. Since then, many terrestrial observations have confirmed the COBE result and extended our knowledge of the pattern of temperature fluctuations to ever finer angular scales. While the angular resolution of COBE's full-sky maps of the temperature fluctuations was limited to be greater than about 10 degrees, terrestrial observations over small areas on the sky have detected temperature fluctuations with angular scales as small as a few minutes of arc. In 2003, the Wilkinson Microwave Anisotropy Probe (WMAP) satellite provided the results of the pattern of temperature fluctuations over the whole sky to an angular resolution of 0.2–0.8 degrees (see Figure 1.7). Even finer angular resolution is expected from the Planck satellite, which is expected to be launched in 2007 (see Section 1.5)

**Figure 1.7.** An all-sky map of the temperature of the cosmic background radiation. The average temperature is 2.725 ± 0.002 K. Indicated are regions that are hotter or colder than the average. Typical temperature fluctuations are a few parts in 100 000. The pattern of anisotropies in the temperature of the microwave background shown here is a direct picture of conditions in the Universe 400 000 years after the bang. From the pattern of temperature fluctuations we can deduce the pattern of mass density fluctuations that served as the primordial seeds for the growth by gravitational instability of galaxies and other large-scale structure in the Universe. (Image courtesy of the NASA/WMAP Science Team.)

There is a wealth of information in the pattern of temperature fluctuations. Associated with the temperature fluctuations are fluctuations in the mass density around the time of recombination. During and after recombination the amplitude of the density inhomogeneities began to grow because of gravitational instability. Pressure from photons, acting counter to the force of gravity due to matter, set up oscillations, whose frequency spectrum can be predicted. The precise form of the oscillation spectrum depends upon almost all cosmological parameters: the expansion rate of the Universe, the spatial curvature of the Universe, the densities of baryonic and nonbaryonic matter, and the amount of dark energy. Thus, while measurements of CMB anisotropies provide detailed constraints about the Universe, at the same time, extracting a unique set of cosmological parameters is unfortunately not possible.

Measurements of the CMB anisotropies yield a value of the baryon density in excellent agreement with Big-Bang-nucleosynthesis results (and completely independent of them). In addition, combined with measurements of the large-scale structure of the Universe, the WMAP results yield a Hubble constant of 71 km $s^{-1}$ $Mpc^{-1}$ (almost identical to the value reported by the Hubble Key Project), and a Universe with the present mass–energy fraction of one-third matter and two-thirds dark energy, consistent with many other independent recent measurements. The results are also consistent with a Universe of flat spatial geometry.

## 1.2.4 The abundances of the light elements

One of the major successes of the Big-Bang theory is the agreement between observations of the present light-element abundances and the predictions of the abundances of elements formed in the Big Bang. The visible matter in the Universe is observed to be composed mainly of hydrogen (about 75%), helium-4 (about 25%), and only trace amounts of deuterium, helium-3 (a lighter isotope of helium), lithium, and all of the other elements in the periodic table up through carbon, oxygen, iron (those ingredients vital to our existence on Earth). A triumph of modern astrophysics is an understanding of the abundances of the chemical elements.

The basic idea is that about three minutes after the bang the primordial neutrons and protons were processed into hydrogen, helium, and lithium. Elements heavier than lithium were formed during stellar evolution. Earlier than about a minute after the bang the temperature of the Universe was too high for the existence of nuclei; later than a few minutes after the bang the temperature and density were too low for nuclear reactions to occur and Big-Bang nucleosynthesis came to an end, leading to a "freeze-out" of the element abundances. Between one and several minutes after the bang the density and temperature were conducive to the formation of these elements via a series of nuclear reactions.

The exact abundances of the elements produced during Big-Bang nucleosynthesis may be reliably calculated (see Figure 1.8). The most important cosmological parameter is the ratio of photons to baryons during nucleosynthesis. This ratio can be expressed in terms of the present baryon density of the Universe. The predicted abundances for these light elements, which vary by nine orders of magnitude, agree extremely well with the observed abundances for a present baryon density of about 3–4 $\times 10^{31}$ g $cm^{-3}$. Recent advances in theoretical modeling and observational astronomy suggest a primordial mass fraction of helium-4 of 24%–25%, a primordial abundance of deuterium relative to hydrogen of a few times $10^{-5}$ by number, and a primordial abundance of lithium-7 relative to hydrogen of a few times $10^{-10}$ by number.

One important consequence of Big-Bang nucleosynthesis is knowledge of the present baryon density. Expressed in terms of the fraction of the critical density,

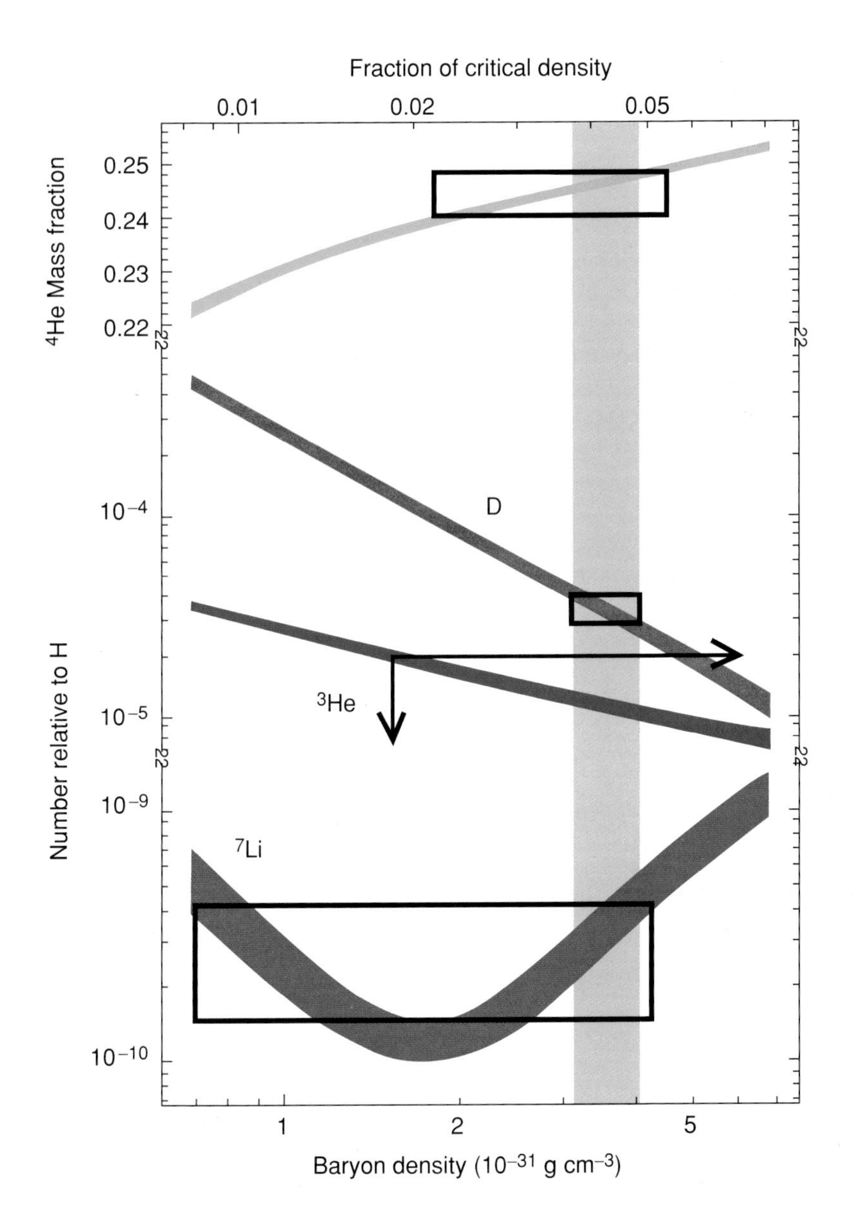


**Figure 1.8.** The abundances of the light elements hydrogen (deuterium), helium (helium-3), and lithium-7 as functions of the present baryon density are indicated by the curves. The widths of the curves reflect residual uncertainties in nuclear reactions and the neutron half life. The inferred primordial values are indicated by the boxes. Values of the present baryon density where theory and observations agree are indicated by the vertical shaded region.

$\Omega_{\mathrm{BARYONS}} \simeq 0.04$, which is much smaller than the observed total mass fraction. This is one of the strongest pieces of evidence that the dark matter is nonbaryonic.

## 1.3 Modern developments in cosmology

The evidence for the expansion of the Universe, the success of the theory of Big-Bang nucleosynthesis, and the existence of the cosmic microwave background radiation provide a solid basis for the standard hot-Big-Bang cosmology, which provides a very successful description of the expanding Universe within the theory of general relativity. We turn now to more recent developments in cosmology: the increasing observational evidence for dark matter and evidence for a new form of dark energy; and the theory of inflation, which appears to address many of the issues not explained by standard hot-Big-Bang cosmology.

### 1.3.1 Dark matter

#### Baryonic dark matter

As described in Section 1.2.4, the very strong constraint on the total amount of baryonic matter from theoretical considerations of Big-Bang nucleosynthesis limits the baryonic-matter contribution to about 5% of the overall density of the Universe. Thus, only about 5% of the overall mass–energy density of the Universe can be accounted for by ordinary matter (composed mainly of neutrons and protons, and collectively referred to as baryons). Of this 5%, only about 1% of this matter is observed in luminous form. There are many possibilities for the remaining "dark" baryonic matter: black holes, other dark and dense remnants of past stars, very faint stars, planets, rocks, and warm or hot ionized gas. Over the past decade, extensive searches for the additional 4% of unseen baryonic material have been carried out. It appears that most of this additional mass is warm gas associated with groups of galaxies, whereas the remainder is located within galaxies, being composed both of stars and of cold (atomic and molecular) gas. The total baryon density, including the dark baryons, is consistent with the total baryon density deduced from observations of the fluctuations in the microwave background radiation.

#### Nonbaryonic dark matter

While 4% of the overall density may be in dark baryons, other evidence overwhelmingly points to the existence of additional (nonbaryonic) dark matter, comprising about 25% of the overall total density. (Until the discovery of dark energy, about 95% of the overall matter density was suspected to reside in nonbaryonic dark matter.) The evidence for nonbaryonic dark matter has mounted over time as several independent and increasingly higher-precision measures have indicated the existence of nonbaryonic dark matter.

One of the ways to infer the existence of dark matter is from observations of distant galaxies located behind massive clusters of galaxies. The images of these galaxies can be drawn out into arc-like features, and their brightnesses amplified, due to gravitational lensing effects (Figure 1.9). Measurements of the velocities of stars within galaxies and small satellite galaxies in the outer haloes of the Milky Way, and motions of galaxies within clusters, also indicate the presence of significant amounts of unseen

**Figure 1.9.** A HST image of a gravitationally lensed cluster, A2218. Prominent arcs are visible due to the distortion and magnification of background objects by the massive cluster. Such gravitationally lensed objects provide a means of studying more distant objects, too faint to be otherwise observed, as well as for determining the mass distribution within the cluster. (Credit: W. Couch (UNSW) and NASA.)

matter. Without the unseen dark matter, these stars and satellites would no longer remain bound. Temperature measurements of the X-ray gas (which makes up most of the ordinary matter within clusters) yield another estimate of the dark-matter mass in clusters. In addition, the amount of gas can also be determined by measuring the scattering of cosmic-microwave-background photons off the hot gas in clusters: the hot gas is not the dark matter, but measures the depth of the gravitational potential resulting from the dark matter.

Nonbaryonic particles can be characterized in terms of whether they are fast or slow – that is, semi-relativistic (hot) or very nonrelativistic (cold). An example of a semi-relativistic dark-matter candidate is the neutrino, which was recently found to have a nonzero mass. However, theoretical considerations limit the contribution of neutrinos to the overall matter density. If there are too many hot, dark-matter particles in the early Universe, then these fast-moving particles will destroy the smaller-density irregularities (the seeds of future structure) by diffusion. In this case, the first objects to form in the Universe would be the largest structures; the superclusters and smaller objects such as galaxies would form later by fragmentation. This would predict that there should be very little structure at the current time. Such a picture is inconsistent with results of extensive supercomputer studies following the growth of galaxies from initial fluctuations. In addition, using space-based and large ground-based telescopes, galaxies have been observed back to extremely high redshifts and large look-back times, before superclusters apparently had had sufficient time to assemble. Nonbaryonic dark matter comprised predominantly of hot dark matter is therefore ruled out. Thus, most of the nonbaryonic matter in the Universe is inferred to be cold dark matter.

The most plausible explanation for dark matter is that it is comprised of some new, undetected type of elementary particle that survived annihilation in the early Universe in sufficient numbers to dominate the present rest-mass density of the Universe. It is hoped that the dark matter will eventually be detected and discovered either in direct searches for the relic dark matter, or by producing the dark-matter particles in accelerator experiments.

Nonbaryonic dark-matter candidates are classified as thermal relics (WIMPs – weakly interacting massive particles) or nonthermal relics (such as axions and WIMPZILLAs).

### *Thermal relics/WIMPs*

A thermal relic, or WIMP, is a hypothetical stable species of elementary particle that was once in thermodynamic equilibrium in the early Universe. So long as the particle remained in equilibrium its relative abundance (its abundance relative to photons) was simply determined by the ratio of the mass of the particle to the temperature of the Universe (it is commonly assumed that there is no chemical potential associated with the new species). Eventually the particle species dropped out of equilibrium when the rate for processes that created or destroyed the particle became much smaller than the expansion rate of the Universe. After the particle species dropped out of equilibrium, its relative abundance decreased only due to the expansion of the Universe.

**Hot thermal relics (neutrinos)** Once believed to be massless, neutrinos have very recently been shown to have a small mass, about one ten-millionth of the mass of the electron. Given their weak interactions, they fall into the category of nonbaryonic dark matter. There are three species of neutrino: the electron, tau, and muon neutrinos. The Japanese experiment Super-Kamiokande and the Canadian Sudbury Neutrino Observatory (SNO) recently found evidence that neutrino species are transformed into other neutrino species as they travel. From these measurements, it is inferred that the neutrino indeed has mass (and there is evidence for additional physics beyond the

Standard Model of particle physics, which links the strong, weak, and electromagnetic interactions – see Chapter 4).

Neutrinos are an example of a thermal relic. They were once in thermodynamic equilibrium with the rest of the Universe. When the temperature of the Universe dropped below about $10^{10}$ K, the rate of creation and annihilation of neutrinos became much less than the expansion rate of the Universe at that time, and the relative abundance of neutrinos "froze out." Since the thermal energy associated with a temperature of $10^{10}$ K ($E = k_B T \simeq 1$ MeV) is much greater than the rest-mass of neutrinos, neutrinos froze out when they were relativistic and had an abundance comparable to that of photons. Thus, the Big-Bang model predicts a neutrino background similar to the photon background. Of course the feebleness of neutrinos' interactions renders the neutrino background seemingly undetectable.

The present contribution to the mass density of a neutrino species with mass larger than the present temperature of the Universe depends on the mass of the neutrino: $\Omega_\nu h^2 \simeq m_\nu/90$ eV. Using the Hubble Key Project value of the Hubble constant and the neutrino mass indicated by neutrino-oscillation experiments (0.1 eV), the neutrino contribution to $\Omega$ is less than 1%. Although the mass of the neutrino is tiny, these particles are very abundant, and their total mass contribution rivals that in luminous stars. However, their contribution is insufficient to explain the total amount of observed dark matter. This should be expected, since neutrinos would be hot dark matter, which cannot be the bulk of the dark matter.

**Cold thermal relics** Neutrinos froze out when they were relativistic. WIMPs that freeze out when they are nonrelativistic are called *cold* thermal relics. The present contribution to $\Omega$ for a cold thermal relic is largely determined by the WIMP's annihilation cross-section. The more strongly the particle species interacts, the longer it will stay in equilibrium. The longer it remains in equilibrium, the lower the freeze-out temperature. If the particle species is in equilibrium at temperatures below the mass of the particle, its relative abundance will be suppressed by the Boltzmann factor, $\exp(-M/T)$, where $M$ is the mass of the species and $T$ is the temperature of the Universe. Thus, a large annihilation cross-section results in a lower freeze-out temperature, which translates into a lower relative abundance. Demanding that $\Omega$ for the cold thermal relic can account for the observed dark-matter density results in an annihilation cross-section somewhat smaller than "weak" annihilation cross-sections. Therefore, any stable particle with annihilation cross-section comparable to weak cross-sections is a candidate for a WIMP.

The leading candidate for a cold thermal relic is the lightest supersymmetric particle. A very promising idea from particle physics is that there is a new symmetry in nature, called supersymmetry (SUSY – see Chapter 4). SUSY would provide an explanation for the observed hierarchy between the weak scale (about 100 GeV) and the Planck scale ($10^{19}$ GeV). SUSY is also a fundamental ingredient to superstring theory (Section 1.5.4). SUSY requires that for every fundamental particle of matter and radiation there is a corresponding supersymmetric partner, as yet not directly observed. SUSY particles are expected to have masses larger than those of the observed superpartners and would be expected to decay very rapidly. The exception is the lightest supersymmetric particle, which would be stable as the result of a conserved parity. Since the lightest SUSY particle is expected to be stable and have a weak annihilation cross-section, it should still exist in copious amounts in the Universe, thus providing an excellent candidate for dark matter.

Since dark matter interacts only weakly, dark-matter particles would be inefficient in dissipation of energy in gravitational collapse and would not form tightly bound objects. Thus, the weakly interacting massive particles would be only very loosely associated with the luminous matter in the Milky Way and other galaxies, and would form a dark-matter halo surrounding galaxies and galaxy clusters.

*Nonthermal relics*

The axion is a particle that would result from the breaking of a symmetry invoked to explain the lack of strong charge–parity (CP) violation. The axion mass is expected to be much lower than those of WIMPs, about $10^{-3}$–$10^{-5}$ eV. Axions would be produced in the early Universe around the epoch of the quark–hadron transition. If the axion scenario is correct, then axions should be abundant in the Universe in such numbers that they would be a serious contender for the dark matter in spite of their small mass.

Another possible source of dark matter is ultra-high-mass particles or "WIMPZILLAs." Such particles may have been generated during the end of inflation (see Section 1.3), perhaps due to the changing gravitational field. With masses in the range $10^{12}$–$10^{16}$ GeV, WIMPZILLAs would have only very weakly interacted with the plasma in the early Universe and never reached thermal equilibrium when they froze out, unlike the lighter WIMPs. To contribute significantly to the total density of the Universe, these nonthermal particles must be stable over long periods, comparable to the age of the Universe or greater. If they are unstable but have a lifetime well in excess of the age of the Universe, the rare decay of WIMPZILLAs may create ultra-high-energy cosmic rays, measurements of which can provide constraints on the WIMPZILLA contribution to dark matter in the Universe.

## 1.3.2 Dark energy

If the Universe is dominated by matter or radiation, if gravity is attractive, and if it is the only force acting on large scales, then the expansion velocity of the Universe is expected to slow over time. It was therefore a surprise when measurements of distant supernovae failed to reveal a deceleration in the expansion of the Universe. Supernovae, the result of thermonuclear explosions of stars, are so bright that they can briefly rival the brightnesses of the galaxies in which they reside, and hence can be observed to great distances. The availability of sensitive, wide-area CCDs opened up the possibility of searches for large samples of faint and distant supernovae. Observations by two independent groups found that supernovae at high redshifts are fainter in comparison with those observed nearby than expected in cosmological models in which the expansion velocity slows over time (see Figure 1.10). It is of course possible that astrophysical effects could conspire to make the supernovae at higher redshifts fainter than expected on the basis of observing their local counterparts. Differing amounts or types of dust could make the distant supernovae appear to be fainter, as could the effects of gravitational lensing; different chemical compositions, evolution of the supernovae, and other effects could also plausibly cause differences in intrinsic luminosities of the supernova population. However, many groups have searched for these potential systematic effects, but have been unable to find any evidence for them. Moreover, recent observations by the WMAP satellite, based on entirely independent physics, yield results entirely consistent with the supernova data.

The implication of this result is that the Universe is now accelerating in its expansion. Within the FRW cosmology, an acceleration of the scale factor is possible if the energy density and pressure satisfy the inequality $\rho + 3p < 0$. Since both matter and radiation have positive pressure, the conditions for acceleration suggest the existence of another component of mass–energy density. The energy density and pressure associated with a cosmological constant are related by $p = -\rho$, so a cosmological term that contributes positive $\rho$ will result in $\rho + 3p < 0$.

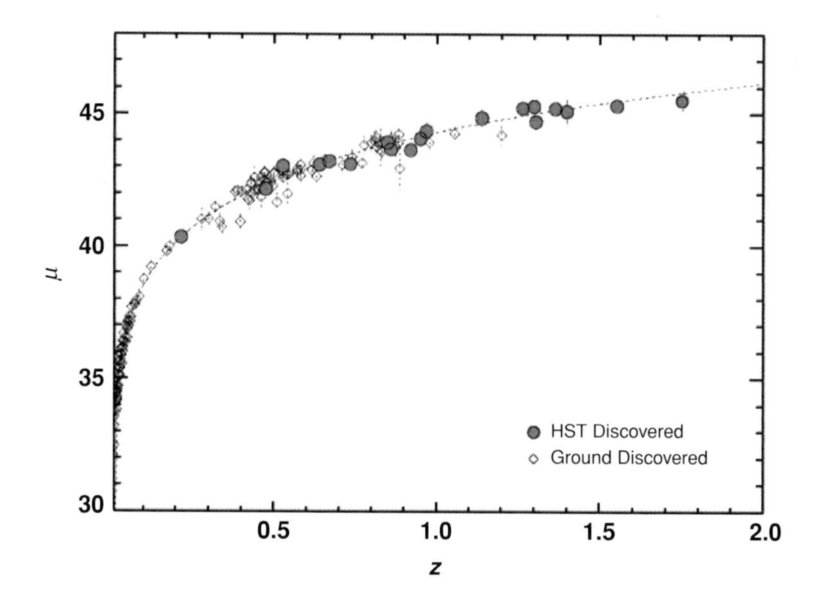


**Figure 1.10.** A Hubble diagram for type-Ia supernovae observed from HST and ground-based telescopes. Plotted is the distance modulus ($\mu = 5 \log d - 5$, where $d$ is the distance in megaparsecs) as a function of redshift. The model fit is for $\Omega_M = 0.3$ and $\Omega_\Lambda = 0.7$, providing evidence for a recent acceleration. The data are also consistent with a transition to deceleration at higher redshifts (courtesy of A. Riess *et al.*, 2004).

The observational results are consistent with a Universe in which one-third of the overall density is in the form of matter (ordinary plus dark), and two-thirds is in some different form, say dark energy with a large, negative pressure. Dark energy does not cluster gravitationally with matter (see Box 1.6). The supernova data provide evidence for a change in the Hubble parameter over time, which is consistent with an accelerating Universe; the WMAP data provide evidence for a missing energy component, in quantitative agreement with the interpretation of the supernova data. Dark energy currently has no explanation: it presents one of the greatest theoretical challenges in cosmology and particle physics.

### 1.3.3 Inflation

The standard FRW cosmology is a homogeneous and isotropic cosmological model. But the standard cosmological model is incapable of explaining why the Universe should be homogeneous and isotropic. The issue is the existence of *causal horizons* in the standard Big-Bang model. Today we observe regions of the Universe that were until recently causally disconnected according to the standard FRW cosmology. Why should the temperature, density, and properties of causally disconnected regions be so remarkably similar? The standard FRW cosmology also describes a Universe in which the local geometry can have either positive curvature (so the geometry is that of a three-sphere), negative curvature (with geometry described by a three-hyperboloid), or zero spatial curvature. Observations are consistent with a spatially flat Universe. The standard FRW cosmology cannot explain why the Universe is so spatially flat. These issues are explained very naturally in a theory called inflation.

Inflationary theory is based upon ideas of modern elementary-particle physics. It predicts that, early in the history of the Universe, the energy density of the Universe was dominated by vacuum energy, which caused a very rapid increase in size of the scale factor $a(t)$. The inflation of the scale factor meant that a small, smooth spatial region of the Universe expanded exponentially to encompass a volume that would grow to become larger today than the size of our observable Universe. In the process of expansion, the spatial geometry became flat.

### BOX 1.6 DARK ENERGY

Dark energy has properties very different from those of dark matter. The influence of dark matter on the expansion rate of the Universe is governed by familiar (attractive) gravity alone. The influence of dark energy on the expansion rate of the Universe depends on the dark-energy density *and* the dark-energy pressure.

In general relativity there is a second critical equation describing the expansion of the Universe:

$$\ddot{a}/a = -4\pi G \sum_i (\rho_i + 3p_i)\,,$$

where the sum is over the various contributions to the mass–energy density of the Universe. According to this equation, both energy and pressure (not just mass as in the Newtonian case) govern the dynamics of the Universe. It also allows the possibility of negative pressure, resulting in an effective repulsive gravity.

Any component of the mass-energy density can be parameterized by its ratio of pressure to energy density, $w = p/\rho$. For ordinary matter $w = 0$, for radiation $w = \frac{1}{3}$, and for the cosmological constant $w = -1$. The effect on $\ddot{a}/a$ of an individual component is $-4\pi G\rho_i(1 + 3w_i)$. If $w_i < -\frac{1}{3}$ that component will drive an acceleration (positive $\ddot{a}$) of the Universe.

In the lambda-CDM model the expansion rate of the Universe is given by

$$H^2(z)/H_0^2 = [\Omega_M(1+z)^3 + \Omega_\Lambda(1+z)^{3(1+w)}],$$

where here $w$ is associated with the dark energy. Under the assumption of a flat Universe, the current observations of distant supernovae and measurements by the WMAP satellite are consistent with a cosmological model in which $\Omega_M \simeq 0.3$, $\Omega_\Lambda \simeq 0.7$, and $w = -1$. The observations are inconsistent with cosmological models without dark energy.

There are at least two major challenges to theoretical understanding of the dark energy. The first is the sheer magnitude of the dark-energy component. Calculating the expected energy of the quantum vacuum leads to a discrepancy with the observed value of no less than 55 orders of magnitude. The second glaring question is that of why we coincidentally appear to be living during the epoch when the dynamics of the expansion are only now becoming dominated by the dark energy. Addressing these issues presents immense challenges, but their solution promises to lead to a much deeper understanding both of cosmology and of particle physics.

The source of the vacuum energy that drove inflation is unknown: it is usually modeled as the potential-energy density associated with a new scalar field called the inflaton. After inflation the energy density must be extracted from the inflaton to initiate the radiation-dominated era. The process of extraction of the vacuum energy and the conversion into radiation is another unknown associated with inflation. Connecting the inflaton field to some dynamical degree of freedom of very-high-energy physics is one of the goals of early-Universe cosmology.

Although inflation is believed to occur very early in the history of the Universe (much less than one second after the bang), we can study the dynamics of inflation because events during inflation left a detectable imprint on the present Universe in the form of the observed primordial temperature fluctuations, the observed primordial density fluctuations, and a yet-to-be-observed background of primordial gravitational waves. The primordial temperature and density fluctuations result from quantum fluctuations in the inflaton field during inflation. In simple models for inflation, the amplitude and spectrum of the primordial fluctuations depend on the nature of the inflaton potential. The amplitude and spectrum of the gravitational-wave background result from quantum fluctuations in the gravitational field itself during inflation. The spectrum and amplitude of the gravitational-wave background also depend on the nature of the inflaton potential.

These primordial density fluctuations became the seeds that grew as a result of gravitational instability, and gave rise to the galaxies, clusters, and large-scale structure observed today. Thus, inflation, in combination with gravitational instability, naturally provides a way to explain both the smoothness on large scales and the detailed structure on smaller scales.

Inflation predicts a spatially flat Universe, a nearly power-law distribution of primordial density and temperature fluctuations, and a background of gravitational waves. It is not yet possible to measure the gravitational-wave radiation from the early Universe, but measurements of the cosmic microwave background to date are consistent both with flatness and with a nearly scale-invariant density- and temperature-fluctuation spectrum (Section 1.2.3). Upcoming space-based gravitational-wave missions, as well as polarization measurements of the cosmic microwave background radiation, may reveal the gravitational-wave background from the early Universe and provide a stronger test of inflation.

## 1.4 The Universe after reionization

### 1.4.1 The reionization history of the Universe

At a redshift $z \sim 1100$, the Universe cooled to a temperature low enough to allow electrons and protons to combine. At this point, most of the matter in the Universe rapidly transitioned from being fully ionized to being completely neutral. As structure developed due to gravitational instability, the first massive stars formed, along with their parent galaxies harboring black holes in their nuclei, and the radiation emitted from them is thought to have reionized the Universe.

Observations of bright quasars (Section 1.4.3) offer a means of probing the evolution of the intergalactic medium and seeing directly when the epoch of reionization occurred. The spectra of quasars, now extending to redshifts greater than 6, reveal Lyman-$\alpha$ absorption lines due to the presence of clouds of neutral hydrogen along the line of sight. The Lyman-$\alpha$ transition is due to resonant scattering, and is sufficiently strong that even a relatively low column density of neutral hydrogen will cause significant absorption. At high redshifts, the resulting optical depth to Lyman-$\alpha$ is expected to be so high that no flux shortward of Lyman-$\alpha$ will be present, resulting in what is referred to as the "Gunn–Peterson trough." After reionization the light from quasars at lower redshifts and later times is free to travel unabsorbed. If reionization occurs rapidly, then at some characteristic redshift the clearing of the "trough" would mark the epoch of reionization. However, surveys of quasars out to redshifts $z \sim 6$ indicate that the fraction of neutral hydrogen increases systematically, but not abruptly, from a redshift beyond $z \sim 3$, out to the current limit of quasar surveys at $z \sim 6.4$. Independent results from the WMAP satellite suggest that reionization may have begun even earlier, at a redshift of $17 \pm 5$. These combined results indicate that reionization occurred over an extended period of time. Models suggest that massive-star formation in the early Universe occurred at $z \sim 10$–$30$. Such stars are expected to be of very low heavy-element abundance and very massive, and they may be the source of energy required for the observed reionization.

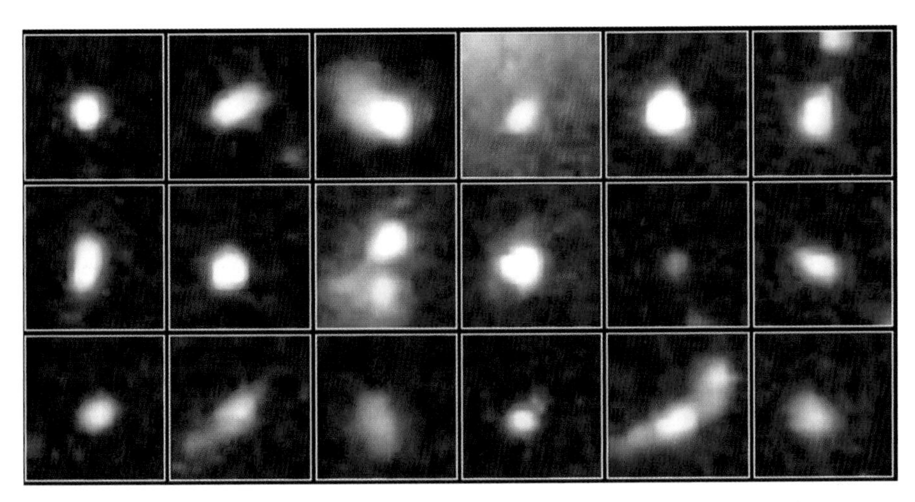

**Figure 1.11.** HST Wide Field and Planetary Camera 2 images of Lyman-$\alpha$-emission objects at a redshift $z \sim 2.4$. These objects may be merging as part of the process of galaxy assembly. In CDM models, galaxy formation proceeds by the hierarchical merging of small objects, which are expected to be more numerous at higher redshifts. (Credit: R. Windhorst, Arizona State University and NASA.)

## 1.4.2 Galaxy formation and evolution

Very little is known empirically about the so-called "dark ages," the epoch between a redshift of about 1100 (the surface of last scattering of the CMB), and a redshift of $z \sim 6.4$, the limit out to which the most distant quasars have been discovered. A tremendous amount of astrophysics remains to be learned: the details of the first stars, clusters, and active galaxy nuclei to be formed; establishing which were the first luminous objects to form (stars, active galactic nuclei, black holes) and how they affected their environment; the formation of galaxies; the initial distribution of stellar masses (the initial mass function); the star-formation history; chemical evolution, cooling, and heating processes; the buildup of the Hubble sequence of galaxies; and the ionization history of the Universe.

Deep images of the sky obtained by the Hubble Space Telescope (HST), in combination with spectra obtained with ground-based telescopes, have identified an epoch during which galaxies underwent significant bursts of star formation a few billion years after the Big Bang, at redshifts of 1–3 (e.g., Figure 1.11). Many of the galaxies identified at these redshifts are high emitters of UV radiation, and actively star-forming, but they might not be representative of the general population of galaxies at these redshifts, which might not be visible at UV wavelengths. There are current searches for older and redder galaxies at high redshifts, the signature of even earlier star formation in the Universe. Actively star-forming galaxies beyond a redshift of 5 are now being identified on the basis of the detection of the redshifted Lyman-$\alpha$ line at restframe 1216 Å, and amplification by gravitational lensing is allowing a glimpse of galaxies possibly beyond a redshift of 7.

Observing directly the first light in the Universe, the formation of the first stars, and the early formation of galaxies must await higher resolution and sensitivity. Plans are under way for large-aperture (20–100-m-class) telescopes on the ground, as well as a 6-m-class telescope in space, a collaboration of the National Aeronautics and Space Administration (NASA) and European Space Agency (ESA): the James Webb Space Telescope (JWST) (see Section 1.5.5).

## 1.4.3 Active galactic nuclei

The radiation output from the nuclei of some galaxies can exceed those of normal galaxies by factors of hundreds (see Chapter 3). Quasars, the most energetic active galactic nuclei (AGN), have now been found in the Sloan survey out to redshifts over 6, with the peak of quasar activity having occurred at a redshift of about 2.5. Strong evidence has accumulated over the past four decades that suggests that the engine that powers the various kinds of observed active nuclei is black holes with masses ranging from $10^6$ to $10^9$ solar masses. Most likely these black holes grew by mergers and the accretion of gas onto high-density regions of cold dark matter.

With HST observations it has become apparent that these supermassive black holes are found in the nuclei of all galaxies so far studied, and that there is tight correlation between the mass of the black hole and both the velocity dispersion and the luminosity of the spheroid of the host galaxy. These observations have important implications for understanding the dynamics of spheroid assembly and its relation to black-hole formation. The fundamental basis for the correlation between mass and velocity dispersion, and the details of the formation of such supermassive objects at high redshift, remain to be understood.

### 1.4.4 Gamma-ray bursts

These extremely luminous objects, some of which have now been established to be at cosmological distances, emit most of their energy ($10^{51}$–$10^{53}$ ergs) at gamma-ray frequencies, and over only a very short duration of a few seconds (see Chapter 3). Observed gamma-ray bursts (GRBs) range in duration from a few microseconds to a few hundred seconds, but the physical nature of these sources appears to fall into two classes, with a dividing line at about two seconds. There is currently no information on the distances and luminosities of the short-duration bursts. Measurements of redshifts of GRBs with durations greater than two seconds have recently established that they reside in normal, star-forming galaxies at high redshifts. The field has advanced rapidly with the capability of satellites such as BeppoSAX and HETE to measure the positions of these sources to high accuracy and on a rapid timescale, allowing followup at other wavelengths. The afterglows of GRBs span a very broad range from X-ray, through optical, to radio frequencies. In some cases, optical spectra have not only allowed redshifts to be obtained, but also revealed a connection between GRBs and supernovae. Because GRBs are so luminous, exceeding the luminosity of supernovae or active galaxies, they potentially offer a unique means of studying massive-star formation in the early, obscured Universe if they can be discovered in the redshift range of 6 to 20. NASA has two upcoming missions to study GRBs: SWIFT and GLAST.

## 1.5 The future of cosmology

### 1.5.1 Detecting dark matter

For cold thermal relics there are considerable complications in calculating the exact present abundance and scattering cross-sections of the relic dark matter. Detailed model calculations must take into account questions such as the spin dependence of the WIMP–nucleus interactions, the angular-momentum dependence of the annihilation cross-section, and whether the WIMP can "co-annihilate" with any other particle species. Nevertheless, in all models a cold, thermal, relic, dark-matter particle should have feeble interactions with normal matter. Thus, a cold thermal relic might be detected in sufficiently sensitive experiments.

The local density of dark matter in our neighborhood of the Milky Way is estimated to be about 0.3 GeV $c^{-2}$ cm$^{-3}$, so the number density of cold thermal dark-matter

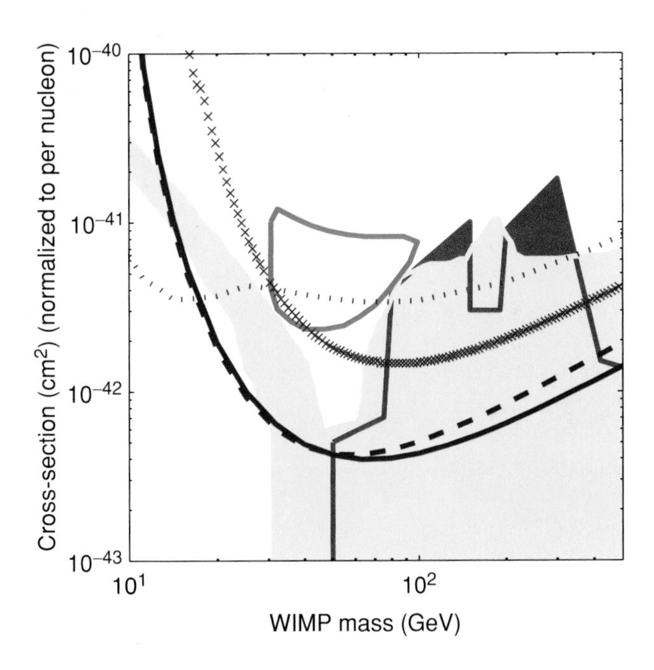


**Figure 1.12.** New constraints from the Cryogenic Dark Matter Search (CDMS) in the Soudan mine. Plotted is nucleon cross-section versus WIMP mass. The parameter space above the solid curve is excluded at the 90% confidence level. These results are inconsistent with the green closed ($3\sigma$) contour from the earlier published results from the DAMA experiment, using radiopure NaI. The CDMS experiment utilizes independent germanium and silicon detectors: two 250-g Ge and two 100-g Si devices, cooled to temperatures below 50 mK.

relics of mass $M$ should be $n = 0.3(\text{GeV } c^{-2}/M)$ cm$^{-3}$. Using the local Galactic rotation velocity of 220 km s$^{-1}$ as the velocity of the WIMPs, the local flux of WIMPs would be approximately 66 000 (GeV $c^{-2}/M$) cm$^{-2}$ s$^{-1}$. WIMP masses in the range 10 GeV to 10 TeV should produce nuclear recoils of energy 1–100 keV. The experimental challenge is detecting the very small recoil energies from a rare WIMP–nucleus collision. Separating the event from the background contamination of cosmic rays and natural radioactivity is an immense problem.

Several cryogenic experiments seeking to measure the recoil energy expected from the collisions between WIMPs and the nuclei of atoms in large (~100–400 g) germanium and silicon detectors are under way in the USA and in Europe. The results are typically given in terms of an excluded region on the mass-scattering cross-section plane (see Figure 1.12). The WIMP mass is expected to be in the range of 10 GeV to 10 TeV. The expected event rates for these interactions are extremely low, between 0.001 and 1 events per kilogram per day. A new generation of such experiments is being carried out in deep underground mines to reduce the background events, but to date no WIMPs have been detected.

Another avenue for WIMP detection is the search for the products of present-day WIMP annihilations. Although universal WIMP annihilations ceased when the WIMPs froze out in the early Universe, today WIMPs may have an anomalously large concentration at the center of the Earth, at the center of the Sun, or in our Galactic center. The annihilation of WIMPs at those locations may produce a characteristic signal such as high-energy neutrinos from the center of the Earth, the center of the Sun, and the Galactic center, or a flux of antiprotons, positrons, or gamma rays from the Galactic center. To date, no unambiguous signal has been seen. However, new experiments will begin in the near future with the prospect of indirect detection of dark matter.

Finally, if WIMPs were present in the early Universe, they should also be produced in the collisions at high-energy particle accelerators. Production and detection of WIMPs in high-energy physics experiments is challenging because of the expected large mass and feeble interactions of the WIMPs.

If the dark matter is a cold thermal relic, then eventually direct detection experiments will become sensitive enough to detect WIMPs if they are there, eventually indirect detection observations will have enough discrimination to separate any possible WIMP signature from background contamination, and eventually they will be produced and detected in high-energy colliders. The WIMP hypothesis will be either confirmed or disproved experimentally.

Axions can be converted into photons in the presence of a strong magnetic field. The axion couples to two photons, so the strength of the $a\gamma\gamma$ coupling is the important parameter of axion models relevant for axion searches. Axion searches are conducted by placing microwave cavities within strong magnetic fields, and tuning the cavity to various frequencies, and searching for the conversion of background axions into photons. Axion-search experiments are only just starting to probe the expected range of the $a\gamma\gamma$ coupling. Again, axion-search experiments will eventually confirm or refute the hypothesis that the dark matter is axions.

### 1.5.2 Connecting dark energy with fundamental physics

Today the vacuum energy is approximately $\rho_\Lambda \simeq 10^{-29}$ g cm$^{-3}$. This may be expressed in terms of an energy scale as $\rho_\Lambda \simeq (3 \times 10^{-3}\ \mathrm{eV})^4$ or in terms of a length scale as $\rho_\Lambda \simeq (8 \times 10^{-3}\ \mathrm{cm})^{-4}$. There are many possible contributions to the energy density of the vacuum; the problem is to understand why the vacuum energy today is so small.

The vacuum energy should result from many contributions.

1. There may be a "bare" cosmological constant, i.e., a fundamental term in the gravitational action as originally introduced by Einstein. There is no way to estimate the size of this contribution.
2. The vacuum of quantum field theory should include the zero-point energies associated with vacuum fluctuations. Quantum fields may be thought of as an infinite number of harmonic oscillators in momentum space. The zero-point energy of such an infinite collection will be infinite. It is usually argued that there should be some physical cutoff of the high-momentum modes of the field, and the vacuum energy will be finite, of order $k_{\mathrm{MAX}}^4$, where $k_{\mathrm{MAX}}$ is the imposed cutoff at a maximum wavevector. Proposed cutoffs are the Planck scale, $10^{28}$ eV, and the supersymmetry-breaking scale, believed to be about $10^{12}$ eV. The problem is the incredibly small value of $k_{\mathrm{MAX}}$ required from observations: $3 \times 10^{-3}$ eV.
3. In theories with fundamental scalar fields, the minimum of the scalar field potential may have a vacuum energy density associated with it. For instance, in the Glashow–Weinberg–Salam theory the broken and unbroken symmetry phases differ by a potential-energy difference of approximately $10^{11}$ eV. Symmetry breaking associated with a grand unified theory would be associated with a potential-energy difference of as much as $10^{25}$ eV.
4. Nonzero vacuum expectation values of fermion bilinear terms $\bar{\psi}\psi$, such as those associated with dynamical chiral-symmetry breaking in the strong interactions, would contribute to the vacuum energy. For the strong interactions the difference between the symmetric and broken phase would be about $3 \times 10^8$ eV.

The cosmological constant receives contributions from all these seemingly unrelated sources, some with a positive contribution, some negative. In the absence of an unknown symmetry causing the vacuum energy to vanish, there is no known reason for the cosmological constant to be so small.

Since there are so many sources of vacuum energy, presumably the solution to the cosmological-constant problem must await a unified theory of everything.

## 1.5.3 Connecting inflation with fundamental physics

Primordial inflation is an epoch of the very early Universe when the expansion rate of the Universe, $H$, was nearly constant. If $H$ is constant, the scale factor increases exponentially with time; a near-constant $H$ results in a near-exponential increase in the scale factor. Such a behavior requires that the energy density must have been nearly constant during inflation. Primordial inflation must eventually end, and the energy density driving inflation must be converted into radiation to commence the radiation-dominated era. The usual picture is of a slowly decreasing expansion rate until the end of inflation, then a conversion of the residual energy density driving inflation into radiation.

The nature of the energy density of the Universe during inflation is unknown. The energy during the inflationary epoch is usually associated with the potential-energy density of a scalar field, the inflaton. The idea is that during inflation the inflaton field is homogeneous, starting from some value away from the value at the minimum of the inflaton potential. As the scalar field evolves toward the minimum of its potential (assumed to have zero potential-energy density) the expansion rate of the Universe slowly decreases. During the evolution of the inflaton, the potential energy decreases and the kinetic energy decreases. Eventually inflation will end when the potential-energy density no longer dominates.

The nature of the inflaton potential and the mechanism for converting the inflaton energy into radiation are completely unknown. The energy and mass scales associated with inflation are believed to be well above the energy scales of the standard model of particle physics. Presumably the inflaton, if it exists, must be part of some fundamental theory like grand unified theories, supersymmetry, or string theory. Connecting the inflaton with fundamental physics is one of the outstanding challenges of early-Universe cosmology.

Fortunately, the signature of inflation is encoded in the primordial density and temperature fluctuations, as well as a background of gravitational waves produced during inflation. The amplitude and spectrum of the density/temperature fluctuations and the gravitational-wave spectrum depend on the exact nature of the inflaton potential (or equivalently the exact nature of the evolution of the expansion rate during inflation).

A challenge for observational cosmology is the precise determination of the nature of the primordial spectra. The primordial density/temperature fluctuations have been revealed by precision CBR measurements and large-scale structure surveys (Sections 1.2.2 and 1.2.3). The next decade will see even more precise measurements of the density/temperature fluctuations. The crucial missing piece is the determination of background gravitational waves. In addition to the possibility of direct detection of the cosmological background gravitational waves, a gravitational-wave background leaves a distinctive imprint on the polarization pattern of the CBR (Section 1.3.3). The detection of a cosmological wave background, either directly or indirectly, would provide important information about the nature of inflation.

## 1.5.4 String cosmology

While inflation solves a number of problems, there are many remaining questions, perhaps the most fundamental of which is that of what powered inflation in the first place. The merging of quantum mechanics and general relativity has remained unsolved, suggesting that there is additional physics beyond the theory of general relativity. Superstring theory (or M-theory) appears a promising route for addressing these problems (see Chapter 5). In string or M-theory, elementary particles result from

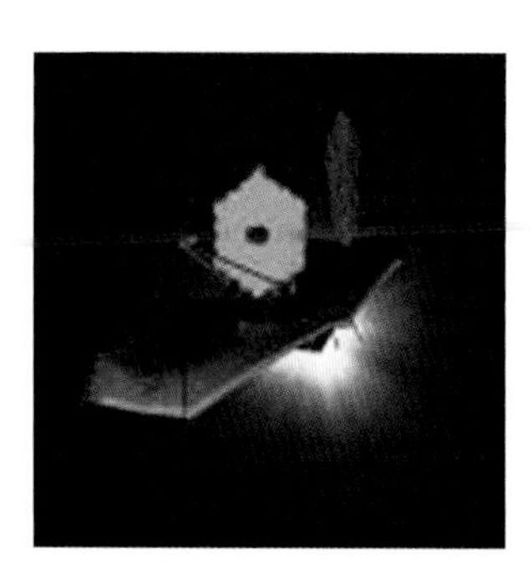

**Figure 1.13.** The James Webb Space Telescope, a 6-m-class near-infrared optimized telescope to be launched by NASA in 2011.

the vibrational modes of tiny strings, approximately a Planck length in scale, $10^{-33}$ cm. The theory predicts that there should be additional spatial dimensions beyond those in general relativity. These extra dimensions are hidden, either because they are small, or perhaps because current experiments have no means of detecting their existence. An idea in string theory with implications for cosmology is that the electromagnetic, weak, and strong interactions are confined to a three-dimensional "brane" in a higher-dimensional space known as the "bulk," but gravity can propagate through the bulk. These ideas raise the possibility that there may be new effects that could leave a cosmological signature, or that laboratory experiments may reveal an entirely new behavior of gravity on small scales. An intriguing new idea to emerge from brane cosmology is that the Big Bang might have been the collision of two such sheets and that such collisions might happen in cycles.

## 1.5.5 New observational tools

In addition to the clear future challenges for theory, an array of accelerator and laboratory experiments, and new space- and ground-based facilities are planned in the coming decades to address many of the cosmological questions we have discussed. A sample of these new initiatives includes, but is not limited to the following.

The James Webb Space Telescope (JWST) A NASA 6-m-class telescope, optimized for infrared wavelengths (1–28 μm), with primary-science drivers to detect light from the first stars in the Universe and illuminate the "dark ages," and to observe directly the reionization of the Universe, the evolution of galaxies, and star formation. It is scheduled for launch in 2011 (see Figure 1.13).

The Giant Segmented Mirror Telescope (GSMT) A ground-based complement to the JWST. The GSMT is a 20–30-m-class optical/near-infrared telescope designed for faint-object spectroscopy using adaptive optics. It will be used to study the star formation in the early Universe, the evolution of black holes, dark matter, and dark energy (see Figure 1.14).

The Large Synoptic Survey Telescope (LSST) A ground-based survey telescope with an effective collecting area of about 6.5 m$^2$, very rapid sky-coverage capability, and a field of view of 3 degrees on the sky. One of its primary-science goals is the study of dark matter through weak gravitational lensing, and it will also produce large numbers of type-Ia supernova candidates (that can be followed spectroscopically by other facilities) useful for studying dark energy.

The Joint Dark Energy Mission (JDEM) A 2-m-class space telescope dedicated to the study of dark energy. One design, the Supernova Acceleration Probe, aims to discover about 2000 type-Ia supernovae, over the redshift range 0 to 2, with very high photometric precision. It would also enable precision weak-lensing studies.

Planck A European Space Agency (ESA) mission for high-angular-resolution, high-sensitivity all-sky measurements of CMB anisotropies to a precision of $\Delta T/T \simeq 10^{-6}$, with the aim of measuring cosmological parameters (e.g., spatial curvature, the Hubble constant) to precisions of a few percent. It is planned for launch in 2007.

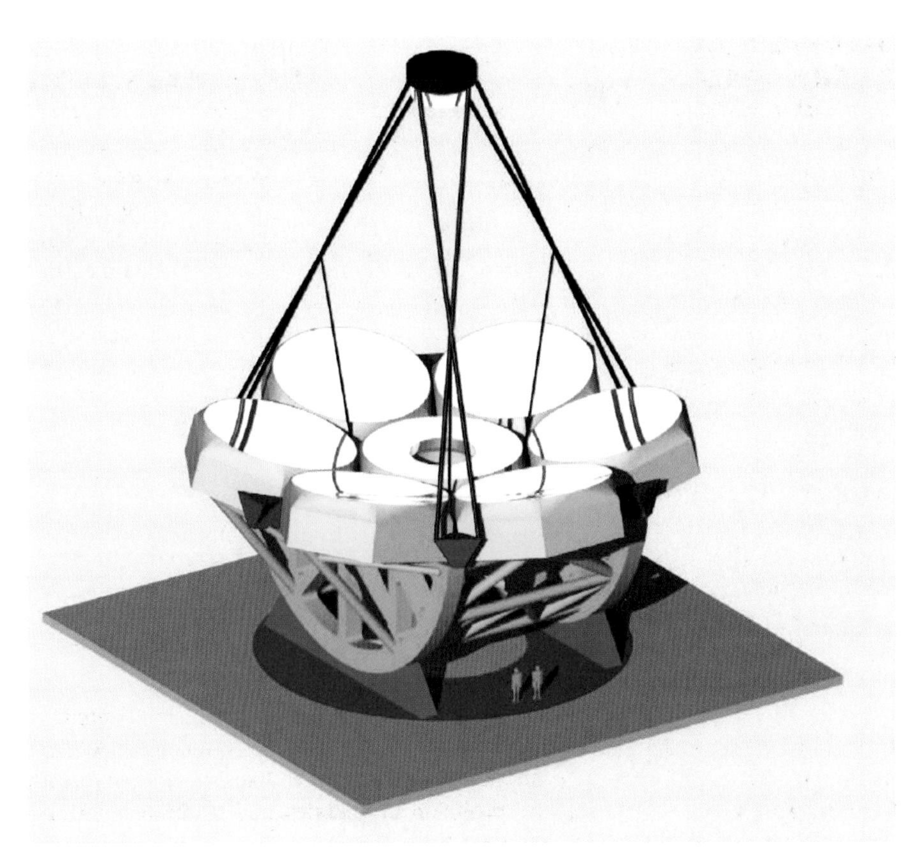

**Figure 1.14.** The Giant Magellan Telescope, a 21.5-m ground-based telescope currently under design for completion in the next decade. Seven 8.4-m mirrors reside on a common mount. Adaptive optics will correct for the turbulence in the Earth's atmosphere.

The Large Hadron Collider (LHC) A collider experiment currently under construction at CERN, occupying the 27-mile tunnel originally built for the Large Electron–Positron Collider (LEP), and aimed at achieving the TeV energies needed for detecting supersymmetric particles and the Higgs boson. It is planned for operations in 2007.

## 1.6 Summary

It is not simply that the Universe appears to be accelerating, but progress in cosmology during the latter half of the twentieth century has accelerated rapidly also. Most satisfying is the emergence of a physical model, motivated by particle physics, which connects the earliest moments of the Universe to the observed Universe today. As we have also seen, while there is understanding of a broad framework, there are still fundamental pieces missing from this emerging new cosmological framework. It is very likely that the 95% of the Universe that is in a dark form will have some surprises yet to reveal. In this sense alone, the discovery potential for cosmology remains bright. Further understanding of the nature of matter, space, and time holds the promise of addressing some of the most fundamental questions in cosmology, indeed some of the deepest questions that humanity has posed.

## FURTHER READING


1. S. Dodelson, *Modern Cosmology*, New York Academic Press, 2003.
2. W. L. Freedman and M. S. Turner, *Measuring and Understanding the Universe*, Reviews of Modern Physics Colloquia, **75** (4), 2003, p. 1433.
3. B. Greene, *The Elegant Universe*, New York, W. W. Norton and Co., 1999.
4. E. W. Kolb and M. S. Turner, *The Early Universe*, New York, Addison-Wesley, 1990.
5. A. R. Liddle and D. H. Lyth, *Cosmological Inflation and Large-Scale Structure*, Cambridge, Cambridge University Press, 2000.
6. M. Rees, *Before the Beginning: Our Universe and Others*, New York, Addison-Wesley, 1997.


## Wendy L. Freedman

Astronomer Wendy Freedman is the Crawford H. Greenewalt Director of the Carnegie Observatories in Pasadena, California. She is a native of Toronto, Canada, and she received her doctorate in astronomy and astrophysics from the University of Toronto in 1984. She was awarded a Carnegie Fellowship at the Observatories in 1984, joined the permanent faculty in 1987, and was appointed director in 2003. She was awarded the Marc Aaronson Lectureship and prize in 1994, and was selected as an American Physical Society Centennial Lecturer in 1999. In 2000 she received the McGovern Award for her work on cosmology, and was elected a Fellow of the American Academy of Arts and Science. She was awarded the American Philosophical Society's Magellanic Prize in 2002 and was elected to the US National Academy of Sciences in 2003.

Her principal research interests are in observational cosmology. Dr. Freedman was a principal investigator for a team of thirty astronomers who carried out the Hubble Key Project to measure the current expansion rate of the Universe. Her current research interests are directed at measuring the past expansion rate of the Universe and characterizing the nature of the dark energy which is causing the Universe to speed up its expansion.

## Edward W. Kolb

Known to most as "Rocky," Edward W. Kolb is a founding head of the NASA Fermilab Astrophysics Group at the Fermi National Accelerator Laboratory and a Professor of Astronomy and Astrophysics at the University of Chicago. He is the Director of the Particle Astrophysics Center at Fermilab.

He received a Ph.D. in physics from the University of Texas, and did postdoctoral research at the California Institute of Technology and Los Alamos National Laboratory where he was the J. Robert Oppenheimer Research Fellow. He has served on editorial boards of several international scientific journals as well as *Astronomy* magazine.

He is a Fellow of the American Academy of Arts and Sciences and a Fellow of the American Physical Society. He was the recipient of the 2003 Oersted Medal of the American Association of Physics Teachers and the 1993 Quantrell Prize for teaching excellence at the University of Chicago. His book for the general public, *Blind Watchers of the Sky*, received the 1996 Emme Award of the American Aeronautical Society. His research field is the application of elementary-particle physics to the very early Universe. In addition to having published over 200 scientific papers, he is a co-author of *The Early Universe*, the standard textbook on particle physics and cosmology.

# Gravity

Ronald Adler

## 2.1 The attraction of gravity

Gravity attracts. It attracts every body in the Universe to every other, and it has attracted the interest of physicists for centuries. It was the first fundamental force to be understood mathematically in Isaac Newton's action-at-a-distance theory, it is a center of current attention in Albert Einstein's general relativity theory, and it promises to be the last force to be fully understood and integrated with the rest of physics.

After centuries of success, Newton's theory was finally replaced by Einstein's theory, which describes gravity at a deeper level as due to curvature of spacetime. General relativity is widely considered to be the most elegant physical theory and one of the most profound. It has allowed the study and understanding of gravitational phenomena ranging from laboratory scale to the cosmological scale – the entire Universe; but many mysteries remain, especially in cosmology.

An impressive understanding of the other fundamental forces of nature, electromagnetism and the weak and strong nuclear forces, is now embodied in the Standard Model of elementary particles. However, these other forces are understood in terms of quantum field theory and are not geometric in the manner of general relativity, so gravity remains apart. Many physicists believe that the final phase of understanding gravity will be to include quantum effects and form a union of general relativity and the Standard Model. We would then understand all the forces and spacetime on a fundamental quantum level. This is proving to be quite a task.

## 2.2 Some extraordinary properties of gravity

Gravity is astoundingly weak. This might not be obvious when we struggle to climb a mountain, but it becomes more apparent if we remember that the entire mass of the Earth, about $6 \times 10^{24}$ kg, is pulling on our bodies, and yet we are able to overcome gravity and climb thousands of meters to the summit. In more objective terms the gravitational force between an electron and proton in a hydrogen atom is about $2 \times 10^{39}$ times weaker than the electrical force. The only reason that gravity is relevant at all is that it is always attractive, most bodies have little unbalanced electrical charge, and the much stronger nuclear forces are negligible over distances larger than nuclear size. As a result it is gravity that holds us on the Earth, and it is gravity that dominates the Universe on the cosmological scale.

*The New Physics for the Twenty-First Century*, ed. Gordon Fraser.

In both Newtonian and general relativity theory the gravitational constant $G$ is an arbitrary parameter, so these theories cannot explain the weakness of gravity, and we must look deeper. At the extreme low end of the distance scale it appears, somewhat paradoxically, that gravity again dominates. This can be understood from the quantum uncertainty principle, whereby small distances are associated with large momenta and energy; since energy is a source of gravity, it is gravity that dominates at sufficiently small distances. The scale at which this happens is about $10^{-35}$ m, about $10^{20}$ times smaller than a nucleus. This is the Planck scale, a center of current attention for theorists.

An extraordinary dynamical property of gravity is the universality of free fall, which was first understood by Galileo Galilei: diverse objects fall at the same rate in a gravitational field. This is now usually referred to as the equivalence principle, or more precisely the weak equivalence principle. According to Newtonian theory the fall is universal because the gravitational force on a body is proportional to its mass, while its resistance to acceleration is inversely proportional to its mass, so the mass cancels out of the equation of motion; it is an ad-hoc explanation. According to general relativity a body follows a generalized straight line or geodesic in spacetime, independently of its internal structure, so the universality of free fall is simply and beautifully explained.

## 2.3 General relativity and curved spacetime

Einstein knew from the beginning that Newtonian gravity was not consistent with his special relativity, according to which no physical effect can exceed the speed of light, $c$. Newtonian gravity is an action-at-a-distance force, implying infinite propagation speed. In developing his theory Einstein took a geometric viewpoint and built on an analogy: spacetime with gravity is to spacetime without gravity as a curved surface (such as a sphere) is to a flat surface (such as a sheet of paper).

In the early nineteenth century Carl Friedrich Gauss used the concept of a metric to study two-dimensional surfaces, and Georg Friedrich Bernard Riemann later generalized the ideas of Gauss to more abstract spaces and higher dimensions. To locate a point in an $n$-dimensional space requires $n$ coordinates. A coordinate system is only an arbitrary set of $n$ labels for the points of the space. A coordinate need not even have the dimension of a distance; angles make good coordinates. The metric is the mathematical object that relates coordinate differences to physical distances. The simplest example is a one-dimensional space or line, where a point is labeled with a single coordinate $u$ that increases along the line. The distance $ds$ between two nearby points on the line is not in general equal to the coordinate difference $du$: the metric function $g$ relates these via $ds^2 = g\, du^2$. For a two-dimensional space with coordinates $u$ and $v$ the relation is nearly as simple. The distance is determined by four functions $g_{ij}$, called the metric array or metric tensor or simply metric, via

$$ds^2 = g_{11}\, du^2 + g_{22}\, dv^2 + g_{12}\, du\, dv + g_{21}\, dv\, du \qquad \text{(line element in two-dimensions)} \qquad (2.1)$$

Since $du\, dv = dv\, du$ we always take $g_{12} = g_{21}$: the metric is symmetric. In the special case $g_{11} = g_{22} = 1$ and $g_{12} = g_{21} = 0$ this is Pythagoras' theorem. The physical distance, together with its mathematical form (2.1), is also called the line element. Generalization to a general Riemann space such as four-dimensional spacetime is in principle straightforward. To describe gravity we need a four-by-four metric for spacetime.

For two-dimensional surfaces the idea of curvature is simple and intuitive; a sphere and a plane surface are clearly intrinsically different. However, for a general

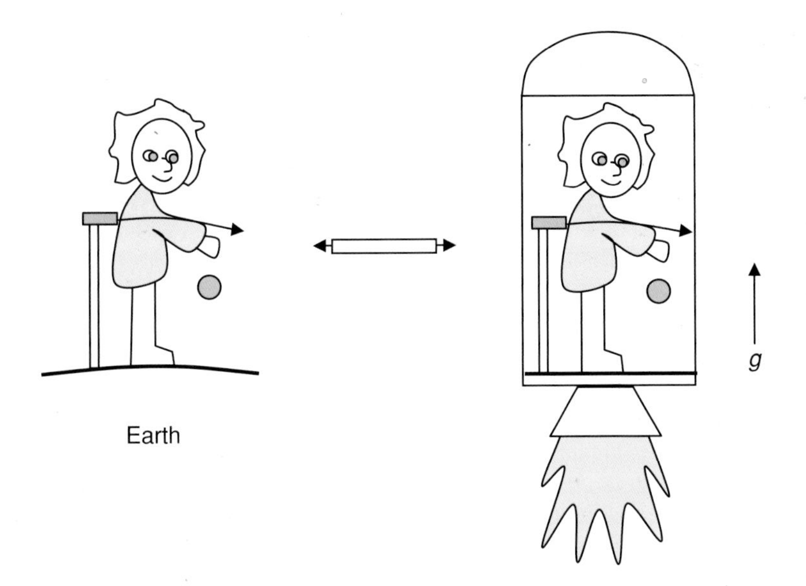


**Figure 2.1.** Einstein realizes that gravity and acceleration are equivalent in a profound way. His "happiest thought" is that a falling person would not feel the effects of gravity.

$n$-dimensional Riemann space curvature is described by an array of $n^4$ functions called the Riemann curvature tensor. Tensors, arrays of numbers or functions, are fundamental mathematical objects in Riemannian geometry. In general relativity the curvature tensor signals the presence of a gravitational field; physically it is the analog of tidal forces in Newtonian theory.

Einstein's great idea, that gravity is curvature of spacetime, is based on the equivalence principle. He used elevators in his thought experiments, but we will use a lab in a rocket ship. See Figure 2.1. All objects are acted on by gravity and fall to the floor with the same acceleration in the Earth lab; in the rocket ship the lab floor accelerates and catches up with the objects, so they appear to accelerate toward the floor at the same rate. From inside the lab one cannot tell from such experiments whether one is in a gravitational field or in an accelerating lab: they are equivalent. There is also a converse of this equivalence shown in Figure 2.1. If we let the Earth lab fall freely then objects inside will accelerate along with it and appear to float, just like in the rocket lab with the motor turned off. We can think of gravity as being transformed away in the freely falling lab. Einstein called this insight his "happiest thought."

An interesting prediction of general relativity follows from the equivalence principle. See again Figure 2.1. A beam of light sent across the lab in the rocket ship will appear to fall downward as the lab accelerates upward past it. Of course, since the speed of light is so large it will only have time to fall a small amount. We therefore expect that light should also fall or be deflected in a gravitational field. However, the deflection calculated using just the equivalence principle is only half the total obtained from the full theory.

How can we use the equivalence principle to understand gravitational fields such as that of the Earth? As Stephen Hawking whimsically points out, we certainly cannot think of the Earth as accelerating outward since it is not observed to become larger. Einstein showed that we obtain the same effect if spacetime is curved. For bodies moving slowly in a weak gravitational field, such as the Earth's, it is the warping of time that produces the effect of gravity, while the warping of space is less important. Thus we may say that the Earth's gravity is due mainly to the fact that a clock near the surface runs slightly slower than a more distant clock, even though the amount of slowing is only a few parts in $10^8$.

According to general relativity, matter, energy, and stress cause gravity by curving spacetime. Bodies in a gravitational field do not feel a force, but instead move along paths called geodesics, which are as straight as they can be in the curved geometry. Because of the curvature the paths wiggle about as if there were a force present.

Einstein found the fundamental field equations of general relativity in 1915, as did the mathematician David Hilbert. They are

$$G_{\mu\nu} = (8\pi G/c^4)T_{\mu\nu} \qquad \text{(Einstein field equations)} \tag{2.2}$$

These relate the Einstein tensor $G_{\mu\nu}$, which is constructed from the metric tensor and its derivatives, to the sources of gravity, which are described by the stress-energy tensor $T_{\mu\nu}$. They are nonlinear differential equations for the metric. The nonlinearity means that two solutions added together do not give another solution, unlike in Newtonian gravitational theory where we may sum up the fields of point masses to give rather general solutions. Physically this is because the gravitational field itself has energy and is thus a source of more gravity. Nonlinearity means that the strong fields associated with neutron stars and black holes are intrinsically more complex and interesting than the weak fields which occur in the Solar System.

Einstein referred to his field equations as "damnably difficult" and solved them using approximations. In 1916, almost immediately after the discovery of general relativity, Karl Schwarzschild obtained an exact solution for the field exterior to any spherically symmetric body such as the Sun. Much later, in 1963, a solution was found by Roy P. Kerr for a rotating body that is not spherically symmetric. These solutions are the bases for the study of black holes.

## 2.4 Weak gravity and experimental tests

For a weak gravitational field spacetime is only slightly curved and the metric differs only slightly from the Lorentz metric of special relativity, if we use Cartesian coordinates. If we retain only the first-order deviation of the metric from that of special relativity the mathematics is greatly simplified.

If we also assume that the field varies slowly with time and that all bodies move slowly compared with light, then general relativity reduces to Newtonian theory in a beautiful way. See Box 2.1. Time is warped by a small amount $2\phi/c^2$, where $\phi$ is the Newtonian potential, which is equal to $-GM/r$ for spherical bodies. Space is warped by an equal and opposite amount. Thus the Newtonian potential completely determines the metric. Warping of time means that a clock runs slower in the potential by a fractional amount $\phi/c^2$, so light emitted by an atom in the potential is redshifted by this amount relative to infinity. The redshift between two points in the potential is equal to the difference of the potential over $c^2$; for two points separated by a small vertical distance $\Delta r$ in the Earth's field it is about $(g/c^2)\Delta r$, where $g$ is the acceleration due to gravity. In the Solar System the maximum warping of spacetime occurs at the surface of the Sun, where it is only about $10^{-6}$.

The geodesic or generalized straight-line path of a body in spacetime is also an extremum, that is the longest distance between two points in spacetime. This means that a body follows the path in spacetime for which a clock on it reads the maximum elapsed time. It turns out from the algebra of the geodesic equation that the warping of time is most important in determining the path of a body moving slowly at $v$; warping of space is less important by a factor of $v^2/c^2$. On combining these facts we may summarize Newtonian gravity in two remarkably simple statements.

1. A clock in a gravitational field runs slower by a fractional amount $\phi/c^2$.
2. A body acted on only by gravity, such as a planet, follows the path that maximizes proper time as measured by a clock on the body.

BOX 2.1 THE NEWTONIAN LIMIT AND EDDINGTON'S PARAMETERS

In the limit of weak fields and low velocities the metric differs only slightly from the Lorentz metric, and is

$$ds^2 = (1 + 2\phi/c^2)c^2\,dt^2 - (1 - 2\phi/c^2)(dx^2 + dy^2 + dz^2), \qquad \text{(B1)}$$

where $\phi$ is the Newtonian potential. This metric can be used to understand many phenomena, including the redshift.

The Sun is nearly spherically symmetric and has a Newtonian potential $\phi = -G\,M_s/r$. Eddington wrote the metric of the Solar System in a parameterized form that is more general and accurate than the Newtonian limit, and describes the geometry of the Solar System for a rather wide range of theories. His metric is similar to (B1) but contains an additional nonlinear term,

$$\begin{aligned} ds^2 = {} & \{1 - 2G\,M_s/(rc^2) + 2\beta[G\,M_s/(rc^2)]^2\}c^2\,dt^2 \\ & - [1 + 2\gamma\,G\,M_s/(rc^2)](dx^2 + dy^2 + dz^2). \qquad \text{(B2)} \end{aligned}$$

In general relativity $\beta = \gamma = 1$. The term with $\beta$ describes nonlinear warping of time, while the $\gamma$ term describes warping of space. The mass of the Sun is actually measured using the first term in the time warping, so it contains no arbitrary parameter.

One may also study the limit of general relativity when the field is weak but for any velocity of the bodies involved. This linearized theory resembles classical electrodynamics and contains analogs of electric and gravitational fields appropriately called gravitoelectric and gravitomagnetic fields. Gravitoelectric fields are produced by the mere presence of mass–energy in analogy with electric charge, while gravitomagnetic fields are produced by the motion of mass–energy in analogy with electric current. While the existence of the gravitomagnetic field has not yet been directly verified experimentally, it may be solidly inferred from the experimentally verified gravitoelectric field and Lorentz invariance. The Gravity Probe-B (GP-B) spacecraft, launched in April 2004 (see Figure 2.2), has as one goal the direct detection of the gravitomagnetic field of the Earth. Since the prediction of gravitomagnetism is so definitive, a negative result from GP-B would be a major catastrophe.

Linearized general relativity theory gives insight into some peculiar facts about gravity. If we study the gravitational field produced by a perfect fluid (characterized only by its energy density $\rho$ and pressure $p$), we find that the source of the field is the combination $\rho + 3p$, not just $\rho$ as in Newtonian theory. Pressure produces gravity. Since the pressure in most ordinary fluids is much less than the energy density this is usually of no consequence, but for a gas of very hot material or photons the pressure is one-third the density, $p = \rho/3$, so pressure is an important source. For example it cannot be neglected in neutron stars, and is one reason why a neutron star of sufficient mass must collapse to a black hole.

An extraordinary situation occurs if the pressure is negative. This may sound peculiar, but for the vacuum defined as the ground state of a quantum field the pressure is the negative of the energy density, $p = -\rho$. The source strength is then negative, $\rho + 3p = -2\rho$, so elements of the fluid repel each other. Recent cosmological observations indicate that the energy of the actual vacuum is indeed not zero and appears to be the dominant stuff of the Universe, as we will discuss in Section 2.7.

**Figure 2.2.** April 2004 – the launch of the Gravity Probe-B (GP-B) spacecraft, which aims to detect the gravitomagnetic field of the Earth. (Photo credit: Dimitri Kalligas.)

Experimental tests of general relativity were discussed by Clifford Will in the previous volume of *The New Physics*, so we will mention only some new developments and give a brief summary.

We can divide gravity experiments into those that test fundamental principles and those that test the dynamics of individual theories such as general relativity. The fundamental principles include the invariance of spacetime under translations and rotations and changes of inertial frame, and most notably the equivalence principle. The equivalence principle was roughly tested in the lab by Galileo and Newton, and has been verified independently by Robert H. Dicke and by Eric Adelberger to better than one part in $10^{11}$.

A satellite space test of the equivalence principle (STEP), planned for the next decade, is expected to achieve a sensitivity of about one part in $10^{18}$, a particularly interesting number. String theory suggests the existence of a long-range scalar field, called the dilaton, which leads to a modification of general relativity in which the equivalence principle is violated by a small amount, estimated by some theorists to be about $10^{-18}$, thus perhaps detectable by STEP. (On naive dimensional grounds we might expect a violation of the equivalence principle at a level given by the ratio of the rest energy of an ordinary nucleon of matter to the Planck energy, which is about 10 GeV/ $10^{19}$ GeV $= 10^{-18}$.) STEP might thus turn out to be a probe of quantum-gravity effects.

Measurements of the redshift also test fundamental principles since the redshift may be derived from the equivalence principle. The most accurate measurement to date is that of Robert Vessot and Martin Levine, who in 1976 verified the theoretical prediction to one part in $10^4$ using a hydrogen maser clock in a rocket flown to about $10^4$ km.

Most dynamical tests of general relativity are based on a formalism invented by Arthur S. Eddington, and later developed by Kenneth Nordtvedt Jr. and Clifford Will into the parameterized post-Newtonian (PPN) formalism. Eddington wrote the metric for a weak spherically symmetric gravitational field such as that of the Sun in a rather general form with only two parameters, $\beta$ describing the nonlinearity in time warping and $\gamma$ describing space warping. See Box 2.1. General relativity predicts that $\beta$ and $\gamma$ are both equal to 1, and we can think of most experimental tests as measuring $\beta$ and $\gamma$.

The anomalous perihelion shift of Mercury is proportional to $(2 + 2\gamma - \beta)/3$, which is equal to 1 in general relativity. This is the only "classic" test that involves nonlinear effects. It has been measured with radar over a period of about 30 years, and is in agreement with relativity to within about one part in $10^3$.

The deflection of starlight by the Sun, another "classic" test, is proportional to $(1 + \gamma)/2$, which is equal to 1 in general relativity. This has been measured accurately for radio waves from quasars using arrays of radio telescopes. The measurements are in agreement with relativity to within about a few parts in $10^4$.

Light is also slowed by a gravitational field by an amount proportional to the same combination $(1 + \gamma)/2$. Irwin Shapiro has measured the delay using radar ranging to planets and spacecraft; a typical delay is some hundreds of microseconds. His measurements agree with relativity to within about one part in $10^3$.

From these measurements the Eddington parameters predicted by general relativity have accurately been verified: $\gamma$ is 1 to within about $10^{-4}$ and $\beta$ is 1 to within about $10^{-3}$.

The GP-B spacecraft will soon measure the precession of a gyroscope in Earth orbit. The precession due to the Earth's gravitoelectric field is predicted to be 6.6 arcseconds per year and that due to the Earth's gravitomagnetic field is 0.042 arcseconds per year (analogous to the precession of a magnetic dipole in a magnetic field). The two precessions are orthogonal. This should give the best measurement yet of $\gamma$, to within about $10^{-5}$, and a verification of the gravitomagnetic field to within about 1%.

In summary, general relativity is now accurately tested for weak fields in the Solar System. The binary pulsar discussed in the next section has provided confirmation of the Solar-System tests. However, there is little accurate confirmation for strong fields such as those found near neutron stars, or for larger distance scales such as galactic.

## 2.5 Gravitational radiation and the binary pulsar

Classical electrodynamics predicts that an accelerated charge emits electromagnetic radiation, and in the same way linearized general relativity theory predicts that an accelerated mass emits gravitational radiation. The gravitational waves that we expect to see on the Earth from astronomical sources should involve incredibly small distortions of spacetime, less than about $10^{-20}$, so the linearized theory is an excellent approximation to study their detection. However, they may be produced by violent processes such as collisions of neutron stars and black holes, which involve very strong fields; these calculations can be more difficult.

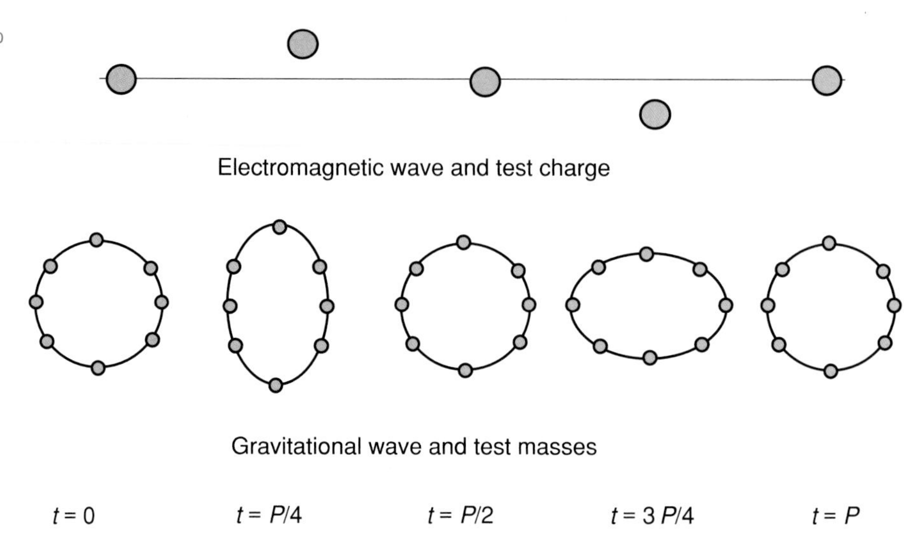


**Figure 2.3.** Motions imparted to a test charge by an electromagnetic wave and to a ring of test masses by a gravitational wave, shown over one period *P*.

Gravitational-wave theory is similar to electromagnetic-wave theory. Electromagnetic plane waves are transverse in that the electric and magnetic fields are perpendicular to each other and to the direction of propagation. A charged particle acted on by the wave is accelerated perpendicular to the propagation direction, as in Figure 2.3. Gravitational plane waves are also transverse and move at the speed of light; but the wave produces a relative force between point masses that is proportional to their separation, with the motion of a ring of free particles shown in Figure 2.3.

Detection of gravitational waves is difficult because they produce extremely small motions, but the principle of operation of actual detectors is nearly as simple as illustrated in Figure 2.3. Since the waves cause relative motions of free particles, they exert a force and will subject any solid object to a strain exactly like a Newtonian tidal force. Early detectors were large metal cylinders designed to respond resonantly, like a bell, to such a force. More sensitive modern detectors use interferometry to make precise distance measurements. The Laser Interferometer Gravitational-wave Observatory (LIGO) is a pair of interferometers about 4 km in size, in which beams of laser light measure the difference between the lengths of the legs induced by a gravitational wave. The light bounces back and forth hundreds of times between the mirrors to give an effective length much greater than 4 km. With a strain of about $10^{-20}$, equal to the amplitude of the wave, the effective displacement is only about $10^{-14}$ m, the size of a nucleus. Such a measurement is possible only with careful suppression of seismic, thermal, and shot noise. (Shot noise is due to the laser light being quantized.) Figure 2.4 shows the sensitivity expected in three generations of LIGO, and the strain and frequencies expected from some astrophysical sources. LIGO I will soon take data. In a few years a second-generation LIGO II with improved sensitivity, about $10^{-23}$, may make the first definitive detection of gravitational waves. Within about a decade at least three other interferometric detectors should also be built, in Europe, Japan, and Australia.

A system of three satellites called the Laser Interferometer Space Antenna (LISA) is planned for about 2010. It would be a triangular system of satellites in Solar orbit forming an interferometer with $5 \times 10^6$ km legs, able to detect strains of as little as $10^{-23}$. It would operate at frequencies as low as $10^{-4}$ Hz, where many interesting sources are expected. See Figure 2.4.

Interesting astrophysical sources of gravitational radiation include violent events such as supernovae and gamma-ray bursts, stellar-mass objects falling into large black holes, diverse binary stars, nonsymmetric pulsars, and coalescing neutron stars and

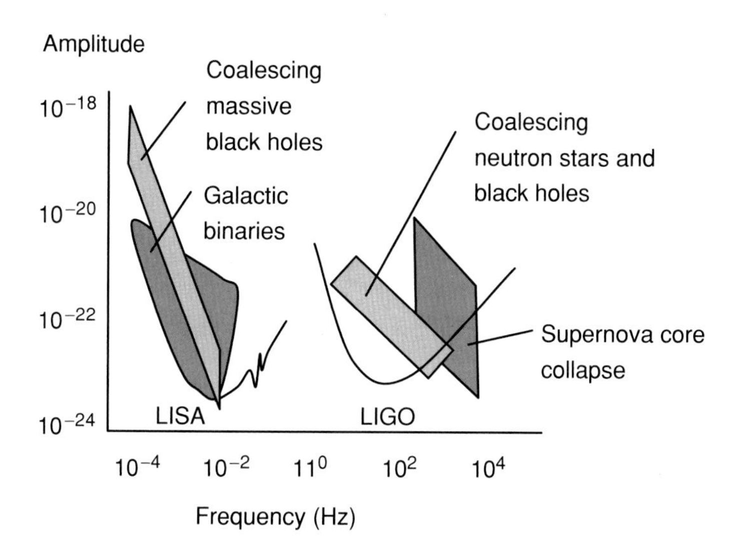


**Figure 2.4.** The sensitivities of LIGO and LISA to gravitational waves, and some expected sources.

black holes. One source of note is the collision of two supermassive black holes, with $10^6$ to $10^9$ Solar masses. Most radio galaxies are believed to contain such supermassive black holes, which absorb material around their equator and then eject it at relativistic speed in jets from the poles. Radio emission results from the collision of the jets with interstellar matter. Some radio sources have been observed to have four lobes, which may be the result of collisions that would generate prodigious amounts of gravitational radiation and may occur with a frequency as high as one per year.

Calculation of radiation from such strong sources generally requires a combination of analytical approximations and numerical work to calculate the wave magnitude and spectrum. The spectra range from periodic emissions by orbiting stars to short pulses from transient events such as black-hole collisions. Knowledge of the spectrum is important for detecting the waves.

Although gravitational radiation has not been detected directly, there is convincing evidence for its existence. The binary pulsar PSR 1913 + 16, discovered by Russell Hulse and Joseph Taylor in 1974, is a pulsar in orbit about a condensed companion, probably also a neutron star. The orbital period is about 8 h, the orbital radius is about $10^6$ km, the masses of both bodies are about 1.4 Solar masses, and the pulse period is about 59 ms. Orbital velocities are about 300 km $s^{-1}$, and the system is clean in that it has few non-gravitational complications. Timing of its pulses to an accuracy of about one part in $10^{10}$ has been used to derive an amazing amount of information about the system. Its orbital period is obtained from the Doppler shift of the pulses as the pulsar moves toward and away from us, which in turn gives the periastron advance, which is over 4 degrees per year. Many of the relativistic effects observed in the Solar System have also been measured for the binary pulsar, and all are consistent with general relativity. Moreover, over the course of three decades the period has decreased as predicted by relativity for energy loss to gravitational radiation. The data agree with theory to well within about 1%. See Figure 2.5.

Most relativists consider the binary-pulsar observations to be convincing evidence for gravitational radiation. Nonetheless, direct detection is still of the utmost importance because it should open a new observational window that might be as important as the windows opened by neutrino astronomy and gamma-ray astronomy, which have totally changed our view of the Universe. Gravitational radiation might let us observe such things as the interiors of exploding supernovae and gamma-ray bursters, and could make it possible for us to see directly into the early stages of the Universe, which are not accessible with light.

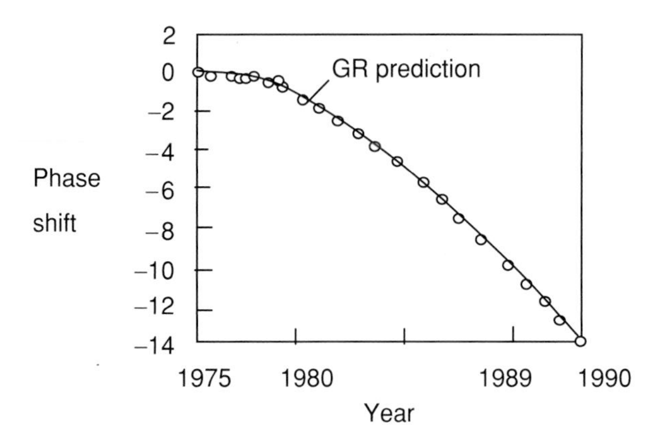


**Figure 2.5.** The period of the binary pulsar has decreased, in accord with general relativity, by energy loss to gravitational radiation.

## 2.6 Strong gravity, neutron stars, and black holes

Ordinary stars like the Sun produce quite small distortions in the spacetime geometry. As noted in Section 2.4 the distortion for the Solar System is approximately $GM/c^2r$. In terms of the geometric mass, $m = GM/c^2$, we may write this as $m/r$. (See Box 2.2.) The geometric mass of the Sun is about 1.47 km, and the distortion is greatest at the Solar surface, where it is only about $10^{-6}$. However, neutron stars have about a Solar mass packed into a radius of about 10 km, so the spacetime distortion is of order 0.1. To study such dense objects we need the exact solution of Schwarzschild given in Box 2.2, while for rapidly spinning dense bodies we need the exact solution found by Kerr.

Neutron stars were first suggested as theoretical possibilities in the 1930s by Fritz Zwicky and Walter Baade, but were taken seriously only after the 1967 discovery of pulsars, which are spinning neutron stars. They occur as an end product of a supernova explosion, when the matter in the stellar core is compressed to very high densities. Under sufficient pressure the electrons and nuclei in the core undergo inverse beta decay, wherein the protons absorb electrons and become neutrons with the emission of neutrinos. The resulting neutron-rich stars have interior densities of about $2 \times 10^{14}$ g $\mathrm{cm}^{-3}$, well above that of nuclei. Analysis of these stars requires relativistic structure equations and an equation of state giving the pressure in the neutron fluid as a function of density, for densities beyond those measured experimentally. Despite this the theory is considered trustworthy since the neutrons form a well-understood degenerate Fermi fluid.

An important and surprising theoretical prediction is a maximum mass for a neutron star, above which it cannot be stable. This limit is similar to the famous white-dwarf mass limit of about 1.4 Solar masses obtained by Subrahmanyan Chandrasekhar. One reason for neutron-star instability above the limit is that, as we noted before, pressure in the star contributes to the downward force of gravity. The value of the limit depends on the theoretical equation of state used, but is roughly 4 Solar masses. Observed neutron-star masses cluster around 1.4 Solar masses, and none is much greater. For example the binary pulsar mentioned earlier consists of a pulsar and a neutron-star companion, both with about 1.4 Solar masses. Many X-ray sources are believed to be powered by the strong gravitational fields of neutron stars.

Strange things happen if the core of an imploding supernova is greater than the mass limit for a neutron star, for there is no stable end state for the collapse. In 1937 J. Robert Oppenheimer and Hartland Snyder predicted that the core would collapse indefinitely, and that viewed from the outside the surface would asymptotically approach the Schwarzschild radius, which is twice the geometric mass, $R_s = 2m = 2GM/c^2$. At that radius the redshift is infinite and no light can escape. (See Box 2.2.)

BOX 2.2 THE SCHWARZSCHILD SOLUTION

Schwarzschild assumed spherical symmetry and with clever arguments reduced the mathematical problem to finding only two unknown functions from the field equations. His solution is

$$ds^2 = (1 - 2m/r)c^2\,dt^2 - dr^2/(1 - 2m/r) - r^2(d\theta^2 + \sin^2\theta\,d\varphi^2). \tag{B3}$$

Here $m$ is a constant of integration called the geometric mass, which must be determined from physical considerations. For large $r$ this is of course the metric of special relativity expressed in spherical coordinates. It looks similar to the Newtonian limit (Box 2.1), but the radial coordinate is different. The coordinates used in the Newtonian limit and Eddington metric are called isotropic since the space part is proportional to that of flat three-space, with no preferred direction. By transforming the Schwarzschild metric to isotropic coordinates and comparing with the Newtonian limit and Eddington metric one may identify $m = GM/c^2$ and find that $\beta = \gamma = 1$.

As we approach the Schwarzschild radius $R_s = 2m$ from above the time coefficient in (B3) goes to zero, so light emitted from this radius is infinitely redshifted and cannot be seen. Moreover, the coefficient of the radial term becomes infinite, so a very small change in $r$ corresponds to a very large change in the physical distance. Worse yet, for $r$ less than $R_s$ the signs of the time and radial terms change, so the coordinates $t$ and $r$ no longer correspond to time and space. For astrophysics problems only the outside of the black hole is relevant, so Schwarzschild's coordinates serve well, but for theoretical studies of the interior other coordinates are used.

They called this asymptotic state a frozen star, but it is now known as a black hole, a name invented by John Archibald Wheeler.

The prediction of Oppenheimer and Snyder was not taken seriously until the 1970s when astrophysical X-ray sources were discovered. Some of the sources appeared to be very compact objects with masses too large to be neutron stars, and black holes were the only known candidates. For example the source GRO J 1655-40 has $7.02 \pm 0.22$ Solar masses. Many X-ray sources are at present believed to be stellar-mass black holes. Moreover, most galaxies are believed to contain supermassive black holes, and quasars and radio galaxies are interpreted in terms of matter falling into black holes and emitting radiation.

Although astronomical observations are limited to black-hole exteriors, theorists are greatly interested in their surface regions and interiors. To an outside observer the surface of a collapsing star asymptotically approaches the Schwarzschild radius with a timescale of about $10^{-5}$ seconds for a typical star. Since the star is filled with burnt-out stellar matter, the interior is not described by the Schwarzschild solution, which is valid only for the empty exterior space. Thus there is no singularity inside the star. Figure 2.6(a) shows the collapse as "seen" by a distant outside observer.

Things are more exciting for an observer sitting on the surface of the collapsing star because he measures a different time. For him the surface of the star passes through the Schwarzschild radius and continues on to the origin, where the density becomes infinite and a spacetime singularity forms. The infinite tidal forces at the singularity would tear apart any object. At the Schwarzschild radius the surface observer's time differs by an infinite amount from that of a distant outside observer. After he passes the Schwarzschild radius he enters a region of spacetime that is simply not accessible to the outside observer – the black-hole interior. The collapse he "sees" is shown in Figure 2.6(b).

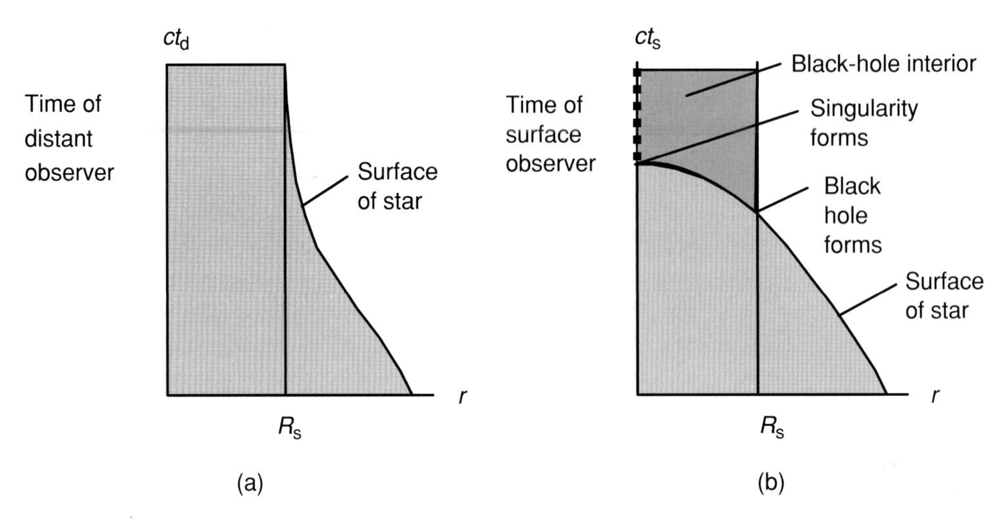


**Figure 2.6.** (a) As seen by a distant observer (with time $t_d$) the surface of a collapsing star asymptotically approaches the Schwarzschild radius. (b) As seen by an observer on its surface (with time $t_s$) it collapses through the Schwarzschild radius to zero size and a singularity.

Despite this strangeness a surface observer doing local physics experiments in a small lab falling past the Schwarzschild radius would not notice anything peculiar – so long as he ignored the outside world! This is guaranteed by the equivalence principle. In a sense conditions near a black hole are ordinary on a small scale.

There is a nice geometric way to view the black-hole surface. The light cone is a surface in spacetime along which light moves; it is also called a null surface since the line element is zero along it. (If we suppress one space dimension it looks like a cone; see Figure 2.7.) The statement that no particle moves faster than light implies that particles move within the light cone and cannot pass outward through it. A black-hole surface is also a null surface, and the local light cones are tangential to it. As a result particles cannot pass outward through the black-hole surface, any more than they can move faster than light. Light itself cannot pass outward through the surface, and at best can move along it. The surface acts as a one-way membrane or horizon, and outside observers cannot see inside.

A spherical black-hole surface is both an infinite-redshift surface and a null surface or horizon, but these surfaces are not always the same. In a rotating black hole, described by the Kerr solution, they are different. There is a gravitomagnetic field in the Kerr exterior, which indicates that the field is caused by moving mass–energy in the interior, although no exact solution for the interior is known. The Kerr solution is believed to be the general final asymptotic state for uncharged collapsing stars

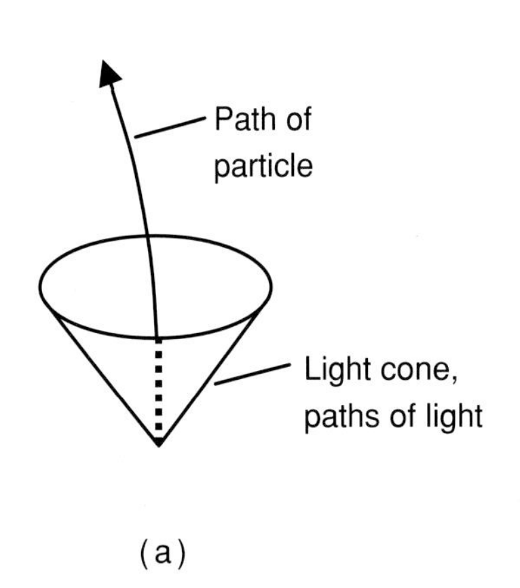


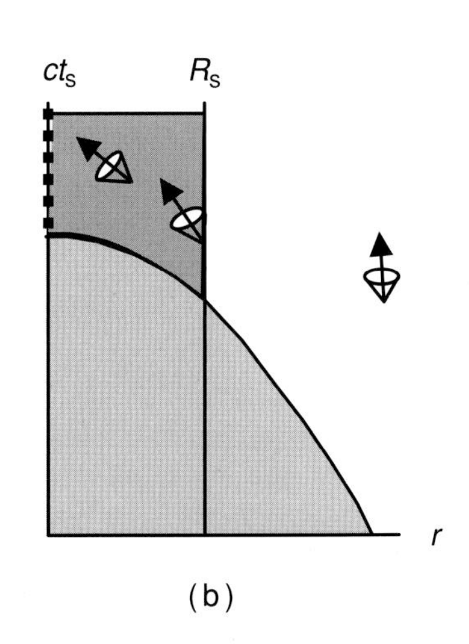


**Figure 2.7.** (a) Paths of light in spacetime determine the light cone. (b) Particles must move within the light cone. Light cones are tangential to the black-hole surface.

with angular momentum. The infinite-redshift surface is nonspherical and outside the horizon, which is the true black-hole spherical surface. The region between is called the ergosphere, and has odd properties. For example a freely falling particle in the ergosphere can have negative total energy, and can send information to the outside world.

The singularity inside a black hole is bad news to theorists. It can produce paradoxical effects such as a flux of unlimited energy. But it is inside a horizon and cannot affect the exterior world, which is thereby protected. Penrose and Hawking have conjectured a "cosmic censorship" principle, that all singularities are similarly hidden by a horizon. Other theorists believe that singularities would not occur in a consistent quantum-gravity theory, making cosmic censorship unnecessary and irrelevant. There is as yet no general agreement.

An "eternal" black hole is a theoretical object described by the Schwarzschild geometry everywhere in spacetime. It is a region of empty space, containing a singularity, that has existed forever and was not formed in the past. Part of the spacetime is a black-hole interior, part is a time-reversed black hole called a white hole, and part is a second exterior region. Eternal black holes have been used in amusing theoretical studies involving "wormholes," which connect the exterior regions, and might allow time travel. Real black holes that formed by collapse of matter in the past should not have such extra regions, and there is no reason to think that eternal black holes actually exist.

Black holes were not taken seriously until a few decades ago, but may possibly be the dominant form of matter in the Universe. It has been suggested that the dark matter, which appears to make up over 80% of the "cold matter" in the Universe, is composed of small black-hole remnants with a mass of about $10^{-8}$ kg. (See Section 2.7.) If this is true then they dominate the matter in the Universe, and the ordinary matter that physicists study in the lab is only a small contaminant.

## 2.7 Gravity and the Universe, dark matter and dark energy

General relativity was the first theory to allow a science of cosmology. Soon after discovering general relativity in 1915, Einstein developed a cosmological model. It is a well-known story that he added to the left-hand side of the field equations a cosmological term $\Lambda g_{\mu\nu}$, representing a large-scale repulsion, in order to obtain a static model of the Universe; $\Lambda$ is called the cosmological constant. With the discovery of the expansion of the Universe by Edwin Hubble a decade or so later, the static model became obsolete, and the idea of an expanding Big-Bang Universe has dominated cosmology ever since. Although the cosmological term was motivated by the desire for a static Universe, its existence must finally be decided by observations. We will discuss later how $\Lambda$ can be interpreted as the energy of the vacuum.

In cosmology galaxies or clusters of galaxies are looked upon as the particles of the material or fluid that fills the Universe. This material appears to be roughly homogeneous and isotropic on the cosmological scale of some 10 billion light years, although there is much clumping on a smaller scale. Because the material of the Universe is homogeneous and isotropic it follows that the space should be also, which simplifies the mathematics. Essentially only three types of space are possible. These are illustrated by the two-dimensional analogs shown in Figure 2.8: (1) a flat infinite sheet; (2) the surface of a sphere; and (3) the surface of a pseudo-sphere, which is like the center region of a saddle at every point. (It behaves mathematically like a sphere with imaginary radius.) These three spaces are called flat, positively curved or hyperspherical, and negatively curved or pseudo-hyperspherical, and are often labeled by the curvature

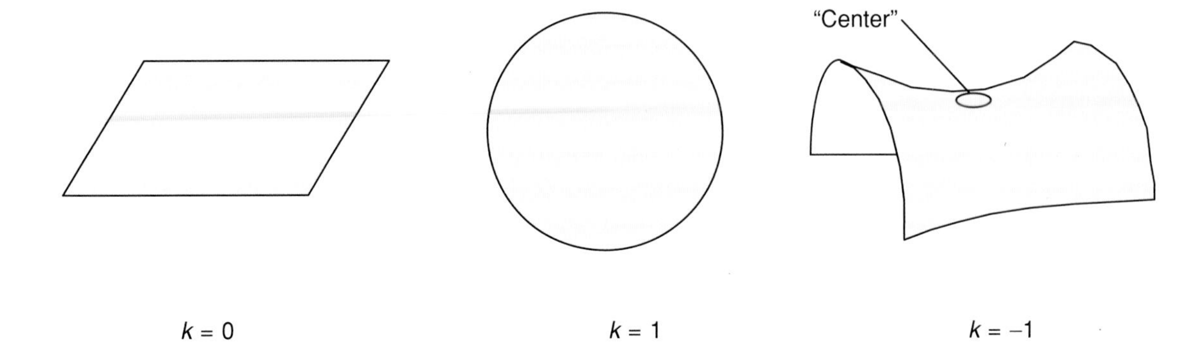


**Figure 2.8.** Two-dimensional analogs of the FRW spaces. All are homogeneous and isotropic. The pseudo-sphere is like the circled center region everywhere.

parameter $k = 0$, $k = 1$, or $k = -1$. The spacetimes used in cosmology are one of these three spaces with a universal time simply added on. They are called FRW spaces, after Alexander Friedmann, who first used them, and H. P. Robertson and A. G. Walker who developed them some years later. See Box 2.3.

A hyperspherical geometry has philosophical appeal. It provides an answer to the old question "Is the Universe infinite, and if so, how can we conceive of anything physical being infinite, and if not, what lies beyond it?" The hypersphere, like its two-dimensional analog, curves back on itself and has no boundary, but it is finite in extent and has a finite volume.

Expansion of the Universe is easy to visualize by picturing grid lines on the analogous two-dimensional surfaces, with galaxies attached at intersections. These move apart as the surface expands. In the case of the hypersphere the radius becomes larger.

**BOX 2.3 THE STANDARD MODEL OF COSMOLOGY**

The FRW metric has the form

$$ds^2 = c^2\,dt^2 - a(t)^2[dr^2/(1-kr^2) - r^2(d\theta^2 + \sin^2\theta\,d\varphi^2)]. \tag{B4}$$

The scale function $a(t)$ has the dimension of a length, the radial coordinate $r$ is dimensionless, and $t$ is a universal time. If this metric is substituted into the Einstein field equations, two fairly simple differential equations for the scale function result,

$$-\left(\frac{8\pi G}{c^4}\right)\rho = \Lambda - 3\left(\frac{k}{a^2} + \frac{\dot{a}^2}{c^2a^2}\right), \quad \left(\frac{8\pi G}{c^4}\right)p = \Lambda - \left(\frac{k}{a^2} + \frac{\dot{a}^2}{c^2a^2} + \frac{2\ddot{a}}{c^2a}\right). \tag{B5}$$

These must be augmented by an equation of state relating the pressure and density in the fluid. If this is the linear relation $p = w\rho$ then a solution may be written implicitly in the form of an integral,

$$ct = \int_0^a da\,[(D/a)^{3w+1} + \Lambda a^2/3 - k]^{-1/2} \quad (D = \text{constant of integration}) \tag{B6}$$

For the values $w = 0$ and $k = 0$ favored at present, this leads to

$$a(t) = \left(\frac{3D}{\Lambda}\right)^{1/3}\left[\sinh\left(\frac{\sqrt{3\Lambda}}{2}ct\right)\right]^{2/3}, \tag{B7}$$

which is plotted in Figure 2.9; it begins at zero for $t = 0$, the Big Bang, and finally increases exponentially for late times, the accelerating Universe. (Note that for $k = 0$ the scale function is not directly observable.)

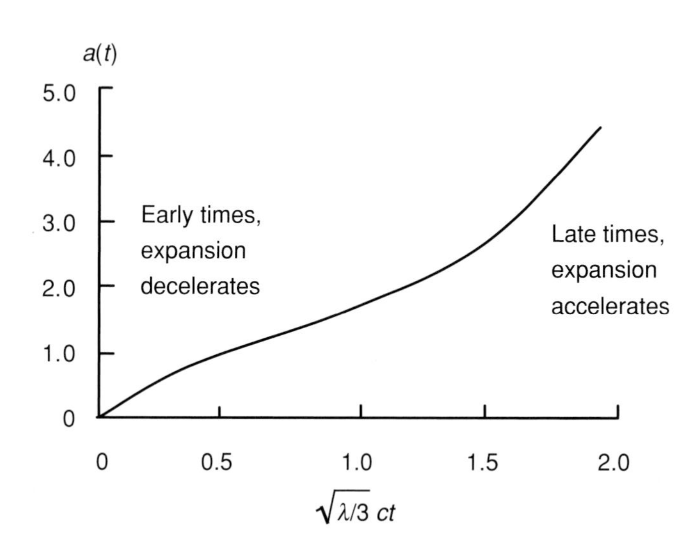


**Figure 2.9.** The scale function for the flat Universe favored at present, with a cosmological constant and "cold" dark matter.

This illustrates the meaning of coordinates in relativity: galaxies remain at rest in the coordinate grid system but move apart physically. Note that for $k = 0$ the three-space is flat, but since the scale increases with time the four-dimensional spacetime is curved.

With the use of the FRW metric the mathematical cosmological problem is relatively simple. See Box 2.3. There is only one unknown function to be found from the field equations, the scale function $a(t)$. This allows us to calculate many observables. To find it we must choose a value for the curvature parameter, and a value for the cosmological constant $\Lambda$. The cosmic material is usually described as a perfect fluid having an energy density $\rho$ and a pressure $p$, with a relation between them called the equation of state. A popular choice is the linear relation $p = w\rho$. This describes "cold matter" of negligible pressure with $w = 0$, "hot matter" or radiation with $w = \frac{1}{3}$, and the "vacuum" with $w = -1$. Anything that behaves as if $w < -\frac{1}{3}$, so that gravity is repulsive (see Section 2.4), is generally called dark energy, but we limit ourselves here to the most popular case of a vacuum with $w = -1$ (see Section 2.11).

For the linear equation of state the solution for the scale function may be expressed in terms of an integral (see Box 2.3). One notable solution is that found by Willem de Sitter in 1917 for a universe empty of matter except for the cosmological constant; this has an exponential scale function, and is now believed to describe the early Universe during inflation and also, with a much smaller cosmological constant, an approximation to the present Universe.

Observational cosmology relates the scale function $a(t)$ and parameters like $k$, $\Lambda$, and $w$ to the real Universe. For example the Hubble law for the recession velocity of galaxies is $v = HL$, where $L$ is the distance to the galaxy and $H$ is called the Hubble "constant." $H$ actually varies slowly with time and is related to the scale function by $H = \dot{a}/a$, where $\dot{a} = da/dt$. Observation gives a value of $H = 71 \pm 2$ km s$^{-1}$ Mpc$^{-1}$ in the units favored by astronomers; $H^{-1}$ is a rough measure of the lifetime of the Universe, some 10 billion years. The curvature parameter $k$ is related to the energy density of all the contents of the Universe, including the energy of the vacuum: if the density is greater than a critical value defined as $\rho = 3H^2c^2/8\pi G \cong 8.0 \times 10^{-10}$ J m$^{-3}$ then $k = 1$ and the Universe is a hypersphere; if it is equal, then $k = 0$ and the Universe is spatially flat; if it is less, then $k = -1$ and the Universe is a pseudo-hypersphere. Measurements of the density, deviations from linearity in the Hubble expansion (from recent supernova data), and measurements of the cosmic microwave background (from the Wilkinson microwave-anisotropy probe WMAP) have led astronomers to favor a spatially flat

Universe ($k = 0$), a nonzero cosmological constant (or vacuum energy density), and a small value of $w \approx 0$ for the matter in the Universe. With these parameters the scale function is the simple analytical expression in (B7), see Box 2.3.

The measurements imply two unexpected and astounding things. First, most of the matter in the Universe is not the ordinary visible matter in stars and interstellar dust and gas but invisible matter that interacts only weakly with ordinary matter. It is called dark matter. Secondly, the energy of the vacuum is not zero as intuition suggests, but instead appears to be the dominant energy stuff of the Universe. The vacuum energy (a dark energy) appears to make up about 73% of the energy density of the Universe, dark matter makes up about 23%, and the ordinary matter that physicists have studied for centuries is only about 4%. The total density is about equal to the critical density, which is consistent with a flat or $k = 0$ Universe.

Probably the clearest evidence for dark matter comes from observing the rotation velocities of stars and gas in the outer regions of galaxies. The velocity for a circular orbit around the center of a galaxy is a measure of the mass $M(r)$ interior to the orbit, and is given by $v^2 = GM(r)/r$. Observations indicate that the mass does not decrease rapidly outside the central bulge, as the luminous matter does, but much more slowly, and the total mass of a typical galaxy appears to be as much as ten times that of the luminous matter.

There has been no direct lab detection of dark-matter particles yet. The leading candidate is the neutralino, a supersymmetric partner of the photon suggested by theory; it is the lightest supersymmetric particle and thus stable. (See Chapter 4.) Remnants of evaporated black holes, with a mass of about $10^{-8}$ kg, are another attractive candidate.

Alternatively the observations may be a signal that our laws of gravity or dynamics should be modified for large distances or small accelerations. Newtonian mechanics and gravity are tested mainly on the scale of the Solar System, so a modification on a galactic scale is reasonable. In the approach called modified Newtonian dynamics (MOND), Newton's second law is modified to $F = ma^2/a_0$ for small accelerations $a \ll a_0$. This implies that rotation velocities should become constant far from galactic centers, approaching $v^4 = GMa_0$. This fits galactic rotation curves rather well with $a_0 \approx 10^{-10}$ m s$^{-2}$. MOND is a phenomenological idea with little theoretical justification at present, but is interesting because of its simplicity. Moreover, the characteristic acceleration is comparable to the speed of light divided by the lifetime of the Universe, $c/T \approx 7 \times 10^{-10}$ m s$^{-2}$, which may be a coincidence or a hint of some significance. There has also been an anomaly observed in the motion of the two spacecraft Pioneer 10 and Pioneer 11, which are now about 60 AU from the Sun. The two have anomalous accelerations of $8.09 \pm 0.2 \times 10^{-10}$ m s$^{-2}$ and $8.56 \pm 0.15 \times 10^{-10}$ m s$^{-2}$, comparable to $a_0$. No definitive explanation has been offered for this anomaly.

Another alternative to dark matter is that of shadow universes, which is based on a theory of extra dimensions and "branes," that is membranes in more than two dimensions. (See Chapter 5.) The idea is that we live on a brane, along with all physical phenomena that we observe, except gravity, which operates in more dimensions. If there were other universes on nearby branes we would interact with them only by gravity, and ordinary matter in those universes should appear to us as dark matter. The idea that gravity is manifested in more dimensions than other forces is testable in the lab, and we will discuss it later.

It is curious that the energy density of the vacuum is about twice that of the dark matter, because vacuum energy density is constant in time, while the dark-matter density should decrease from a very large value to a very small value as the Universe expands. Do we just happen to live at a time when they are only a factor of two apart? This may be an accident or a clue to something much deeper.

Cosmology is in an odd state at present. The general properties are in accord with the standard model of cosmology, but the basic stuff of the Universe, the dark matter and dark energy, have not been detected in the lab, and are not at all understood. The cosmological dark-energy observations are quite new but the results of different measurements are consistent and rather convincing.

## 2.8 Gravity and the early Universe

In earlier times the Universe was quite different than the star-filled one we see now. If we extrapolate backward in time using the physics at our disposal, we arrive at the following sketch of its history. About $3 \times 10^5$ years after the hot Big Bang, atoms like hydrogen condensed out of a previously existing hot plasma at a temperature of about $10^4$ K or $kT \approx 1$ eV. This left photons to decouple from matter and propagate almost freely, redshifted in the expanding Universe, to become the cosmic microwave background (CMB) radiation that we now detect at 2.7 K. The CMB is thus a photo of the Universe at the time of decoupling. Before that, a minute or so after the Big Bang, light nuclei like helium condensed out of a plasma of hot nucleons at about $kT \approx 1$ MeV. Before that, a microsecond or so after the Big Bang, the nucleons condensed out of a plasma of quarks and gluons at $kT \approx 1$ GeV. Before the time of the quark–gluon plasma, at $kT \gg 10^3$ GeV, we have no direct experimental lab evidence and must use indirect evidence and speculations such as grand unified theories (GUTs) of particles and theories of inflation. Finally, approaching the Planck time of about $10^{-43}$ seconds and a temperature $kT = 10^{19}$ GeV we have nearly total freedom to speculate.

The image presented by the CMB is strikingly uniform, with a temperature variation of only about one part in $10^5$ across the entire sky, so the Universe was extremely isotropic at decoupling time. This leads to the following so-called horizon puzzle. If the early Universe was indeed dominated by very hot matter and radiation, then the cosmological equations give a scale function $a(t)$ proportional to $\sqrt{t}$. This is zero at $t = 0$ and has an infinite derivative, so the initial expansion would have been so rapid that fairly nearby points would have separated faster than the speed of light. This leads to no paradox in itself, since it does not mean that an object could outrun light; but it does mean that fairly nearby parts of the Universe could never have been in causal or thermal contact before decoupling time. In fact a horizon would have separated regions of the Universe that are now more than a degree or so apart as viewed from Earth. How then could such separated parts have had so nearly the same temperature at the time of decoupling? This is the horizon puzzle.

A favored solution to the horizon puzzle is to develop a theory that gives a different scale function. For example an exponential scale function first increases slowly but later expands rapidly, and there is no horizon. One way to have exponential expansion, called inflation, is to suppose that the dominant material of the early Universe had an equation of state approximately like the vacuum, $p = -\rho$, with enormous energy density. Inflation theories usually achieve this by filling the Universe with one or several scalar fields, called inflatons. There are many versions of inflation but it generally happens well after the Planck time and lasts about $10^{-35}$ seconds, during which time the Universe expands by a very large amount, typically $e^{60} \approx 10^{26}$. Inflation theories make predictions about the spatial distribution of the CMB that are now being vigorously tested. (See Chapter 1.)

The present density of the Universe appears to be within a factor of two of the critical density. For a universe containing vacuum energy and matter with a linear equation of state the ratio can be calculated theoretically to be $\{1 - k/[(D/a)^{3w+1} + \Lambda a^2/3]\}^{-1}$, where $D$ is an arbitrary constant of integration. This approaches unity as the scale function $a(t)$ approaches zero for early times, assuming only that $3w + 1 > 1$, which is true for ordinary matter. Thus the density should have been very nearly equal to the critical density in the early Universe. The fact that it is roughly equal now means either that the present scale function is much smaller than $D$ or that the curvature parameter $k = 0$ (a flat universe). Some theorists favor $k = 0$ as the more natural explanation. The great inflationary expansion makes the Universe behave *as if* it were flat.

Cosmogenesis, the very beginning of the Universe, is an epistemologically murky problem. We will mention only one approach, based on work of Hawking and James Hartle. It is clearly hard to say what initial or boundary conditions are appropriate for the beginning of the Universe. Indeed, before the Planck time of about $10^{-43}$ seconds, it can be argued that there was no clearly defined direction in time, due to the quantum nature of spacetime, which we discuss in the next section. Hartle and Hawking have suggested that the boundary condition should be the simplest one possible, that in a sense there is no boundary. This means that the geometry of the Universe was initially a four-dimensional hypersphere of Planck scale radius $R$, having no boundary and analogous to the three-dimensional hypersphere of FRW cosmology. All of the coordinates were spatial, so time did not exist. But quantum fluctuations in such a small space would be very large, large enough that components of the metric could change sign. This is mathematically equivalent to making the relevant coordinate imaginary and singling it out as a time coordinate. In terms of that special time coordinate the Universe would have expanded to freeze in the fluctuation that caused time to begin. In terms of the imaginary time coordinate the Universe would have been an exponentially expanding FRW universe with positive curvature parameter $k = 1$, containing vacuum energy as its only ingredient. This model appears to be viable for the earliest times.

## 2.9 Gravity and the quantum, the Planck scale

Most of the ideas discussed so far are founded on a reasonable mixture of theory and observation, but we now enter more speculative areas. Electromagnetism and the weak and strong nuclear forces are now understood in terms of the standard model of particles, a quantum theory. General relativity is a classical theory since it does not contain Planck's constant, and is based on the concept of spacetime distances. Gravity stands apart. In non-gravitational physics spacetime is the arena in which events take place, but in general relativity it is the very stuff of the physics. Moreover, when we combine some quantum ideas of measurement with the concept of gravity as spacetime geometry the classical spacetime continuum concept breaks down.

In relativity it is implicitly assumed that spacetime distances can be measured to any desired accuracy. We will do a thought experiment, as Einstein was fond of, using only basic concepts of quantum theory and gravity, to see how the concept of classical spacetime interval fails and at what distance. (We will ignore factors of 2 and $2\pi$ and deal with orders of magnitude, in the spirit of the uncertainty principle.) We want to measure the distance $L$ between two points in space, A and B, so we will send a light signal from A to B and back and measure the time $T$ it takes; the distance is $L = cT/2$. See Figure 2.10. Our clock need not be at A but must be observable from A, so a macroscopic clock is acceptable, and we assume that it is arbitrarily accurate. Nevertheless, inescapable limits on the accuracy of the measurement are imposed by the wave nature and quantization of light and the geometric nature of gravity.

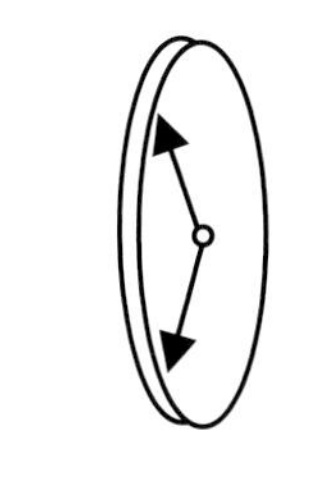
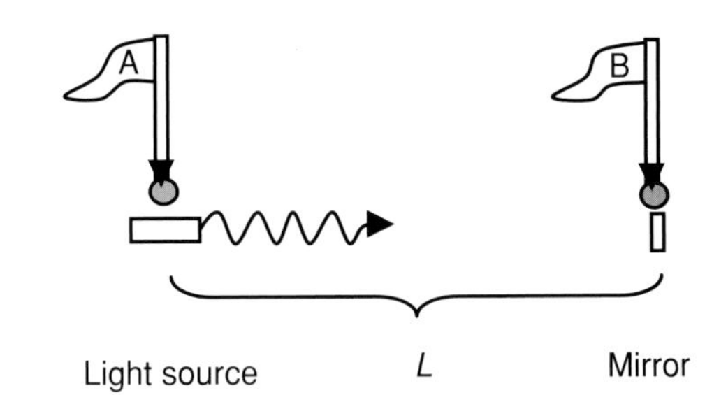


**Figure 2.10.** Measuring the distance between A and B with light, using a source, a mirror, and a clock, which may be large but must be visible from A.

First, light is a wave so there is uncertainty in $L$ of about the wavelength of the light. Secondly, light is quantized so it must be composed of at least one photon and have energy of at least $E = h\nu$. Lastly, the photon has energy, so it causes a gravitational field in the region between A and B, which is manifested by a distortion of the space that causes a further uncertainty in $L$. The result is a minimum uncertainty of about $L_{\mathrm{P}} \equiv \sqrt{G\hbar/c^3} = 1.6 \times 10^{-35}$ m, called the Planck distance. See Box 2.4. For an accurate measurement we must limit ourselves to much larger distances. Note that the Planck distance is the only distance that can be formed from the fundamental constants $G$, $c$, and $\hbar$. Uncertainty in distance measurement may be interpreted much like the uncertainty principle in quantum mechanics: the act of measuring the distance makes it uncertain by at least $L_{\mathrm{P}}$.

A similar phenomenon occurs in particle physics. If we try to confine an electron to a region smaller than its Compton wavelength, $\hbar/(2m_{\mathrm{e}}c)$, we find that, due to the uncertainty principle, electron–positron pairs are created, so the single-particle description breaks down and we must use a many-body quantum field description. By analogy we may think of an attempt to measure spacetime at the Planck scale as creating and destroying spacetime, so we must use a quantum description. Some theorists, notably Wheeler, visualize spacetime at the Planck scale as wildly fluctuating *quantum foam*, without even a well-defined signature or topology. This is the kind of initial state we mentioned previously for the no-boundary model of cosmogenesis.

Similar reasoning leads to a generalization of the uncertainty principle. Imagine measuring the position of a particle with light. The wave nature of light implies an uncertainty $\Delta x \approx \lambda$. A photon has momentum $p = h/\lambda$ and will generally impart a significant fraction to the particle, causing a momentum uncertainty of about $\Delta p \approx \hbar/\lambda$. This gives the usual uncertainty principle $\Delta x\, \Delta p \approx \hbar$. But there will also be a gravitational effect since the photon has energy. The gravitational field of the photon

### BOX 2.4 LIMIT ON DISTANCE MEASUREMENTS

We estimate how accurately we can measure a distance $L$ with light of wavelength $\lambda$. Figure 2.10 shows the thought experiment, with light sent from A to B and back in time $T = 2L/c$. First, since light is a wave there is an uncertainty of about a wavelength $\lambda$. Secondly, a single photon has energy $E = h\nu = hc/\lambda$ and thus an effective mass of $m_{\mathrm{eff}} = E/c^2 = \hbar/(\lambda c)$ spread over $L$. Finally, the photon produces a potential $\phi \approx G\, m_{\mathrm{eff}}/L$ in the region between A and B, and therefore a fractional distortion of the metric of about $\phi/c^2 \approx G\hbar/(L\, c^3\lambda)$. (See Box 2.3.) On putting these together we find the total uncertainty,

$$\Delta L \approx \lambda + L\phi/c^2 = \lambda + G\hbar/(c^3\lambda) = \lambda + L_{\mathrm{P}}^2/\lambda. \qquad \text{(B8)}$$

The minimum of this occurs for a wavelength equal to the Planck distance, and gives a minimum uncertainty of about the Planck distance.

must be proportional to its energy (and thus its momentum) and also to $G$, that is $\Delta x_{\text{gr}} \propto G\Delta p$. The only distance we can form from this and $c$ and $\hbar$ is $\Delta x_{\text{gr}} \approx G\,\Delta p/c^3 = L_{\text{P}}^2(\Delta p/\hbar)$. On adding this to the usual uncertainty principle we obtain

$$\Delta x \approx \hbar/\Delta p + L_{\text{P}}^2(\Delta p/\hbar) \qquad \text{(Generalized uncertainty principle)} \tag{2.3}$$

The minimum of this is about equal to the Planck distance, which suggests that smaller distances might not have physical meaning.

The Planck distance is extremely small, 20 orders of magnitude smaller than a nucleus. On a log scale humans are far closer to the size of the Universe than to the Planck distance. Associated with the Planck distance are Planck units of time, energy, and mass,

$$\begin{aligned} L_{\text{P}} &= \sqrt{G\hbar/c^3} = 1.6 \times 10^{-35}\,\text{m}, \qquad & t_{\text{P}} &= L_{\text{P}}/c = \sqrt{G\hbar/c^5} = 0.54 \times 10^{-43}\,\text{s}, \\ m_{\text{P}} &= \hbar/(cL_{\text{P}}) = 2.2 \times 10^{-8}\,\text{kg}, \qquad & E_{\text{P}} &= m_{\text{P}}c^2 = 1.2 \times 10^{10}\,\text{GeV} \qquad \text{(Planck scale)} \end{aligned} \tag{2.4}$$

Whereas the Planck distance is far less than the nucleon size, the Planck energy is much larger than that of present-day particle accelerators, which is about $10^3$ GeV or 1 TeV.

Gravity becomes comparable in strength to the other fundamental forces when the interaction energy approaches the Planck energy, as we may show. The gravitational force between two masses $m$ is $Gm^2/r^2$, while the electric force between two electron charges is $e^2/r^2$. These are equal for $m^2 \approx e^2/G$. But $e^2 \approx \hbar c/137$, so to within a few orders of magnitude equality occurs for $m^2 \approx \hbar c/G$, which is the Planck mass. The same is roughly true for the other forces.

It is widely believed that quantum electrodynamics (QED) and quantum chromodynamics (QCD) are really low-energy effective field theories that break down at some high energy such as the Planck energy, and must be replaced by a more complete theory. This is to be expected since gravity becomes too strong to neglect, so the complete theory must include gravity. Thus the Planck scale serves as a natural cutoff for the infamous divergences of QED and QCD, which should not occur in the complete theory.

## 2.10 Black holes and quantum theory

For distances much larger than the Planck scale it is reasonable to use thermodynamics and quantum theory in a classical curved background spacetime. This leads to some interesting results – and questions.

Jacob Bekenstein began the study of black-hole thermodynamics in 1973 when he suggested that a black hole has an intrinsic entropy proportional to its area $A_{\text{BH}} = 4\pi R_{\text{s}}^2 = 4\pi(2GM/c^2)^2$. This seems to be a consistent extension of the idea of entropy since the total black-hole area increases in collisions, but only up to a point. Hawking did not initially believe black-hole thermodynamics since nothing can escape from a black hole, whereas a normal body at nonzero temperature radiates energy as photons. But when he applied quantum field theory in the spacetime of a black hole he discovered to his surprise that a black hole should produce a flux of radiation characteristic of a black body. The radiation originates *near* the surface, and thus can escape to the external world.

We can obtain Hawking's result heuristically using the uncertainty principle. QED predicts the existence of vacuum fluctuations, wherein virtual pairs of electrons and photons materialize from the vacuum, then annihilate so that energy and momentum are conserved. These are described by Feynman diagrams such as those in Figure 2.11(a).

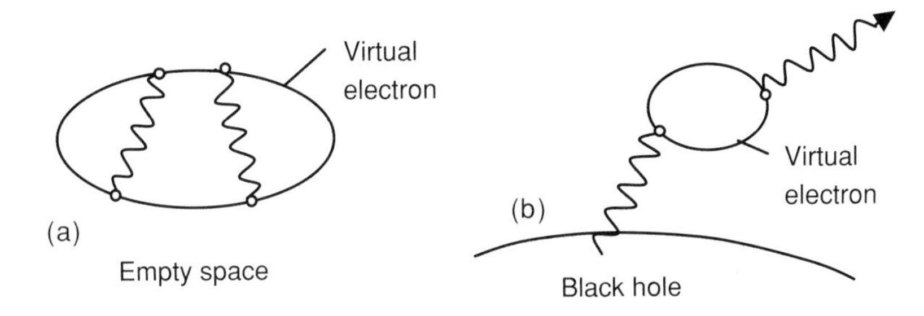


Figure 2.11. Vacuum quantum fluctuations in flat space and near a black hole.

Near a black hole processes like that in Figure 2.11(b) are also possible, in which one negative-energy photon is absorbed by the black hole and a positive-energy one is emitted to the exterior. The emitted photon energy is stolen from the mass of the black hole. An estimate of the emitted photon energy can be obtained from the uncertainty principle. For a small particle near a black-hole surface the field lines appear to radiate from near the center of the black hole, implying an intrinsic position uncertainty of about the Schwarzschild radius. (This is reasonable on dimensional grounds.) The uncertainty principle thus implies an intrinsic photon momentum uncertainty $\Delta p \approx \hbar/R_s = \hbar c^2/(2GM)$ and associated energy uncertainty of $E \approx \hbar c^3/(2GM)$. There is no heuristic way to show that the spectrum of the emitted photons is thermal, but if we assume that it is then the temperature must be roughly $kT_{\rm BH} \approx \hbar c^3/(2GM)$. Hawking's exact result, and the associated entropy defined by $d S_{\rm BH} = d(Mc^2)/(kT_{\rm BH})$, are

$$kT_{\rm BH} = \hbar c^3/(8\pi GM), \qquad S_{\rm BH} = 4\pi GM^2/(\hbar c) = 4\pi (M/m_{\rm P})^2 = A/\left(4L_{\rm P}^2\right). \tag{2.5}$$

Hawking's celebrated discovery put black-hole thermodynamics on firmer ground, but it is peculiar in that the black-hole temperature is inversely proportional to mass. Thus a black hole in a radiation environment with an ambient temperature will absorb radiation if its own temperature is smaller, and will radiate if it is greater. If it radiates it will lose mass and its temperature will increase, and so forth in a runaway process. If sufficiently small black holes existed we should detect bursts of radiation from them with a very distinctive spectrum. Alas we do not see such bursts, which could be because the theory is wrong, or because not enough small black holes exist to give detectable radiation.

As an evaporating black hole approaches the Planck size the theory cannot tell us what happens next. Some theorists expect that it should radiate completely away to vacuum since there is nothing to prevent this. Others expect that it must stop radiating or it would become smaller than the minimum Planck size allowed by the generalized uncertainty principle. Black-hole remnants of Planck size and mass are an interesting candidate for the dark matter. Unfortunately, direct experimental detection would be quite difficult since they interact only gravitationally, and due to their large Planck-scale mass they would have a very low number density.

Black-hole evaporation leads to an amusing information puzzle. In one sense an uncharged spherical black hole is very simple since it is entirely described by its mass, despite the fact that it is the final state of collapse of very many different systems. It is thus a statistical system in which a macrostate corresponds to many microstates. Its entropy should be given by Boltzmann's equation in terms of the number of microstates $n$ that correspond to the single macrostate, $S = \log n$. For example, the entropy of a small 1-g black hole, according to (2.5), is about $3 \times 10^{10}$, and the corresponding number of microstates is about $\exp(3 \times 10^{10}) \approx 10^{10^{10}}$, a very large number indeed. According to basic information theory the entropy of a black hole is a measure of the amount of information that has flowed into it. But as the black hole radiates energy it loses mass and about one unit of entropy for each photon it radiates. We may ask where the information has gone at the end point of evaporation; if the black hole radiates entirely away there is no information, and if it leaves a remnant there should be very little information left. In either case it appears that a great deal of information has been lost.

Associated with this is a question of quantum predictability, which is most easily seen looking backward in time. Suppose that a black hole is formed by the evolution of a quantum state evolving via the Schrödinger equation. In principle we should be able to calculate the initial state by running time backwards, but if the final state is a statistical mixture of photons and vacuum or a remnant we clearly cannot do this. Stated somewhat more technically, a pure initial state should not be able to evolve into a statistical or mixed state, which the final state appears to be.

The entropy of a black hole is proportional to its area, which is peculiar since entropy is proportional to volume for most other systems. Bekenstein showed that the black-hole entropy is an upper bound on the entropy of any system enclosed by the same area $A$. To see this suppose the contrary, namely that a system has a larger entropy; we could add energy and entropy to it until it collapses into a black hole with entropy larger than the Bekenstein value, which is a contradiction. Some take this to be a hint of something very deep, called the holographic principle. Since entropy is a measure of missing information, it appears that any region of space can in some sense be described in terms of a two-dimensional surrounding area instead of its three-dimensional volume. The situation is analogous to holography in optics, in which the interference pattern on a two-dimensional film surface produced by an object allows us to view the object from a range of directions; the two-dimensional film encodes information about the three-dimensional structure of the object. Interesting speculations on the appropriate degrees of freedom and structure of spacetime have been stimulated by this analogy.

## 2.11 Much ado about nothing, the cosmological constant

We now come to what is probably the biggest embarrassment in theoretical physics – the cosmological-constant problem. We may view the cosmological term $\Lambda g_{uv}$ as geometric and put it on the left-hand side of the field equations, as Einstein did. But we may also move it to the right-hand side of the equations and view it as a source of gravity. It then represents the stress energy when no ordinary matter is present, that is empty space or vacuum. Specifically, the energy density of the vacuum is related to the cosmological constant by $\Lambda = (8\pi G/c^4)\rho_{\text{vac}}$. Intuition might suggest that the energy density of the vacuum should be zero, since it is "nothing," but the observations mentioned in Section 2.7 indicate that vacuum energy or dark energy constitutes about 73% of the stuff in the Universe. (This is not detectable on the scale of the Solar System, so it is consistent with observations.) The value of the vacuum energy density is estimated to be about $6 \times 10^{-10}$ J m$^{-3}$, or 4 GeV m$^{-3}$.

Viewed as a stress-energy tensor, the cosmological term corresponds to a perfect fluid with the rather odd negative pressure $p = -\rho$. It is not hard to see how this comes about with a thought experiment. Place vacuum with density $\rho$ in a closed cylinder, with *ordinary* zero-density vacuum outside, as in Figure 2.12. Now slowly move the piston out a distance $dx$, to expand the volume by $A dx$. Since the vacuum energy density is constant the energy in the cylinder increases by $\rho A dx$. The increase in energy comes from work done on the gas by the piston, so the pressure of the interior vacuum on the cylinder must be negative. The work done is $pAdx$ so we see that $p = -\rho$.

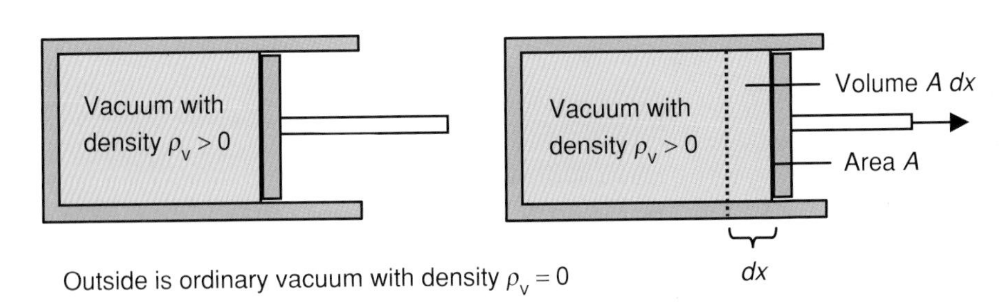


**Figure 2.12.** An experiment giving the vacuum equation of state as $p = -\rho$.

When we use quantum theory to estimate the energy density of the vacuum we encounter big trouble. In QED the ground-state energy density of the electromagnetic field is infinite, that is a divergent integral. Moreover, the divergence is not logarithmic like the more familiar divergences, but is quadratic and more difficult to handle. Almost any but the simplest of Feynman diagrams with no loops leads to a divergence, which must be removed by renormalization, that is by subtracting out the infinity on the basis of the physical argument that it occurs because the theory is not valid at short distances or high energies. Thus it might seem that we should simply cut off the divergent integral for the vacuum energy density at some reasonable value, such as the Planck energy, to obtain a finite and reasonable result. On dimensional grounds the result of this must be about the Planck energy divided by the Planck volume, $E_P/L_P^3$. This agrees with the actual calculation up to a factor of $1/(4\pi)$, but the observational value is about 4 GeV $m^{-3}$, whereas the Planck density is of order $10^{19}$ GeV$/(10^{-3}$ m$)^3 = 10^{124}$ GeV $m^{-3}$. These differ by an incredible 124 or so orders of magnitude, which is the cosmological-constant problem in its simplest form.

The problem is particularly embarrassing because the energy cutoff needed to naively resolve it is absurdly small. Since the Planck distance and energy are related by $L_P = \hbar c/E_P$ the Planck density can be written as $E_P^4/(\hbar c)^3$. To reduce this by 124 orders of magnitude requires that the cutoff energy be reduced from the Planck energy of $10^{19}$ GeV by about 31 orders of magnitude, to less than 1 eV, which is less than the energy of visible light and thus absurd. The problem is not a high-energy problem that we can defer until we understand the Planck scale and quantum gravity. It should be understandable now.

There has been much work done on the cosmological-constant problem, but surprisingly few claims of a solution. None is generally accepted. It is widely considered to be the biggest problem in fundamental physics because it is such a large discrepancy and casts doubt on our understanding of the basic concepts of quantum fields and geometric gravity.

## 2.12 Quantum-gravity theory

The conventional wisdom is that the main goal of theoretical physics is a complete quantum theory of particles and forces, including gravity and spacetime. There have been interesting attempts in this direction, but there is little empirical data to guide us since we cannot do experiments at the Planck scale.

A deep division has long existed in how physicists view spacetime. Relativists generally view it as the fundamental dynamical stuff of the Universe, whereas particle physicists generally view it as an arena that is unchanged by the events within it. Feynman began with the latter view and developed his own approach to quantum gravity in flat space, which we may call perturbative quantum gravity (PQG). This is a close analog of QED, and involves a spin-2 graviton field represented by a second-rank tensor instead of the spin-1 photon field represented by a 4-vector potential. The

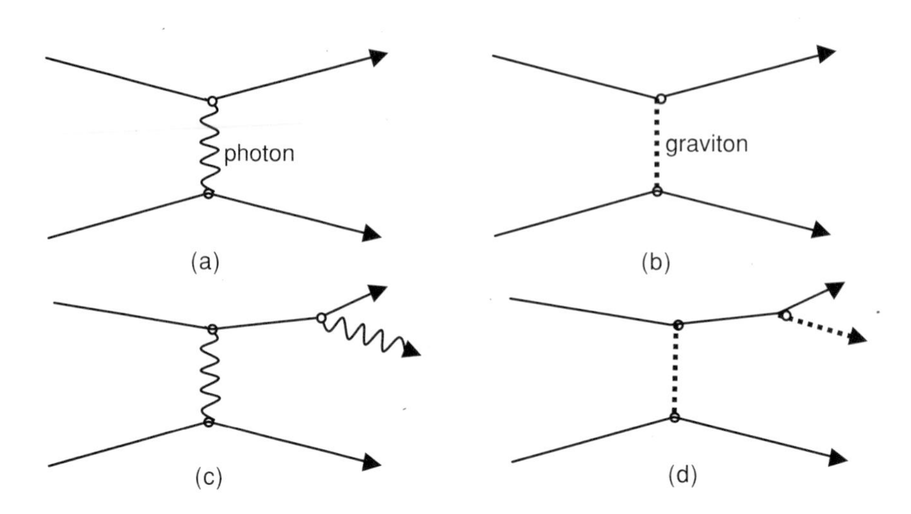


**Figure 2.13.** Feynman diagrams for (a) scattering of electrons with a virtual photon, (b) scattering of masses with a virtual graviton, (c) production of a real photon in electron scattering, and (d) production of a real graviton in mass scattering.

graviton is coupled to conserved energy momentum in the same way as the photon is coupled to conserved electric current. PQG leads to Feynman diagrams for processes involving gravitons, analogous to those of QED for photons, as in Figure 2.13. The predictions of PQG for effects like the deflection of light agree with general relativity. One amusing result is that a body in a weak graviton field behaves *as if* it were in a slightly curved space; thus the geometric interpretation of gravity is suggested instead of postulated. There are significant technical differences between PQG and QED for diagrams containing loops. The recipes to handle the divergent integrals are more complicated, and an infinite number of renormalization parameters is needed, but the theory is predictive up to about the Planck scale.

The more or less orthodox view is that as we approach the Planck scale all forces become comparable in strength, so that none can be ignored. Present theories are only approximate effective field theories that should be replaced by a unified theory, which is valid at the Planck scale and takes account of quantum effects on spacetime. Presumably such a theory would automatically explain the weakness of gravity.

We mentioned previously the breakdown of the classical spacetime continuum at the Planck scale. Let's look briefly at one quantum view of spacetime. In the Schrödinger picture of quantum fields a wave *functional* $\psi[\phi(\vec{x})]$ gives the relative amplitude that the field has the configuration $\phi(\vec{x})$. This functional is the direct analog of the wave function $\psi(\vec{x})$ in quantum mechanics, which is the amplitude that a particle is at position $\vec{x}$. We may specify any configuration of the field, and the wave functional gives the amplitude at which it may be observed. For the gravitational field we can set up a formalism in which spacetime geometry can be viewed as a 3-space geometry evolving in time. The wave functional is the amplitude for finding any 3-space geometry at time $t$, and can be written as $\psi[g^{(3)}_{\mu\nu}(\vec{x}), t]$. Naturally the classically expected geometry should have a large amplitude, but geometries with wild distortions like black holes and wormholes and the wrong signature can also occur. In general the shorter the time over which they occur the wilder such fluctuations can be. This is the sort of behavior we mentioned for the no-boundary model of cosmogenesis.

In the standard Copenhagen interpretation of quantum mechanics both a small quantum world of amplitudes and a large classical world of measuring instruments are necessary. The absolute square of the wave function is the relative probability of observing a particle at $\vec{x}$. For a small region of spacetime we can in principle use the same sort of interpretation of the wave functional, but a problem arises for a large system, especially the Universe containing ourselves and our measuring instruments. We know that a wave-function description of macroscopic objects is not generally appropriate due to decoherence effects, which wash out superposition and interference. It is thus doubtful that we should think of the entire present Universe in probabilistic quantum terms.

For the small early Universe a probability description remains problematic. Is an ensemble of universes merely a theoretical construct, or should we believe in the existence of a multitude of real alternative universes (a multiverse), or is this simply not the correct viewpoint? Such questions make quantum cosmology interesting. Indeed, this is further motivation to consider new ideas and experiments about decoherence and the relation of the quantum and classical worlds.

Certainly the best-known contender for a unified quantum theory of all forces is string theory, or the generalization called M theory. Any version of string theory appears to contain the germ of geometric gravity theory automatically, and moreover seems to generally predict a scalar field component to gravity called the dilaton. As mentioned previously, this might give rise to a violation of the equivalence principle and thus be testable in the STEP experiment. Very few ideas related to quantum gravity are testable.

Other notable approaches to quantum gravity result from mathematically reformulating the Einstein equations so that their structure is more amenable to quantization. These include the loop gravity of Abhay Ashtekhar and collaborators and the twistor gravity of Roger Penrose. A particularly nice feature of these is that a spacetime continuum need not be assumed a priori, but a quantum spacetime may be constructed, for example from mathematical objects called spin networks.

## 2.13 And now for something completely different . . . about spacetime

There is a famous story that Wolfgang Pauli once remarked that a certain theory was crazy, but not crazy enough. Quite crazy ideas may be needed if we are to understand spacetime and gravity at the fundamental level.

We first mention the so-called hierarchy question, of why the Planck scale is so remote from that of the nucleons of ordinary matter. We understand matter up to an energy of about 1 TeV. Beyond that there is little experimental data, and it is widely assumed that no drastically new things occur, except perhaps for the appearance of supersymmetric particles, up to the GUT scale at about $10^{16}$ GeV or the Planck scale at about $10^{19}$ GeV. On a logarithmic scale the GUT and Planck scales are very roughly the same. The gap between 1 TeV and the Planck energy is called the desert, on the presumption that it is devoid of new phenomena.

Some theorists have suggested that the desert might not exist. Since gravity is so weak, it has not been measured for distances below a centimeter or so, and it is possible that gravity becomes as strong as the other forces at a distance much larger than the Planck distance. If gravity operates in more than three space dimensions (say $3 + n$), as suggested by string theory, then the force at small distances is proportional to the inverse $(2 + n)$th power of distance. See Box 2.5. An analogy is shown in Figure 2.14; our perceived world is represented by the surface, and the space in which gravity operates by the bulk. Close to the mass the lines of force spread out in the bulk, but far away they spread out in the lower-dimensional surface. In $3 + n$ dimensions the strength of gravity grows more rapidly with decreasing distance (or increasing energy) than it would in three dimensions, lowering the unification energy at which all forces become equal. We can choose the number of extra dimensions $n$ and the crossover distance $R$, at which the force changes power behavior, so that the unification energy is somewhat

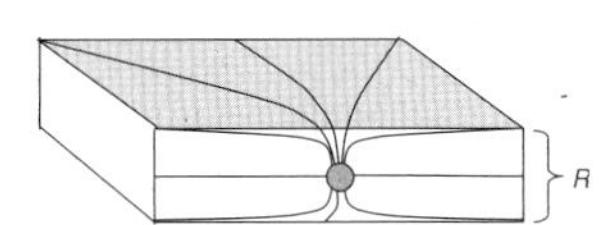


**Figure 2.14.** In this analogy field lines spread out like $1/r^2$ in the bulk near the body and like $1/r$ on the surface far away from the body.

BOX 2.5 MORE DIMENSIONS FOR GRAVITY

The gravitational force law in three space dimensions, far from a point mass, is Newton's inverse square law, but near the point mass it is assumed to become of inverse $2 + n$ power,

$$F_g = G m_1 m_2 / r^2 \quad \text{for } r \gg R, \qquad F_g = G m_1 m_2 R^n / r^{2+n} \quad \text{for } r \ll R. \tag{B9}$$

$R$ is roughly the size of the extra space dimensions, as in Figure 2.14. The unification energy at which the gravitational force becomes equal to the electroweak and strong forces is easy to estimate. The non-gravitational forces are very roughly given by

$$F_{ng} \approx e^2/r^2 \approx \hbar c \alpha / r^2. \tag{B10}$$

The dimensionless coupling parameter $\alpha$ is within a few orders of magnitude of unity for all the non-gravitational forces well above 1 TeV. Equality of the gravitational and non-gravitational forces for two particles of effective mass $m_1 \approx m_2 \approx E_u/c^2$ thus occurs when

$$G(E_u/c^2)^2(R/r)^n \approx \hbar c. \tag{B11}$$

We may write Newton's constant in terms of the Planck energy $E_P$ as $G = \hbar c^5/E_P^2$ and relate the distance $r$ being probed to the probe energy by the standard relation $r \approx \hbar c/E_u$. This gives the desired expression for $R$,

$$(\hbar c/R)^n = (E_u/E_P)^2 E_u^n \quad \text{so} \quad R = (E_P/E_u)^{2/n}(\hbar c/E_u). \tag{B12}$$

If the unification energy is equal to the Planck energy then the size $R$ is the Planck distance, independently of the dimension $n$. Reasonable values for $n$ and $R$ are given in the main text.

above 1 TeV, thus eliminating the desert. Some examples are, very roughly, as follows: if $n = 1$ (four space dimensions) then we must choose $R \approx 10^{13}$ m, which is trivially ruled out by everyday experience; if $n = 2$ (five space dimensions) then $R \approx 10^{-3}$ m, which is a possibility that has motivated new gravitational experiments in this range; if $n = 3$ (six space dimensions) then $R \approx 10^{-9}$ m, which is about atomic size and of interest. Remarkably, the modification to gravity would have a negligible effect on atomic spectra.

According to this picture there should, in our observable world, be an apparent nonconservation of energy and momentum since they can be transferred by gravity to the bulk. Such effects should be detectable in particle experiments at the Large Hadron Collider (LHC) scheduled to begin experiments in a few years at energies well above 1 TeV. We have previously noted the possibility that the cosmological dark matter may be ordinary matter in a parallel or shadow universe which acts on our Universe only by gravity.

Gravity is generally considered to be fundamental, but this can be questioned, especially in view of its extreme weakness. It may be only a classical large-distance manifestation of something more fundamental, analogous to the Van der Waals force between molecules, which is a weak residual effect of the electromagnetic force. Andrei Sakharov suggested that fundamental fields such as electromagnetism cause, in their ground states, a stress in the vacuum that is perceived as gravity. The weakness of gravity is due to the large Planck cutoff energy for the divergent integrals that arise. The rather qualitative suggestion of Sakharov has led to further work in the field, now called induced gravity.

A somewhat similar idea, in spirit, is that general relativity emerges as an approximation from a more fundamental spacetime substructure. To some condensed-matter theorists general relativity has the objectionable property that there is often no global time coordinate for the Schrödinger evolution of a wave function. (In special relativity the proper time in any inertial frame serves this purpose.) This motivates abandoning general relativity as a fundamental theory in favor of a theory with a global time, and probably a corresponding special restframe. Robert B. Laughlin and collaborators have emphasized that the behavior of light approaching a black-hole surface is reminiscent of that of a sound wave approaching a phase-discontinuity surface; the sound wave slows down and loses its identity as a wave. They therefore suggest that a black-hole interior may be vacuum in a phase different from that of the exterior, one with larger energy density, analogous to a condensed quantum fluid. The black-hole surface would be a phase boundary, not the simple infinitely thin horizon of general relativity. We can think of the vacuum in this view as a quantum-fluid aether.

The idea of emergent general relativity has other attractive features. There would be no singularities inside black holes, and probably none anywhere. Black holes would not have the Hawking temperature or radiate, and lab experiments on analog quantum fluids might provide clues to their behavior, such as distinctive observable properties like reflection and frequency shifts of radiation falling on their surface.

There are also big problems with this approach. The biggest is consistency with observation, since special and general relativity have been verified extensively and accurately. In particular, the problem of explaining the constancy of the velocity of light in the original classical aether theory reappears.

Humans have made good use of the concepts of space and time. Our survival in the real world from prehistory to the present has depended on understanding where things are and when events happen. Immanuel Kant insisted that space and time are necessary mental constructs, innate to the human mind. But we should recognize that the description of physical spacetime by a mathematical continuum is only a model. It has served well from the scale of the cosmos down to subnuclear size, but there is no guarantee that it will continue to be useful for arbitrarily small scales. The divergences of quantum field theories may already indicate that a better mathematical model of physical spacetime is necessary, or perhaps that the very concept of spacetime is flawed.

Two opposing metaphysical views of space and time have influenced physical theories. One view is that space and time have independent reality and would exist in the absence of objects and events. The opposite view is that space and time are nothing but the relation between objects and events, so there would be no spacetime without objects and events. The relational view leads us to ponder how we actually measure spacetime and question whether it serves a useful purpose at a small scale where there are no longer distinguishable objects and events. Indeed, it is clear that for subatomic phenomena we generally measure momenta and only infer spacetime properties using our models and Fourier transforms.

Many ingenious attempts to understand spacetime at the Planck scale have been made: viewing it as the argument of a wave functional, a manifestation of the ten dimensions of string theory, constructed from loops or spin networks, a quantum fluid, the eigenstate of a quantum operator, and many more. These are rather reminiscent of attempts to understand the aether in the later half of the nineteenth century. Aether was then considered necessary for the propagation of electromagnetic waves, but finally proved to be unobservable and extraneous, that is effectively nonexistent. It may be that Planck-scale spacetime similarly does not exist. Perhaps the best we can do is to understand the behavior of momenta and cross-sections at the Planck scale, in the spirit of an S-matrix theory of momenta like that of the 1960s, with spacetime emerging only as a larger-scale approximation.

## WEBSITES

For references and links to educational and research material, see http://math.ucr.edu/home/bacz/relativity.html.
For information and a review on supernova data and the dark energy there is a website for the Supernova Acceleration Probe (SNAP), a proposed satellite experiment, at SNAP.LBL.GOV. For extensive information on the cosmic microwave background and cosmological parameters see the WMAP site at MAP.GSFC.GOV.

## FURTHER READING


1. P. Davies (ed.). *The New Physics*, Cambridge, Cambridge University Press, 1988.
2. S. W. Hawking, *A Brief History of Time*, New York, Bantam Books, 1988.
3. S. W. Hawking, *The Universe in a Nutshell*, New York, Bantam Books, 2001.

Both Hawking books are readable and charming, but Hawking often does not distinguish between ideas that are verified or mainstream and his own speculations.

4. A. Zee, *An Old Man's Toy: Gravity at Work and Play in Einstein's Universe*, New York, Macmillan, 1989.

Very readable with a good discussion of the equivalence principle.

5. M. Riordan and D. Schramm, *The Shadows of Creation: Dark Matter and the Structure of the Universe*, New York, W. H. Freeman and Co., 1991.
6. L. Dixon (ed.), *Gravity From the Hubble Length to the Planck Length: Proceedings of the XXVI SLAC Summer Institute on Particle Physics*, Stanford, California, Stanford Linear Accelerator Center (SLAC report 538), 1998.
7. B. F. Schutz, *A First Course in General Relativity*, Cambridge, Cambridge University Press, 1999.
8. N. Arkani-Hamed, S. Dimopoulos, and G. Dvali, *Phys. Lett. B* **429** (1998) 263.
9. R. J. Adler, Pisin Chen, and D. Santiago, The generalized uncertainty principle and black hole remnants, *General Relativity and Gravitation*, **33** (2001) 2101.
10. G. Chapline, E. Hohlfeld, R. B. Laughlin, and D. I. Santiago, Quantum phase transitions and the breakdown of classical general relativity, *Phil. Mag. B* **81** (2001) 235.


# Ronald Adler

After having obtained his B.S. in physics from Carnegie Mellon University in 1959 and a Ph.D. in physics from Stanford in 1965, Ronald J. Adler moved in turn to the University of Washington, Seattle, the University of Colorado, Boulder, the Virginia Polytechnic Institute, the American University, Washington, and the Institute for Advanced Study, Princeton. From 1974 until 1977 he was Professor of Physics at the Federal University, Pernambuco, Recife, Brazil. In 1988 he became Professor of Physics (adjunct) at San Francisco State University. After having been a Visitor at the Stanford Linear Accelerator Center, in 1995 he became Visiting Professor of Physics at Stanford University, and had meanwhile joined the Stanford University Gravity Probe B theory group.

On the industrial side, he has worked as a Mathematical Analyst at Decision Focus Inc., Palo Alto, and as a Senior Scientist at Lockheed's Palo Alto Research Laboratory. He has contributed on relativity to many books, encyclopedias, physics journals, reports, and conference talks.

# 3 The new astronomy

Arnon Dar

## 3.1 Introduction

The great discoveries in physics and the technological breakthroughs in the twentieth century have completely revolutionized astronomy – the observational study of the physical Universe beyond Earth and its theoretical understanding. These great discoveries included special relativity, general relativity, quantum mechanics, atomic structure, and nuclear structure, together with the elementary particles and their unified interactions. The technological developments of the twentieth century which had the greatest impact on observational astronomy included microelectronics, microdetectors, computers, and space-age technologies. They allowed astronomical observations deep into space with unprecedented resolution and sensitivity. The New Physics, together with these observations, led by the end of the twentieth century to an amazing understanding of an extremely complex Universe that contains more than $10^{21}$ stars in more than 100 billion galaxies with enormous variety, diverse environments, and complex evolutions. Nevertheless, astronomy, one of the oldest sciences, is still one of the most rapidly developing. This is because many fundamental questions related to the origin of our physical Universe, to its contents, to its laws, and to the existence of life in it are still unanswered. They may be answered as science progresses, new technologies for high-resolution observations are exploited, and new fundamental theories are developed and tested. In this chapter, we give a brief account of our present knowledge of the physical Universe, our current understanding of it, and our major observational endeavors to widen this knowledge and understanding.

## 3.2 Advances in observational astronomy

Until the invention of the optical telescope for military purposes at the beginning of the seventeenth century, astronomical observations were made with the naked eye. The Universe observable from planet Earth included only five other planets – Mercury, Venus, Mars, Jupiter, and Saturn – orbiting the Sun and a few thousand more distant stars. The invention of the telescope dramatically increased the horizon of the observable Universe, the number of observable stars, and the resolving power of observations. The subsequent development and construction of large reflecting telescopes with the use of photographic plates for imaging and photometry and of prisms for spectroscopy led, by the end of the third decade of the past century, to more discoveries: the Solar

*The New Physics for the Twenty-First Century*, ed. Gordon Fraser.

System contains another three planets – Uranus, Neptune, and Pluto; the Sun is but one star on the outskirts of the Milky Way, an island of about 100 billion stars; the Milky Way is but one island in an enormously larger observable Universe, which contains more than ten billion such islands of stars, called *galaxies*; and the Universe looks roughly similar in all directions and is expanding, so all galaxies appear to be receding from each other at a rate proportional to their separation – the law discovered by Edwin Hubble in 1929.

By the middle of the twentieth century, observations also confirmed that the microscopic laws of physics seem to be the same everywhere and determine the behavior of the Universe at all scales and times. These emphasized the two major achievements in theoretical astronomy in the second half of the twentieth century – the standard stellar-evolution theory and the standard Big-Bang model of cosmic evolution. Not only could these theories explain the main astronomical observations, but they also made predictions that were verified, sometimes accidentally, by improved observations and by new observational techniques.

A major advance in ground-based observations after World War II was the development of radio telescopes, which benefited from the wartime development of radar. This marked the birth of radio astronomy and led to the accidental discovery by Arno A. Penzias and Robert W. Wilson in 1964 of the microwave background radiation from the Big Bang (whose existence had been predicted by George Gamow in 1946) and to the discovery in 1967 by Jocelyn Bell and Antony Hewish of *radio pulsars* – highly magnetized *neutron stars*, born in *supernova explosions* following the evolution of massive stars.

Another major advance in observational astronomy, which benefited from the arms race and the technological development during the Cold War, has been the ability to place telescopes in space to observe those wavelengths of electromagnetic radiation which do not penetrate the Earth's atmosphere. These include almost all wavelengths except the visible and the radio bands. The deployment of such telescopes led to major discoveries. The Cosmic Background Explorer (COBE) launched in 1990 measured the microwave background radiation spectrum with unprecedented accuracy and discovered the predicted tiny spatial fluctuations in its temperature associated with the beginning of structure formation in the Universe. The X-ray satellites GINGA, ASCA, and ROSAT launched in the 1980s, RXTE and Beppo-Sax launched in the 1990s, and the X-ray observatories Chandra and XMM Newton launched in 2000 discovered a variety of X-ray sources such as *X-ray binaries*, stellar-mass *black holes*, isolated X-ray pulsars and *anomalous pulsars*, active galactic nuclei with *supermassive black holes*, and the cosmic X-ray background, and provided important information on the diffuse cosmic X-ray background as well as on diffuse X-ray sources such as *supernova remnants*, the interstellar space in galaxies, and the intergalactic space in *clusters* and *groups of galaxies*.

The Compton Gamma Ray Observatory has discovered and clarified the nature of a variety of Galactic and extragalactic gamma-ray sources, provided the first evidence that *gamma-ray bursts* are of cosmological origin (this was confirmed in 1997 by the localization by the Beppo-SAX satellite of gamma-ray bursts and ensuing ground-based observations), and provided valuable information on the diffuse gamma-ray background radiation. However, a telescope in space can be important even in the visible and near-visible part of the spectrum, as was demonstrated by the Hubble Space Telescope. It allows us to make observations free of the blurring caused by atmospheric conditions and to reach an angular resolution near the theoretical limit, $\sim\lambda/D$, where $\lambda$ is the wavelength and D is the diameter of the telescope aperture. In the future, adaptive optics may overcome, to a large extent, atmospheric blurring and distortions due to weight in ground-based optical telescopes (Figure 3.1). Nonetheless, space-based

**Figure 3.1.** The Hubble Space Telescope (HST), with its solar arrays backlit against the black background of space. The 2.4-m-diameter HST was the first and flagship mission of NASA's Great Observatories program. From its orbit above the Earth's atmosphere, it opened an important new window on the heavens, observing in visible, near-ultraviolet, and near-infrared wavelengths. (Courtesy of STScI/NASA.)

telescopes will still be needed for observations in wavelength bands other than optical and radio such as the far-infrared, far-ultraviolet, X-ray, and gamma-ray bands.

Past experience has taught us that whenever a new window in the electromagnetic spectrum was opened for observational astronomy, this led to major astronomical discoveries. Thus, although the major observational efforts in astronomy are still based on viewing the Universe through electromagnetic radiation (Figure 3.2), astronomers are constantly looking for new observational windows. At present the new windows include ultra-high-energy cosmic rays, neutrinos, and gravitational waves. The observational efforts in these windows are still mainly exploratory. They are strongly motivated by evidence that the most energetic phenomena in the Universe are also copious sources of high-energy cosmic rays, neutrinos, and gravitational waves.

Cosmic rays, discovered by Victor Hess in 1911, are high-energy particles, mostly atomic nuclei and electrons, that constantly bombard the Earth's atmosphere from outer space. Their energies extend up to at least  eV per particle. Their energy spectrum and composition provide valuable astrophysical information on their sources and the properties of space along their trajectories. However, despite intensive investigations throughout the twentieth century, their origin is still not known; nor is the origin of their energy spectrum and chemical composition fully understood. The arrival directions of most cosmic rays are made isotropic by the magnetic fields in the Galactic and intergalactic space that they travel through, and so they do not point to their sources. Only the arrival direction of the most energetic ones, whose energy is above eV, can point to their sources, if they are Galactic, or to their extragalactic sources if they have a higher energy and come from relatively nearby extragalactic sources. The mean free path of ultra-high-energy cosmic-ray nuclei and electrons in the electromagnetic background radiation is much shorter than the typical distance to their putative

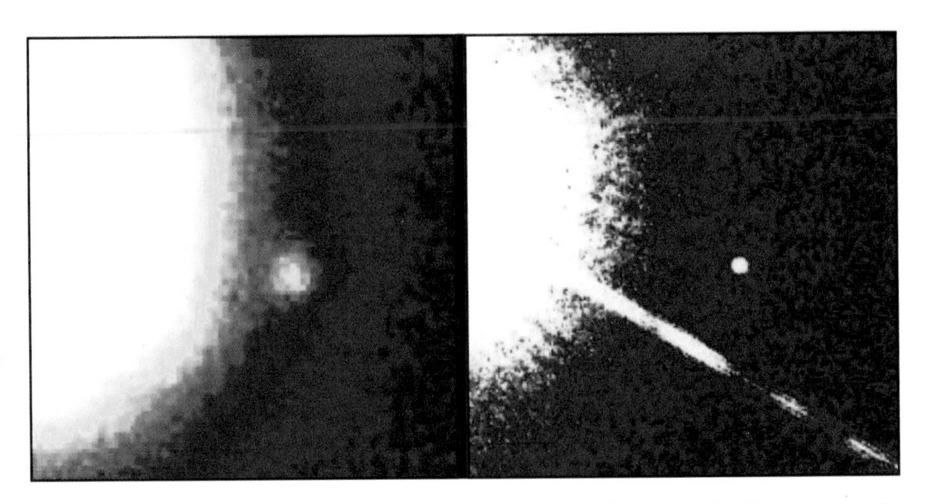

**Figure 3.2.** Astronomical objects come in a wide range of sizes. The Brown Dwarf Gliese 229B is a small companion to the red star Gliese 229, 19 light-years from Earth in the constellation Lepus. Estimated to be 20–50 times the mass of Jupiter, it is too massive and hot to be classified as a planet, but too small and cool to shine like a star. Brown dwarfs form in the same way as stars do, by condensing out of a cloud of hydrogen gas. However, they do not accumulate enough mass to generate the high temperatures needed to sustain nuclear fusion in their core, which is the mechanism that makes stars shine. Instead brown dwarfs shine in the same way as gas-giant planets like Jupiter radiate energy, through gravitational contraction, and are difficult to observe. The left-hand side shows the discovery of Gliese 229B by the Palomar telescope and the right-hand side shows its resolved image obtained by the Hubble Space Telescope. (Courtesy of S. Kulkarni, D. Golimowski, and NASA.)

extragalactic sources, namely quasars and active galactic nuclei. Preliminary results from large ground-based extensive air-shower arrays, which detect them and measure their energies through their atmospheric showers, did not provide clear evidence for their Galactic or extragalactic origin. This is due to poor statistics and difficulties in energy calibration at the highest energies. These may be solved with the new giant ground-based air-shower arrays such as the Auger project, or with space-based telescopes such as FUSE that will detect fluorescent light from extensive atmospheric air showers and the Čerenkov light from their ground impact.

Neutrinos are weakly interacting particles whose cross-sections are very small except at very high energies. This makes them very penetrating. Neutrino emission from optically thick astrophysical objects can be used to view their interiors, which are invisible from the electromagnetic radiation emitted from their surfaces. However, the very weak interactions of neutrinos also make their detection very difficult, in particular in the high-background environment created by cosmic-ray interactions in the atmosphere and by radioactive decays in the ground and in detector materials. To reduce background, neutrino telescopes are usually placed deep underground (Super-Kamiokande, SNO), in tunnels under high mountains (Baksan, Macro), underwater (Baikal, Antares, Nestor), or under ice (Amanda). At high energies, the large mass around the detector is used also to convert neutrinos into charged particles that are detected electromagnetically or acoustically by the telescopes. So far, neutrino telescopes have succeeded in detecting and measuring solar neutrinos and neutrinos from the nearby supernova 1987A in the large Magellanic cloud. These results provided valuable astrophysical information and signaled the birth of neutrino astronomy (Figure 3.3).

Einstein's theory of gravity, general relativity, predicts the existence of gravitational waves. The first gravitational antennas to detect gravitational waves from astrophysical sources were built in the 1960s but did not have sufficient sensitivity. The design and development of more sensitive interferometric gravitational-wave detectors began in the mid 1970s and early 1980s, finally leading to the large LIGO and VIRGO projects, as well as others of smaller dimensions (GEO-600 and TAMA), and to the future space-based interferometric antennas LISA. In parallel, cryogenic resonant detectors were designed and constructed in several laboratories, and in the mid 1990s the ultracryogenic antennas NAUTILUS, AURIGA, ALLEGRO, and NIOBE were developed. In spite of many years of endeavor, gravitational waves have proved elusive. However, the discovery of the binary radio pulsar PSR 1913+16 by R. Hulse and J. Taylor in 1974 and their long-term high-precision monitoring of the decay of its orbital motion, behaving exactly like that expected from gravitational-wave emission, is compelling, although indirect, evidence for the emission of gravitational waves.

**Figure 3.3.** A view of the Sudbury Neutrino Observatory (SNO) during assembly. The detector, 2200 m underground in the Creighton mine near Sudbury, Ontario, Canada, is a Čerenkov light detector using 1000 tonnes of heavy water in a 12-m-diameter acrylic vessel. Neutrinos react with the heavy water to produce flashes of light, which are detected by an array of 9600 photomultiplier tubes mounted on a geodesic support structure surrounding the heavy-water vessel. The detector is immersed in light (normal) water in a 30-m barrel-shaped cavity. The rock above shields the detector from cosmic rays. SNO succeeded in detecting the ("missing") solar neutrinos which on their way to Earth transform ("oscillate") into other neutrino species, providing new insights into the properties of neutrinos and the mechanisms at work deep inside the Sun. (Courtesy of SNO.)

What impact on our understanding of the Universe did this enormous progress in observational astronomy have? Basically, it confirmed our two fundamental theories of stellar and cosmic evolution, and helped us to refine, improve, and expand these theories to better describe our Universe.

## 3.3 The Standard Model of cosmic evolution

The cosmos, the grand stage where stellar evolution takes place, is also evolving. Cosmic evolution (see Chapter 1) is described quite well by the inflationary Big-Bang model – the standard model of cosmology (the name "Big Bang" was originally, and facetiously, invented by Fred Hoyle). It is based on four observationally motivated assumptions.

- The space, time, and mass of the Universe are related by Einstein's general relativity theory.
- The Universe is expanding from a homogeneous, isotropic, very hot, and dense initial state.
- In its very early stage the Universe's spatial dimensions were inflated at an apparently superluminal speed.
- The elementary constituents of matter in the Universe and their interactions are well described by the Standard Model of particle physics (see Chapter 4).

The hot-Big-Bang model yields testable predictions that are in good agreement with observations. The age of the Universe, which can be calculated from its expansion rate and its matter and energy density to be around 14 billion years, agrees well with the age inferred by radioactive dating of the oldest matter in the Universe, or from the age of the oldest stars. The relative abundances of the light elements (hydrogen, deuterium, helium, and lithium), which were produced during the first few minutes during which the Universe cooled down by expansion to a temperature below about a billion degrees Kelvin, agree with those observed. The observed relic electromagnetic radiation from the Big Bang, which fills the whole Universe, has exactly the properties predicted by the theory (the right temperature, almost perfect isotropy over the sky, and tiny fluctuations due to the beginning of large-scale structure formation in the Universe), as does the relic neutrino radiation from the Big Bang, whose existence and properties can be inferred from the observed relative abundances of the light elements.

Despite the spectacular success of the Standard Model of cosmic evolution, there are still major puzzles concerning the origin of the Universe, its matter and energy content, and how they affect the formation of structure in it. Most of the matter in the Universe is not the stuff that stars and planets are made of, and is dark – not emitting electromagnetic radiation – and we do not know what it is. Most of the current energy density in the Universe seems to be antigravitating, accelerating its expansion. It is not known, either, what this mysterious "dark energy" is. However, although dark matter and dark energy play a dominant role in structure formation and probably determine when and how such structure formation and later stellar formation do begin, there is no evidence that they play a significant role in stellar evolution. Although not completely understood, it is believed that after matter and radiation decoupled, sufficiently large density fluctuations decoupled from the cosmic expansion, grew through gravitational attraction and accumulation of more matter, contracted, cooled, and finally formed protogalaxies with dense molecular clouds inside them – the main sites of stellar evolution.

## 3.4 The standard theory of stellar evolution

Stars form mainly inside dense molecular clouds of typically $10^5$ Solar masses which are shielded from the galactic ionizing radiation. For reasons that are not fully understood, but which may have to do with collisions of molecular clouds, or with passage of shock waves through molecular clouds as the clouds pass through spiral structure in galaxies, or with magnetic–gravitational instabilities, the dense core of a molecular cloud begins to collapse under its self-gravity and to fragment into clumps 0.1 parsec in size and of 10–50 Solar masses. These clumps continue to collapse under their own gravity. Angular-momentum conservation turns their irregular shape into a rotating disk around a denser central region that becomes a *protostar* whose mass, density, and temperature are increased rapidly by infalling matter. During the initial collapse, the star is transparent to its thermal radiation and the collapse proceeds fairly quickly. As the clump becomes more dense, it becomes opaque to its own radiation. Its central temperature rises and, after a few million years, thermonuclear fusion of hydrogen nuclei into helium nuclei begins in its core. The thermonuclear energy release increases further the central temperature and accelerates the rate of thermonuclear energy release. The increase in kinetic and radiation pressure produces a strong stellar wind that stops the infall of new mass. The protostar then becomes a young star with a fixed mass, which determines its future evolution. It is usually surrounded by a massive, opaque disk that gradually accretes onto the stellar surface. Energy is radiated both from the disk at infrared wavelengths and from the regions where material falls onto the star at optical and ultraviolet wavelengths. Somehow a fraction of the material accreted onto the star is ejected in highly collimated jets perpendicular to the disk plane. Mass accumulation through gravitational attraction and energy dissipation in the circumstellar disk probably lead to the formation of a *planetary system* around the young star. The radiation and stellar winds from the clusters of young stars in the cores of the molecular clouds heat the surrounding gas and form H II (ionized hydrogen), which is blown away, and the stars form a cluster where they continue their evolution (Figure 3.4).

The evolution of a star basically depends on its mass. For most of its life, a star that is mainly made of the hydrogen and helium that were formed in the Big Bang is in quasistatic hydrostatic equilibrium with its self-gravity balanced by internal pressure. In this stage, essentially, every intrinsic property is determined by its mass. The length of time the star can live by burning hydrogen (the available fuel) and its luminosity (how quickly it uses that fuel) depend on its mass. The mass of a star also determines what other elements it can use as nuclear fuel, and as what kind of stellar remnant it will end up. This is one of the great achievements of stellar-evolution theory: given the initial mass of a star, one can write its entire life story in amazing detail. The life history of a typical star is called the *Main Sequence.*

A star's life is a constant battle against gravity, the force that tries to compress it, and the pressure support that resists. That battle begins as soon as a molecular cloud begins to collapse into a protostar. During the Main Sequence, the star achieves stability by gas pressure. Collapse is prevented because, if the star were to compress slightly, it would become hotter and denser and the rate of nuclear fusion would increase, causing it to expand again. But when the hydrogen in the core is used up, the core must resume its original collapse. This collapse converts gravitational energy into heat. Because this heat is transferred to the outer layers of the star, the core can't use it to maintain pressure support and the star contracts. Meanwhile, that heat which is released to the outer layers has important consequences. It heats the hydrogen just outside the core to high enough temperatures to undergo fusion. This causes the luminosity to increase. The hydrogen burns in a shell around the helium core, making the outer layers or "envelope" of the star expand. The radius increases so much that the surface temperature actually decreases, and the star becomes a *red giant.* The stellar mass is now distributed over a much larger volume, and has a much smaller density (Figure 3.6).

All this time the core continues to contract and heat up, until finally it becomes hot enough to fuse helium into carbon by a reaction called the triple-alpha process. Carbon can also fuse with helium to form oxygen. At this stage, the contraction of the core stops: the star has a helium-burning core surrounded by a hydrogen-burning shell. Eventually, of course, even the helium in the core becomes exhausted. This happens after only about 100 million years, because helium burning is less efficient than hydrogen burning. Now the core is mostly carbon and oxygen (CO) and once

**Figure 3.4.** A star incubator: hydrogen gas and dust pillars protrude from the interior wall of a dark molecular cloud in the Eagle Nebula, 6500 light-years away. The tallest pillar (left) is about four light-years high. (Courtesy of J. Hester, P. Scowen, and NASA.)

again continues its collapse. The star produces energy by burning helium in a shell around the core and hydrogen in a shell around that shell.

Stars smaller than about five Solar masses never become hot enough to achieve fusion in their CO cores at the end of their red-giant phase. The collapse of the core therefore continues until a new kind of pressure is able to support the star against gravity. This is electron-degeneracy pressure, a consequence of the Pauli exclusion principle in quantum mechanics which does not allow two electrons to occupy the same quantum state. When electrons are compressed into smaller volumes they must occupy higher and higher quantum energy levels, where they move with higher and higher momentum and produce a higher pressure. Even at absolute zero (0 K), the electrons produce a pressure that increases with density and becomes important in very dense matter. The radius of a core of about one Solar mass that is supported by electron-degeneracy pressure is about that of planet Earth, and then its mean density is about one ton per cubic centimeter. In the final stages of a low-mass star's life, the outer layers of the star become unstable and are ejected into space, forming a planetary nebula. Fusion in the hydrogen and helium shells stops and the core becomes exposed as a *white-dwarf* star (Figure 3), slowly cooling because it is no longer generating energy. Since degeneracy pressure is insensitive to temperature, the white dwarf does not contract as it cools, eventually becoming a cold *black dwarf.*

A white dwarf whose mass exceeds about 1.4 Solar masses is no longer able to support itself against gravity by electron-degeneracy pressure and continues its collapse until a new source of pressure is able to stop the collapse. This limit, known as the Chandrasekhar mass limit, was derived by Subrahmanyan Chandrasekhar from simple quantum-mechanical considerations. This theoretical prediction is strongly supported by observations: no white dwarf with a mass greater than 1.4 Solar masses has been found. In fact, the vast majority of white dwarfs we see are nowhere close to the Chandrasekhar limit. It seems that stars even of 2–5 Solar masses are able to lose enough mass during their planetary-nebula stage to leave behind white dwarfs of less than a Solar mass.

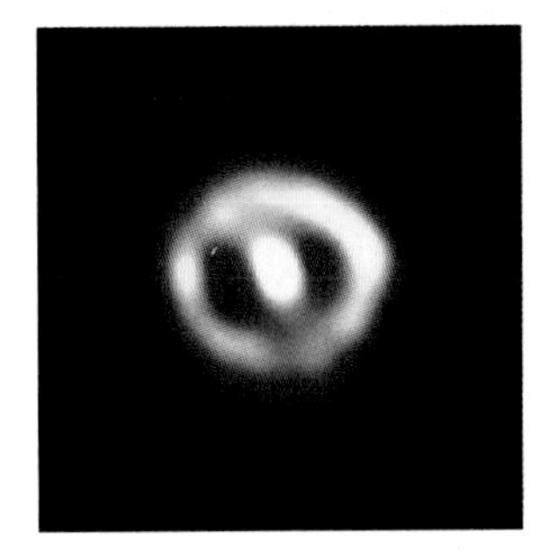

**Figure 3.5.** Nova Cygni 1992 is a thermonuclear explosion seen by the Hubble Space Telescope on February 19, 1992 on the surface of a white dwarf in a double star system 10 000 light-years away in the constellation of Cygnus. The ring is the edge of a bubble of hot gas blasted into space by the nova. (Courtesy of F. Paresce, R. Jedrzejewski, and NASA/ESA.)

Once created, isolated white dwarfs simply cool and fade away. However, a white dwarf in a close binary system can accrete material from the companion star and slowly increase its mass. Such a companion may be a red giant or red supergiant whose outermost layers are held by a very weak gravitational force. The details of the mass transfer are complex and depend on a variety of factors: the proximity of the two stars to each other, their masses, the extent of the secondary star's envelope, and the presence of magnetic fields. However, such mass-transfer systems seem to share one common feature – the formation of an accretion disk around the compact star. The matter pulled off the secondary star does not fall directly onto the compact star. Instead, it orbits around, forming a disk. Magnetic turbulence and friction slow the matter's orbital motion, which causes the matter to spiral through the disk or fall down directly onto the surface of the compact star. The inspiralling of the matter toward the white dwarf releases large amounts of gravitational energy and heats the accretion disk. Because of the large gravitational potential on the surface of a white dwarf, the accreted material may accumulate, reach nuclear-ignition conditions and trigger a thermonuclear explosion and mass ejection that are accompanied by an intense brightening. The star becomes visible across large interstellar distances hence the name *nova*, meaning "new." Mass transfer and accretion then resume until another nova outburst occurs. In a *dwarf nova*, for reasons that are not well understood, the matter accumulates in the accretion disk until an instability develops and the entire accretion disk crashes down onto the white dwarf. There is no nuclear explosion but large amounts of gravitational energy are released. The released energy heats the white-dwarf surface and temporarily brightens the star, less than in classical novae. Within weeks to months, a new accretion disk forms and in its turn becomes unstable, and the outburst is repeated.

If the accreted mass does not explode, the white dwarf may go over the Chandrasekhar mass limit and the sudden collapse of the star can cause runaway fusion of its carbon and oxygen. The star becomes a huge thermonuclear bomb, a *type-I supernova* as bright as a billion stars, and can be seen from cosmological distances. It is thought that this kind of supernova was observed in 1572 by Tycho Brahe and in 1604 by Johannes Kepler (Figure 3.7).

The nuclear reactions that occur during a type-I supernova explosion can produce substantial quantities of heavier elements. Carbon and oxygen can fuse to form silicon, and silicon nuclei can fuse to form nickel. This is one of the ways that heavy elements can be synthesized and ejected into the interstellar medium, where new stars, planets, and life can be born.

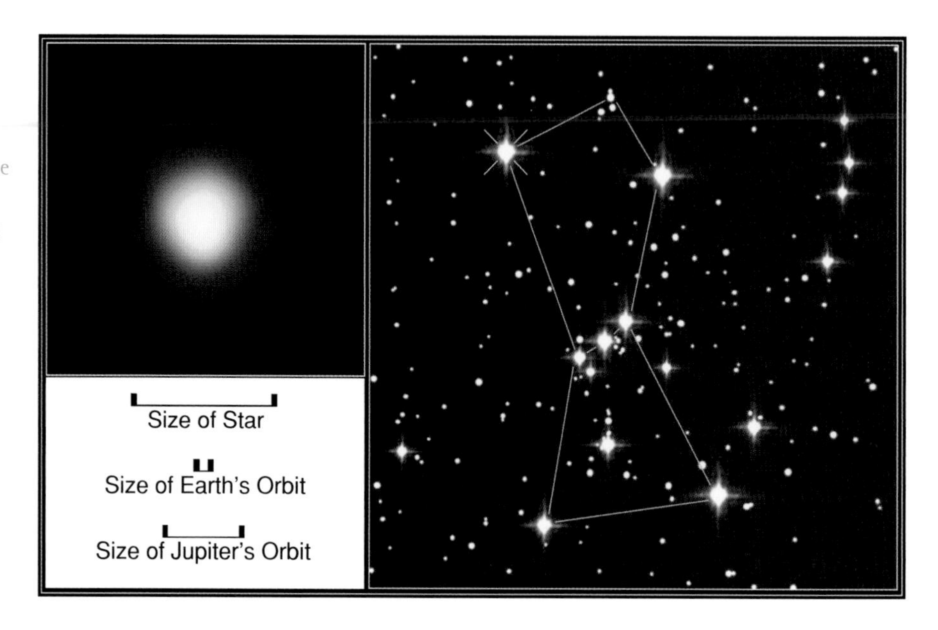


**Figure 3.6.** Betelgeuse is a red supergiant star marking the shoulder of the constellation Orion. Betelgeuse is so huge that, if it replaced the Sun at the center of our Solar System, its outer atmosphere would extend past the orbit of Jupiter. This image taken by the Hubble Space Telescope reveals a huge ultraviolet atmosphere with a mysterious hot spot on the stellar surface, more than ten times the diameter of Earth. (Courtesy of A. Dupree, R. Gilliland, NASA, and ESA.)

**Figure 3.7.** The Crab Nebula is the remnant of the supernova explosion seen in 1054 AD (see also Figure 3.9). This is how it appears today as seen by the Palomar ground-based optical telescope. In the explosion, most of the mass of the short-lived massive progenitor star was ejected at speeds of thousands of kilometers per second, sweeping up all the interstellar gas in its path as it rushed outward from the explosion site. The deceleration of the ejected matter is the result of the uneven concentration of gas it encounters in the depths of space. Thus the nebula appears irregular, with beautiful wisps, knots, and filaments. (Courtesy of J. Hester, P. Scowen, and NASA.)

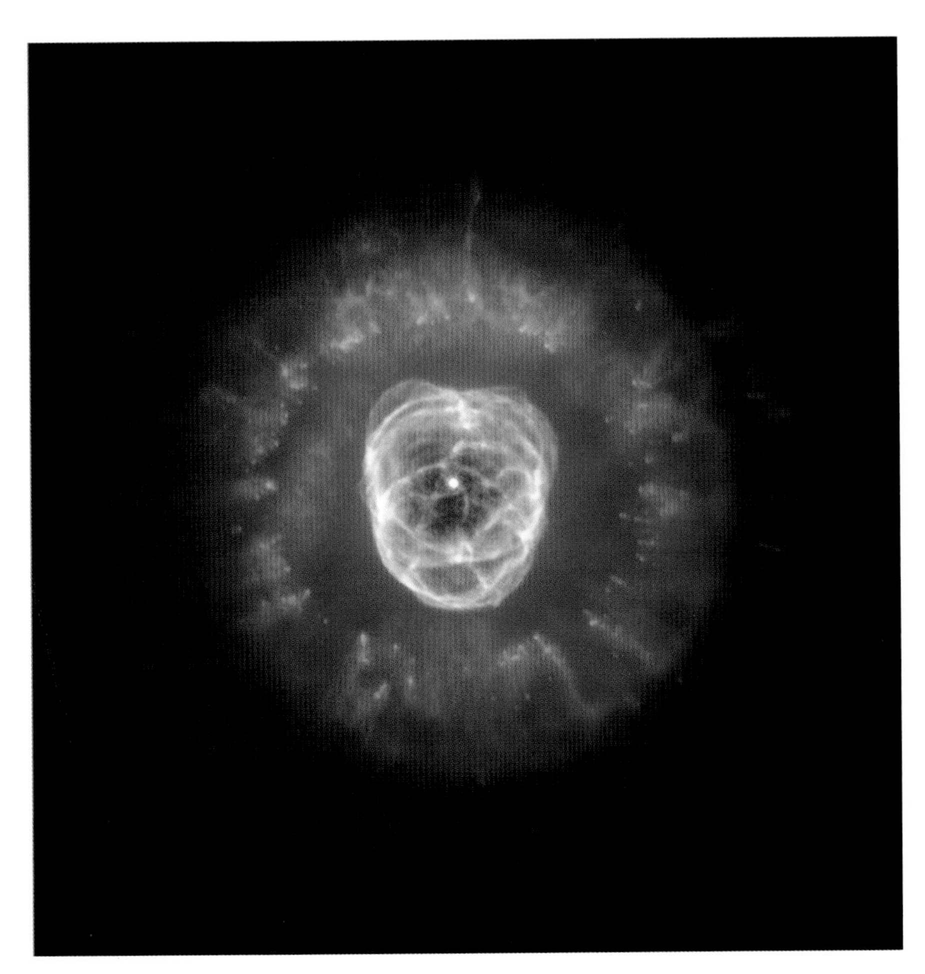

**Figure 3.8.** The planetary nebula nicknamed the Eskimo is the spectacular death throes of a Sun-like star after it has depleted its nuclear fuel – hydrogen and helium – and begins blasting off layers of material. This wind of gas and dust blowing out of the dying star evolves into a "planetary nebula." This term came about not because of any real association with planets, but because, when viewed in early telescopes, these objects resembled the disks of planets. (Courtesy of NASA, ESA A. Fruchter, and the ERO Team.)

Stars more massive than about five Solar masses don't stop with helium burning. Hydrostatic equilibrium predicts that, for a star supported by gas pressure, the larger the mass, the higher the central temperature must be to support that mass. Stars more massive than eight Solar masses have high enough temperatures in their cores to fuse helium and develop a CO core. The core becomes hot enough to fuse carbon into neon, and oxygen into sulfur and silicon. Finally, silicon is fused into iron. Every time a heavier element is made, it sinks to the center of the star, where it eventually becomes hot enough to undergo fusion. The ash of yesterday's burning becomes the fuel for today's. The result is that the star has an "onion-skin" structure, with hydrogen on the outside, inner shells of helium, carbon/oxygen, oxygen/neon/magnesium, and sulfur/silicon, and an iron core at the very center (Figure 3.8).

Fusion occurs in all shells simultaneously, but since iron is the most strongly bound nucleus, the iron core cannot generate energy by nuclear fusion and begins to shrink. In less than a day, silicon burning in the dying star produces so much iron that the iron core exceeds the Chandrasekhar limit and collapses. In the collapse, the Fermi energy of the degenerate electrons increases beyond the energy threshold for electron capture by protons, which become neutrons. This reduces the electron-degeneracy pressure further and accelerates the collapse. In less than a second, the core collapses into a ball of neutrons only few tens of kilometers in radius – a proto neutron star. The gravitational collapse is stopped by the increasing neutron and electron-degeneracy pressure in the proto neutron star while the proto neutron star is cooling by neutrino emission and contracting. The enormous release of gravitational energy in the core,

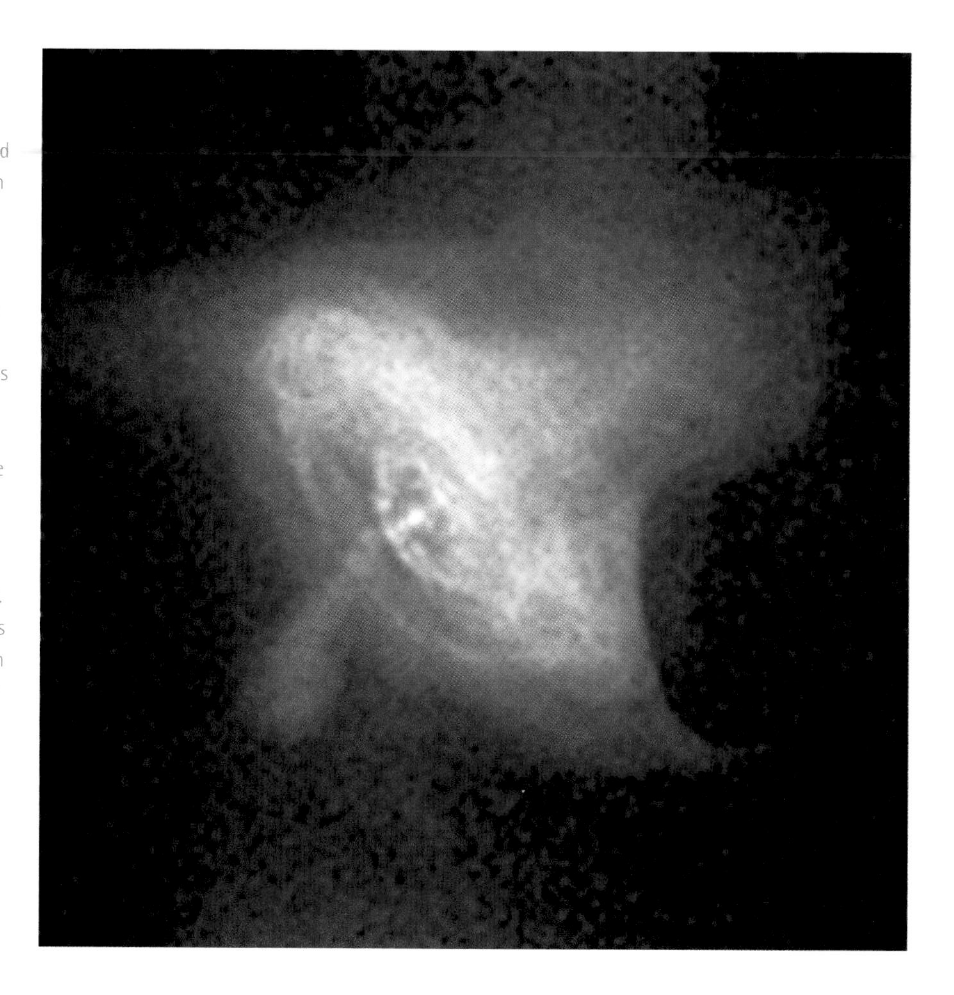

**Figure 3.9.** A rapidly spinning neutron star, or pulsar, that emits bursts of radiation 30 times a second, that was formed in the supernova explosion seen on Earth in 1054 AD in the constellation Taurus (see also Figure 3.7). The X-ray image taken by NASA's Chandra X-ray Observatory shows the central pulsar surrounded by tilted rings of high-energy particles that appear to have been thrown more than a light-year from the pulsar. Perpendicular to the rings, jet-like structures produced by high-energy particles blast out of the pulsar. The diameter of the inner ring is about one light-year, more than 1000 times the diameter of our Solar System. The X-rays are produced by high-energy particles spiraling around magnetic-field lines in the nebula. The nebula's bell shape could be due to the way this huge magnetized bubble was produced, or to its interaction with clouds of gas and dust. (Courtesy of NASA/CXC/SAO.)

of the order of 20% of the rest-mass energy of the Sun, is transported out, mostly by the neutrinos. A small fraction of the gravitational energy released is deposited in the stellar envelope by the escaping neutrinos and by strong shocks that somehow blow away the outer layers of the star, leaving finally a *neutron star* and a supernova shell that expands into the interstellar medium (Figure 3.9).

If the star is very massive and the energy deposition in the layers beyond the core is not sufficient to impart escape velocities to them, part of the blown-off matter may fall back on the neutron star. As the mass of the neutron star increases, its radius decreases, and when it becomes smaller than the limiting Schwarzschild radius (see Chapter 2), the neutron star collapses to a *black hole* from which matter and radiation cannot escape.

If the neutron star is born in a close binary system and accretes matter from a companion star, far more energy is released by mass accretion onto it than if the accreting star were a white dwarf, because of its smaller radius. The result is an emission of more energetic radiation, namely X-rays. An X-ray-emitting neutron star and its close stellar companion are called an *X-ray binary*. In some X-ray binaries, the accreted matter periodically undergoes brief episodes of explosive nuclear burning. These explosions heat the surface still further and produce bursts of X-ray emission. Within minutes, the explosion has run its course and the surface cools, ending the X-ray burst. Within one to several hours, fresh matter is accreted onto the neutron star and another nuclear explosion is set off, again accompanied by a burst of X-rays.

In a few systems, the accreting matter is briefly and repeatedly held back in the accretion disk before it crashes onto the neutron star. Each crash is accompanied by intense heating and a burst of X-ray emission. The temporary holding back of matter in the accretion disk is reminiscent of dwarf novae, in which the compact star is not a neutron star, but a white dwarf. Binary X-ray systems that exhibit repeated bursts of X-rays – either from nuclear explosions or owing to the release of gravitational energy when matter crashes onto a neutron star – are called *X-ray bursters*.

Despite the impressive success of the standard stellar-evolution theory, there is still a long list of astrophysical systems and phenomena that are not yet fully understood, and many unsolved puzzles. This list includes systems such as pulsars and black holes; phenomena such as supernova explosions, mass accretion, and jet ejection in compact binary systems; and puzzles such as anomalous X-ray pulsars, soft-gamma-ray repeaters, gamma-ray bursts, and the origin of high-energy cosmic rays. At present, most of the theoretical research in stellar evolution is focused on these topics. Special observational and theoretical endeavors are also focused on the search for extra-Solar planets that may be life-supporting and on the theory of formation and evolution of planetary systems around Main-Sequence stars.

## 3.5 The frontiers of astrophysics

### 3.5.1 Pulsars and anomalous pulsars

Pulsars are rapidly rotating, magnetized neutron stars with a magnetic moment misaligned with their spin axis that are born in supernova explosions. They are believed to spin down by emission of magnetic-dipole radiation over millions of years. They were first detected in the radio band, but many emit also in the optical, X-ray, or gamma-ray bands. How and where their emissions take place are still not known. Some of these pulsars are anomalous. In particular, they contain small groups of X-ray pulsars and gamma-ray bursters that are relatively very young but, unlike young radio pulsars, already spin slowly and cannot be powered by their rotational energy. Among the alternative sources that may power them are the energy stored in an ultra-strong magnetic field, a fossil accretion disk formed by fall-back supernova material, or a phase transition to strange quark matter. Ongoing observations with the X-ray observatories Chandra and XMM Newton, and future observations with these observatories and with many other X-ray and gamma-ray telescopes under construction, are expected to shed more light on these enigmatic objects.

### 3.5.2 Microquasars

Isolated black holes cannot be seen directly because nothing escapes from them. However, if a supernova explosion produces a black hole in a close binary system, radiation can be emitted from the companion star orbiting the invisible black hole, from the hot accretion disk surrounding the black hole, and from the highly relativistic bipolar jets that seem to be launched every time matter that had been accumulated in the accretion disk falls suddenly onto the black hole. Such highly relativistic bipolar jets from close binary systems that contain an accreting black hole were first discovered in our Galaxy by F. Mirabel and L. Rodriguez in 1994. They were named microquasars because they seem to be the stellar-mass analogs of the accreting massive black holes in quasars and active galactic nuclei. Because of their relative proximity, the dozen or so microquasars that have been discovered so far in our Galaxy offer a "laboratory" for studying the

**Figure 3.10.** An image of bipolar jets from a new star, HH 30, taken in 2000 with the Wide Field and Planetary Camera aboard the Hubble Space Telescope. It shows an edge-on disk at the bottom of the image, which blocks light from a flattened cloud of dust split into two halves by the disk around the central star. This is not itself visible – but the reflection of the star's light by dust above and below the plane of the disk is clearly seen. The disk's diameter is 450 times the Earth–Sun distance. (Courtesy of NASA, A. Watson, K. Stapelfeldt, J. Krist, and C. Burrows.)

much more distant massive black holes and the spacetime environments near black holes. They have become the target of intensive observational and theoretical investigations (Figure 3.10).

### 3.5.3 Gamma-ray bursts

Gamma-ray bursts (GRBs) are short but very intense flashes of soft gamma rays occurring at a rate of one or two per day over the whole sky. They were discovered in 1967 by the Vela military satellites that were launched in order to monitor the compliance of the Soviet Union with the nuclear test ban treaty. For a long time their nature and origin was a mystery. The Compton Gamma Ray Observatory, launched in 1991, established an isotropic sky distribution and an intensity distribution that were consistent only with a cosmological origin. Direct confirmation for their cosmological origin came only in 1997 when the Beppo-SAX satellite discovered that gamma-ray bursts have an X-ray afterglow that lasts for days and weeks after a burst, and provided sufficiently fast and accurate localization for their detection in the optical and radio bands as well. Absorption lines in their optical afterglows and emission lines from their host galaxies were used to determine their distance and show their cosmological origin. So far observations have established that the relatively long GRBs originate in star-formation regions in distant galaxies, that they are beamed emissions, and that some of them, or perhaps all of them, are associated with supernova explosions. The nature of the progenitors of both short and long GRBs and their production mechanism are still not known. Nevertheless, the large intensities of radiation from GRBs in various bands of the electromagnetic spectrum, their large distances, and their association with star formation make them very promising beacons for studying the history of star formation and of the interstellar and intergalactic space with advanced telescopes.

### 3.5.4 Accretion disks and jets

Mass accretion is the accumulation of matter onto an object from its environment due to the gravitational pull of the object. Accreting objects are numerous in the Universe and are of very different sizes and appearances. They include $10^6$–$10^9$-Solar-mass black holes in active galactic nuclei, white dwarfs, neutron stars, and stellar black holes in X-ray binaries, protostars, and protoplanets. Because of angular-momentum conservation, the accreted material usually forms a disk around the central object, from where it is accreted onto the object. Accretion disks seem to be present around most of these accreting objects and power their emission. Such accretion disks give rise to a variety of complex quasi-steady and transient phenomena. Mass accretion on rotating (Kerr) black holes is the most efficient known astrophysical mechanism of energy release. It is believed to be the power source of active galactic nuclei, quasars, and microquasars, and, perhaps, gamma-ray bursts. Highly collimated flows with large velocities – jets – are often observed near accreting compact objects. Correlations between emission from the accretion disk and that from the jet provide evidence that the jets are launched directly from the disks. It is suspected that magnetic rotation instabilities in the accretion disk play a major role in mass accretion and bipolar jet ejection, but the exact mechanism of mass accretion and launching of bipolar jets from accretion disks is still not understood.

### 3.5.5 Cosmic accelerators

Radio, X-ray, and gamma-ray observations show that high-energy cosmic rays are present and must be constantly injected in the interstellar space in galaxies, in the intergalactic space in clusters and groups of galaxies, and in the radio lobes of quasars and active galactic nuclei (AGN). They also provide strong limits on cosmic-ray density in the intergalactic space and evidence that at least very-high-energy cosmic-ray electrons are accelerated in supernova remnants, in gamma-ray bursts, in relativistic jets from compact stellar objects such as pulsars and microquasars, in the powerful jets from massive black holes in quasars, AGN, and radio galaxies, and in the lobes formed when these jets stop and expand. A complete theory of cosmic-ray acceleration in these objects is still lacking. Despite intensive investigations extending over the entire twentieth century, the origin of the extra-Solar cosmic rays observed near Earth is still not known, nor are their energy spectrum and chemical composition fully understood. At energies above the threshold for photopion production on the microwave background radiation, their mean free path in space is much shorter than the typical distance to their potential cosmic sources, quasars and AGN. Thus, after the discovery of the cosmic microwave background radiation, it was pointed out by K. Greisen, and independently by G. Zatsepin and V. Kuzmin, that, if the ultra-high-energy cosmic rays are of extragalactic origin, their flux should be strongly suppressed above the threshold energy for their absorption in the intergalactic space, the so-called "GZK cutoff." Preliminary results from large ground-based extensive air-shower arrays, which detect them and measure their energies through their atmospheric showers, do not provide clear evidence for their Galactic or extragalactic origin. This is due to poor statistics and difficulties in energy calibration at the highest energies, which may be overcome with the new giant ground-based air-shower arrays such as the Auger project or with space-based telescopes such as FUSE that are intended to detect fluorescent light from extensive atmospheric air showers and Čerenkov light from their ground impact.

## 3.6 Physics beyond the standard models

Astrophysics and cosmology can be used to test the standard models of physics under conditions and over distance scales not accessible to laboratory experiments. Most of the astrophysical observations are in good agreement with the standard models. Thus, supernova explosions, stellar evolution, primordial nucleosynthesis, and cosmic background radiations have been used to derive strong limits on particle physics beyond its Standard Model. General relativity has been tested successfully in the weak-field limit over distances from a few astronomical units up to galactic and cosmological distances by gravitational-lensing phenomena, and in the strong-field limit by observations of binary pulsars. However, there are observations that seem to need physics beyond the standard models. They include the baryon asymmetry of the Universe, the dark-matter problem, the dark-energy problem, the origin of the Big Bang, the inflation phase of the early Universe, and neutrino phenomena (see Chapter 4). In particular, the Sudbury Neutrino Observatory has shown that neutrinos from the Sun change their flavor on their way to Earth – a possibility that was first suggested by A. Gribov and B. Pontecorvo in 1968. This change of flavor solves the 30-year-old Solar-neutrino problem – the discrepancy between the electron-neutrino flux from the Sun measured in various Solar-neutrino experiments (the Homestake chlorine experiment, Kamiokande, SAGE – Soviet–American Gallium Experiment, GALLEX – European gallium experiment, and

Super-Kamiokande) and that expected from the Solar luminosity. Although it was clearly demonstrated before by all Solar-neutrino experiments, and by experiments with neutrinos produced by cosmic rays in the atmosphere, that neutrino-flavor oscillations can explain the missing neutrinos in all these experiments, it was the first time that the appearance of the missing neutrinos with a different flavor was detected. Neutrino oscillations require that neutrinos of different flavors have nonzero mass difference, i.e., they cannot all have the same mass, for example the zero masses that were initially adopted (for simplicity) in the Standard Model of the elementary particles. This breakthrough discovery has important consequences not only for particle physics but also for cosmology and for astrophysical processes where neutrinos play an important role.

## 3.7 Distant planets and astrobiology

One of the major recent breakthroughs in astronomy was the discovery of an extra-Solar planet orbiting a nearby star by D. Queloz and M. Mayor in 1995. Since then, more than 100 extra-Solar planets orbiting nearby Main-Sequence stars have been discovered. This number is now reaching the point at which robust statistical inferences about trends within the sample will become possible. All the discoveries used the "classical methods" for detection of extra-Solar planets (ultra-precise radial-velocity measurements of Main-Sequence stars, astrometry, sensitive photometry to measure transits, and direct imaging). Thus, the discoveries were limited to nearby stars since they rely on light from the planet or from the parent star. Moreover, these classical methods are sensitive enough only when the planets are heavier than Jupiter, but not for Earth-like planets. Gravitational lensing of light from stars by planets of nearer stars is now being used to search for "terrestrial" planets near Solar-mass stars. Within a decade or two new space telescopes and infrared space interferometry will be able to examine the atmospheres and the conditions on such planets, and show whether they can support life on their surfaces. Astrobiology seeks to understand the origin of the building blocks of life; how and where these biogenic compounds formed and how they combined to create life; how life affects, and is affected by, the environment from which it arose; and whether and how life expands beyond its planet of origin. This interdisciplinary field combining astronomy, chemistry, and biology is still in its infancy but is bound to grow quickly because it tries to solve one of the most fascinating puzzles in science.

## 3.8 Concluding remarks

Physicists have been impressively successful in reducing almost everything to simple fundamental physical laws. The dream of astrophysicists has been to start from these fundamental laws and reconstruct the whole Universe. The early Universe, perhaps, was simple; but the present Universe, when examined in great detail, appears to be infinitely more complex. It is beyond belief that science will be able to reconstruct this complexity from basic principles, but, perhaps, it is not unreasonable to expect that science will be able to provide answers to some of the most fundamental questions in astrophysics and cosmology, such as the following.

- How did the Universe begin?
- How did the laws of physics come about?
- Are there extra spacetime dimensions?
- Is Einstein's general relativity the full story of gravity?

- What is the origin of the baryon asymmetry of the Universe?
- What is dark matter?
- Is there dark energy and, if so, what is it?
- Are there other states of matter at ultra-high density?
- How did life begin and how does it end?
- Are we alone in the cosmos?

## FURTHER READING


1. S. P. Maran (ed.), *The Astronomy and Astrophysics Encyclopedia*, New York, Van Nostrand Reinhold, 1992.
2. G. O. Abell, D. Morrison, and S. C. Wolf, *The Physical Universe*, Philadelphia, Saunders College Publishing, 1991.
3. J. M. Pasachoff and A. V. Filippenko, *The Cosmos: Astronomy in the New Millennium*, Pacific Grove, Brooks Cole, 2000.
4. J. Silk, *The Big-Bang*, New York, W. W. Freeman, 2001.
5. D. S. Clayton, *Principles of Stellar Evolution and Nucleosynthesis*, Chicago, University of Chicago Press, 1983.
6. S. L. Shapiro and S. A. Teukolsky, *Black Holes, White Dwarfs and Neutron Stars. The Physics of Compact Objects*, New York, Wiley, 1983.
7. M. S. Longair, *High Energy Astrophysics*, Cambridge, Cambridge University Press, 2000.


# Arnon Dar

Arnon Dar, Professor of Physics at the Department of Physics and Asher Space Research Institute of the Technion, the Israel Institute of Technology, Haifa, received his Ph.D. in 1964 from the Hebrew University in Jerusalem for research in nuclear physics. He then worked on high-energy particle reactions and on the quark model of elementary particles at the Weizmann Institute of Science at Rehovot and at MIT. In the late 1960s and the 1970s, he applied the quark model to high-energy elementary particles, nuclei, and interactions of cosmic rays with atomic nuclei, while working at the Technion, MIT, the University of Paris at Orsay, and Imperial College, London. In the late 1970s he became interested in neutrino oscillations, atmospheric neutrinos, and neutrino astronomy. Since the early 1980s his main research interest has been particle astrophysics and cosmology, particularly in astrophysical and cosmological tests of the standard particle-physics model and of general relativity. These have included astrophysical and cosmic puzzles, such as solar neutrinos, the origin of cosmic rays, dark matter, and gamma-ray bursts. This research was done at the Technion, the University of Pennsylvania, the Institute of Advanced Studies in Princeton, NASA's Goddard Space Flight Center, the Institute for Astronomy, Cambridge, UK, and the CERN European research center in Geneva. His most recent work has been on cosmic accelerators, gamma-ray bursts, supernovae, and strange-quark stars.

# 4 Particles and the Standard Model

Chris Quigg

## 4.1 Origins of particle physics

Within the lifetime of my grandparents, there lived distinguished scientists who did not believe in atoms. Within the lifetime of my children, there lived distinguished scientists who did not believe in quarks. Although we can trace the notion of fundamental constituents of matter – minimal parts – to the ancients, the experimental reality of the atom is a profoundly modern achievement. The experimental reality of the quark is more modern still.

Through the end of the nineteenth century, controversy seethed over whether atoms were real material bodies or merely convenient computational fictions. The law of multiple proportions, the indivisibility of the elements, and the kinetic theory of gases supported the notion of real atoms, but a reasonable person could resist because no one had ever seen an atom. One of the founders of physical chemistry, Wilhelm Ostwald, wrote influential chemistry textbooks that made no use of atoms. The physicist, philosopher, and psychologist Ernst Mach likened "artificial and hypothetical atoms and molecules" to algebraic symbols – tokens, devoid of physical reality – that could be manipulated to answer questions about Nature.

Atoms became irresistibly real when they began to come apart, with J. J. Thomson's discovery of the electron at the Cavendish Laboratory in 1897. In the end, the atomists won not because they could see atoms – atoms are far too small to see in visible light – but because they learned how to determine the size and weight of a single atom. Even under conditions of macroscopic tranquility – perfect equilibrium and constant temperature – microscopic particles suspended in a liquid are perpetually in erratic motion. At the Sorbonne in 1908, Jean Perrin established that the wild-wandering "Brownian" movement results from the relentless buffeting of the suspended grains by agitated molecules of the surrounding medium (Figure 4.1). Verifying a relation for the rate of wandering derived by Albert Einstein, he deduced the mass of an individual molecule. In demonstrating the mechanical effects of tiny atoms and molecules, Perrin effectively ended skepticism about their physical reality. Ostwald announced his conversion in 1909, the year he won the Nobel Prize. Mach went to his grave in 1916, doggedly fighting a futile rearguard action.

Though the battle was won, Perrin still felt compelled to defend the atomic hypothesis in the preface to his popular 1913 book, *Atoms*:

*The New Physics for the Twenty-First Century*, ed. Gordon Fraser.

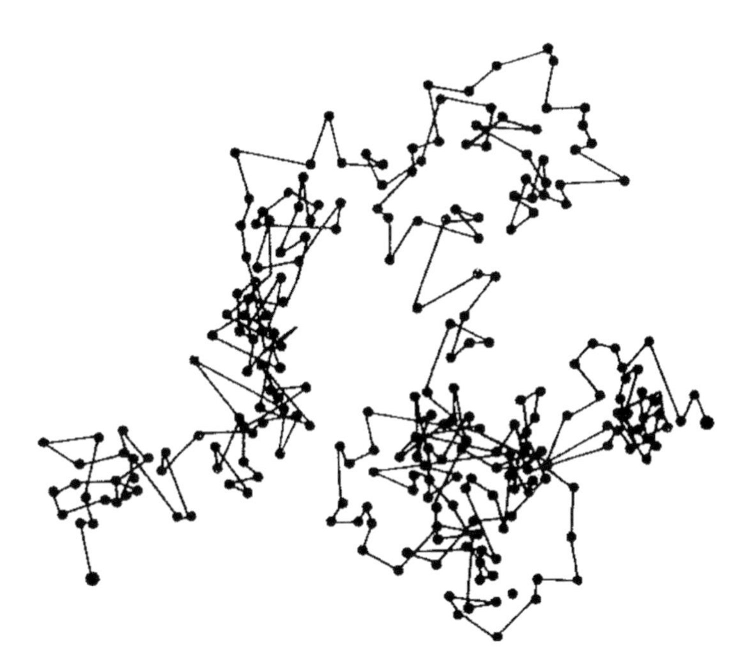

**Figure 4.1.** Positions of a mote in Brownian motion at thirty-second intervals. (From Jean Perrin, *Les atomes*, 1913).

Studying a machine, we don't just think about the parts we can see, which are the only objective reality we can establish short of taking the machine apart. We observe the visible parts as best we can, but we also try to guess what hidden gears and levers might explain the machine's movements.

To divine in this way the existence or the properties of objects that we haven't yet experienced directly – *to explain a complicated visible by a simple invisible* – that is the kind of intuitive intelligence to which, thanks to men such as Dalton or Boltzmann, we owe the doctrine of atoms.

A century after Perrin, we have learned to extend our senses so we can visualize – and even manipulate – individual atoms. A quantum ranging instrument called the scanning tunneling microscope surveys surfaces at fraction-of-an-atom resolution, creating remarkable topographic maps such as the example in Figure 4.2.

The experimental establishment of the atomic hypothesis had revolutionary consequences for science. Richard Feynman opined that the statement richest with useful information about the physical world is ". . . that all things are made of atoms – little particles that move around in perpetual motion, attracting each other when they are a little distance apart, but repelling upon being squeezed into one another." Atoms link

**Figure 4.2.** A scanning-tunneling-microscope image of a gallium arsenide 1–1–0 surface. The protrusions are the arsenic atoms; a missing atom in the arsenic lattice is apparent. (Department of Synchrotron Radiation Research, Lund University, Sweden).

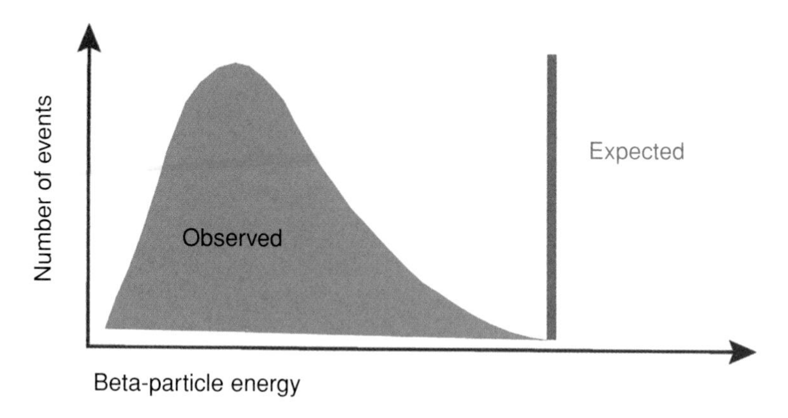


**Figure 4.3.** Expectations and reality for the beta-decay spectrum. The continuous spectrum of electron energies baffled physicists for many years.

heat and motion and the properties of gases through kinetic theory, bringing together mechanics and thermodynamics. The new reality of atoms led in short order to questions of atomic structure and to the invention of quantum mechanics, joining chemical reactions and the properties of diverse substances with the behavior of individual atoms under the influence of electromagnetism.

The discovery of the electron and the prospect of peering within the atom renewed the ancient quest for the ultimate constituents of matter and the interactions among them. But a hint soon developed that physicists would have to cast a wider net if they hoped to understand why the world is the way it is. That hint – the neutrino – emerged from a puzzle in 1914 that led to an idea in 1930 that was confirmed by an experiment in 1956.

Neutrinos are tiny subatomic particles that carry no electric charge, have almost no mass, move nearly at the speed of light, and hardly interact at all. They are among the most abundant particles in the Universe. As you read this page, inside your body are more than ten million relic neutrinos left over from the Big Bang. Each second, about a hundred million million neutrinos made in the Sun pass through you. In one tick of the clock, about a thousand neutrinos made by cosmic-ray interactions in Earth's atmosphere traverse your body. Other neutrinos reach us from natural sources, including radioactive decays of elements inside the Earth, and artificial sources, such as nuclear reactors.

There was a neutrino puzzle even before physicists knew there was a neutrino. One variety of natural and artificial radioactivity is nuclear beta decay, observed as the spontaneous transmutation of one element into its periodic-table next-door neighbor by the emission of an electron, or beta particle. An example is the transformation of lead into bismuth, $^{214}\mathrm{Pb}_{82} \rightarrow {}^{214}\mathrm{Bi}_{83} + \beta^-$. According to the principle of energy and momentum conservation, the beta particle should have a definite energy at the value indicated by the spike in Figure 4.3. James Chadwick observed something quite unforeseen. While working at Berlin's Imperial Physical Technical Institute in 1914, he measured a continuum of electron (beta) energies less than the expected value, as shown in Figure 4.3.

What could account for the missing energy? Might it mean that energy and momentum are not uniformly conserved in subatomic events? (The grand master of the quantum theory, Niels Bohr, was willing to entertain this possibility.) The riddle of the continuous beta spectrum gnawed at physicists for years. On December 4, 1930, theorist Wolfgang Pauli was moved to address an open letter to his colleagues attending a radioactivity congress in Tübingen. Pauli remained in Zurich, where he was obliged to attend a student ball. "Dear Radioactive Ladies and Gentlemen," Pauli begins, "I have hit upon a desperate remedy regarding . . . the continuous beta-spectrum . . ." The conventional picture of beta decay was incomplete, Pauli asserted. He speculated that an unknown and unseen particle – a very penetrating, neutral particle of vanishingly small mass – might be emitted in beta decay, along with the electron and transformed nucleus. Pauli's hypothetical particle would escape undetected from any known apparatus, taking with it some energy, which would appear to be missing. In reality, the balance of energy and momentum would be restored.

To conjure a new particle was indeed a "desperate remedy" and a radical departure from the prevailing theoretical style. But in its way Pauli's invention was very conservative, for it preserved the principle of energy and momentum conservation and with it the notion that the laws of physics are invariant under translations in space and time. Inspired by Pauli's speculation, Enrico Fermi formulated a theory of the weak interaction that – with one modification – is still in use to describe beta decay. It remained to prove the existence of Pauli's stealth particle, which Fermi named "neutrino," the little neutral one.

Seeing the neutrino interact had to await dramatic advances in technology. Detecting a particle as penetrating as the neutrino required a large target and a copious source of neutrinos. In 1953, Clyde Cowan and Fred Reines exposed 300 liters of liquid scintillator – about $10^{28}$ protons – to the emanations outside a fission reactor at the Hanford Engineering Works in the state of Washington. According to Fermi's theory, a reactor should emit an intense flux of (anti)neutrinos, and it might be possible to capture a few examples of inverse beta decay: antineutrino + proton → neutron + antielectron (positron). When the initial runs at Hanford were suggestive but inconclusive, Cowan and Reines moved their team to a more powerful reactor at the Savannah River nuclear plant in South Carolina. There they made the definitive observation of inverse beta decay – of neutrino interactions – in 1956.

Recently, far more massive detectors have recorded the interactions of neutrinos produced along with heat and light by nuclear burning in the Sun. Figure 4.4 shows the neutrino sky as observed over 500 days by the Super-Kamiokande Detector (Super-K) installed one kilometer deep in the Koizumi mine in Japan. Super-K consists of eleven thousand photomultiplier tubes that view fifty *million* liters of pure water to detect cones of light radiated by fast-moving charged particles. It records individual neutrino interactions (neutrino + neutron → proton + electron) in real time, and determines the neutrino direction from the electron's path. Super-K has demonstrated that the brightest object in the neutrino sky is the Sun, which proves that nuclear fusion powers our star.

As we will see shortly, the properties of neutrinos are one of the most active – and most puzzling – areas of particle physics today. But I have chosen to highlight the neutrino episode as a crucial antecedent of modern particle physics because it represents two important departures in the history of our subject.

First, the neutrino was the first particle conjectured out of respect for an established conservation law, or symmetry principle. Symmetries of the laws of Nature have taken on such central importance in the way we look at physical phenomena that nearly all the particles we are searching for today owe their hypothetical existence to a symmetry or a hidden symmetry. The neutrino was accepted – provisionally – and used fruitfully long before it was detected directly.

Second, the neutrino is not a constituent of everyday matter, yet we cannot understand the properties of matter or the origin of an everyday phenomenon such as sunshine without it. Particle physics aspires not just to find the smallest bits of ordinary matter. Our goal is to discover all the stuff – matter, energy, maybe more – the Universe is made of; to explore Nature's grand stage of space and time; and to understand – as simply and comprehensively as possible – how the Universe works.

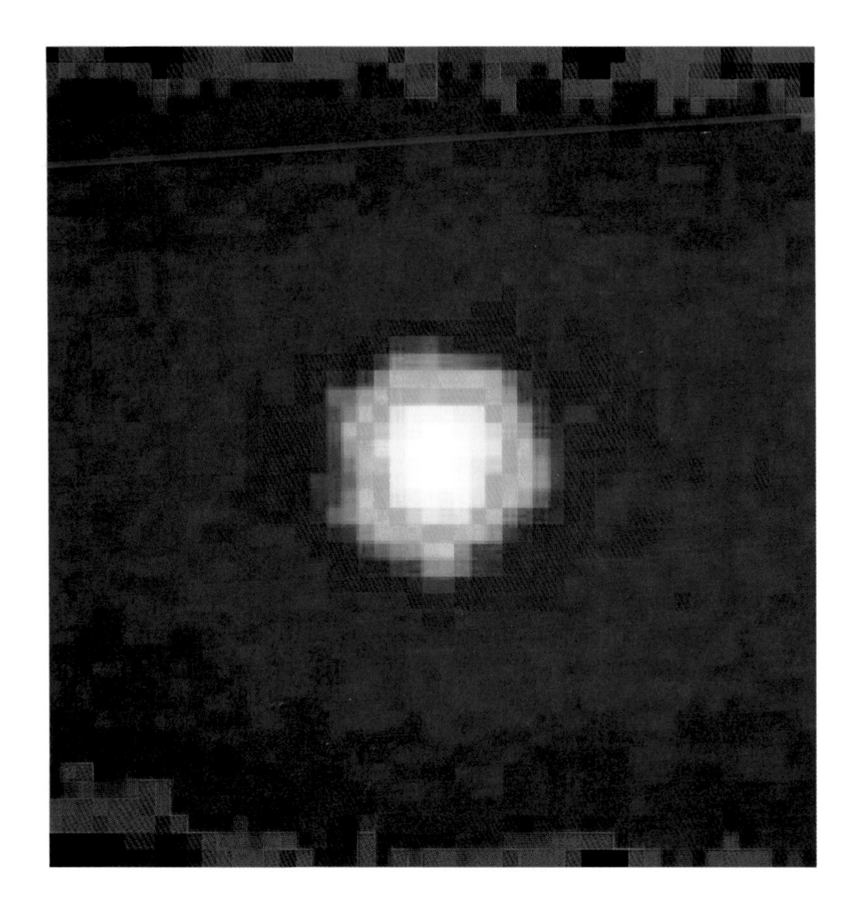

**Figure 4.4.** The first sky map made by the Super-Kamiokande experiment, showing that the Sun (center) is the brightest object in the neutrino sky. The direction to the neutrino source is calculated from the direction of the electrons produced when a neutrino interacts in the water of the detector. Scattering effects blur the electron directions, so each pixel in this image is about twice the Sun's actual size on the sky. (Courtesy H. Sobel.)

## 4.2 The Standard Model

The distinguished Princeton mathematical physicist Arthur Wightman wrote in the thirteenth edition of the *Encyclopaedia Britannica* (1968) that "Running through the theoretical speculation since World War II has been the idea that the observed particles are not really elementary but merely the states of some underlying simple dynamical system. Such speculations had not led very far by the 1960s." The fifteen years that followed saw a revolution in the prevailing view of the structure of matter and the forces of Nature. We learned that a fundamental description of matter must be based on the idea that the strongly interacting particles, or hadrons (the proton, pion, and hundreds more), are composed of quarks. Together with leptons, such as the electron and neutrino, quarks seem to be elementary particles – indivisible and structureless at the current limits of resolution. We learned too that the strong, weak, and electromagnetic interactions of the quarks and leptons are consequences of symmetries that we observe in Nature, and established a deep connection between the weak interactions and electromagnetism. The gauge theories that describe all three interactions have a common mathematical structure that seems to bespeak a common origin. The new synthesis of quarks and leptons and gauge interactions describes experimental observations so well that it has been designated the Standard Model of particle physics, where "standard" is to be taken as a reliable reference and a point of departure for a still more comprehensive understanding.

I will begin by describing the main elements of the Standard Model and some of the insights we have derived from it. My aim is not to give a comprehensive review

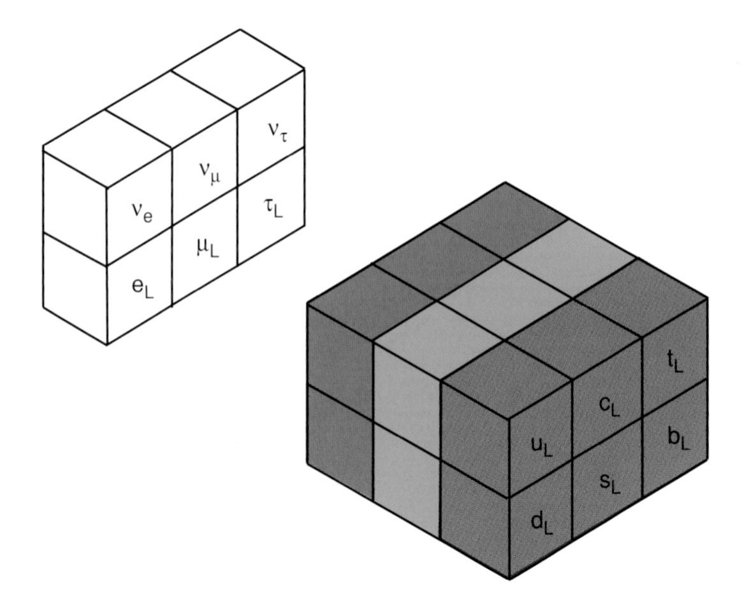


Figure 4.5. Three "generations" of leptons (electron, muon, tau, and their associated neutrinos) and three of quarks (up/down, charm/strange, top/bottom). Quarks carry a color charge and are affected by the strong interactions. Leptons do not carry color and are not affected by the strong interactions. The quarks and charged leptons feel the effects of electromagnetism. The charged-current weak interaction acts on the left-handed (L) quarks and leptons.

or justification of the Standard Model, but to lead us to a discussion of the unfinished business of the Standard Model and the new physics on our threshold.

The leptons are the fundamental particles that are not affected by the strong force that binds atomic nuclei. Three leptons, the electron, muon, and tau, differ in mass, but all carry electric charge of $-1$. The other three leptons are the electrically neutral neutrinos, which are nearly massless. Each lepton has two spin states, so they are called spin-$\frac{1}{2}$ particles. All of the leptons are observed directly in the laboratory. The weak interaction responsible for beta decay changes a charged lepton into its neutrino partner.

We owe the idea of quarks to Murray Gell-Mann (then at Caltech) and George Zweig (then at CERN, the European Laboratory for Particle Physics), who sought in 1964 to bring order to a burgeoning catalog of hadrons. They argued that the strongly interacting particles were composite objects, each a combination of a small number of elementary particles that Gell-Mann named quarks. Like the leptons, the quarks are spin-$\frac{1}{2}$ particles, but quarks are affected by the strong interaction. Each quark carries a fraction of the electron's charge: the down, strange, and bottom quarks have a charge of $-\frac{1}{3}$, and the up, charm, and top quarks carry a charge of $+\frac{2}{3}$. The combinations of quarks that form hadrons always add up to integer charges. A proton, for example, is composed of two up quarks and a down quark, for a total charge of $+1$; a neutron is two downs and an up, for a total charge of 0; and mesons are quark–antiquark combinations with total charge $+1$, 0, or $-1$. The weak interaction also mediates transformations among the quarks. According to the quark model of hadrons, the beta decay of a neutron into a proton occurs when the weak interaction changes a down quark in the neutron into an up quark, emitting an electron and an antineutrino in the process. Down to up is a dominant transition, though down to charm and down to top occur with reduced strength. The weak-interaction families are indicated in the representation of the fundamental constituents shown in Figure 4.5.

Many unavailing searches have been mounted for fractionally charged particles, so, in contrast to the leptons, free quarks have never been observed. The case for the existence of quarks has become unassailable, however. The first dramatic supporting evidence came from a MIT-SLAC collaboration led by Jerry Friedman, Henry Kendall, and Dick Taylor at the Stanford Linear Accelerator Center in the late 1960s. Their studies of high-energy electrons scattered by protons indicated that the proton is composed of pointlike electrically charged objects. Subsequent studies of electron and neutrino scattering from nucleons at many laboratories around the world have confirmed and extended the initial discovery and pinned down the properties of the small things within the proton: they have all the properties of quarks.

Any remaining resistance to the quark hypothesis crumbled in 1974 when a team at Brookhaven National Laboratory led by Sam Ting and a team at SLAC led by Burt Richter simultaneously announced the discovery of a remarkable new state: a heavy hadron with a mass of 3.1 GeV with an extraordinarily long lifetime. Within two weeks the discovery of a second state at SLAC showed that the teams had discovered a hadronic atom consisting of a charm quark and charm antiquark. Quarks were real, period! Subsequent experimentation has elaborated a full spectrum of charmonium states; in fact, two new states have been found since 2002. In 1977, Leon Lederman and his collaborators at Fermilab discovered a second hadronic atom called upsilon, with a mass near 10 GeV, this one composed of a bottom quark and bottom antiquark. The charmonium and upsilon systems have been remarkably productive laboratories for testing the new understanding of the strong interactions that emerged from the quark model.

What distinguishes quarks from leptons, and what accounts for the rule that hadrons are made of three quarks or a quark plus an antiquark? The response to both questions is the answer to an early quark-model puzzle. According to Pauli's exclusion principle, which elegantly guides the electron configurations that explain chemical properties in the periodic table, two identical spin-$\frac{1}{2}$ particles cannot occupy the same quantum state. That is, they cannot have the same values of spin and orbital angular momentum. That restriction posed a problem for hadrons such as $\Omega^-$, composed of three strange quarks with a total spin of $\frac{3}{2}$. The three strange quarks in the $\Omega^-$ appear to occupy the same quantum state – which the exclusion principle forbids. The problem is resolved by supposing that the three otherwise identical quarks are distinguished by a hidden quantum number called color, which is the charge on which the strong force acts. Each quark flavor – up or down, charm or strange, top or bottom – comes in three colors – red, green, and blue. Since leptons do not carry color, they do not feel the strong force.

Color charge recasts the rules by which quarks form hadrons: colored states are never seen in isolation. The simplest colorless combinations are color-anticolor, or quark-antiquark, states, and a red quark plus a green quark plus a blue quark. Remarkably, color provides not just a restatement of an empirical rule. As we shall see, it provides the basis for a dynamical theory of the strong interaction that explains the rule. All this would matter little if color remained no more than a fiction for theorists. A host of observables, from the rate at which electrons and positrons annihilate into hadrons to the decay pattern of the W-boson that mediates the weak interaction, measures the number of quark colors to be three!

Establishing the quarks and leptons as the basic bits of matter is only part of our revolution. Our new picture of the fundamental interactions is rooted in Hermann Weyl's efforts in the 1920s to find a geometric basis for both gravitation and electromagnetism. Weyl attempted to unify the two fundamental interactions known in his day by requiring that the laws of Nature be invariant under a change of scale chosen independently at every point in spacetime. The attempt came up short, but his terminology, gauge invariance, has survived. Weyl's hypothesis that symmetries determine interactions was given full expression in the work of Chen Ning Yang and Robert L. Mills in 1954; it has proved amazingly fertile.

If the mathematical expression of the laws of Nature remains unchanged when we transform some characteristic of the physical system in the same way at every point in space and time, we say that the equations display a global symmetry. If the equations remain unchanged when we transform the same characteristic by different amounts at different times and places, we call the theory invariant under a local symmetry. A rough understanding of how local symmetries imply the existence of forces can be found in an idealized rubber disk. Suppose that we impose a local symmetry by demanding that the disk keep its shape when we displace various points within it by different amounts. The displacements stretch the disk and introduce forces between the points. Similarly, in gauge theories the fundamental forces are the response required to preserve a symmetry in the face of local transformations. For the symmetry operations that generate the fundamental interactions, the corresponding force is carried by spin-one gauge bosons, one for each symmetry operation that leaves the theory invariant. The same invariance requirement demands that the gauge bosons have zero mass.

Quantum electrodynamics, or QED, is derived by demanding that the equations describing the motion of a charged particle remain invariant when the phase of the particle's wave function is altered independently at every point in space. Remarkably, that requirement is enough to require the existence of electromagnetism; it also implies the existence of a zero-mass spin-one photon and dictates how the photon interacts with matter.

### 4.2.1 Strong interactions

The theory of the strong interactions, quantum chromodynamics, or QCD, follows from the requirement that the theory of quarks remain invariant if the convention for the quark colors is set independently at each point in space and time. In this case, the SU(3) color symmetry implies eight separate symmetry operations, so eight massless spin-one force carriers emerge, the gluons.

The photon is electrically neutral, so photons do not interact directly among themselves. In contrast, each gluon carries a color charge and an anti-color charge, so the gluons do interact among themselves. That distinction is responsible for an important difference between the two theories – and a most remarkable property of QCD.

In quantum field theory, the strength of an interaction – the effective charge, we may say – is not a fixed number, but depends on the circumstances of the observation. The effect is actually familiar. An electron that enters a medium composed of mobile molecules that have positively and negatively charged ends, for example, will polarize the molecules, drawing the positive ends near while repelling the negative ends. The nearby positive charges screen the electron's charge, making it appear smaller than it would be outside the medium. Only by inspecting the electron at very close range – on a submolecular scale, within the screen of positive charges, can we measure its full charge. That is to say, the effective charge of the electron increases at short distances.

In the quantum vacuum, virtual pairs of electrons and positrons have a fleeting existence. These ephemeral vacuum fluctuations are themselves polarizable, so QED predicts that electric charge is screened; the effective charge is reduced at large distances. Because of the inverse relation between distance and energy, we may say that the electric charge effectively increases from low energies to high energies.

The same sort of screening occurs in QCD, where it is color charge rather than electric charge that is screened. But here, the fact that the gluons carry color adds a crucial twist to the story. By vacuum fluctuations, a quark continuously changes its color by emitting and reabsorbing gluons. In effect, the virtual gluons disperse the quark's color charge through the surrounding space, camouflaging the quark that is the source of the charge. The smaller a region of space surrounding a quark, the smaller will be the proportion of the quark's color it contains. Because the gluons are colored, camouflage – antiscreening – tends to reduce the effective color charge of a quark at short distances.

Overall, the antiscreening effect prevails over screening; as a result, the strong interaction is stronger at long distances than at short distances. Through the uncertainty principle, long distances correspond to low energies and short distances to high energies. Thus, QCD explains what had seemed paradoxical: quarks can be permanently confined and yet act independently when probed at high energies. This profound insight, called asymptotic freedom, emerged in 1973 from calculations carried out by David Gross and Frank Wilczek at Princeton and by David Politzer at Harvard.

### 4.2.2 Weak and electromagnetic interactions

Fermi's weak-interaction theory long served as a successful description of beta decay and other low-energy phenomena, with a single essential revision: the weak interactions are not invariant under mirror reflection (parity), and the charged-current weak interactions apply – in modern language – only to the left-handed quarks and leptons. The observation that parity might be not a symmetry of the weak interactions was made by theorists Tsung Dao Lee and Chen Ning Yang in 1956, when they noticed that parity invariance had been presumed, but never subjected to experimental test, in weak processes. Later that year, C. S. Wu and her collaborators at the National Bureau of Standards demonstrated that parity is maximally violated in the beta decay of $^{60}$Co. Though deep in its implications, the modification to Fermi's theory was mathematically simple. But Fermi's was an effective theory; when applied at very high energies or used to calculate quantum corrections, it made no sense. The development of a proper theory of the weak interactions was a central part of the Standard-Model revolution.

The modern replacement for Fermi's theory is the electroweak theory developed through the 1960s by Sheldon Glashow, Abdus Salam, and Steven Weinberg. As its name implies, it embodies a deep connection between the weak and electromagnetic interactions. It is derived from the up–down or neutrino–electron symmetry of the familiar weak interaction plus a phase symmetry analogous to the one that is the basis of QED. There is one little problem. We have seen that gauge theories imply massless gauge bosons, whose influence extends to great distances. That is very well for electromagnetism, but the range of the weak interactions is known to be less than about $10^{-15}$ centimeters, which implies that the mass of its force carrier must be roughly 100 GeV.

That massive gauge bosons might be accommodated in the gauge-theory paradigm is suggested by the phenomenon of superconductivity (see Chapter 8). Superconductivity consists of two miraculous results. The first, better known, is the ability of many substances to carry electrical currents with no resistance at all, when cooled to low temperatures. The second, no less astonishing, is that magnetic fields do not penetrate into superconductors. This effect, discovered in 1933 by Walther Meissner and Robert Ochsenfeld, is the basis for the levitation of a superconductor above a magnet, supported by a cushion of expelled magnetic field lines. The interpretation is that, although it is a gauge boson, the photon acquires a mass within the superconducting medium, where circumstances conspire to hide the gauge symmetry.

Here, then, is a model for the weak bosons. If the vacuum state of the Universe does not express the full electroweak symmetry, the gauge bosons that correspond to the hidden symmetries will acquire mass. The electroweak theory then yields four gauge bosons: a massless photon that couples to electric charge and reproduces the predictions of QED, two massive charged gauge bosons, $W^+$ and $W^-$, which reproduce the successes of the old effective theory of weak interactions; and a massive gauge boson, $Z^0$, that mediates weak processes that do not change the charge of the particle acted upon.

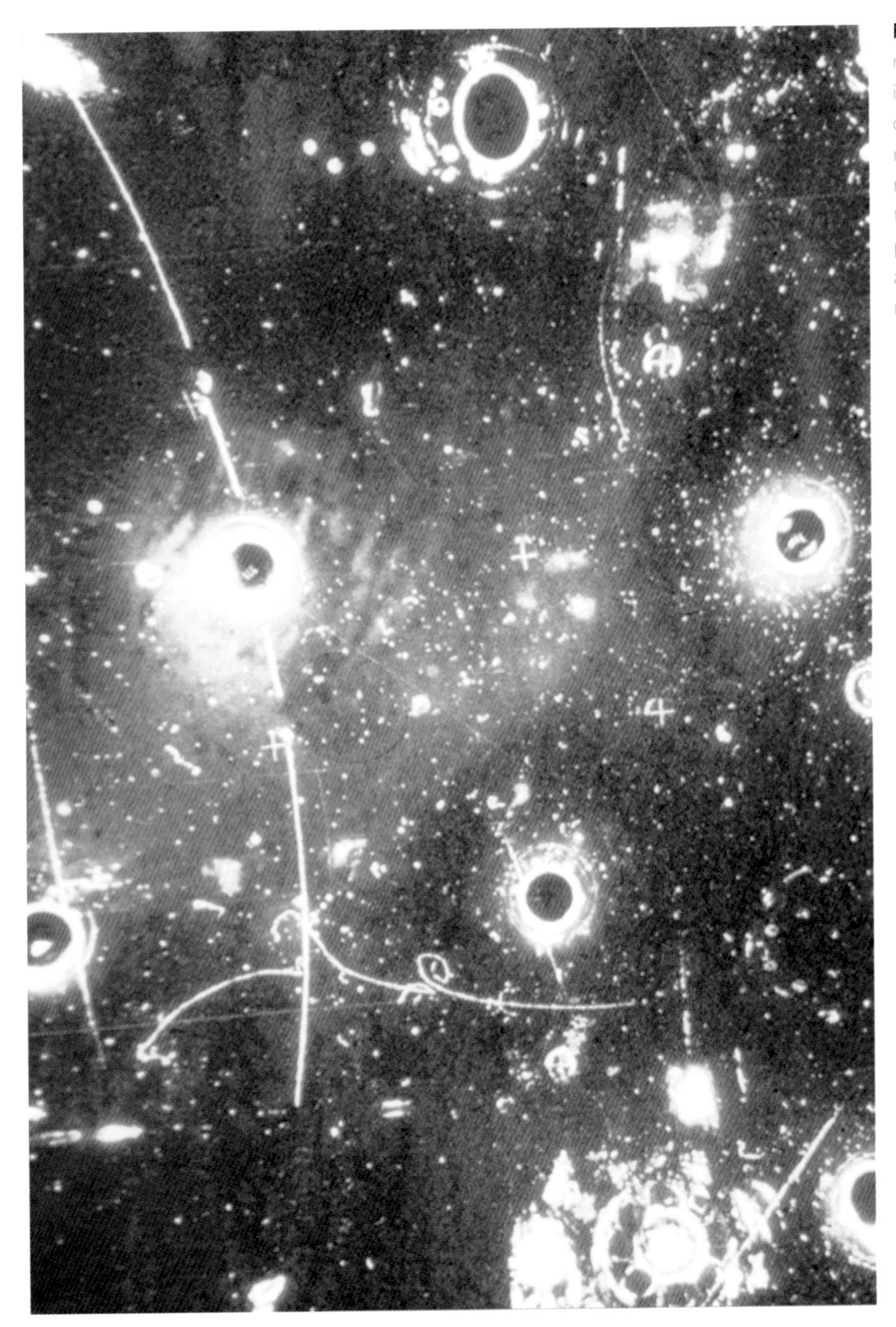

**Figure 4.6.** The first neutral-current event, observed in the Gargamelle bubble chamber at CERN. A beam of muon neutrinos enters from the right. The showering horizontal track near the bottom of the photo is an electron recoiling after being struck by a neutrino. (Photo credit: CERN.)

Such "weak-neutral-current" processes had never been observed; they are predictions of the electroweak theory. In 1973, André Lagarrigue and colleagues discovered neutral-current processes in the Gargamelle bubble chamber at CERN. The first example, a muon-neutrino scattering from an electron, is shown in Figure 4.6. Today, many observations confirm the essential unity of the weak and electromagnetic interactions.

Discovering the gauge bosons themselves required a heroic effort that is one of the proudest achievements of modern particle physics. In a triumph of accelerator art, CERN scientists led by Carlo Rubbia and Simon van der Meer refitted the Super Proton Synchrotron as a proton–antiproton collider capable of reaching energies at which

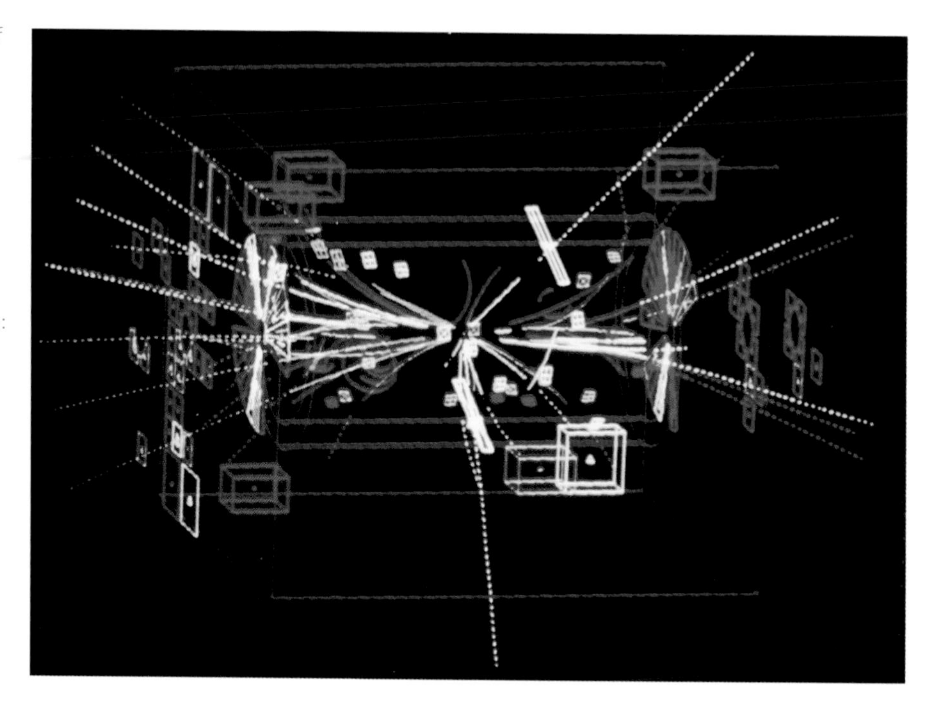

**Figure 4.7.** An early example of a $Z^0$ boson, observed in the UA1 detector at CERN. Proton and antiproton beams collide at the center of the detector, with a center-of-mass energy of 540 GeV. The white dashed lines at bottom and top right represent an electron and positron, decay products of the $Z^0$. (Photo credit: CERN.)

the weak bosons would be produced. International collaborations led by Rubbia and Pierre Darriulat constructed two elaborate detectors that recorded W and Z bosons in 1983 – and, incidentally, changed the landscape of high-energy experiments forever. The properties of the gauge bosons are precisely in line with the expectations of the electroweak theory. An early example of a $Z^0$ boson created in proton–antiproton collisions is shown in Figure 4.7.

### 4.2.3 Hiding electroweak symmetry

Unlike Fermi's theory, the electroweak theory does make sense as a quantum field theory that applies up to extremely high energies. One issue remains: what is the agent that conspires to hide the electroweak symmetry? We do not know, and finding out is at the top of the experimental agenda. The electroweak theory itself tells us that we will find the answer if we are able to conduct a thorough investigation of physics on the 1-TeV scale. That ambitious undertaking is the main business of the Large Hadron Collider under construction at CERN. The simplest guess for the agent of electroweak-symmetry breaking, based on a very close analogy with the Meissner effect, is a massive spin-zero particle known as the Higgs boson, about which I have more to say in a later section.

Without a mechanism to hide the electroweak symmetry, the character of the physical world would be profoundly changed. The quarks and leptons would remain massless, but QCD would still confine the quarks into color-singlet hadrons. Because most of the mass of the protons and other hadrons arises from the energy stored in binding the quarks in a small volume, the proton mass would be little changed if the quarks were massless. In the real world, the neutron has a slightly greater mass than the proton; that mass difference determines the pattern of beta decays and the nature of the stable nuclei in Nature. The neutron mass is greater than the proton mass because the down quark is more massive than the up quark. (Remember that a proton is made of two ups and a down, the neutron of two downs and an up.) Without the small quark-mass difference to tip the balance, the proton would outweigh the neutron because of the self-energy associated with its electric charge. The pattern of beta decay would be entirely different from the one we know, and the lightest nucleus would be one neutron; there would be no hydrogen atom.

If the electroweak symmetry were unbroken, the weak gauge bosons would be massless, so beta decay would proceed at the rapid clip of an electromagnetic process. QCD modifies this conclusion in a very interesting way, as we can see by considering a world with one set of massless quarks, up and down. For vanishing quark masses, QCD has an exact chiral symmetry – an up–down symmetry that acts separately on the left-handed and right-handed quarks. At low energies, the strong interactions become strong, and quark–antiquark condensates appear in the vacuum, establishing communication between the left-handed and right-handed quarks. The symmetry is reduced: in place of the separate up–down symmetries for left-handed and right-handed quarks, only the familiar up–down isospin symmetry remains. The breaking of chiral symmetry in turn hides the electroweak symmetry and gives masses to the W and Z, leaving the photon massless. The masses that the weak bosons would acquire in this way are some 2500 times smaller than we observe, and beta-decay rates would be about ten million times faster than in the real world.

I do not know of an exhaustive analysis of Big-Bang nucleosynthesis in this fictional world, but it seems likely that some light elements would be produced in the early Universe. However, even if some nuclei are produced and survive, they would not form atoms as we know them. The vanishing electron mass means that the Bohr radius of an atom would be infinite. A world without compact atoms would be a world without chemistry and without stable composite structures like the solids and liquids we know. *When we search for the mechanism of electroweak-symmetry breaking, we are seeking to understand why the world is the way it is.* This is one of the deepest questions humans have ever pursued, and it is coming within the reach of particle physics.

## 4.3 A decade of discovery past

After the heady days of the 1970s and 1980s when the main elements of the Standard Model fell into place, many particle physicists look upon the 1990s as an era of consolidation, rather than revolution. It is true that no lightning bolt of new phenomena radically incompatible with the Standard Model has struck, and no theoretical worldview richer and more comprehensive than the Standard Model has been vindicated. However, the achievements are by any measure impressive, and experimental surprises have pointed in new directions for both theory and experiment. Taken together, the notable accomplishments that follow point to a very lively decade ahead. Perhaps we will look back to see that we have been in the midst of a revolution all along.

- Experiments in laboratories around the world have elevated the electroweak theory to a law of Nature that holds over a staggering range of distances, ranging from subnuclear ($10^{-17}$ cm) to galactic ($10^{22}$ cm). The electroweak theory offers a new conception of two of Nature's fundamental interactions, ascribing them to a common underlying symmetry principle. It joins electromagnetism with the weak interactions in a single quantum field theory. The electroweak symmetry is reflected in the delicate cancellation among contributions to the reaction electron + positron → $W^+ + W^-$ illustrated in Figure 4.8. Dozens of measurements have tested the agreement between theory and experiment at the level of one part in a thousand.

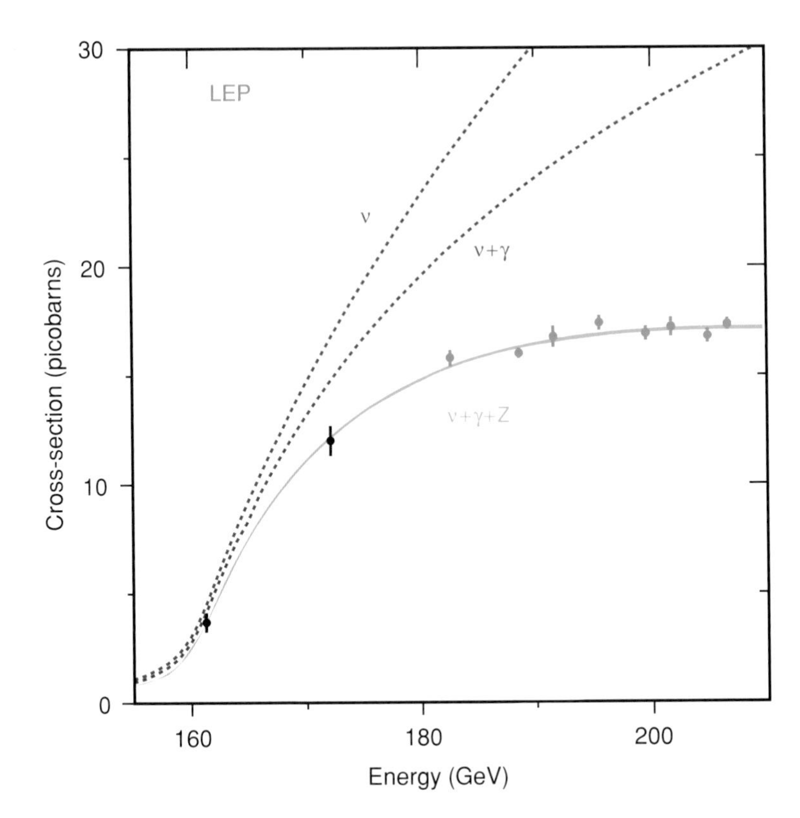


Figure 4.8. The electroweak symmetry balances the contributions of neutrino, photon, and $Z^0$-boson exchange to the reaction electron + positron → $W^+ + W^-$. The individual contributions lead to reaction rates that grow rapidly with energy. (Courtesy LEP Electroweak Working Group.)

- Even as the Higgs boson has eluded detection, experiments may have observed its influence, beyond its presumed role as the agent that hides electroweak symmetry. Measurements carried out using the Large Electron Positron Collider (LEP) at CERN, the European Laboratory for Particle Physics, and other instruments are sensitive not just to the symmetry of the electroweak theory, but also to quantum corrections that are subtly affected by the Higgs boson. The evidence strongly hints (see Figure 4.9) that the Higgs boson must be present, with mass less than about 250 GeV.
- The observation that neutrinos born as one flavor oscillate during propagation into other flavors establishes that neutrinos have mass.
- The CDF and D0 Experiments at Fermilab discovered the top quark in proton–antiproton collisions. (A specimen from the D0 detector is shown in Figure 4.10.) Long anticipated as the partner of the bottom quark, top stands out because its mass, about 174 GeV, is some forty times the b-quark mass. In an early success for the electroweak theory, precision measurements detected the top quark's virtual quantum effects and predicted its great mass.
- Experiments of all sorts and many theoretical approaches have contributed to the elaboration and understanding of quantum chromodynamics in settings ranging from B-meson decays to electron–proton scattering, from top-quark production to heavy-ion collisions.
- The NA48 Experiment at CERN and the KTeV Experiment at Fermilab established that particle–antiparticle symmetry is violated not only in quantum-mechanical mixing, but also in the decays of neutral K-mesons.
- The Belle Experiment at KEK, the Japanese high-energy physics laboratory, and the BaBar experiment at the Stanford Linear Accelerator Center detected differences between the decays of $B^0$-mesons and anti-$B^0$-mesons, as predicted by the Standard Model. This was the first observation of CP violation outside

the $K^0$-anti-$K^0$ complex, for which James Cronin, Val Fitch, and collaborators detected a particle–antiparticle asymmetry in 1964. (CP symmetry – for charge conjugation [matter–antimatter exchange] times parity [mirror reflection] – implies that a state in the matter world looks identical to the mirror image of the corresponding state in the antiworld.)

- Measurements of high-redshift supernovae offer strong evidence that the Universe is expanding at an accelerating rate (see Chapter 1). A remarkable

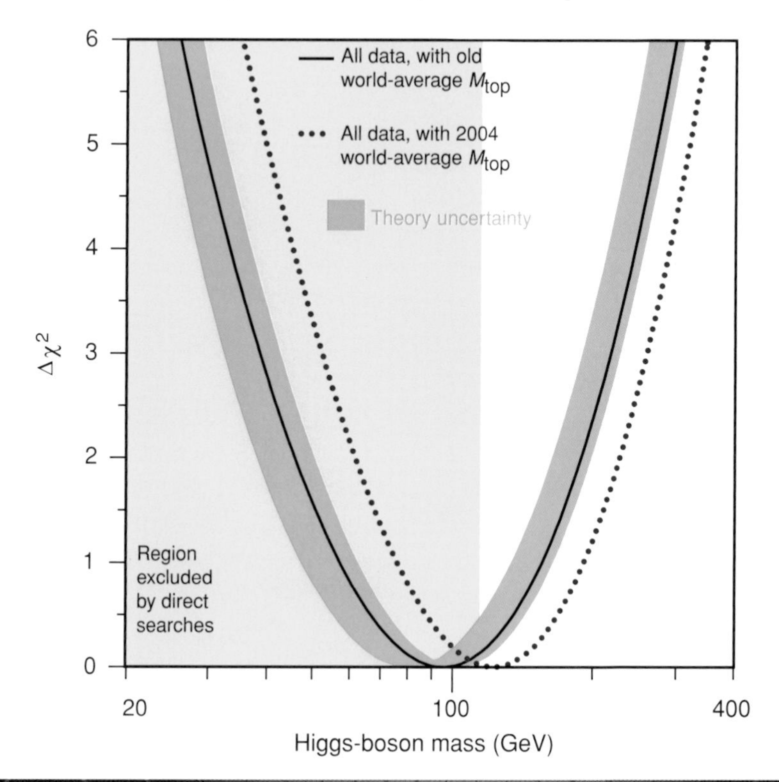


**Figure 4.9.** Electroweak-theory fits to a Universe of two dozen observables demand the existence of a Higgs boson with mass smaller than about 250 GeV (blue dots). The best fit corresponds to the minimum value of the statistical measure chi-squared. Direct searches for the Higgs boson carried out at CERN's Large Electron–Positron Collider exclude (yellow region) a Higgs-boson mass up to 114 GeV. (Courtesy LEP Electroweak Working Group.)

**Figure 4.10.** A pair of top quarks reconstructed in the D0 Experiment at Fermilab. Each top quark decays into a bottom quark and a W boson. In this end view, one W-boson decays into a muon (turquoise, top right) and a neutrino (pink, right). The other decays into a quark and antiquark that materialize in the two jets on the left. The small jet at the top and the jet containing a muon (turquoise) at bottom right are characteristic of bottom quarks. (Fermilab image.)

concordance of those measurements with maps of the cosmic microwave background indicates that we live in a flat Universe dominated by dark matter and an unidentified form of "dark" energy that drives the cosmic expansion.
- The DONUT Experiment at Fermilab detected for the first time the interactions of tau neutrinos, recording the last of the known quarks and leptons.
- Scattering experiments at the highest energies yet attained (proton–antiproton collisions approaching 2000 GeV at the Fermilab Tevatron and electron–positron collisions above 200 GeV at the LEP) reveal quarks and leptons to be structureless at a resolution of about $10^{-17}$ cm.

## 4.4 The great questions

When reflecting on the future agenda of particle physics, I find it useful to proceed from broad scientific themes to the specific questions – the minute particulars – that will illuminate them, and then to instruments and technology development. Here it will suffice to cite the main themes and develop them briefly.

### 4.4.1 Elementarity

Are the quarks and leptons structureless, or will we find that they are composite particles with internal structures that help us understand the properties of the individual quarks and leptons?

### 4.4.2 Symmetry

One of the most powerful lessons of the modern synthesis of particle physics is that (local) symmetries prescribe interactions. Our investigation of symmetry must address which gauge symmetries exist (and, eventually, why). We have learned to seek symmetry in the laws of Nature, not necessarily in the consequences of those laws. Accordingly, we must understand how the symmetries are hidden from us in the world we inhabit. The most urgent problem in particle physics is to complete our understanding of electroweak symmetry breaking by exploring the 1-TeV scale. In the immediate future, this is the business of experiments at the Tevatron Collider and the Large Hadron Collider.

### 4.4.3 Unity

In the sense of developing coherent explanations that apply not to one individual phenomenon in isolation, but to many phenomena in common, unity is central to all of physics, and indeed to all of science. At this moment in particle physics, our quest for unity takes several forms.

First is the fascinating possibility that all the interactions we encounter have a common origin and thus a common strength at suitably high energy. Second is the imperative that quantum corrections respect the symmetries that underlie the electroweak theory, which urges us to treat quarks and leptons together, not as completely independent species. Both of these ideas are embodied in unified theories of the strong, weak, and electromagnetic interactions, which imply the existence of still other forces – to complete the grander gauge group of the unified theory – including interactions that change quarks into leptons.

The third aspect of unity is the idea that the traditional distinction between force particles and constituents might give way to a common understanding of all the particles. The gluons of quantum chromodynamics carry color charge and interact among themselves, so we can imagine quarkless strongly interacting matter in the form of glueballs. Supersymmetry, which relates fermions and bosons, offers another challenge to the separation between messengers and constituents. Finally, we desire to reconcile the pervasive outsider, gravity, and the forces that prevail in the quantum world of our everyday laboratory experience.

### 4.4.4 Identity

We do not understand the physics that sets quark masses and mixings. Although we are testing the idea that a complex phase in the quark-mixing matrix lies behind the observed CP violation, we do not know what determines that phase. The accumulating evidence for neutrino oscillations presents us with a new embodiment of these puzzles in the lepton sector. At bottom, the question of identity is very simple to state: what makes an electron an electron, and a top quark a top quark?

### 4.4.5 Topography

Ordinary experience tells us that spacetime has $3 + 1$ dimensions. Could our perceptions be limited? If there are additional, hidden, dimensions of spacetime, what is their size and shape? Part of the vision of string theory (see Chapter 5) is that excitations of the curled-up dimensions determine the particle spectrum. We have recognized recently that Planck-scale compactification is not – according to what we can establish – obligatory, and that current experiment and observation admit the possibility of dimensions not navigated by the strong, weak, and electromagnetic interactions that are almost palpably large. A whole range of new experiments will help us explore the fabric of space and time, in ways we didn't expect just a few years ago.

## 4.5 The Higgs boson

The many successes of the electroweak theory and its expansive range of applicability are most impressive, but our understanding of the theory is incomplete. We have not yet learned what differentiates electromagnetism from the weak interactions – what endows the W and Z with great masses while leaving the photon massless. The agent of electroweak-symmetry breaking represents a novel fundamental interaction whose effects are felt at an energy of a few hundred GeV. *We do not know the nature of the new force.*

It is commonplace in physics to find that symmetries we observe in Nature's laws are not displayed openly in the expressions of those laws. A liquid is a disordered collection of atoms or molecules held together by electromagnetism that looks the same from every vantage point, reflecting the fact that the laws of electromagnetism are indifferent to direction. A crystal is an ordered collection of the same atoms or molecules, held together by the same electromagnetism, but it does not look the same from every vantage point. Instead, a crystal displays ranks and files and columns that single out preferred directions. The rotational symmetry of electromagnetism is hidden in the regular structure of a crystal. Symmetry means disorder – even if, at first look, that statement may seem to run counter to Western notions of ornamentation.

The central challenge in particle physics today is to understand the nature of the mysterious new force that hides the symmetry between the weak and electromagnetic interactions. Is it a fundamental force of a new character? Or a new gauge interaction, perhaps acting on undiscovered constituents? Or a residual force that emerges from strong dynamics among the weak gauge bosons? Or is electroweak-symmetry breaking the work of the strong interaction, QCD? Theorists have explored examples of all these ideas, but we do not yet know which path Nature has taken.

The simplest guess goes back to theoretical work by Peter Higgs and others in the 1960s. According to this picture, we owe the diversity of the everyday world to a vacuum state that prefers not a particular direction in ordinary space, but a particular particle composition. The vacuum of quantum field theory is an agitated broth in which many kinds of virtual particles wink in and out of existence. Interchanging the identities of different species does not change the laws of physics. At high temperatures – as in the disordered liquid – the symmetry of the laws of physics is manifest in the equal populations of all species; but at low temperatures, one species – chosen at random – prevails, and condenses out in great numbers, concealing the symmetry.

We can picture the condensate of Higgs particles as a viscous medium that selectively resists the motion of other particles through it. In the electroweak theory, the drag on the W and Z particles caused by their interactions with the Higgs condensate gives masses to the weak-force particles. Interactions with the condensate could also give rise to the masses of the constituent particles – quarks and leptons – that compose ordinary matter. Today's version of the electroweak theory shows how this could come about, but does not predict what the mass of the electron or the top quark or the others should be.

When we melt a crystal, the rotational symmetry is restored in the disordered liquid that results. In the same manner, if we could heat up the vacuum enough, we could see the symmetry restored: all particles would become massless and interchangeable. For the present, it is beyond our human means to heat even a small volume of space to the energy of 1 TeV ($10^{12}$ electron volts) – the temperature of $10^{16}$ K – that would be required. We can hope to succeed on a smaller scale, by exciting the Higgs condensate to see how it responds. The minimal quantum-world response is the Higgs boson: an electrically neutral particle with zero spin.

The electroweak theory predicts that the Higgs boson will have a mass, but it cannot predict what that mass should be. Consistency arguments require that it weigh less than 1 TeV. Measurements sensitive to the virtual influence of the Higgs boson suggest (see Figure 4.9) that the mass of the Standard-Model Higgs boson should lie below about 200 GeV. Given an assumption about the Higgs boson's mass, we can say enough about the Higgs boson's properties – about how it could be produced and how it would decay – to guide the search.

The most telling searches have been carried out in experiments investigating electron–positron annihilations at energies approaching 210 GeV at the LEP collider in Geneva. The quarry of the LEP experimenters was a Higgs boson produced in association with the Z. A Higgs boson accessible at the LEP would decay into a b (bottom) quark and a b antiquark, and have a total width (inverse lifetime) smaller than 10 MeV. Figure 4.11 shows how the decay pattern of a Standard-Model Higgs boson depends on its mass.

In 2000, the final year of LEP experimentation, accelerator scientists and experimenters made heroic efforts to stretch their discovery reach to the highest Higgs mass possible. At the highest energies explored, a few tantalizing four-jet events showed the earmarks of Higgs + Z production. A statistical analysis showed a slight preference for a Higgs-boson mass near 116 GeV, but not enough to establish a discovery. Future experiments will tell whether the LEP events were Higgs bosons or merely unusual background events. The LEP observations do place an experimental lower bound on the Higgs-boson mass: the Standard-Model Higgs particle must weigh more than about 114 GeV.

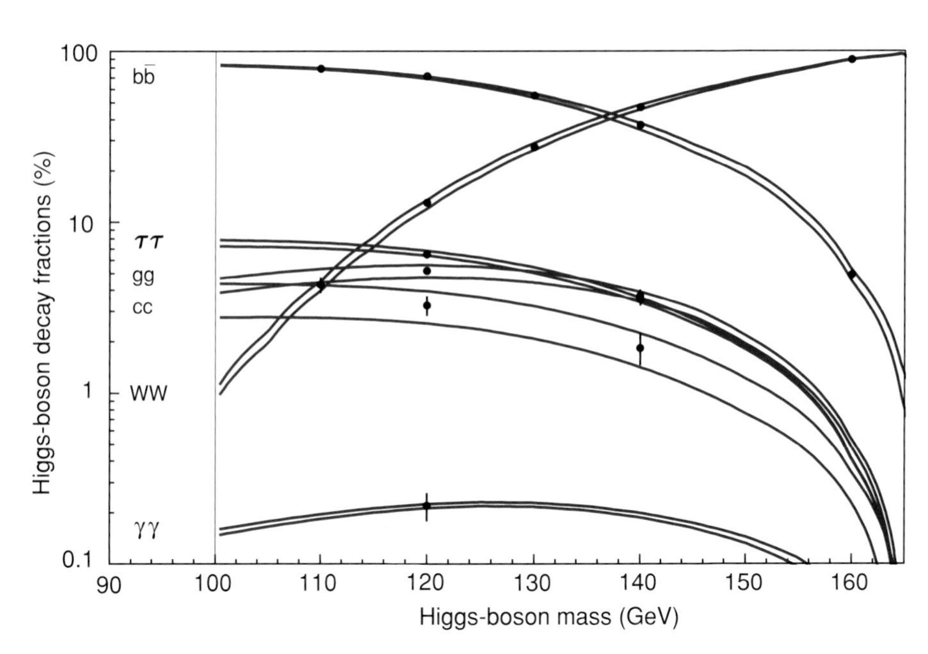


**Figure 4.11.** Probabilities for the decay of a Standard-Model Higgs boson into various products as a function of the Higgs-boson mass. Present-day theoretical uncertainties are indicated by the widths of the bands. Simulated data points indicate the precision of measurements that will be achieved in experiments at an electron–positron linear collider. (Courtesy Marco Battaglia.)

Experiments at Fermilab's Tevatron, where 1-TeV protons collide with 1-TeV antiprotons, may be able to extend the search, looking for Higgs + W or Higgs + Z events. Our best hope for the discovery of the Higgs boson lies in experiments at the Large Hadron Collider (LHC) at CERN, where two 7-TeV beams of protons will collide head-on. When the LHC is commissioned around the year 2007, it will enable us to study collisions among quarks at energies approaching 1 TeV. A thorough exploration of the 1-TeV energy scale will determine the mechanism by which the electroweak symmetry is hidden and teach us what makes the W and Z particles massive.

Once a particle that seems to be the Higgs boson is discovered, a host of new questions will come into play. Is there one Higgs boson or several? Is the Higgs boson the giver of mass not only to the weak W and Z bosons, but also to the quarks and leptons? How does the Higgs boson interact with itself? What determines the mass of the Higgs boson? To explore the new land of the Higgs boson in more ways than the LHC can do alone, physicists are planning a TeV linear (electron–positron) collider. The simulated data points in Figure 4.11 show how precisely a linear collider will map the decay pattern of a Standard-Model Higgs boson.

Our inability to predict the mass of the Higgs boson is one of the reasons many physicists believe that the standard electroweak theory cannot be the whole story. As we search for the Higgs boson, we are searching for extensions that make the electroweak theory more coherent and more predictive. Supersymmetry – which entails several Higgs bosons – associates new particles with all the known quarks and leptons and force particles. Dynamical symmetry breaking interprets the Higgs boson as a composite particle whose properties we may hope to compute once we understand its constituents and their interactions. These ideas and more will be put to the test as experiments begin to explore the 1-TeV scale.

## 4.6 Unified theories

With the Standard Model in hand, what more remains to be understood? If quantum chromodynamics and the electroweak theory are both correct descriptions of Nature, are they complete? Our discussion of the Higgs boson has exposed some limits to the predictive power of the electroweak theory, and others will appear when we turn to the origins of mass. In fact, many observations are explained only in part, or not at all, by the separate gauge theories of the strong and electroweak interactions. It is natural to look for a more comprehensive theory to make sense of them.

Quarks and leptons are spin-$\frac{1}{2}$ particles, structureless at the current limits of resolution. Moreover, both quarks and leptons appear in weak-interaction pairs – up and down, neutrino and electron, and so on – and the weak interactions of quarks have the same strength as the weak interactions of leptons. Quantum corrections respect the symmetry on which the electroweak theory is based only if each pair of quarks is accompanied by a pair of leptons – the pattern we see in Nature. All of these observations prompt the following question: how are quarks and leptons related?

Can the three distinct coupling parameters of the Standard Model be reduced to two or one? We have already noted that the coupling strengths determined by experiment are not constants; they depend on the energy scale (or distance scale) of the determination. The electromagnetic coupling strength increases toward short distances or high energies. The strong coupling behaves oppositely – it decreases at high energies. Might there be a unification energy at which all the couplings, suitably defined, coincide?

The proton's charge exactly balances the electron's charge. Why? In other terms, what fixes the charge of the electron to be precisely three times the charge of the down quark? Why does the sum of the charges of all the distinct quarks and leptons in a generation vanish?

Such questions lead us to take seriously a unification of quarks and leptons, and impel us to search for a more complete electroweak unification or a "grand" unification of the strong, weak, and electromagnetic interactions.

In 1974, Jogesh Pati and Abdus Salam put forward the evocative notion of lepton number as the fourth color, complementing the red, green, and blue labels for the strong interaction charge. Quarks and leptons would then fit together in a single extended family, or multiplet. One meaning of a symmetry is that any member of a multiplet can be transformed into any other; the agent of change is a gauge boson, or force particle. The first implication of quark–lepton unification is therefore that there must be transitions that interchange quarks and leptons. The physical consequences depend on the details of the unification scheme, but proton decay is a natural, and extremely interesting, outcome.

The simplest example of a unified theory that encompasses QCD and the electroweak theory is the SU(5) model developed in 1974 by Howard Georgi and Sheldon Glashow. The Standard Model entails twelve gauge bosons: eight gluons that mediate the strong (QCD) interaction, the $W^+$ and $W^-$ responsible for familiar weak interactions such as beta decay, the $Z^0$ that drives weak neutral-current interactions, and the photon of electromagnetism. Recall that the weak neutral current was the price – or reward – of electroweak unification. Enlarging the symmetry to SU(5) adds twelve more interactions carried by the leptoquark gauge bosons $X^{\pm 4/3}$ and $Y^{\pm 1/3}$ that mediate transitions between quarks and leptons, including electron $\rightarrow$ antidown + $X^{-4/3}$ and up $\rightarrow$ positron + $Y^{-1/3}$.

The leptoquark bosons can induce the proton to disintegrate into a positron and a neutral pion, and by other paths. If X and Y were massless, like the photon or the gluons, protons would decay at a very rapid rate, contrary to experience. A mechanism similar to the one that hides electroweak symmetry, and gives masses to the weak-force carriers, can endow X and Y with masses roughly equal to the unification scale. That prolongs the proton lifetime to approximately $10^{30}$ years – far longer than the age of the Universe of a bit more than $10^{10}$ years.

We can't wait forever to see a single proton decay, but we can watch many protons for a much shorter time. Since one cubic centimeter of water contains Avogadro's

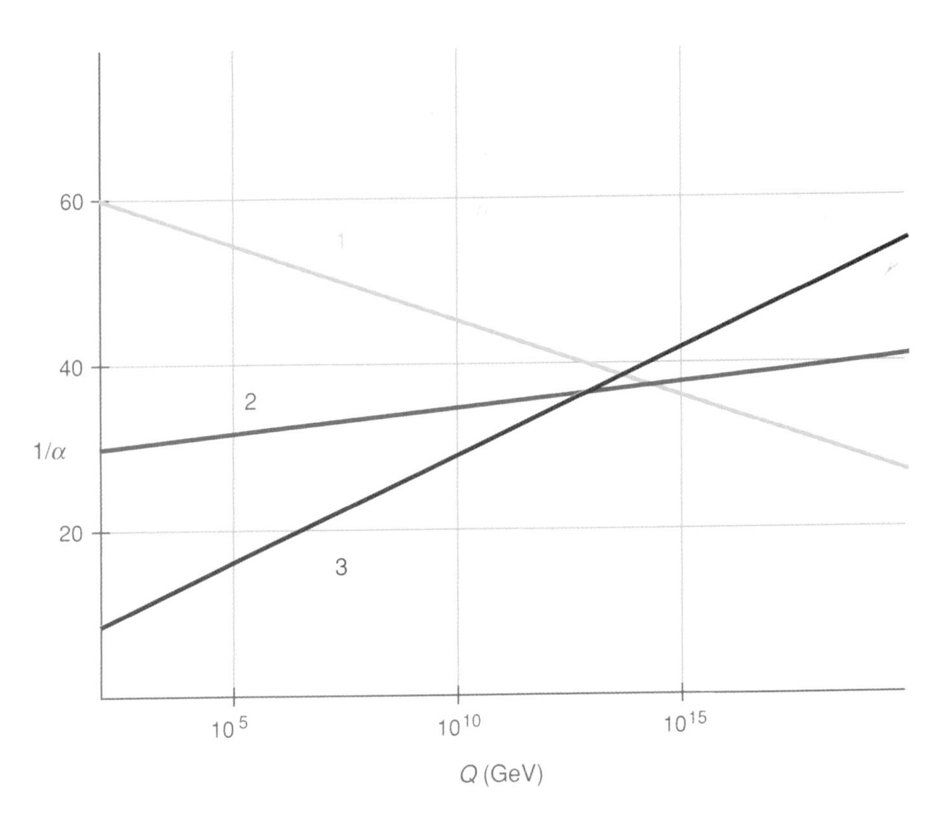


**Figure 4.12.** Inverse coupling strengths of the three component interactions of the Standard Model are extrapolated from values measured in accelerator experiments toward higher energies. The strong (QCD) coupling is labeled 3; the charged-current weak coupling is labeled 2; the (appropriately scaled) weak-hypercharge coupling is labeled 1. According to the SU(5) unified theory, they should coincide at some high energy.

number ($6 \times 10^{23}$ nucleons), a year's careful scrutiny of a cube ten meters on a side ($6 \times 10^{32}$ nucleons) is the magnitude of effort required to reach the decay rate suggested by the SU(5) theory. The target set by SU(5) motivated the construction of a number of experiments in underground laboratories, culminating in the Super-Kamiokande detector that has made such signal contributions to neutrino physics. No examples of proton decay have been observed; depending on the decay mode, the proton lifetime is inferred to be longer than about $10^{31}$ to $10^{33}$ years, surpassing the lifetime predicted by the SU(5) theory. The search continues, with physicists developing plans for detectors ten times as massive as Super-K.

The SU(5) unified theory makes precise the relationship among the three separate interactions of the Standard Model: the SU(3) color gauge theory QCD, the SU(2) weak isospin force, and the U(1) weak hypercharge force. If the unified theory is correct, the coupling strengths of these three interactions – properly normalized – should come together at high energy. Experiments have determined the three couplings with high precision at laboratory energies. If no surprises are encountered over many orders of magnitude in energy – a big if that we should treat with deference – we can use our knowledge of the Standard Model to project the values of the couplings at energies far beyond those accessible to experiment.

The projections take a particularly simple form for the inverses of the coupling strengths, which are shown in Figure 4.12. The trends are encouraging: the coupling strengths are quite different at low energies, but approach one another as the energy increases. However, they do not quite meet in a single point. Put another way, if we begin with a common coupling strength at a unification energy near $10^{14}$ GeV, the SU(5) theory predicts low-energy coupling strengths that come close to, but don't quite equal, the observed values.

The near miss of coupling-strength unification and the failed prediction for the proton lifetime rule out SU(5) as the correct unified theory, unless some new phenomena

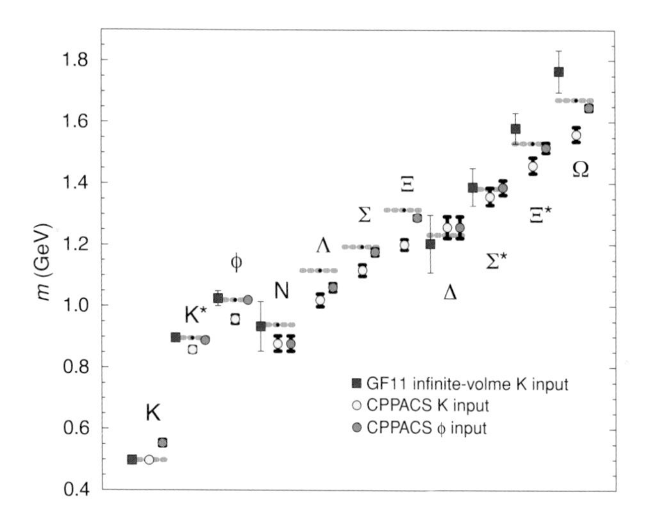


**Figure 4.13.** Calculating particle masses. Final results of the CP-PACS Collaboration's light-hadron spectrum in the valence-quark approximation. Experimental values (horizontal lines) and earlier results from the GF11 Collaboration are plotted for comparison.

intervene between the laboratory scale of about 100 GeV and the unification scale. Most particle physicists nevertheless find the notion of unified theories extraordinarily attractive. The requirement mentioned above, namely that quark pairs must be matched with lepton pairs, seems to me a very powerful message from Nature to build a unified theory. If SU(5) is not the right way, at least it has shown us that there can be a way.

## 4.7 Origins of mass

The origins of mass suffuse everything in the world around us, for mass determines the range of forces and sets the scale of all the structures we see in Nature. Not long ago, physicists regarded questions about the value of the proton mass or electron mass as metaphysical – beyond the reach of science. The properties of the atom's constituents seemed givens set by Nature, not outcomes to be understood. Thanks to the Standard Model of particle physics, we now understand that such questions are indeed scientific, and some answers are already in hand.

Quantum chromodynamics, the theory of the strong interactions, teaches us that the dominant contribution to the masses of the proton, rho meson, and other composite, strongly interacting particles is not the masses of the quarks of which they are constituted, but the energy stored up in confining a color-singlet configuration of quarks in a tiny volume. Computing hadron properties has required the development of large-scale computer simulations of QCD on a spacetime lattice. In 1993, theorists at IBM's Watson Research Center used the purpose-built GF11 supercomputer to make the first precise evaluation of the hadron spectrum, approximating mesons as quark–antiquark states and baryons as three-quark states. Their simulations yielded masses that agree with experiment to within about ten percent. More recently, members of the CP-PACS (Computational Physics by Parallel Array Computer Systems) project at the University of Tsukuba have embarked on an ambitious program that will soon lead to a full calculation, including the effects of a quark–antiquark sea of virtual particles within the hadrons. The valence-quark calculations are summarized in Figure 4.13. The gross features of the light-hadron spectrum are reproduced very well, but, in detail, the computed spectrum deviates systematically from experiment. Those deviations seem to be resolved by the inclusion of quark–antiquark pairs.

This is a remarkable achievement, both for the understanding in principle that hadron masses arise from the interactions of quarks and gluons and for the quantitative success. To put it another way: because most of the luminous matter in the

Universe is in the form of neutrons and protons, *quantum chromodynamics explains nearly all the visible mass of the Universe.*

The Higgs boson is an essential part of the analogy to the Meissner effect in superconductivity that yields an excellent understanding of the masses of the weak-interaction force carriers $W^{\pm}$ and $Z^0$. The weak-boson masses are set by the electroweak scale and a single parameter that may be taken from experiment or derived from a unified theory.

A key aspect of the problem of identity is the origin of the masses of quarks and leptons. Interactions with the Higgs field may also generate quark and charged-lepton masses; however, our understanding of these masses is considerably more primitive. (For neutrinos, which may be their own antiparticles, there are still more possibilities for new physics to enter.) The Standard-Model response is tantalizing but not completely satisfying. The theory shows how masses can arise as consequences of electroweak-symmetry breaking, but it does not permit us to calculate the masses. In the electroweak theory, the natural mass scale is the electroweak scale, about 175 GeV, but the specific value of each quark or lepton mass is set by a separate coupling strength. The values of these mass parameters are vastly different for different particles: the value for the top quark is 1, whereas for the electron it is about 0.000003. Inasmuch as we do not know how to calculate the mass parameters, I believe that *we should consider the sources of all these fermion masses as physics beyond the Standard Model.*

The mass parameters also determine the relative importance of weak-interaction transitions that cross family lines: the up quark couples dominantly to down, less to strange, and still less to bottom. If they are complex, rather than real, numbers, the mass parameters induce the subtle asymmetry between the behavior of matter and antimatter known as CP (for charge-conjugation–parity) violation. The CP-violating effects observed in the decays of K-mesons and B-mesons give every indication of arising from the mixing among quark families. According to my convention that the sources of all quark and lepton masses relate to physics beyond the Standard Model, what we call "Standard-Model CP violation" may itself be a window on new physics.

What accounts for the range and values of the quark and lepton mass parameters? The Standard Model has no response. Unified theories of the strong, weak, and electromagnetic interactions can relate some quark and lepton masses, but have not yet given us a comprehensive understanding. It is possible that the masses, family mixings, and quantum numbers of the quarks and leptons reflect some sort of internal structure. Models based on subconstituents have so far proved unrewarding. Or perhaps the quark and lepton properties carry messages from other dimensions of spacetime. In string theory, particle properties are set by the vibrations of tiny bits of string in spatial dimensions beyond the familiar three. Theorists must work to turn this assertion into an unambiguous calculation.

For experiment and theory together, an essential goal is to refine the problem of identity – to clarify the question so we will recognize the answer when we see it.

## 4.8 Neutrino oscillations

If neutrinos have masses, then a neutrino of definite flavor $\nu_a$, produced in association with a charged lepton of flavor a, need not be a particle with definite mass. It can instead be a quantum-mechanical mixture of particles with definite masses. Each of these components of definite mass is represented in quantum mechanics by a traveling wave. When $\nu_a$ is born, the peaks of the waves corresponding to its different mass components line up and the waves add up to a resultant with the properties of $\nu_a$. If $\nu_a$ has a certain mean momentum, the waves corresponding to different masses evolve with different frequencies. After the neutrino has propagated for some time, the peaks of the different component waves will have fallen out of phase, and the sum will no longer correspond to a state of definite flavor. In this way, for example, a neutrino born as $\nu_\mu$ may become an admixture of $\nu_\mu$ and $\nu_\tau$. Upon interacting in a detector it produces not the expected muon, but a tau lepton, so we conclude that the original $\nu_\mu$ has spontaneously changed into $\nu_\tau$. If we consider the simplest case of two neutrino species, we find that the probability of the neutrino changing flavor is proportional to $\sin^2[\triangle m^2 L/(4E)]$, where $\triangle m^2$ is the difference between the masses squared of the two neutrinos with definite mass, $L$ is the distance the neutrino travels between birth and interaction, and $E$ is the neutrino's energy. The phenomenon is called neutrino oscillation, because the probability of the neutrino changing flavors oscillates with the distance traveled. Because neutrino oscillation occurs only if neutrinos have distinct masses, the observation of neutrino oscillation is immediate proof that at least one neutrino has mass.

Several recent experiments have proved that neutrinos do change flavor during propagation. The first striking evidence came in 1998 from the behavior of atmospheric neutrinos observed in the Super-Kamiokande detector, shown in Figure 4.14. Since the Earth is transparent to neutrinos, an underground detector can compare the flux of downward-going neutrinos produced by cosmic-ray interactions in the atmosphere just above the detector with the flux of upward-going neutrinos produced in the atmosphere on the other side of the Earth. The upward-going neutrinos may travel as far as 13 000 kilometers before interaction, whereas the downward-going neutrinos travel only a few tens of kilometers. Super-Kamiokande investigators have reported that the flux of muon neutrinos from below the horizon is only about half the flux of muon neutrinos from above the horizon. Since the neutrino-production rates are uniform all around the Earth, it follows that something is happening to the neutrinos that travel a long distance from birth to Super-K. The most graceful interpretation is that the upward-going muon neutrinos are oscillating into another species, with oscillation into tau neutrinos the path favored by measurements. New experiments using accelerator-produced muon-neutrino beams directed through the Earth to distant detectors aim to refine the Super-K observations, observe the expected oscillatory behavior, and search for tau leptons produced by the well-traveled neutrinos.

In 2001, the Sudbury Neutrino Observatory in Ontario provided elegant evidence that electron neutrinos produced in the Sun arrive as muon neutrinos and tau neutrinos after their eight-minute trip to Earth. Their detector is a smaller version of Super-Kamiokande filled with "heavy water" in which deuterium (hydrogen with an extra neutron in its nucleus) replaces ordinary hydrogen. They have measured three different reactions that have different sensitivities to the fluxes of electron, muon, and tau neutrinos. The charged-current reaction, $\nu_e$ + deuteron $\rightarrow$ electron + proton + proton is induced only by electron neutrinos. Neutrino–electron scattering and the neutral-current reaction neutrino + deuteron $\rightarrow$ neutrino + proton + neutron are induced by all three neutrino flavors. Taken together, the results shown in Figure 4.15 establish that about two-thirds of the electron neutrinos produced in the Sun appear on Earth as a mixture of muon and tau neutrinos. Moreover, the total flux of Solar neutrinos is just what is predicted by the theoretical "standard Solar model" of John Bahcall and his collaborators.

A third piece of evidence that neutrinos change flavors still awaits confirmation. Investigators at the Liquid Scintillator Neutrino Detector (LSND) at Los Alamos National Laboratory reported in 1995 that electron antineutrinos appeared in a beam derived from the decay of positively charged muons, which yield muon antineutrinos and electron neutrinos, but no electron antineutrinos. They interpreted their results as evidence that muon antineutrinos oscillate into electron antineutrinos. A new experiment

**Figure 4.14.** The Super-Kamiokande neutrino detector, shown here during filling, consists of a cylinder 40 meters in diameter and 40 meters high that holds 50 000 tons of pure water. Many of the 11 200 phototubes that view the water are visible. (Kamioka Observatory, ICRR (Institute for Cosmic Ray Research), The University of Tokyo.)

at Fermilab, MiniBooNE, which uses neutrinos produced by an 8-GeV proton beam, is taking data that will test the LSND report. Taken together with the evidence from atmospheric and solar neutrinos, the squared neutrino-mass difference inferred by LSND would require the existence of a fourth neutrino flavor. Only three neutrino flavors fit with the observed generations of quarks and leptons, and we know from other measurements that only three neutrino flavors have normal weak interactions. Therefore a new kind of neutrino – perhaps a "sterile" neutrino immune to the weak interactions – is required to fit the LSND data.

In a few short years, the study of neutrino oscillations has exploded from patient searches to become one of the liveliest areas of particle physics. What we are learning

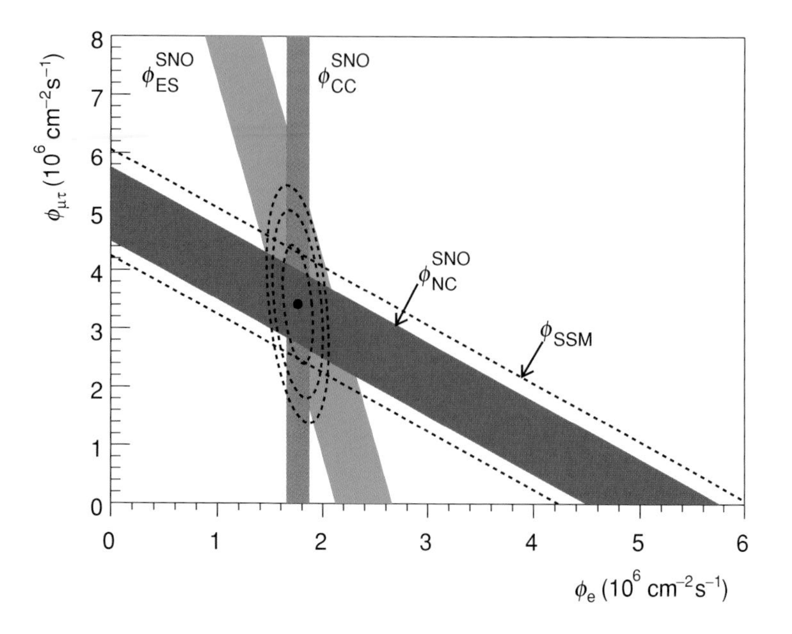


**Figure 4.15.** Measurements of three neutrino-induced reactions at the Sudbury Neutrino Observatory determine the flux of electron neutrinos that arrive from the Sun and the flux of muon and tau neutrinos.

has implications for particle physics itself, but also for nuclear astrophysics and for cosmology. The observation of neutrino oscillations raises many burning questions. How many flavors of neutrinos are there? If the number is greater than three, what is the nature of the new neutrinos? What are the neutrino masses? Is the neutrino its own antiparticle? Do neutrinos play a role in developing the observed excess of matter over antimatter in the Universe? What accounts for neutrino masses? Is it the same – still mysterious – mechanism as that which accounts for quark and charged-lepton masses, or will neutrino mass open a new window on physics beyond the Standard Model? It is a rich menu that will keep experimenters, theorists, and machine builders occupied for years to come.

## 4.9 Supersymmetry

Supersymmetry is the name physicists have given to a possible new symmetry of Nature, hitherto unobserved, that underlies one of the most audacious efforts to build a theory more comprehensive than the Standard Model. The new symmetry relates fermions – particles with half-integer spin – to bosons – particles with integer spin. Within the context of quantum field theory, the most general symmetry is the combination of relativistic invariance plus internal symmetries such as those we have used to derive gauge theories plus supersymmetry. Even if we had not found appealing consequences of a supersymmetric world, that statement would be reason enough for theorists to explore the idea of supersymmetry, because mathematical inventions have a way of proving useful for physics.

Roughly speaking, supersymmetry requires that each known particle should have a superpartner with spin offset by $\frac{1}{2}$, and it relates the interactions of particles with those of their superpartners. Rich as it is, the known spectrum of quarks and leptons and gauge bosons contains no candidates for superpartners. If supersymmetry is correct, the spectrum of particles would be doubled. The electron would have a spin-zero partner, the scalar electron or selectron; each quark would have a spin-zero partner called a squark, and each gauge boson would have a spin-$\frac{1}{2}$ partner, generically called a gaugino. Each of these superpartners would have the same mass as its ordinary partner, if

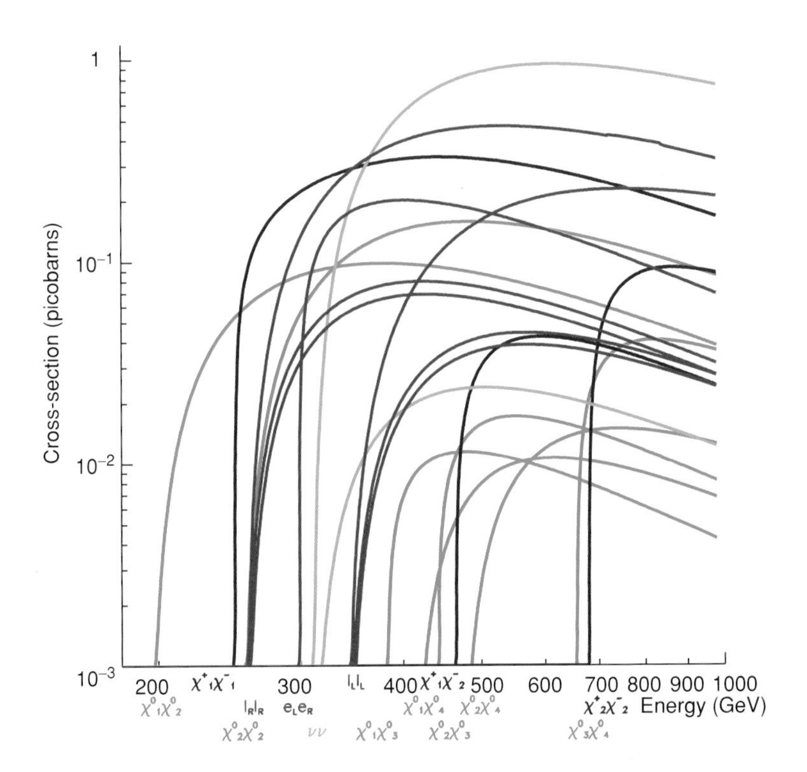


**Figure 4.16.** The thicket of thresholds for supersymmetric-particle production that might be encountered at an electron–positron linear collider. The spectrum of superpartners depicted is one of many possibilities theorists have explored. (Courtesy of Grahame Blair.)

supersymmetry were exact. That is obviously not the case, so if our world is supersymmetric, supersymmetry cannot be exact. Mapping the spectrum of superpartners in detail would help to uncover the mechanism that hides supersymmetry.

So far, supersymmetry may seem like a complication without any benefits, but the potential benefits make a very impressive list. In many supersymmetric models, the superpartners carry a supersymmetric quantum number, called R-parity, that means they must be produced in pairs. The lightest supersymmetric particle (LSP) must therefore be stable; it has no way to shed its super-ness and decay into ordinary particles. Relic LSPs left over from the Big Bang therefore could account for the dark matter of the Universe. We can imagine that a complete supersymmetric model would predict that the Universe contains dark matter, how much, and of what kind. Treating supersymmetry as a local symmetry leads to a quantum theory of gravity that reproduces Einstein's gravity in the classical limit.

The Higgs-boson mass, we have seen, must lie below about 1 TeV. There is no mechanism in the Standard Model to keep the Higgs mass within that range, particularly when quantum corrections can link it with much larger scales such as the unification scale or the Planck scale that defines the energy at which gravity must be added to the mix. Supersymmetry, with its correlation between fermions and bosons, stabilizes the Higgs-boson mass below 1 TeV, provided that the superpartners themselves appear with masses around 1 TeV. If that is so, the discovery of a host of superpartners is just around the corner. Figure 4.16 shows the production rates for a succession of superpartners in electron–positron collisions. The forest of nearby thresholds certainly would qualify as an embarrassment of experimental riches!

The presence of so many particles, in addition to the Standard-Model roster, changes the energy dependence of the coupling constants. It is very suggestive that the coupling strengths of the SU(3), SU(2), and U(1) gauge interactions coincide at about $10^{16}$ GeV, as shown in Figure 4.17.

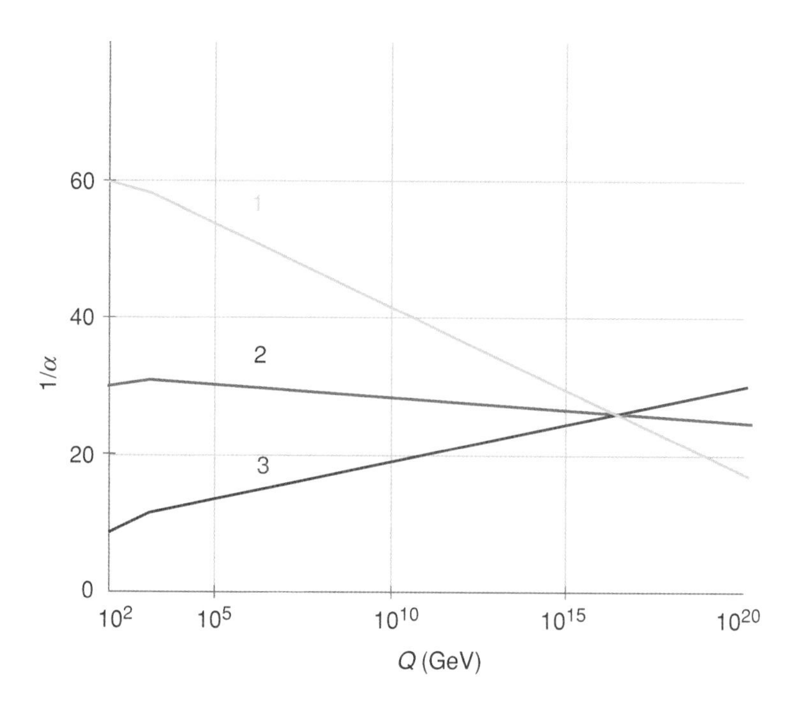


**Figure 4.17.** Inverse coupling strengths of the three component interactions of the Standard Model are extrapolated from values measured in accelerator experiments toward higher energies, assuming that a full spectrum of supersymmetric particles appears around 1 TeV. In this supersymmetric version of the SU(5) unified theory, the couplings coincide near $10^{16}$ GeV. Compare this with Figure 4.12.

The discovery of supersymmetry would represent not only a magnificent experimental achievement but also a great triumph for esthetic theoretical reasoning. It would mark the opening of new dimensions – not new dimensions of ordinary spacetime, but quantum dimensions – and it would supply formidable encouragement for the ambitious string-theory program.

## 4.10 Gravity and particle physics

Gravity is by far the weakest of the known fundamental interactions – strong, weak, electromagnetic, and gravitational – so it is entirely natural to neglect gravity in most particle-physics applications. The gravitational attraction between two protons is some thirty-five orders of magnitude feebler than their mutual electrostatic repulsion, for example. Yet we cannot put gravity entirely out of our minds, because gravity is universal: all objects that carry energy attract one another.

Recent measurements of supernovae at high redshift imply that the Universe contains a cosmologically significant amount of a curious form of energy that accelerates the expansion of the Universe (see Chapter 1). That discovery has called new attention to a longstanding mystery: why is empty space so nearly weightless? This question, which physicists call the vacuum-energy problem or the cosmological-constant problem, shows that something essential must be missing in our understanding of the electroweak theory.

Astronomical observations show that the energy density of space can be no more than about $10^{-29}$ grams per cubic centimeter, in everyday units. That stupendously tiny amount would still be enough to constitute seventy percent of the matter and energy of the Universe. What the nature and origin of the "dark energy" could be is a defining question for cosmology today, and demands an explanation from particle physics.

Particle physics has been ready with an answer, but it is an understatement to say that it isn't the answer we might hope to hear. According to our present understanding, the pervasive Higgs field of the electroweak theory fills all of space with a uniform energy density that is at least $10^{+25}$ grams per cubic centimeter, an amount very much greater than astronomy tells us is there. If free space were as heavy as the electroweak theory suggests, the Universe would have collapsed back on itself instants after the Big Bang.

We do not know what to make of this fifty-four-orders-of-magnitude mismatch, but we know that the mismatch is even greater – by more than a hundred orders of magnitude – in unified theories of the strong, weak, and electromagnetic interactions. Will the resolution require a new principle, or might a proper quantum theory of gravity, in combination with the other interactions, resolve the puzzle of the cosmological constant? That new principle might be supersymmetry, and the proper theory might be string theory, but for the moment there is no established explanation for the vanishing of the cosmological constant – nor for a vacuum energy that is not quite zero. Many of us feel that the vacuum-energy problem must be an important clue – but to what?

Evidence for dark energy is recent, but we have known about the vacuum-energy problem since the 1970s. Over that span, we have developed the electroweak theory into a precise predictive tool, verified its qualitative consequences, and tested its predictions at the level of one part in a thousand. It is perfectly acceptable in science – indeed, it is often essential – to put hard problems aside, in the expectation that we will return to them at the right moment. What is important is never to forget that the problems are there, even if we do not allow them to paralyze us.

Encouraged by our growing understanding of the strong, weak, and electromagnetic interactions, we have begun to ask why gravity is so weak. This question motivates us to face up to the implicit assumptions we have made about gravity. Newton's inverse-square force law is one of the great triumphs of human intelligence: no deviations have been observed from astronomical distances down to small, but still macroscopic, distances. Elegant experiments using torsion balances or microcantilevers have confirmed that gravitation closely follows the Newtonian force law down to distances on the order of a hundredth of a centimeter, which corresponds to an energy scale of only about $10^{-12}$ GeV! (At shorter distances, corresponding to higher energies, the constraints on deviations from the inverse-square law deteriorate rapidly.) In contrast, we have tested quantum chromodynamics and the electroweak theory down to distances shorter than $10^{-17}$ cm, or energies of about 1000 GeV. That is, we have scrutinized the Standard-Model interactions over a range of fifteen orders of magnitude in energy or distance where we have no experimental information about gravity. No experimental result prevents us from considering changes to gravity even on a small but macroscopic scale.

"What is the dimensionality of spacetime?" is a delicious scientific question that tests our preconceptions and unspoken assumptions. It is given immediacy by the impressive developments in string theory. For its internal consistency, string theory requires an additional six or seven space dimensions, beyond the $3 + 1$ dimensions of everyday experience. Unlike the three space dimensions we know, the extra dimensions are curled up in small circles of circumference $L$, so that a traveler who moves a distance $L$ in an extra dimension returns to the starting place. Until recently, theorists presumed that the extra dimensions must be compactified on the Planck scale, where classical concepts of space and time break down, with radius $R \approx 10^{-35}$ m. The experimental evidence is remarkably unconstraining; so far as we know, extra dimensions could be large compared with the normal scales of particle physics.

One way to change the force law is to imagine that gravity propagates freely throughout an enlarged spacetime that contains extra spatial dimensions beyond the known three (see Chapter 2). To respect the stronger constraints on the behavior of the Standard-Model interactions, we suppose that all the other forces are confined to our

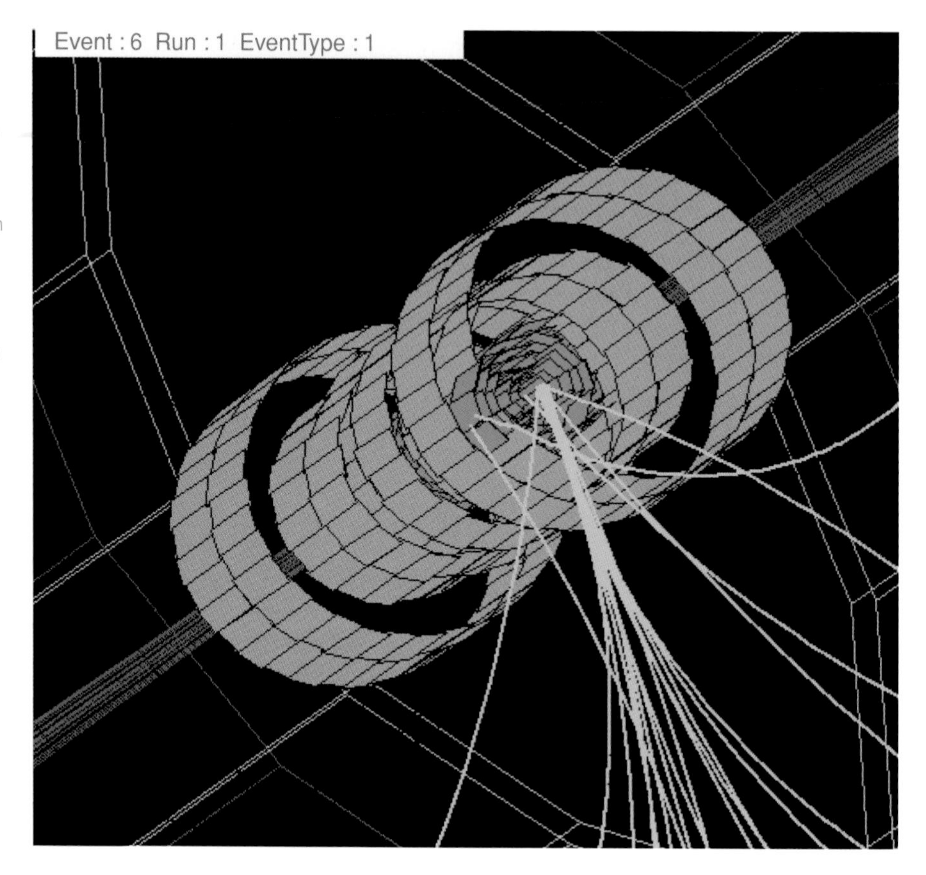


**Figure 4.18.** Looking for the graviton. Simulation in the CDF detector of a missing-energy event that might signal graviton emission in proton–antiproton collisions at 2 TeV. No detected particle balances the momentum of the jet of particles at the bottom right. (Maria Spiropulu, using CDF simulation, reconstruction, and visualization tools.)

three dimensions. What difference do extra dimensions make? At separations larger than the scale of extra dimensions, gravity obeys the Newtonian force law. At shorter separations, gravity – unlike the other forces – spreads out in all the spatial dimensions, and the gravitational force gains an extra inverse power of the separation for each extra dimension. On this picture, gravity is not intrinsically weak; it is as strong as electromagnetism at small distance scales, but appears weak at the relatively large distances we have studied because its effects are dispersed throughout the volume of extra dimensions. The larger or more numerous the extra dimensions, the nearer is the unification of the strength of gravity with the other fundamental forces.

The possibility of large extra dimensions changes the way we think about the influence of gravity on particle-physics experiments, and makes it conceivable that accelerator experiments might someday probe gravity. Two examples will illustrate the new opportunities.

In a high-energy collision, any particle can radiate a graviton into the extra dimensions. Since it couples only gravitationally, the graviton sails off into the extra-dimensional beyond without registering in the detector. Although the probability of individual processes of this kind is tiny, the phenomenon may be observable because a graviton may excite the extra dimensions to ring in a gigantic number of modes. The sheer number of possibilities may compensate for the improbability of each particular possibility. Because the graviton carries away energy that is unrecorded by the detector, the experimental signature is missing energy. Figure 4.18 shows the simulated gravitational excitation of extra dimensions in the CDF experiment at Fermilab: a single jet of particles produced in proton–antiproton collisions at 2 TeV carries energy transverse to the direction of the colliding beams that is not balanced by anything visible to the detector. Searches have begun at the Fermilab Tevatron and will continue at the CERN Large Hadron Collider.

If the size of extra dimensions is close to the distances around $10^{-19}$ m now being probed in collider experiments, some theorists have speculated that tiny black holes might form in high-energy collisions. Concentrating the required vast energy in a tiny volume is implausible in conventional three-dimensional gravity, because gravity is so weak. If the extra dimensions do exist, and if gravity is intrinsically strong, then two high-energy particles might pass close enough to trigger gravitational collapse. The resulting microscopic black holes would evaporate explosively into large numbers of quarks and gluons that might be seen as spectacular cosmic-ray showers or hedgehog events in collider experiments.

Irrespective of whether any of these dramatic signatures shows itself, gravity is here to stay as part of particle physics.

## 4.11 New tools for particle physics

Theoretical speculation and synthesis is valuable and necessary, but we cannot advance without new observations. The experimental clues needed to answer today's central questions can come from experiments at high-energy accelerators, experiments at low-energy accelerators and nuclear reactors, experiments with found beams, and deductions from astrophysical measurements. Past experience, our intuition, and the current state of theory all point to an indispensable role for accelerator experiments.

The opportunities for accelerator science and technology are multifaceted and challenging, and offer rich rewards for particle physics.

One line of attack consists in refining known technologies to accelerate and collide the traditional stable, charged projectiles – electrons, protons, and their antiparticles – pushing the frontiers of energy, sensitivity, and precise control. The new instruments might include brighter proton sources to drive high-intensity beams of neutrinos or kaons; very-high-luminosity electron–positron storage rings to produce huge samples of B-mesons, charmed particles, or tau leptons; cost-effective proton–proton colliders beyond the LHC at CERN; and linear electron–positron colliders.

The energy lost through radiation on each turn around a storage ring is a barrier to attaining very high energies in electron–positron colliders. Colliding the beams made in two linear accelerators placed head-to-head avoids the radiation problem, at the price of offering only one chance for particles in the two beams to collide. Four decades of development, including the successful demonstration of the linear-collider concept at Stanford in the 1990s, have established the feasibility of a next-generation linear collider with a collision rate matched to the needs of experiments. A worldwide consensus supports the timely construction of a TeV-scale linear collider as an international project. The scientific opportunities – to study the nature of the mysterious new force that hides the electroweak symmetry, to search for the dark matter of the Universe, and more – are rich and urgent. The benefits of a linear collider in conversation with the Large Hadron Collider would be very great.

Discoveries at the Large Hadron Collider could point to the need to explore energies well above the 1-TeV scale. A very large proton–proton collider is the one machine we could build to reach to much greater energies with technology now in hand. A center-of-mass energy exceeding 100 TeV – seven times the energy of the LHC – seems well within reach. Cost reduction is the imperative driving research and development activities.

A second approach entails the development of exotic acceleration technologies for standard particles: electrons, protons, and their antiparticles. We don't yet know what instruments might result from research into new acceleration methods, but it is easy to imagine dramatic new possibilities for particle physics, condensed-matter physics, applied science, medical diagnostics and therapies, and manufacturing, as well as a multitude of security applications.

A third path involves creating nonstandard beams for use in accelerators and colliders. Muon-storage rings for neutrino factories, muon colliders, and photon–photon colliders each would bring remarkable new possibilities for experiment.

The muon's 2.2-microsecond lifetime is a formidable challenge for muon colliders, but it presents an opportunity for the construction of a novel neutrino source – a very rich source of high-energy electron neutrinos. The muon's decay yields a muon neutrino and an electron antineutrino (or the other way around, depending on the muon's charge), and no tau neutrinos. Storing a millimole of muons per year in a racetrack would produce neutrino beams of unprecedented intensity along the straightaways. These beams would enable oscillation studies over a wide range of distances and energies, and make possible studies of neutrino interactions on thin targets rather than the massive-target detectors now required. Experiments now beginning are aimed at producing and gathering large numbers of muons into intense, controlled beams that can be contained in a storage ring. If this development succeeds, it would revolutionize neutrino experiments.

Finally, let us note the continuing importance of enabling technologies: developing or domesticating new materials, new construction methods, new instrumentation, and new active controls. To a very great extent, the progress of particle physics has been paced by progress in accelerator science and technology. A continuing commitment to accelerator research and development will ensure a vigorous intellectual life for accelerator science and lead to important new tools for particle physics and beyond.

## 4.12 Changing viewpoints

Poets speak of attaining some distance from the world – finding some strangeness – to express the truths that make a rock stony or a love sublime. Like the poet, the physicist seeks not merely to name the properties of Nature, but to plumb their essence. To truly understand the world around us, to fully delight in its marvels, we must step outside everyday experience and look at the world through new eyes.

Some say that we have ranged far enough, that looking further from the everyday world toward the Higgs boson, or supersymmetry, or unified theories, or superstrings will teach us nothing about our world. I share neither their conclusion nor the self-assurance that prompts them to speak for Nature, for the essence of science is doubt, curiosity, and humility. It is a comfort of sorts to know that such stay-at-home voices have been with us for a long time. We read in Jean Perrin's 1926 Nobel Lecture,

> Particularly at the end of the last century, certain scholars considered that since appearances on our scale were in the end the only important ones for us, there was no point in seeking what might exist in an inaccessible domain. I find it very difficult to understand this point of view since what is inaccessible today may become accessible tomorrow (as has happened by the invention of the microscope), and also because coherent assumptions on what is still invisible may increase our understanding of the visible.

In the event, physicists have known since the 1920s that to explain why a table is solid, or why a metal gleams, we must explore the atomic and molecular structure of matter. That realm is ruled not by the customs of everyday life, but by the laws of quantum mechanics. It is hard to imagine that any amount of contemplating the properties of matter at human dimensions would have led to these insights – insights that today govern not only our understanding of the world, but much of commerce.

The emergence of the Standard Model of particle physics and the glimpses it gives us of the understanding that may lie beyond are part of a larger change that we are living through in the way humans think about their world. Indeed, the great lesson of science in the twentieth century is that the human scale – our notion of size and time – is not privileged for understanding Nature, even on the human scale. The recognition that we need to leave our familiar surroundings the better to understand them has been building since the birth of quantum mechanics. As it emerges whole, fully formed, in our unified theories and equations that map physical understanding from one energy scale to another, the notion seems to me both profound and irresistible. It is all the more potent for the parallels in biology, cosmology, and ecology – parallels that range to scales both larger and smaller than the human scale.

I find it fully appropriate to compare this change in perception with great scientific revolutions of the past. I believe that, in time, our new understanding of scales – our new sense of proportion – will be considered as important as the enlightened sense of place we owe to Copernicus, Galileo, and Newton.

## 4.13 The path before us

Wonderful opportunities await particle physics, with new instruments and experiments poised to explore the frontiers of high energy, infinitesimal distances, and exquisite rarity. We look forward to the Large Hadron Collider at CERN to explore the 1-TeV scale (extending efforts at the LEP and the Tevatron to unravel the nature of electroweak-symmetry breaking) and many initiatives to develop our understanding of the problem of identity: what makes an electron an electron and a top quark a top quark. Here I have in mind the work of the B factories and the hadron colliders on CP violation and the weak interactions of the b quark; the wonderfully sensitive experiments around the world on CP violation and rare decays of kaons; the prospect of definitive accelerator experiments on neutrino oscillations and the nature of the neutrinos; and a host of new experiments on the sensitivity frontier. We might even learn to read experimental results for clues about the dimensionality of spacetime.

If we are inventive enough, we may be able to follow this rich menu with the physics opportunities offered by a linear electron–positron collider and by new-technology neutrino beams based on muon-storage rings. I expect a remarkable flowering of experimental particle physics, and of theoretical physics that engages with experiment.

Experiments that use natural sources also hold great promise. We suspect that the detection of proton decay is only a few orders of magnitude away in sensitivity. Dark-matter searches and astronomical observations should help to tell us what kinds of matter and energy make up the Universe. The areas already under development include gravity-wave detectors, neutrino telescopes, measurements of the cosmic microwave background, cosmic-ray observatories, gamma-ray astronomy, and large-scale optical surveys. Indeed, the whole complex of experiments, observations, and interpretations that we call astro-/cosmo-/particle physics should enjoy a golden age.

Our theories of the fundamental particles and the interactions among them are in a very provocative state. The Standard Model summarizes a simple and coherent understanding of an unprecedented range of natural phenomena, but our new understanding raises captivating new questions. In search of answers, we have made far-reaching speculations about the Universe that may lead to revolutionary changes in our perception of the physical world, and our place in it. Truly, we are entering a remarkable age of exploration and new physics!

## FURTHER READING


1. *The Particle Adventure*, an interactive tour of quarks, neutrinos, antimatter, extra dimensions, dark matter, accelerators, and particle detectors, http://particleadventure.org.
2. Interactions.org, a communication resource from the world's particle-physics laboratories, http://interactions.org.
3. *Quarks Unbound*, from the American Physical Society, http://www.aps.org/units/dpf/quarks_unbound/.
4. Stanford Linear Accelerator Center's Virtual Visitor Center, http://www2.slac.stanford.edu/vvc/.
5. Fermi National Accelerator Laboratory's *Inquiring Minds*, http://www.fnal.gov/pub/inquiring/index.html.
6. *What Is CERN?* from the European Laboratory for Particle Physics, http://public.web.cern.ch.
7. *Anatomy of a Detector*, from Fermilab, http://quarknet.fnal.gov/run2/boudreau.shtml.
8. S. Weinberg, *The Discovery of Subatomic Particles*, 2nd edn., Cambridge, Cambridge University Press, 2003.
9. F. Close, M. Marten, and C. Sutton, *The Particle Odyssey: A Journey to the Heart of Matter*, New York, Oxford University Press, 2002.
10. G. Fraser, E. Lillestøl, and I. Sellevåg, *The Search for Infinity: Solving the Mysteries of the Universe*, New York, Checkmark Books, 1995.
11. G.'t Hooft, *In Search of the Ultimate Building Blocks*, Cambridge, Cambridge University Press, 1996.
12. A. Pais, *Inward Bound: Of Matter and Forces in the Physical World*, Oxford, Oxford University Press, 1988.


## Chris Quigg

Fermilab theorist Chris Quigg is internationally known for his studies of heavy quarks and his insights into particle interactions at ultra-high energies. He has held numerous visiting appointments in the USA and Europe, and has lectured at summer schools around the world.

He has been a member of the Fermi National Accelerator Laboratory staff since 1974, where he was for ten years head of the Laboratory's Theoretical Physics Department. He served as deputy director of the Superconducting Super Collider Central Design Group in Berkeley from 1987 to 1989.

Professor Quigg received his Bachelor of Science degree in 1966 from Yale University and his Ph.D. from the University of California at Berkeley in 1970. He is a Fellow of the American Association for the Advancement of Science and of the American Physical Society, and was awarded an Alfred P. Sloan Research Fellowship. He has written a celebrated textbook on particle physics, and edited the *Annual Review of Nuclear and Particle Science* for a decade.

Chris Quigg's theoretical research has always been strongly influenced by experimental results and possibilities. His interests span many topics in particle physics, from hadron structure through ultra-high-energy neutrino interactions. His work on electroweak-symmetry breaking and supercollider physics highlighted the importance of the 1-TeV scale. In recent years, he has devoted much energy to helping to define the future of particle physics – and the new accelerators that will take us there.

# 5 Superstring theory

Michael B. Green

## 5.1 Introduction

In the course of the twentieth century fundamental notions concerning the physical world were radically altered by a succession of unifying insights. In the early part of the century quantum theory provided the framework for understanding the structure of the atom, its nucleus, and the "elementary" particles of which it is made. This understanding, combined with the special theory of relativity, led, by the early 1970s, to a comprehensive account of three of the fundamental forces – the weak, strong, and electromagnetic forces – in the "Standard Model." Meanwhile, Einstein's 1915 general theory of relativity replaced Newton's theory of gravity, leading to a unification of the force of gravity with spacetime geometry, and explanations of the evolution of the Universe from the time of the Big Bang and of weird astrophysical phenomena, such as black holes. Yet, with each advance ever-deeper questions were posed and remained unanswered. What determines the species of observed elementary particles, such as the electron, neutrinos, quarks, etc., and what principle leads to a unified description of all of them? How can quantum theory be consistent with general relativity? What is the meaning of space and time at extremely small distances, where quantum theory plays a dominant role? How did the Universe begin and how will it end?

An optimist should regard such unresolved puzzles as the seeds from which further insights will arise. This spirit of incompleteness is a distinguishing feature of this chapter. Here we will see that many, if not all, of these deep questions might be resolved by string theory. Although there is no single mathematical equation that summarizes string theory – indeed, it is not yet a complete theory – it has many intriguing interrelations with some of the most modern areas of mathematics. Its compelling qualities suggest that string theory has the potential to overcome key problems present in previous physical theories and to overcome them in a manner that is surprisingly simple and at the same time very novel.

A word of caution – string theory is sometimes referred to as a "Theory of Everything," which is a somewhat immodest description. More realistically, string theory has the potential to unify all the physical forces, including gravity, in a manner that is consistent with quantum mechanics. This is a lot, but almost certainly new questions will arise as the old ones are answered, which will surely modify what is meant by "Everything."

*The New Physics for the Twenty-First Century*, ed. Gordon Fraser.
Published by Cambridge University Press. 

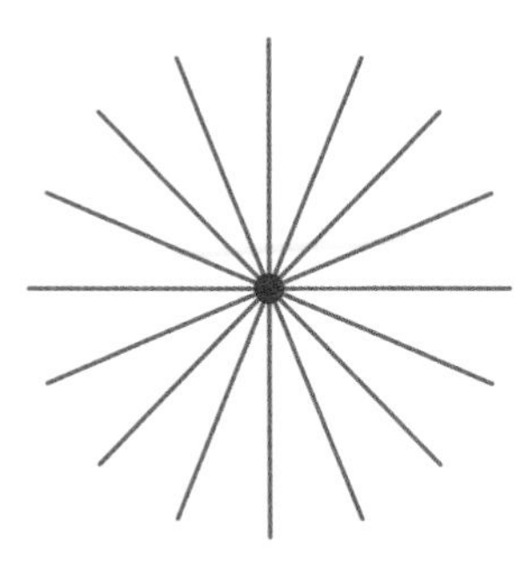

**Figure 5.1.** Gravitational and electric field lines around a point particle converge to give an infinite field at the position of the particle.

## 5.2 Problems with points

At the dawn of the twentieth century the understanding of the laws of physics was in a state of confusion. The discovery of the electron in 1897 by J. J. Thomson had been the first direct confirmation that matter is corpuscular on tiny distance scales. But what is an electron made of? Atoms contain electrons, but what is the structure of the atom?

It rapidly became clear that, despite the triumphs of nineteenth-century physics, some of the most straightforward questions could not even be addressed by the theories then current. Even the most respectable and longstanding theories had manifest inconsistencies. For example, within Newton's theory of gravity, which had been sacrosanct for over two hundred years, the concept of a pointlike particle does not make sense. The problem is that the density of the gravitational energy increases without limit close to the point – the gravitational energy becomes infinite at the location of the particle. But the concept of infinite energy is meaningless. In similar fashion the laws of electricity and magnetism, which were so elegantly encapsulated in Maxwell's theory of electromagnetism in the 1860s, give rise to infinite energy in the electric field surrounding a point particle that carries electric charge.

Therefore, despite the fact that the electron did appear to Thomson to be a point (as it still does with today's far more accurate experiments) these theoretical issues immediately prompted Abraham, Lorentz, and others to invent models in which the electron was a sphere of very small but nonzero radius. This indeed avoided the problem of infinite energy but substituted other equally puzzling inconsistencies. These are intimately related to questions addressed in Einstein's special theory of relativity, which appeared on the scene in 1905 and provided a brilliant resolution of a closely related and even more pressing problem. The problem that Einstein addressed was the profound conflict between Maxwell's theory of electromagnetism and Newton's equations that had long described the dynamics of moving objects. Special relativity postulates that no information can travel faster than the speed of light. Einstein argued that this requires abandoning the age-old notion of time as a universal property that has the same value for all objects – a viewpoint that dated back to Galileo and earlier. According to special relativity space and time are viewed as related qualities – time is somewhat like the three spatial dimensions, making up four-dimensional *spacetime*. The word *somewhat* hides the subtle differences which permit time to progress in only one direction (everything ages and nothing grows younger) whereas a spatial distance can be either positive or negative. Later, in his general theory of relativity, Einstein took this unification of space and time to new heights, as we will discuss later.

Once special relativity had arrived, it became obvious why there had been such profound problems in constructing a theory in which particles are pointlike. In particular, special relativity does not allow objects of nonzero size to be rigid – after all, when a rigid object is moved by pushing one point on it, the whole object moves – all the points on the body move instantaneously and therefore respond to the push faster than the speed of light, which is forbidden. In order to construct a theory of a particle of nonzero size it would be essential to take special relativity into account. The extended object could not be rigid and the passage of time as measured on each separate point on the object would be different – time is relative!

In the aftermath of 1905 it might have been logical for physicists to have constructed theories of extended particles that do satisfy the very stringent conditions imposed by special relativity. If they had done this they would have discovered string theory – but that was not to happen for over sixty years. In fact, much of the mathematical apparatus necessary to describe the structure of an extended string-like particle was developed in the intervening period. Furthermore, only after the development of both quantum theory and general relativity was it possible to understand the structure and physical implications of string theory.

## 5.3 Special relativity, quantum theory, and the Standard Model

The fact that the classical physical laws governing point particles are inconsistent soon became a subsidiary issue, submerged under other rapidly accumulating experimental and theoretical advances. Twenty-five years after Planck had initiated quantum theory, Heisenberg formulated the complete theory of quantum mechanics in 1925, with Schrödinger's version following in 1926. An essential property of quantum theory is the uncertainty principle, which implies that certain "common-sense" qualities that are normally ascribed to everyday objects do not make sense in the microworld. For example, it is not possible to simultaneously specify the precise position and velocity of a particle. According to quantum theory the uncertainty (or inaccuracy) in the position of an object is, at best, inversely proportional to the uncertainty in its momentum (or velocity). Roughly speaking, its position is fuzzy and there is some probability of the particle being at any particular position. If the particle is specified to be at a precise position in space then the uncertainty in its velocity is infinite, which makes no sense. Although this smearing out of the position of a particle makes the problems with infinities less severe than in the classical theory, nevertheless infinite quantities remain and cannot be ignored.

These problems are even more profound in the extension of quantum mechanics invented by Dirac around 1928 which incorporates the law of special relativity. Dirac realized that material particles, such as the electron, could be described in a manner that is consistent with both quantum theory and special relativity only if antiparticles exist. This means that for every species of particle there has to be another species of particle of the same mass (the antiparticle) that carries charge of the opposite sign. In this way Dirac predicted the existence of the antielectron, called the positron, which was discovered in 1932. Now remember another feature of special relativity, namely that the mass of a particle is just a special form of energy and that energy and mass can be converted into each other. A particle and its antiparticle can therefore annihilate each other, leaving a bunch of energy equivalent to the masses of the two particles – energy that may be in the form of electromagnetic radiation, or photons. Conversely, it is also possible for a particle and its antiparticle to be created in a collision of highly energetic particles. The energy of the colliding particles (such as cosmic rays hitting the atoms in the atmosphere) is converted into the mass of the electron and the positron.

Even stranger is the fact that particles and antiparticles can appear spontaneously from "nothing" by virtue of quantum fluctuations implied by the uncertainty principle. This means that, if a specific small region of empty space of a certain size is studied with some apparatus, there will be an uncertainty in the momentum inside that region that is inversely proportional to its size. But this leads to an uncertainty in the energy inside the region. So the energy inside the region is nonvanishing – there are fluctuations of energy that increase as the size of the region is reduced. If the region in question is as small as $10^{-15}$ m the fluctuations are large enough to create an electron and a positron from nowhere, which annihilate a short time later. As a result the notion of empty space, or the vacuum, is meaningless when special relativity is combined with quantum mechanics! Although space may be empty in an average sense, it is more accurate to picture it as full of a sea of particles and antiparticles at very small

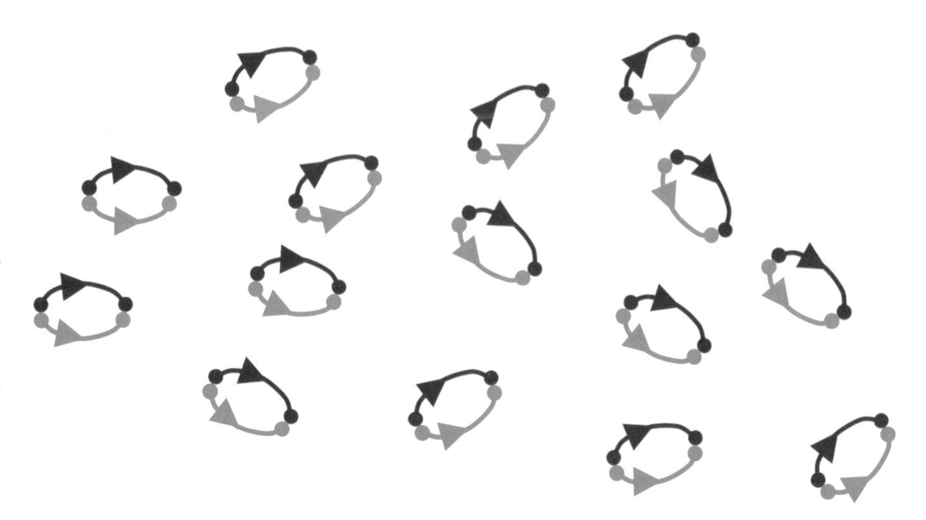

**Figure 5.2.** The vacuum can be viewed as a sea of particles and antiparticles being created and very quickly being destroyed (their trajectories are represented by red and blue arrows). This effect alters the orbits of electrons in an atom as they collide with the particles created from the vacuum. Measurements of the spectrum of hydrogen confirm this picture to a very high accuracy.

distance scales, illustrated in Figure 5.2. These ideas are synthesized in the structure of quantum field theory, which was pioneered by Dirac and others in the 1930s. The fact that arbitrary numbers of particles permeate space makes quantum field theory a subject replete with technical problems.

These problems are most important at ultra-short distances – according to the uncertainty principle the smaller the distance the greater the energy fluctuations. These fluctuations, which create particles of increasingly high mass, lead to new infinities in quantum field theory. Nevertheless, methods of making sensible approximations were found and the work of Schwinger, Tomonaga, and Feynman in the 1940s led to spectacular agreement between theoretical ideas and experimental observations. Almost everything that is known about explicit properties of quantum field theory is by means of such approximations, which are encoded in "Feynman diagrams," such as those shown in Figure 5.3. This involves expanding the physical property under study in a series of powers of the constant, $g$, which determines the strength of the force. In the case of quantum electrodynamics $g$ is proportional to the electric charge of the electron and is approximately equal to 1/137.

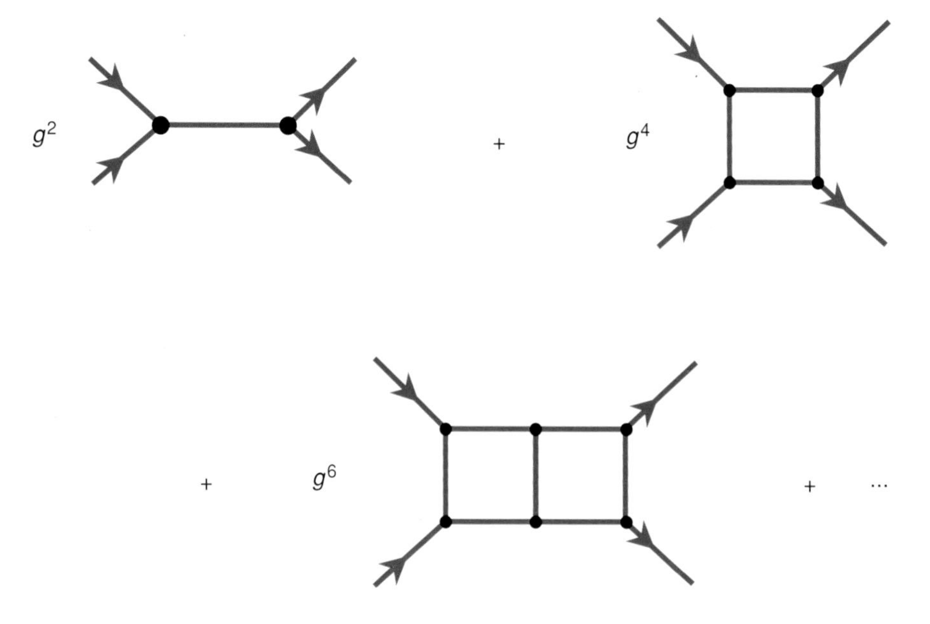


**Figure 5.3.** Feynman diagrams represent the scattering of particles as they collide. The successive diagrams have a greater number of loops and are suppressed by powers of $g$, where $g$ is a small parameter that determines the strength of the force. The number of different diagrams associated with a given power of $g$ escalates dramatically as that power increases.

This kind of approximation in which a physical quantity is expressed as a power series in a small parameter is known as a "perturbative" approximation and is very common in a wide variety of physical theories. In the case of quantum electrodynamics, the quantum version of Maxwell's theory of electromagnetism, the series of Feynman diagrams is astonishingly accurate after very few terms in the expansion and produces predictions that are the most accurate in the whole of science!

The evolution of these ideas, together with a massive accumulation of experimental data, eventually led to the unification of the theory of the electromagnetic and weak forces by Glashow, Salam, and Weinberg in the electroweak theory in 1967. The structure of the electroweak theory is based on a theory of the type first discovered by Yang and Mills, which is a mathematically elegant generalization of Maxwell's nineteenth-century theory of electromagnetism. Although Yang–Mills theory, also known as non-Abelian gauge theory, was formulated in 1954, it was only at the end of the 1960s that this kind of theory found its place in theoretical physics. Particular problems remained in describing the strong force, which were finally overcome in 1973 with the formulation of quantum chromodynamics (QCD), which is another example of a Yang–Mills field theory. The combination of Yang–Mills theories that describes all the known elementary particles and the combined strong, weak, and electromagnetic forces became known as the Standard Model (see Chapter 4).

The Standard Model has an importance within particle physics that is analogous to that of the periodic table of the elements in late-nineteenth-century chemistry. Recall that Mendeleev showed how the many apparently unrelated chemical elements can be arranged into a pattern – the periodic table – according to their properties, such as the kinds of chemical forces they exert. This pattern has symmetries that related the known elements and led to predictions of further elements and their forces. Only much later, once the structure of the atom had been uncovered, was it understood that all of the elements and the variety of chemical forces could be explained from one fundamental principle – the principle that determines the structure of atoms. In much the same way the Standard Model fits the known elementary particles together into patterns with particular symmetries according to the forces the particles exert – the strong, weak, and electromagnetic forces. The particles of the Standard Model are shown in Figure 5.4. These symmetries indicate which particles are needed in order to complete the table.

We see here the all-pervasive role that the notion of symmetry plays in theoretical physics. An important feature of quantum field theory is that particles and forces are no longer to be viewed as distinct properties of matter – each kind of force is characterized by the presence of certain particles. For example, when charged particles, such as electrons, are accelerated they emit electromagnetic radiation, which is itself described as a stream of particles, or photons. Similarly, the weak force is associated with the so-called W and Z particles (discovered experimentally in 1983) and the strong force with the "gluons" (which are responsible for the binding of quarks within the proton and neutron).

The field theories of the Standard Model again suffer from infinities due to short-distance quantum fluctuations. As with quantum electrodynamics, these can be eliminated in a consistent and sensible manner. This procedure for eliminating infinite quantities is known as "renormalization" and it makes sense so long as the theory does not pretend to describe the gravitational force and is therefore not a complete theory. Detailed measurements of atomic structure as well as of properties of elementary particles have confirmed in great detail the picture of the vacuum as a seething collection of energy fluctuations.

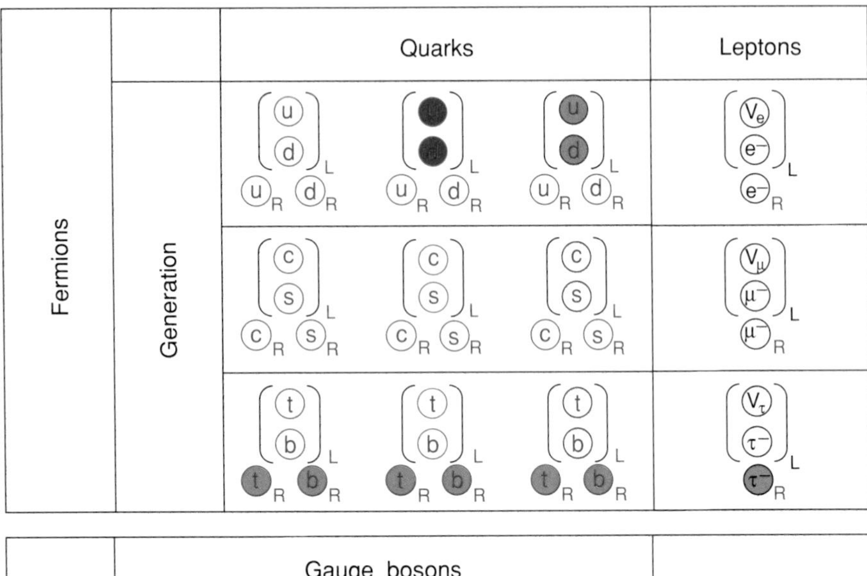


**Figure 5.4.** The elementary particles that enter into the Standard Model. The fermions (electron, quarks, neutrinos, and others) are arranged in a pattern of singlets, pairs, and triplets that indicates their charges under the electroweak and strong forces. The bosons consist of the gauge bosons that transmit the electroweak and strong forces and the Higgs particle (which has yet to be discovered).

Although the Standard Model gives an economical description of observed particle physics and forces, it has two obvious major problems. The first of these is an esthetic problem: the Standard Model does not give a unified explanation of the forces. In particular, it does not relate the electroweak force to the strong force – it simply adds them together in a manner that fits experiment but is not dictated by any theoretical principle. The model has a great deal of additional arbitrariness. Although it makes very accurate predictions concerning the properties of elementary particles, many parameters, such as the species of particles together with their charges and masses, have to be input into the model in an arbitrary manner. There are at least 26 arbitrary parameters of this kind that need to be fed into the Standard Model before it can make any predictions. These are precisely the parameters that a truly unified theory should predict. The second major problem with the Standard Model is that it is obviously not the whole story since it omits the most familiar force of all; the force of gravity! The superficial excuse for this omission is that gravity is by far the weakest force and is not measurable in particle experiments in accelerator laboratories. However, not only is gravity predominant in determining structure on astronomical scales, but also it is the determining factor in the inability of quantum field theory to describe physics on the shortest imaginable distance scales. Any truly unified theory has to come to grips with the most profound problem in twentieth-century theoretical physics – Einstein's theory of gravity is apparently inconsistent with quantum theory.

In other words, although the Standard Model gives a very impressive account of the observed elementary particles and the non-gravitational forces, it does not account for the force of gravity, which necessarily requires a new understanding of the nature of spacetime within quantum theory. Hence the Standard Model is in no sense a unified theory.

## 5.4 Gravity, spacetime, and quantum theory

The force that is left out of the Standard Model was the subject of Einstein's remarkable extension of the special theory of relativity to the general theory, which was finalized in 1915. This overturned Newton's theory of gravity and in the process led to a synthesis of geometry and physics. In the general theory the force of gravity is attributed to the effect of the curvature of spacetime (see Chapter 2). The meaning of curvature is most familiar from the everyday notion of curved surfaces, such as the surface of a ball, which is two-dimensional. However, this idea extends naturally to the notion of curvature in the four dimensions of spacetime. This is difficult to conceptualize without mathematical abstraction, but suffice it to say that within general relativity the more massive an object is, the greater the curvature of space around it. An extreme situation arises when the density of an object reaches a critical value, beyond which a black hole forms. This is a region in which the object is surrounded by an "event horizon" – a region inside of which the geometric properties of space and time are quite different from those outside. Within the classical laws of physics (neglecting quantum theory) nothing can escape from inside the horizon – even a beam of light is trapped by the force of gravity and cannot escape from a black hole.

There is good evidence that large black holes exist in many, maybe most, galaxies. However, when these ideas are combined with the quantum properties of matter on microscopic scales the situation becomes even weirder. To begin with, as Hawking realized in the mid 1970s, a black hole is no longer black once quantum effects are taken into account. Quantum fluctuations allow matter to emerge from the hole in the form of radiation – "Hawking radiation." It is as if the hole were a hot radiating object. As the radiation is emitted, the hole evaporates. In his original treatment Hawking posed a very deep question – the radiation emitted from a black hole bears no apparent relation to the matter that went into the formation of the black hole in the first instance. The hole might have been formed by infalling protons but will emit protons and neutrons and many other types of particles in an uncorrelated manner. All information about the nature of the infalling matter is apparently wiped out in the process of black-hole formation and its subsequent evaporation. This contradicts one of the main precepts of quantum theory, which does not permit such loss of information. There is one important gap in the argument, which is that it assumes quantum theory can be applied to all the forces *except* gravity, which was not treated quantum mechanically. The modern challenge is to discover how this paradox may be resolved in the framework of a more complete theory that treats all of the forces, including gravity, in a quantum-mechanical manner. As we will see, in recent years string theory has provided insight into how quantum mechanics is to be reconciled with Hawking radiation.

Attempting to describe the force of gravity in a quantum-mechanical manner leads to the most profound problems of all. A superficial indication of these problems is that the renormalization program does not work for quantum gravity – in the case of gravity the problematic infinities cannot be eliminated, so the theory appears to make no sense. The underlying origin of these problems can be seen from a simple intuitive combination of the quantum-mechanical uncertainty principle with the fact that general relativity leads to the presence of black holes. As we saw, the uncertainty principle leads to a description in which there are large fluctuations in energy within any small region of space. An extreme situation arises when this principle is applied to extraordinarily small regions. The energy fluctuations in a region bounded by a box whose size is about the "Planck length," which is $10^{-35}$ m, are so large that a tiny black hole forms spontaneously within that region and quickly evaporates. This tiny length scale is equal to $G_N h/c^3$, where $c$ is the speed of light, $G_N$ is Newton's constant that determines the strength of the gravitational force, and $h$ is Planck's constant, which is the constant that is fundamental to quantum theory. The average lifetime of such black holes is the "Planck time" of around $10^{-42}$ s, which is the time taken by light to traverse a Planck length. Roughly (very roughly!) speaking, this means that everywhere in "empty" space and at all times there are microscopic black holes appearing and disappearing. Since the nature of space and time inside a black hole is radically different from that outside, this means that all notions of classical spacetime will have to be jettisoned in describing the quantum theory of gravity. Determining the nature of quantum spacetime has long been the aim of any quantum theory of gravity and is the central objective of the radical directions in which string theory is currently heading.

Before addressing these modern directions, we shall see how string theory evolved in the most unexpected manner from altogether different considerations.

## 5.5 String theory – precursors and early developments

String theory developed in a long and curious manner. We have seen that as long ago as 1905 there were good reasons to contemplate theories of fundamental particles of nonzero size. However, with few exceptions, subsequent developments were based on the notion of pointlike particles, leading eventually to the very impressive successes of quantum field theory and culminating in the Standard Model in the early 1970s. It is interesting that the neglected attempts to describe extended particles by Born and Infeld in 1934 and by Dirac in 1962 play a central role in modern developments within string theory.

Despite the astonishing successes of quantum electrodynamics, for a long period in the late 1950s and 1960s quantum field theory was in the doldrums. This was a period in which a wealth of experimental information was accumulated from large particle accelerators concerning the mysterious nature of the strong force, which seemed impossible to describe in terms of quantum field theory. This is the force responsible for binding protons and neutrons into the nucleus of the atom. On a more microscopic level, it is the force that holds three quarks together inside each individual proton or neutron. Thus, the proton and neutron are now viewed as composite objects, whereas the various species of quarks inside them are considered as fundamental as the electron. Similarly, mesons are composite particles that contain a quark and an antiquark. Experimental observations showed that the strong force has the extraordinary property that it binds the quarks inside the proton, neutron, and other particles so strongly that they can never be separated from each other – the further a quark moves from its companion quarks, the greater the force pulling it back. Quarks are said to be "confined" and isolated quarks do not exist. This mysterious property seemed impossible to account for within the precepts of quantum field theory until the advent of QCD in 1973.

Despite the eventual success of QCD in describing the strong force, the apparent failure of quantum field theory in the 1960s was the immediate motivation for the novel ideas that were to lead to string theory. Most significantly, following the many attempts to explain the behavior of particles at very high energies, in 1968 Veneziano made an inspired guess for the mathematical formula for the "scattering amplitude" of mesons. This is the expression that determines the manner in which particles such

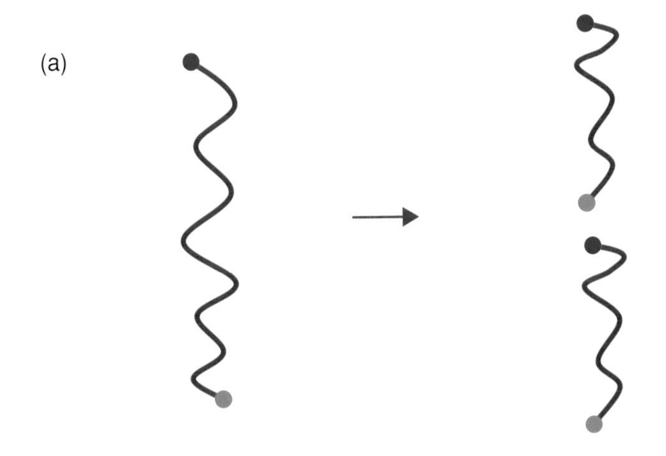


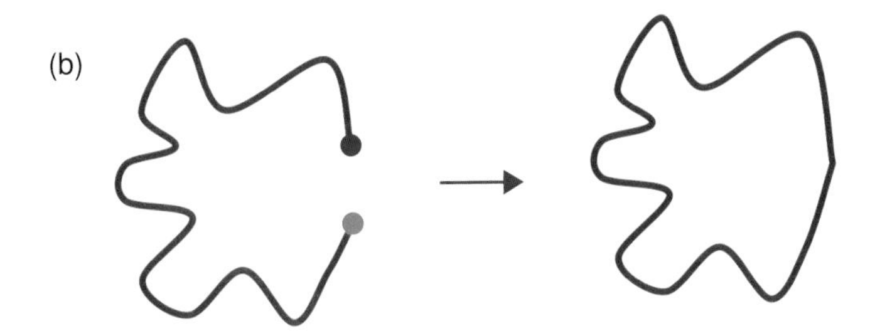


**Figure 5.5.** A string representing a meson with a quark at one end and an antiquark at the other. It can break into two strings (a), each of which also has a quark and an antiquark at its ends. Its end-points can also join to form a closed string (b), which is observed as a particle called a glueball.

as mesons behave when they collide and which contains essentially all the information about the properties of the particles. It turned out, after a series of imaginative insights, that Veneziano's expression is actually the scattering amplitude that comes from considering the collisions of particles that have the structure of "relativistic" strings!

In this perversely circuitous manner, theoretical physicists finally came upon the kinds of theories that describe particles as strings. The original idea was that the quark and antiquark inside any meson are tied together by a string. As shown in Figure 5.5, the string can break into a pair of strings only by creating a quark and an antiquark so that each string in the pair again has a quark at one end and an antiquark at the other. Cutting the string does not create isolated quarks. This is analogous to the well-known fact that cutting a bar magnet in half does not create isolated magnetic charges. In QCD such a string should arise as a consequence of properties of the gluons that bind the quarks together.

The most obvious distinction between a string-like object and a point particle is that a string can vibrate. When a string vibrates, the energy of the string depends on the frequency of the vibration – the higher the frequency the greater the energy. However, one of the more famous consequences of special relativity is that energy is equivalent to mass. This means that when a string vibrates in a particular harmonic it behaves like a particle of a given mass, with the mass increasing as the frequency of the harmonic increases. But a string has an infinite number of harmonics – it produces notes of ever-increasing frequency or ever-decreasing wavelength. In this way it follows that the vibrations of a string describe a wealth of particles with ever-increasing masses.

The masses of these particles, or "states" of the string, are measured in multiples of the string's tension. In the original string theory these states were supposed to be identified with the many different species of mesons.

Of course, there are very important differences between the behavior of these relativistic strings and that of those more familiar in stringed musical instruments. A familiar violin string is made of massive atoms and special relativity is utterly irrelevant for describing its motion. On the other hand, a relativistic string is massless unless it is stretched – the masses of its excited states arise entirely from the energy involved in stretching it. In this case the points along the string move at speeds approaching the speed of light, so the effects of special relativity are crucial. According to special relativity, the passage of time is different for each point along the string – this is a key concept and one that makes the mathematical consistency of string theory quite remarkable. Many false attempts to construct theories of extended particles consistent with relativity were made, but with little success before the advent of string theory. The way in which string theory resolves this fundamental problem is by embodying an infinite extension of gauge invariance that lies behind Maxwell's theory of electromagnetism and non-Abelian gauge theory. This rich underlying mathematical structure is one of the most attractive features of the theory.

However, when studied in detail in the 1970s, it turned out that this initial string theory had terrible problems. Although it was possible to account for some of the general features of the mesons by assuming them to be small strings, there were striking inconsistencies. One notable problem was that the theory made sense only if spacetime were 26-dimensional, requiring the presence of 22 extra space dimensions! As we will see later, extra dimensions in spacetime had been postulated in theories of gravity soon after Einstein formulated his general theory of relativity, but they are not sensible in a theory of the strong force, which was the supposed purpose of the original string theory. An even more profound problem with that theory is that one of its states is a "tachyon" – a particle that travels faster than the speed of light. The presence of a tachyon in any quantum theory indicates a disaster – the vacuum, empty space, in such a theory is unstable and so the theory makes no sense. For these reasons, by the mid 1970s work on string theory had virtually been abandoned. This was in part due to rapid progress on a number of fronts in conventional quantum field theory.

## 5.6 Strings unify the forces

Some of the inconsistencies of the original string theory remain unresolved to this day. However, it was understood how to extend string theory in order to incorporate a new symmetry, known as "supersymmetry," which has long been postulated to be a property of the fundamental physical laws. Such theories are known as superstring theories. Before describing them in more detail, it is important to emphasize the importance of supersymmetry. This novel kind of symmetry was invented for purely theoretical reasons in the 1970s in order to provide unification of two broad classes of particles that are not unified in the original Standard Model. These classes of particles, known as bosons and fermions, are distinguished by their spin. The quantum notion of a spinning particle is the analog of the classical notion, except that quantum spin is always measured in integer or half-integer amounts, whereas a non-quantum object has arbitrary amounts of spin, or angular momentum. Bosons are particles with integer spin (such as the photon and the W and other particles) whereas fermions are particles with half-integer spin (such as the quarks and the electron and other particles). The fact that fermions satisfy Pauli's "exclusion principle," whereas bosons do not, means

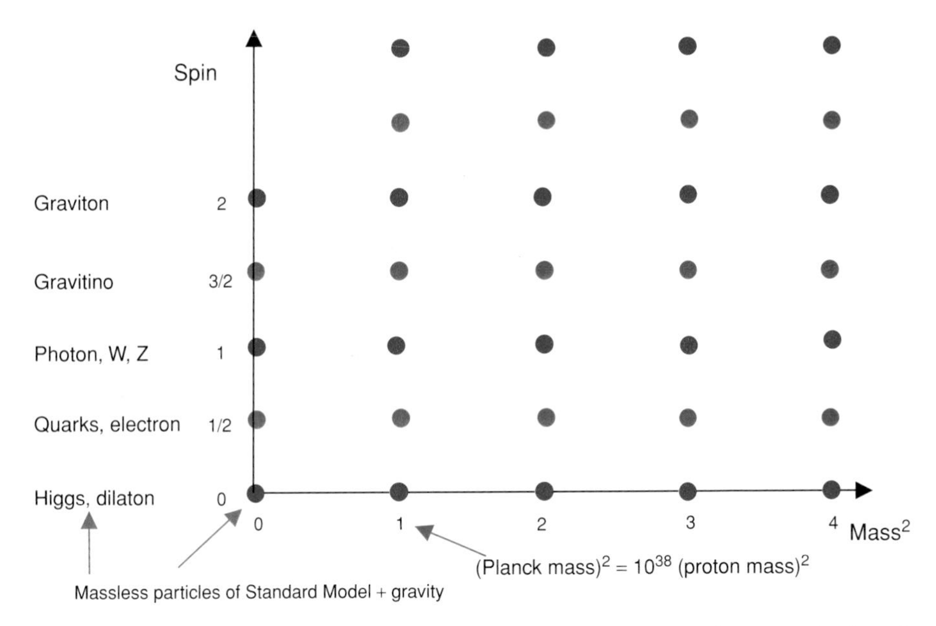


**Figure 5.6.** The states described by a superstring. The masses of the states satisfy the equation (mass)$^2$ = multiple of string tension. The vertical axis represents the spin of each state. States with half-integer spin are fermions (red filled circle as on graph), whereas those with integer spin are bosons (blue filled circle as on graph). Remarkably, the massless states correspond precisely to the kinds of particles in the Standard Model (spin-0 scalars, spin-$\frac{1}{2}$ quarks, neutrinos, and other leptons, spin-1 photon and vector bosons) together with the spin-2 graviton and the spin-$\frac{3}{2}$ gravitino, which is connected with supersymmetry.

that they have dramatically different properties, so it was a theoretical triumph to find a symmetry that relates them to each other. Supersymmetry was originally invented for purely theoretical reasons, partly motivated by the 1971 extension of string theory by Ramond, Neveu, and Schwarz that included fermions as well as the bosons that are present in the original string theory. In the intervening period supersymmetry has become widely accepted as one of the main ingredients that should enter into the description of particle physics that unifies the physical forces at energy scales that lie well beyond those tested by the Standard Model. It now seems that almost any imaginable unification of the forces will necessarily require supersymmetry. One very special feature of supersymmetry in quantum theories is that there are important mutual cancellations of the effects of bosons and fermions. For a time it was hoped that the infinities that we discussed earlier would cancel out in such theories, but this does not happen within conventional field theory. However, of particular relevance here is the fact that supersymmetry is a necessary ingredient in formulating a consistent theory of strings, which is free of the problematic infinities. The experimental verification of the predictions of supersymmetry would be a triumph for this web of theoretical ideas.

The inclusion of supersymmetry within string theory leads to a theory with a remarkable spectrum of states, which is illustrated in Figure 5.6. To begin with, there is no longer a tachyon state, so superstrings are stable. Furthermore, and most impressively, the massless states of the superstring correspond to just the particles that are needed in order to describe all the forces of the Standard Model together with general relativity, which describes the force of gravity. For example, there are massless states of spin 1, which might correspond to the photon, W and Z bosons, and the gluons of QCD, as well as massless spin-$\frac{1}{2}$ states corresponding to the electron, quarks, neutrinos, and other particles. These are just the kinds of particles that enter into the Standard Model. In addition, there is also a massless spin-2 state, which corresponds to the graviton. This is the particle that transmits the gravitational force and is present in any theory of quantum gravity. Its interpretation as a graviton in string theory was suggested in 1974 by Scherk, Schwarz, and Yoneya. These states all arise in the string spectrum as if by magic – no-one actually asked for them to be there. In this manner it turns out that superstring theory contains all the ingredients necessary for a theory of all the known physical forces, including gravity! In addition to all this, in string theory the description of physics at very small length scales is radically altered by the absence of pointlike particles. Such a change of short-distance physics is necessary in order to avoid the terrible problems encountered at short distances in quantum field theory based on pointlike particles – the terrible problems with infinities of all previous models of quantum gravity are absent from string theory.

During the period 1980–1984 John Schwarz and I constructed superstring theories that possessed these exceptional properties. We were particularly excited by our discovery in 1981 that the problematic infinities that had plagued all previous theories of quantum gravity were absent from certain superstring theories. This signaled that such theories modify short-distance physics in a radical manner that makes them consistent. However, an equally daunting problem arises in any realistic theory of quantum gravity due to the occurrence of "quantum anomalies." This refers to a breakdown of certain sacred conservation laws, such as the conservation of energy and conservation of electric charge. These conservation laws are necessary ingredients, without which any theory would be inconsistent, but their breakdown seemed to be unavoidable in any quantum theory. We discovered that very special versions of superstring theory consistently evade these anomalies. These were versions that contain specific Yang–Mills symmetries associated with the very large mathematical symmetries known as SO(32) or E8 $\times$ E8 (the nomenclature is technical, but it indicates huge underlying symmetries that generalize those of the Standard Model). This subtle manner in which superstring theory avoids anomalies finally provided impetus for an explosion of interest in the theory as the basis of a unified theory of forces and particles. Since 1984 string theory has developed in many directions.

### 5.6.1 Interactions between strings

The way in which strings interact with each other as they move is represented by "stringy Feynman diagrams," which are generalizations of conventional Feynman diagrams of theories based on pointlike particles. Just as with conventional theories of pointlike particles, this series of diagrams represents an approximation, known as the "perturbative" approximation, to the theory. Although it treats the strings according to quantum theory, this approximation is one in which the effects of quantum theory on the spacetime in which the string is moving can be ignored. As we saw earlier, a quantum theory of gravity must ultimately treat spacetime, as well as the particles and forces, in a quantum manner, so it will be important to progress beyond the perturbative approximation. Nevertheless, this approximation already elucidates many unusual features of string theory. For example, Figure 5.7 indicates that a string necessarily interacts with the spacetime through which it moves. The fluctuations of the string world-sheet automatically describe closed strings disappearing into empty space. A consequence of such interactions between the string-like particle and spacetime can be consistent only if Einstein's theory of general relativity is satisfied at distance scales

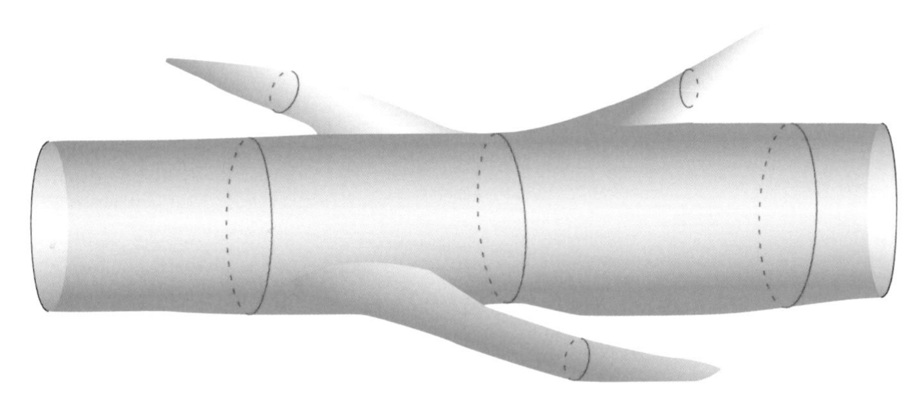

**Figure 5.7.** A single string sweeps out a world-sheet as it moves through spacetime. The fluctuations of this sheet include configurations in which a closed string is emitted into empty space (the vacuum) or is absorbed from the vacuum. In this way the string is sensitive to the nature of the vacuum.

much larger than the string scale. In other words, Einstein's theory can be derived from the requirement that string theory is consistent! With conventional theories of pointlike particles there are no constraints of this kind.

As we saw earlier, when two strings meet they can join by touching at a point, or a single string can split into two when two points on the string touch. Two open strings can join when their end-points touch or a single open string can split into two. With this geometrically appealing picture of how string-like particles interact it is simple to build the sequence of stringy Feynman diagrams that describe the scattering of string-like particles. For example, the diagrams in Figure 5.8 represent the collision of two closed strings. Although this series of approximations is analogous to the Feynman series, the string diagrams have some very special features that radically change their interpretation. In contrast to the situation with Feynman diagrams of conventional quantum field theory, where there are very many separate diagrams at each order in the expansion in powers of $g$, in string theory there is a single diagram at each order. The most surprising feature of the string Feynman diagrams is that they do not possess any short-distance infinities of the kind that has plagued all previous theories that incorporate the force of gravity. Recall that in theories with pointlike particles the problem of infinities is first encountered in the evaluation of the Feynman diagram with one loop. The infinities arise because of the presence of nodes at the places where two particles meet inside the diagram. The corresponding string diagram is the smooth surface with one loop in Figure 5.8. This has no nodes and therefore no possibility of producing the same problematic infinities. Although there are other potential

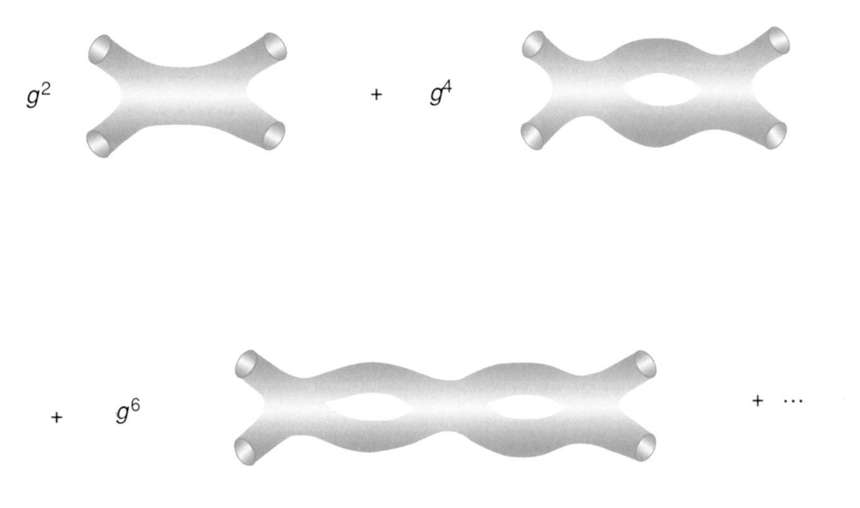


**Figure 5.8.** Feynman diagrams for closed strings. The incoming strings interact with each other by joining together or splitting apart. The figure illustrates the leading approximation and the first two corrections. The surfaces spanned by the strings are smooth, unlike the corresponding Feynman diagrams of conventional pointlike theories shown in Figure 5.3.

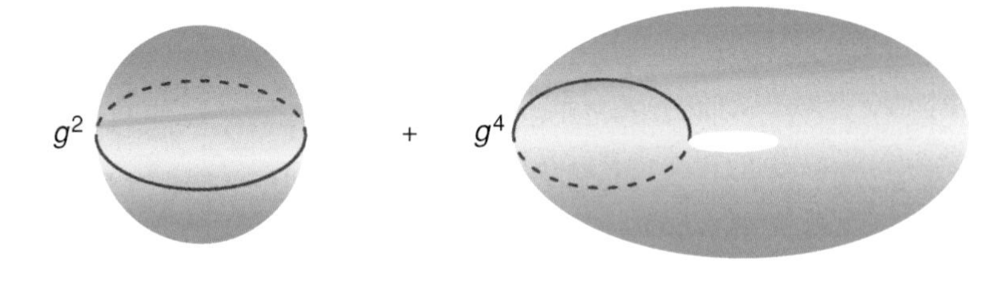


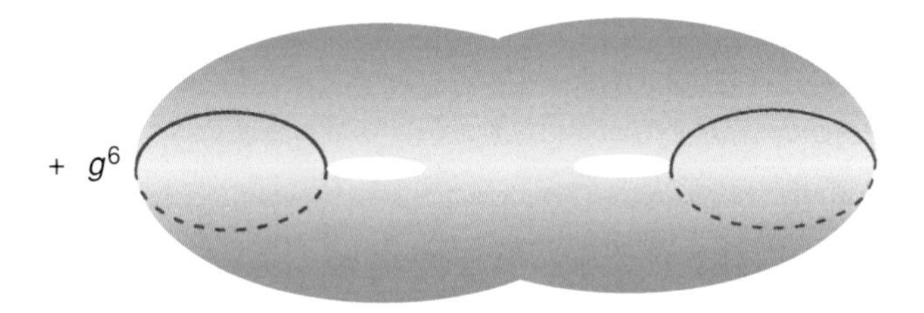


**Figure 5.9.** The topological classification of string-theory Feynman diagrams corresponds to the classification of the topology of two-dimensional surfaces, which is an elegant branch of mathematics. The sequence of Feynman diagrams of Figure 5.8 is equivalent to a sequence of tori with increasing numbers of holes.

divergences arising from this diagram, these are absent from superstring theory by virtue of the cancellations between fermions and bosons embodied in supersymmetry. The theory provides for the first time a finite, well-defined scheme for calculating quantum effects of gravity.

In fact, the classification of the various string-theory diagrams is simply related to an elegant branch of mathematics that studies the embeddings of surfaces, such as the two-dimensional world-sheets spanned by strings. This proceeds by imagining that the surfaces are rubber sheets that can be distorted. It is clear that the first diagram of Figure 5.8 can be deformed into a sphere, as shown in Figure 5.9. The second diagram in the series is the one-loop diagram that can be deformed into a donut shape, or torus, and so on for diagrams with more holes.

However, it should be stressed that string theory as described above cannot pretend to be a complete theory, but rather is an approximation, analogous to the Feynman diagrams that approximate quantum field theories based on pointlike particles. In the case of string theory, the perturbative approximation corresponds closely to the picture described above, in which string-like particles move through spacetime, where the properties of space and time are assumed to be classical and not altered by quantum mechanics. Normally, the perturbative approximation to a theory is deduced by approximating some exact, or nonperturbative, formulation of the theory. Bizarrely, in the case of string theory the perturbation approximation is all we have – the underlying "nonperturbative" theory to which this approximates is not yet known.

A peculiar feature of superstring theory is that it makes sense only in ten spacetime dimensions – which is six more than the dimensions that are apparent from ordinary observation. As we will see shortly, in a theory that contains gravity, extra dimensions can curl up and become unobservably small. By the end of 1984 there were five apparently different versions of ten-dimensional superstring theory. Two of these, called type IIA and type IIB, contained only closed strings and seemed to be unable to account for the physical forces, since they do not contain Yang–Mills gauge particles of the kind needed that enter into the Standard Model. In addition, there were two so-called heterotic-string theories called $H_{SO(32)}$ and $H_{E8 \times E8}$. The subscripts

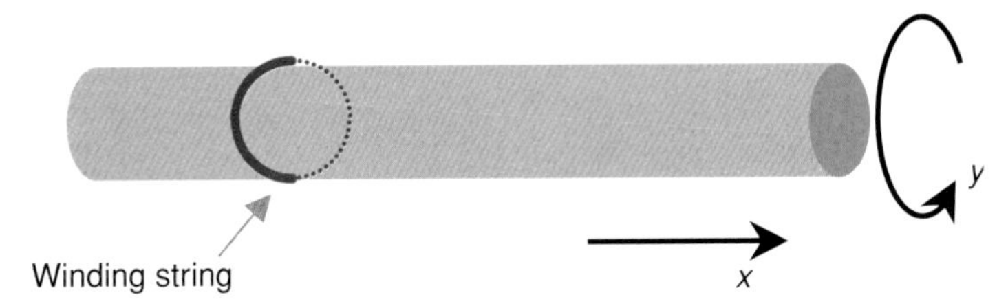


Figure 5.10. A dimension of space can curl up to form a circle. At distance scales large relative to the radius of the circle the space effectively has one dimension fewer. Note that a string can wind around the circle but a point cannot, which leads to a rich new structure in string theory.

indicate that these theories contain Yang–Mills symmetries associated with the very large mathematical symmetries known as SO(32) and E8 × E8 that were known to be consistent with the absence of anomalies. The last of the five ten-dimensional string theories is the one that contains both open strings and closed strings that is known as the type-I theory. An important feature here is that the massless states of an open string describe a spin-1 state, which is something like a photon or a Yang–Mills gauge particle. The open-string part of the theory therefore generalizes Yang–Mills theory. However, the massless states of a closed string include a spin-2 state, which is the graviton associated with general relativity. The closed-string part of the theory therefore generalizes general relativity. The fact that the end-points of an open string can join to form a closed string, as in Figure 5.5, indicates a unification of Yang–Mills theory and gravity within string theory that is not apparent in any conventional theory based on point particles.

The unifying properties of perturbative superstring theory quickly led to an attractive scenario for finally explaining the origin of the forces and elementary particles that enter into the Standard Model. Most of the early work in this area concentrated on the $H_{E8 \times E8}$ superstring theory, which comes closest to explaining the Standard Model. This required an understanding of why the six extra spatial dimensions are not observed. The possibility of extra space dimensions in a theory of gravity has been considered since the work of Kaluza and Klein in the early years after Einstein's formulation of general relativity. Kaluza, followed by Klein, realized that a theory of gravity in five spacetime dimensions, where the extra spatial dimension is curled up into an extremely small circle, would look like a theory in four spacetime dimensions since nothing of normal size would sense the presence of the extra dimension (see Figure 5.10). This approximately four-dimensional theory would, however, describe gravity unified with electromagnetism. The original Kaluza–Klein proposal was inadequate for a number of reasons – not least because it does not account for the weak and strong forces. Nevertheless, this phenomenon whereby the presence of very small extra space dimensions leads to a unification of forces generalizes naturally in superstring theory. Here there are six extra dimensions that must be curled up into a very small six-dimensional "ball," leaving a theory with four normal spacetime dimensions. There are many possible ways in which this can happen, although severe restrictions are imposed by requiring the consistency of the four-dimensional theory. Each distinct "shape" of the six-dimensional space leads to a distinct four-dimensional description of forces and particles. In fact, many important physical features do not depend on the precise shape of the extra dimensions, but rather on their "topology." Topology is the most basic geometric feature of any space, which remains unchanged when the shape of the space is changed. For example, a donut has a single hole in it even when it is squashed – it is said to have "genus 1," whereas a double-holed donut has "genus 2," and so on. There are many more sophisticated measures of topology than the number of holes, but they are all designated by integers. In string theory such integers correspond to properties such as the number of generations of elementary particles in the Standard Model. In conventional particle physics based on quantum field theory different choices for such integers would correspond to different theories, but here they arise from different configurations within one theory. A particularly important class of such curled-up six-dimensional spaces is known as Calabi–Yau spaces. These mathematical spaces, which were originally conjectured to exist by the mathematician Calabi in 1957 and proved to exist by Yau in 1977, play a dominant role in the study of string theory in four dimensions. Not only are these issues of relevance to finding out how superstring theory might fit observational data, but also the study of such spaces has a close connection with some of the most interesting recent ideas in geometry.

It is disappointing that there is no known theoretical principle that selects one of these many possible configurations as the one that describes observed physics. But it is impressive that there are solutions to superstring theory that are tantalizingly close to describing the Standard Model. This endeavor of describing experimental data in terms of superstrings developed rapidly in the late 1980s. Even though it is by no means complete, its lasting legacy has been to indicate novel ways in which the Standard Model might be embedded in a larger framework that includes gravity and unifies all of the forces. Today almost all attempts to unify the forces are based on string theory, which is the only game in town since it is the only framework within which quantum effects of gravity can be estimated.

However, the description of physics based on the perturbative approximation to superstring theory has conspicuous limits and the true meaning of the theory will be apparent only once the underlying nonperturbative formulation has been understood. Some of these limitations are quite obvious. For example, supersymmetry is a symmetry that is most certainly not present in Nature in a directly observable way. Assuming that supersymmetry is a property of Nature, and one that will, we hope, be discovered before too long, it must be a "broken" symmetry. That is, it is a symmetry of the equations of the theory but not of their solutions. Among other things, unbroken supersymmetry in string theory leads to the statement that all the observed elementary particles – the quarks, leptons, etc. – should be massless! This is not as absurd as it may sound. Although particles such as the W and Z bosons and the top quark have masses hundreds of times the mass of the proton, the natural mass scale of any theory of quantum gravity is not the proton mass but the Planck mass. This is a mass about $10^{19}$ times the mass of the proton and is the mass scale of the microscopic black holes that are implied by the uncertainty principle discussed earlier. In comparison with the Planck mass all the observed particle masses are indeed tiny and might be taken to be zero to a first approximation. However, it is these "tiny" nonzero masses that we need to understand in order to understand the nature of matter in our Universe and so it is crucial to understand how supersymmetry is broken. Since supersymmetry is guaranteed not to be broken in any perturbation approximation, this is something that can be understood only in a more complete description of the theory. Furthermore, many cosmological problems and the analysis of properties of objects such as black holes require deeper nonperturbative insight.

## 5.7 Beyond the naive perturbative approximation

The fact that the simplest approximation to string theory describes the forces and particles in such an elegant and consistent manner led to a widespread belief that there is a deep and very new physical principle at the heart of the theory. However, it also emphasized the profound importance of deciphering the underlying formulation of the theory beyond its perturbative approximation, which is essential for discovering the meaning of the concept of quantum spacetime. A series of transformations in the understanding of these issues has occurred since the early 1990s.

The essential feature of this shift in understanding was the realization that the fundamental string-like particles in terms of which string theory was originally formulated are not the only particles in the theory. In fact, string theory also contains a plethora of other extended particles that are not seen in the perturbative approximation. This situation is also well known in standard quantum field theories. There, only the pointlike fundamental particles arise in perturbation theory. However, such theories often contain other kinds of particles, known generically as solitons. These are stable, or nearly stable, solutions of the theory that represent massive particles. Familiar examples are black holes in general relativity, and magnetic monopoles in certain versions of Yang–Mills theory. Solitons are traditionally considered to be complicated composite objects in field theory rather than fundamental point particles. However, many recent field-theory results make it clear that when nonperturbative quantum effects are included the solitonic particles are as fundamental as the point particles.

In string theory there are solitonic particles that may have extensions in several directions. These are generalizations of black holes. For example, there is a membrane-like soliton that, just like a soap bubble, is two-dimensional. More generally, there are solutions that are extended in $p$ directions, which are collectively known as $p$-branes (by analogy with membranes). Since string theory requires nine dimensions it is possible to have $p$-branes where $p$ can be as high as nine. These $p$-branes are generalizations of a black hole, which is a $p = 0$ solution – a black hole is localized around a point. Included among these solitonic solutions of string theory is the $p = 1$ case, which is a string-like particle, and the $p = 2$ case, which is a membrane-like particle like a soap bubble. As we will see later the $p = 3$ case, which describes an object with three space dimensions, may be of special significance in describing our Universe. Higher-dimensional objects also arise in the theory.

This means that there is a wealth of other particles in string theory beyond those described by strings. Although it appears from the original perturbative viewpoint that the strings are in some sense fundamental, in the complete theory there is no reason to believe that $p$-branes are less important. The study of the properties of these objects has revealed a wealth of interconnections between superstring theories. Early on, in the mid 1980s, there appeared to be a number of different theories with distinct properties – in one way of counting there were five distinct theories. However, it is now known that these are all different approximations to the same underlying theory. The fundamental strings of one of these theories arise as solitonic solutions of another of the theories. This can happen in a manner that depends crucially on geometric properties of the theory, that is difficult to imagine although a simple example can be visualized. Consider a situation in which one dimension of space is curled up into a circle (which was originally considered by Kaluza and Klein). A membrane, which is an object that is extended in two dimensions, can wrap around that direction to form a long tube. When the circumference of the tube is very small it will appear to be extended in one direction only and looks like a string. Generalized arguments of this sort involve wrapping $p$-branes around extra dimensions. It turns out that all the apparently different superstring theories are really different approximations to the same theory. The fact that an object that appears as a membrane in one approximation is the fundamental string in another approximation is a consequence

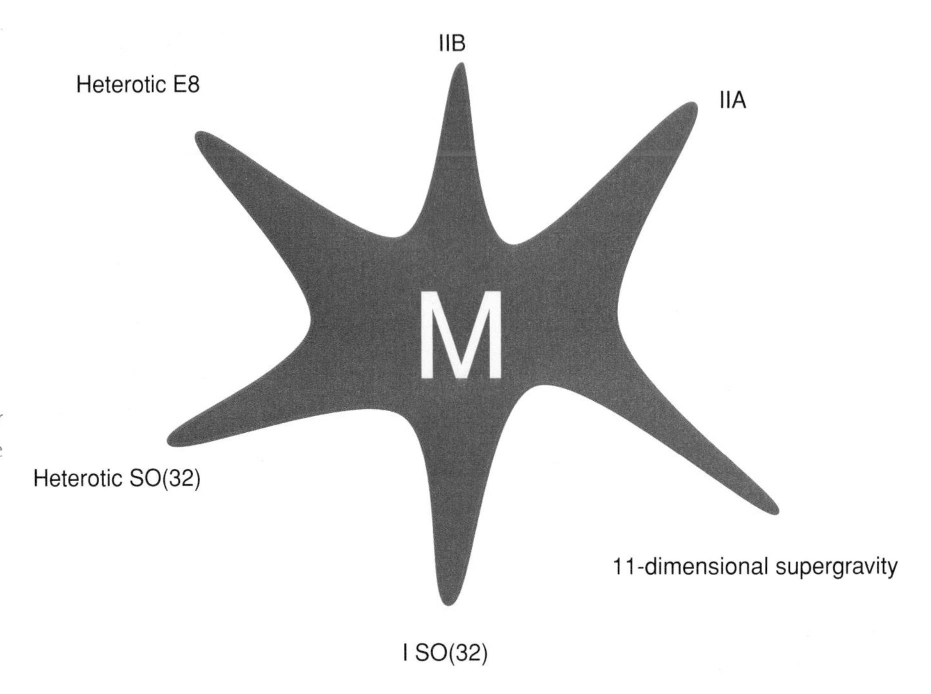


**Figure 5.11.** The currently understood approximations to string theory are represented by points at the boundary of this diagram. The five apparently different string theories are connected to each other when various parameters are varied. Furthermore, string theory is connected to 11-dimensional supergravity in the low-energy limit. No reliable method is known for describing the interior points of the diagram. The whole diagram is often referred to as the "moduli" space of M-theory.

of a profound set of mathematical symmetries of the equations, known as "duality" symmetries.

We have seen that superstring theory reduces at low energies to some supersymmetric version of general relativity, or supergravity in ten spacetime dimensions. For a long time this appeared to exclude any relation between string theory and the most symmetric version of supergravity that had been discovered by Cremmer, Julia, and Scherk in 1978 and which exists in *eleven* spacetime dimensions! However, an extension of the duality arguments that relate all the superstring theories to each other also shows that they are related to 11-dimensional supergravity. Since this is not a string theory at all, this connection is not seen within the perturbative approximation to string theory. Recall that the quantity $g$, which is the charge that determines the strength of the inter-string forces, has to be small for the perturbative approximation to make sense. We now know that when $g$ is chosen to be large it is identified with the circumference of a circular extra dimension – the eleventh spacetime dimension. The fact that a dimension of space can be described as the value of a charge is a strange and novel feature of nonperturbative extension of string theory. The problem is that there is not yet a satisfactory way of evaluating the quantum effects within 11-dimensional supergravity – it is a conventional quantum field theory that possesses all the problematic features of quantum general relativity.

This set of interrelated ideas that connect all string theories to each other, as well as to 11-dimensional supergravity, is the subject of intense study and has tentatively been given the name "M-theory." Through a series of sophisticated guesses and the application of a great deal of modern mathematics a picture has emerged of M-theory as an underlying theoretical framework out of which all the apparently different string theories emerge as distinct approximations, in addition to 11-dimensional supergravity. The diagram in Figure 5.11 represents these relationships. For the moment there is no consensus regarding the correct mathematical language which lies at the heart of the theory – current research is aimed at finding a description of physics at interior points of the diagram that are far from the regions in which known approximation procedures are reliable. Nevertheless, these investigations have uncovered some imaginative new

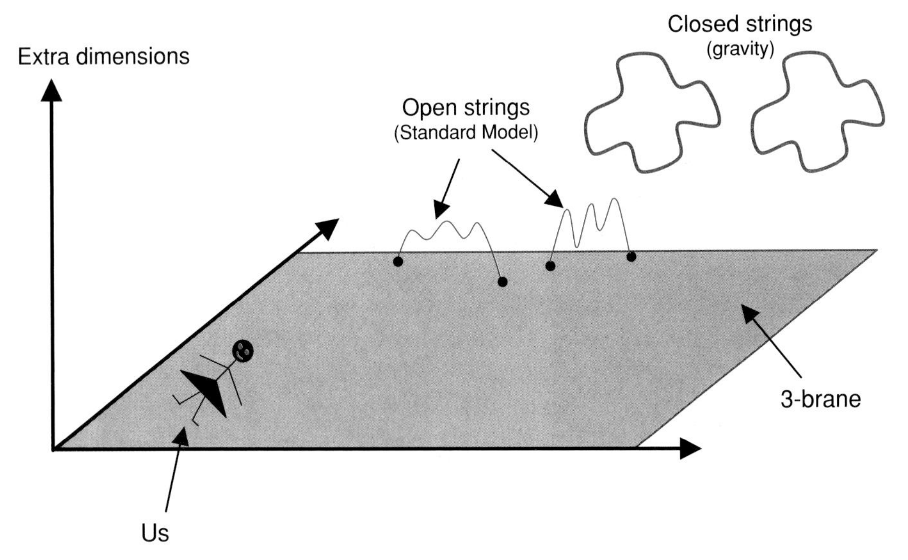


**Figure 5.12.** A 3-brane universe is embedded in the ten-dimensional spacetime of string theory. The six extra dimensions would have to be small, but nevertheless might be measurable. String theory gives a natural explanation for why the non-gravitational forces are confined to the four spacetime dimensions of the 3-brane while the force of gravity permeates all ten spacetime dimensions.

possibilities for the physical implications of string theory that are not apparent from its original perturbative formulation.

## 5.8 A Universe made of string

The developments of the last few years have led to insights into the nature of the geometry of spacetime at the shortest distance scales implied by string theory. Although the picture is still very incomplete, fascinating connections are being made, which relate traditional theories to string theory in unexpected ways.

### 5.8.1 Life on a 3-brane

The existence of higher-dimensional objects, or $p$-branes, within string theory suggests a novel description of our physical Universe. In particular, since we live in (approximately) three space dimensions it has been suggested that the Universe might be a 3-brane embedded in ten space dimensions. The analogy with the four-dimensional spacetime of the observed Universe is quite profound. According to string theory the end-points of open strings are tethered to the 3-brane as the strings move around so that these strings are restricted to four spacetime dimensions and cannot move off in the six extra dimensions transverse to the 3-brane. This is shown in Figure 5.12. Since open strings carry Yang–Mills forces, such as those of the Standard Model, this means that the Universe appears to be four-dimensional if gravity is ignored. However, closed strings, which carry the force of gravity, can move through the whole ten-dimensional spacetime and therefore the force of gravity senses all the directions.

Since the non-gravitational laws have been tested very thoroughly, it is known with great accuracy that they should work only in four spacetime dimensions. However, the closed strings that carry the force of gravity can move in all dimensions, including the six extra ones that are perpendicular to the 3-brane. The force of gravity has been measured much less accurately than the other forces since it is so much weaker, so there is a possibility that the extra dimensions in which gravity is effective may be ridiculously larger than previously anticipated (see Chapter 2). In fact, there is an outside possibility that they could be as large as 1 mm! This is to be compared with the previously expected size of extra dimensions, which was around $10^{-34}$ m. Even a size of around $10^{-14}$ m would be such a huge distance scale that it would overturn previously held convictions. Among other things, the string tension would turn out to be very much smaller than expected and it would be much easier to detect the excited states of the string.

### 5.8.2 Quantum properties of black holes

In 1974 Hawking showed that black holes emit radiation when the quantum properties of matter are taken into account, so they are not really black. Shortly after, he concluded that the outgoing radiation is unrelated to the matter that formed the hole originally – this leads to a loss of information, which is incompatible with a basic tenet of quantum theory. This led to his dramatic statement that quantum mechanics breaks down in the presence of black holes. However, in the absence of a consistent treatment of quantum gravity this conclusion has always been questionable. With the advent of string theory there is at last a consistent framework for describing the quantum properties of black holes. Although it has so far been possible to discuss only particular examples of black holes that carry special charges, it is apparent that they have a completely consistent quantum interpretation. It has been possible to construct these special black holes explicitly from microscopic quantum states consisting of collections of $D$-branes and strings. Remarkably, the properties of such black holes, as well as the way they absorb and emit radiation, are completely in accord with quantum mechanics. There are still important unresolved issues as to how to generalize these arguments to arbitrary black holes, but the picture looks very promising.

### 5.8.3 The holographic Universe

The understanding of how $D$-branes behave has provided a concrete illustration of some novel mathematical ideas that should be relevant to understanding the quantum geometry of spacetime. Among these, the area of *noncommutative* geometry seems to be particularly relevant. The key distinction between this and the classical geometric ideas that underpin general relativity is that noncommutative geometry is based on the premise that the coordinates that label distances in space do not "commute" with each other. Any two numbers, call them $x$ and $y$, have the property that they do commute, which means that the product $x$ times $y$ is equal to $y$ times $x$. These numbers may represent the distance between two separated objects along two directions. However, in noncommutative geometry the product of two distances in distinct directions depends on the order in which the two distances are multiplied, so $x$ times $y$ and $y$ times $x$ are *not* equal. This means that, at the most fundamental level, the coordinates $x$, $y$, and $z$ that label distances in different directions cannot be identified with ordinary numbers! This novel set of ideas radically alters the notion of spacetime at small distances – for example, the area of a square can no longer be defined as the product of the length of its sides. It has close parallels to the way in which the uncertainty principle overthrew the notion that properties of particles such as velocity and position can be assigned simultaneously. Roughly speaking, this means that within a quantum description of spacetime there is a limit to the size of a volume within which it makes sense to think of space as a continuous set of points – this limiting size is around the size of a cube whose edges are a Planck length or $10^{-35}$ m. This means that space and time are granular on the Planck scale – an effect that is sometimes called "spacetime foam," a concept that has yet to be defined in a very precise manner.

Closely related to this is a wonderfully counter-intuitive conjecture that says that in any theory of quantum gravity the physical behavior of matter within any given volume of space is determined by physical laws on the *boundary* of that volume. In other words, at a fundamental level the laws of physics that determine properties of our four-spacetime-dimensional Universe are formulated in one dimension fewer. Furthermore, because of the quantum nature of spacetime, these three spacetime dimensions are again granular, or foamy, at the Planck scale. This idea was first suggested to be a property of quantum gravity by 't Hooft in 1993 using rather general arguments based on properties of microscopic black holes and an indication of how it might be realized in string theory was pointed out by Susskind in 1995. This property is reminiscent of the phenomenon of holography in optics – three-dimensional optical images can be encoded on a two-dimensional plate. However, holography in quantum gravity is much more profound since it implies that it is not simply the optical image of reality that is encoded on a surface surrounding the object but rather the reality of the object itself! This is obviously very difficult to conceive of in a direct way, but studies of several very special examples of string theory in particular spacetimes, pioneered by Maldacena, have led to concrete illustrations of the idea.

## 5.8.4 Cosmology and string theory

Although string theory emerged from the study of elementary particles of the kind that are seen in experiments at large accelerators such as CERN (Geneva), SLAC (Stanford), and Fermilab (Chicago), the natural scale of energy in the theory is the string scale. This is generally expected to be close to the Planck energy scale, which is enormously higher than can be reached in any conceivable accelerator. For this reason there will probably always be a problem in predicting the physics observed in terrestrial experiments directly from string theory. It is a little like trying to predict the motion of water waves in terms of the particles inside the molecules of water. Although we know that water is made of $H_2O$ at the molecular level, this is not useful for understanding water waves, which are hugely bigger than water molecules. However, in the earliest moments after the Big Bang the Universe was very small and extremely hot so the physics of super-high-energy scales was predominant. String theory should therefore have a lot to say about physics in the very early Universe. According to Einstein's theory of general relativity a Big Bang taking place at an initial time is an inevitable condition for explaining the subsequent evolution of the expanding Universe. The actual moment of the Big Bang is known as the "initial singularity" and is a pathology of the equations. However, when combined with other assumptions about the early state of matter, this has led to a very impressive set of cosmological predictions that are confirmed by observation. The hope is that these assumptions about the early Universe will be explained by string theory and that the theory will radically alter the nature of the initial Big Bang in surprising ways. The most radical possibility is that the physics described by string theory when matter is at super-high density is so different from that of conventional theories that there are *no* singularities in the theory. This hope is based on intuition but is far from being proved.

Perhaps the single most puzzling experimental fact that might be explained by any theory that purports to describe the physical Universe is the fantastically small value of the "cosmological constant." This is the quantity that enters into Einstein's equations that determines how curved the Universe would be if it contained no material particles. This is known to be incredibly small – about $10^{-120}$ in natural units, which is by far the most accurately known quantity in the whole of science and deserves explanation. Yet the natural value for the cosmological constant in any conventional particle theory is at least 80 orders of magnitude larger than this, which is a gigantic error! However, no previous theory has incorporated gravity in a consistent manner, so this problem has not been taken very seriously. The challenge of explaining the near vanishing of the cosmological constant is a stimulating source of ideas. Added to this, recent observations suggest that the rate of expansion of the Universe is accelerating, which indicates that something like 75% of the energy in the Universe is "dark energy" as distinct from the other 25% that is carried by matter. One explanation of this dark energy is that the cosmological constant is tiny but nonzero and positive (see Chapter 1).

## 5.9 Prospects

It is essential that, no matter how mathematically elegant a physical theory is, it should explain observations and make predictions for future experiments. It might be argued that string theory has so far failed such a test since it has not yet made very detailed connections with experiment. As we have seen, a major problem is that the theory naturally describes physics at energies hugely greater than those accessible on Earth. Ultimately, the tests of the theory are likely to come from cosmological observations that detect the state of the Universe during the first moments after the Big Bang. Developments during that era were dominated by extremely high-energy phenomena, which can be described only by a theory such as string theory that unifies all the forces.

Although there is still no detailed understanding of how to make predictions from string theory, several general features have emerged. Foremost among these is the fact that the structure of the theory provides a natural explanation for the existence of gravity and the structure of general relativity – even if Einstein got there first! Furthermore, string theory provides an elegant mathematical framework for a unified theory of all the forces and particles. Supersymmetry seems to play an important role in the theory and its experimental discovery would undoubtedly be of great importance. Indeed, the discovery of supersymmetry would involve more than simply the discovery of some new particles; it would also imply very novel extensions of the meaning of the dimensions of space and time.

There is currently a fundamental divergence of views among physicists as to whether a theory of physics should be able to explain the values of parameters such as the cosmological constant. One point of view is that the laws of physics allow the formation of universes in which the cosmological constant can have a wide range of values. However, unless the value of this constant of our Universe has approximately the tiny value that is observed, our Universe would never have evolved in a way that would have led to the galaxies, stars, and, ultimately, to planet Earth. So, evolution of human beings would not have been possible. Since we have evolved, and are thinking about physics, the cosmological constant of our Universe has to be incredibly small. This is one example of an "anthropic" argument that says that, although all kinds of universes might have evolved from the Big Bang, the fact that we are here strongly restricts the Universe in which we live. From this viewpoint, we do not need to calculate the actual value of the cosmological constant but only to show that the observed value is not excluded by the theory. Some estimates suggest that there may be a very large number of solutions of string theory (at least as many as $10^{200}$), of which a small fraction (but a number that may be as large as $10^{10}$) might have a small positive cosmological constant and approximately describe the Standard Model, which is consistent with a Universe that supports life. Therefore, according to the anthropic argument, one of these solutions describes the Universe we live in. The question of which quantities in the physical Universe are ones we can expect to calculate and which ones have an anthropic origin provokes arguments but is far from being resolved within, or outside, of the framework of string theory.

It should be borne in mind that, since string theory is profoundly different from preceding theories, it is likely to have observational implications that are difficult to foresee in its present incomplete state. Optimistically, a lesson may be provided by the history of general relativity. When Einstein finally formulated general relativity in 1915 he realized that it explained small peculiarities of the observed orbit of the planet Mercury. Although these were already well known to be paradoxical, nobody had previously considered this to be an indication of the need to overturn Newton's venerable law of gravity.

In spite of the present incompleteness of its formulation and the lack of direct experimental evidence, to its practitioners string theory has a life of its own. It combines quantum theory with the kinds of forces that are seen in experiments in a manner that makes use of, and extends, all of the theoretical ideas that have evolved over the past fifty years and it does this in a mathematically compelling manner. Whatever else may emerge from string theory, it is likely to be a source of fascination for many years to come.

## FURTHER READING


1. B. Greene, *The Elegant Universe: Superstrings, Hidden Dimensions, and the Quest for the Ultimate Theory*, New York, W. W. Norton, 1999.
2. P. C. W. Davies and J. Brown (eds.), *Superstrings: A Theory of Everything?*, New York, Cambridge University Press, 1988.
3. M. B. Green, Superstrings, *Scient. Am.* **255**, September 1986, 3.
4. N. Arkani-Hamed, S. Dimopoulos, and G. Dvali, The universe's unseen dimensions, *Scient. Am.* **283**, August 2000, 62.
5. S. Muhki, The theory of strings: an introduction, *Current Sci.* Textbooks: **77** (1999) 1624.
6. http://www.sukidog.com/jpierre/strings/.
7. http://superstringtheory.com.
8. M. B. Green, J. H. Schwarz, and E. Witten, *Superstring Theory*, New York, Cambridge University Press, 1987.
9. J. G. Polchinski, *String Theory*, New York, Cambridge University Press, 1998.
10. B. Zwiebach, *An Introduction to String Theory*, New York, Cambridge University Press, 2004.


## Michael B. Green

London-born Michael Green studied physics at Cambridge University where he obtained his Ph.D. in 1970. After a period of postdoctoral research at the Institute for Advanced Study in Princeton, and at Cambridge and Oxford Universities, he became a lecturer at Queen Mary College, University of London in 1978. Between 1979 and 1984 he worked extensively with John Schwarz (of Caltech) on the formulation of superstring theory. Following the explosion of interest in this work, he was promoted to professor in 1987, elected to the Royal Society in 1989, and awarded the Dirac Medal of the International Centre for Theoretical Physics in Trieste. He became the John Humphrey Plummer Professor of Theoretical Physics at Cambridge University in 1993. He was awarded the Dannie Heinemann Prize of the American Physical Society in 2002 and the Dirac Medal and prize of the Institute of Physics in 2004.

# II Quantum matter

# 6 Manipulating atoms with photons

Claude Cohen-Tannoudji and Jean Dalibard

## 6.1 Introduction

Electromagnetic interactions play a central role in low-energy physics, chemistry, and biology. They are responsible for the cohesion of atoms and molecules and are at the origin of the emission and absorption of light by such systems. They can be described in terms of absorptions and emissions of *photons* by charged particles or by systems of charged particles like atoms and molecules. Photons are the energy quanta associated with a light beam. Since the discoveries of Planck and Einstein at the beginning of the last century, we know that a plane light wave with frequency $\nu$, propagating along a direction defined by the unit vector $\boldsymbol{u}$, can also be considered as a beam of photons with energy $E = h\nu$ and linear momentum $\boldsymbol{p} = (h\nu/c)\boldsymbol{u}$. We shall see later on that these photons also have an angular momentum along $\boldsymbol{u}$ depending on the polarization of the associated light wave.

Conservation laws are very useful for understanding the consequences of atom–photon interactions. They express that the total energy, the total linear momentum, and the total angular momentum are conserved when the atom emits or absorbs a photon. Consider for example the conservation of the total energy. Quantum mechanics tells us that the energy of an atom cannot take any value. It is quantized, the possible values of the energy forming a discrete set $E_a$, $E_b$, $E_c$, ... In an emission process, the atom goes from an upper energy level $E_b$ to a lower one $E_a$ and emits a photon with energy $h\nu$. Conservation of the total energy requires

$$E_b - E_a = h\nu. \tag{6.1}$$

The energy lost by the atom on going from $E_b$ to $E_a$ is carried away by the photon.

According to Equation (6.1), the only possible frequencies emitted by an atom are those corresponding to the energy differences between pairs of energy levels of this atom. This important result means that light is an essential source of information on the atomic world. By measuring the frequencies emitted or absorbed by an atom, it is possible to determine the differences $E_b - E_a$ and thus to obtain the energy diagram of this atom. This is what is called *spectroscopy*. High-resolution spectroscopy provides very useful information on the internal dynamics of the atom. Furthermore, each atom

*The New Physics for the Twenty-First Century*, ed. Gordon Fraser.

has its own spectrum. The frequencies emitted by a hydrogen atom are different from those emitted by a sodium or a rubidium atom. The spectrum of frequencies emitted by an atom is in some way its fingerprint. It is thus possible to collect information on the constituents of different types of media by observing the light originating from these media.

During the past few decades, it has been realized that light is not only a source of information on atoms, but also a tool that can be used to act on them, to manipulate them, to control their various degrees of freedom. These methods are also based on conservation laws and use the transfer of angular and linear momentum from photons to atoms. With the development of laser sources, this research field has expanded considerably during the past few years. Methods for polarizing atoms, trapping them, and cooling them to very low temperatures have been developed. New perspectives have been opened in various domains such as atomic clocks, atomic interferometry, and Bose–Einstein condensation. The purpose of this chapter is to review the physical processes which are the basis of this research field, and to present some of its most remarkable applications. Other applications are also described in Chapters 7, 9, and 11.

Two types of degrees of freedom have to be considered for an atom: (i) the internal degrees of freedom, such as the electronic configuration or the spin polarization, in the center-of-mass reference frame; and (ii) the external degrees of freedom, i.e. the position and the momentum of the center of mass of the atom. In Section 6.2 we present the basic concepts used in the control of the internal degrees of freedom. We then turn to the control of the external motion of an atom using an electromagnetic field. We show how one can trap atoms (Section 6.3) and cool them (Section 6.4). Finally, we review in Section 6.5 a few important applications of cold atoms.

## 6.2 Manipulation of the internal state of an atom

### 6.2.1 Angular momentum of atoms and photons

Atoms are like spinning tops. They have an internal angular momentum $J$. Like most physical quantities, the projection $J_z$ of $J$ along the $z$ axis is quantized: $J_z = M\hbar$, where $\hbar = h/(2\pi)$ and $M$ is an integer or half-integer, positive or negative.

Consider the simple case of a spin-$\frac{1}{2}$ atom. The *quantum number* $J$ characterizing the angular momentum is $J = \frac{1}{2}$ and there are two possible values of the magnetic quantum number $M$:

$$M = +\frac{1}{2}\text{: spin up } \Uparrow; \qquad M = -\frac{1}{2}\text{: spin down } \Downarrow$$

At room temperature and in a low magnetic field, the Boltzmann factor $\exp[-\Delta E\,/\,(k_B T)]$ corresponding to the energy splitting $\Delta E$ between the two states is very close to 1 ($k_B$ is the Boltzmann constant). The populations of the two spin states are nearly equal and the spin polarization is negligible.

Photons have also an angular momentum $J_z$, which depends on their polarization (see Figure 6.1). For a right-circular polarization with respect to the $z$ axis (called $\sigma_+$ polarization), $J_z = +\hbar$. For a left-circular polarization with respect to the $z$ axis (called $\sigma_-$ polarization), $J_z = -\hbar$. Finally, for a linear polarization parallel to the $z$ axis (called $\pi$ polarization), $J_z = 0$.

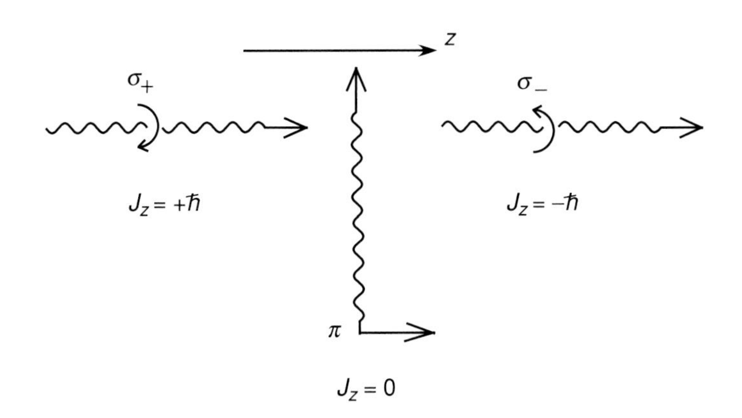


**Figure 6.1.** Three different states of polarization of a photon and corresponding values of $J_z$.

## Polarization selection rules

When an atom absorbs a photon, it gains the angular momentum of the absorbed photon and its magnetic quantum number changes from $M_g$ to $M_e$. The change $(M_e - M_g)\hbar$ of the atomic angular momentum must be equal to the angular momentum of the absorbed photon. It follows that $M_e - M_g$ must be equal to $+1$ after the absorption of a $\sigma_+$-polarized photon, to $-1$ after the absorption of a $\sigma_-$-polarized photon, and to 0 after the absorption of a $\pi$-polarized photon. These *polarization selection rules* result from the conservation of the total angular momentum and express the clear connection which exists between the variation $M_e - M_g$ of the magnetic quantum number of an atom absorbing a photon and the polarization of this photon (see Figure 6.2).

### 6.2.2 Optical pumping

Optical pumping, which was developed in the early 1950s by Alfred Kastler and Jean Brossel, is our first example of manipulation of atoms with light. To explain optical pumping, let us consider the simple case of a transition connecting a ground state g with an angular momentum $J_g = \frac{1}{2}$ to an excited state e with an angular momentum $J_e = \frac{1}{2}$, so that there are two ground-state Zeeman sublevels $g_{+1/2}$ and $g_{-1/2}$ and two excited Zeeman sublevels $e_{+1/2}$ and $e_{-1/2}$ (see Figure 6.3). If one excites such an atom

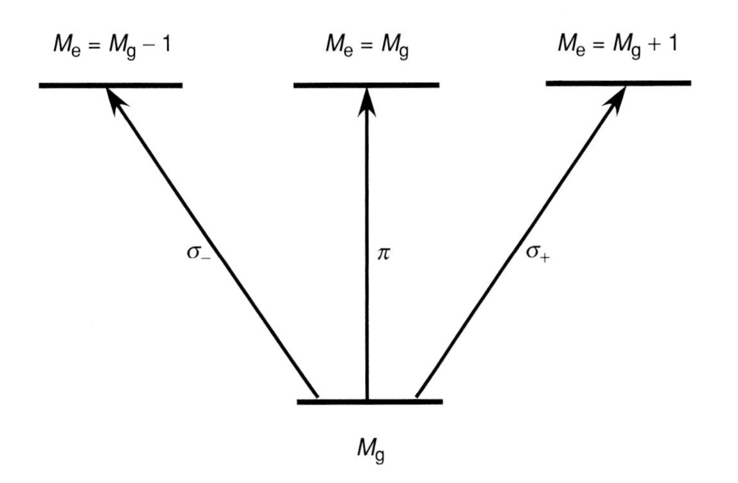


**Figure 6.2.** Polarization selection rules resulting from the conservation of the total angular momentum after the absorption of a photon by an atom.

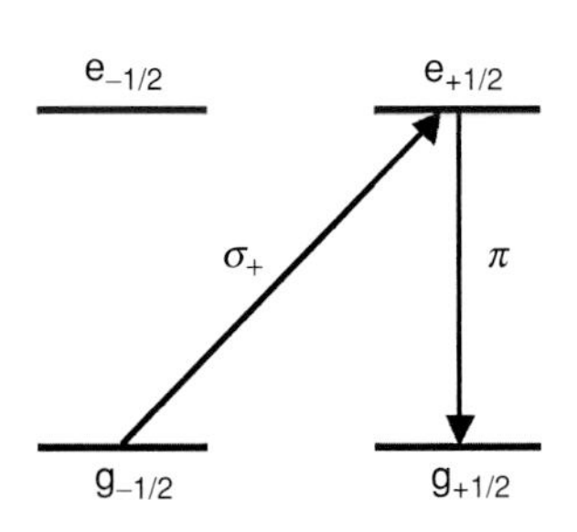


**Figure 6.3.** The principle of optical pumping for a $1/2 \to 1/2$ transition. Atoms are transferred from $g_{-1/2}$ to $g_{+1/2}$ by an *optical pumping cycle*, which consists of an absorption of a $\sigma_+$-polarized photon followed by the spontaneous emission of a $\pi$-polarized photon.

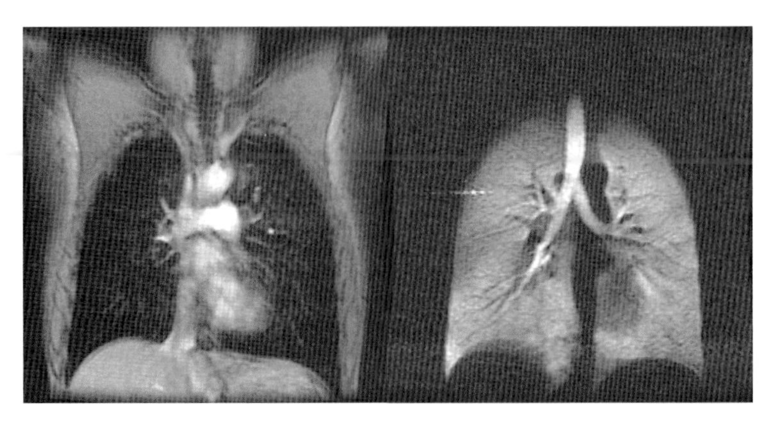

**Figure 6.4.** Magnetic-resonance imaging of the chest of a patient. Left: image obtained with ordinary proton-based resonance. Right: image obtained with gaseous-helium-based resonance. The patient has inhaled a mixture of air and helium-3, and the latter has been polarized using optical pumping. (Photographs courtesy of *Physics World*. Figure taken from the article of G. Allan Johnson, Laurence Hedlund, and James MacFall, November 1998.)

with $\sigma_+$-polarized light, one drives only the transition $g_{-1/2} \to e_{+1/2}$ because this is the only transition corresponding to the selection rule $M_e - M_g = 1$ associated with a $\sigma_+$ excitation (see Section 6.2.1). Once the atom has been excited to $e_{+1/2}$, it can fall back by spontaneous emission either into $g_{-1/2}$, in which case it can repeat the same cycle, or into $g_{+1/2}$ by emission of a $\pi$-polarized photon. In the last case, the atom remains trapped in $g_{+1/2}$ because there is no $\sigma_+$ transition starting from $g_{+1/2}$ (see Figure 6.3). This gives rise to an optical-pumping cycle transferring atoms from $g_{-1/2}$ to $g_{+1/2}$ through $e_{+1/2}$. During such an absorption–spontaneous-emission cycle, the angular momentum of the impinging photons has been transferred to the atoms, which thus become polarized.

It clearly appears in Figure 6.3 that atoms absorb resonant light only if they are in $g_{-1/2}$. If they are in $g_{+1/2}$, they cannot absorb light because there is no $\sigma_+$ transition starting from $g_{+1/2}$. This means that any transfer of atoms from $g_{+1/2}$ to $g_{-1/2}$ that would be induced by a resonant radio-frequency (RF) field or by a relaxation process can be detected by monitoring the amount of light absorbed with a $\sigma_+$ polarization.

A first interesting feature of these optical methods is that they provide a very efficient scheme for polarizing atoms at room temperature and in a low magnetic field. Secondly, they have very high sensitivity. A single RF transition between the two ground-state sublevels is detected by the subsequent absorption or emission of an optical photon, and it is much easier to detect an optical photon than a RF photon because it has a much higher energy. Finally, these optical methods allow one to study and to investigate non-equilibrium situations. Atoms are removed from their thermodynamic equilibrium by optical pumping. By observing the temporal variations of the absorbed or emitted light, one can study how the system returns to equilibrium.

## Magnetic-resonance imaging with optical pumping

Optical pumping has recently found an interesting application for imaging the human body. One prepares a sample of polarized gaseous helium-3, using optical pumping

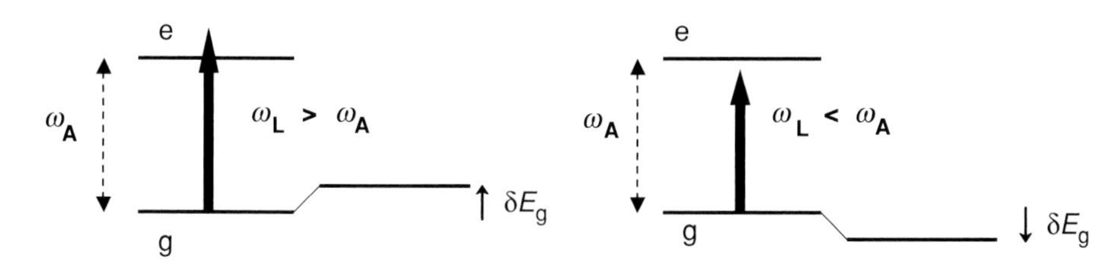


**Figure 6.5.** Light shift of the ground state g of an atom produced by a non-resonant light excitation detuned to the red side of the atomic transition (right) or to the blue side (left).

with a laser. This gas is inhaled by patients (this is harmless!) and it is used to perform magnetic-resonance imaging (MRI) of the cavities in their lungs (Figure 6.4). Current proton-based MRI provides information only on solid or liquid parts of the human body such as the muscles, the brain, and the blood (the left-hand part of Figure 6.4). The use of gaseous polarized helium with high degrees of spin polarization provides MRI signals strong enough to be detected even with a system as dilute as a gas. These signals allow internal spaces of the body, such as the cavities of the lung, to be visualized at unprecedented resolutions (the right-hand part of Figure 6.4). The use of this polarized gas is a promising tool for improving our understanding of lung physiology and function.

## 6.2.3 Light broadening and light shifts

The interaction of an atom with an electromagnetic field perturbs atomic energy levels. This perturbation exists even in the absence of any incident light beam, in the "vacuum" of photons. Atomic energy levels are shifted. The interpretation of this effect, which was discovered by Willis Lamb in 1947, has stimulated the development of quantum electrodynamics, which is the prototype of modern quantum field theories. It can be interpreted as due to "virtual" emissions and reabsorptions of photons by the atom. Atomic excited states are also broadened, the corresponding width $\Gamma$ being called the natural width of the atomic excited state e. In fact, $\Gamma$ is the rate at which an excited atom spontaneously emits a photon and falls into the ground state g. It can be also written $\Gamma = 1/\tau_R$, where $\tau_R$ is called the radiative lifetime of the excited state, i.e. the mean time after which an atom leaves the excited state by spontaneous emission.

Irradiation with light introduces another perturbation for the ground state of an atom, which can be also described, at low enough intensities, as a broadening $\Gamma'$ and a shift $\delta E_g$ of the atomic ground state g. Both quantities depend on the intensity of light $I_L$ and on the detuning $\Delta = \omega_L - \omega_A$ between the frequency of light $\omega_L$ and the atomic frequency $\omega_A$. The broadening $\Gamma'$ is the rate at which photons are scattered from the incident beam by the atom. The shift $\delta E_g$ is the energy displacement of the ground level, as a result of virtual absorptions and stimulated emissions of photons by the atom within the light-beam mode; it is called the light shift or AC Stark shift. It was predicted and observed by Jean-Pierre Barrat and one of the authors of this article (C.C.-T.) in the early 1960s. At resonance ($\Delta = 0$), $\delta E_g = 0$ and $\Gamma'$ takes its maximum value. For large enough detunings, $\delta E_g$ is much larger than $\hbar\Gamma'$. One can show that $\delta E_g$ is then proportional to $I_L/\Delta$. The light shift produced by irradiation with nonresonant light is thus proportional to the intensity of light and inversely proportional to the detuning. In particular, it has the same sign as the detuning (see Figure 6.5).

## 6.3 Electromagnetic forces and trapping

The most well-known action of an electromagnetic field on a particle is the Lorentz force. Consider a particle with charge $q$ moving with velocity $\boldsymbol{v}$ in an electric field $\boldsymbol{E}$ and a magnetic field $\boldsymbol{B}$. The resulting force $\boldsymbol{F} = q(\boldsymbol{E} + \boldsymbol{v} \times \boldsymbol{B})$ is used in all electronic devices. For neutral particles, such as neutrons, atoms, and molecules, the Lorentz force is zero. However, one can still act on these particles by using the interaction of their electric dipole moment $\boldsymbol{D}$ or magnetic dipole moment $\boldsymbol{\mu}$ with a gradient of electric or magnetic field. The interaction energy of the dipole with the field is $-\boldsymbol{D}\cdot\boldsymbol{E}$ or $-\boldsymbol{\mu}\cdot\boldsymbol{B}$, and it constitutes a potential energy for the motion of the center of mass of the particle. Depending on the relative orientation between the dipole and the field, the resulting force is directed toward regions of large or small field.

In the following we shall first address the possibility of trapping charged particles with electromagnetic fields (Section 6.3.1). We shall then turn to neutral particles and we shall discuss separately the case of magnetic forces (Section 6.3.2) and electric (Section 6.3.3) forces. The physical concepts involved are quite different since one deals with a permanent dipole moment in the first case, and an induced moment in the latter. We shall finally turn to the radiation-pressure force (Section 6.3.4), and present two spectacular applications of this force to atom manipulation: the deceleration of an atomic beam and the magneto-optical trap.

### 6.3.1 Trapping of charged particles

At first sight, the simplest trap for charged particles should consist in a pure electrostatic field $\boldsymbol{E}(\boldsymbol{r})$ such that the electrostatic force $\boldsymbol{F} = q\boldsymbol{E}$ in the vicinity of a given point O would be a restoring force. This is unfortunately forbidden by the Gauss equation for electrostatics $\boldsymbol{\nabla}\cdot\boldsymbol{E} = 0$, which entails $\boldsymbol{\nabla}\cdot\boldsymbol{F} = 0$. This result, which makes it impossible to have a force pointing inwards for a sphere centered at any point O, is known as the Earnshaw theorem. We present hereafter two possibilities for how to circumvent this theorem. The first one takes advantage of the time dependence of the applied electric field. The second possibility uses a combination of electric and magnetic fields.

#### The Paul trap

The principle of this trap, which was invented by Wolfgang Paul in Bonn in the mid 1950s, places the particle to be trapped in an electric field that is rapidly oscillating at frequency $\Omega$: $\boldsymbol{E}(\boldsymbol{r}, t) = \boldsymbol{\mathcal{E}}(\boldsymbol{r}) \cos(\Omega t)$, the amplitude $\mathcal{E}(\boldsymbol{r})$ vanishing at the center of the trap O. The motion of the particle can then be decomposed into a fast oscillation at frequency $\Omega$ (the micro-motion) superimposed upon a motion with a slower characteristic frequency. After averaging over a time period $2\pi/\Omega$, the kinetic energy associated with the micro-motion is $U(\boldsymbol{r}) = q^2\mathcal{E}^2(\boldsymbol{r})/(4m\Omega^2)$. This kinetic energy plays the role of a potential energy for the slow motion. It is null at O, since $\mathcal{E}$ vanishes at this point, and positive everywhere else. One thus achieves in this way a potential well centered on O, which confines the particle. The consistency of the above treatment is ensured by choosing the field amplitude $\mathcal{E}$ such that the oscillation frequency of the particle in the potential $U(\boldsymbol{r})$ is indeed much smaller than the fast frequency $\Omega$.

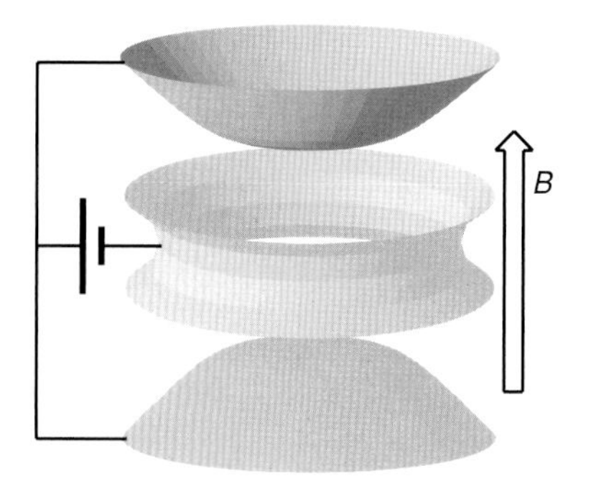


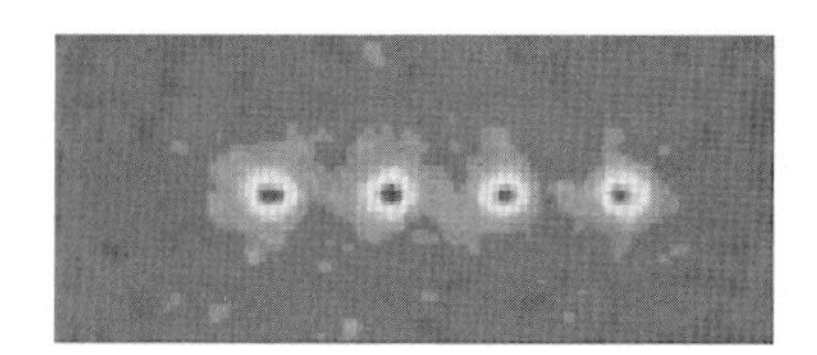

**Figure 6.6.** Trapping charged particles. Left: the scheme of electrodes used in a Penning trap, with a vertical magnetic field. Right: a chain of four calcium ions confined in a linear Paul trap; the distance between two ions is 15 μm. (Photograph by courtesy of Rainer Blatt, Innsbruck.)

### The Penning trap

The Penning trap is formed by the superposition of a quadrupole electrostatic potential $V(\boldsymbol{r})$ and a uniform magnetic field $\boldsymbol{B}$ parallel to the $z$ axis (Figure 6.6, left-hand side). The electrostatic potential ensures that trapping along the $z$ direction occurs, and it expels the particle in the perpendicular $x$–$y$ plane; it can be written $V(\boldsymbol{r}) = \kappa(2z^2 - x^2 - y^2)$, where $\kappa$ is a constant. The Newtonian equations of motion for the particle are linear in $\boldsymbol{r}$ and $\boldsymbol{v}$; hence they can be solved exactly. For $qB > (8\kappa m)^{1/2}$, one finds that the motion in the $x$–$y$ plane is stabilized by the magnetic field, while the trapping along the $z$ axis by the quadrupole electric field is not affected. One thus achieves a stable three-dimensional confinement of the particles.

### Applications

Both Paul and Penning traps are extensively used in modern physics. They are often associated with efficient cooling of the particles using either quasiresonant laser light (see Section 6.4.3) or resonant coupling with a damped electrical circuit. They allow a precise measurement of the cyclotron frequency $qB/m$ of the trapped particle. By facilitating comparison of this ratio for a single electron and a single positron, or for a single proton and a single antiproton, Penning traps have allowed tests of the symmetry between matter and antimatter with unprecedented precision. One can also use this trap to measure accurately the gyromagnetic ratio of a particle, i.e. the ratio between the cyclotron frequency and the Larmor frequency, characterizing the evolution of the magnetic moment. Using a single trapped electron, Hans Dehmelt and his group have employed this system to test quantum electrodynamics at the $10^{-12}$ level of accuracy. Single ions trapped in Penning or Paul traps are used for ultra-high-resolution spectroscopy and metrology. It is also possible to trap large assemblies of ions in these traps and to study in great detail the macroscopic behavior of these plasmas. Last but not least, a string of a few ions trapped in a Paul trap is considered a very promising system with which to implement the basic concepts of the new field of quantum information (Figure 6.6, right-hand side – see also Chapter 11).

## 6.3.2 The magnetic dipole force

One of the most celebrated experiments performed in the early days of quantum mechanics was the Stern–Gerlach experiment. In that experiment, performed in 1921, a beam of silver atoms was sent into a region with a large magnetic gradient. Otto

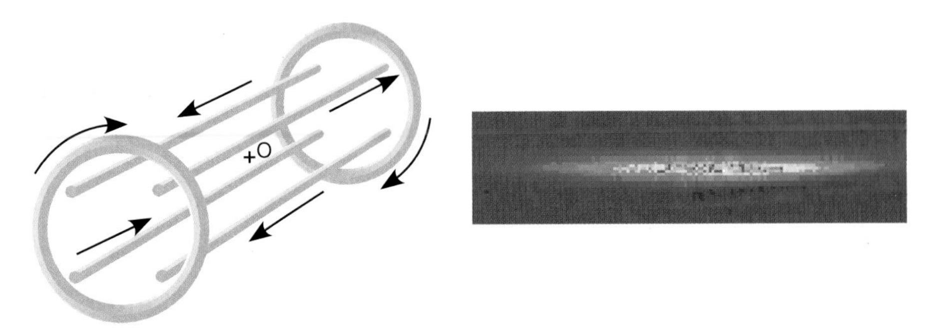


**Figure 6.7.** Magnetic trapping of neutral atoms. Left: a Ioffe–Pritchard trap, consisting of four linear conductors and two circular coils. The arrows indicate the current direction in each conductor. The modulus of the magnetic field has a nonzero local minimum at the center of symmetry O of the system. Atoms with a magnetic moment anti-parallel to the local magnetic field are confined around O. Right: a photograph of $10^7$ cesium atoms confined in a Ioffe–Pritchard trap. The image of the cigar-shaped atom cloud has been obtained by recording the absorption of a short resonant laser pulse and making the image of the shadow of the atom cloud onto a CCD camera. The temperature is of the order of 10 microkelvins. (Photograph: ENS.)

Stern and Walter Gerlach observed that the beam was split into two components, one being deflected toward the region of large magnetic field, the other being deflected away from this region.

The modern interpretation of this experiment is straightforward within the framework that we outlined in Section 6.2. The silver atoms have a spin $J = \frac{1}{2}$, and they possess a magnetic moment $\boldsymbol{\mu}$ proportional to their angular momentum $\boldsymbol{\mu} = \gamma \boldsymbol{J}$, where $\gamma$ is a constant. Therefore the projection of the magnetic moment along the axis of the magnetic field can take only two values, $\pm\hbar\gamma/2$, corresponding to an orientation of $\boldsymbol{\mu}$ either parallel or anti-parallel to $\boldsymbol{B}$. The magnetic-interaction energy is then $-\boldsymbol{\mu}\cdot\boldsymbol{B} = \mp\mu B$. Atoms with $\boldsymbol{\mu}$ parallel to $\boldsymbol{B}$ (energy $-\mu B$) are attracted by the high-field region, whereas atoms with $\boldsymbol{\mu}$ anti-parallel to $\boldsymbol{B}$ are deflected toward the low-field region. The magnetic force can be quite large in practice. Consider a hydrogen atom in its ground state; $\mu$ is the Bohr magneton, of the order of $10^{-23}$ J T$^{-1}$. In the vicinity of a strong permanent magnet, the gradient is $\sim$10 T m$^{-1}$, hence a force 6000 times larger than gravity.

## Magnetic trapping of neutral atoms

The magnetic force is now widely used to trap neutral particles such as atoms, neutrons, and molecules. Static magnetic traps are centered around a point O where the amplitude $B$ of the magnetic field is minimum. Atoms prepared with a magnetic moment $\boldsymbol{\mu}$ anti-parallel with $\boldsymbol{B}$ (*low-field seekers*) have a magnetic energy $\mu B(\boldsymbol{r})$ and they feel a restoring force toward O. It is not possible to achieve a local maximum of $B$ (except on the surface of a conductor). Therefore one cannot form a stable magnetic trap for *high-field seekers*, i.e. atoms with $\boldsymbol{\mu}$ parallel to $\boldsymbol{B}$.

The first magnetic trap for neutrons was demonstrated by Wolfgang Paul and his group in 1975. William Phillips and his team at the NIST, Gaithersburg, observed the first magnetically trapped atomic gas in 1985. Nowadays the most commonly used magnetic trap is the Ioffe–Pritchard trap, which ensures confinement around a location O where the field $B_0$ is nonzero, typically from 0.1 to 1 mT (Figure 6.7). This ensures that the Larmor frequency characterizing the evolution of $\mu$ is large, on the order of 1–10 MHz. Since the oscillation frequency of the atom in the trap is usually much smaller (a few hundred hertz only), this allows the magnetic moment $\mu$ to adjust adiabatically to the direction of $B$ during the displacement of the atom in the trap: an atom initially prepared in a low-field-seeking state will remain in this state during the course of its evolution.

Magnetic traps are simple to design and to build, and atoms can be stored for some time (several minutes) at very low temperatures (microkelvins), without any appreciable heating. In fact the lifetime of an atom in a magnetic trap is mostly determined by the quality of the vacuum in the chamber containing the trap. Indeed, the trap depth is relatively low, so a collision with a molecule from the background gas ejects the atom out of the trap. Magnetic traps have played a key role in the achievement of Bose–Einstein condensation with atomic gases. They are also used to trap atomic fermionic species and molecules.

### 6.3.3 The electric dipole force

#### Permanent dipole moments: molecules

For a system with a permanent electric dipole $D$ such as a hetero-molecule (CO, $NH_3$, $H_2O$), one can transpose the reasoning given above for a magnetic dipole. When the molecule is placed in an electric field, the projection $D_z$ of its dipole moment in the electric-field direction is quantized. It can take $2J + 1$ values, where $J$ is the angular momentum of the molecular state under consideration; the electrostatic energy $-D \cdot E = -D_z E$ gives rise to $2J + 1$ potential-energy surfaces. If the molecular beam propagates in a region where the electric field is inhomogeneous, a different force corresponds to each surface and the beam is split into $2J + 1$ components.

This electric dipole force is used in many devices, such as the ammonia maser, for which it is at the basis of the preparation of the population inversion. A recent spectacular application of this force has been developed by the group of Gerard Meijer in Nijmegen. It consists of decelerating a pulsed beam of molecules using electric-field gradients. The beam is sent into a region of increasing field, so that molecules with $D$ anti-parallel to $E$ are slowed down. With a maximum field of $10^7\ \mathrm{V\,m^{-1}}$ the decrease in kinetic energy is on the order of $k_B \times 1\ \mathrm{K}$. The electric field is then switched off as soon as the pulsed beam of molecules has reached the location where this field is large. Using a carefully designed stack of electrodes, one repeats this operation a large number of times over the total length of the beam (typically 1 m) and the pulsed beam of molecules can be brought nearly to rest.

#### Induced dipole moments: atoms

For atoms, the permanent dipole moment is null, as a consequence of the symmetry of the physical interactions at the origin of the atom's stability. However, it is still possible to act on them using an electric-field gradient, through an *induced* electric dipole moment. When an atom is placed in a static electric field $E$, it acquires a dipole moment $D = \alpha_0 E$, where $\alpha_0$ is the static polarizability. For simplicity we shall assume here that $\alpha_0$ is a scalar, although it may also have a tensorial part.

The potential energy of an atom in an electric field is $W = -\alpha_0 E^2/2$ and the corresponding force is $F = -\nabla W = \alpha_0 \nabla(E^2)/2$. For an atom in its ground state g, $\alpha_0$ is positive. Therefore the atom is always attracted to the regions where the electric field is the largest. The potential energy $W$ is nothing but the shift of the relevant atomic internal state induced by the electric field. It is calculated here at the second order of perturbation theory, assuming a linear response of the atom with respect to the field. We neglect saturation effects, which is valid as long as the applied electric field is small compared with the inner field of the atom created by the nucleus.

This analysis can be generalized to the case of a time-dependent electric field. Consider a field oscillating with the angular frequency $\omega_L$. The static polarizability must then be replaced by the dynamic polarizability $\alpha(\omega_L)$. When $\omega_L$ is much smaller than the relevant atomic Bohr frequencies $\omega_A$ of the atom, then $\alpha(\omega_L) \simeq \alpha_0$. This is the case in many experiments in which one manipulates ground-state alkali atoms (Bohr frequencies $\omega_A \sim 3 \times 10^{15}\,s^{-1}$) with very-far-detuned laser light, such as the radiation from a $CO_2$ laser ($\omega_L \sim 2 \times 10^{14}\,s^{-1}$).

## The resonant dipole force

A very important practical case concerns an atom in its ground internal state, which is irradiated with a laser wave whose frequency $\omega_L$ is comparable to the atomic Bohr frequency $\omega_A$ corresponding to the resonance transition g $\leftrightarrow$ e. In this case the dipole potential $W(\boldsymbol{r}) = -\alpha(\omega_L)E^2(\boldsymbol{r})/2$ is nothing but the light shift $\delta E_g$ of the ground state that we derived in Section 6.2.3. We recall that the sign of the light shift, and hence the direction of the dipole force, depends on the sign of the detuning to resonance $\Delta = \omega_L - \omega_A$. When $\Delta$ is negative, the result is qualitatively the same as for a static field; the atom is attracted to the region of large laser intensities. On the contrary, when $\Delta$ is positive, the force on the atom tends to push it away from high-intensity regions. The dipole force is nonzero only if the light intensity is spatially inhomogeneous. One can show that it can be interpreted as resulting from a redistribution of photons between the various plane waves forming the laser wave in absorption–stimulated-emission cycles.

When the laser intensity is increased to a large value, the atom spends a significant time in the excited internal state e. In this case the preceding expression for the dipole potential must be modified. A convenient point of view on the system is obtained through the *dressed-atom* formalism, in which one deals with the energy levels of the combined system "atom + laser photons." Two types of dressed states are found, connecting respectively to the ground and to the excited atomic states when the laser intensity tends to zero. The forces associated with the two dressed states are opposite. Since spontaneous-emission processes cause random jumps between the two types of dressed states, the atomic motion is stochastic, with an instantaneous force oscillating back and forth between two opposite values in a random way. Such a dressed-atom picture provides a simple interpretation of the mean value and of the fluctuations of dipole forces.

## Dipole traps for neutral atoms

One of the most spectacular uses of the resonant dipole force is the possibility of trapping atoms around local maxima or minima of the laser intensity. The first laser trap was demonstrated in 1985 at Bell Labs, by the group of Steven Chu and Arthur Ashkin,

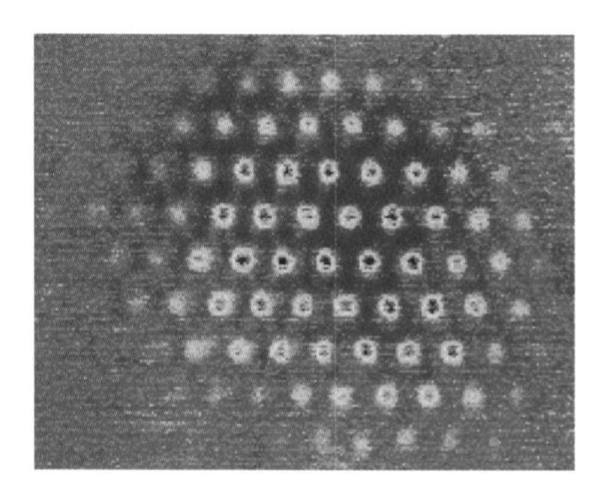

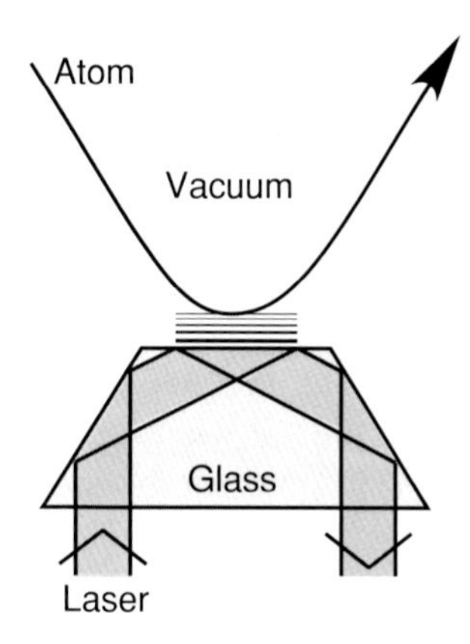


**Figure 6.8.** Manipulation of atoms using the resonant dipole force. Left: a photograph of cesium atoms captured in a hexagonal optical lattice with a period of 29 μm. Each spot contains $10^4$ atoms. (Picture courtesy of C. Salomon, ENS Paris.) Right: an atom mirror formed by an evanescent wave propagating at the surface of a glass prism. For a positive detuning $\Delta$ of the laser beam with respect to the atomic resonance, the atoms are repelled from the high-intensity region. If their incident kinetic energy is low enough, they bounce elastically on the light sheet without touching the glass surface.

using a single focused traveling laser wave. Atoms were accumulated at the vicinity of the focal point of the light wave. Later on, several other traps were investigated, such as hollow tubes used as atom guides. The research on dipole traps has led to another spectacular development, *optical tweezers*. The object being trapped is no longer a single atom, but a micrometer-sized dielectric sphere. It can be attached to objects of biological interest, such as a DNA molecule, and it allows the microscopic manipulation of these objects.

## Optical lattices

Optical lattices are formed by the periodic modulation of the light intensity in a laser standing wave. Depending on the sign of the detuning $\Delta$, the atoms accumulate at the nodes or the antinodes of the standing wave (Figure 6.8, left-hand side). Optical lattices, which were initially studied by the groups of Gilbert Grynberg at the Ecole Normale Supérieure (Paris) and William Phillips at the NIST, have led to several spectacular developments, from both theoretical and experimental points of view. The tunneling between adjacent wells plays a significant role in the dynamics of the atoms, and these lattices constitute model systems for studying quantum transport in a periodic potential. As an example, the group of Christophe Salomon has shown that atoms submitted to a constant force in addition to the lattice force undergo periodic Bloch oscillations, instead of being uniformly accelerated as in the absence of a lattice. Also, the team of Theodor Hänsch and Immanuel Bloch at Munich has observed the superfluid–insulator Mott transition for an ultra-cold gas placed in an optical lattice. The superfluid phase corresponds to a Bose–Einstein condensate, in which each atom is delocalized over the whole lattice. The isolating phase is obtained by increasing the lattice depth so that the tunneling between adjacent wells is reduced. The repulsion between atoms then favors a situation in which the number of atoms at each lattice node is fixed.

### Atomic mirrors

For a positive detuning of the laser wave with respect to the atomic frequency, the dipole force repels the atoms from the high-intensity region. It is thus possible to create a potential barrier on which the atoms can be elastically reflected. Following a suggestion by Richard Cook and Richard Hill, several groups have used an evanescent wave propagating at the surface of a glass prism to form an atomic mirror (Figure 6.8, right-hand side). The incident atoms arrive on the vacuum side and feel the repulsive dipole force as they enter the evanescent wave. If their incident kinetic energy is smaller than the potential barrier created by the light, atoms turn back before touching the glass.

In practice, with a laser intensity of 1 W focused on a surface of the order of 1 $\text{mm}^2$, an atom can be reflected if the component of its velocity normal to the mirror is lower than a few meters per second. Such atomic mirrors are therefore well suited for manipulating laser-cooled atoms (see Section 6.4). They constitute very useful components for the development of atomic optics. Using a curved dielectric surface, one can focus or defocus an atomic beam. Using an evanescent wave whose intensity is modulated in time, one can make a vibrating mirror. The corresponding modulated Doppler shift introduces a frequency modulation of the reflected de Broglie wave (Section 6.5).

## 6.3.4 The radiation-pressure force

### Recoil of an atom emitting or absorbing a photon

Consider an atom in an excited electronic state e, with its center of mass initially at rest. At a certain time, the atom emits a photon and drops to its electronic ground state g. The total momentum of the system is initially zero and it is conserved throughout the whole process. Therefore, in the final state, since the emitted photon carries away the momentum $\hbar \boldsymbol{k}$, the atom recoils with momentum $-\hbar \boldsymbol{k}$. This recoil phenomenon also occurs when an atom absorbs a photon. Consider the atom in its ground state g and with its center of mass initially at rest; suppose that a photon with wavevector $\boldsymbol{k}$ is sent toward this atom. If the atom absorbs the photon, it jumps to the excited state and recoils with the momentum $\hbar \boldsymbol{k}$.

To the change $\hbar k$ of the atom's momentum there corresponds a change $v_{\text{rec}} = \hbar k/m$ of the atom's velocity, where $m$ is the atom's mass. For a hydrogen atom absorbing or emitting a photon on the Lyman-$\alpha$ line (2p $\rightarrow$ 1s transition), this recoil velocity is 3 $\text{m}\,\text{s}^{-1}$. For a sodium atom, a photon emitted or absorbed on its resonance line (wavelength 590 nm) corresponds to a change in velocity of $v_{\text{rec}} = 3\ \text{cm}\,\text{s}^{-1}$. These are very low velocities compared with those of atoms or molecules at room temperature, which are on the order of several hundreds of meters per second. This explains why the changes in velocity due to recoil effects have most of the time been neglected in the past. However, as we see below, the repetition of these changes in velocity can lead to large forces.

### The radiation pressure in a resonant light wave

Consider an atom placed in a traveling laser wave with wavevector $\boldsymbol{k}$. We assume that the laser frequency $\omega_{\text{L}}$ is resonant with the atomic transition g $\leftrightarrow$ e at frequency $\omega_{\text{A}}$. The atom then undergoes a succession of *fluorescence cycles*. The atom initially

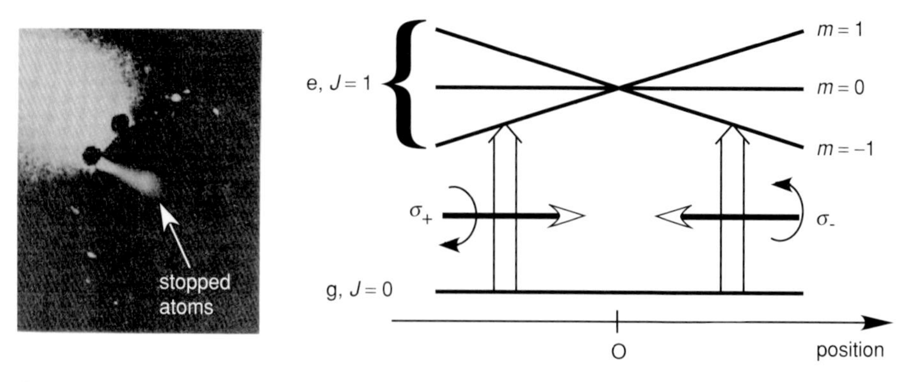


**Figure 6.9.** Manipulation of atoms using the radiation-pressure force. Left: a photograph of a beam of sodium atoms stopped by the radiation pressure of a counter-propagating laser beam. (Photograph courtesy of W. D. Phillips, NIST Gaithersburg.) Right: the principle of the magneto-optical trap. The two counter-propagating laser waves have the same intensity and the same frequency, and they are respectively $\sigma_+$- and $\sigma_-$-polarized. In the presence of a magnetic field, the balance between the two radiation-pressure forces is broken; using a gradient of magnetic field, one achieves a situation in which atoms feel a restoring force toward the center O.

in its ground state absorbs a photon from the laser beam and gains momentum $\hbar \mathbf{k}$. After a time on the order of the radiative lifetime $\tau_R$ of the electronic excited state e, the atom decays back to the ground state by emitting spontaneously a photon. The direction of emission of this fluorescence photon is random; the symmetry properties of spontaneous emission are such that the probabilities of the photon being emitted in two opposite directions are equal. Therefore the change of momentum in a spontaneous emission averages out to zero. It follows that, in a fluorescence cycle, the average variation of the atomic velocity is related only to the absorption process and is equal to $\hbar \mathbf{k}/m$.

The repetition rate of these cycles is limited only by the lifetime $\tau_R$ of the excited state e. Since $\tau_R$ is on the order of $10^{-8}$ s, about $10^8$ (one hundred million!) fluorescence cycles can take place per second. During each cycle, the velocity of the atom changes on average by an amount $v_{rec} \sim 1$ cm s$^{-1}$. Being repeated 100 million times per second, this produces a change in velocity per second 100 million times larger than the recoil velocity, corresponding to an acceleration or deceleration on the order of $10^6$ m s$^{-2}$. Radiation-pressure forces are therefore $10^5$ times larger than the force due to gravity!

## Stopping an atomic beam

This considerable radiation-pressure force makes it possible to stop an atomic beam. Consider a sodium atomic beam coming out of an oven at a temperature of 500 K, corresponding to an average speed of 1 km s$^{-1}$. We irradiate this atomic beam with a counter-propagating resonant laser beam, so that the radiation-pressure force slows the atoms down. If the available distance is large enough, the atoms may even stop and return in the opposite direction: an atom with an initial velocity of 1 km s$^{-1}$ subjected to a deceleration of $10^6$ m s$^{-2}$ is brought to rest in one millisecond. During this deceleration time it travels 50 cm only, which makes such an atomic decelerator very practicable (Figure 6.9, left-hand side).

A complication in the deceleration process originates from the Doppler effect. As the atom's velocity $v$ changes, the laser frequency in the atomic frame $\tilde{\omega}_L = \omega_L - kv$

also changes and the resonance condition $\tilde{\omega}_L = \omega_A$ is no longer satisfied. In order to circumvent this problem, several solutions have been proposed and demonstrated: use of an inhomogeneous magnetic field so that the Zeeman effect also changes the atomic resonance frequency $\omega_A$ as the atom progresses in the decelerator, chirping of the laser frequency during the deceleration process, use of a broadband laser, etc. The first atoms stopped by radiation pressure were observed in 1984 by the groups of William Phillips (NIST) and John Hall (JILA, Boulder).

### The magneto-optical trap

The radiation-pressure force can be used to trap neutral atoms quite efficiently. The trap is based upon the imbalance between the opposite radiation-pressure forces created by two counter-propagating laser waves. The imbalance is made position-dependent through a spatially dependent Zeeman shift produced by a magnetic-field gradient. The principle of the trap takes advantage of both the linear and the angular momenta carried by the photons. For simplicity we present its principle for a one-dimensional configuration, as was first suggested by one of the authors (J. D.) in 1986; we assume that the angular momenta of the ground g and excited e internal levels involved in the trapping are respectively $J_g = 0$ and $J_e = 1$ (Figure 6.9, right-hand side). The two counter-propagating waves have the same negative detuning $\Delta$ ($\omega_L < \omega_A$) and they have opposite circular polarizations; they are thus in resonance with the atom at different places. At the center O of the trap, the magnetic field is zero. By symmetry the two radiation-pressure forces have the same magnitude and opposite directions. They balance each other and an atom at O feels no net force. Consider now an atom to the left of O. The laser wave coming from the left, which is $\sigma_+$-polarized, is closer to resonance with the allowed transition g $\leftrightarrow$ e, $m = +1$ than it is for an atom at O. The radiation pressure created by this wave is therefore increased with respect to its value at O. Conversely, the radiation-pressure force created by the wave coming from the right is decreased with respect to its value at O. Indeed, the wave is $\sigma_-$-polarized and it is further from resonance with the transition g $\leftrightarrow$ e, $m = -1$ than it is at O. Therefore the net force for an atom to the left of O is pointing toward O. For an atom located to the right of O, the reverse phenomenon occurs: the radiation-pressure force created by the wave coming from the right now dominates, so the resulting force also points toward O. One therefore achieves a stable trapping around O.

Such a scheme can be extended to three dimensions, as was first demonstrated in a collaboration between the groups at MIT and Bell Labs, and it leads to a robust, large, and deep trap called a *magneto-optical trap*. It has a large velocity-capture range and it can be used for trapping atoms in a cell filled with a low-pressure vapor, as was shown by the JILA Boulder group. Furthermore, the nonzero value of the detuning provides cooling of the trapped atoms, along the lines that will be discussed in the next section.

## 6.4 Cooling of atoms

The velocity distribution of an ensemble of atoms is characterized by the mean velocity and the velocity dispersion around the mean value. In physics, temperature is associated with this velocity spread, i.e. with the disordered motion of the atoms. The hotter the temperature of the medium, the higher the velocity dispersion of its constituents. To cool a system, this velocity spread has to be reduced.

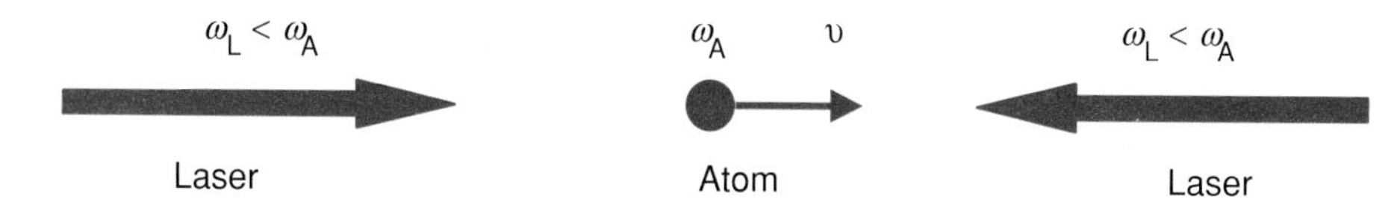


**Figure 6.10.** Doppler cooling in one dimension, resulting from the Doppler-induced imbalance between the radiation-pressure forces of two counter-propagating laser waves. The laser detuning is negative ($\omega_L < \omega_A$).

## 6.4.1 Doppler cooling

The simplest cooling scheme uses the Doppler effect and was first suggested in 1975 by Theodor Hänsch and Arthur Schawlow for free atoms and by David Wineland and Hans Dehmelt for trapped ions. The concept is basically simple; we explain it for free atoms, for which case it is very reminiscent of the principle of the magneto-optical trap discussed previously. Consider an atom being irradiated by two counter-propagating laser waves (Figure 6.10). These two laser waves have the same intensity and the same frequency $\omega_L$ slightly detuned below the atomic frequency $\omega_A$. For an atom at rest with zero velocity, there is no Doppler effect. The two laser waves have then the same apparent frequency. The forces being exerted have the same value with opposite signs; they balance each other and no net force is exerted on the atom. For an atom moving to the right with a velocity $v$, the frequency of the counter-propagating beam seems higher because of the Doppler effect. The wave becomes closer to resonance, more photons are absorbed, and the force created by this beam increases. Conversely, the apparent frequency of the co-propagating wave is reduced because of the Doppler effect and becomes farther from resonance. Fewer photons are absorbed and the force decreases. For a moving atom, the two radiation-pressure forces no longer balance each other. The force opposite to the atomic velocity finally prevails and the atom is thus subjected to a nonzero net force opposing its velocity. For a small velocity $v$, this net force can be written as $F = -\alpha v$, where $\alpha$ is a friction coefficient. The atomic velocity is damped out by this force and tends to zero, as if the atom were moving in a sticky medium. This laser configuration is called *optical molasses*.

### The limit of Doppler cooling

The Doppler friction responsible for the cooling is necessarily accompanied by fluctuations due to the fluorescence photons which are spontaneously emitted in random directions and at random times. Each emission process communicates to the atom a random recoil momentum $\hbar k$, which is responsible for a momentum diffusion described by a diffusion coefficient $D$. As in usual Brownian motion, competition between friction and diffusion leads to a steady state, with an equilibrium temperature proportional to $D/\alpha$. A detailed analysis shows that the equilibrium temperature obtained with such a scheme is always larger than a certain limit $T_D$, called the Doppler limit. This limit is given by $k_B T_D = \hbar\Gamma/2$, where $\Gamma$ is the natural width of the excited state. It is reached for a detuning $\Delta = \omega_L - \omega_A = -\Gamma/2$, and its value is on the order of 100 μK for alkali atoms. In fact, when the measurements became precise enough, the group of William Phillips showed in 1988 that the temperature in optical molasses was much lower than had been expected. This indicated that other laser-cooling mechanisms, more powerful than Doppler cooling, are operating. They were identified in 1998 by the Paris and Stanford groups. We describe one of them in the next subsection, namely the Sisyphus cooling mechanism, which was proposed by the authors of this chapter.

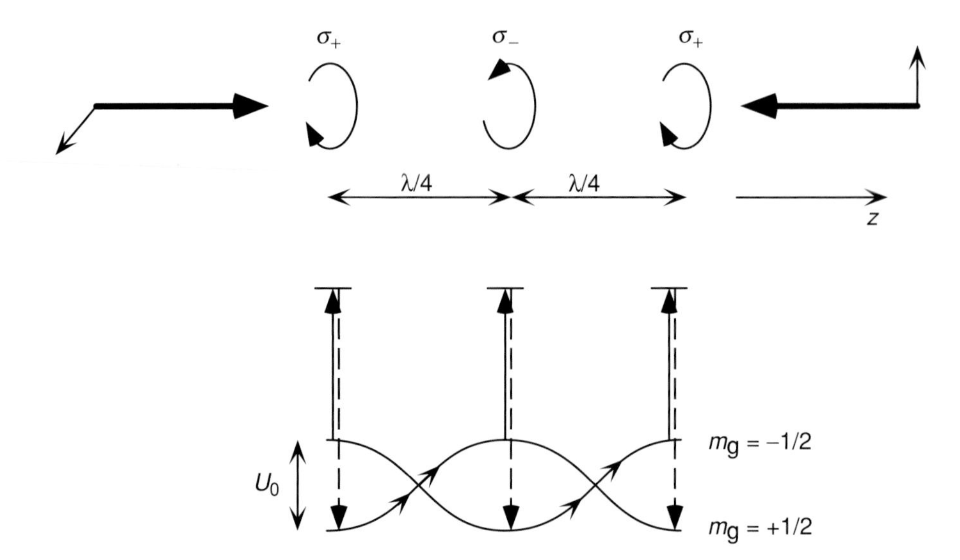


**Figure 6.11.** One-dimensional Sisyphus cooling. The laser configuration is formed by two counter-propagating waves along the $z$ axis with orthogonal linear polarizations. The polarization of the resulting field is spatially modulated with a period $\lambda/2$. For an atom with two ground Zeeman sublevels $m_{\mathrm{g}} = \pm\frac{1}{2}$, the spatial modulation of the laser polarization results in correlated spatial modulations of the light shifts of these two sublevels and of the optical-pumping rates between them. Because of these correlations, a moving atom runs up potential hills more frequently than down.

## 6.4.2 Sisyphus cooling

The ground level g of most atoms, in particular alkali atoms, has a nonzero angular momentum $J_{\mathrm{g}}$. This level is thus composed of several Zeeman sublevels. Since the detuning used in laser-cooling experiments is not large compared with $\Gamma$, both differential light shifts and optical-pumping transitions exist for the various Zeeman sublevels of the ground state. Furthermore, the laser polarization and the laser intensity vary in general in space, so the light shifts and optical-pumping rates are position-dependent. We show now, with a simple one-dimensional example, how the combination of these various effects can lead to a very efficient cooling mechanism.

Consider the laser configuration of Figure 6.11, consisting of two counter-propagating plane waves along the $z$ axis, with orthogonal linear polarizations and with the same frequency and the same intensity. Because the phase shift between the two waves varies linearly with $z$, the polarization of the total field changes from $\sigma_+$ to $\sigma_-$ and vice versa every $\lambda/4$. In between, it is elliptical or linear. We address here the simple case in which the atomic ground state has angular momentum $J_{\mathrm{g}} = \frac{1}{2}$. The two Zeeman sublevels $m_{\mathrm{g}} = \pm\frac{1}{2}$ undergo different light shifts, depending on the laser polarization, so the Zeeman degeneracy in zero magnetic field is removed. This gives the energy diagram of Figure 6.11, showing spatial modulations of the Zeeman splitting between the two sublevels with a period $\lambda/2$.

If the detuning $\Delta$ is not very large compared with $\Gamma$, there are also real absorptions of photons by the atom, followed by spontaneous emission, which give rise to optical-pumping transfers between the two sublevels, whose direction depends on the polarization: $m_{\mathrm{g}} = -\frac{1}{2} \longrightarrow m_{\mathrm{g}} = +\frac{1}{2}$ for a $\sigma_+$ polarization, $m_{\mathrm{g}} = +\frac{1}{2} \longrightarrow m_{\mathrm{g}} = -\frac{1}{2}$ for a $\sigma_-$ polarization. Here also, the spatial modulation of the laser polarization results in a spatial modulation of the optical-pumping rates with a period of $\lambda/2$.

The two spatial modulations of light shifts and optical-pumping rates are of course correlated because they are due to the same cause, the spatial modulation of the light polarization. These correlations clearly appear in Figure 6.11. With the proper sign of the detuning, optical pumping always transfers atoms from the higher Zeeman sublevel to the lower one. Suppose now that the atom is moving to the right, starting from the bottom of a valley, for example in the state $m_g = +\frac{1}{2}$ at a place where the polarization is $\sigma_+$. Because of the finite value of the optical-pumping time, there is a time lag between the dynamics of internal and external variables. The atom can climb up the potential hill before absorbing a photon. It then reaches the top of the hill, where it has the maximum probability of being optically pumped in the other sublevel, i.e. at the bottom of a valley, and so on.

Like Sisyphus in Greek mythology, who was always rolling a stone up the slope, the atom is running up potential hills more frequently than down. When it climbs a potential hill, its kinetic energy is transformed into potential energy. Dissipation then occurs by emission of light, since the spontaneously emitted photon has an energy higher than that of the absorbed laser photon. After each Sisyphus cycle, the total energy $E$ of the atom decreases by an amount on the order of $U_0$, where $U_0$ is the depth of the optical potential wells of Figure 6.11. When $E$ becomes smaller than $U_0$, the atom remains trapped in the potential wells.

### Limits of Sisyphus cooling

The previous discussion shows that Sisyphus cooling leads to temperatures $T_{\mathrm{Sis}}$ such that $k_B T_{\mathrm{Sis}} \simeq U_0$. We have seen in Section 6.2.3 that the light shift $U_0$ is proportional to $I_L/\Delta$. Such a dependence of $T_{\mathrm{Sis}}$ on the laser intensity $I_L$ and on the detuning $\Delta$ has been checked experimentally.

At low intensity, the light shift is much smaller than $\hbar\Gamma$. This explains why Sisyphus cooling leads to temperatures much lower than those achievable with Doppler cooling. One cannot, however, decrease the laser intensity to an arbitrarily low value. The previous discussion ignores the recoils due to the spontaneously emitted photons. Each recoil increases the kinetic energy of the atom by an amount on the order of $E_R$, where

$$E_R = \hbar^2 k^2/(2M) \tag{6.2}$$

is the recoil energy of an atom absorbing or emitting a single photon. When $U_0$ becomes on the order of or smaller than $E_R$, the cooling due to Sisyphus cooling becomes weaker than the heating due to the recoil, and Sisyphus cooling no longer works. This shows that the lowest temperatures which can be achieved with such a scheme are on the order of a few $E_R/k_B$. This is on the order of a few microkelvins for heavy atoms such as rubidium and cesium. This result is confirmed by a full quantum theory of Sisyphus cooling and is in good agreement with experimental results.

Under the optimal conditions of Sisyphus cooling, atoms become so cold that they are trapped in the few lowest quantum vibrational levels of each potential well, more precisely the lowest allowed energy bands of this periodic potential. This is an example of the optical lattices discussed in Section 6.3.3. The steady state corresponds to an antiferromagnetic ordering, since two adjacent potential wells correspond to opposite spin polarizations.

## 6.4.3 Sub-recoil cooling

In Doppler cooling and Sisyphus cooling, fluorescence cycles never cease. Since the random recoil $\hbar k$ communicated to the atom by the spontaneously emitted photons

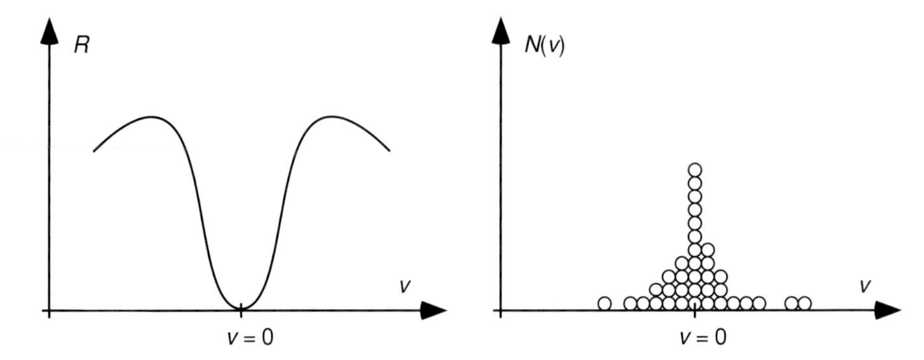


**Figure 6.12.** Sub-recoil cooling. The random walk in velocity space is characterized by a jump rate $R$ vanishing at $v = 0$. As a result, atoms that fall within a small interval around $v = 0$ remain trapped there for a long time and accumulate.

cannot be controlled, it seems impossible to reduce the atomic momentum spread $\delta p$ below a value corresponding to the photon momentum $\hbar k$. The condition $\delta p = \hbar k$ defines the *single-photon recoil limit*, the effective recoil temperature being set as $k_B T_R/2 = E_R$. The value of $T_R$ ranges from a few hundred nanokelvins for heavy alkalis to a few microkelvins for a light atom such as metastable helium, irradiated on its resonance line $2\,^3S \leftrightarrow 2\,^3P$.

## Sub-recoil cooling of free particles

It is possible to circumvent the recoil limit and to reach temperatures $T$ lower than $T_R$. The basic idea is to create a situation in which the photon-absorption rate $\Gamma'$, which is also the jump rate $R$ of the atomic random walk in velocity space, depends on the atomic velocity $v = p/M$ and vanishes for $v = 0$ (Figure 6.12). For an atom with zero velocity, the absorption of light is quenched. Consequently, there is no spontaneous re-emission and no associated random recoil. In this way one protects ultra-slow atoms (with $v \simeq 0$ ) from the "bad" effects of the light. On the contrary, atoms with $v \neq 0$ can absorb and re-emit light. In such absorption–spontaneous-emission cycles, their velocities change in a random way and the corresponding random walk in $v$-space can transfer atoms from the $v \neq 0$ absorbing states into the $v \simeq 0$ dark states, where they remain trapped and accumulate.

Up to now, two sub-recoil cooling schemes have been proposed and demonstrated. In the first one, called *velocity-selective coherent population trapping* (VSCPT), which was investigated by the Paris group in 1988, the vanishing of $R(v)$ for $v = 0$ is achieved by using quantum interference between different absorption amplitudes, which becomes fully destructive when the velocity of the atom vanishes. The second one, called Raman cooling, which was investigated by the Stanford group in 1992, uses appropriate sequences of stimulated Raman and optical-pumping pulses for tailoring the desired shape of $R(v)$. Using these two schemes, it has been possible to cool atoms down to a few nanokelvins.

## Sideband cooling of trapped ions

The states $v \simeq 0$ of Figure 6.12 are sometimes called "dark states" because an atom in these states does not absorb light. Dark-state cooling also exists for ions and is called

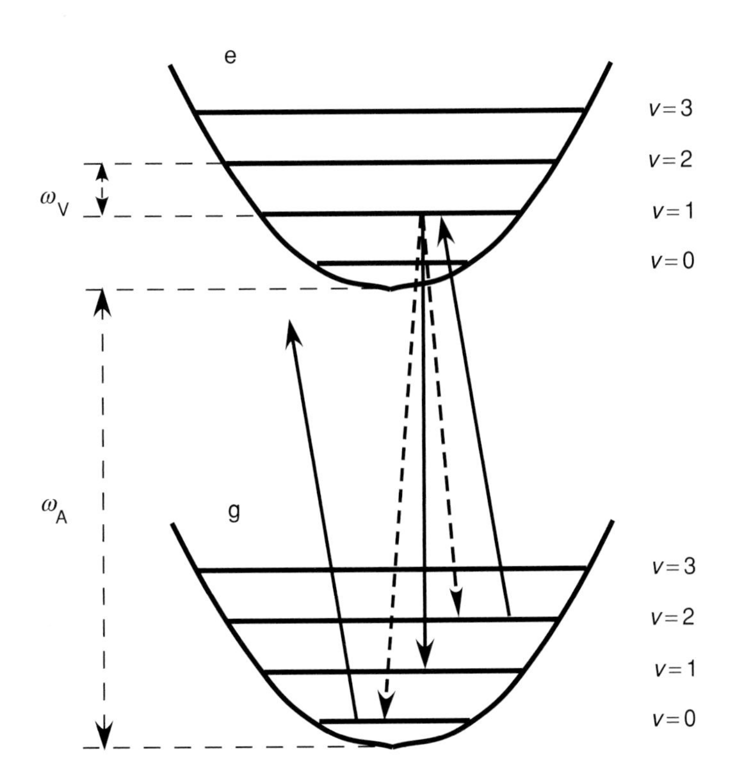


**Figure 6.13.** Sideband cooling. A laser excitation at frequency $\omega_A - \omega_V$ excites selectively transitions g, $v \longrightarrow$ e, $v-1$, where $v$ is the vibrational quantum number, if the natural width $\Gamma$ of the excited state is small relative to the vibration frequency $\omega_V$. The most intense spontaneous transitions bringing the ion back to the ground state obey the selection rule $\Delta v = 0$, so $v$ decreases after such a cycle. When the ion reaches the ground state g, 0, it remains trapped there because there are no transitions at frequency $\omega_A - \omega_V$ that can be excited from this state.

*sideband cooling* (see Figure 6.13). Consider an ion trapped in a parabolic potential well. The vibrational motion of the center of mass of this ion is quantized. The corresponding levels are labeled by a vibrational quantum number $v = 0, 1, 2, \ldots$ and the splitting between two adjacent levels is equal to $\hbar\omega_V$, where $\omega_V$ is the vibrational frequency of the ion in the parabolic potential well. The motion of the center of mass is due to the external electric and magnetic forces acting on the charge of the ion and is, to a very good approximation, fully decoupled from the internal motion of the electrons. It follows that the parabolic potential well and the vibrational levels in the ground electronic state g and in the electronic excited state e are the same (Figure 6.13). The absorption spectrum of the ion therefore consists of a set of discrete frequencies $\omega_A \pm n\omega_V$, where $\omega_A$ is the frequency of the electronic transition, and where $n = 0, \pm 1, \pm 2, \ldots$ We have a central component at frequency $\omega_A$ and a series of "sidebands" at frequencies $\omega_A \pm \omega_V, \omega_A \pm 2\omega_V, \ldots$ We suppose here that these lines are well resolved, i.e. that the natural width $\Gamma$ of the excited state, which is also the width of the absorption lines, is small compared with their frequency spacing $\omega_V$: $\Gamma \ll \omega_V$.

The principle of sideband cooling is to excite the ion with laser light tuned at the lower sideband frequency $\omega_A - \omega_V$. One excites in this way the transitions g, $v \longrightarrow$ e, $v - 1$. For example, Figure 6.13 shows the excitation of the transition g, $2 \longrightarrow$ e, 1. After such an excitation in e, $v - 1$ the ion falls to the ground state by spontaneous emission of a photon. One can show that the most probable transition obeys the selection rule $\Delta v = 0$. For the example of Figure 6.13, the ion falls preferentially into g, 1. There are also much weaker transitions corresponding to $\Delta v = \pm 1$, the ion falling into g, 0 and g, 2. It is thus clear that the cycle consisting of the excitation by a lower sideband followed by spontaneous emission decreases the vibrational quantum number $v$. After a few such cycles the ion reaches the vibrational ground state g, $v = 0$. It is then trapped in this state because there is no possible resonant excitation from g, 0 with laser light of frequency $\omega_A - \omega_V$. The transition with the lowest frequency from g, 0 is the transition g, $0 \longrightarrow$ e, 0 with frequency $\omega_A$ (see Figure 6.13). The state

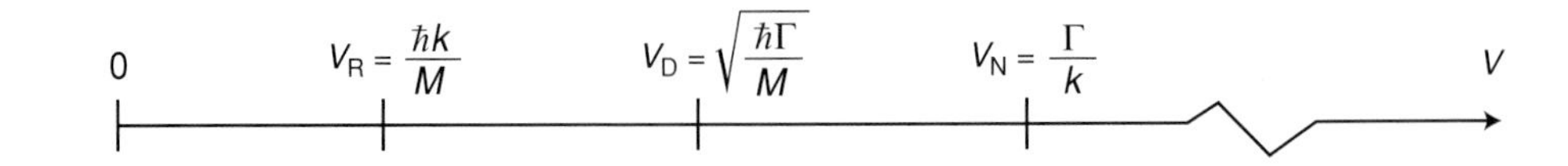


**Figure 6.14.** A few characteristic velocities associated with the various cooling schemes.

g, 0 is therefore a dark state and, after a few cycles, the ion is put into this state. Sideband cooling is a very convenient way for preparing a single ion in the vibrational ground state. Strictly speaking, the population of the first vibrational excited state is not exactly equal to zero, because of the nonresonant excitation of the transitions $\Delta v = 0$ by the laser light at frequency $\omega_A - \omega_V$, but, if $\Gamma$ is small enough relative to $\omega_V$, this population is negligible.

### Velocity scales for laser cooling

To conclude this section, we show in Figure 6.14 a few characteristic velocities given by the previous analysis and appearing in the velocity scale. The first, $v_R = \hbar k/M$, is the recoil velocity. The second, $v_D = \sqrt{\hbar\Gamma/M}$, is such that $Mv_D^2 = \hbar\Gamma$ and thus gives the velocity dispersion which can be achieved by Doppler cooling. The last, $v_N = \Gamma/k$, satisfies $kv_N = \Gamma$, which means that the Doppler effect associated with $v_N$ is equal to the natural width $\Gamma$. It gives therefore the velocity spread of the atoms which can be efficiently excited by a monochromatic light. It is easy to check that $v_N/v_D$ and $v_D/v_R$ are both equal to $\sqrt{\hbar\Gamma/E_R}$, where $E_R = \hbar^2k^2/(2M)$ is the recoil energy. For most allowed transitions, we have $\hbar\Gamma \gg E_R$, so that $v_R \ll v_D \ll v_N$ (the $v$ scale in Figure 6.14 is not linear). This shows the advantage of laser cooling. Laser spectroscopy with a monochromatic laser beam produces lines whose widths expressed in velocity units cannot be smaller than $v_N$. Doppler cooling reduces the velocity dispersion of the atoms to a much lower value, which, however, cannot be smaller than $v_D$. Sisyphus cooling reduces this lower limit to a few times $v_R$. Finally, sub-recoil cooling allows one to go below $v_R$.

## 6.5 Applications of ultra-cold atoms

Ultra-cold atoms move with very small velocities. This opens new possibilities for basic research and applications. Firstly, ultra-cold atoms can be kept a much longer time in the observation zone than can thermal atoms, which leave this zone very rapidly because of their high speed. This lengthening of the observation time considerably increases the precision of the measurements. Fundamental theories can be tested with a higher accuracy. Better atomic clocks can be built. Secondly, we know since the work of Louis de Broglie that a wave is associated with each material particle, the so-called de Broglie wave. The wave–particle duality, which was established initially for light, applies also to matter. The wavelength $\lambda_{dB}$ of this wave, called the *de Broglie wavelength*, is given, in the nonrelativistic limit, by the formula $\lambda_{dB} = h/(Mv)$, where $h$ is the Planck constant, $M$ the mass of the particle, and $v$ its velocity. Very small values of $v$ thus correspond to large values of $\lambda_{dB}$, which means that the wave aspects of atoms will be easier to observe with ultra-cold atoms than with thermal atoms. We now illustrate these general considerations by a few examples.

### 6.5.1 Atomic clocks

The principle of an atomic clock is sketched in Figure 6.15 (on the left-hand side). An oscillator (usually a quartz oscillator) drives a microwave source and its frequency

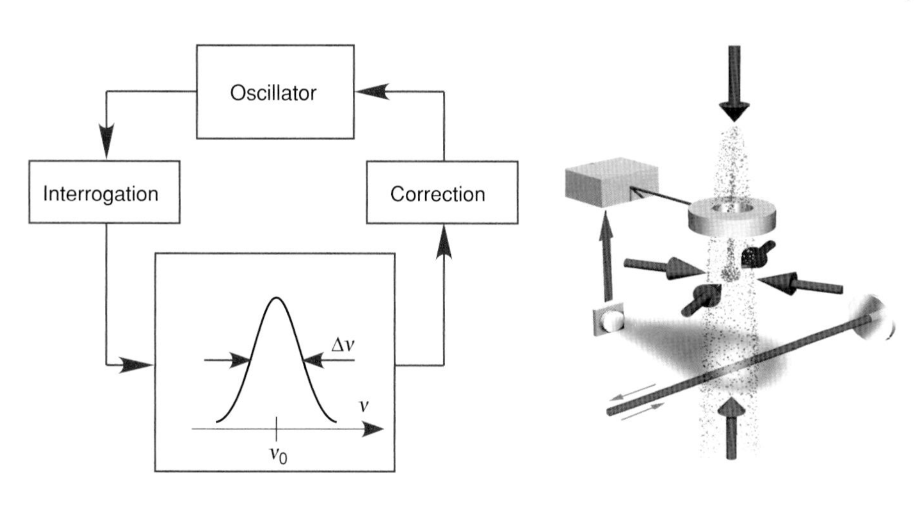


**Figure 6.15.** Atomic clocks. Left: the principle of an atomic clock. Right: an atomic fountain

is scanned through an atomic resonance line. This resonance line is centered at the frequency $\nu_0$ of a transition connecting two sublevels a and b of the ground state of an atom and has a width $\Delta\nu$. A servo loop maintains the frequency of the oscillator at the center of the atomic line. In this way, the frequency of the oscillator is locked at a value determined by an atomic frequency and is the same for all observers. Most atomic clocks use cesium atoms. The transition connecting the two hyperfine sublevels a and b of the ground state of this atom is used to define the unit of time: the second. By convention, the second corresponds to 9 192 631 770 periods of oscillations $\nu_0^{-1}$. In standard atomic clocks, atoms from a thermal cesium beam pass through two microwave cavities fed by the same oscillator. The average velocity of the atoms is several hundred m s$^{-1}$; the distance between the two cavities is on the order of 1 m. The microwave resonance line exhibits *Ramsey interference fringes*. The width $\Delta\nu$ of the central component of the signal varies as $1/T$, where $T$ is the time of flight of the atoms from one cavity to another: the larger $T$, the narrower the central line. For the longest devices, $T$ can reach 10 ms, leading to values of $\Delta\nu$ on the order of 100 Hz.

Much narrower Ramsey fringes, with sub-hertz linewidths can be obtained in the so-called Zacharias *atomic fountain* (see Figure 6.15, right-hand side). Atoms are captured in a magneto-optical trap and laser-cooled before being launched upward by a laser pulse through a microwave cavity. Because of gravity they are decelerated, so they return and fall back, passing a second time through the cavity. Atoms therefore experience two coherent microwave pulses when they pass through the cavity, the first time on their way up, the second time on their way down. The time interval between the two pulses can now be on the order of 1 s, i.e. about two orders of magnitude longer than with usual clocks. Atomic fountains have been realized for sodium by Steven Chu's group in Stanford and for cesium by the group of André Clairon and Christophe Salomon in Paris. A short-term relative frequency stability of $4 \times 10^{-14}\tau^{-1/2}$, where $\tau$ is the integration time, has recently been measured for a 1-m-high cesium fountain. This stability reaches now the fundamental quantum noise induced by the measurement process: it varies as $N^{-1/2}$, where $N$ is the number of atoms detected. The long-term stability of $6 \times 10^{-16}$ is most likely limited by the hydrogen maser which is used as a reference source. The real fountain stability, which will be more precisely determined by beating the signals of two fountain clocks, is expected to reach $\Delta\nu/\nu \sim 10^{-16}$ for a one-day integration time. In addition to the stability, another very important property of a frequency standard is its accuracy. Because of the very low velocities in a fountain device, many systematic shifts are strongly reduced and can be evaluated with great precision. With an accuracy of $2 \times 10^{-15}$, the Paris fountain is at present the most accurate primary standard. Improvement of this accuracy by a factor of ten is expected

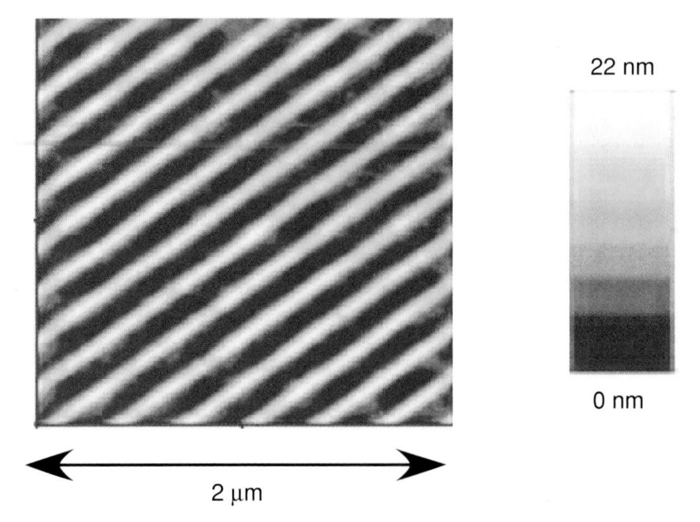


**Figure 6.16.** Atomic-force-microscopy image of lines of chromium atoms channeled by a laser standing wave and deposited onto a silicon substrate. The width of the lines is 63 nm and the period is 213 nm. (Photograph courtesy of T. Pfau and J. Mlynek.)

in the near future. In addition, cold atomic clocks designed for a reduced-gravity environment are currently being built and tested, in order to increase the observation time beyond 1 s. These clocks should operate in space in the relatively near future.

Atomic clocks working with ultra-cold atoms can of course provide an improvement of the Global Positioning System. They could also be used for basic studies. A first line of research consists in building two fountain clocks, one with cesium and one with rubidium atoms, in order to measure with high accuracy the ratio of the hyperfine frequencies of these two atoms. Because of relativistic corrections, the hyperfine frequency is a function of $Z\alpha$, where $\alpha$ is the fine-structure constant and $Z$ is the atomic number. Since $Z$ is not the same for cesium and rubidium, the ratio of the two hyperfine frequencies depends on $\alpha$. By making several measurements of this ratio over long periods of time, one could check cosmological models predicting a variation of $\alpha$ with time. The present upper limit for $\dot{\alpha}/\alpha$ in laboratory tests could be improved by two orders of magnitude. Another interesting test would be to measure with a higher accuracy the gravitational redshift and the gravitational delay of an electromagnetic wave passing near a large mass (the Shapiro effect).

## 6.5.2 Atomic optics and interferometry

### Atomic lithography

The possibility of controlling the transverse degrees of freedom of an atomic beam with laser light opens interesting perspectives in the domain of lithography. One uses the resonant dipole force created by a laser to guide the atoms of a beam and deposit them onto a substrate, where they form the desired pattern. Using, for example, a standing laser wave orthogonal to the beam axis to channel the atoms at the nodes or antinodes of the wave (depending on the sign of the detuning $\Delta$, see Section 6.3.3), several groups have succeeded in depositing regular atomic patterns (see Figure 6.16). The typical length scale of these patterns is a few tens of nanometers, which makes this technique competitive with other processes of nanolithography. Efforts are currently being made to adapt this technique to atoms of technological interest (indium and gallium).

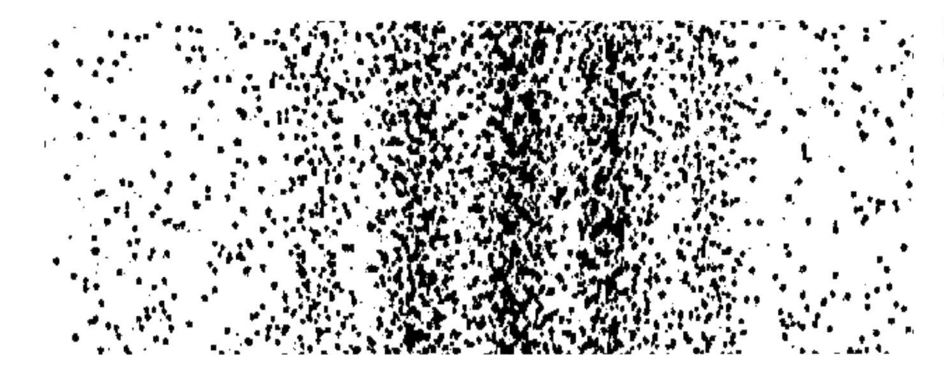

**Figure 6.17.** Young fringes observed with metastable neon atoms. (Photograph courtesy of F. Shimizu.)

## Young-slit interferometers

Because of the large value which can be achieved for atomic de Broglie wavelengths, a new field of research, atomic interferometry, has developed considerably during the past few years. It consists of extending to atomic de Broglie waves the various experiments which were previously performed with electromagnetic waves. For example, Young fringes have been observed in the laboratory of Fujio Shimizu in Tokyo by releasing a cloud of cold atoms in a metastable state above a screen pierced with two slits. The impact of the atoms on a detection plate is then observed, giving clear evidence of the wave–particle duality. Each atom gives rise to a localized impact on the detection plate. This is the particle aspect. But, at the same time, the spatial distribution of the impacts is not uniform (see Figure 6.17). It exhibits dark and bright fringes, which are nothing but the Young fringes of the de Broglie waves associated with the atoms. Each atom is therefore at the same time a particle and a wave, the wave aspect giving the probability of observing the particle at a given place.

## Ramsey–Bordé interferometers

The existence of internal atomic levels brings an important degree of freedom for the design and application of atomic interferometers. The general scheme of an interferometer using this degree of freedom is represented in Figure 6.18. The atoms are modeled by a two-level system, g being the ground state and e an excited state. They interact with two pairs of laser beams, perpendicular to the atomic beam and separated in space. We suppose that spontaneous-emission processes are negligible during the whole interaction time. An atom initially in g with a momentum $p_z = 0$ in the direction $z$ of the laser beams emerges from the first interaction zone in a coherent linear superposition of g, $p_z = 0$ and e, $p_z = \hbar k$ because of the transfer of linear momentum in the absorption process. This explains the appearance of two distinct trajectories differing in both internal and external quantum numbers in the region between the two lasers of the first pair. After the second interaction zone, there is a certain amplitude that the state e, $p_z = \hbar k$ is transformed into g, $p_z = 0$ by a stimulated-emission process while the state g, $p_z = 0$ remains unaffected (see the left-hand side of Figure 6.18). Another possibility is that the state e, $p_z = \hbar k$ remains unaffected while the state g, $p_z = 0$ is transformed into e, $p_z = \hbar k$ (see the right-hand side of Figure 6.18). Finally, the interaction with the second pair of laser beams propagating in the opposite direction can close the two paths of each interferometer. Two relevant interference diagrams therefore occur in such a scheme. In the first one (on the left-hand side of Figure 6.18), the atoms are in the ground state in the central zone between the two pairs of beams.

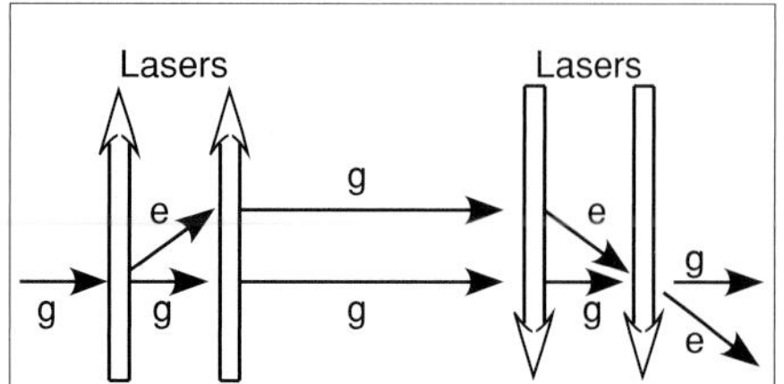


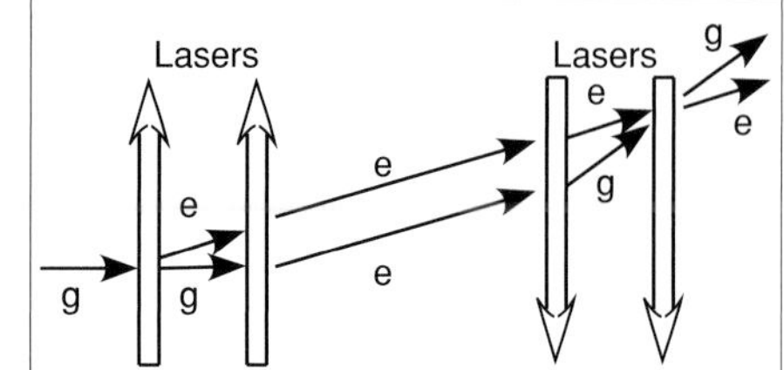


**Figure 6.18.** The two interferometers arising in the Ramsey–Bordé type geometry. Two-level atoms interact with two pairs of laser waves, which induce a coherent coupling between the two internal levels g and e. The horizontal axis may represent time as well as space. The vertical axis is used to represent the recoil of an atom after the absorption or the stimulated emission of a photon. Depending on the experimental conditions, this interferometer can be used to measure rotations perpendicular to the plane of the figure or accelerations. If one adjusts the frequency of the lasers to maximize the interference signal, it constitutes the prototype of an optical clock.

In the second interferometer (on the right), the atoms are in the excited state in the central zone. Note that the four interactions are separated here in space. The scheme can be easily transposed to a situation in which these four interactions are separated in time, the atom being initially in g at rest.

Depending on the geometry of the experiment, the two interferometers of Figure 6.18 are sensitive to the frequency of the exciting lasers (thus forming an atomic clock in the optical domain), to the acceleration of the system, or to its rotation (a gyroscope). In all cases, cold atoms have brought an important increase in sensitivity, thanks to their small velocity. Consider for example the case of the detection of a rotation around an axis perpendicular to the plane of the interferometer. The measurement is based on the Sagnac effect: the rotation modifies the difference in length between the two interfering paths of Figure 6.18. The best demonstrated sensitivity of atom interferometers to rotation is $6 \times 10^{-10}$ rad s$^{-1}$ for an integration time $T = 1$ s, and the sensitivity improves as $\sqrt{T}$. This result, obtained by the group of Mark Kasevich, is comparable to the best optical gyroscopes. One can show that the sensitivity of these atomic gyroscopes is higher than the sensitivity of laser gyroscopes with the same area between the two arms of the interferometer by a factor that can be as large as $mc^2/(h\nu)$, where $m$ is the mass of the atom and $\nu$ the frequency of the photon, a factor that can reach values on the order of $10^{11}$.

These atomic interferometers can also be used to measure fundamental constants accurately. For example, the measurement of the interference pattern as a function of the frequency of the lasers reveals two different resonances for the two interferometers shown in Figure 6.18. The difference in frequency between the two resonances is related to the recoil shift $\hbar k^2/m$, where $k$ is the wavevector of the light and $m$ the mass of the atom. The measurement of this recoil shift, which has been performed by Steven Chu's group at Stanford, can be combined with the measurement of the Rydberg constant and the ratio between the proton and the electron masses $m_p/m_e$. It then yields a value of the fine-structure constant $\alpha$ with a relative accuracy of $10^{-8}$. The precision of this method, whose advantage is that it does not depend on quantum-electrodynamics calculations, is comparable to those of the other most accurate methods.

## 6.6 Concluding remarks

The manipulation of atomic particles by electromagnetic fields has led to spectacular new results during the past two decades. The combination of the trapping and cooling methods described in this chapter allows the temperature of atomic gases to be decreased by several orders of magnitude and to reach the sub-microkelvin region. Conversely, the thermal wavelength $\lambda_T$ of the particles is increased by four orders of magnitude with respect to its room-temperature value. It reaches values on the order of an optical wavelength when the atoms are cooled at the recoil temperature.

For these low temperatures and large wavelengths, the quantum features of the motion of the atomic center of mass become essential. For example, an assembly of cold atoms placed in an optical lattice is a model system for the study of quantum transport in a periodic potential. This allows one to draw profound and useful analogies with condensed-matter physics, where one deals with the motion of electrons or holes in the periodic potential existing in a crystal. Another field of application of these large wavelengths is atomic interferometry, which is now becoming a common tool for realizing ultra-sensitive sensors of acceleration or rotation.

Systems of cold atoms have played an important role in the development of theoretical approaches to quantum dissipation based either on a master equation for the density operator of the system, or on a stochastic evolution of its state vector. The study of sub-recoil laser cooling has also brought some interesting connections with modern statistical-physics problems, by pointing out the possible emergence of Lévy flights in the dynamics of these atoms.

When the spatial density $\rho$ of the laser-cooled atoms increases, collective processes occur. The formation of molecules assisted by light, or *photoassociation*, has been a very fruitful theme of research whereby new information could be collected from these ultra-cold molecular systems. Conversely this formation of molecules in laser-cooled atomic gases limits the achievable spatial density. This has so far prevented one from reaching the threshold for quantum degeneracy ($\rho\lambda^3 \geq 1$) with purely optical cooling: Bose–Einstein condensation of a bosonic-atom gas is observed only when a final step of evaporative cooling is used, after an initial pre-cooling provided by the optical methods described above (see Chapter 7).

To summarize, the manipulation of atoms with light is nowadays a tool that is encountered in most atomic physics laboratories. It is a key step in the production of quantum degenerate gases, both for bosonic and for fermionic species – see Chapter 7. It plays a central role in modern metrology devices, and is at the heart of the new generation of atomic clocks. It is also an essential element for the practical implementation of quantum information concepts (see Chapter 10). Thanks to the development of miniaturized systems, cold atoms can be used in very diverse environments, on Earth or even in space: the time provided by a cold atomic clock will soon be available from the International Space Station!

## FURTHER READING

1. A more detailed presentation can be found in the texts of the three Nobel lectures of 1997: S. Chu, *Rev. Mod. Phys.* **70** (1998) 685, C. Cohen-Tannoudji, *Rev. Mod. Phys.* **70** (1998), 707, and W. D. Phillips, *Rev. Mod. Phys.* **70** (1998) 721.
2. C. S. Adams and E. Riis, Laser cooling and trapping of neutral atoms, *Progr. Quantum Electron.* 21 (1997).
3. E. Arimondo, W. D. Phillips, and F. Strumia (eds), *Laser Manipulation of Atoms and Ions*, Proceedings of the 1991 Varenna Summer School, Amsterdam, North Holland, 1992.
4. H. Metcalf and P. van der Straten, *Laser Cooling and Trapping* Springer, 1999.
5. [5] C. Cohen-Tannoudji, J. Dupont-Roc, and G. Grynberg, *Atom–Photon Interactions – Basic Processes and Applications*, New York, Wiley, 1992.
6. M. O. Scully and M. S. Zubairy, *Quantum Optics*, Cambridge, Cambridge University Press, 1997.

## Claude Cohen-Tannoudji

Born in Constantine, Algeria (then part of France), Claude Cohen-Tannoudji studied physics at the Ecole Normale Supérieure (ENS) in Paris, where he went on to do diploma work under Alfred Kastler. After military service, he returned to the ENS for further research under Kastler and Jean Brossel on atom–photon interactions. He was awarded his Ph.D. in 1962 and obtained a teaching position at the University of Paris. In 1973 he was appointed Professor of Atomic and Molecular Physics at the Collège de France, where he formed an influential group investigating new methods of laser trapping and cooling. These went on to open up the fruitful study of physics in the microkelvin and nanokelvin range. He has played an important part in the teaching of modern quantum physics in France, and with his collaborators has written standard books on quantum mechanics, quantum electrodynamics, quantum optics, and Lévy statistics, together with about 200 theoretical and experimental papers on various problems of atomic physics and quantum optics. In 1997, he shared the Nobel Prize for Physics with Steven Chu and William D. Phillips for his work on the development of methods to cool and trap atoms with laser light.

## Jean Dalibard

Following his studies at the Ecole Normale Supérieure (ENS) in Paris, Jean Dalibard began his research career at the ENS Kastler–Brossel laboratory, where his thesis work on the laser cooling of atoms was under the supervision of Claude Cohen-Tannoudji. As well as making important contributions to the physics of atoms and photons and to quantum optics, he has also worked with Alain Aspect on experiments to test fundamental quantum mechanics. More recently he has been investigating the detailed properties of Bose–Einstein condensates.

He is research director at the French Centre National de la Recherche Scientifique and Professor at the Ecole Polytechnique, Palaiseau. He was invited to the US National Institute of Standards and Technology in 1991 and has also taught at Innsbruck, the Technion in Haifa, Tokyo, and Connecticut. He was awarded the Gustave Ribaud Prize of the French Academy of Science in 1987, its Mergier Bourdeix Prize in 1992, and the Jean Ricard Prize of the French Physical Society in 2000.

# 7 The quantum world of ultra-cold atoms

William Phillips and Christopher Foot

## 7.1 Introduction

This chapter describes remarkable new experiments that have been made possible through recent advances in laser cooling of atoms. Cooling of atoms in a gas by laser light is in itself surprising and this technique is important in a range of applications. In particular the experiments described here take atoms that have been laser cooled and then use other methods to reach even lower temperatures, at which wonderful and fascinating quantum effects occur. As well as revealing new phenomena, these experiments explore the deep analogies between the roles played by light waves in lasers and matter waves. These techniques are widely accessible and enable skilled researchers in laboratories all over the world to explore exciting new physics. The key physics and techniques of laser cooling are described in depth in Chapter 6.

## 7.2 What is "temperature"?

We use the words "cold" and "hot" every day; for example to refer to the temperature of the air outside. But what do "cold" and "hot" mean for the air's molecules? Broadly speaking, we can say that atoms in a hot gas move faster than those in a cold gas and that cooling, by taking energy out of the gas, slows the atoms down. In this description of the gas as a collection of atoms behaving like billiard balls, or little hard spheres, the lowest possible temperature occurs when the atoms stop moving but, as we shall see, this picture does not give an accurate description at very low energies, for which quantum effects become important. In this chapter we will look at the fascinating properties of atomic gases cooled to temperatures within one millionth of a degree Kelvin from absolute zero. We shall use the term "ultra-cold" to refer to such extremely low temperatures that have only recently been achieved in experiments on atomic gases. The techniques used to cool liquids and solids have a longer history and have made available a different set of possibilities (see Chapter 8). Note that most of this chapter refers to gases in which the individual particles are atoms, but the general statements about temperature scales apply equally well to gases of molecules, e.g. air, whose major constituents, oxygen and nitrogen, exist as diatomic molecules.

*The New Physics for the Twenty-First Century*, ed. Gordon Fraser.
Published by Cambridge University Press. 

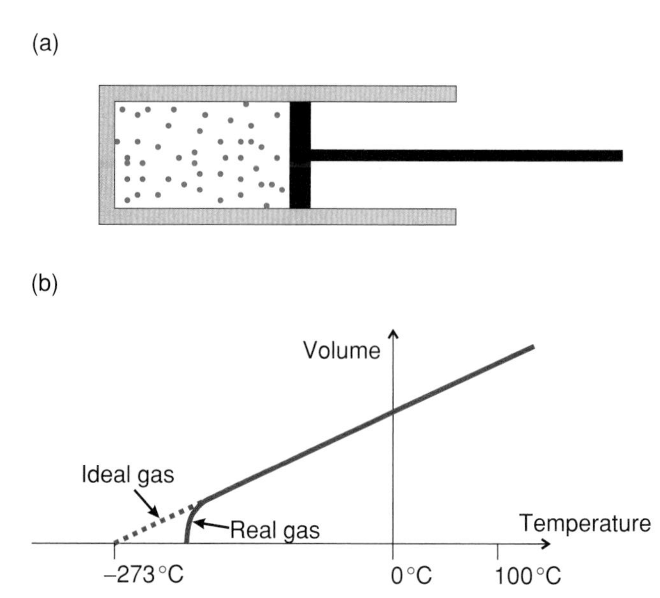


**Figure 7.1.** (a) A cylinder with a piston contains a sample of gas – this apparatus resembles a bicycle pump but, rather than changing the volume of the gas by exerting pressure on it, as when you pump up a flat tyre, we consider the situation in which the temperature of the gas is changed, by heating or cooling the piston cylinder, at a constant pressure, i.e. the piston is allowed to move freely. (b) The results of a constant-pressure experiment show how the volume of the gas changes as its temperature varies. Gases undergo much greater thermal expansion, or contraction, than solids or liquids. The data points shown in the range 0–100 °C indicate typical results for a gas and extrapolation of this straight line to lower temperatures suggests that the gas contracts to zero volume at about -273 °C. (See also the later section in this chapter on quantum effects.)

Before we proceed to describe how to cool atoms, we should first discuss what temperature means and the concept of an absolute zero of temperature. Historically people defined various practical scales of temperature by measurements using thermometers, e.g. the commonly used mercury thermometer contains liquid mercury that expands along a glass capillary tube as it heats up. To calibrate the thermometer it is placed in water at freezing point, i.e. a mixture of ice and water, and a mark is made at the position of the mercury that will be zero on the scale. The thermometer is then placed in boiling water (at normal atmospheric pressure) and another mark put on the scale; dividing the distance between the two marks into 100 equal graduations gives the Celsius scale. With this instrument we can assign a temperature in the usual way. For example, to determine the temperature of the air outdoors, we wait until the heat flow into, or out of, the thermometer brings it to the same temperature as the surrounding air; the reading on the instrument then gives the temperature. This practical definition of temperature is somewhat arbitrary and both the Celsius and Fahrenheit scales (for which a different arbitrary zero is chosen) are used by weather forecasters in different parts of the world. Physicists prefer a more fundamental scale of temperature defined through consideration of the thermal expansion of a gas, rather than of a liquid like mercury. Consider the experiment shown in Figure 7.1(a) in which a cylinder with a piston contains a sample of the gas at a constant pressure and the temperature of the gas is changed. Figure 7.1(b) shows a plot obtained in such an experiment. Extrapolation of the straight line to lower temperatures suggests that a gas has zero volume at about −273 °C, which is defined as the zero of the Kelvin temperature scale.

In this chapter we will look at recent experiments that have cooled atomic vapors to within 1/1 000 000th of a degree from this absolute zero (less than a microkelvin). These ultra-cold atomic vapors have a very low pressure, so the atoms remain far apart, thus preventing, or at least slowing down, the processes by which atoms stick together to form molecules, or condense into a liquid. Real gases at close to atmospheric pressure follow curves like the dotted line in Figure 7.1(b); as the volume decreases the particles come closer together and the attractive force between molecules becomes stronger, until eventually the gas condenses into a liquid, e.g. nitrogen liquefies at −196 °C, equivalent to 77 K. (This effect can be strikingly demonstrated by putting an ordinary rubber balloon, filled with air, into a bucket of liquid nitrogen – with suitable safety precautions – when the balloon is taken out it is as flat as a pancake, but it expands back to its original size as the air warms up.) All gases follow the same behavior under conditions where they are far from becoming a liquid, or solid, e.g. for nitrogen at room temperature, but for steam the temperature needs to be much greater than 100 °C, at atmospheric pressure. Gases obey this universal behavior no matter whether their constituent particles are molecules or individual atoms, and in this general discussion statements about "molecules" in gases also apply to atomic vapors that are the main theme of this chapter.

From such experiments, physicists deduced the behavior of a perfect gas, i.e. the theoretical ideal case in which the interactions are negligible in comparison with the thermal energy of the moving molecules and the particles collide with each other like perfectly elastic billiard balls. At constant pressure, the volume of such a perfect gas would contract to zero at the absolute zero of temperature and all the particles would be stationary at a single point in space. Obviously, this cannot happen in real gases because the molecules have a finite size and interact with each other. It is impossible even in principle because of quantum effects: Heisenberg's uncertainty principle states that we cannot know simultaneously the precise values of the momentum and position of particles; a gas that has zero energy and volume would violate the principle since the particles would have zero momentum and their position would be precisely known. Later in this chapter, we will see that the description of a gas in terms of classical particles ceases to be valid at low energies, for which the wavelike nature of particles becomes important – such wave–particle duality is a well-established feature in quantum mechanics and leads to an uncertainty in the positions of individual atoms as illustrated in Figure 7.2.

For many years, the gas that is closest to the ideal was considered to be helium, because the interactions between helium atoms are so weak that it remains gaseous down to 4 K. Below about 2 K the quantum effects due to the wavelike nature of the atoms cause liquid helium to become a superfluid and it exhibits the range of fascinating phenomena described in Chapter 8. We shall see that dilute atomic gases are also superfluids with many intriguing properties. In the 1980s it was realized that theoretically hydrogen atoms in a strong magnetic field should remain a gas right down to absolute zero because the field keeps all the magnetic moments of the atoms aligned in the same direction (the magnetic moment of a hydrogen atom arises mostly from its single electron). Hence the effect of the magnetic field is to keep the spins of all the atoms in the same direction (anti-parallel to the direction of their magnetic moments), and this polarization of the spins prevents the hydrogen atoms combining with each other to give molecules – hydrogen molecules form when the electrons on the two atoms that come together have spins in opposite directions: spin up and spin down. (A similar situation arises for the two electrons in the ground state of the helium

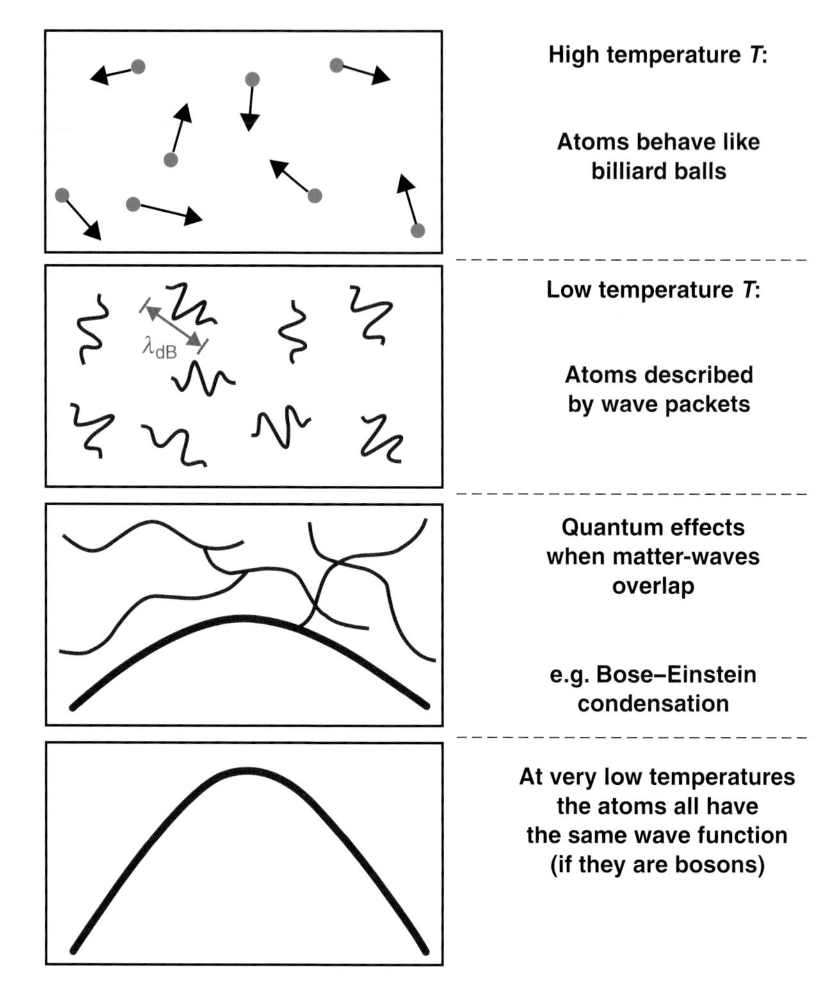


Figure 7.2. A pictorial representation of atoms in a gas at various temperatures. At high temperatures the atoms behave like particles bouncing about, such as billiard balls. As the temperature decreases the wavelength associated with each atom increases until eventually they overlap with each other; in this regime the atoms can no longer be considered as separate entities (individual particles) and quantum effects occur. (The de Broglie wavelength of a particle is introduced later and in Chapter 6. It represents approximately the size of the region of space in which there is a high probability of finding the particle.) Adapted from a picture produced by Wolfgang Ketterle's group at MIT.

atom – giving a stable electronic configuration in which the two electrons have opposite spins so that they may both enter the lowest atomic energy level.)

Alkali-metal atoms such as sodium and rubidium in magnetic traps share this special property of spin-polarized hydrogen, to some extent. Unlike the case of hydrogen, the formation of alkali molecules is not completely inhibited; however, the process occurs so slowly that sodium and rubidium atoms can exist for many seconds as a dilute gas at near zero temperature. (Experiments in which an ultra-cold gas of alkali-metal atoms has been converted entirely into a molecular gas are described later.) The stabilization by a magnetic field that allows vapors of alkali metals to exist at such low temperatures is really very surprising and few people appreciated that it was possible until after the pioneering experiments had been carried out. A crucial factor in the success of the work with alkali metals has been laser cooling of the atoms and a brief summary of this powerful new technique is given in the following section.

## 7.3 Laser cooling

At room temperature atoms in a gas whiz around at a speed comparable to the speed of sound in air, 345 m $s^{-1}$. This is not a coincidence but comes from the fact that sound is transmitted by molecules knocking into one another. Chapter 6 describes the methods

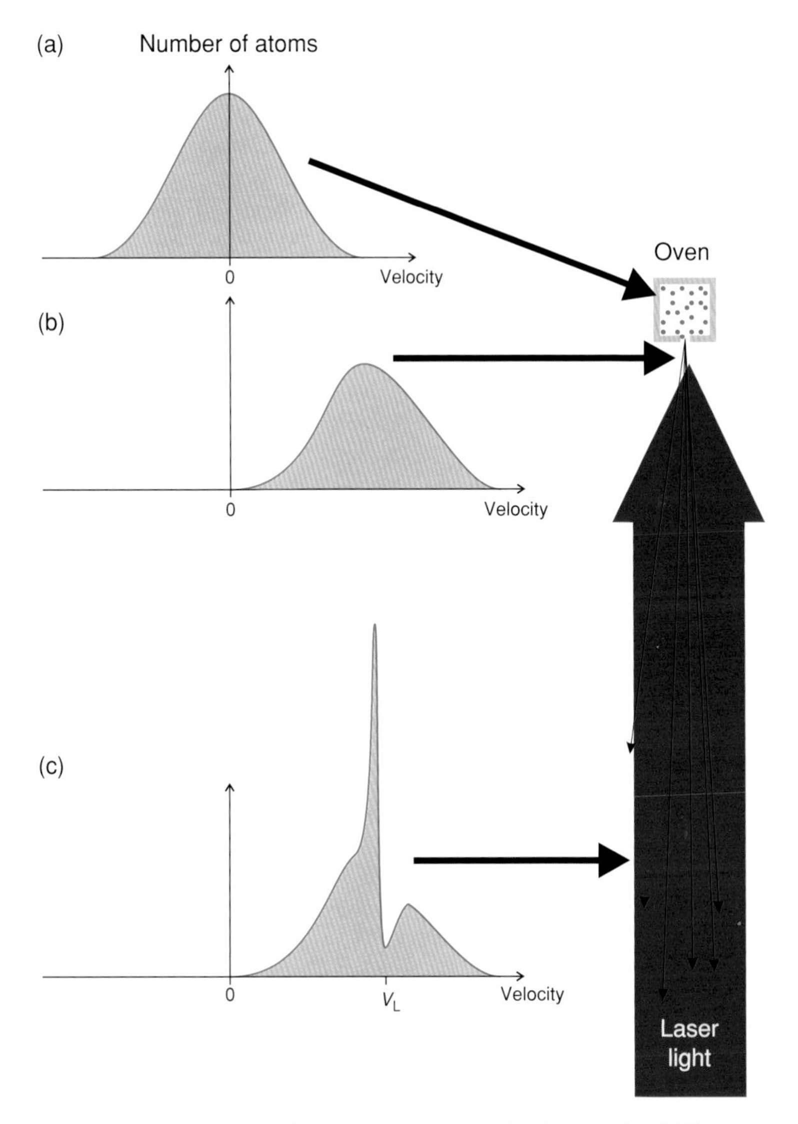


**Figure 7.3.** How the velocity distribution of atoms changes during laser cooling. (a) The number of atoms with a given velocity in a gas, along a particular direction, has the bell-shaped distribution shown. There are equal numbers of atoms with positive and negative velocities and the average velocity is zero. (b) An atomic beam, created by allowing some atoms to escape through a hole in the wall of a container of gas, has a one-sided velocity distribution. Once atoms have passed through the exit hole they travel in straight lines, with their original velocity, provided that the region outside the container has a good vacuum so there are negligibly few collisions. All the atoms move in approximately the same direction in this atomic beam. (c) A strong laser beam in the direction opposite to the atom's motion changes the velocities of some of the atoms. Laser light of a particular frequency interacts strongly with atoms that have velocities within a certain narrow range. These atoms absorb light and decelerate until their velocity is no longer within the narrow range where light is absorbed. Effectively the light "burns a hole" in the velocity distribution around $V_L$.

used to slow these atoms from speeds comparable to that of a jet plane down to a snail's pace. We shall not repeat the full story here, but Figure 7.3 briefly summarizes some of the important points.

The absorption of the radiation depends on the velocity of the atoms because of the Doppler effect, and this allows control of atomic velocities by laser light. The Doppler effect is familiar from the everyday example of the change in pitch of the sound from a passing car, or train, i.e. the noise is shifted to a higher frequency as the vehicle approaches and then to a lower frequency as it moves away from the listener. Of more relevance to atoms is the Doppler effect for light, e.g. in astronomy the light emitted by distant stars has a redshift because they are moving away from observers on Earth. Figure 7.3(c) shows that, because of the Doppler effect, light at a single frequency from a laser affects atoms only over a narrow range of velocities; the light exerts a force on these atoms so their velocity decreases until they no longer interact with the light. The light does not affect atoms with other velocities, so in the resulting distribution the atoms pile up on the edge of the hole (on the lower-velocity side). In order for the light to interact with most of the atoms in an atomic beam the experiments must compensate for the changing Doppler shift as the atoms slow down. This is achieved either by changing the laser frequency with a magnetic field to track the atoms as they slow down, or by altering the atomic resonance frequency using the Zeeman effect of an applied magnetic field (as described in Chapter 6). In this way alkali-metal atoms such as sodium and rubidium from an oven (at above room temperature) are slowed from velocities of hundreds of m $s^{-1}$ down to a few m $s^{-1}$ over a distance of about 1 m, which is a reasonable length for an apparatus. Such atoms, or indeed any atoms with velocities less than a few tens of m $s^{-1}$ produced by other methods, can be cooled further by another technique that uses counter-propagating laser beams at the same frequency (see Chapter 6).

As described in Chapter 6 by Cohen-Tannoudji and Dalibard, early experiments on the optical-molasses technique found that it worked much better than expected! Atoms were cooled to much lower temperatures than had been predicted by simple calculations based on the Doppler effect and a two-level atom. Even with the more subtle laser-cooling process in optical molasses, atoms do not go below the so-called recoil limit to the temperature. The recoil limit arises from the spontaneous photons that the atoms emit as they decay back down to their ground state after absorption of a photon from the laser beam. Recoil from each spontaneously emitted photon gives the atom a kick in a random direction and changes the atomic velocity by an amount called the recoil velocity (that can easily be calculated from conservation of momentum). These random kicks prevent the atom from coming entirely to rest and a simple theory of this process in terms of a random walk (of the velocity) shows that the atom's velocity cannot be reduced below the recoil velocity – an intuitively reasonable result since the velocity changes in steps of this recoil velocity. (Fluctuations in the impulses received from the absorbed photons also contribute to the heating.) At the recoil limit sodium atoms have a temperature of only a few microkelvins.

The spontaneous emission also limits the density of atoms that can be achieved in laser-cooling experiments, e.g. when the cloud of cold atoms becomes so large and dense that a photon emitted by an atom near the center of the cloud is absorbed by other atoms on its way out; the scattering of this photon within the cloud gives additional kicks to other atoms and hence leads to a radiation force pushing the cloud apart (plus additional heating). An analogous situation occurs in stars, where the outward radiation pressure counterbalances the inward gravitational force to stop the stellar cloud collapsing. People have found various ways to cool atoms below the recoil limit and also to overcome the limitations to the temperature and density related to scattering of light within the cloud; however, it has turned out that those methods are not essential for the experiments that we shall describe here because the temperature at which atoms can be trapped by magnetic fields is easily within the reach of laser cooling. Once the atoms are trapped the simple but very effective method of evaporative cooling works extremely well, as described later.

## 7.4 Magnetic trapping

We cannot put laser-cooled atoms into a conventional container such as the cylinder shown in Figure 7.1 since there is no way of making the container cold enough, and if it were cold enough the atoms would stick to the walls. Physicists have long used magnetic forces to deflect and guide atomic beams and these techniques have been adapted to build a "magnetic bottle" for cold atoms that holds the atoms close to the center of an ultra-high-vacuum chamber; this keeps the atoms isolated from the environment and so acts like a vacuum flask used both for hot coffee and also in a more sophisticated form in the laboratory to store cryogenic liquids, e.g. liquid nitrogen and liquid helium. Even the strongest gradient of the magnetic field that can be produced in the laboratory, however, gives only a weak force on atoms. Therefore these "magnetic bottles," or magnetic traps as they are more commonly called, cannot confine atoms at room temperature and it was not until the advent of laser cooling that magnetic traps became important – immediately following the first laser-cooling experiments at the NIST, Gaithersburg the experimenters showed that the cold sodium atoms could be held in a simple magnetic trap. In contrast, ions (atoms that have lost, or gained, electrons to become electrically charged) experience much stronger electrostatic forces, at moderate electric fields, that enable hot ions to be trapped and then cooled afterwards. Such ion-trapping techniques are important in atomic physics and in quantum computation, see Chapter 11.

We shall attempt to explain the principles of magnetic trapping using only "well-known" properties of magnets; however, the reader who has taken an introductory course in electromagnetism may find that writing down some equations makes things easier. An atom with a magnetic moment can be represented as a small bar magnet that can move around freely and we shall consider how to confine this free magnet with larger magnets arranged at fixed positions, as illustrated in Figure 7.4.

None of the arrangements shown in Figure 7.4 is stable. They are analogous to trying to balance a pencil upright on its point; once it starts to topple it falls faster and faster. Consideration of these cases gives physical insight into the problem, in a non-mathematical way, but you still might think that there is some cunning arrangement that will work. It can be proved that there is not. The equations of electromagnetism show that there is no stable equilibrium for a (freely moving) magnetic dipole in a static magnetic field.

A scientific toy that furnishes a more realistic analog of magnetic trapping of atoms is the Levitron shown in Figure 7.5. In this device, a small magnet floats above a larger magnet in the base of the device. The upward force arises from the repulsion of the two magnets, as in Figure 7.4(c), but in the Levitron the floating magnet does not turn over because it spins about its axis like a gyroscope – recall that unstable equilibrium is like trying to balance a pencil upright on its point, but a spinning top stays upright (while it rotates sufficiently fast). The floating magnet is in the shape of a ring, or a disk, with an axle through the center that is used to set it spinning. Atoms have spin that has similar properties to the angular momentum of a spinning top, but there are some important differences between a classical spinning object and the quantum-mechanical systems. The spin of a quantum object takes only values that are integer, or half-integer, multiples of $\hbar$ (Planck's constant), and this quantization means that the angular momentum of atoms in a magnetic trap stays fixed, i.e. the discreteness of the allowed values does not allow atoms to "slow down." (The spinning magnet in the Levitron slows down because of the air resistance until the rate of rotation drops below some critical value and the system is no longer stable.) The orientation (angle) of the spin angular momentum relative to a magnetic field is quantized, as

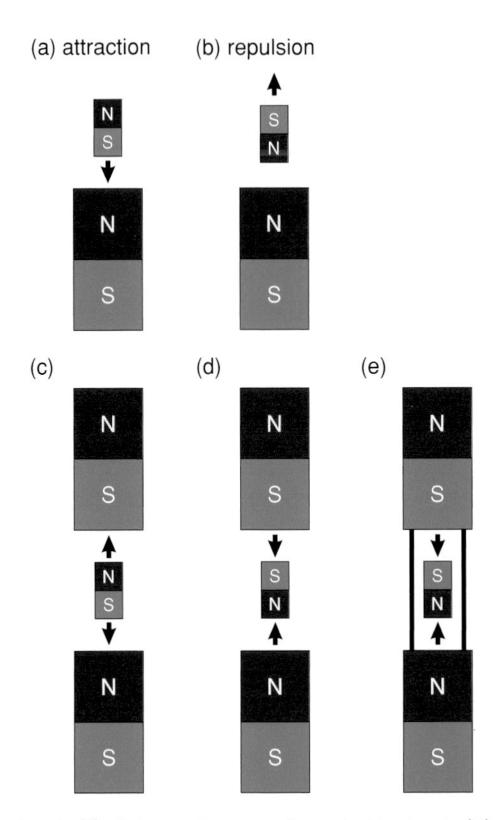


**Figure 7.4.** (a) The South pole (S) of the small magnet faces the North pole (N) of a larger magnet so there is a strong attractive force that pulls the small magnet into contact with the larger one. (b) When like poles face each other, S and S (or N and N), you might expect that the repulsion would cause the magnets to fly apart. Actually, if you try this you will find that the small bar magnet twists around so that the situation becomes like that in (a). Neither (a) nor (b) keeps the small magnet in a fixed position (with the magnets not in contact), but a more symmetrical arrangement might seem more promising. (c) The two big magnets have equal size (and strength) so that midway between them the small bar magnet experiences equal and opposite forces on each end. These forces balance each other; however, this does not lead to a stable equilibrium. Any small displacement from the midpoint toward one of the larger magnets leads to stronger attraction that increases the displacement further in the same direction, so the bar magnet crashes into one of the larger magnets as in case (a). (d) Another case, is that in which the small bar magnet has the opposite orientation to that in (c), so that for small displacements from the mid-point there is a repulsive force that pushes the bar magnet out from between the others (or the bar magnet twists around). (e) Stable equilibrium can be obtained with a magnet that slides inside a tube. In this apparatus the magnet cannot twist around or move sideways and the magnetic confinement works in one direction only (along the tube). When the tube is vertical, only the lower magnet is needed to provide an upward force that balances the downward force of gravity, and this arrangement whereby a small ring magnet "floats" above the lower magnet is used in a common "scientific toy." (Such devices are often constructed using a ring magnet that slides along a rod through its center, instead of being inside a tube.)

well as its magnitude, and only certain orientations are allowed, e.g. spin up and spin down for spin $\frac{1}{2}$ such a for the electron. Atoms may have larger values of angular momentum when the contributions from all the electrons (outside closed shells) and the nuclei of the atoms are added together, which means that more orientations are possible. Such complications, however, do not affect the basic principles explained here for the simple case in which the spin equals $\frac{1}{2}$ (in units of $\hbar$). The important point is

**Figure 7.5.** In the Levitron (R), a ring magnet is levitated above a larger magnet in the base of the device. The upward force arises from the repulsion of the two magnets; the floating magnet does not turn over because it is spinning about its axis so that it behaves like a gyroscope. (Photo courtesy of Graham Quelch.) Levitron (R) is a registered trademark of Fascinations.

that the atom's spin, and its magnetic moment, does not flip over, but stays at a fixed angle relative to the magnetic field (as for the spinning Levitron). Most atom traps use coils of wire carrying high currents rather than permanent magnets, as shown in Figure 7.6, since this allows the magnetic field to be switched on and off rapidly.

The trapped atoms undergo harmonic oscillation (like that of a pendulum) and the frequency of this sinusoidal motion is independent of the amplitude. In this motion, the atoms slow down as they move away from the center of the trap and come to rest at the extremity of the oscillation, where all the energy has the form of potential energy; the atoms then start to move back toward the center of the trap. Atoms with the highest energy make the largest excursions from the center, i.e. they go the furthest up the sides of the trap. These general features apply to the three-dimensional motion in magnetic traps where the atoms oscillate at different frequencies in different directions; in the Ioffe–Pritchard trap atoms oscillate more slowly along the axis than in the direction perpendicular to the four straight wires.

## 7.5 Evaporative cooling

Evaporative cooling of atoms in a magnetic trap operates on the same principle as the cooling of a cup of coffee. The coffee cools down because the hot water molecules that escape as steam carry away energy from the liquid in the cup. In experiments with magnetic traps the most energetic atoms are allowed to escape by reducing the depth of the trap as illustrated in Figure 7.7. Atoms that escape carry away an above-average amount of energy, so the remaining atoms have less average energy than the initial cloud.

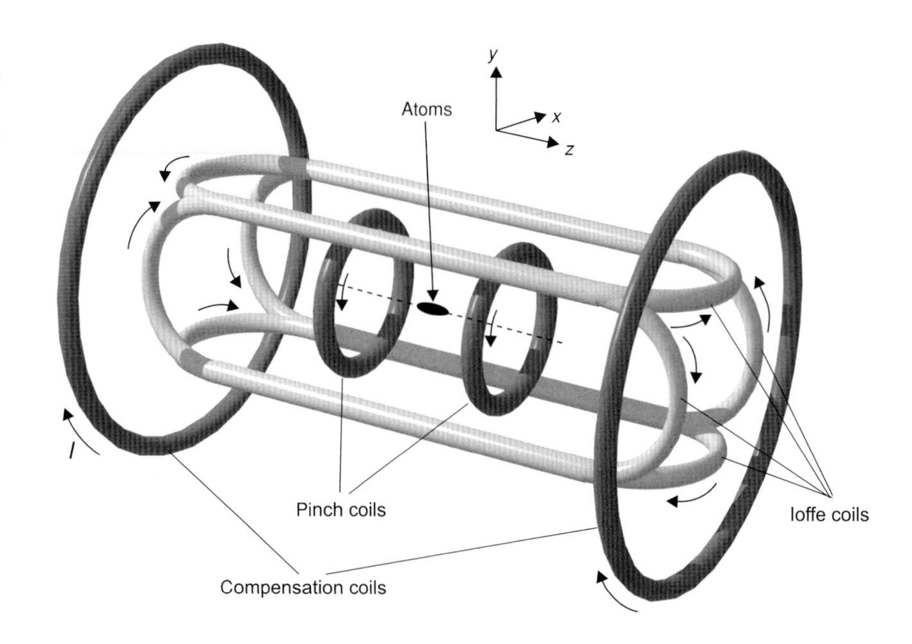


**Figure 7.6.** A magnetic trap. The four straight sections of wire create a magnetic field that is zero along the central line of the wires. (In each wire the current flows in the direction opposite to that of adjacent wires, as indicated in the figure.) Atoms feel a force pushing them toward this central line if their magnetic moments have the correct orientation with respect to the magnetic field, and this configuration acts like a "magnetic tube" that confines the atoms. This "tube" is sealed off at both ends of the trap by the field produced by two circular coils with currents in the same direction. These circular "pinch" coils act like the magnets in Figure 7.4(e) that repel the atom. This configuration is called a Ioffe–Pritchard trap and the confined atoms congregate in a long cigar-shaped cloud along the axis. (Image courtesy of K. Dieckman, AMOLF [Institute for Atomic and Molecular Physics], Amsterdam.)

Evaporation is described in the caption to Figure 7.7 as if it were carried out as a series of steps followed by periods to allow the atoms to re-establish equilibrium, which is convenient for explaining the principle of the technique. In practice, the trap depth is reduced in an almost continuous series of small steps that continually cut away atoms at the outside of the trapped cloud. This so-called "forced" evaporation can be likened to blowing on a cup of coffee to cool it down faster. In the first experiment on Bose–Einstein condensation, described below, the final sample contained only 2000 rubidium atoms but they could still be readily observed. More recent experiments typically start with more than a billion ($10^9$) atoms of rubidium, or sodium, in the magnetic trap at a temperature of tens of microkelvins and more than a million atoms remain after evaporation to temperatures below a microkelvin. Alkali-metal vapors have been evaporatively cooled to temperatures of less than a nanokelvin, but it is not generally necessary to go to such low temperatures in order to see quantum effects.

## 7.6 Quantum effects

Quantum effects have a very different nature from the classical behavior of large objects and trying to explain such effects strains everyday language – the correct language to use is mathematics, but we shall not resort to that here. Generally speaking, we think of atoms as particles at high temperatures, but atoms with low energies behave like waves. In the early days of quantum mechanics this was called wave–particle duality, and Louis de Broglie introduced the idea that the waves associated with an object have a wavelength $\lambda$ given by $\lambda = h/p$; here $h$ is Planck's constant and $p$ is the momentum. For particles the momentum equals mass times velocity, $p = mv$. This de Broglie relation also applies to photons, which can be regarded as the particle-like manifestation of light.

As already mentioned, at room temperature the atoms in a gas behave like billiard balls bouncing off the walls of the container and colliding with each other. This description allows accurate calculation of the relation among the pressure, volume, and temperature for a gas. The properties of real gases differ slightly from those of an ideal gas of hard elastic spheres because atoms have more complicated interactions. (The force

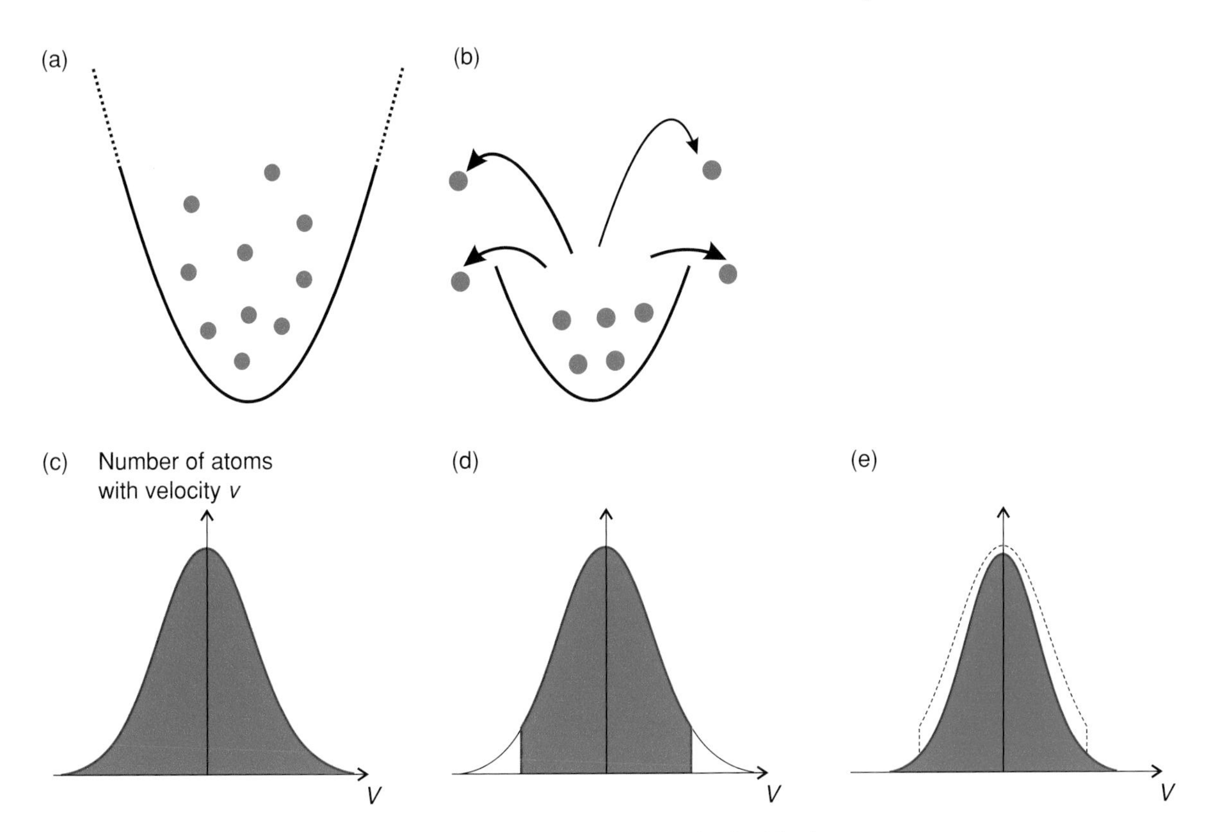


**Figure 7.7.** Evaporative cooling reduces the temperatures of trapped atoms by the following steps. (a) The trapped atoms are initially contained in a deep trap. (b) A reduction of the trap depth allows atoms with higher than average energy to escape. (c) Initially the atoms have a distribution of velocities with the typical shape for a gas. Note, however, that the spread of velocities of the trapped atoms is small because magnetic traps are not strong enough to hold uncooled atoms and this figure is on a much smaller scale than Figure 7.3. (d) A sudden reduction of the trap depth chops off the wings of the distribution – the atoms that escape are those that have the highest velocity at the center and go furthest up the sides of the potential well. A sharp edge in the velocity distribution soon smoothes out as the atoms collide with each other and the wings of the distribution grow again to re-establish the equilibrium distribution but with a lower temperature than before the evaporation. The mean energy per atom is the same in (d) and (e); however, temperature has a well-defined meaning only for systems that are in thermal equilibrium; therefore the distribution in (d) cannot be assigned a temperature. In a trap with a finite depth there will always be a few atoms with exceptionally high energy that escape; however, the rate of cooling becomes very slow when the mean energy of the gas becomes much less than the depth of the potential. Therefore, to make evaporation occur quickly, the depth of the well is reduced again. In most experiments, the magnetic potential remains constant and radio-frequency radiation is used to induce the atomic spin to flip (change direction) at a certain distance from the center of the trap – these atoms then fly out of the trap. This removal of the atoms that have a resonant interaction with the radio-frequency radiation gives a very precise process for cutting away atoms from the outside of the cloud.

between two atoms is attractive when they are far apart but repulsive at short range.) As the temperature decreases the average de Broglie wavelength of the atoms increases; this wavelength characterizes the size of the region within which there is a significant probability of finding an atom, so we can regard the particles as becoming fuzzy or delocalized. At a sufficiently low temperature the quantum fuzziness of the atoms

### BOX 7.1 INTERNAL AND EXTERNAL DEGREES OF FREEDOM

The internal degrees of freedom of atoms are associated with the motion of the electrons around the nucleus, which in a simple quantum model are treated like planets orbiting around the Sun, leading to the well-known energy levels and shell structure of atoms. However, to describe the behavior of a gas of atoms we can, for the most part, ignore their internal structure and just consider them as particles that move and collide with each other. The temperature depends on how the atoms as a whole move, i.e. their center-of-mass motion. The electrons within the atoms might not be in the lowest internal energy state, but this does not matter so long as that internal state does not change. This assumption breaks down at extremely high temperatures, such as at the surface of the Sun, where the thermal energy is so high that electrons break off the atoms to create a plasma of free electrons and ions. At room temperature, and below, collisions involve far less energy than is required to excite the atomic electrons, so evaporative cooling has no effect on the internal structure of the atoms. (The electrons keep whizzing rapidly around the nucleus in the same way, no matter whether the atoms in the gas are hot or cold.) Thus, for the quantum effects considered in this chapter, the atoms move in exactly the same way as "elementary" or indivisible particles of the same mass (and spin); the fact that atoms are really composite particles, made up of protons, electrons, and neutrons, has no influence.

In quantum mechanics a particle confined in a potential well, such as an atom in a magnetic trap, can have only certain energies, i.e. the energy is quantized. The allowed values of this energy shown in Figure 7.8 correspond to wave functions that are standing waves in the potential well. These resemble the standing waves on a string, but in this case the spread of the wave functions increases with energy; this feature of the quantum-mechanical description corresponds to the classical behavior that higher-energy atoms go further up the "sides" of the harmonic potential. For this shape of potential the allowed energy levels have equal spacing between them as indicated. (For a real string with a fixed length, as on a musical instrument such as a violin, the modes of oscillation that correspond to the fundamental frequency and higher-order harmonics lead to a different spacing of the energy levels that is applicable for particles confined in a box with rigid walls.) In a magnetically trapped gas, that is hotter than the critical temperature at which quantum effects occur, the atoms are spread over many high-lying levels as shown in Figure 7.8. As the gas is cooled the atoms sink down into lower energy levels of the trap, as shown in Figure 7.8. Quantum effects become important when more than one atom tries to crowd into the same level. This criterion in terms of the occupation of energy levels turns out to be equivalent to the condition that the wavepackets overlap, and the two descriptions in terms of (a) energy levels (and the associated wave functions) and (b) atoms as particles (with an associated de Broglie wavelength) are two facets of the same underlying behavior.

"overlaps" and the atoms undergo a "quantum identity crisis," that is, we can no longer distinguish individual atoms from each other. These are the conditions for which we enter the *quantum world of ultra-cold atoms* where the wavelike nature of the atoms must be considered. This description in terms of fuzzy atoms shows the conditions for which "something" happens, and was indicated pictorially in Figure 7.2, but to understand in more detail what happens, and why, we need to consider the wave functions and the energy levels associated with the motion of the atom's center of mass. The distinction between the atom's external degrees of freedom (momentum and position) and the internal structure is discussed in Box 7.1.

The way that identical particles arrange themselves in the accessible energy levels of a quantum system is intimately related to their spin. Two types of particles exist in nature: bosons whose spin is an integer multiple of $\hbar$ (Planck's constant over $2\pi$),

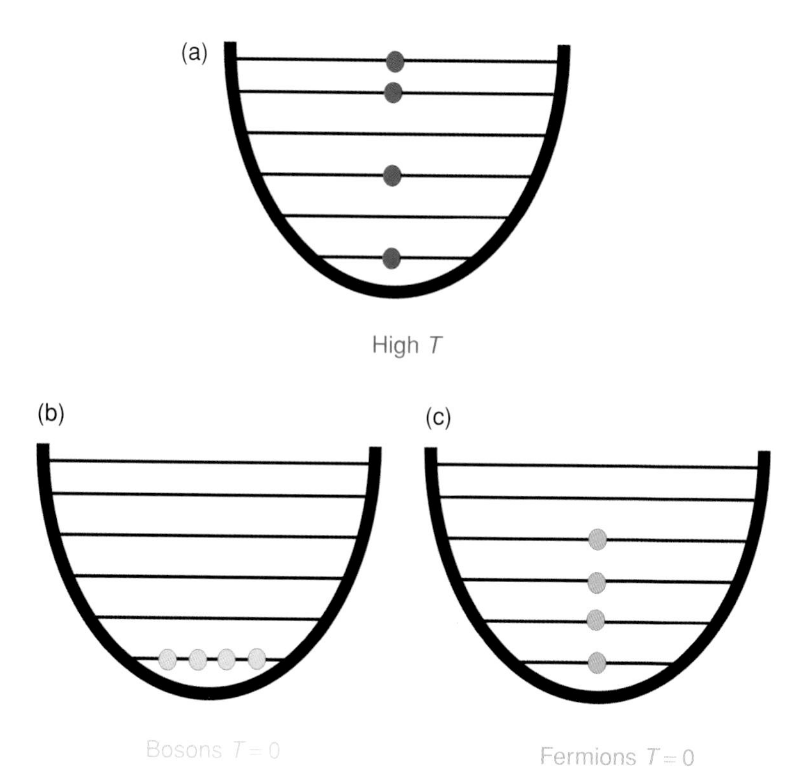


**Figure 7.8.** The way that atoms distribute themselves over the energy levels of a system, in this case a magnetic trap, depends on the temperature. (a) This shows atoms distributed over many energy levels corresponding to temperatures hotter than that for quantum effects (the energy levels are sparsely populated so quantum effects are negligible). Both (b) and (c) show systems at lower temperatures (near to absolute zero) at which the occupation of the lower levels approaches unity and quantum effects become important: (b) shows the case of bosons, for which the atoms all avalanche into the lowest level, whereas (c) shows that, in contrast, for fermions there is only one atom in each level.

e.g. 0, $\hbar$, $2\hbar$, . . . etc. and fermions, with spins of $\frac{1}{2}\hbar$, $\frac{3}{2}\hbar$, $\frac{5}{2}\hbar$ . . . etc. The major breakthrough in work with ultra-cold gases was made with atoms that are bosons and we shall consider them first. The contrasting behavior of fermions will be described later. Identical bosons tend to congregate in the same quantum state, because the probability of bosonic particles going into a given state is greater if there are already some particles of the same species in that state. At heart, this gregarious behavior arises because of constructive interference of the matter waves. Thus, when a gas of bosons is cooled to a critical temperature near which the number of atoms in the lowest level becomes greater than unity, then more atoms are encouraged to go into that state. This increases the probability of atoms going into that state still further, so this process rapidly builds up into an avalanche. The sudden accumulation of atoms in the lowest level is a "phase transition" called Bose–Einstein condensation (BEC). This phase transition arises because of the quantum properties of bosons and it is of a completely different nature from other types of phase transition, such as the condensation of steam into (liquid) water as it cools. Phase changes of a substance from gas to liquid, or from liquid to solid, arise because of the attractive interactions between the atoms, or molecules. In contrast, Bose condensation is a phase transition from a classical gas to a quantum gas and in this process a large population of atoms goes into the quantum level of lowest energy (as shown in Figure 7.8(b)). We can draw a picturesque analogy between atoms undergoing BEC and people searching for the place where a party is to be held; after a few of them have found the right place, randomly, the rest rapidly join them, if they are close by. In some ways this also applies to experimental research groups in this area: initially there were a few groups and now many others have joined them to explore the fascinating physics of ultra-cold atoms.

The phenomenon of BEC was predicted by Einstein in the early days of quantum mechanics, on the basis of theoretical work by the Indian physicist Satyendranath

Bose (see Chapter 9). It was recognized as an important theoretical idea, but people did not believe that it could be observed in the laboratory because it was expected that any gas would undergo "ordinary" condensation into a liquid at higher temperatures than the phase transition to a BEC. They were wrong. In 1995, two separate experimental groups succeeded in making a Bose–Einstein condensate in a dilute gas as described in the following section. The key factor in this breakthrough was the extremely low temperatures produced by evaporative cooling that enabled BEC to be achieved while the gas remained at a low density. For such a very dilute gas the problems caused by interactions between the atoms are not as bad as people had assumed. In subsequent sections we shall look at the exciting range of new experiments made possible by BEC

### 7.6.1 Bose–Einstein condensation

In the first experiment, Eric Cornell, Carl Wieman, and their team at the Joint Institute for Laboratory Astrophysics (JILA) in Boulder cooled the rubidium atoms in their magnetic trap by evaporation and observed clear evidence for BEC as shown in Figure 7.9.

In Figure 7.9(c), almost all the atoms are in the ground state and, in some sense, this is as "cold" as can be obtained in this trap. The elliptical shape of the peak gives another clear signature of Bose condensation. This characteristic shape arises from the expansion of the wave function when the trapping potential is switched off, whereas the thermal cloud expands equally in all directions.

In the same year, a team at the Massachusetts Institute of Technology (MIT) led by Wolfgang Ketterle also created a Bose–Einstein condensate, but using sodium atoms rather than rubidium. In these first experiments neither team used the Ioffe trap, described above, but later this type of trap was used at MIT to create a large Bose–Einstein condensate of over a million sodium atoms. This configuration enabled the MIT team to carry out a remarkable experiment in which they created two condensates at the same time. As explained previously, the atoms in a Ioffe trap form a long cigar-shaped cloud. The MIT team chopped the cloud into two pieces, each of half the original length, using a sheet of laser light. The light exerted a repulsive dipole force on the atoms that pushed the atoms out of the region of high intensity. (This light was not absorbed by the atoms because it had a frequency far away from the atomic resonance frequency.) After the sheet of light and magnetic trap had been turned off, the two independent condensates expanded so that they overlapped with each other, and the image of atoms shown in Figure 7.10 was recorded. The two condensates interfered to give fringes. Someone with experience of optical experiments might be surprised that interference occurs between two condensates that act like independent sources of matter waves with an arbitrary phase difference between them. Indeed, this issue was hotly debated, with some people claiming that, because two separately created independent condensates have no phase relationship, there would be no interference. In the experiment they did see interference, just as two beams from independent lasers (or other coherent sources) interfere.

Each time the experiment with atoms was repeated, however, the position of the fringes changed, because it depends on the difference between the phases of the two condensates formed in that particular run. Thus, if the number of atoms at each position is averaged over many experimental runs, the fringes would wash out. This experiment demonstrates that a Bose condensate is a source of coherent matter waves, just like a laser is a source of coherent light. Further details of matter-wave interferometry in atomic beams are given in Chapter 6, but let us explore the parallels between BEC and

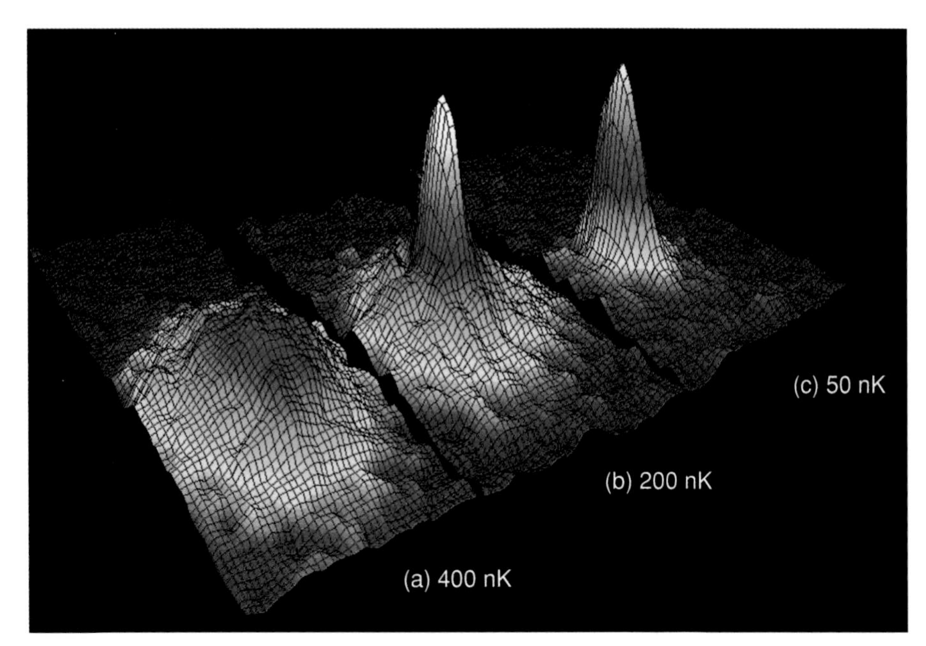


Figure 7.9. These three images of the atoms were obtained after turning off the magnetic trap to allow the atoms to expand freely. Some time later a laser beam was flashed on to illuminate the atoms and record an image of the cloud on a video camera. The size and shape of the expanded cloud depends mostly on the initial velocity distribution, or more accurately the energy in the case of the Bose–Einstein condensate (the initial size is negligible in comparison – 100 nanokelvin (nK) is equivalent to 0.1 microkelvin). (a) A cloud released from the trap at 400 nK, which is above the temperature for quantum effects. Thermal equilibrium ensures that these atoms have the same velocity in all directions and so after expansion they give a spherical cloud (referred to as a thermal cloud). (b) A narrow peak of cold atoms emerges out of the thermal cloud at a temperature of 200 nK. The peak corresponds to the atoms that have Bose-condensed into the lowest energy level of the trap. (c) Below 50 nK the sample is an almost pure condensate. The distinct changes in the size and shape that occur over a relatively narrow temperature range are a clear indication of a phase transition. (Image courtesy of Eric Cornell and Carl Wieman.)

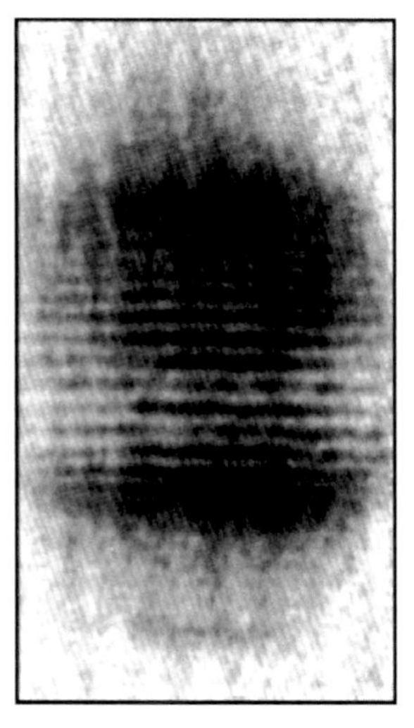

Figure 7.10. This interference pattern was created by two overlapping Bose–Einstein condensates. Atoms are observed at certain places and not at other positions. It might seem strange that atoms arriving at the same place "cancel each other out," but this behavior is expected when the atoms are considered as matter waves. The atoms that do not appear at positions where the matter waves interfere destructively are redistributed to positions of constructive interference. Similar interference patterns occur in optics, e.g. Young's famous double-slit experiment, in which light diffracted from two closely spaced slits overlaps to give straight fringes. (Image courtesy of Wolfgang Ketterle.)

lasers a little more here. The random photons emitted by an ordinary light bulb can be compared to the atoms in a gas moving in random directions; the coherent light from a laser, for which all the photons are "in step," has similarities to a condensate in which all the atoms occupy the ground state of the trap.

## 7.6.2 The atomic laser

The word laser is an acronym for light amplification by stimulated emission of radiation. (Actually most lasers are not operated as amplifiers with an input and output beam but as stand-alone oscillators; however, the acronym with "o" instead of "a" has less appeal!) In stimulated emission (see Figure 7.11(b)), an incident photon stimulates an atom in an excited state to make a transition down to its ground electronic state and emit a photon in phase with the original photon. This process amplifies a light wave that propagates through a medium with more atoms in the excited level than in the ground state. (The techniques of manipulating atoms with photons are described in Chapter 6.)

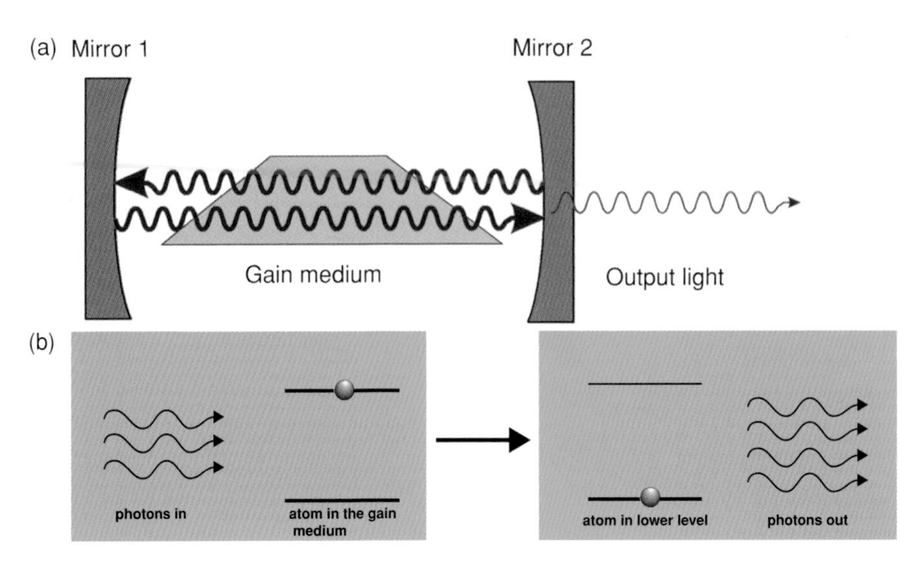


Figure 7.11. (a) A schematic diagram of a simple laser. The light reflected back and forth between the two highly reflecting mirrors is amplified as it passes through a "gain medium." This is a medium in which there is a population inversion, i.e. more of the atoms are in an excited state, in which their electrons are in an upper level, than in the lower level (which is opposite to the situation when there is no excitation of the laser gain medium). This makes the process in which a photon stimulates the emission of another photon more probable than absorption of the incoming photon, (b) In the process of stimulated emission, an incident photon stimulates an atom in an excited state to make a transition to its ground electronic state and emit a photon in phase with the original one. Here, three incoming photons become four, with the extra photon being a copy of the original ones, with the same frequency and phase. This amplification leads to a build-up of coherent light between the two mirrors; one of them is slightly less reflective than the other, so that some light escapes to give a laser beam.

In stimulated emission the photons behave like bosons, i.e. photons go into a state where there are other photons and, as in BEC, this leads to a rapid buildup of photons in a particular state. There are, however, profound differences between photons and atoms: radiation disappears as the temperature is reduced toward zero whereas atoms do not. For example, physicists consider thermal radiation in a box at a certain temperature as a "gas of photons" and this radiation exchanges energy with the walls of its enclosure by the absorption and emission of photons. (To promote this process the walls should not be reflective; ideally they should be completely absorbing at all wavelengths, i.e. "perfectly black," but this is not crucial.) In a box with a temperature close to absolute zero the energy density of the radiation is negligible. Magnetic traps are used to confine the atoms specifically because they do not have walls on which ultra-cold atoms can stick. In a laser, the photons are not in the lowest energy state but populate a higher energy level as shown in Figure 7.11(a). One of the mirrors in the laser cavity allows some light out of the cavity to form a useful source of light. Similarly a coherent beam of matter waves is made by allowing atoms to spill out of the trap as shown in Figure 7.12. Atoms pour out over the lower lip of the trap, like water out of a jug, to form a continuous beam of matter waves. The condensate in the trap is described by a wave function with a well-defined phase and consequently the matter waves coupled out of the trap are coherent like the light from a laser. The fact that the condensate within the trap has a definite phase leads to the Bose gas being a superfluid, as described later. In the following sections we pursue the analogy between the Bose condensate and optical experiments with intense, coherent beams of light produced by lasers.

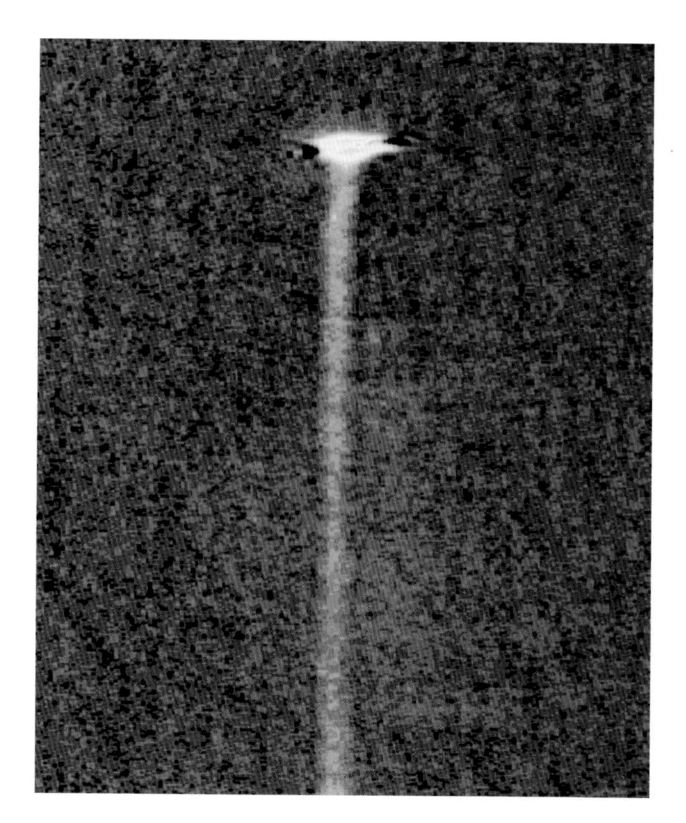

**Figure 7.12.** An atom laser is created by extracting cold atoms from a Bose–Einstein condensate in a magnetic trap (the blob of atoms at the top of the picture). The atoms are extracted by the same mechanism as that used for evaporation but with a deeper cut that removes atoms from the condensate itself rather than those on the edges. The atoms fall downwards because of gravity in a continuous beam of matter waves. (Courtesy of N. Smith, W. Heathcote, and E. Nugent, Oxford.)

## 7.7 Diffraction of atoms by a standing-wave light field and the Bragg interferometer

In the previous section, we saw that two condensates released from a double-well trap produce interference fringes in a similar way to Young's classic double-slit experiment with light (but with the interesting feature that the fringes are visible even when the phases of the two condensates are independent of each other). Experiments that are analogs of the diffraction of light from multiple slits, as in a diffraction grating, have also been performed. The diffraction pattern shown in Figure 7.13 was obtained by trapping atoms in a one-dimensional array of regularly spaced potential wells formed with a standing wave of laser light.

Three-dimensional arrays of regularly spaced potential wells are created using three orthogonal standing waves of light, i.e. counter-propagating pairs of laser beams along the $x$, $y$, and $z$ axes, respectively. These periodic structures are called optical lattices; their potential surface in a cross-sectional plane looks like an "egg-box." Atoms in such optical lattices have fascinating properties that can be deduced by observation of the diffraction patterns after the atoms have been released, as described in Chapter 9.

### 7.7.1 Bragg diffraction of atoms

The one-dimensional optical lattices that act like diffraction gratings are produced by the interference of two laser beams having the same frequency. The interaction of the atoms with these lattices gives diffraction patterns, as shown in Figure 7.13. The multiple diffraction spots correspond to many different momentum states of the atoms.

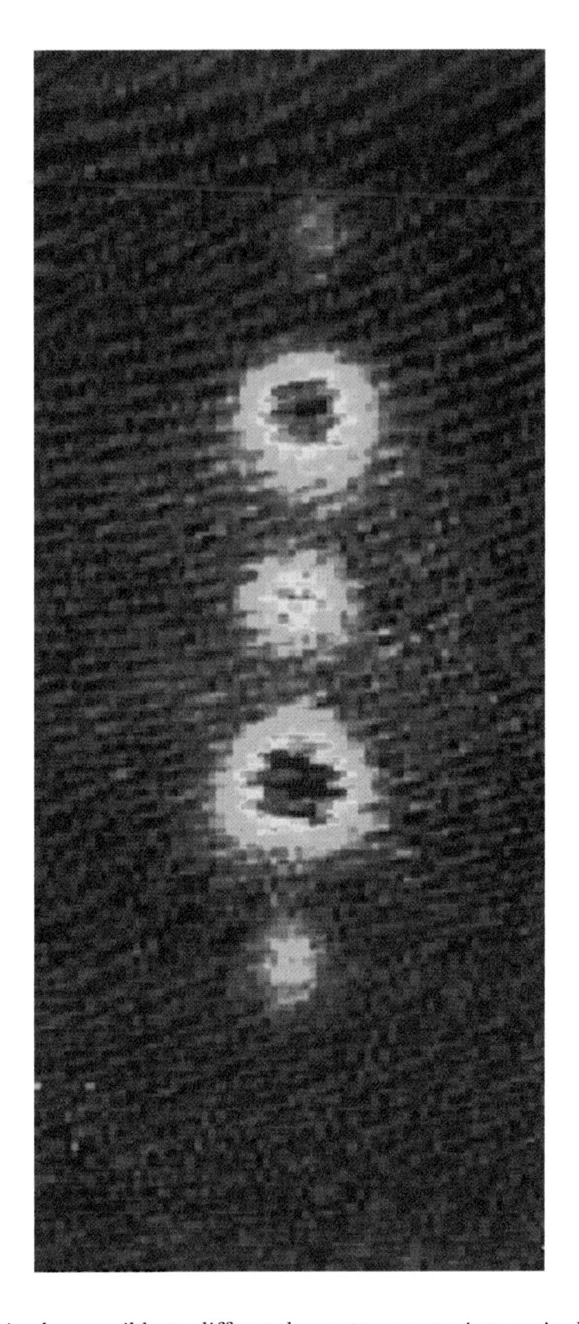

**Figure 7.13.** To create this diffraction pattern ultra-cold atoms from a Bose–Einstein condensate were loaded into an array of potential wells created by a standing wave of light – the "hills and valleys" are formed by the same potential as that which leads to the dipole force described in Chapter 6. When the atoms are released from the wells they expand and overlap, just like light waves propagating from a diffraction grating to a distant observation plane. Sharp maxima occur at positions where the matter waves interfere constructively. Unlike the case with just two wells, as in Figure 7.10, where a diffraction pattern occurs regardless of phase difference, diffraction from many wells requires that the atoms have the same phase (or a uniform phase gradient) across all the wells. This can be achieved by loading from a BEC. (Courtesy of R. Godun, V. Boyer, D. Cassettari, and G. Smirne, Oxford.)

By contrast, it is also possible to diffract the matter waves into a single direction, or momentum state, as explained here. The interference of two laser beams with different frequencies produces a moving optical lattice, i.e. the potential wells produced by the light move with a velocity relative to the laboratory frame of reference. (An observer moving with the lattice would see the two laser beams Doppler shifted to have the same frequency in the moving frame; this is similar to the situation in the optical-molasses technique mentioned in Section 6.4.) Atoms diffract from this moving lattice in an analogous way to the Bragg diffraction of X-rays from the lattice planes of a crystal: the incident beam of X-rays must hit the crystal at a specific angle in order to be diffracted. This Bragg condition on the angle corresponds to a condition on the frequency difference $\delta f$ between the laser beams that is required for Bragg diffraction of the atoms in a particular direction. The diffraction of atoms from a standing wave can be regarded as a scattering process in which atoms in a Bose condensate absorb photons from one of the laser beams and are stimulated to emit into the other beam. This process changes the atom's momentum by the difference in the momentum of the two photons, and the change in kinetic energy of the atoms must also be equal to the difference in energy $h \times \delta f$ of the two photons. This exchange of momentum and energy between atoms and photons is what we call Bragg diffraction of atoms by light; it uses a long, smooth pulse of light, so that the atoms are Bragg diffracted into a single new momentum state. (A short pulse would lead to multiple momentum states, or diffraction spots, as in Figure 7.13.)

Bragg diffraction preserves the coherence of the matter waves for a condensate. That is, if the atoms in the original condensate all have the same de Broglie wavelength, the diffracted atoms also all have the same de Broglie wavelength (but not the same wavelength as the condensate). This Bragg diffraction is analogous to bouncing a laser beam from a mirror or a beam-splitter in optics: all, or part of, the incident beam of light is sent in a different direction from the original laser beam. In Bragg diffraction of atoms by laser beams, we can change the momentum of all, or part of, the condensate by varying the strength of those laser beams (or "Bragg beams"), to realize the action of a beam-splitter or a mirror for matter waves.

Having elements like beam-splitters and mirrors, we can make an interferometer for atoms – a device that allows us to split a coherent sample of atoms (from a Bose–Einstein condensate), send the two components along different paths, and then recombine them to observe their wave interference. Figure 7.14 shows one way in which this is done.

## 7.8 Nonlinear atom optics: four-wave mixing

We have mentioned the fact that a Bose–Einstein condensate is an atom-wave analog to the light field in a laser. Both the laser and the Bose–Einstein condensate are coherent, in that all the photons or atoms have the same wavelength, and intense, in that there are lots of photons or atoms. It is these two shared characteristics of coherence and intensity that make condensates very useful as sources of matter waves and allow us to use them in ways similar to how lasers are used as sources of light.

One of the first truly new things to be done with lasers, shortly after their appearance around 1960, was the demonstration of "nonlinear" optics. Transparent optical materials like glass have an index of refraction whose value is the ratio of the speed of light in a vacuum to the speed of light in the material. Normally, for low light intensities, this refractive index does not depend on the intensity, and we say the material is "linear." But for high intensities, all materials are nonlinear – their index depends on the intensity. This feature has some remarkable effects, but only with the advent of lasers and their high intensity (and coherence) was it possible to see these effects. In 1961 physicists took pure red light from a ruby laser, passed it through a transparent quartz crystal, and found that some of the emerging light was ultraviolet, at exactly half the wavelength (double the frequency) of the incoming red light. The nonlinearity of the quartz had partially rectified the oscillating electric field of the red light, producing a small field oscillating at double the frequency – the ultraviolet light. Strange as it seems, we can do much the same thing with atom waves. We do not double the energy of the atoms as in the 1961 experiment with photons. Instead, we combine three atom waves together to produce a new, fourth wave, in a process known as "four-wave mixing" that is illustrated in Figure 7.15.

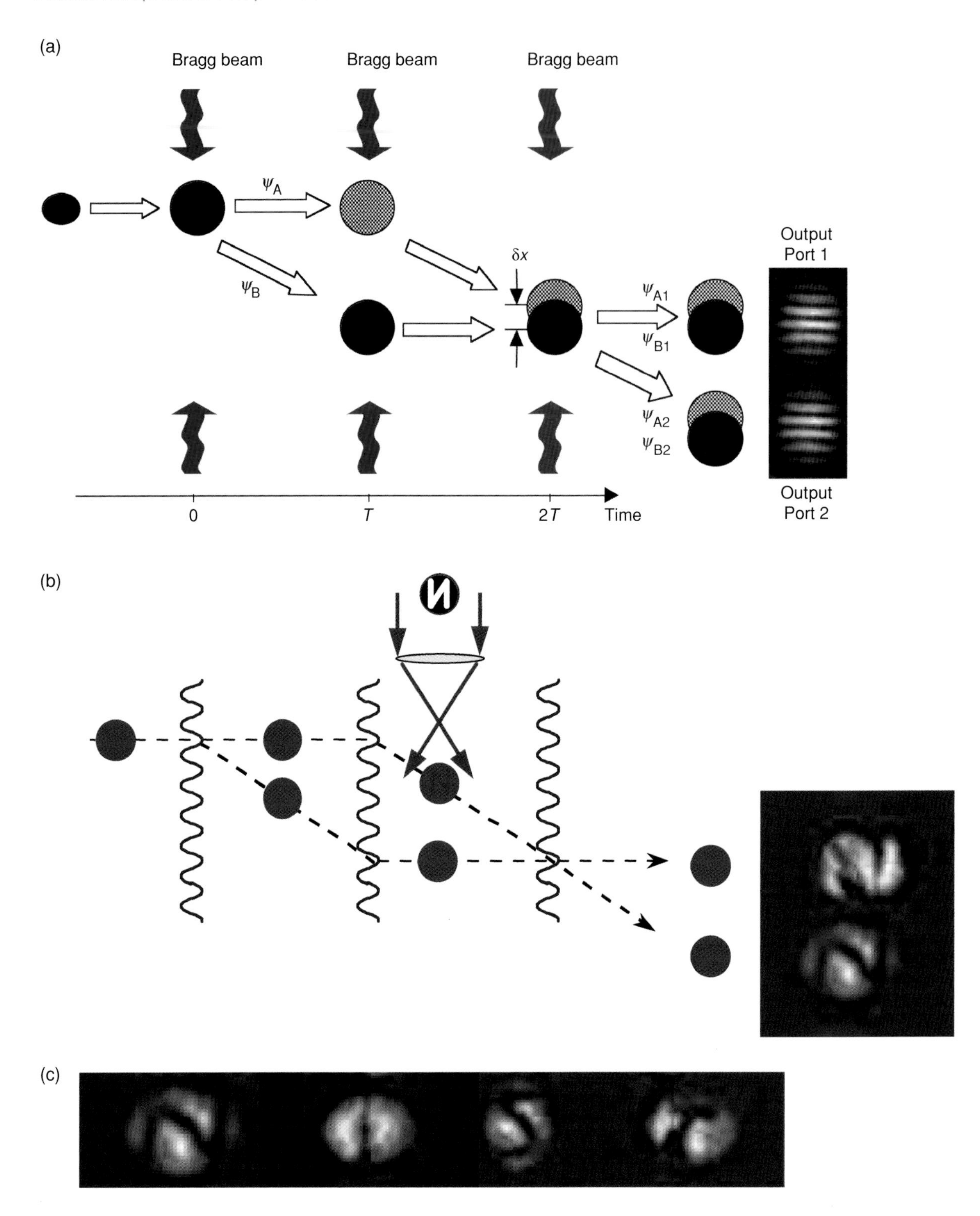
(a)
Bragg beam
Bragg beam
Bragg beam
$\psi_A$
$\psi_B$
$\delta x$
$\psi_{A1}$
$\psi_{B1}$
$\psi_{A2}$
$\psi_{B2}$
Output Port 1
Output Port 2
0
$T$
$2T$
Time
(b)
(c)

## 7.9 Controllable collisions between ultra-cold atoms

At very low temperatures the collisions between atoms have a simple form, and for some particular atomic states have the interesting feature that the interactions change from being effectively repulsive to being attractive as the applied magnetic field is varied. As mentioned previously, these interactions between atoms are not the cause of Bose condensation, but when the condensate nucleates there is an increase in density so that, although the cloud remains a dilute gas, the interactions between the atoms determine the properties of the condensate. Some of these characteristics are examined in this section (see also the review paper "Quantum encounters of the cold kind" in the suggestions for further reading).

For ultra-cold atoms the scattering does not depend on the detailed form of the force between the atoms, as we can see from the following analogy. An optical microscope cannot resolve details of the structure of an object smaller than the wavelength of the light (that illuminates the object). Similarly, in a collision between ultra-cold atoms that have a de Broglie wavelength much longer than the range of the interaction, we cannot tell anything about the detailed nature of that interaction. The nature of ultra-cold collisions is shown vividly in Figure 7.16.

An extraordinary feature of ultra-cold atoms is that the strength of the interactions (scattering length) can often be controlled by varying the magnitude of an applied magnetic field. There are states with this special feature in both sodium and rubidium. Generally in experiments with these states the Bose condensate is formed at a magnetic field for which the scattering length is positive, which corresponds to an effective repulsive interaction. The scattering length can be changed to a negative value by altering the applied magnetic field so that the interactions between atoms become effectively attractive. For a strongly attractive interaction the condensate collapses – a phenomenon that has poetically been called a "bosonova" because of the parallel between the behavior of the Bose gas and a supernova. The implosion of the condensate leads to a large increase in density and a large rate of inelastic collisions that lead to the ejection of a "hot" gas at a temperature of only a few hundred microkelvins. The

←

**Figure 7.14.** (a) To make an interferometer, a BEC that is initially at rest, at time $t = 0$, is exposed to a Bragg pulse that splits it into equal parts, one that remains at rest and the other that moves with twice $v_{rec}$, the change in atomic velocity caused by the recoil from the momentum of two photons. (This velocity change is about 6 cm $s^{-1}$ for the sodium atoms used in these experiments.) A time $T$ later (when $t = T$), the two clouds of atoms have separated, and another pulse, twice as long as the first, acts like a mirror, giving the zero-velocity atoms a kick of 6 cm $s^{-1}$ and bringing the atoms moving at 6 cm $s^{-1}$ to rest. At time $t = 2T$, the two clouds of atoms again overlap and a final beam-splitter pulse puts half of the atoms from each of the two paths, upper and lower, into the zero-velocity and 6-cm $s^{-1}$ velocity states. The phase difference between the two paths determines whether they add constructively or destructively; the final result can be all the atoms at rest, all moving at 6 cm $s^{-1}$, or anything in between. Many things, like the phase of the Bragg beams, interactions between atoms in the clouds or the application of another beam of light to atoms in one of the paths, change the phase and therefore the nature of the interference. (b) In a whimsical application of this Bragg-interferometry technique, researchers focused a laser beam containing the pattern of an alphabetic letter onto the atom cloud in just one path of the interferometer. This light caused a change of phase of the matter waves so that, when the clouds recombined, the letter showed up as constructive interference in one of the arms, i.e. the letter "N" in the case shown and a reverse "N" in the other output port. The researchers from the National Institute of Standards and Technology were able to write the acronym of their laboratory onto the quantum phase of their atoms and the string of four outputs spelling out "NIST" is shown in (c). (Images courtesy of the Laser Cooling Group, NIST, Gaithersburg.)

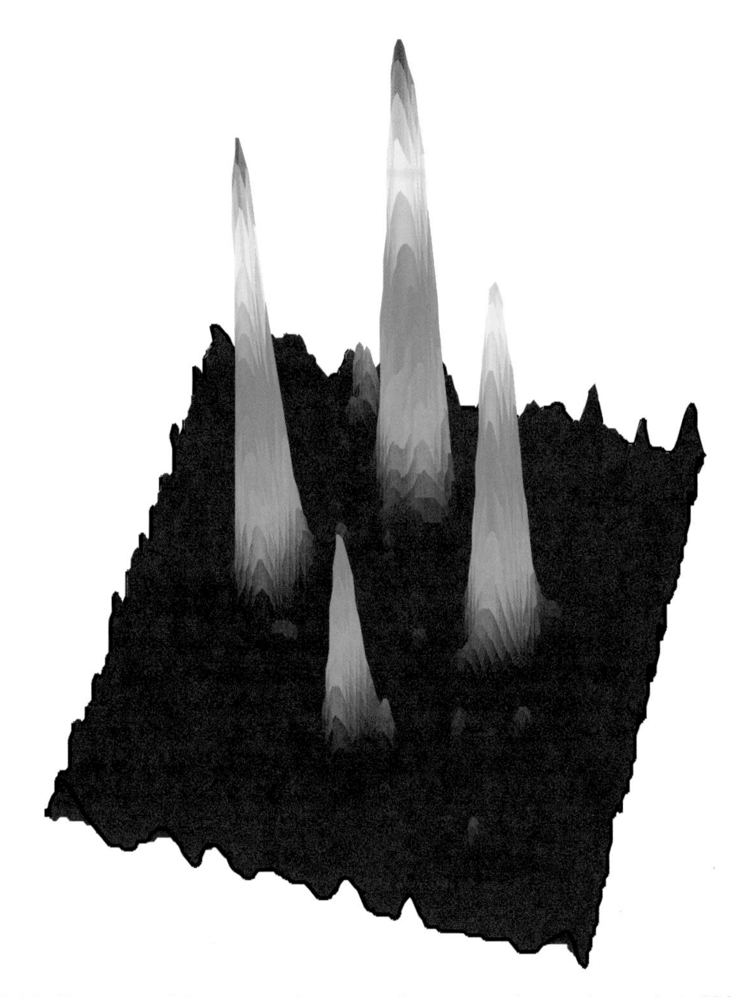

**Figure 7.15.** The process of Bragg scattering creates three waves of atoms from a single BEC: a Bragg pulse splits off one cloud of atoms from the original BEC as illustrated in Figure 7.14, then a second, different Bragg pulse creates another cloud of atoms with different velocities. Together with the original BEC, this makes three clouds, all with different atomic momenta (and thus three different de Broglie wavelengths) and all overlapping at the position of the original BEC. These mix together, the atoms themselves acting as the "material" with an index of refraction. Beams of ultra-cold atoms interact with each other through the collisions between the atoms that are discussed in a later section, whereas in free space two beams of light simply pass through each other without effect. The four peaks shown each correspond to a different atomic velocity (or momentum): the three larger peaks are the three original velocities created from a single BEC, and the smaller one is the fourth atom wave generated by the nonlinearity caused by the atomic collisions. One way to think about the creation of this fourth wave is to consider two atoms colliding in the presence of a third atom (where each of these atoms is in a different one of the three original waves). Because the atoms are bosons, the third wave makes it more likely that the two atoms will bounce off each other in such a way that one of them goes in the same direction as the third atom – the collision is stimulated to have this outcome (with a high probability) by the presence of lots of atoms in the same beam as the third one. Conservation of energy and momentum dictate that the other atom of the colliding pair must go in the direction of the new, fourth wave. (Images courtesy of the Laser Cooling Group, NIST, Gaithersburg.)

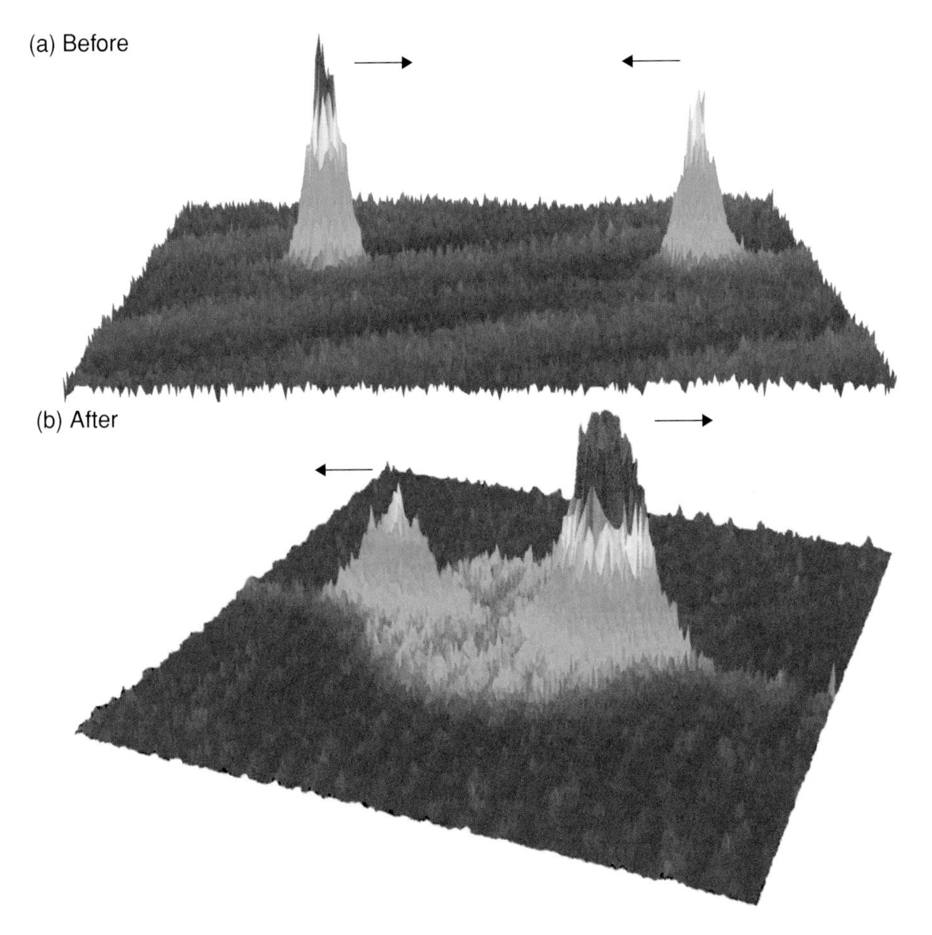


**Figure 7.16.** (a) This image shows two separate condensates of rubidium atoms at either end of a Ioffe trap. These condensates were projected toward each other at a low velocity; as they passed through each other some atoms collided and scattered elastically to create the sphere of outgoing atoms visible after the collision (b). This figure shows that, at low energies, the scattering is the same in all directions (isotropic), so we can deduce only one piece of information from these data, namely the overall strength of the interaction, which is proportional to the probability that the atom is scattered. Although we do not need to go into details of the quantum-mechanical theory of scattering, it is useful to mention one technical term: the scattering length is the radius of a hard sphere that gives exactly the same scattering as a given potential – this single parameter does not tell us the shape of an actual potential but only the overall strength of the scattering. In this regime the atoms behave like small billiard balls that bounce off each other without loss of energy (elastically). The rate of inelastic collision processes in which one or both of the atoms is lost from the trap must be small. To probe the interaction in more detail needs higher-energy, shorter-wavelength atoms. Indeed, the de Broglie wavelength at which the scattering stops being isotropic gives an indication of the range of the interaction. (Courtesy of Jook Walraven, AMOLF [Institute for Atomic and Molecular Physics], Amsterdam.)

attractive interactions also lead to the formation of ultra-cold molecules. More details of this process are given later.

Collisions between the atoms, including molecule formation, prevented Bose condensation of hydrogen atoms for the high densities required at the temperatures that could be achieved by cryogenic techniques. Such temperatures are higher than those produced by the combination of laser and evaporative cooling techniques used for the alkali metals; however, BEC of hydrogen was eventually achieved by the team of Daniel Kleppner and Thomas Greytak at MIT. They transferred a cryogenically cooled sample of spin-polarized atoms into a magnetic trap and then used evaporative cooling. Their experiment did not use laser cooling, but they used a laser system to detect that Bose condensation had taken place; not by imaging the condensate directly as in the experiments with alkali metals, but instead by measuring the absorption of a laser beam that excited a narrow two-photon transition in the atoms – the increase in density as the atoms go into the Bose condensate caused the atoms to collide more frequently, which leads to a measurable frequency shift of the spectral line, giving a clear signature of Bose condensation.

A frequency shift caused by collisions also arises in other precision measurements, e.g. atomic fountain clocks (see Chapter 6). Time is defined in terms of the hyperfine frequency of the cesium atom and to measure this frequency precisely and accurately laser-cooled atoms are launched upward to form the fountain. Unfortunately cesium atoms have exceptionally strong interactions between them (they have a scattering length 50 times larger than that of rubidium) and this sets a severe limit on the density of atoms than can be used without perturbing the transition by an unacceptable amount. In this section we have described effects of interactions between bosons; more recent ground-breaking experiments have used fermionic atoms to observe new physics.

## 7.10 Fermi degeneracy

So far we have concentrated on quantum effects seen when a Bose gas reaches quantum degeneracy at extremely low temperatures, but laser cooling, magnetic trapping, and evaporation have been carried out with isotopes of lithium and potassium that are composite fermions (the isotopes with relative atomic masses of 6 and 40, respectively). Fermions do not undergo a phase transition as they enter the quantum regime, rather they simply fill up all the accessible energy levels. For reasons that are very deep in quantum mechanics two fermions cannot occupy the same state. This is usually expressed as the Pauli exclusion principle and it leads to the shell structure of atoms that underlies the periodic table of the elements – if this principle did not operate then all electrons in atoms could go into the lowest orbit around the nucleus. In fact, two electrons (but no more than two) can occupy the same orbital if they have different spin states, spin up and spin down. Similar behavior occurs in experiments with trapped atoms that are composite fermions – at low temperatures these atoms stack up in the allowed energy levels to give the distinctive distribution of the atoms over the energy levels shown in Figure 7.8(c). Other examples of where the fermionic behavior manifests itself are electrons in metals, and also the He-3 isotope in "traditional" low-temperature physics as described in Chapter 8. (Helium has two isotopes: He-4 is a boson and He-3 is a fermion.)

The filled levels in Figure 7.8(c) are known as the Fermi sea and at the top of the sea is the Fermi surface where the occupation drops from 1 particle per level to 0. However, it turns out to be difficult to cool fermions when they are at low temperatures for two reasons, both of which arise from the Pauli exclusion principle. When two fermions collide they are unlikely to change their energies because there are few vacant energy levels into which they can go; this is called Pauli blocking. The second factor that impedes cooling of fermions is that two identical fermions cannot be at the same place at the same time; this implies that fermionic atoms, that are in the same state, remain spatially separated and therefore their scattering is strongly inhibited. This problem has been circumvented in experiments by using either a mixture of different spin states, e.g. the potassium isotope with a relative atomic mass of 40 has a spin of $\frac{9}{2}$

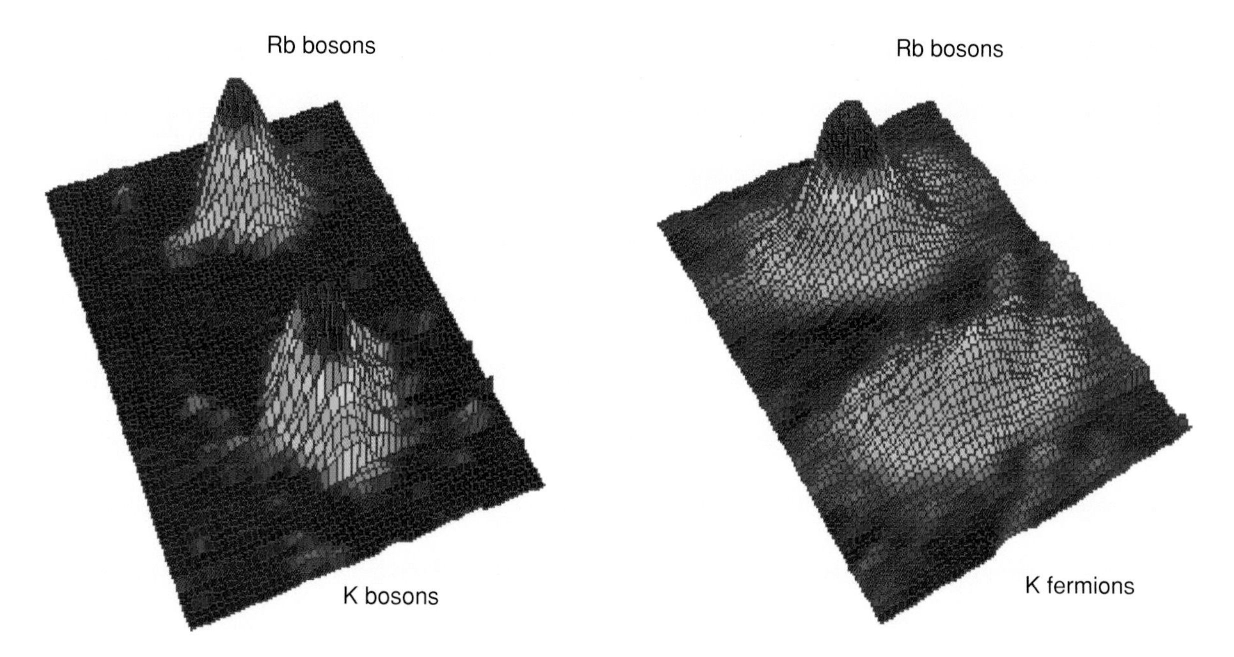


Figure 7.17. The different quantum behaviors of fermions and bosons affect their distribution in a magnetic trap. In this experiment, both rubidium (Rb) and potassium (K) atoms were confined in the same magnetic trap. The rubidium atoms are bosons and at low temperatures undergo Bose–Einstein condensation, which gives a characteristic peak as in Figure 7.9. A similar peak appears in the distribution of the potassium atoms when the experimenters used a bosonic isotope of potassium. A fermionic potassium isotope gives rise to a more spread-out distribution because the fermionic atoms do not go into the same state, as illustrated in Figure 7.8(c) (Courtesy of Massimo Inguscio, LENS, University of Florence.)

and can be magnetically trapped in several orientations, or a mixture of fermions with atoms of a completely different species, e.g. potassium and rubidium atoms as shown in Figure 7.17. This figure shows the difference between fermions and bosons very clearly. A key objective of the work on fermions is to reach the regime where there is predicted to be a transition to a state like that in a superconductor (see Chapter 8).

We have already mentioned that the tendency of atoms in cold dense gases to form molecules presented a barrier to reaching the quantum regime – this was circumvented by working at extremely low density and consequently extremely low temperatures. Once an ultra-cold quantum gas of various species of atoms had been attained, researchers were able to perform experiments to study the effects of varying the range of the interactions between the atoms by changing the applied magnetic field; in particular the behavior at certain magnetic-field strengths for which the range of the interatomic interactions becomes very large (so-called "Feshbach resonances"). In this section we shall primarily consider the very interesting physics that arises near these resonances in the case of fermionic atoms (which is quite different from the case of bosons described earlier).

Sweeping the magnetic-field strength through a region where the interactions are resonantly enhanced converts a large fraction of the atoms into molecules – this transformation has been particularly successful in experiments with the fermionic isotopes of lithium and potassium (referred to above). The diatomic molecule formed from two fermions is a boson because two spin-$\frac{1}{2}$ particles combine to give a total spin that is an integer; in the experiments that have been carried out with magnetically trapped atoms whose spins are parallel, the two atoms with spin-$\frac{1}{2}$ combine to give a total spin of 1 (in units of Planck's constant, $\hbar$). If the spins are anti-parallel then the composite particle has a total spin of 0, which is also an integer. The relative weakness of the interactions between two fermionic atoms with aligned spins means that they combine to give a "long-range" molecule, i.e. one in which the two atoms are much further apart than in "typical" molecules such as those in gases at room temperature and pressure, e.g. the nitrogen molecules in air. In a gas at microkelvin temperatures these large molecules do not break apart when they knock into each other because they have low kinetic energy and hence undergo very gentle collisions. Paradoxically, the large size

(a)

(b)

of these ultra-cold molecules makes it less likely that they change their quantum state during a collision, and this has proved to be a great advantage in experiments.

This cannot be explained from a classical point of view; broadly speaking, it arises because, when the atoms have a large separation, the wave functions of the various possible states are not similar and have little overlap with each other, giving a smaller rate of transitions than between (more tightly bound) states that have similar wave functions. (The overlap between molecular wave functions is also an underlying feature of the Franck–Condon principle that determines which transitions between quantum levels have the highest probability in electronic transitions.) The accumulation of many molecules in the same quantum state has allowed BEC of molecules to be achieved. The transformation of a gas of fermionic atoms into a pure Bose–Einstein condensate of molecules is accompanied by a strong change in behavior, e.g. the size and distribution of the atoms; this can be detected, as in the experiments on atoms, by releasing the molecular condensate from the confining potential and allowing it to expand. The images of the expanded cloud look very similar to those shown in Figure 7.9. The creation of a molecular Bose–Einstein condensate starting from a cloud of fermionic atoms was a truly surprising result bearing in mind how difficult it had been to attain BEC of atoms in the first place, and the experiments on molecules have led on to further remarkable work.

## 7.11 Conclusions

The techniques of laser cooling and evaporation in magnetic traps have opened the door to the world of ultra-cold atoms and this chapter gives a glimpse of the fascinating phenomena that have been seen so far. These methods are now widely used in physics laboratories throughout the world. After this major experimental breakthrough the properties of the Bose–Einstein condensate were studied and it is now well understood; indeed, the expectation that the properties of such dilute gases would correspond closely to theoretical predictions was a motivation for the quest for BEC in the first place. In these experiments, optical systems similar to those in a microscope and highly sensitive cameras are used to obtain the high-quality images of the clouds of ultra-cold atoms shown in this chapter. As well as producing beautiful pictures, this ability to visualize the atoms directly has enabled the field to develop very rapidly. There are many remaining physics challenges regarding systems with mixtures of different spin states and also regarding fermionic systems; however, the field is now sufficiently developed for the important technological applications of Bose condensates to be considered, e.g. the application of atomic interferometry to measure the gravitational acceleration of atoms, and to measure rotation in navigational

**Figure 7.18.** Part (a) shows an analogy between a molecular Bose–Einstein condensate and couples dancing – each couple represents a molecule formed from two identical fermionic atoms. Making the interaction between the atoms weaker (by changing the applied magnetic field strength) gives a gas comprised of loosely associated pairs of atoms as illustrated in (b); this quantum gas closely resembles the situation in a superconductor where the electrons in a metal (or other type of solid) form so-called Cooper pairs. Conventional superconductors are described by the theory of Bardeen, Cooper, and Schrieffer (BCS theory); with ultra-cold atoms it is now possible to study the crossover from the BEC to the BCS regime in a carefully controlled way and this will elucidate important features of superconductivity. It is hoped that this new approach may give important insights into phenomena such as high-temperature superconductivity. (Figure by Markus Greiner, JILA.)

gyroscopes. Four-wave mixing was described as an example of the laser-like properties of the matter waves in the condensate and there are other related phenomena yet to be studied. Futuristic dreams also include applications to quantum information processing in which the quantum information is stored in the spin of atoms in an optical lattice. Large arrays can be prepared in a state with one atom per lattice site by means of a quantum phase transition (see Chapter 9) and these represent a very promising physical basis for future work that is only just starting to be explored.

## FURTHER READING


1. C. J. Pethick and H. Smith, *Bose–Einstein Condensation in Dilute Gases*, Cambridge, Cambridge University Press, 2001.
2. L. P. Pitaevskii and S. Stringari, *Bose–Einstein Condensation*, Oxford, Oxford University Press, 2003.
3. C. J. Foot, *Atomic Physics*, Oxford, Oxford University Press, 2004.
4. J. R. Anglin and W. Ketterle, Bose–Einstein condensation of atomic gases, *Nature* **416** (2002) 211.
5. S. L. Rolston and W. D. Phillips, Nonlinear and quantum atom optics, *Nature* **416** (2002) 219.
6. K. Burnett, P. S. Julienne, P. D. Lett, E. Tiesinga, and C. J. Williams, Quantum encounters of the cold kind, *Nature* **416** (2002) 225.
7. Nobel website: http://www.nobel.se/. This site includes links to research groups.


# Disclaimer

Certain commercial equipment, instruments, and materials are identified in this chapter. Such identification is not intended to imply recommendation or endorsement by the National Institute of Standards and Technology, nor is it intended to imply that the materials and equipment identified are necessarily the best available for this purpose.

# William D. Phillips

After studying physics at Juniata College, Huntingdon, Pennsylvania, William Phillips began research for his Ph.D. at MIT under Daniel Kleppner, making a precision measurement of the magnetic moment of the proton in water. During this time, tunable dye lasers became available, opening up the study of laser-excited atomic collisions and providing a second Ph.D. research topic. During postdoctoral work at MIT under a Chaim Weizmann Fellowship, he turned to the search for Bose–Einstein condensation in spin-polarized hydrogen.

In 1978 he moved to the National Bureau of Standards (subsequently the National Institute of Standards and Technology – NIST), making precision measurements. But his experience with lasers and atomic physics opened up new possibilities. These crucial developments in laser cooling went on to attain temperatures much lower than had been expected, effectively stopping atoms in their path and opening up new paths to Bose–Einstein-condensation physics with alkali vapors. For these achievements, he shared the 1997 Nobel Physics Prize with Steven Chu and Claude Cohen-Tannoudji.

In 1998 he was awarded the American Physical Society's Arthur L. Schawlow Prize in Laser Science. He is a Fellow of the American Physical Society, the Optical Society of America, the American Academy of Arts and Sciences, and the US National Academy of Sciences.

## Christopher Foot

After having begun his physics career with an honours degree and doctorate from the University of Oxford, Christopher Foot spent several years working on high-resolution laser spectroscopy of atomic hydrogen at Stanford University, supported for part of that time by a Lindemann Trust Fellowship. He returned to Oxford as a Junior Research Fellow at Jesus College and started research on laser cooling and trapping of atoms. He continued this work after he became a University Research Fellow supported by the Royal Society and he also collaborated with the National Physical Laboratory on the development of a cesium atomic-fountain clock. In 1991 he became a University Lecturer and Tutorial Fellow at St. Peter's College, Oxford. Professor Foot is a Fellow of the American Physical Society. His current research interests include the study of the superfluid properties of Bose–Einstein-condensed atoms and the experimental realization of schemes for direct quantum simulation of quantum-mechanical systems, such as spins arranged on a lattice, using neutral atoms in an array of optical traps.

# 8 Superfluids

Henry Hall

## 8.1 What is a superfluid?

The word "superfluid" was coined to describe a qualitatively different state of a fluid that can occur at low temperatures, in which the resistance to flow is identically zero, so that flow round a closed path lasts for ever – a persistent current. Superfluidity can occur either for uncharged particles such as helium atoms or for charged particles such as the electrons in a metal. In the latter case the flow constitutes an electric current and we have a superconductor. Since an electric current is accompanied by a magnetic field, it is much easier to demonstrate the presence of a persistent current in a superconductor than in a neutral superfluid.

In this chapter I shall describe the properties of superfluids, starting with the simplest and working up to more complicated examples. But in this introductory section I shall depart from the historical order even further by turning to the last page of the detective story so as to catch a glimpse of the conclusion from almost a century of experimental and theoretical research; we shall then have an idea of where we are heading.

The conclusion is that superfluidity is more than just flow without resistance. It is an example of a transition to a more ordered state as the temperature is lowered, like the transition to ferromagnetism. At such a transition new macroscopically measurable quantities appear: in the case of a ferromagnet, the spontaneous magnetization. Such new measurable quantities are known as the *order parameter* of the low-temperature phase. The new quantities in the case of a superfluid are more subtle and more surprising: the amplitude and phase of the de Broglie wave associated with the motion of the superfluid particles. A superfluid can therefore exhibit quantum-mechanical effects on a macroscopic scale.

In the quantum mechanics of a single particle it is often stated that the only measurable property of the wave function is $|\Psi|^2$; the phase $\varphi$ is not observable. But for $N$ particles a form of the Heisenberg uncertainty principle states that

$$\Delta N\,\Delta\varphi \gtrsim 1. \tag{8.1}$$

It is thus possible to construct a minimum-uncertainty wavepacket for $N$ particles in which the fractional uncertainty in particle number $\Delta N/N$ and the uncertainty in phase $\Delta\varphi$ (in radians) are both of order $1/\sqrt{N}$. This is essentially what happens in superfluids, and also in laser light and in Bose-condensed atomic gases.

*The New Physics for the Twenty-First Century*, ed. Gordon Fraser.

## 8.2 Quantum fluids

Quite apart from its superfluidity, liquid helium is unlike other liquids in that it remains liquid down to the lowest attainable temperatures at its vapor pressure; a pressure of 25 atmospheres is required to solidify it at the absolute zero of temperature. This is itself an indication of the importance of quantum mechanics in this system, specifically the familiar uncertainty relation between position $x$ and momentum $p = mv$,

$$\Delta p \, \Delta x \gtrsim \hbar, \tag{8.2}$$

where $\hbar$ is Planck's constant divided by $2\pi$. Equation (8.2) shows that a particle cannot be localized too strongly in space without large fluctuations in momentum and consequently high energy, known as zero-point energy. Therefore helium atoms prefer to move freely past each other, rather than be localized on a regular lattice, until they are compressed to a density at which the short-range repulsion between them prevents them from changing places. This effect is similar for the light isotope helium-3, except that the lighter atomic mass leads to a rather higher solidification pressure, about 30 atmospheres.

However, in another way the two isotopes are profoundly different. Quantum particles are *indistinguishable* because particles of the same type are strictly identical and so cannot be labeled. Consequently $|\Psi|^2$ for $N$ particles cannot change if any two of them are exchanged. This leaves two possibilities: in the exchange either $\Psi \to \Psi$ or $\Psi \to -\Psi$. If $\Psi$ is unchanged, the particles are said to obey Bose–Einstein statistics and are known as bosons; if $\Psi$ changes sign they are said to obey Fermi–Dirac statistics and are known as fermions. Whether a particle is a boson or a fermion turns out to depend on another property: quantum particles can have an intrinsic angular momentum as if they were spinning about their axis. This spin is measured in units of $\hbar$, and relativistic arguments show that particles with integral spin (including zero) are bosons, whereas particles with half-integral spin are fermions. For compound particles such as helium atoms we just count how many fermions are exchanged when the atoms are exchanged. If the number is odd then $\Psi$ changes sign and the atom is a fermion; if it is even the atom is a boson. Since electrons, protons, and neutrons are all spin-$\frac{1}{2}$ fermions, we find that helium-4 is a boson and helium-3 is a fermion. We shall see that helium-3 becomes superfluid only at a temperature 1000 times lower than helium-4, so it is a good guess that this is somehow connected with the difference between fermions and bosons.

What is that difference? If two particles *that are in the same state* are exchanged $\Psi$ cannot change; this is possible for fermions only if $\Psi \equiv 0$. In other words, the probability of having two fermions in the same state is zero; this is the Pauli exclusion principle. At high enough temperatures the particles will be so energetic that it is unlikely that two will want to be in the same quantum state anyway, so the difference between bosons and fermions is unimportant. But at absolute zero all the particles in a Bose gas can go into the lowest-energy state, whereas for a Fermi gas each particle must be in a different state from all the others because of the exclusion principle, so that all states up to an energy $E_F$, called the Fermi energy, are occupied and all higher-energy states are unoccupied. The lowest-energy states of gases of bosons and fermions are thus completely different.

We can estimate a characteristic temperature $T_0$ below which the difference between bosons and fermions matters by equating the average separation $(V/N)^{1/3}$ of $N$ particles in volume $V$ to the de Broglie wavelength $\lambda = h/(mv)$ of a particle of thermal kinetic energy $\frac{1}{2}mv^2 = \frac{3}{2}k_B T_0$, where $k_B$ is Boltzmann's constant. The result is

$$T_0 = \frac{h^2}{3mk_B}\left(\frac{N}{V}\right)^{2/3}. \tag{8.3}$$

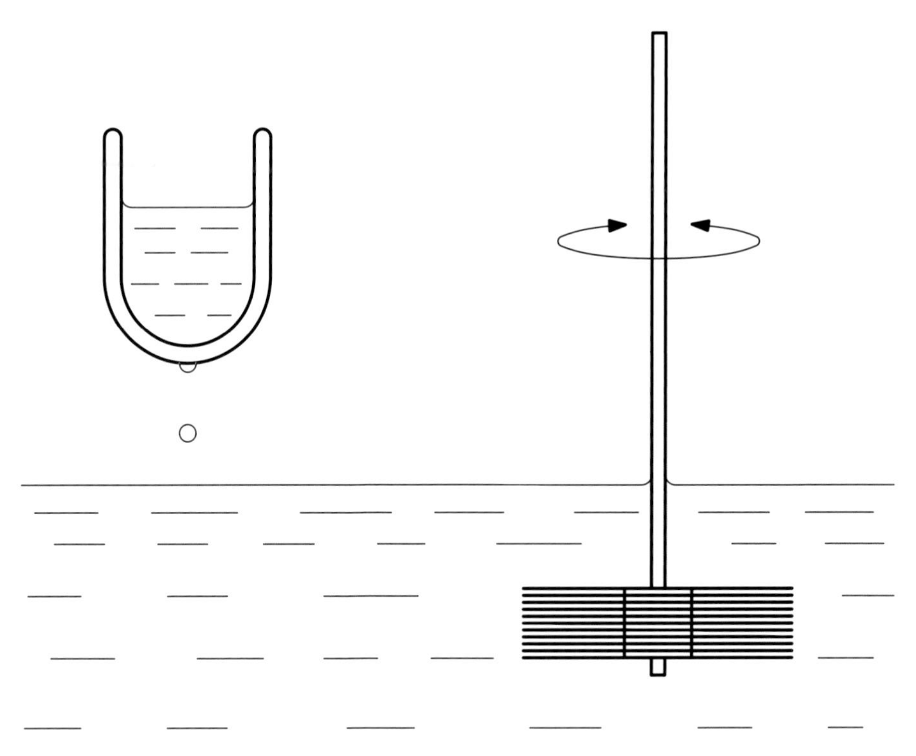

**Figure 8.1.** A schematic diagram illustrating two key early flow experiments on helium II. On the left, liquid flows through the adsorbed surface film to drip off the bottom of a beaker, indicating negligibly small viscosity. On the right, Andronikashvili's experiment with torsional oscillations of a pile of disks. The damping of the oscillations indicates a viscosity of the same order as that of helium I, but the period of the oscillations depends on temperature and shows that the fraction of the liquid that is dragged round by the disks decreases from unity at $T_\lambda$ to zero at absolute zero.

Much below $T_0$ the de Broglie wavelength is much larger than the particle separation and we have a quantum fluid; much above $T_0$ the behavior is almost classical. If we put numbers appropriate for liquid helium into Equation (8.3) we find $T_0 \sim 3\,\mathrm{K}$; for electrons in a typical metal with one free electron per atom we find $T_0 \sim E_\mathrm{F}/k_\mathrm{B} \sim 2 \times 10^4\,\mathrm{K}$. Conduction electrons in a metal are therefore very much a quantum fluid even at room temperature.

## 8.3 Liquid helium-4

The existence of a phase transition in liquid helium-4 at 2.17 K was first noticed as a minor anomaly in the density, which was found to be accompanied by a sharp peak in the heat capacity shaped like a Greek letter lambda. The transition was therefore called the $\lambda$-point and the transition temperature $T_\lambda$. The high-temperature phase was named helium I and the low-temperature phase helium II. The remarkable properties of helium II that led to the invention of the word "superfluid" were discovered during the 1930s.

Figure 8.1 illustrates a paradox discovered in the early flow experiments. Helium II can flow without resistance through very narrow channels, which is dramatically illustrated by its ability to empty a beaker by flowing out of it via an adsorbed film only about 100 atoms thick. On the other hand, motion of a solid body through the liquid seemed to be resisted by viscosity much as in helium I. A particularly clear illustration of this, which indicated the resolution of the paradox, was the classic experiment of Andronikashvili in 1946, suggested by the Russian theorist Landau (Figure 8.2). Andronikashvili studied the period and damping of torsional oscillations of a stack of closely spaced disks. The damping suggested the presence of much the same viscosity in helium II as in helium I, but the period was temperature-dependent below

**Figure 8.2.** The Soviet theorist Lev Landau (Nobel Prize 1962) outside the Institute for Physical Problems in Moscow in 1937. This photograph and Figures 8.19 and 8.20 were taken by the late David Shoenberg, who was the author's research supervisor; *his* research supervisor was Piotr Kapitza (Figure 8.19).

$T_\lambda$, tending toward the period in vacuum at the lowest temperatures. Since the period indicates the inertia of the liquid that is being dragged round by the disks, it seems that a decreasing fraction of the liquid moves with the disks as the temperature is reduced below $T_\lambda$.

These observations suggest a *two-fluid model* of helium II, in which the total density $\rho$ of the liquid is divided into a normal part $\rho_n$ that behaves like any other liquid and a superfluid part $\rho_s$ that flows without resistance. The proportions change smoothly from all normal fluid at $T_\lambda$ to all superfluid at absolute zero. A theoretical basis for this model had been provided in 1941 by Landau's idea of thermal excitations against a background of the quantum-mechanical ground state of the liquid. According to this theory all the entropy of the liquid would be associated with the normal fluid, so thermal effects of superflow were to be expected. A striking example is shown in Figure 8.3, where the superfluid component is drawn toward a source of heat so strongly that a fountain is created. Thermal effects are shown more quantitatively in the phenomenon of second sound, a wavelike propagation of temperature. In ordinary

**Figure 8.3.** The fountain effect. When heat is applied via the spiral wire heater liquid is drawn in through the plug of fine powder at the bottom with such force that it shoots out of the top. In the original discovery by Allen and Jones in 1938 the accidental source of heat was light from a torch; for this photograph heat radiation was removed by a blue filter. This photograph and Figure 8.7 are stills from movies made by the late J. F. Allen. (Courtesy of J. F. Allen, University of St. Andrews.)

systems temperature variations are rapidly attenuated: a meter below the surface of the Earth there is hardly any annual temperature variation. But in helium II temperature can propagate as a wave by counterflow of the normal fluid and superfluid, so that compressions of normal fluid correspond to rarefactions of superfluid. Second sound propagates with a temperature-dependent speed of about 20 $\mathrm{m\,s^{-1}}$, an order of magnitude slower than ordinary sound, in which the two fluids move together to create a pressure wave moving at about 240 $\mathrm{m\,s^{-1}}$.

The experimental verification of the two-fluid model does suggest that we should take seriously an obvious similarity to the phenomenon of Bose condensation in an ideal gas of bosons. Below a temperature of order $T_0$ (Equation (8.3)) a finite fraction of all the gas atoms is in the state of lowest energy, and this fraction increases on cooling until it reaches unity at absolute zero. The phenomenon has been demonstrated and studied extensively over the last few years with clouds of atoms in a magnetic trap. It is tempting to identify the superfluid fraction of helium II with the fraction of Bose-condensed atoms, since these have a common wave function; but this would be quite wrong. The fraction of atoms with zero momentum in helium II can be estimated from neutron-scattering experiments. If the neutrons are sufficiently energetic that they scatter essentially off single atoms, the momentum distribution of the atoms off which they scatter can be found from a "billiard-ball" calculation. The conclusion is that only about 10% of the atoms are in the zero-momentum state at absolute zero, although the liquid is 100% superfluid. We conclude that, because of the stronger interactions in a liquid, the ground-state wave function of liquid helium is more complicated than a simple Bose condensate. But the properties of the superfluid do suggest that the concept of a macroscopic wave function is applicable to the whole superfluid fraction. This implies that, whatever the ground-state wave function is, it is highly correlated, probably involving correlated pairs of atoms with zero mean momentum, and perhaps quartets also. The order parameter in the form of a macroscopic wave function is thus robust against the introduction of interactions, in a way that the condensate fraction is not. Indeed, the ideal Bose gas is a somewhat pathological limit, in that it is infinitely compressible below the condensation temperature: squeezing it just puts more atoms into the ground state.

A more extreme departure from an ideal Bose gas is solid helium-4, but even this system has very recently been shown by Kim and Chan to display superfluid behavior, albeit with a superfluid fraction of only about 1% at $T = 0$, and presumably an even smaller Bose-condensate fraction (see item 12 of Further reading). The possibility of such an effect was first suggested by Leggett in 1970. The essential difference between a fluid (liquid or gas) and a solid is that in a fluid the probability of finding an atom is spatially uniform, whereas in a solid the probability of finding an atom is high near the lattice sites and very low in between. It is argued that this very low probability between sites limits the amplitude of any macroscopic wave function, and hence the maximum superfluid fraction. The property of superfluidity is thus remarkably reluctant to disappear entirely.

The key piece of evidence to demonstrate superfluidity is the existence of persistent currents. It should be possible to set up a persistent current in a toroidal container by rotating the torus about its axis while cooling through $T_\lambda$. Once the system is in the superfluid state, the container can be brought to rest (along with the normal fluid), and the superfluid component should continue to rotate. John Reppy realized that the angular momentum of a persistent current could be measured non-destructively by detecting gyroscopic effects. Those who have played with a toy gyroscope will be aware of its peculiar response: if you push it one way it moves in a perpendicular direction in an effort to keep its angular momentum constant. Reppy's superfluid gyroscope, shown in Figure 8.4, uses this effect to measure the angular momentum of a persistent current in helium II; the experiment is very delicate because of the small amount of angular momentum involved.

A particularly important result of the experiments of Reppy and Depatie in 1964 came from investigating the effect of a slow change of temperature (below $T_\lambda$) on the persistent angular momentum. It decreased as the temperature was raised, in line with the variation of $\rho_s$, and *increased reversibly when the temperature was lowered.* Raising the temperature above $T_\lambda$ destroyed the persistent current, but it could be raised to a few millikelvin below $T_\lambda$ without ill effect. Thus, upon a slow change of temperature, angular momentum of the supercurrent is *not* conserved but the circulating velocity is. In thermal equilibrium excitations are continually being created and destroyed, so $\rho_n$ fluctuates about its mean value. The experiment shows that these fluctuations occur at constant superfluid velocity $v_s$. The angular momentum of the persistent current can change because the thermal excitations serve to exchange momentum with the walls of the container.

If the velocity of a persistent current is not too far below the maximum that is stable, the angular momentum does decay at an observable rate at constant temperature, but in a very characteristic way. If 1% of the angular momentum is lost in an hour, then another 1% is lost in 10 hours, another 1% in 100 hours, another 1% in 1000 hours, and so on. This decay as the logarithm of the time is indicative of fluctuation over a potential barrier. It shows that persistent currents are metastable – the ground state of no flow is lower in energy. A persistent current substantially smaller than the maximum that can be created does essentially last for ever.

To understand persistent currents in terms of a macroscopic wave function we use the fact that there must be an integral number of wavelengths of the de Broglie wave

**Figure 8.4.** Reppy's superfluid gyroscope. The persistent current is originally set up by cooling through $T_\lambda$ with the axis of the toroidal container vertical, and the apparatus rotating about this axis. Rotation is then stopped and the axis of the torus is slowly turned horizontal as shown, by means of the worm gear on the left, so that further slow rotation about a vertical axis forces the trapped angular momentum to change direction (precession). This produces a twist of the horizontal torsion fiber from which the torus is suspended, which is sensed by the block with electrical connections in front of the torus. The torus is packed with fine powder because superflow is stable to higher velocities in narrow channels. (Photo courtesy of J. D. Reppy.)

in the circumference of the annulus, or, alternatively, the change in phase must be an integral multiple of $2\pi$. The change in phase $d\varphi$ due to a displacement $dl$ normal to the wavefront is given by

$$\frac{d\varphi}{dl} = \frac{2\pi}{\lambda} = \frac{mv_s}{\hbar}, \tag{8.4}$$

so integration round the annulus gives

$$\oint d\varphi = \oint \frac{mv_s}{\hbar} \cdot dl = 2\pi n. \tag{8.5}$$

The dot in Equation (8.5) is a reminder that $dl$ is a component of displacement parallel to $v_s$. Equivalently, and more conveniently, we can take the component of $v_s$ parallel to the displacement $dl$. Equation (8.5) can be rearranged to give

$$\oint v_s \cdot dl = n\frac{h}{m}. \tag{8.6}$$

This integral is known in hydrodynamics as the circulation, and we see that it is quantized in units of $\kappa = h/m = 9.98 \times 10^{-8}\ \mathrm{m^2\,s^{-1}}$. The most convincing argument for this result was given by Feynman (Figure 8.5).

Figure 8.5. Richard Feynman speaking at Kyoto in 1985. His seminal work on superfluidity (see item 1 of Further Reading) was a brief episode in a career of innovation in most areas of theoretical physics. (Photo *CERN Courier*.)

The quantization of circulation was demonstrated experimentally and the value of the quantum confirmed by Vinen in 1961. He used a particular form of toroidal container, a cylinder with a fine wire on the axis. Circulation could be trapped on the wire by cooling through $T_\lambda$ while rotating the cylinder, and then stopping the rotation. This circulation was then measured by studying the transverse vibrations of the wire. The measurement makes use of the mechanism that produces lift on an aircraft wing, known as the Magnus effect: when a moving solid body has a circulation around it, there is a transverse force on it proportional to the circulation. In Vinen's experiment the Magnus force caused a precession of the plane of vibration of the wire. Vinen demonstrated the trapping of a single quantum of circulation. In later experiments two and three quanta were trapped.

The quantization condition (8.6) explains the stability of persistent currents: there is no continuous process that will change the value of $n$; it is a topologically stable quantum number. However, what happens if we apply Equation (8.6) to a simply connected volume of liquid? There is then no hole to prevent our integration contour shrinking continuously to zero, so something terrible must happen if $n \neq 0$. Consider $n = 1$. When our integration is round a small circle of radius $r$, $v_s = \hbar/(mr)$; when $r$ is of atomic dimensions $v_s$ is on the order of the speed of sound and macroscopic ideas break down. There must be a node in the wave function and a region of atomic dimensions where superfluidity is suppressed. A little thought shows that such nodes must form lines in the liquid: what we have is a vortex line, as seen when emptying the bath, but very much weaker. Nucleation of such a vortex and passing it across the flow channel is one possible mechanism for the decay of persistent currents. Vortices with $n > 1$ are probably not important because their energy is proportional to $n^2$, so it is preferable to have $n$ single quantum vortices.

Quantized vortices solve the problem of how helium II can achieve thermal equilibrium with a rotating container. In the absence of constraints the lowest-energy state of any system in equilibrium with a rotating container is rotation in synchronism with the container. A body rotating with angular velocity $\Omega$ has velocity $r\Omega$ at radius $r$, so the circulation round a circle of radius $r$ is $2\pi r^2\Omega$. In the absence of vortices the superfluid component of helium II cannot achieve this in a simply connected container and will not rotate; only the normal fluid can follow the motion of the container. However, this circulation can be very closely approximated if we insert into the superfluid a uniform array of single quantum vortices at a density of $2\Omega/\kappa$ lines per unit area, and this gives the lowest-energy state that can be achieved subject to the constraint of quantized circulation. Even at quite modest rates of rotation there are a lot of vortices: rotation at $1\ \mathrm{rad\,s^{-1}}$ produces $20\ \mathrm{lines\,mm^{-2}}$.

The first evidence for the existence of this vortex array was obtained from experiments on second sound by Vinen and the author in 1956. They found an additional attenuation of second sound proportional to the angular velocity in uniformly rotating

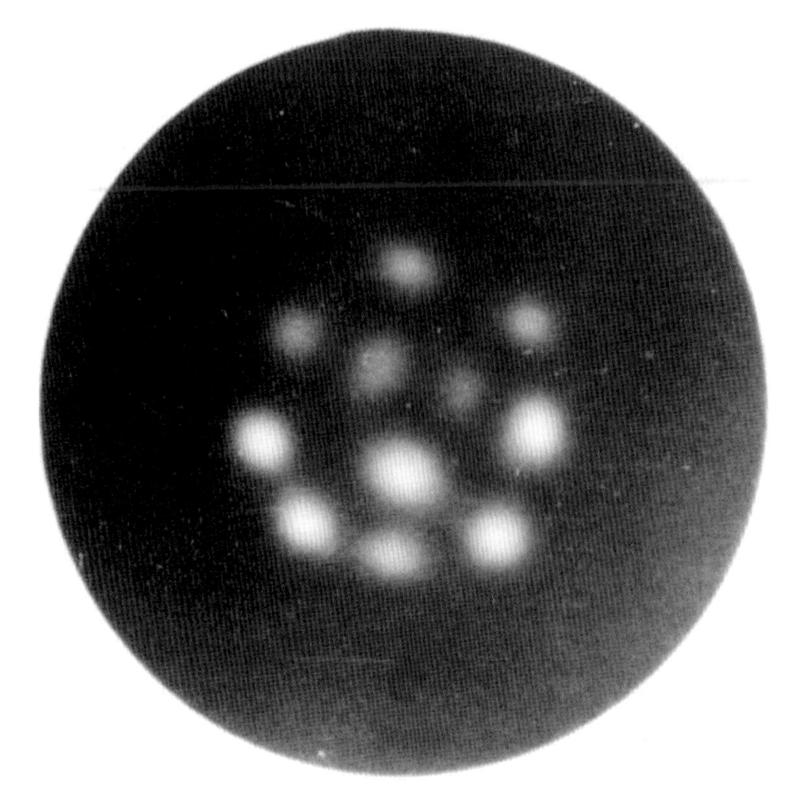

**Figure 8.6.** An image of an array of 11 vortices in rotating helium II obtained by decorating the vortex cores with negative ions. (Image courtesy of R. E. Packard.)

helium II. This could be explained by invoking a resistive force opposing the counterflow of normal fluid and superfluid due to the scattering of thermal excitations by vortices. A vortex density proportional to the angular velocity neatly explained the observations. More direct evidence was obtained by Packard and colleagues, making use of the fact that negative ions (bubbles containing an electron) are attracted to the core of a vortex (which is a sort of linear bubble). They decorated the vortex cores with electrons, and then applied an electric field to pull the electrons out through the liquid surface and focused them magnetically onto a fluorescent screen, which could be photographed. This gave an image of the points on the liquid surface from which the electrons had emerged. The first evidence of discrete vortices was obtained in 1974, but it took until 1979 to reduce vibration to the point at which an array stationary in the rotating frame could be photographed, an example of which is shown in Figure 8.6. Each vortex moves with the velocity field due to all the others, and thus follows the rotation of the container. It is somewhat easier to see vortices in the Bose condensate of trapped atomic gases; they are sufficiently close to an ideal Bose gas to be highly compressible, so there are substantial density variations associated with a vortex.

## 8.4 Superconductivity

Superconductivity was discovered by Kamerlingh Onnes (Nobel Prize 1913) in 1911, not long after his first liquefaction of helium. He observed the resistance of mercury drop to an unmeasurably small value in a very narrow temperature interval around 4.2 K. For a material with identically zero electrical resistance, Maxwell's equations of electromagnetism show that the magnetic field in the interior of the sample cannot change:

$$\frac{dB}{dt} = 0. \tag{8.7}$$

Figure 8.7. A magnet floating over a superconducting tin dish on a cushion of magnetic lines of force, produced by the expulsion of flux from the superconductor by the Meissner effect. The concave upper surface of the superconductor enables the magnet to float in a laterally stable position. (Courtesy of J. F. Allen, University of St. Andrews.)

If the external field changes, screening currents are induced in a thin surface layer so as to keep the interior field constant. It was not until 1933 that Meissner and Ochsenfeld showed that a superconductor is more than a perfect conductor: for a pure annealed sample Equation (8.7) should be replaced by the stronger condition

$$B = 0. \tag{8.8}$$

When a superconducting metal is cooled through the transition in a magnetic field the field is expelled from the interior – the Meissner effect. It costs energy to expel the field, but this is more than compensated by the energy gained by going into the superconducting state, provided that the field is not too large. This argument suggests that a sufficiently large magnetic field will destroy superconductivity, which is indeed what is observed.

A particularly striking demonstration of the Meissner effect is the floating magnet, Figure 8.7. This is usually demonstrated by lowering a magnet over a dish that is already superconducting, but this merely demonstrates zero resistance. However, with a thick dish of pure tin the magnet can be rested on the dish before it is cooled below $T_c$, and will then lift off the dish as the tin becomes superconducting and expels magnetic field: this is the Meissner effect. Although the currents that levitate the magnet are in the upper surface of the superconductor, this demonstration of the Meissner effect does not work if the dish is made too thin. There is then not enough energy released in the superconducting transition to raise the magnet against gravity. Instead, concentrated bundles of flux penetrate the tin and hold it locally normal. (Note the simplicity and visual impact of this demonstration compared with the difficulty of demonstrating persistent currents in a neutral superfluid.)

These properties are clearly similar to those of superfluid helium, so we might expect the order parameter of the superconducting phase to be a macroscopic wave function also. But for helium we related this macroscopic wave function to the phenomenon of Bose condensation. How can we obtain a similar result for fermions? The answer is to combine them together in pairs so that they look like bosons. The theoretical breakthrough came in 1956 when Leon Cooper showed that with any attractive interaction a Fermi gas is unstable with respect to the formation of bound pairs with equal and opposite momenta and opposite spins, now known as Cooper pairs. The pairs form with

**Figure 8.8.** John Bardeen, Leon Cooper, and John Schrieffer (Nobel Prize 1972) in alphabetical order, with Cooper symbolically pointing the way, at the opening of the IBM Yorktown Heights laboratory. (Photo courtesy of Sir Brian Pippard.)

zero relative orbital angular momentum, $L = 0$, known as an s-wave state. Bardeen, Cooper, and Schrieffer (BCS; Figure 8.8) showed in 1957 how to construct a superconducting ground-state wave function in which all the pairs had the same center-of-mass momentum – zero in the ground state, nonzero in a current-carrying state. The attraction mechanism that they proposed involved lattice vibrations or phonons, which had been suggested by the observed dependence of $T_c$ on isotopic mass. It can be visualized by considering the attraction between two people on a bed: each makes a depression in the mattress that the other tends to roll into.

It is important to notice that a Cooper pair is not really a boson, because excited states of Cooper pairs do not exist. The pairing is a collective phenomenon, and the lowest-energy excitation of the system is breaking up a Cooper pair, which requires an energy of order $k_B T_c$. The fact that all the pairs necessarily have the same mean momentum is sufficient to ensure the existence of a macroscopic wave function as an order parameter. An important aspect of the BCS wave function is the existence of a coherence length $\xi$, which can be thought of as the size of a Cooper pair. Uncertainty-principle arguments show that $\xi$ is of order $T_0/T_c$ times the atomic spacing, and thus typically of order 1 μm. The coherence length is a measure of the rigidity of the superconducting wave function: it is energetically unfavorable to change it on a shorter length scale than this.

We now enquire whether a quantization condition like Equation (8.6) exists for a superconducting ring. An important change for a particle of charge $q$ is that the momentum that relates to the de Broglie wavelength is not just $mv$ but $mv + qA$, where $A$ is a vector potential related to the magnetic flux $\Phi$ by

$$\oint A \cdot dl = \Phi, \tag{8.9}$$

in which the integral is around the perimeter of the area through which the flux passes. A good explanation of how this electromagnetic contribution to the momentum comes about can be found in the second volume of Feynman's *Lectures on Physics*. Equation (8.5) is now replaced by

$$\oint d\varphi = \oint \frac{mv_s + qA}{\hbar} \cdot dl = 2\pi n. \tag{8.10}$$

Since currents flow only in the surface of a superconductor, for a path of integration around the ring away from the surface we find that

$$\oint A \cdot dl = n\frac{h}{q} = n\Phi_0, \tag{8.11}$$

where $\Phi_0$ is known as the flux quantum. Experiments to determine the flux quantum were first done in 1961 by Deaver and Fairbank and by Doll and Näbauer. They measured the flux trapped by a thin layer of superconductor deposited on either a quartz fiber or a copper wire, and found that Equation (8.11) was satisfied with $q = 2e$, as would be expected if the supercurrent were carried by Cooper pairs of electrons.

In view of Equation (8.11) it is reasonable to ask why quantized flux lines do not penetrate a superconductor as vortices do helium II? Why does the Meissner effect happen? The answer is that there is a surface energy at a boundary between normal and superconducting regions proportional to the difference between the coherence length $\xi$ and the penetration depth of screening currents $\lambda$. For most pure superconductors $\xi > \lambda$, and it is energetically unfavorable for flux to penetrate, so the Meissner effect occurs. But for many alloys, and also pure niobium, $\xi < \lambda$, so the surface energy is negative and it pays for normal and superconducting regions to be as finely divided as possible, limited only by the quantization condition (8.11). Such materials are known as type-II superconductors; a flux lattice in a type-II superconductor is shown in Figure 8.9. These materials are of great technical importance because they can carry supercurrents in much higher magnetic fields than can type-I superconductors. It is important for applications that the flux lines should be pinned by metallurgical defects so that they cannot move, because moving flux lines induce an EMF, so the resistance is then not zero.

In a neutral superfluid the supercurrents are not confined to a surface layer, so superfluid helium can be regarded as the limit $\lambda \to \infty$ and classified as extreme type II.

A useful feature of metals is that it is possible to make a *tunnel junction*, through which electrons can pass by quantum mechanical tunneling, by interposing a thin layer of insulator between two layers of metal. Measurements of tunneling between a superconductor and a normal metal provided the most direct evidence for an energy gap in the available states for an electron in a superconductor, representing the energy required to break up a Cooper pair. Josephson (Figure 8.10) predicted in 1962, while still a graduate student, that for a sufficiently thin insulating layer it should be possible for Cooper pairs to tunnel between two pieces of superconductor. The basic idea is that the macroscopic wave function of the Cooper pairs penetrates into the insulator as an evanescent wave in much the same way as a light wave penetrates into vacuum in the phenomenon of total internal reflection. Josephson predicted two effects: a steady supercurrent could flow through the junction according to

$$I = I_0 \sin(\delta\varphi), \tag{8.12}$$

where $\delta\varphi$ is the difference in phase across the junction; and with a potential difference $\delta V$ across the junction an alternating current should flow at a frequency $\nu$ given by

$$h\nu = 2e\,\delta V. \tag{8.13}$$

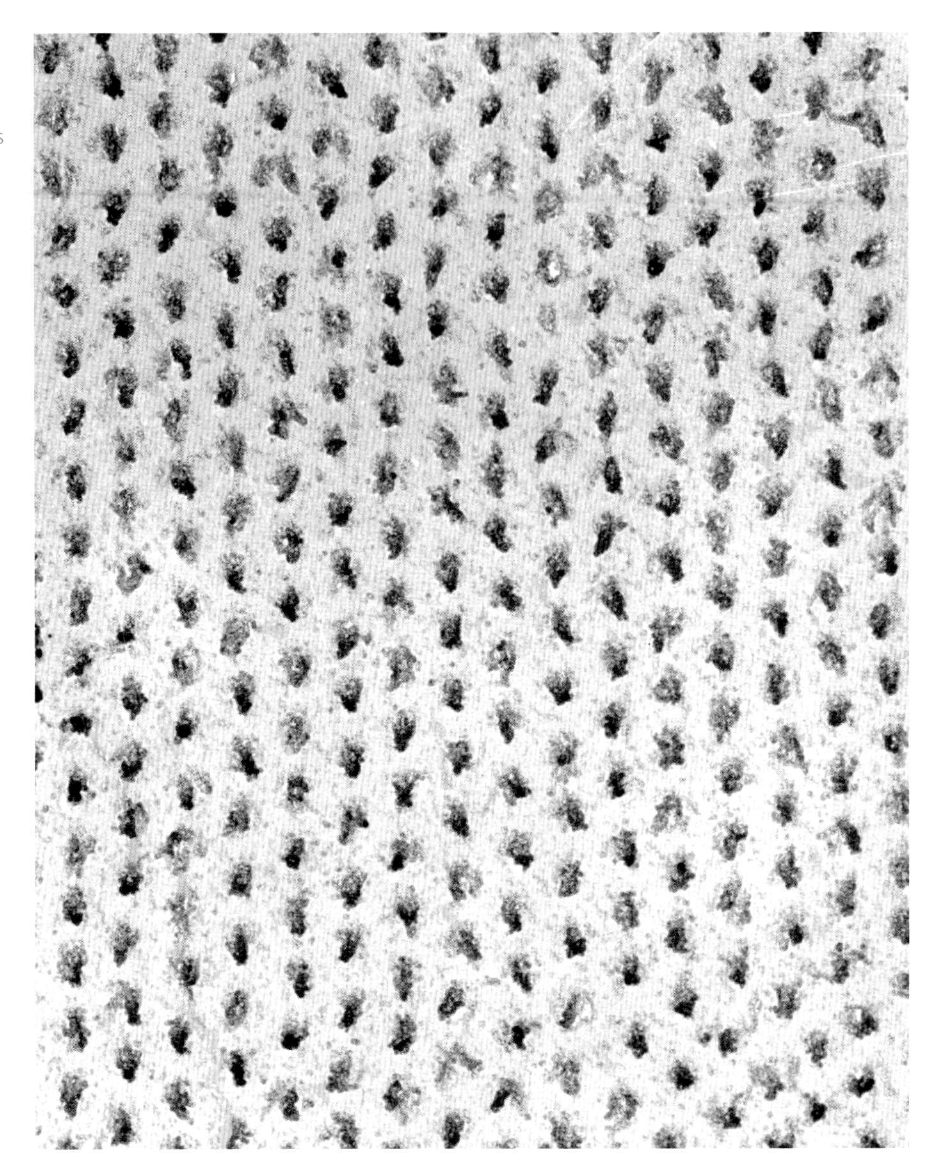

**Figure 8.9.** A photograph of flux lines in a single crystal of a lead–indium alloy, made by sprinkling colloidal iron particles onto the sample, making a replica of the surface, and examining it under an electron microscope. The separation of the flux lines is about 0.6 μm. (Image courtesy of U. Essmann.)

Equation (8.13) is now the basis of the most precise standard of voltage used at national standards laboratories.

Now consider the device shown in Figure 8.11. A superconducting ring is interrupted at two points by Josephson tunnel junctions and provided with input and output connections, so that current from an external circuit flows through the two junctions in parallel. Essentially no phase differences can arise in the bulk superconductor, so the phase integral around the ring (Equation (8.10)) is dominated by the two junctions. For zero flux through the ring, or an integral number of flux quanta, we can have equal phase differences across the two junctions, because they cancel out when integrated round the ring; hence a current up to $2I_0$ can flow through the external circuit. But for a half-integral number of flux quanta we require a phase integral of $\pi$ round the ring. This can be achieved only with a phase difference of $+\pi/2$ at one junction and $-\pi/2$ at the other. Current $I_0$ then flows in opposite directions through the two junctions, so that there is a current $I_0$ around the ring but no current can flow through the external circuit. We have constructed an interferometer for de Broglie waves, a

**Figure 8.10.** Brian Josephson (Nobel Prize 1973) at about the time that he predicted the tunneling of Cooper pairs. This was actually the second Josephson effect. His first prediction, made while still an undergraduate, concerned an effect of temperature differences on experiments using the Mössbauer effect to measure the gravitational redshift. (Photographic Archives of the Cavendish Laboratory.)

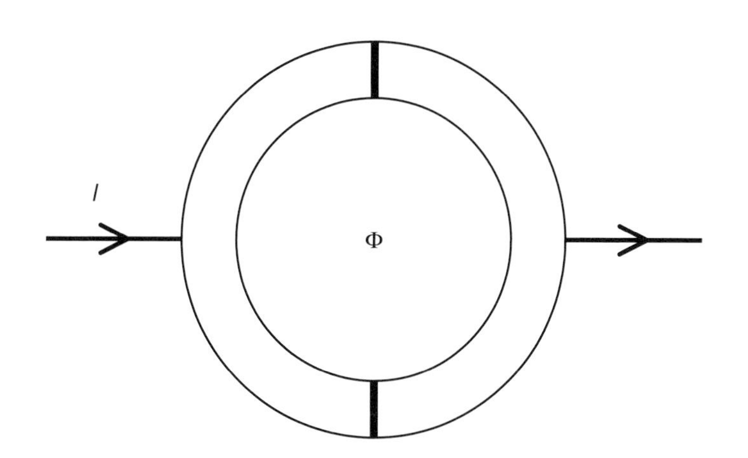


**Figure 8.11.** A superconducting circuit divides into two branches, each containing a Josephson junction. When a magnetic flux Φ exists between the two paths it introduces a phase difference between them, so interference effects occur.

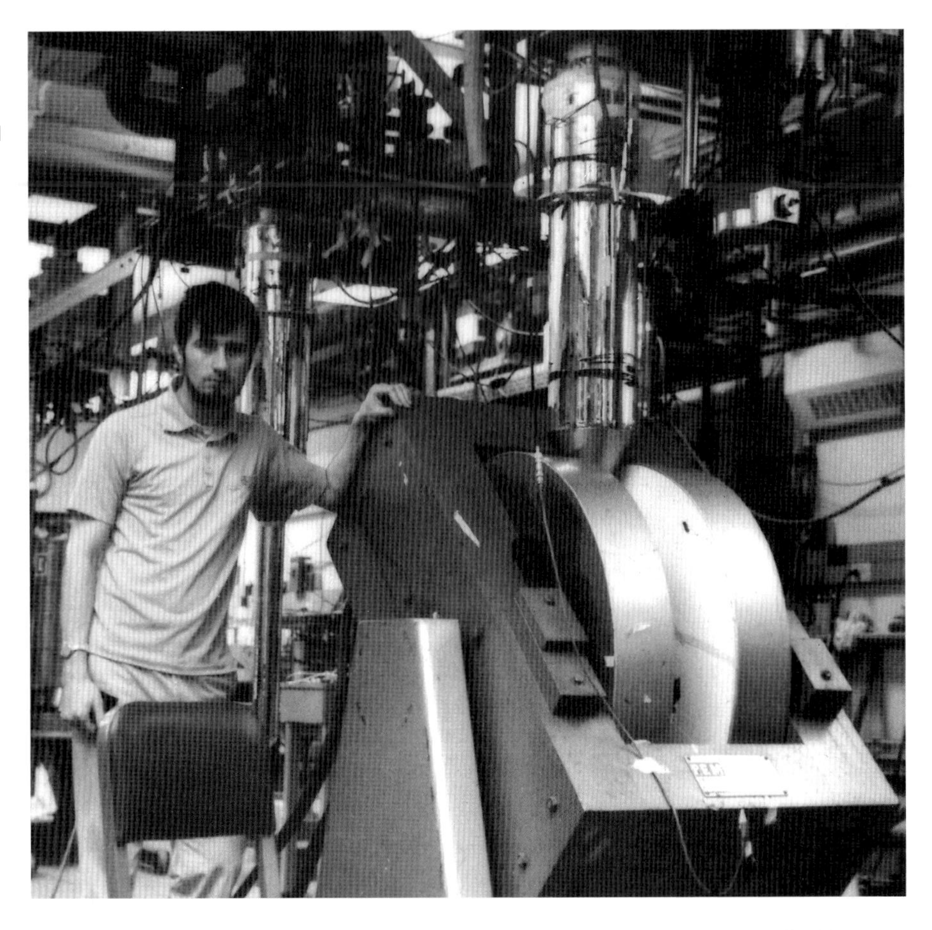

**Figure 8.12.** Graduate student Douglas Osheroff with the apparatus he used to discover the superfluid transition in liquid helium-3. (Photo courtesy of D. D. Osheroff.)

superconducting quantum-interference device or SQUID. Various types of SQUID are now used in high-precision electrical and magnetic measurements.

## 8.5 Liquid helium-3: an anisotropic superfluid

Once the BCS theory of superconductivity had been established it was confidently expected that liquid helium-3 would become superfluid at a low enough temperature by the formation of Cooper pairs of fermionic helium-3 atoms. It was recognized that for the interaction to be effectively attractive the pairs would have to form with nonzero orbital angular momentum, so as to keep the atoms apart and avoid the hard-core repulsion; orbital angular momenta $L = 1$ (known as p-wave) and $L = 2$ (known as d-wave) were considered likely candidates. Because of the requirement for the wave function to change sign when fermions are exchanged, anti-parallel spins ($S = 0$) are possible only for even orbital angular momentum, zero for the superconductors considered so far. For odd orbital angular momentum a triplet spin state with total spin $S = 1$ is required.

When a phase transition in liquid helium-3 was discovered at Cornell in 1972 by Osheroff (Figure 8.12), Richardson, and Lee (Nobel Prize 1996), the first thing to be noticed was a change in the nuclear-magnetic-resonance (NMR) properties; the demonstration of superfluidity came some time later. The phenomenon of nuclear magnetic resonance arises because nuclei with spin have a magnetic moment parallel to their angular momentum. Gyroscopic effects mean that the forces tending to align the magnetic moment with an external magnetic field instead cause it to precess around the

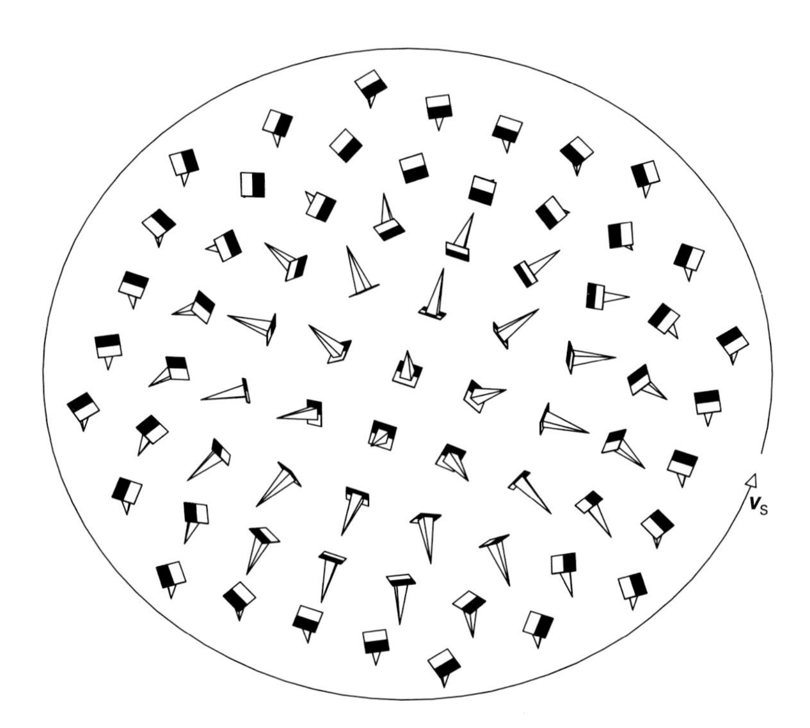


**Figure 8.13.** An example of a textural vortex in $^3$He-A. The cones indicate the local direction of $\hat{\ell}$ and the black and white halves of the base enable rotation about $\hat{\ell}$ to be followed. Note that turning $\hat{\ell}$ smoothly from up at the center to down at the perimeter produces a rotation of $4\pi$ about $\hat{\ell}$ as we go round the perimeter, and therefore two quanta of circulation.

field direction at a rate proportional to the field, given by the Larmor frequency. The system thus exhibits a resonant response at this frequency, which is modified by other forces tending to orient the nuclei. The continued existence of NMR in liquid helium-3 well below $T_c$ was a strong indication of spin-triplet pairing, since singlet pairs ($S = 0$) have no magnetic moment. Particularly significant was a large shift of NMR frequency as the liquid cooled below $T_c$ at the melting pressure, which Leggett (Nobel Prize 2003) soon showed could be explained by an ordering of the relative orientation of spin and orbital angular momenta leading to large amplification of the effects of the very weak interaction between nuclear magnetic moments. One mechanism for producing such ordering would be the formation of Cooper pairs.

The transition temperature $T_c$ varies from 2.7 mK at the melting pressure to 1.0 mK at zero pressure. It is interesting to note that $T_c/T_0$ is of the same order as for superconductors and hence the coherence length $\xi$ is similar. There are two principal superfluid phases: the A-phase at pressures above 21 bar and temperatures near $T_c$; and the B-phase at low temperatures. The phases differ in the nature of the ordering of the spin and orbital angular momenta of the Cooper pairs, but both phases have $L = 1$ and $S = 1$.

The A-phase combines the properties of a nematic liquid crystal and a superfluid, and has unique properties as a result of having both types of ordering at once. The liquid-crystal aspect of the order parameter is represented by a vector $\hat{\ell}$, which is the direction of the orbital angular momentum of the Cooper pairs; in a nematic this special direction would be the orientation of a long-chain molecule. However, $\hat{\ell}$ is more than just a direction, it is more like the $\hat{z}$ axis of a Cartesian coordinate system, because rotation around the direction $\hat{\ell}$ also has a physical significance: it represents a change in phase of the superfluid wave function. Thus, as we move through the liquid, rotations of $\hat{\ell}$ produce a liquid-crystal texture and rotations around $\hat{\ell}$ produce superflow.

To see how texture and superflow interact, consider the example shown in Figure 8.13. The figure shows a perspective view of a plane in the liquid, with the local direction of $\hat{\ell}$ indicated by cones. We assume an essentially two-dimensional texture, so that there is no variation perpendicular to the plane shown. The texture illustrated is formed by starting from $\hat{\ell}$ up in the center and, as we move out radially from the center in any direction, rotating $\hat{\ell}$ smoothly about an axis in the plane perpendicular to the radius, until it is pointing down at the edge of the diagram. Figure 8.13 shows that the result of this process is to produce a rotation of the order parameter by $4\pi$ around $\hat{\ell}$ as we go round the perimeter of the diagram, and thus a $4\pi$ change in the phase of the superfluid wave function. There are therefore two quanta of circulation round the perimeter of the diagram, with the quantum given by Equation (8.6) with $m$ the mass of a Cooper pair, twice the mass of a $^3$He atom. Note, however, that the circulation increases continuously from zero as we move radially outward; a general expression for the circulation includes an integral of the texture along the path of integration.

This example shows that it is possible to manipulate textures in $^3$He-A so as to produce vortices without a singular core. Such vortices are much more easily produced than singular vortices. More generally, a texture can evolve in time so as to destroy superflow. This effect is specific to $^3$He-A, because in $^3$He-B the phase of the wave function does not couple to other degrees of freedom of the order parameter. Gyroscopic experiments on persistent currents in superfluid helium-3 do indeed show that the largest persistent currents that can be generated in the A-phase are very much smaller than those in the B-phase, which is consistent with these ideas. It is likely that the residual stability of superflow in the A-phase is due to a boundary condition that restricts the evolution with time of the texture. The vector $\hat{\ell}$ has to be normal to a solid surface, and therefore the surface texture can evolve only by the motion of singularities whereby $\hat{\ell}$ changes from pointing into the boundary to pointing out of it.

When we consider the spin degrees of freedom of the order parameter it becomes apparent that we have something resembling a "two-superfluid model": spin up and spin down behave as independent superfluids, each with their own phase variable. This is fairly obvious for the A-phase, in which only Cooper pairs with spin projection $S_z = +1$ and $S_z = -1$ are formed. In the B-phase pairs with $S_z = 0$ are also formed, but it turns out that there are still only two independent phase variables. These are usually taken as $\varphi$, associated with the motion of all spin states together, and $\varphi_{\mathrm{sp}}$, associated with the counterflow of up spin and down spin. Gradients of $\varphi$ lead to a flow of matter, whereas gradients of $\varphi_{\mathrm{sp}}$ cause spin supercurrents that transfer magnetization from one place to another. Consequently, spin waves exist in both superfluid phases.

In the B-phase, as we have already remarked, mass superflow is uncoupled from the other degrees of freedom of the order parameter. There is no unique angular-momentum direction for all the Cooper pairs, as in the A-phase, but instead the directions of the spin and orbital angular momenta of each pair are related to the directions of the linear momenta of the atoms making up the pair. The ordering of the angular momenta of the B-phase pairs is closely related to a state in which the total angular momentum of each pair $J = L + S = 0$, corresponding to anti-parallel spin and orbital angular momenta. It differs from this in that the spin and orbital coordinates are rotated relative to each other by an angle of $104^\circ$ about a direction $\hat{n}$ which is *the same for all pairs*. This non-obvious adjustment serves to minimize the energy of the interaction between nuclear magnetic moments, which becomes significant because of the coherent addition of contributions from all the pairs.

The importance of this "magic angle" can be seen by considering a particularly intriguing NMR experiment shown in Figure 8.14, which was first performed by the Moscow group in 1985. In a sample situated in a steady magnetic field that is slightly larger at the bottom than at the top, the spins are set into free precession at a fairly large tipping angle to the magnetic field. Initially the precession is more rapid at the bottom where the field is larger, so a gradient of the precession angle, which corresponds to $\varphi_{\mathrm{sp}}$, arises and a spin supercurrent flows so as to transfer magnetization from the lower

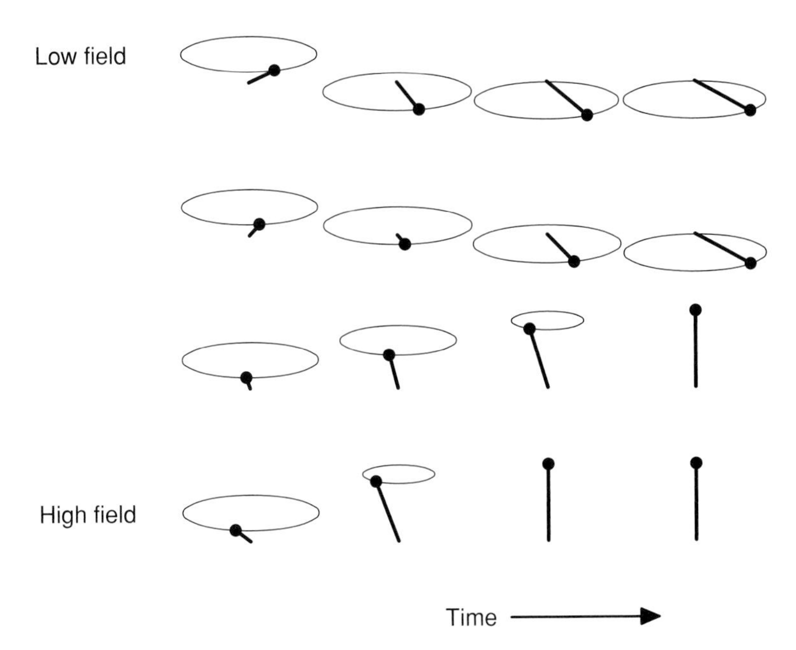


Figure 8.14. Diagrams showing the precession of nuclear magnetization around the field direction during the formation of a homogeneously precessing domain (HPD). Initially the magnetization precesses more rapidly in the higher-field region at the bottom, but the resulting gradient in precession angle creates a spin supercurrent that transfers magnetization downwards until the magnetization is aligned with the field in the lower part of the sample and precessing coherently in the upper part at a tipping angle slightly greater than 104°, which adjusts itself so as to keep the rate of precession constant as the field decreases toward the top.

field at the top to the higher field at the bottom. The tipping angle thus increases at the top and decreases at the bottom. Eventually the sample divides into two domains. In the lower domain the magnetization is aligned with the field and in equilibrium. The NMR signal now comes only from the upper domain, in which the tipping angle is slightly greater than 104° and the spins all precess coherently at the Larmor frequency appropriate to the boundary between the domains. This coherent precession is possible because a tipping angle greater than 104° forces the system to move away from the minimum energy for the magnetic interaction between nuclei, and this raises the NMR frequency for a given field. Consequently, spins in the upper domain can adjust their NMR frequency to match the Larmor frequency at the domain boundary by increasing their tipping angle slightly beyond 104°, and this situation will be stabilized by spin supercurrents. This upper domain is known as the homogeneously precessing domain (HPD).

As the free precession decays and the magnetization returns to thermal equilibrium, the domain boundary moves up the sample until only the lower domain is left. Instead of allowing the HPD to decay, it is possible to maintain it by radio-frequency excitation, as in a normal continuous-wave NMR experiment. Indeed, the domain boundary can be moved up and down at will by adjusting the excitation frequency to match the Larmor frequency at the desired position.

We have already seen that a vortex without a core is possible in the A-phase because of the coupling between flow and textures. In the B-phase mass superflow is uncoupled from the other degrees of freedom, so a core on the scale of the coherence length $\xi$ is required. It does not have to be a normal core, however. There are possible components of the spin and orbital degrees of freedom of the order parameter that have unit angular momentum without involving the phase singularity, and are therefore allowed to be finite on the axis of the vortex. Examples are Cooper pairs with $L_z = +1$ and $S_z = 0$ (like in the A-phase), and pairs with $S_z = +1$ and $L_z = 0$. In general the strengths of various components of the order parameter vary radially within a vortex. Much experimental and theoretical work has been done on vortices in both superfluid phases, mainly at the rotating-cryostat laboratory in Helsinki. The type of vortex seen depends on the speed of rotation, pressure, temperature, and magnetic field. Evidence has been found for even more exotic structures, such as vortices with a non-circular core in the B-phase and vortex sheets in the A-phase. All this work is of interest in connection with a rather less experimentally accessible system, namely rotating neutron stars or pulsars. The neutrons are believed to form a p-wave superfluid of a different type from helium-3, and irregularities in the rotation are believed to be associated with the slipping of vortices pinned in the crust of the star.

The Josephson effect should also exist for a neutral superfluid, provided that a suitable weak link between two reservoirs of liquid can be constructed. One possibility is a hole of dimensions small compared with the coherence length. This is hardly feasible in helium-4, where the coherence length is of the order of atomic dimensions except extremely close to the $\lambda$-point, and phenomena resembling the Josephson effect seen there usually involve the passage of vortices across an aperture rather than tunneling. But it should be feasible for helium-3. One problem with a single hole is that it has very little current-carrying capacity compared with a typical tunnel junction between two layers of metal. The successful observation of oscillatory Josephson currents (Equation (8.13)) at Berkeley in 1997 therefore involved two volumes of helium-3 separated by a silicon nitride window 50 nm thick containing an array of $65 \times 65$ holes of diameter 100 nm spaced 3 $\mu$m apart.

For a neutral system Equation (8.13) is replaced by

$$\kappa \nu = \delta p / \rho, \tag{8.14}$$

where $\kappa$ is the quantum of circulation, $\delta p$ is the pressure difference across the weak link, and $\rho$ is the density of the fluid. In the experiment, flow is driven through the weak link by an elastic diaphragm, the displacement of which gives an indication of $\delta p$. To start the flow a voltage is applied to the diaphragm to produce an initial pressure difference; the displacement is then followed as it relaxes to a new equilibrium. An example of the last part of a transient is shown in Figure 8.15. The first part of the trace shows oscillations at a frequency given by Equation (8.14) and the second part shows oscillations about zero pressure difference. Because of the sine term in Equation (8.12) the behavior is like that of a rigid pendulum. If given a little energy it oscillates about equilibrium, but if given sufficient energy it goes "over the top" and rotates continuously, faster and faster as it is given more energy. The early part of the transient is too noisy for Josephson oscillations to be clearly seen, but if the signal is applied to earphones they can be clearly heard as a descending tone: a remarkable demonstration of the ability of the human ear to extract signal from noise – the cocktail-party effect.

Even more information can be extracted from the oscillations about zero pressure difference. The current is proportional to the rate of change of diaphragm displacement, and the phase can be calculated from an integral of the displacement with respect to time. Equation (8.12) can thus be checked by experiment. The relationship between current and phase is indeed found to be sinusoidal close to $T_c$, but at lower temperatures it tends toward a sawtooth. This is not surprising, because the coherence length is largest near $T_c$ and is not very much greater than the size of the holes at the lowest temperatures.

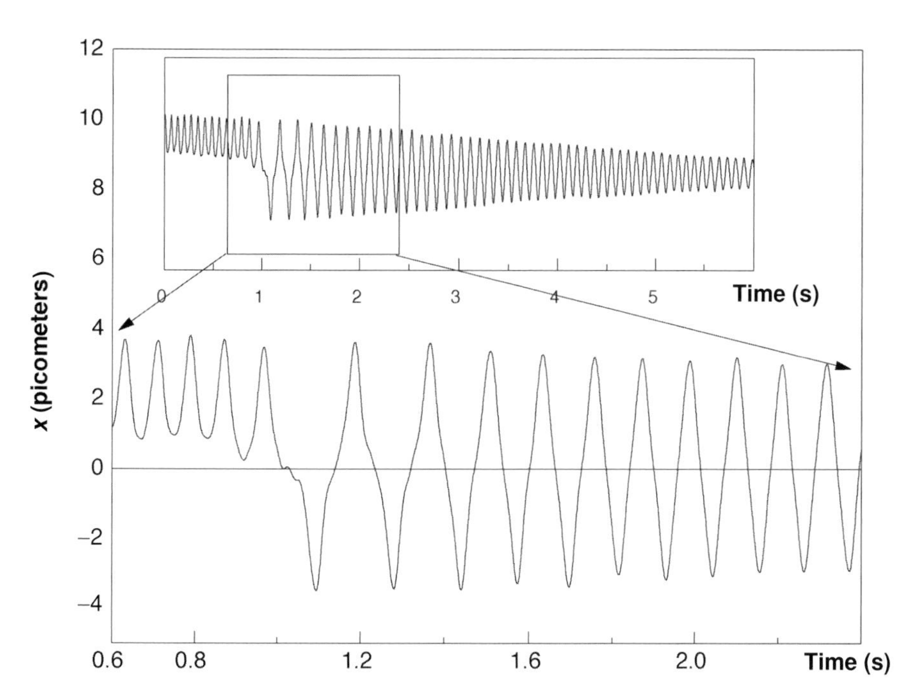


**Figure 8.15.** Flow through a weak link between two volumes of $^3$He-B monitored by the displacement of the diaphragm which drives the flow. The last few Josephson oscillations in the first part of the trace are followed by pendulum oscillations. Reproduced with permission from J. C. Davis and R. E. Packard *Rev. Mod. Phys.* 74, 741–773 (2002).

One way of looking at a phase transition to a more ordered state, such as the superfluid transition, is to say that certain symmetries are spontaneously broken. In the case of superfluid helium-3 the symmetries are gauge symmetry (associated with the phase variable characteristic of all superfluids), three-dimensional rotations in space, and three-dimensional rotations in spin space. These symmetries are of a similar degree of complexity to the symmetries that are believed to have been broken in the early history of the Universe, as the grand unified interaction evolved toward the separate interactions we observe today. It has been suggested that some of the symmetry-breaking transitions in the early Universe may have led to the production of cosmic strings. The analog in liquid helium-3 would be the production of vortices as the system cools from the "grand unified" normal state.

Experiments that can be regarded as Big-Bang simulations have been done at Lancaster, Grenoble, and Helsinki. At temperatures close to 100 μK there are very few excitations in thermal equilibrium; excitations produced when energy is deposited can then be sensitively detected by the resulting damping of a vibrating wire. When a neutron is absorbed it deposits enough energy locally to raise a small region above $T_c$, creating a "mini Big Bang." Evidence has been found for the production of vortices in these experiments.

## 8.6 Exotic superconductors

The meeting of the American Physical Society held in New York on the evening of March 19, 1987 has been called "the Woodstock of Physics." Following the discovery of superconductivity at 38 K in a layered cuprate compound by Bednorz and Müller the previous year, the frenzy was such that by the time of that meeting the highest known superconducting transition temperature had already passed 90 K. Bednorz and Müller were awarded the 1987 Nobel Prize in Physics.

Since 1987 a lot of effort has gone into the development of applications and also into trying to understand the basic physics. I shall concentrate here on the nature of

**Figure 8.16.** The $CuO_2$ sheet common to all cuprate superconductors. Filled circles represent copper and open circles oxygen. In some compounds the pattern is slightly rectangular, one direction being stretched by about 2%.

the order parameter, continuing the theme of this chapter. The mechanism producing the pairing is still controversial, and the suggestions for further reading are the best starting point for the rapidly developing applications.

An early example of a high-temperature superconductor, much studied and fairly typical, is $YBa_2Cu_3O_{7-\delta}$, known as YBCO for short, which has a $T_c$ of 91 K. The $\delta$ in the formula indicates a departure from stoichiometry that is essential in order to make the material a metal, let alone a superconductor; for $\delta = 0$ it is an antiferromagnetic insulator. Thus, the cuprates have a low carrier density as well as a high $T_c$, so $T_0/T_c$ is only of order 10 and the coherence length $\xi$ is therefore very short. The crystal structure is strongly layered, and all cuprates contain $CuO_2$ planes, shown in Figure 8.16; some cuprates are so anisotropic that they can be considered as superconducting $CuO_2$ sheets coupled by the Josephson effect. For YBCO the grid in Figure 8.16 is rectangular, distorted from square by about 2%. The material is most readily prepared as a sintered ceramic, and it is in this form that it is used to make wires. Quite soon after the original discovery, experiments at Birmingham with a sintered ceramic ring of YBCO showed that flux is quantized in the usual unit, $\Phi_0 = h/(2e) = 2.07 \times 10^{-15}$ Wb. This showed that Cooper pairs were formed. A little later, persistent currents and flux quantization were demonstrated in a similar ring interrupted by a niobium section. Since niobium is an ordinary s-wave superconductor, this showed that YBCO is also a spin-singlet, even-orbital-angular-momentum, superconductor. But is it s-wave or d-wave? The strongly anisotropic crystal structure and the four copper–oxygen bonds in the $CuO_2$ sheets were suggestive of a particular $L = 2$ state known as $d_{x^2-y^2}$. The suffix indicates the variation with angle. At unit radius it is $\cos^2\theta - \sin^2\theta$, like a four-leafed clover with two positive leaves and two negative leaves (see Figure 8.17), and this has the symmetry of the coordination of copper by oxygen in Figure 8.16. But is the wave function of the Cooper pairs like this? It has taken about a decade of experiments on good single crystals to settle this question.

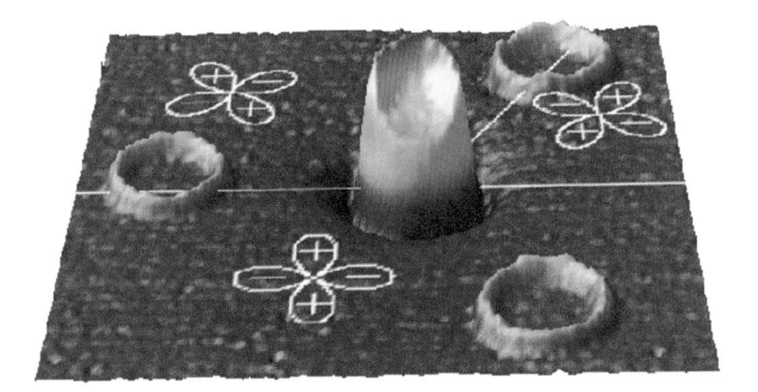

**Figure 8.17.** Three-dimensional rendering of a scanning SQUID microscope image of four YBCO rings, each about 50 μm in diameter, grown on a tricrystal of strontium titanate, cooled and imaged in nominally zero magnetic field. The outer control rings have no flux in them; the central three-junction ring has a half-quantum of flux spontaneously generated in it. The boundaries between the three crystals of the strontium titanate substrate and the orientation of the $d_{x^2-y^2}$ order parameter on each crystal are indicated. (Image courtesy of J. R. Kirtley.)

The anisotropy of the energy gap (the energy required to break a Cooper pair) has been measured by angle-resolved photoemission spectroscopy. In this technique monochromatic synchrotron radiation produces by the photoelectric effect electrons whose energy and momentum are measured. These measurements reveal a sharp minimum in the energy gap, possibly zero, at $\theta = 45^\circ$, which is consistent with $d_{x^2-y^2}$ symmetry. But the magnitude of the energy gap cannot tell us the sign of the order parameter, and therefore cannot distinguish d-wave from s-wave that is highly distorted by crystalline anisotropy.

A phenomenon that is sensitive to the phase of the order parameter is Josephson tunneling between two crystals of different orientations. An idea of what happens can be obtained by examining the order-parameter polar diagrams on the two sides of the junction. If the normal to the interface, which is the dominant direction of tunneling, lies in lobes of the same sign, then there is no phase change across the junction at zero current; but if the normal to the interface lies in lobes of opposite sign, then there is a phase change of $\pi$ at zero current. This is the basis of an ingenious experiment by Tsuei *et al.*, which made use of the fact that YBCO grows epitaxially on strontium titanate. A tricrystal substrate is prepared by fusing together three strontium titanate crystals of chosen orientation and shape. A 120-nm-thick layer of YBCO is then deposited on the polished top surface of the substrate, producing three YBCO crystals in which the $CuO_2$ layers are rotated by different angles around the normal to the surface. Finally, the YBCO is removed except for four 50-μm-diameter rings as shown in Figure 8.17. After cooling this structure down in zero magnetic field, the distribution of magnetic field over the surface is imaged by a scanning SQUID microscope with a sensing coil only 4 μm in diameter. The image shows clearly a half-quantum of flux spontaneously generated in the central ring that straddles all three crystal boundaries, and no flux trapped in the other rings. The outer rings show up in the image because the presence of superconductor under the sensing coil affects its inductance.

Also shown in Figure 8.17 are the grain boundaries between the three crystals and the supposed orientation of a $d_{x^2-y^2}$ order parameter. It can be seen from the rule given above that there should indeed be a phase change of $\pi$ on going round the central ring. That rule is actually oversimplified in that the sign of the order parameter for any crystal is arbitrary. The phase change cannot therefore be assigned to a particular crystal interface, but there is no way of avoiding the total phase change of $\pi$ around

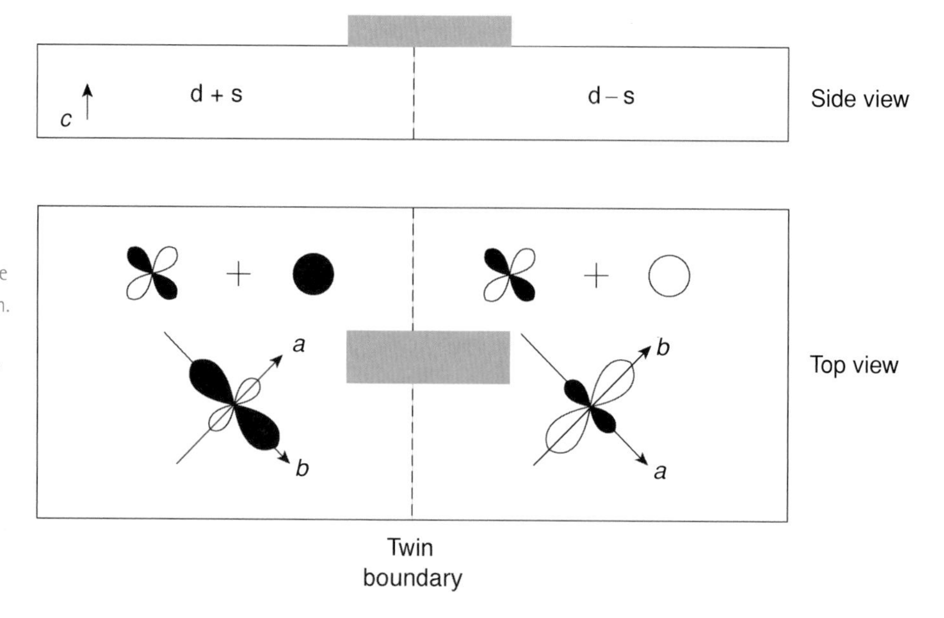


**Figure 8.18.** The experimental geometry for $c$ axis tunneling into lead (gray block) from a single crystal of YBCO with the junction spanning a single twin boundary. Note that the admixture of s-wave into d-wave is such that the large lobes of the hybrid are along the $b$ axis in each crystal of the twin. Reproduced with permission from C. C. Tsuei and J. R. Kirtley *Rev. Mod. Phys.* **72**, 969–1016 (2000).

the central ring. This experiment absolutely rules out s-wave symmetry, for which the order parameter always has the same sign. If the order parameter were $d_{xy}$, which is $d_{x^2-y^2}$ rotated by 45°, the normal to the interface would be along nodes of the order parameter in some cases, so the experiment would probably not work. In any case, the photoemission experiment rules out $d_{xy}$ by showing that the energy-gap minima are at 45° to the crystal axes. The tricrystal experiment shows that those minima are indeed nodes, where the order parameter changes sign.

We have already noted that in an unconstrained single crystal of YBCO the $CuO_2$ lattice is slightly rectangular, being stretched about 2% along the $b$ axis (the axes are labeled by convention). With such a reduced symmetry an admixture of s-wave into d-wave pairing is no longer forbidden, so it is of interest to see whether there is any experimental evidence for such admixture. Kouznetsov *et al.* made use of the fact that YBCO tends to form twinned crystals with the $a$ and $b$ axes exchanged to investigate this, with the arrangement shown in Figure 8.18. An area of lead is deposited over a twin boundary in YBCO and Josephson tunneling into it along the $c$ axis measured. For pure d-wave ordering in YBCO the equal positive and negative lobes have no net overlap with the s-wave in lead, so there should be no current. But with s-wave admixture as shown there should be a net current, with opposite signs for the two elements of the twin. If the sample were ideally constructed these should cancel out. A test for what is happening is to apply a magnetic field in the plane of the sample along the twin boundary. It is found that the current is a minimum at zero field and rises to a maximum when there is half a flux quantum in the twin boundary. This is just what would be expected for s-wave admixture as shown in Figure 8.18. The magnitude of the admixture is much more contentious, because of uncertainty in the strength of coupling between lead and YBCO.

There is a very important distinction to be made between pairing with nonzero angular momentum in superconductors and in liquid helium-3. Normal liquid helium-3 is fully rotationally symmetric, just because it is a liquid, so rotational symmetry in ordinary space and in spin space is spontaneously broken at the superfluid transition. In the cuprates spin-space symmetry is irrelevant for singlet pairing, and rotational symmetry is already broken by the crystal structure. There is no further spontaneous symmetry breaking at the superconducting transition, other than that of the gauge symmetry (leading to a macroscopic phase variable) that is common to all superfluids. The Cooper pairs just adopt a spatial symmetry consistent with the existing symmetry of the crystal lattice, so, in contrast to helium-3, there is no possibility of textures in the spin and orbital part of the order parameter. Superfluid helium-3 therefore remains the example par excellence of complicated spontaneous symmetry breaking on which we can do experiments.

Spurred on by the surprising discovery of the cuprates, other exotic superconductors have been found. An interesting but not very useful one is strontium ruthenate, $Sr_2RuO_4$, which is structurally related to the cuprates. Its transition temperature is a modest 1.5 K, but it is interesting because it has been suggested as a possible example of triplet p-wave pairing, like helium-3. The main evidence for triplet pairing comes from NMR on $^{17}O$, which indicates that there is no change in the magnetic susceptibility of the electrons at $T_c$. A sharp change would be expected for singlet pairing, since singlet pairs are not magnetizable, and this has indeed been seen in YBCO. Strontium ruthenate is a layered structure like the cuprates, and the ordering that has been suggested is a two-dimensional analog of the A-phase, with $L_z = +1$. This would represent an additional spontaneous symmetry breaking at the transition, because the state with $L_z = -1$ has the same energy. If this were the order parameter, domains of $L$ up and $L$ down might arise. The evidence is as yet inconclusive, because phase-sensitive experiments analogous to those on the cuprates have yet to be done.

Another class of superconductors for which p-wave pairing has been suggested is the heavy-fermion intermetallic compounds, which are characterized by a very large effective mass of the charge carriers. These materials tend toward ferromagnetism, whereas the cuprates tend toward antiferromagnetism. $UGe_2$ is a ferromagnet that becomes superconducting under pressure. URhGe is ferromagnetic below 9.5 K and becomes superconducting below 0.26 K at zero pressure. For a ferromagnet the electron energy levels for spin up and spin down are separated, and consequently pairs with equal and opposite momenta can be formed only if the spins are the same. This is a strong argument for triplet pairing, probably p-wave, in these materials, though the issue could be muddied if there is strong interaction between spin and orbital motion. The study of ferromagnetic superconductors is in its infancy; not even the macroscopic behavior is entirely clear yet, though it seems likely that the material will be penetrated by flux lines even in the absence of an applied field. There is a clear role for scanning SQUID microscopy here. In this context it is interesting to note that the ordering proposed for strontium ruthenate is a type of orbital ferromagnetism. However, for none of the heavy-fermion compounds is there experimental evidence of the nature of the pairing as strong as for the cuprates, or even as strong as for strontium ruthenate.

A lasting benefit of the discovery of the cuprates has been to cause people to look more widely for new superconductors and for new types of order parameter. One surprise is that, at this late stage in the history of the subject, the simple binary compound magnesium diboride has been found to be superconducting at 30 K. Understanding these new systems, and particularly the recently discovered coexistence of superconductivity and ferromagnetism, may well require the development of new insights, in much the same way as the understanding of basic superconductivity emerged only long after the initial discovery (see Chapter 9).

## 8.7 A historical epilogue

Our understanding of superfluids has developed over a period of almost a century since the first liquefaction of helium. The development has been at an uneven but accelerating pace, and not always in the expected direction. A key thing to remember is that quantum

**Figure 8.19.** Piotr Kapitza (Nobel Prize 1978) outside his cottage (official residence) in Moscow in 1937, following his forced return from Cambridge in 1934. He proved as willing to stand up to Stalin as he had been to Rutherford. Following the arrest of Landau (Figure 8.2) by the NKVD in 1938, Kapitza wrote to Stalin and Molotov to obtain his release; it took a year.

mechanics was born in 1928, so no fundamental theoretical understanding was possible before then.

It is quite possible that superfluid helium was produced on the day that helium was first liquefied. Onnes was having difficulty in finding evidence of liquid in the glass collecting vessel of his liquefier and so decided to pump on the low-pressure side in an effort to reduce the temperature further. Eventually he found a very calm and barely visible liquid surface near the top of the vessel. The surface of liquid helium is indeed hard to see, because of the small refractive index, but in helium I vigorous boiling usually makes it visible. However, helium II is a superconductor of heat, and so cannot boil; it evaporates only from the surface. The cessation of boiling on cooling through the $\lambda$-point is a conspicuous phenomenon, so it is really surprising that it was not noticed sooner. Even then, it was several years from the observation of the transition to the discovery of the main superfluid properties, largely by Kapitza (Figure 8.19) in Moscow in 1938–41 (he received the Nobel Prize in 1978!).

It is fortunate that Landau was also in Moscow, so that Kapitza's experiments led him to produce the theory of excitations in a ground-state liquid that still stands today as the basis of the two-fluid model, and from which he suggested the Andronikashvili experiment and predicted second sound. Landau based his theory on an attempt to apply quantum mechanics to liquids in general and hence vehemently denied any role for Bose condensation. Several people in the West were convinced of the importance of Bose condensation, but used models that were too gas-like to be quantitative. The matter was almost a cold-war issue for a time. It was resolved by Feynman (see Further reading), who showed that Landau was right about everything except Bose condensation, and invented quantized vortices.

The early attempts to understand superconductivity were dominated by the idea of zero resistance and persistent currents, so it seemed that many different possible equilibrium states were required. The conflict between these ideas and a theorem that the ground state of a system of electrons in the absence of fields carried no current led Bloch to propose, according to Fritz London's account of early ideas, the corollary that "any theory of superconductivity is refutable." Things changed with the surprisingly late discovery of the Meissner effect in 1933, which showed that the equilibrium state was unique. In 1935 the London brothers (Figure 8.20) proposed their phenomenological relation between current and magnetic field, which accounted for the Meissner effect. By 1950 Fritz London had developed this into a clear statement of the concept of a macroscopic wave function, as a target for a microscopic theory. At about this time the impact of people who had worked on radar during World War II led to a large number of microwave experiments on superconductors. Out of this work came Pippard's development of the concept of a coherence length in the early 1950s. Thus was the ground prepared for the appearance of BCS theory in 1957.

Once the BCS theory had appeared, the hunt was on for superfluidity in helium-3. Throughout the 1960s, however, the theorists managed to predict a transition temperature just a little below what was currently attainable experimentally, and many people became discouraged. The eventual discovery at Cornell in 1972 used cooling by the Pomeranchuk method, namely adiabatic compression of a liquid–solid mixture, and the transition was initially attributed to the solid phase. It is important that the variation of pressure with time during the compression was being recorded on an analog chart recorder. The transition may well have been missed in earlier experiments in which discrete points were being recorded by hand. Even modern digital data logging would not be suitable; when you are looking for a brief event at an unknown time in a long time series the memory requirements would be prohibitive. Once the discovery had been made the theorists were poised to work everything out, and with this guidance experimental progress was rapid.

In high-temperature superconductivity progress has been rapid for a different reason: the lure of valuable applications (see suggestions for further reading). The pairing mechanism in these materials is still a matter of controversy, so the story is not quite over yet.

In parallel with this development in our understanding of superfluids there have been amazing developments in the technology of experimentation, which have been a driving force behind that progress. When superconductivity and superfluid helium were discovered experimental technique was still essentially nineteenth century. Electronics began to be used in the 1930s, initially in nuclear physics. The use of mainframe computers on problems that couldn't be tackled otherwise began in the 1950s, but data taking was still entirely manual (or perhaps involved photographing an oscilloscope screen). The development of the dilution refrigerator in the 1960s has been crucial to opening up the temperature region below 1 K, followed by the widespread use of nuclear refrigeration from the mid 1970s. By 1980 data logging and analysis by computer was almost universal, complementing the use of computers for the numerical evaluation of theoretical models. One began to wonder whether everything could be

**Figure 8.20.** The London brothers, who invented the equation relating current to magnetic field that replaces Ohm's law in a superconductor and describes the Meissner effect. Fritz (on the right) developed from this the concept of a macroscopic wave function that is now used to describe all superfluids. Heinz (on the left) invented the principle of the dilution refrigerator, which has become the work-horse of all experiments below 1 K.

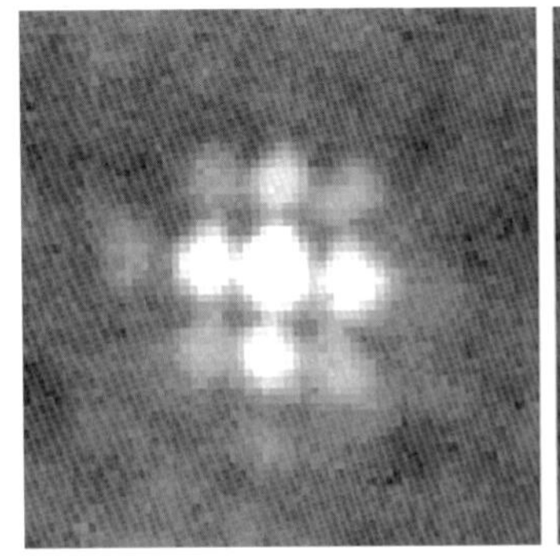

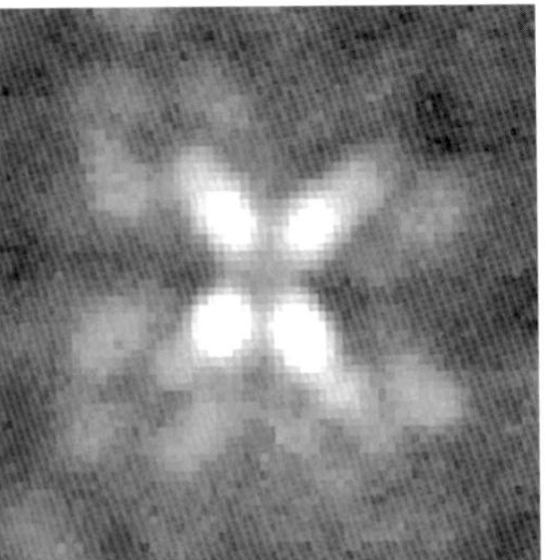

**Figure 8.21.** Nanotechnology in action. Localized impurity states around Ni atoms in superconducting $Bi_2Sr_2CaCu_2O_{8+\delta}$ imaged in differential conductance by scanning tunneling microscopy at 4.2 K. An electron state at +9 mV bias is shown on the left, and a hole state at −9 mV bias on the right. Note the fourfold d-wave symmetry. Each image is about 3.2 nm square. (Images courtesy of J. C. Davis.)

done by sitting at a computer screen. But recently there has been a strong resurgence of the real physical world on which we do experiments. The developments in micrometer and sub-micrometer-scale fabrication and measurement have created qualitatively new possibilities for experiment that would have seemed like science fiction to Kamerlingh Onnes. Some of the first fruits of this have been the demonstration of the Josephson effect in superfluid helium-3, establishing the symmetry of the pairing in cuprate superconductors, and the use of low-temperature scanning tunneling microscopy (Figure 8.21).

I have been fortunate to have lived through much of the above history. This chapter is an attempt to repay the debt I owe to society for having paid me to enjoy myself for half a century.

## FURTHER READING


1. R. P. Feynman, Application of quantum mechanics to liquid helium, *Progress in Low Temperature Physics*, Vol. I, ed. C. J. Gorter, Amsterdam, North-Holland, 1955, p. 17.
   This article, published half a century ago, still says clearly just about everything that is worth saying on the subject of its title. All the stuff about quantized vortices was entirely new at the time.
2. D. R. Tilley and J. Tilley, *Superfluidity and Superconductivity*, 3rd ed., Bristol, Institute of Physics Publishing, 1990.
   A textbook at final-year-undergraduate and first-year-postgraduate level.
3. J. de Nobel, The discovery of superconductivity, *Phys. Today* **49** (September) (1996) 40.
   The accident by which superconductivity was discovered.
4. L. N. Cooper, Bound electron pairs in a degenerate Fermi gas, *Phys. Rev.* **104** (1956) 1189.
   The basic idea behind the BCS theory of superconductivity, nowadays referred to as the Cooper problem.
5. B. D. Josephson, Possible new effects in superconductive tunneling, *Phys. Lett.* **1** (1962) 251.
   A strong contender for the title "Most famous paper to have been rejected by *Physical Review Letters*." An entertaining account of the history associated with this is given by D. G. McDonald, The Nobel laureate versus the graduate student, *Phys. Today* **54** (July) (2001) 46.
6. W. I. Glaberson and K. W. Schwartz, Quantized vortices in superfluid helium-4, *Phys. Today* **40** (February) (1987) 54.
   P. Hakonen and O. V. Lounasmaa, Vortices in rotating superfluid helium-3, *Phys. Today* **40** (February) (1987) 70.
   Two articles in a special issue on liquid and solid helium.
7. O. V. Lounasmaa and G. Pickett, The $^3$He superfluids, *Scient. Am.* **262** (June) (1990) 64.
8. J. Mannhart and P. Chaudhari, High $T_c$ bicrystal grain boundaries, *Phys. Today* **54** (November) (2001) 48.
   Gives an overview of the physics and technology of high-temperature superconductors, including an explanation of how the d-wave character of the pairing can be established.
9. Y. Maeno, T. M. Rice, and M. Sigrist, The intriguing superconductivity of strontium ruthenate, *Phys. Today* **54** (January) (2001) 42.
   A superconductor exhibiting good evidence of p-wave pairing, like $^3$He.
10. J. W. Boag, P. E. Rubinin, and D. Shoenberg (eds.), *Kapitza in Cambridge and Moscow*, Amsterdam, North-Holland, 1990.
    An interesting documentation of Kapitza's extraordinary career.
11. The quantum oracle, *New Scientist* October 12, 2002, p. 28.
    Summarizes for the general reader the book *Universe in A Helium Droplet* (Oxford, Oxford University Press, 2003) by Grigori Volovik. The book expounds for professional physicists analogies between superfluids, especially helium-3, and the evolution of the early Universe. The general reader could attempt the first chapter, "Introduction: GUT and

anti-GUT" for an overview of Volovik's fascinating ideas on possible directions of development for theoretical physics.

12. E. Kim and M. H. W. Chan, Observation of Superflow in solid helium, *Science*, 305(2004), 1941. Presents the experimental evidence for superfluid behavior in bulk solid helium-4. See also the comments by Leggett on p. 1921 of the same issue.

## Henry Hall

Henry Hall is Emeritus Professor of Physics in the University of Manchester, where he conducted research for about forty years. His main area of research is superfluidity, especially helium-3. With W. F. Vinen, he shared the 1963 Simon Prize of the Institute of Physics for pioneering work on quantized vortices in superfluid helium. In 1982 he was elected to the Royal Society, and in 2004 he was awarded the Guthrie Medal and Prize by the Institute of Physics.

# 9 Quantum phase transitions

Subir Sachdev

## 9.1 Introduction

We all observe phase transitions in our daily lives, with hardly a second thought. When we boil water for a cup of tea, we observe that the water is quiescent until it reaches a certain temperature (100 °C), and then bubbles appear vigorously until all the water has turned to steam. Or after an overnight snowfall, we have watched the snow melt away when temperatures rise during the day. The more adventurous among us may have heated an iron magnet above about 760 °C and noted the disappearance of its magnetism.

Familiar and ubiquitous as these and many related phenomena are, a little reflection shows that they are quite mysterious and not easy to understand: indeed, the outlines of a theory did not emerge until the middle of the twentieth century, and, although much has been understood since then, active research continues. Ice and water both consist of molecules of $H_2O$, and we can look up all the physical parameters of a single molecule, and of the interaction between a pair of molecules, in standard reference texts. However, no detailed study of this information prepares us for the dramatic change that occurs at 0 °C. Below 0 °C, the $H_2O$ molecules of ice are arranged in a regular crystalline lattice, and each $H_2O$ molecule hardly strays from its own lattice site. Above 0 °C, we obtain liquid water, in which all the molecules are moving freely throughout the liquid container at high speeds. Why do $10^{23}$ $H_2O$ molecules cooperatively "decide" to become mobile at a certain temperature, leading to the phase transition from ice to water?

We understand these phase transitions by invoking the delicate balance between the diverging interests of the energy, $E$, and the entropy, $S$. The principles of thermodynamics tell us that systems in thermal equilibrium seek to minimize their free energy $F = E - TS$, where $T$ is the temperature measured on the absolute (Kelvin) scale. The energy is determined by the interactions between the $H_2O$ molecules, and this is minimized in the crystalline structure of ice. The entropy, as explained by Boltzmann, is a measure of the degree of "randomness" in a phase; more precisely, it is proportional to the logarithm of the number of microscopic arrangements of $H_2O$ molecules available at a given total energy and volume – the entropy is clearly larger in the liquid water phase. It is now easy to see that, at low $T$, $F = E - TS$ will be smaller in the

*The New Physics for the Twenty-First Century*, ed. Gordon Fraser.

ice phase, whereas at higher $T$ the contributions of the entropy to $F$ become more important, and the free energy of the liquid water phase is lower. The free energies of ice and liquid water cross each other at $0\,^\circ\mathrm{C}$, accounting for the phase transition at this temperature.

So far we have described what are more completely referred to as *thermal* phase transitions, which are caused by the increasing importance of entropy in determining the phase of a system with rising temperature. Let us now turn to the central topic of this chapter, *quantum* phase transitions. Such transitions occur only at the absolute zero of temperature, $T = 0\,\mathrm{K}$, where thermodynamics tells us that the system should be in its lowest energy state (also called the "ground state"). In the simple classical model of $H_2O$ discussed so far, we expect that there is some perfect crystalline arrangement which minimizes the intermolecular interaction energy, and at $T = 0\,\mathrm{K}$ all the $H_2O$ molecules reside at rest on the sites of this lattice. This appears to be a unique quiescent state, so where then is the possibility of a phase transition? The problem with this model is that it is incompatible with the laws of quantum mechanics, which govern the behavior of all the microscopic constituents of matter. In particular, Heisenberg's uncertainty principle tells us that it is not possible to simultaneously specify both the position and the momentum of each molecule. If we do place the molecules precisely at the sites of a perfect crystalline lattice (thus determining their positions), then the momenta of the molecules are completely uncertain – they cannot be at rest, and their kinetic energy will add an unacceptable cost to the energy of this putative ground state. So determining the state of $H_2O$ at $T = 0\,\mathrm{K}$ becomes a delicate matter of optimizing the potential and kinetic energies, while maintaining consistency with Heisenberg's uncertainty principle. As in our earlier discussion of thermal phase transitions, this delicate balance implies that, at least in principle, more than one phase is possible even at $T = 0\,\mathrm{K}$. These phases have distinct macroscopic properties, while containing the same microscopic constituents; they are separated by quantum phase transitions. The key difference from thermal phase transitions is that the fluctuations required by the maximization of entropy at $T > 0\,\mathrm{K}$ have now been replaced by quantum fluctuations demanded by the uncertainty principle. In addition to the familiar ice phase, $H_2O$ exhibits numerous other phases at $T = 0\,\mathrm{K}$ under strong applied pressure, each with a distinct crystal structure and separated from the other phases by quantum phase transitions. We will not describe these transitions here, because they are quite complicated, but focus on simpler examples of quantum-uncertainty-induced transitions.

Our discussion so far would seem to indicate that a quantum phase transition is a theoretical abstraction, which couldn't possibly be of any relevance to an understanding of materials of technological importance. After all, it is impossible to cool any material down to $0\,\mathrm{K}$ in the laboratory, and it takes heroic effort even to get close to it. As we will discuss below, it has become clear during the last decade that these conclusions are quite wrong. Quantum transitions at $T = 0\,\mathrm{K}$ leave characteristic fingerprints in the physical properties of materials at $T > 0\,\mathrm{K}$, and these fingerprints are often clearly visible even at room temperature. A complete understanding of the physical properties of many materials emerges only upon considering the relationship of their $T > 0\,\mathrm{K}$ phases to the distinct ground states and quantum phase transitions at $T = 0\,\mathrm{K}$.

## 9.2 Interacting qubits in the laboratory

We begin our discussion of quantum phase transitions with a simple example. Rather than tackling the full complexity of atomic/molecular potential and kinetic energies, we consider the simplest possible quantum mechanical system, a set of interacting *qubits*. All computers store information as strings of (classical) "bits," which are abstract

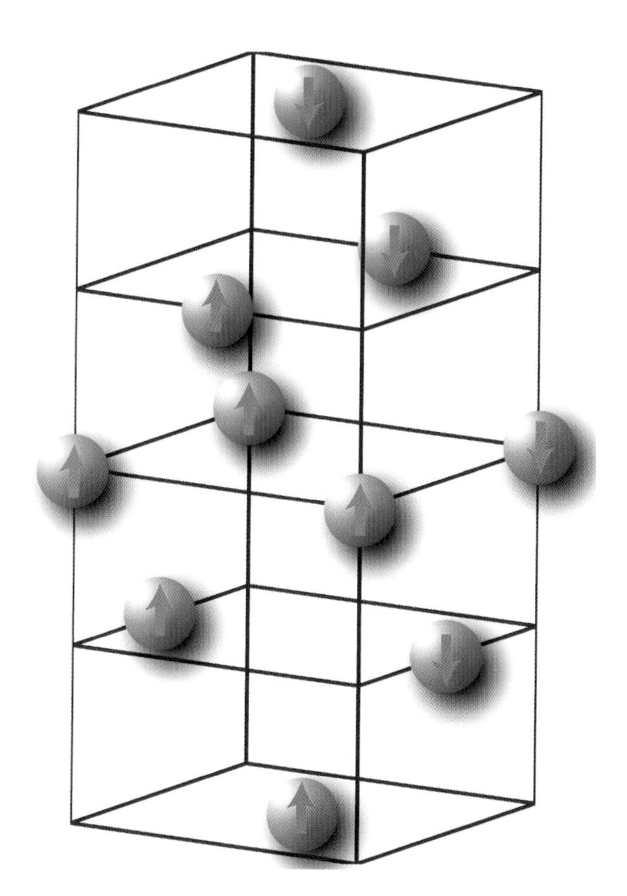

**Figure 9.1.** Locations of the Ho ions in the insulator $LiHoF_4$: the Li and F ions are not shown, and are located in the spaces in between. Each Ho ion has a magnetic moment, which can be oriented "up" (in the $+z$ direction) or "down." This acts as a qubit. The magnetic fields generated by the qubits couple them together to create an interacting qubit system. A quantum phase transition is induced in this system by applying a transverse magnetic field (oriented in the $+x$ direction).

representations of the two possible states of a classical electrical circuit, usually denoted 0 and 1. Similarly, a qubit is the simplest quantum degree of freedom, which can be in only one of two quantum states, which we denote $|\uparrow\rangle$ and $|\downarrow\rangle$. There has been much recent discussion, much of it quite abstract and theoretical, on using coupled qubits to perform computations; here we show that, when many qubits are actually coupled with each other in the laboratory, they undergo a quantum phase transition.

A common physical realization of a qubit is provided by the spin of an electron: each electron spins on its own axis, and its angular momentum about any fixed axis (say $z$) is allowed to take only the values $+\hbar/2$ (in the $|\uparrow\rangle$ state) or $-\hbar/2$ (in the $|\downarrow\rangle$ state); here $\hbar$ is a fundamental constant of Nature, Planck's constant. Associated with this angular momentum is a magnetic moment, $M_z$, which produces a magnetic field that can be measured in the laboratory. Most materials have equal numbers of electrons with spin $+\hbar/2$ and $-\hbar/2$, so the net magnetic moment is zero. However, certain materials (such as iron) have a net excess of electrons with spin $+\hbar/2$ over $-\hbar/2$ (say), endowing them with a macroscopic magnetic moment – this is the explanation for their magnetic properties. Here, we are interested in separating the magnetic qubits on distinct lattice sites and controlling the interactions between them, to allow us to tune the qubits across a quantum phase transition. Iron is not a suitable candidate for this because it is a metal, and its electrons move freely throughout the entire crystal lattice. We need to look at insulators, in which the individual ions have a net magnetic moment. Reasoning in this manner, in 1991 Thomas Rosenbaum at the University of Chicago, Gabriel Aeppli at NEC Research, and their collaborators undertook a series of experiments on the insulator $LiHoF_4$, the simplest of which we shall describe below (see Figure 9.1). In this insulator, the $Li^+$ and $F^-$ ions have equal numbers of up- and down-spin electrons and are nonmagnetic. However, each $Ho^{3+}$ ion has a net magnetic moment. This magnetic moment has contributions both from the electron spin on its own axis and from the orbital motion of each electron around the Ho nucleus, and these motions are strongly coupled to each other via the "spin–orbit" interaction. We do not need to enter into these complexities here because, after the dust settles, the $Ho^{3+}$ ion on site $j$ has only two possible magnetic states, which we denote $|\uparrow\rangle_j$ and $|\downarrow\rangle_j$. Moreover, the crystalline structure also chooses a natural quantization axis, and these states have a magnetic moment, $M_z$, along the preferred crystalline axis. So each $Ho^{3+}$ ion is a perfect physical realization of a qubit.

Before we look at the problem of many coupled qubits, let us dwell a bit more on the physics of an isolated qubit on a single $Ho^{3+}$ ion. One of the fundamental principles of quantum mechanics is the *principle of superposition* of quantum states. For a qubit, this principle implies that its state can be not only $|\uparrow\rangle$ or $|\downarrow\rangle$, but also an arbitrary superposition

$$|\psi\rangle = \alpha|\uparrow\rangle + \beta|\downarrow\rangle, \tag{9.1}$$

where $\alpha$ and $\beta$ are arbitrary complex numbers; the normalization of the state $|\psi\rangle$ requires that $|\alpha|^2 + |\beta|^2 = 1$. The state $|\psi\rangle$ has no classical analog, so it is not possible to provide a picture consistent with our naive classical intuition. An observation of the qubit in state $|\psi\rangle$ will show that it is in state $|\uparrow\rangle$ with probability $|\alpha|^2$ and in state $|\downarrow\rangle$ with probability $|\beta|^2$. But this does not imply that the state $|\psi\rangle$ is a random statistical mixture of the $|\uparrow\rangle$ and the $|\downarrow\rangle$ states – for some measurements there is an additional quantum-interference term that characterizes $|\psi\rangle$ as a true superposition. Crudely speaking, the qubit is both up and down "at the same time." Of particular interest is the state of the qubit with $\alpha = \beta = 1/\sqrt{2}$, which we denote as

$$|\rightarrow\rangle = \frac{1}{\sqrt{2}}(|\uparrow\rangle + |\downarrow\rangle). \tag{9.2}$$

This notation is suggestive: the $|\rightarrow\rangle$ state has the remarkable property that it has a magnetic moment, $M_x$, pointing in the $+x$ direction. So a superposition of a state with magnetic moment up ($|\uparrow\rangle$) and a state with magnetic moment down ($|\downarrow\rangle$) has a magnetic moment pointing right! This is a deep and special property of quantum mechanics, and is ultimately closely linked to Heisenberg's uncertainty principle: states with definite $M_z$ ($|\uparrow\rangle$ or $|\downarrow\rangle$) have uncertain $M_x$, and conversely a state with uncertain $M_z$ ($|\rightarrow\rangle$) has a definite $M_x$. We will also need the state

$$|\leftarrow\rangle = \frac{1}{\sqrt{2}}(|\uparrow\rangle - |\downarrow\rangle), \tag{9.3}$$

which has a magnetic moment pointing in the $-x$ direction. One of the most important properties of a qubit, which has no analog in classical bits, is that the $|\rightarrow\rangle$ and $|\leftarrow\rangle$ states can also be used as a basis for expressing the state of a qubit, and this basis is as legitimate as the $|\uparrow\rangle$ and $|\downarrow\rangle$ states we have used so far. The reader can easily see from (9.1)–(9.3) that the state $|\psi\rangle$ can also be written as a superposition of the $|\rightarrow\rangle$ and $|\leftarrow\rangle$ states:

$$|\psi\rangle = \alpha'|\rightarrow\rangle + \beta'|\leftarrow\rangle \tag{9.4}$$

with $\alpha' = (\alpha + \beta)/\sqrt{2}$ and $\beta' = (\alpha - \beta)/\sqrt{2}$. So we can view $|\psi\rangle$ as a state "fluctuating" between up and down states (as in (9.1)), or as a state "fluctuating" between right and left states (as in (9.4)) – a remarkable fact completely at odds with our classical intuition, but a fundamental property of a qubit.

We are now ready to couple the qubits residing on the Ho atoms in $LiHoF_4$ to each other. This coupling is described by the Hamiltonian, $H$, which is a quantum-mechanical representation of the energy. This Hamiltonian will have two terms, which are the analogs of the kinetic and potential energies of the water molecules we discussed earlier. The quantum phase transition occurs because of a delicate interplay between these energies, and this has a crude parallel to the interplay between $E$ and $S$ near a thermal phase transition. We reiterate, however, that our discussion here of a quantum phase transition is at $T = 0\,\mathrm{K}$, so we are always seeking the quantum-mechanical state with the lowest total energy. We schematically represent the Hamiltonian as

$$H = H_z + gH_x, \tag{9.5}$$

where $H_{z,x}$ are the two announced components of the Hamiltonian, and $g$ is a dimensionless parameter that we will tune to move the qubits across the quantum phase transition – the role of $g$ here will parallel that of $T$ for a thermal phase transition. In the language of the $|\uparrow\rangle_j$, $|\downarrow\rangle_j$ representation of the qubits, the $H_z$ term is a "potential" energy, i.e. it determines the optimum configuration of the $M_z$ magnetic moments which will minimize the energy. The term $H_x$ is a "kinetic" energy, but, unlike in the case of the water molecules, it has a very simple form because there are only two possible states of each qubit. So the only "motion" possible is a flipping of the qubit between the up and down states: it is precisely this up–down flipping, or "quantum tunneling," which is induced by $H_x$. An interesting and fundamental property of $H_{z,x}$, which will become clear from our discussion below, is that the roles of kinetic and potential energies are reversed in the $|{\rightarrow}\rangle_j$, $|{\leftarrow}\rangle_j$ basis of the qubits. In this case, $H_x$ will be a "potential" energy, while the $H_z$ term will induce quantum tunneling between the left and right states, and is thus a "kinetic" energy. So quantum systems have this peculiar property of looking quite different depending upon the choice of observables, but there remains a unique underlying state that has these different physical interpretations.

Let us first discuss the meaning of $H_z$ further – this will give us a good picture of the ground state for $g \ll 1$. In $LiHoF_4$, $H_z$ arises from the "magnetic-dipole" interaction. Each $M_z$ magnetic moment produces a dipolar magnetic-field (much like the familiar magnetic-field patterns around a bar magnet), and this field will tend to align the other magnetic moments parallel to itself. For $LiHoF_4$, these dipolar interactions are optimized for a "ferromagnetic" arrangement of the $M_z$ moments. In other words, the ground state is

$$|\Uparrow\rangle = \ldots|\uparrow\rangle_{j_1}|\uparrow\rangle_{j_2}|\uparrow\rangle_{j_3}|\uparrow\rangle_{j_4}|\uparrow\rangle_{j_5}\cdots \tag{9.6}$$

where the labels $j_1 \ldots$ extend over all the $\sim 10^{23}$ qubits in the lattice. There is nothing in the Hamiltonian or the crystal structure that distinguishes up from down, so another ground state, with the same energy, is

$$|\Downarrow\rangle = \ldots|\downarrow\rangle_{j_1}|\downarrow\rangle_{j_2}|\downarrow\rangle_{j_3}|\downarrow\rangle_{j_4}|\downarrow\rangle_{j_5}\cdots \tag{9.7}$$

The equivalence between (9.6) and (9.7) is therefore related to a *symmetry* between the up and down orientations of the magnetic moments. This last statement needs to be qualified somewhat for any realistic system. While the hypothetical perfect and infinite crystal of $LiHoF_4$ does indeed have an unblemished up–down symmetry, any realistic crystal will always have a slight preference for up or down induced by imperfections and boundaries. Any slight preference is sufficient to break the symmetry, so let us assume that the ground state has been chosen in this manner to be $|\Uparrow\rangle$. This is a simple example of a common phenomenon in physics: a *spontaneously broken symmetry*. The ferromagnetic arrangement of the magnetic moments in such a state will lead to a magnetic field much like that produced by everyday permanent magnets, and this is easy to detect in the laboratory.

The alert reader will have noticed by now that our description above of the states $|\Uparrow\rangle$ and $|\Downarrow\rangle$ surely cannot be the whole story: we have localized each qubit in the up direction, and this must lead to problems with Heisenberg's uncertainty principle. Won't each qubit tunnel to the down state occasionally, just to balance out the uncertainties in $M_z$ and $M_x$? The contribution $H_x$ in (9.5) is one term that performs this tunneling. For $g \ll 1$ it can be shown that our discussion of the $|\Uparrow\rangle$ and $|\Downarrow\rangle$ states is essentially correct, and the ferromagnetic phase is not disrupted by quantum tunneling. However, the story is very different for $g \gg 1$. Now the state of the qubits is determined entirely by the optimization of this tunneling, and this happens when the qubit is equally likely to be in the up or down state. We choose an arbitrary phase in our definition of the qubit states in order to have the tunneling in $H_x$ prefer the state $|\rightarrow\rangle_j$ for all $j$. Thus a large amount of tunneling makes all the qubits point to the "right." This result also suggests a simple way in which the value of $g$ can be tuned in the laboratory: simply apply a magnetic field along the $+x$ or "transverse" direction. This transverse field will enhance the tunneling between the $|\uparrow\rangle_j$ and $|\downarrow\rangle_j$ states of the qubit, and is a powerful tool for "tuning" the strength of quantum fluctuations. In a large transverse field, $g \to \infty$, and then the ground state of $H$ is clearly

$$|\Rightarrow\rangle = \ldots|\rightarrow\rangle_{j_1}|\rightarrow\rangle_{j_2}|\rightarrow\rangle_{j_3}|\rightarrow\rangle_{j_4}|\rightarrow\rangle_{j_5}\ldots \tag{9.8}$$

Note that this state is very different from the $|\Uparrow\rangle$ and $|\Downarrow\rangle$ and, while the relationship (9.2) holds for a single qubit, the analogous relationship for many qubits does *not* hold, i.e.

$$|\Rightarrow\rangle \neq \frac{1}{\sqrt{2}}(|\Uparrow\rangle + |\Downarrow\rangle)\,. \tag{9.9}$$

The correct expression for $|\Rightarrow\rangle$ is derived by inserting (9.2) into (9.8) for each site $j$ and then expanding out the products: one finds a very large number of terms with up- and down-oriented qubits, and only two of these terms appear on the right-hand side of (9.9). The inequality (9.9), and the far more complicated behavior of the many-qubit system, is what allows the appearance of a nontrivial quantum phase transition.

In passing, we note that the state on the right-hand side of (9.9) is often referred to as "Schrödinger's cat." It is a quantum superposition of two macroscopically distinct states (the $|\Uparrow\rangle$ and $|\Downarrow\rangle$ states), much like the quantum superposition of a dead cat and a live cat that Schrödinger speculated about. In practice, such "cat states" are exceedingly hard to create, because (as we have discussed above) even a tiny external perturbation will "collapse" the system's wave function into either $|\Uparrow\rangle$ or $|\Downarrow\rangle$.

We have now assembled all the ingredients necessary to understand the quantum phase transition of the many-qubit system as a function of increasing $g$. For small $g$, as we discussed earlier, the qubits are in the state $|\Uparrow\rangle$ – the most important property of this state is that it has a nonzero value of the average magnetic moment $\langle M_z\rangle$. We expect $\langle M_z\rangle$ to evolve smoothly as the value of $g$ is increased. In contrast, at very large $g$, we have the state $|\Rightarrow\rangle$, in which $\langle M_z\rangle$ is strictly zero, and we expect this to be true for a range of large $g$. Now imagine the functional dependence of $\langle M_z\rangle$ on $g$: it is impossible to connect the two limiting regimes with a smooth function. There must be a singularity at at least one value $g = g_c$ as shown in Figure 9.2, where $\langle M_z\rangle$ first vanishes. This is the location of the quantum phase transition – with increasing $g$, this is the point at which the ferromagnetic moment vanishes, the up–down symmetry of the qubit system is restored, and we reach a "paramagnetic" state. In a very real sense, the complicated many-qubit ground state for $g < g_c$ is similar, and smoothly connected, to the state $|\Uparrow\rangle$ for $g < g_c$, whereas for $g > g_c$ a corresponding satisfactory model is provided by $|\Rightarrow\rangle$. Only at $g = g_c$ do we obtain a truly different "critical" state,

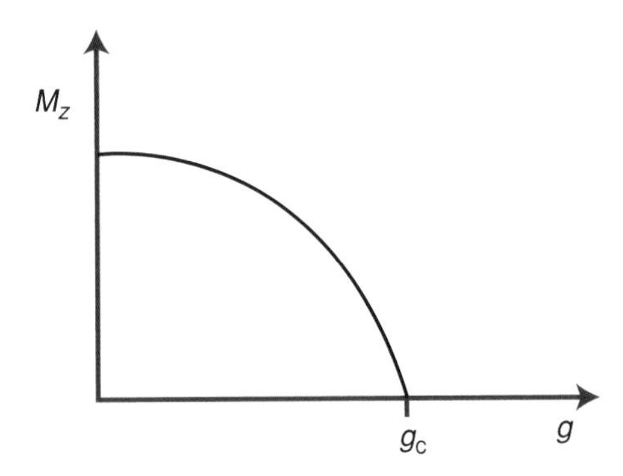


**Figure 9.2.** The dependence of the average magnetic moment along the $z$ axis on the strength of the applied transverse magnetic field at $T = 0$. There is a quantum phase transition at $g = g_c$. For $g < g_c$, the ground state is a ferromagnet, with a wave function qualitatively similar to $|\Uparrow\rangle$ in Equation (9.6) or to $|\Downarrow\rangle$ in Equation (9.7): small imperfections choose between the nearly equivalent possibilities, and thus break the up–down symmetry of the crystal. For $g > g_c$, the ground state is a paramagnet, with a state qualitatively similar to $|\Rightarrow\rangle$ in Equation (9.8).

in a very complex and "*entangled*" qubit arrangement we will not describe in any detail here.

So far we have described the physics of a quantum phase transition at $T = 0$, but this is not of direct relevance to any practical experiment. It is essential to describe the physics at $T > 0$, when the qubits will no longer reside in their lowest-energy state. Describing the structure of the very large number of higher-energy (or "excited") states would seem to be a hopelessly complicated task – there is an exponentially larger number of ways in which the qubits can rearrange themselves. Fortunately, over most of parameter space, a powerful conceptual tool of quantum theory provides a simple and intuitive description: the *quasiparticle*. A quasiparticle appears just like a particle in any experiment: it is a pointlike object that moves while obeying Newton's laws (more precisely, their quantum generalizations). However, rather than being a fundamental degree of freedom in its own right, a quasiparticle emerges from the collective behavior of many strongly coupled quantum degrees of freedom (hence the "quasi-"): it is a "lump" of excited qubits that, when viewed from afar, moves around just like a particle. Moveover, the spectrum of excited states of the qubits can be usefully decomposed into states describing multiple quasiparticles moving around and colliding with each other.

A simple description of the quasiparticle states is also possible. Consider first the limit $g \gg 1$, where the ground state is $|\Rightarrow\rangle$ as in (9.8). To create an excited state, we must flip qubits from right to left, and, because it costs a large amount of energy ($\sim g$) to flip a qubit, let us flip just one at the site $j$; this yields the state

$$|j\rangle = \ldots |\rightarrow\rangle_{j_1} |\rightarrow\rangle_{j_2} |\leftarrow\rangle_{j} |\rightarrow\rangle_{j_3} |\rightarrow\rangle_{j_4} |\rightarrow\rangle_{j_5} \ldots \tag{9.10}$$

The left qubit is just a stationary object at the site $j$ above, but corrections from $H_z$ endow it with all the characteristics of a particle: it becomes mobile and acquires an energy that depends on its velocity (its "dispersion relation"). We can now also create left-oriented qubits on other sites, and these behave much like a gas of particles: the quasiparticles are relatively independent of each other when they are far apart, and they collide and scatter off each other should their paths intersect. Also, for large enough $g$, it should be evident that this quasiparticle picture provides a description of all the lower-energy excited states above the $|\Rightarrow\rangle$ ground state. A somewhat more subtle fact

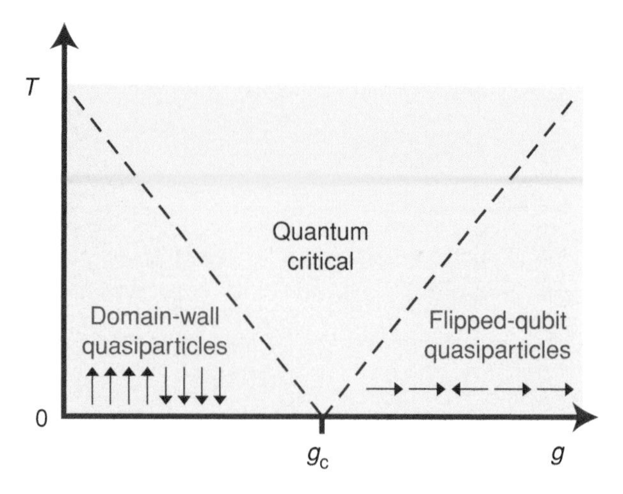


**Figure 9.3.** The phase diagram for a one-dimensional chain of coupled qubits as a function of the parameter $g$ and the absolute temperature, $T$. The ground state at $T = 0$ is as described in Figure 9.2. The flipped-qubit quasiparticles have a state similar to (9.10), while the domain-wall quasiparticles have a state similar to (9.11). No quasiparticle picture works in the quantum-critical region, but we have relaxational dynamics characterized by (9.12).

is that this quasiparticle description works for *all* points, $g > g_c$, on the paramagnetic side of the quantum phase transition. As we lower $g$, each quasiparticle is no longer a single flipped qubit as in (9.10), but becomes a more diffuse object localized around the site $j$; furthermore, there is a lowering of the upper bound on the energy below which the quasiparticle picture applies. Ultimately, as we reach $g = g_c$ from above, the size of the quasiparticle diverges, and the upper bound on the quasiparticle picture reaches the ground-state energy. Only strictly at $g = g_c$ does the quasiparticle picture fail completely,[1] but quasiparticles can be usefully defined and identified at all $g > g_c$.

Let us now turn to a description of the excited states for $g < g_c$. Here again, we find a quasiparticle description, but in terms of an entirely new type of quasiparticle. The description of this quasiparticle is simplest for the case in which the qubits are aligned along a one-dimensional chain, rather than in a three-dimensional crystal, so we restrict our discussion here to this case. In the ground states $|\Uparrow\rangle$ and $|\Downarrow\rangle$, at $g = 0$, the interactions among the magnetic dipoles align the qubits all parallel to each other. The simplest excited state is then the state with exactly one pair of qubits that are anti-aligned, and this leads to a defect (or a "domain wall") between sites $j$ and $j + 1$ separating perfectly aligned qubits (we are numbering the sites consecutively along the chain):

$$|j, j+1\rangle = \ldots |\uparrow\rangle_{j-2} |\uparrow\rangle_{j-1} |\uparrow\rangle_j |\downarrow\rangle_{j+1} |\downarrow\rangle_{j+2} |\downarrow\rangle_{j+3} \cdots \tag{9.11}$$

With a small nonzero value of $g$, the qubits are allowed to tunnel between the up and down states, and it is then not difficult to see that this defect becomes mobile, i.e. it is a quasiparticle. However, there is no simple relationship between this quasiparticle and that in (9.10), and the qubits are organized in very different superposition states. The behavior of the quasiparticle in (9.11) with increasing $g$ parallels that of (9.10) with decreasing $g$: the domain wall becomes increasingly diffuse as $g$ increases, and the quasiparticle description holds only below an energy that goes to zero precisely at $g = g_c$.

Collecting all the information above, we can draw a "phase diagram" of the coupled-qubit model as a function of the transverse field, which is tuned by $g$, and the temperature $T$. This is shown in Figure 9.3. There is a quantum phase transition at the critical point $g = g_c$, $T = 0$, flanked by ferromagnetic and paramagnetic states on its sides. On both sides of the critical point, there is a range of temperatures below which the quasiparticle description is appropriate, and this upper bound in temperature is

[1] A technical aside: for the simple model under consideration here, this statement is strictly true only in spatial dimensions $d < 3$. In $d = 3$, a modified quasiparticle picture can be applied even at $g = g_c$.

related to the upper bound in energy discussed above by Boltzmann's constant $k_B$. Furthermore, there is a "fan" emanating from the quantum critical point where the quasiparticle description is not appropriate and a fundamentally new approach based upon the entangled state at $g = g_c$, and its non-quasiparticle states has to be developed.

The description of this new quantum-critical regime has been the focus of much research in recent years. Some models display a remarkable universality in their properties in this regime, i.e. some observable properties of the qubits are independent of the precise couplings among them. For example, we can ask for the value of the qubit relaxation rate, $\Gamma_R$, which is the inverse time it takes for a perturbation from thermal equilibrium to die away; this rate is given by

$$\Gamma_R = C_Q \frac{k_B T}{\hbar} \tag{9.12}$$

where $\hbar$ is Planck's constant and $C_Q$ is a dimensionless universal number that depends on some known, gross features of the model (such as its spatial dimensionality), but is independent of the details of the couplings in $H$. So, remarkably, a macroscopic collective dynamical property of $10^{23}$ qubits is determined only by the absolute temperature and by fundamental constants of Nature. It is worth noting here that it is only in this quantum-critical region that the qubits are strongly entangled with each other, in a sense similar to that required for quantum computation. The dynamics of the qubits is actually quite "incoherent" here, and overcoming such decoherence effects is one of the major challenges on the road to building a quantum computer in the laboratory.

Extensions of the simple phase diagram in Figure 9.3, and of its physical properties, have been explored in the experiments of Rosenbaum, Aeppli, and collaborators. There is now a good understanding of very clean crystals of $LiHoF_4$. However, crystals that have intentionally been doped with impurities have far more complicated interactions between the qubits, and their study remains an active topic of research.

## 9.3 Squeezing the Bose–Einstein condensate

Prompted by communications with S. N. Bose in 1924 (see Box 9.1), Albert Einstein considered the problem of cooling a gas in a container to very low temperatures. At room temperature, a gas such as helium consists of rapidly moving atoms, and can be visualized as classical billiard balls, which collide with the walls of the container and occasionally with each other. As the temperature is lowered, the atoms slow down, and their quantum-mechanical characteristics become important: de Broglie taught us that each atom is represented by a wave, and the de Broglie wavelength of the helium atoms becomes larger as the temperature is lowered. This has dramatic macroscopic consequences when the wavelength becomes comparable to the typical distance between the atoms. Now we have to think of the atoms as occupying specific quantum states that extend across the entire volume of the container. We are faced with the problem of cataloging all such many-atom states. The atoms are indistinguishable from each other, and quantum mechanics requires that we interpret two states that differ only by the exchange of the position of a pair of atoms not as being two distinct states at all, but rather as components of a single state. Furthermore, if the atoms are "bosons" (any atom with an even total number of electrons, protons, and neutrons is a boson, as is helium) an arbitrary number of them can occupy any single quantum state, i.e. there is no exclusion principle as there is for "fermions" such as electrons. If the temperature is low enough then the many-atom system will search for its lowest-energy state, and for bosons this means that *every* atom will occupy the *same* lowest-energy wavelike quantum state extending across the entire container. A macroscopic number of atoms

BOX 9.1 S. N. BOSE

In 1924, Satyendra Nath Bose was a young reader in physics at Dacca University in present-day Bangladesh. He mailed a six-page article to Albert Einstein entitled "Planck's law and the hypothesis of light quanta." Einstein was already a world-renowned scientist, but he nevertheless made the effort to understand this unsolicited letter from a far-off place: it introduced what we now call Bose statistics, and showed how it could be used to derive the Planck blackbody spectrum from the assumption that light was made of discrete quanta (particles called photons). Einstein was immediately intrigued by this result: he translated the article into German and had it published in the *Zeitschrift für Physik*. He also applied the idea of Bose statistics to particles with mass, and quickly predicted the Bose–Einstein condensate.

occupying a single microscopic state is a Bose–Einstein condensate. Einstein showed that the Bose–Einstein condensate appeared below a critical temperature that is roughly determined by the condition that the de Broglie wavelength of an atom equal the mean atomic spacing.

Unbeknown to Bose and Einstein, the Bose–Einstein condensate had actually already been discovered in the laboratory over a decade before their theoretical work, but several more decades would pass before this connection between theory and experiment was clearly understood – a striking example of how convoluted the progress of science can often be, and how things that seem obvious in retrospect can go unnoticed for a long time. It had been noted by Kammerlingh Onnes that liquid helium at very low temperatures displayed the remarkable property of superfluidity (see Chapter 8): the ability to flow without any appreciable viscosity. However, it was not until 1938 that Fritz London first proposed that superfluidity was linked to the formation of a Bose–Einstein condensate of helium atoms. This proposal was met with skepticism: the Bose–Einstein theory was for an ideal gas of non-interacting bosons, while there was no doubt that the helium atoms interacted strongly with each other. Subsequent theoretical developments have since shown that London was essentially correct: interactions among the bosons do deplete the condensate (not all atoms are in the lowest-energy single-boson state), but the condensate remains intact in the sense that it retains a finite fraction, i.e. a macroscopic number, of atoms.

In the past decade, a beautiful new realization of the Bose–Einstein condensate has been created in the laboratory. In 1995, Eric Cornell, Carl Wieman, and collaborators succeeded in trapping and cooling a gas of rubidium (Rb) atoms to a high enough density and low enough temperature to induce them to form a Bose–Einstein condensate. Unlike the truly macroscopic condensate of $10^{23}$ atoms in liquid helium, this condensate is a much more fragile and delicate object: it exists only in a trap containing $10^{3-6}$ atoms carefully isolated from the world around it and cooled to temperatures extremely close to absolute zero. Indeed, it is not a simple matter to know that atoms have formed a Bose–Einstein condensate. In their experiments, Cornell and Wieman released the atoms from their trap, and then measured the velocities of the escaping atoms: at low enough temperatures, they found a large enhancement in the number of atoms with velocities close to zero, as would be the case for atoms that emerged from the Bose–Einstein condensate. See Chapter 7, Figure 7.9. A more quantitative description of the velocity distribution function is in good agreement with the theory of the interacting Bose gas, so Figure 7.9 is convincing evidence for the formation of a Bose–Einstein condensate of trapped Rb atoms.

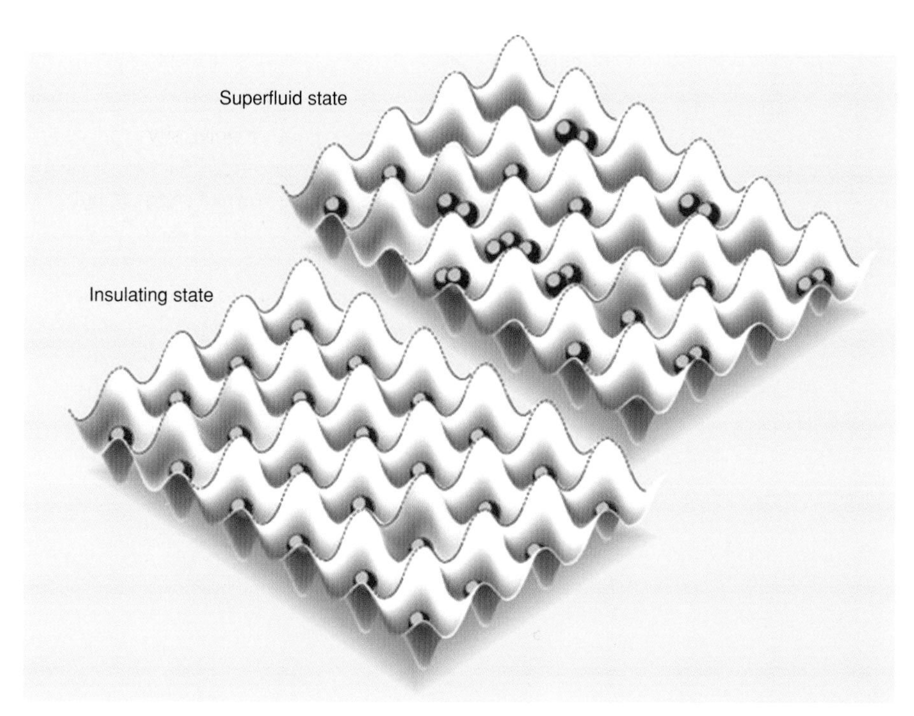


**Figure 9.4.** A two-dimensional section of Rb atoms moving in a periodic potential created by standing waves of laser light. The Rb atoms are like eggs that prefer to reside only at specific sites of an "egg-carton." The superfluid is a superposition of states with strong fluctuations in the number of atoms in any given minimum of the periodic potential, as is expressed in (9.13). The insulator has a fixed number of atoms in each minimum, as expressed in (9.14). (From H. T. C. Stoof, *Nature* **415**, 25 (2002)).

Nothing we have said so far in this section relates to a quantum phase transition: any collection of bosons, when cooled to sufficiently low temperatures, will always reach the same quantum state – a Bose–Einstein condensate. Strong repulsive interactions between the bosons can deplete the condensate, but they never annihilate it, and the system as a whole remains a perfect superfluid. Is there any way to reach a different ground state of bosons and possibly also a quantum phase transition?

The key to the formation of the Bose–Einstein condensate is that some of the atoms move freely throughout the entire system in the same quantum state. This suggests that, if we were able to disrupt the motion of the atoms by a set of microscopic barriers, we may be able to destroy the condensate and reach a new ground state. The precise implementation of this idea was discussed by Sebastian Doniach in 1981 in a slightly different context, and extended to the boson systems of interest here by Matthew Fisher and collaborators in 1989. For the trapped gas of Rb atoms we have discussed here, this idea was demonstrated in a remarkable recent experiment by Immanuel Bloch and collaborators in Munich, so we will discuss the idea using their implementation.

Bloch and collaborators impeded the motion of the Rb bosons by applying a periodic array of barriers within the trap; more precisely, they applied a periodic potential to the Rb atoms. In the two-dimensional view shown in Figure 9.4, we can visualize the Rb atoms as moving across an egg-carton: there is an array of sites at which the Rb atoms prefer to reside in order to lower their potential energy, and they have to go over a barrier to move from any site to its neighbor. In Bloch's laboratory, this periodic potential was created by standing waves of light: along each direction there were two counter-propagating laser beams (and so a total of six lasers) whose interference produces an effective periodic potential for the Rb atoms.

In the presence of a weak periodic potential, Bloch observed only a minor, but significant, change in the state of the Rb atoms. The atoms were released from the trap, as in the experiment of Cornell and Wieman, and then their velocity distribution

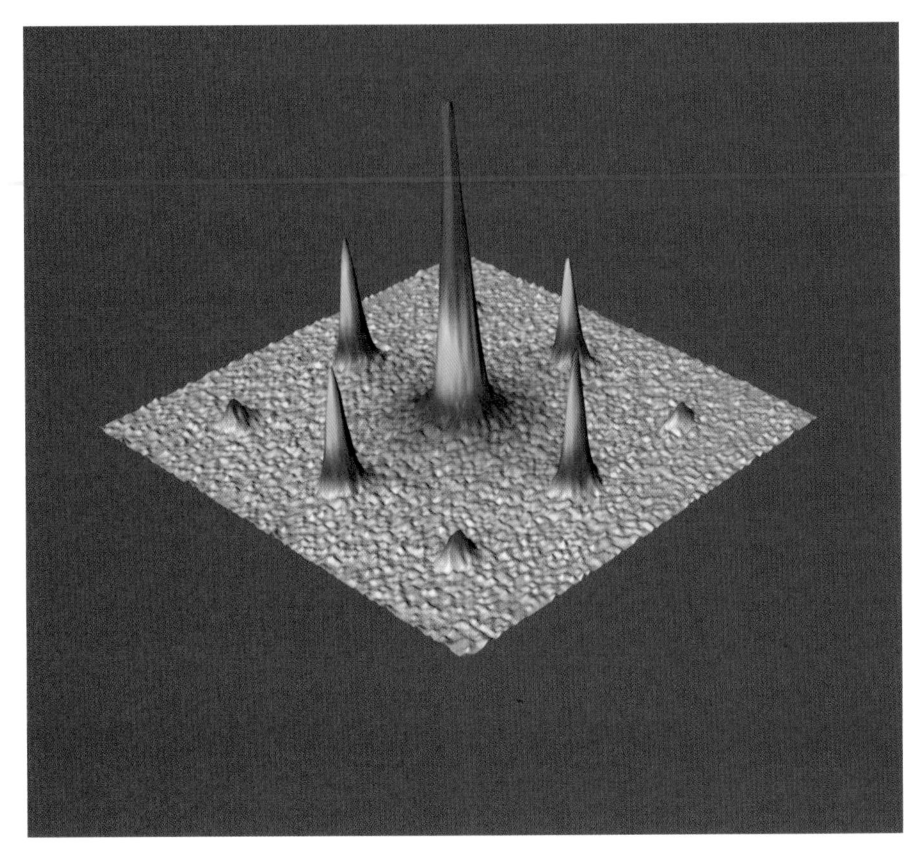

**Figure 9.5.** The velocity distribution function of Rb atoms released from a trap in the presence of a weak periodic potential. The large central peak is the signal of the Bose–Einstein condensate. The satellite peaks result from the diffraction of this condensate off the periodic potential. This signal represents observation of the state $|\mathrm{BEC}\rangle$ in (9.13). Figure courtesy of I. Bloch from M. Greiner, O. Mandel, T. Esslinger, T. W. Hänsch, and I. Bloch, *Nature* **415**, 39 (2002).

function was observed. Bloch observed the distribution shown in Figure 9.5. Note that it has a large peak near zero velocity, as in Figure 7.9, and this is a tell-tale signature of the Bose–Einstein condensate. However, there is also a lattice of satellite peaks, and these record the influence of the periodic potential. Crudely speaking, the quantum state into which the atoms condense has "diffracted" off the period potential, and the satellite peaks in Figure 9.5 represent the diffraction pattern of the periodic potential. So the shape of the condensate has adjusted to the periodic potential, but otherwise the Bose–Einstein condensate retains its integrity: this will change at stronger periodic potentials, as we shall discuss shortly.

In preparation for our description of the quantum phase transition, it is useful to explicitly write down the quantum state of the Bose–Einstein condensate. Let us assume that the atoms are able to occupy only a single quantum state in each minimum of the periodic potential (this is actually an excellent approximation). If atom number $n$ (we number the atoms in some order) occupies the state in the $j$th minimum of the periodic potential (we also number the minima of the periodic potential in some order), we denote this state by $|j\rangle_n$. In the Bose–Einstein condensate, each atom will actually occupy the same state: the state which is a linear superposition of the states in every well. So the Bose–Einstein condensate (BEC) of bosons numbered $n_1, n_2, n_3, \ldots$ in potential wells $j_1, j_2, j_3, \ldots$ is

$$
\begin{aligned}
|\mathrm{BEC}\rangle = \; & \ldots(\cdots|j_1\rangle_{n_1} + |j_2\rangle_{n_1} + |j_3\rangle_{n_1}\cdots) \\
& \times(\cdots|j_1\rangle_{n_2} + |j_2\rangle_{n_2} + |j_3\rangle_{n_2}\cdots) \\
& \times(\cdots|j_1\rangle_{n_3} + |j_2\rangle_{n_3} + |j_3\rangle_{n_3}\cdots)\ldots
\end{aligned}
\qquad (9.13)
$$

Upon expanding out the products above, note that the number of bosons in any given well fluctuates considerably, e.g. one of the terms in the expansion is $\ldots|j_1\rangle_{n_1}|j_1\rangle_{n_2}|j_1\rangle_{n_3}\ldots$ in which all three of the bosons numbered $n_{1,2,3}$ are in the well $j_1$ and none in the wells $j_2$ and $j_3$. The $|\mathrm{BEC}\rangle$ state is a quantum-mechanical superposition of all these states in which the number of bosons in each well fluctuates strongly. This is another way of characterizing the Bose–Einstein condensate, and is indeed the essential property which leads to an understanding of its superfluidity: the strong number fluctuations mean that particles are able to flow across the system with unprecedented ease, and without resistance.

Now turn up the strength of the periodic potential. Theory tells us that there is a critical strength at which there is a quantum phase transition, and beyond this point there is no superfluidity – the Bose–Einstein condensate has disappeared. What has taken its place? As in our discussion in Section 9.2 we can understand the structure of this state by looking at a limiting case: the state should be qualitatively similar to that in the presence of a very strong periodic potential. In such a situation, we expect the tunneling between neighboring minima of the periodic potential to be strongly suppressed. This, combined with the repulsive interactions between the bosons in the same potential well, should lead to a strong suppression of the number fluctuations in the quantum state. An idealized state, with no number fluctuations, has the form

$$\begin{aligned}|I\rangle = \; & \ldots|j_1\rangle_{n_1}|j_2\rangle_{n_2}|j_3\rangle_{n_3}\ldots+\ldots|j_1\rangle_{n_1}|j_2\rangle_{n_3}|j_3\rangle_{n_2}\ldots \\ & +\ldots|j_1\rangle_{n_2}|j_2\rangle_{n_1}|j_3\rangle_{n_3}\ldots+\ldots|j_1\rangle_{n_2}|j_2\rangle_{n_3}|j_3\rangle_{n_1}\ldots \\ & +\ldots|j_1\rangle_{n_3}|j_2\rangle_{n_2}|j_3\rangle_{n_1}\ldots+\ldots|j_1\rangle_{n_3}|j_2\rangle_{n_1}|j_3\rangle_{n_2}\ldots+\ldots \end{aligned} \tag{9.14}$$

Now each site $j_{1,2,3}$ has exactly one of the particles $n_{1,2,3}$ in all the terms. Indeed, any one of the terms above suffices to describe the configuration of the particles in the ground state, because the particles are indistinguishable and we can never tell which particular particle is residing in any given well – we need all the permutations in (9.14) simply to ensure that physical results are independent of our arbitrary numbering of the particles. Note also that all terms in $|I\rangle$ are also present in $|\mathrm{BEC}\rangle$ – the difference between the states is that $|\mathrm{BEC}\rangle$ has many more terms present in its quantum superposition, representing the number fluctuations. In this sense, the inequality between $|\mathrm{BEC}\rangle$ and $|I\rangle$ is similar to the inequality (9.9) between the states on either side of the transition in the coupled-qubit model.

What is the physical interpretation of $|I\rangle$? The complete suppression of number fluctuations now means that the bosons are completely unable to flow across the system. If the bosons were electrically charged, there would be no current flow in response to an applied electric field, i.e. it would be an insulator. Alternatively, in the atom-trapping experiments of Bloch, if we were to "tilt" the optical lattice potential (in two dimensions, we tilt the egg-carton of Figure 9.4), there would be no flow of bosons in the state $|I\rangle$. In contrast, bosons would flow without any resistance in the state $|\mathrm{BEC}\rangle$. The quantum phase transition we have described in this section is therefore a *superfluid–insulator* transition.

It has not yet been possible to observe flows of atoms directly in the experiments by Bloch and collaborators. However, as in the state $|\mathrm{BEC}\rangle$, the velocity distribution function of the atoms in the state $|I\rangle$ can be easily measured, and this is shown in Figure 9.6. Now there is no sharp peak near zero velocity (or at its diffractive images): the absence of a Bose–Einstein condensate means that all the particles have large velocities. This can be understood as a consequence of Heisenberg's uncertainty principle: the particle positions are strongly localized within single potential wells, and so their velocities become very uncertain.

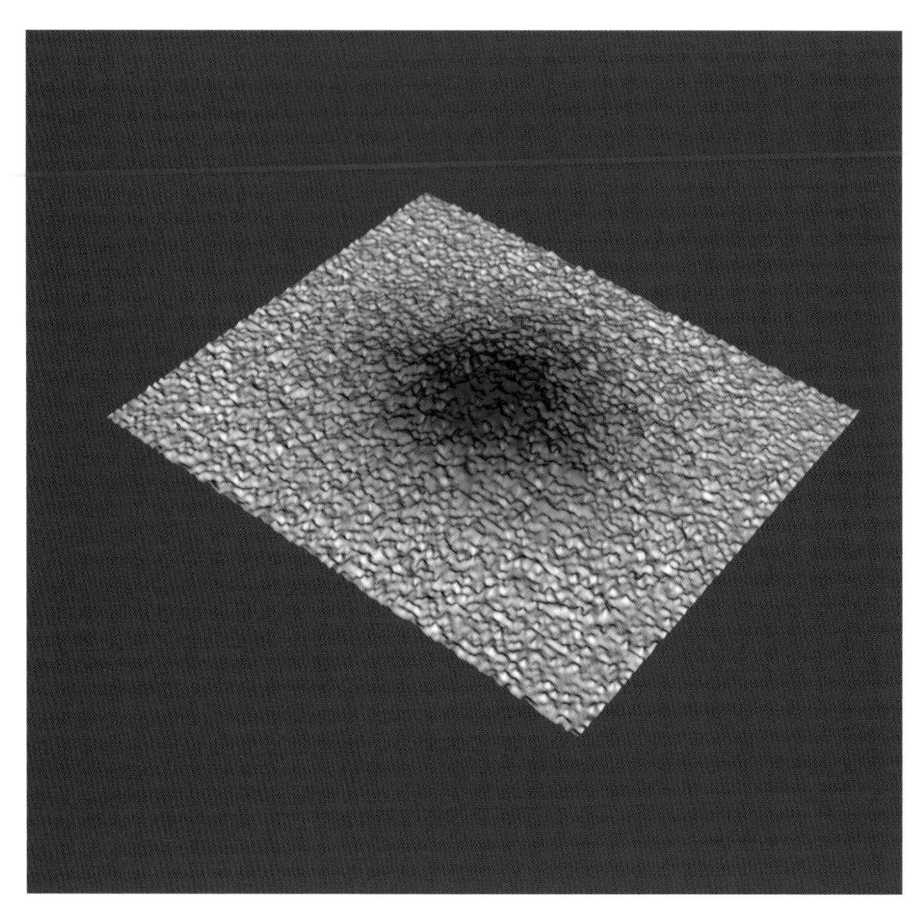

**Figure 9.6.** As in Figure 9.5 but with a stronger periodic potential. Now there are no sharp peaks, indicating the absence of a Bose–Einstein condensate, and the formation of an insulator in the state $|I\rangle$ in (9.14). Figure courtesy of I. Bloch from M. Greiner, O. Mandel, T. Esslinger, T. W. Hänsch, and I. Bloch, *Nature* **415**, 39 (2002).

For future experimental studies, and analogous to our discussion in Figure 9.3 for the coupled qubits, we can now sketch a phase diagram as a function of the strength of the periodic potential and the temperature: this is shown in Figure 9.7. For weak periodic potentials, we have a superfluid ground state: above this, at finite temperature, there are excitations involving superflow over long length scales. At large periodic potentials, we have an insulating ground state: the excitations now involve small number fluctuations between neighboring pairs of sites, which cost a finite energy per number fluctuation. In the intermediate quantum-critical regime, we have behavior

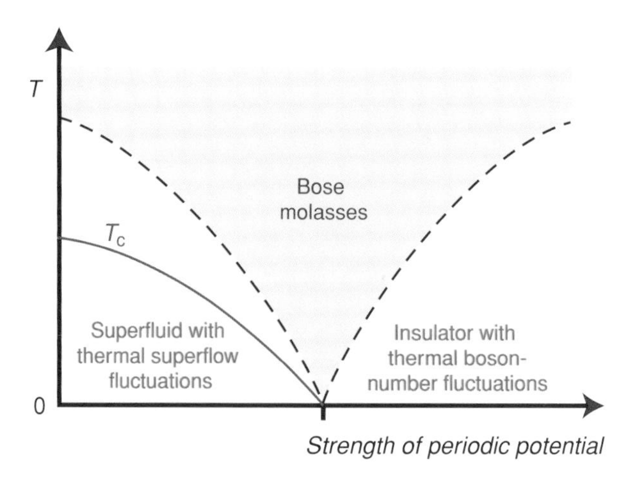


**Figure 9.7.** The phase diagram of a gas of trapped bosonic atoms in the presence of a periodic potential – this diagram is analogous to that for the qubit chain in Figure 9.3. The Bose–Einstein condensate is present below the critical temperature $T_c$, which is the only true phase transition at $T > 0$.

analogous to that in Figure 9.3. The characteristic relaxation rate for boson-number fluctuations obeys an expression analogous to (9.12): the dynamics of the bosons is strongly dissipative (with velocities proportional to applied forces) and so we have christened this yet-to-be-observed regime *Bose molasses.*

## 9.4 The cuprate superconductors

Electricity is the flow of electrons in wires. The wires are made of metals, such as copper, in which each metallic atom "donates" a few of its electrons, and these electrons move freely, in quantum-mechanical wavelike states, through the entire wire. However, single electrons are not bosons, but fermions, which obey the exclusion principle, and so more than one electron cannot occupy the same quantum state to form a Bose–Einstein condensate. The motion of electrons in a metal has a small but finite resistance; this arises from scattering of the electrons off the ever-present defects in the crystalline arrangement of atoms in a metal. Consequently metals are conductors but not *super*conductors.

However, in 1911, Kammerlingh Onnes cooled the metal mercury below 4.2 K and found that its electrical resistance dropped precipitously to an immeasurably small value – he had discovered a superconductor. An explanation of this phenomenon eluded physicists until 1957, when John Bardeen, Leon Cooper, and Robert Schrieffer proposed a theory that again invoked the Bose–Einstein condensate. However the condensate was not of single electrons, but the analog condensate of *pairs* of electrons (now called Cooper pairs): each electron in the metal finds a partner, and the pairs then occupy the same quantum state, which extends across the system. While individual electrons are fermions, pairs of electrons obey Bose statistics at long length scales, and this allows them to form an analog of the Bose–Einstein condensate. Strong fluctuations in the numbers of the Cooper pairs in different regions of the wire are then responsible for superconductivity, just as we found in Section 9.3 that superfluidity was linked to fluctuations in the number of Rb atoms between different minima of the periodic potential.

The phenomenon of superconductivity clearly raises the possibility of many exciting technological applications: resistanceless flow of electricity would have tremendous impact on electrical power transmission, high-frequency electrical circuits behind the wireless communication revolution, and in medical applications like magnetic-resonance imaging (MRI) that require large magnetic fields, to name but a few. Many such applications are already available, and the main obstacle to more widespread usage is the low temperature required to obtain a superconductor. Raising the maximum critical temperature below which superconductivity appears has been a central research goal for physicists since Kammerlingh Onnes carried out his pioneering experiments.

A dramatic improvement in the temperature required for superconductivity came in 1986 in a breakthrough by Georg Bednorz and Alex Müller (see Figure 9.8). These IBM researchers discovered a new series of compounds (the "cuprates") that became superconductors at temperatures as high as 120 K; the highest temperature on record prior to their work had been 15 K. This revolutionary discovery sparked a great deal of theoretical and experimental work, and many important questions remain subject to debate today. Here we will discuss the current understanding of the quantum ground states of these compounds and the role played by quantum phase transitions, and point out some of the open questions.

The best place to begin our discussion is the insulating compound $La_2CuO_4$, whose crystal structure is shown in Figure 9.9. Despite its complexity, most of its electronic properties are controlled by a simple substructure – a single layer of Cu ions that reside

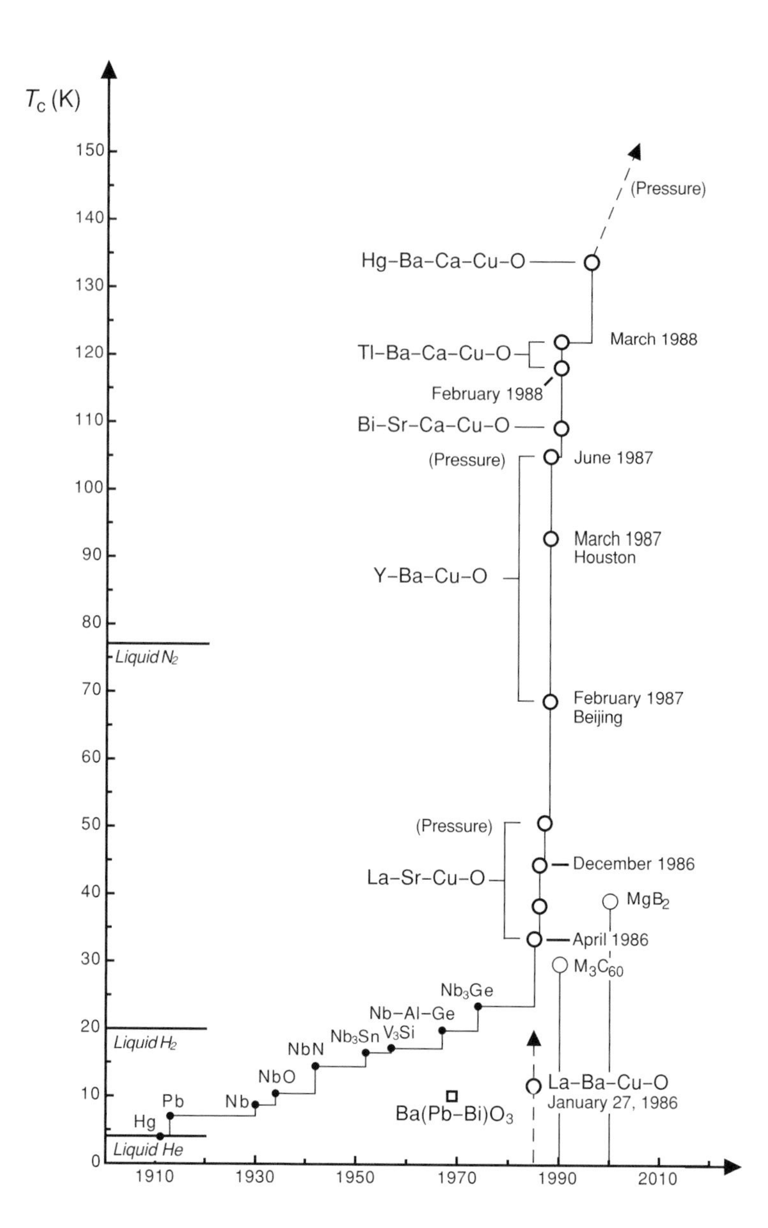


**Figure 9.8.** The history of materials with high critical temperatures ($Tc$) below which they are superconducting, as of early 2003. The cuprate superconductors were discovered on January 27, 1986, and led to a dramatic increase in $T_c$. Since then, apparently unrelated compounds ($MgB_2$) with moderately high $T_c$ have also been discovered. Also indicated are temperatures at which $N_2$ and $H_2$ liquefy – these gases are most commonly used to cool materials. The search for new materials with higher critical temperatures continues, and significant future increases may well appear. Figure courtesy of H. R. Ott.

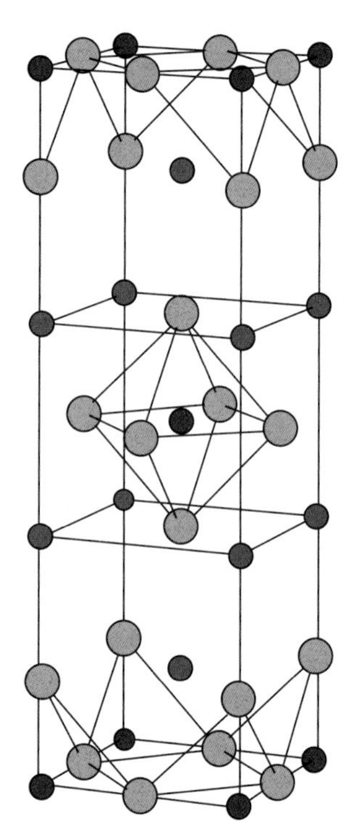

**Figure 9.9.** The crystal structure of the insulator $La_2CuO_4$. Red spheres are Cu, orange are O, and blue are La. The Cu ions are on the vertices of square lattices that extend in the horizontal plane.

on the vertices of a square lattice. The crystal consists of an infinite number of such layers stacked on top of each other, but the coupling between neighboring layers is small enough that we can safely ignore it, and concentrate on a single layer. A further simplification is that we need focus on only a single 3d orbital on each Cu ion. In $La_2CuO_4$ each such orbital has precisely one electron. The Coulomb repulsion between the electrons prevents them from hopping between Cu sites, and this is why $La_2CuO_4$ is a very good insulator. The reader should notice a simple parallel to our discussion of Rb atoms in Section 9.3: there, when the periodic potential was strong enough, each Rb atom was localized in a single well, and the ground state was an insulator. Here, the periodic potential is provided by the existing lattice of positively charged Cu ions, and one electron is trapped on each such ion.

So how do we make this a superconductor? Generalizing the discussion of Section 9.3, the naive answer is that we force $La_2CuO_4$ to undergo an insulator-to-superfluid quantum transition. However, unlike in Section 9.3, we do not have any available tools with which to tune the strength of the periodic potential experienced by the electrons. Bednorz and Müller accomplished the equivalent by chemical substitution: the compound $La_{2-\delta}Sr_{\delta}CuO_4$ has a concentration $\delta$ of trivalent La ions replaced by divalent Sr ions, and this has the effect of removing some electrons from each square lattice of Cu ions, thus creating "holes" in the perfectly insulating configuration of $La_2CuO_4$. With the holes present, it is now possible to transfer electrons from site to site and across the crystal easily, without having to pay the expensive price in energy associated with putting two electrons on the same site. This should be clear from the "egg-carton" model of the insulator in Figure 9.4: the reader can verify that, if a carton is not full of eggs, we can move any egg across the carton by making eggs "hop" between neighboring sites, and without ever putting two eggs on the same site. With holes present, it becomes possible for electrons to move across the crystal, and, when the concentration of holes, $\delta$, is large enough, this material is a "high-temperature" superconductor. We now know that $La_{2-\delta}Sr_{\delta}CuO_4$ undergoes an insulator-to-superfluid quantum phase transition with increasing $\delta$ at the hole concentration $\delta = 0.055$. In some phenomenological aspects, this transition is similar to the superfluid–insulator transition discussed in Section 9.3, but it should be clear from our discussion here that the microscopic interpretation is quite different.

However, there is a complication here that makes the physics much more involved than that of a single insulator–superfluid quantum phase transition. As we noted in Section 9.2, each electron has two possible spin states, with angular momenta $\pm\hbar/2$, and these act like a qubit. It is essential to pay attention also to the magnetic state of these spin-qubits in $La_{2-\delta}Sr_{\delta}CuO_4$. We discussed in Section 9.2 how such coupled qubits could display a quantum phase transition between magnetic and paramagnetic quantum ground states in $LiHoF_4$. The spin-qubits in $La_{2-\delta}Sr_{\delta}CuO_4$ also display a related magnetic quantum phase transition with increasing $\delta$; the details of the magnetic and paramagnetic states here are, however, quite different from those of $LiHoF_4$, and we will describe them further below. So, with increasing $\delta$, a description of the quantum ground state of $La_{2-\delta}Sr_{\delta}CuO_4$ requires not one, but at least two quantum phase transitions. One is the insulator–superfluid transition associated with the motion of the electron charge, while the second is a magnetic transition associated with the electron spin. (Numerous other competing orders have also been proposed for $La_{2-\delta}Sr_{\delta}CuO_4$ and related compounds, and it is likely that the actual situation is even more complicated.) Furthermore, these transitions are not independent of each other, and much of the theoretical difficulty resides in understanding the interplay among the fluctuations of the various transitions.

For now, let us focus on just the state of electron spin-qubits in the insulator $La_2CuO_4$, and discuss the analog of the qubit quantum phase transition in Section 9.2. In this insulator, there is precisely one such qubit on each Cu site. However, unlike the qubit on the Ho site in $LiHoF_4$, there is no preferred orientation of the electron spin, and the qubit of a single Cu ion is free to rotate in all directions. We know from experiments that, in the ground state of $La_2CuO_4$, the "exchange" couplings among the qubits are such that they arrange themselves in an *antiferromagnet*, as shown in Figure 9.10. Unlike the parallel ferromagnetic arrangement of qubits in $LiHoF_4$, now the couplings between the Cu ions prefer that each qubit align itself anti-parallel to its neighbor. The checkerboard arrangement in $La_2CuO_4$ optimizes this requirement, and each qubit has all its neighbors anti-parallel to it. While this arrangement satisfactorily minimizes the potential energy of the spins, the reader will not be surprised to learn that Heisenberg's uncertainty relation implies that all is not perfect: there is an energy cost in localizing each qubit along a definite spin direction, and there are ever-present quantum fluctuations about the quiescent state pictured in Figure 9.10. In $La_2CuO_4$ these fluctuations are not strong enough to destroy the antiferromagnetic order in Figure 9.10, and the resulting checkerboard arrangement of magnetic

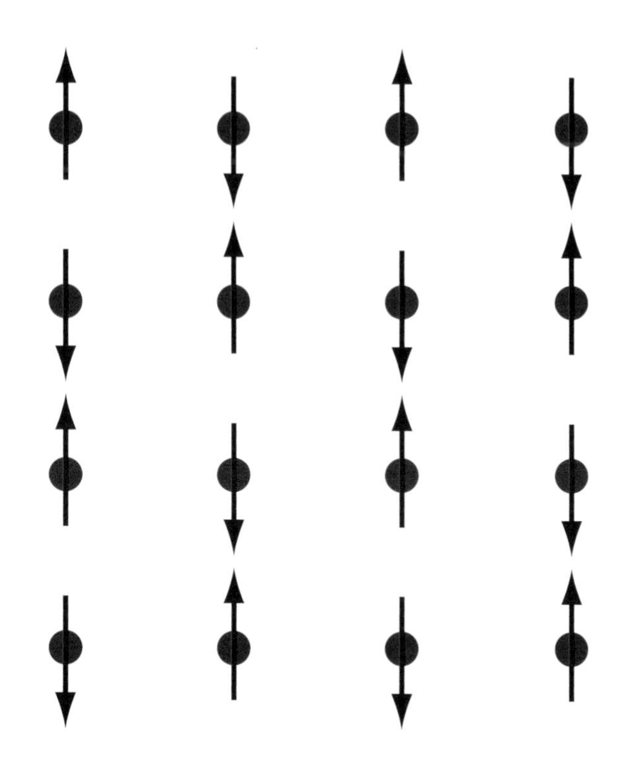

**Figure 9.10.** $La_2CuO_4$, an insulating antiferromagnet. The spin-qubits reside on the Cu sites (red circles), and their average magnetic moments arrange themselves in the checkerboard arrangement shown above. Unlike the qubits in $LiHoF_4$ in Figure 9.1, the spins above are free to rotate in any direction, provided that the spins on neighboring Cu sites retain an anti-parallel orientation.

moments can be, and has been, observed by scattering magnetic neutrons off a $La_2CuO_4$ crystal.

These neutron-scattering experiments can also follow the magnetic state of the spin-qubits with increasing $\delta$. For large enough $\delta$, when the ground state is a superconductor, it is also known that there is no average magnetic moment on any site. So there must be at least one quantum phase transition at some intermediate $\delta$ at which the antiferromagnetic arrangement of magnetic moments in $La_2CuO_4$ is destroyed: this is the antiferromagnet–paramagnet quantum transition. This transition is clearly coupled to the onset of superconductivity in some manner, and it is therefore not straightforward to provide a simple description.

With the insulator–superfluid and antiferromagnet–paramagnet quantum phase transitions in hand, and with each critical point being flanked by two distinct types of quantum order, we can now envisage a phase diagram with at least four phases. These are the antiferromagnetic insulator, the paramagnetic insulator, the antiferromagnetic superfluid, and the paramagnetic superfluid (see Figure 9.11). Of these, $La_2CuO_4$ is the antiferromagnetic insulator, and the high-temperature superconductor at large $\delta$ is the paramagnetic superfluid; note that these states differ in *both* types of quantum order. The actual phase diagram turns out to be even more complex, as there are strong theoretical arguments implying that at least one additional order must exist, and we will briefly review one of these below.

One strategy for navigating this complexity is to imagine that, like in the simpler system studied in Section 9.2, we did indeed have the freedom to modify the couplings between the spin-qubits in undoped $La_2CuO_4$, so that we can tune the system across a quantum phase transition to a paramagnet, while it remains an insulator. This gives us the luxury of studying the antiferromagnet–paramagnet quantum phase transition in a simpler context, and decouples it from the insulator–superfluid transition. It is also reasonable to expect that the insulating paramagnet is a better starting point for understanding the superconducting paramagnet, because these two states differ only by a single type of order, and so can be separated by a single quantum phase transition.

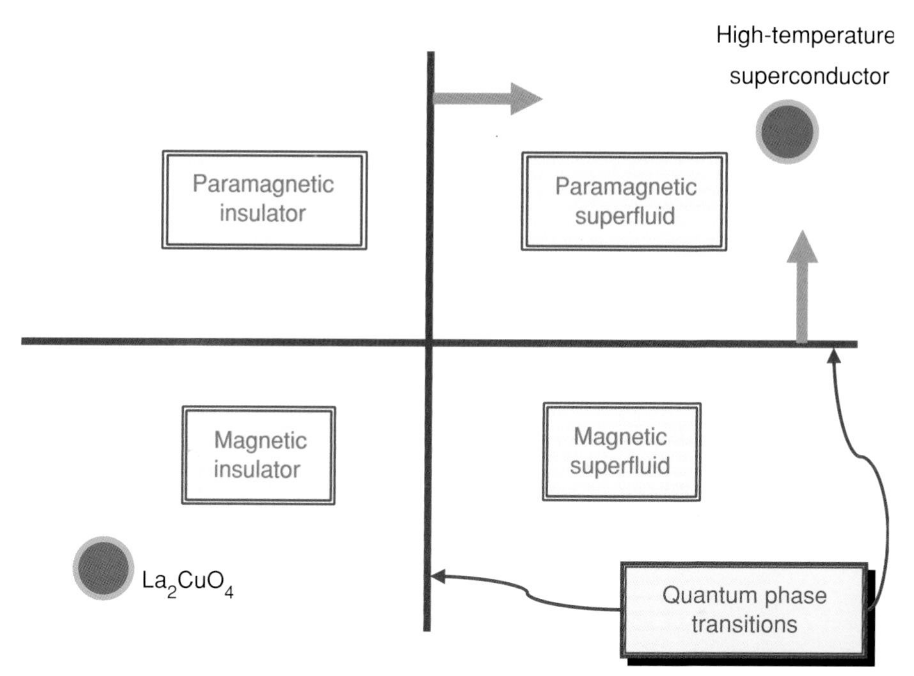


**Figure 9.11.** A minimal schematic theoretical phase diagram for the cuprate superconductors. The axes represent two suitable parameters that can be varied so that the lowest-energy state traverses the quantum phase transitions shown. The insulating magnet, $La_2CuO_4$, is well understood. A theory of the high-temperature superconductor can be developed by understanding the quantum phase transitions, and then moving along the orange arrows.

To study the insulating antiferromagnet–paramagnet quantum phase transition we need to modify the couplings between the qubits on the Cu sites, while retaining the full symmetry of the square lattice – in this manner we perturb the antiferromagnetic ground state in Figure 9.10. As this perturbation is increased, the fluctuations of the qubits about this simple checkerboard arrangement will increase, and the average magnetic moment on each site will decrease. Eventually, we will reach a quantum critical point, beyond which the magnetic moment on every site is precisely zero, i.e. a paramagnetic ground state. What is the nature of this paramagnet? A sketch of the quantum state of this paramagnet, predicted by theory, is shown in Figure 9.12.

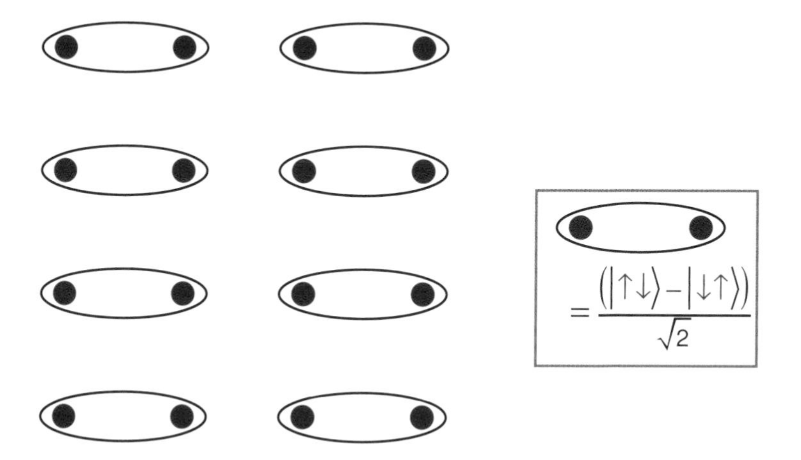


**Figure 9.12.** A paramagnetic insulator obtained by perturbing the antiferromagnet in Figure 9.10. The couplings between the qubits are modified so that quantum fluctuations of the average magnetic moment on each site are enhanced, and the insulator undergoes a quantum phase transition to a state with zero magnetic moment on every site. Each ellipse represents a singlet valence bond as in (9.15) or (9.16). The state above has bond order because the valence bonds are more likely to appear on the links with ellipses shown above than on the other links.

The fundamental object in Figure 9.12 is a *singlet valence bond* between qubits – this valence bond has many similarities to the covalent bond which is the basis of organic chemistry. As in Section 9.2, we represent the states of the electron spin on site $j$ by $|\uparrow\rangle_j$ and $|\downarrow\rangle_j$, and the valence bond between spins on sites $i$ and $j$ is a state in which the spin-qubits are *entangled* as below:

$$|V\rangle_{ij} = \frac{1}{\sqrt{2}}(|\uparrow\rangle_i |\downarrow\rangle_j - |\downarrow\rangle_i |\uparrow\rangle_j). \tag{9.15}$$

Note that, in both states in the superposition in (9.15), the spins are anti-parallel to each other – this minimizes the potential energy of their interaction. However, each qubit fluctuates equally between the up and down states – this serves to appease the requirements of Heisenberg's uncertainty principle, and lowers the total energy of the pair of qubits. Another important property of (9.15) is that it is invariant under rotations in spin space; it is evident that each qubit has no definite magnetic moment along the $z$ direction, but we saw in Section 9.2 that such qubits could have a definite moment along the $x$ direction. That this is not the case here can be seen from the identity

$$|V\rangle_{ij} = \frac{1}{\sqrt{2}}(|\rightarrow\rangle_i |\leftarrow\rangle_j - |\leftarrow\rangle_i |\rightarrow\rangle_j); \tag{9.16}$$

the reader can verify that substituting (9.2) and (9.3) into (9.16) yields precisely (9.15). So the singlet valence bond has precisely the same structure in terms of the $\pm x$ oriented qubits as that of the bond between the $\pm z$ oriented qubits – it is the unique combination of two qubits which does not have *any* preferred orientation in spin space. This is one of the reasons why it is an optimum arrangement of a pair of qubits. However, the disadvantage of a valence bond is that a particular qubit can form a valence bond with only a single partner, and so has to make a choice among its neighbors. In the antiferromagnetic state in Figure 9.11, the arrangement of any pair of spins is not optimal, but the state has the advantage that each spin is anti-parallel to all four of its neighbors. The balance of energies shifts in the state in Figure 9.12, and now the spins all form singlet bonds in the rotationally invariant paramagnet. For the state illustrated in Figure 9.12, the spins have chosen their partners in a regular manner, leading to a crystalline order of the valence bonds. Detailed arguments for these and related orderings of the valence bonds appear in the theoretical works, and some of the physics is briefly discussed in Box 9.2. A significant property of the resulting bond-ordered paramagnet is that, while spin-rotation invariance has been restored, the symmetry of the square lattice has been broken. This *bond order* is an example of the promised third quantum order. The interplay between the spin-rotation and square-lattice symmetries in the antiferromagnet–paramagnet quantum transition makes its theory far more complicated than that for the ferromagnet–paramagnet transition described in Section 9.2.

Tests of these theoretical ideas for the quantum transition between the states in Figures 9.10 and 9.12 have been quite difficult to perform. Most computer studies of models of such antiferromagnets suffer from the "sign" problem, i.e. the interference between different quantum histories of the qubits makes it very difficult to estimate the nature of the quantum state across the transition. In 2002, Anders Sandvik and collaborators succeeded in finding a model for which the sign problem could be conquered. This model displayed a quantum transition between a magnetically ordered state, as in Figure 9.10, and a paramagnet with bond order, as in Figure 9.12.

To summarize, the physics of the spin-qubits residing on the Cu sites is a competition between two very different types of insulating quantum states. The first is the antiferromagnet in Figure 9.10 which is found in $La_2CuO_4$: there is an average magnetic moment on each Cu site, and the "up" and "down" moments are arranged like the

### BOX 9.2 RESONANCE BETWEEN VALENCE BONDS

The first example of "resonance" in chemistry appeared in the planar hexagonal structures proposed in 1872 by Kekulé for the benzene molecule $C_6H_6$ – note that this predates the discovery of quantum mechanics by many years. The structures are represented by the bonding diagrams in Figure 9.13. Of particular interest is (what we now call) the $\pi$-orbital bond between the neighboring C atoms. This bond involves two electrons and its quantum-mechanical state is similar to the valence bond in (9.15) or (9.16) (the main difference is that the electronic charges are not so well localized on the two C atoms, and there are also contributions from states with both electrons on the same C atom). There are two possible arrangements of this bond, as shown in Figure 9.13. With the knowledge of quantum mechanics, and the work of Linus Pauling on the physics of the chemical bond, we know that the bonds tunnel back and forth between these two configurations, and this enhances the stability of the benzene structure. Moreover, the final structure of benzene has a perfect hexagonal symmetry, which is not the case for either of the two constituent structures.

In 1974, Patrick Fazekas and Philip Anderson applied the theory of resonance between valence bonds to quantum antiferromagnets; specific applications to the high-temperature superconductors were proposed in 1987 by G. Baskaran and Anderson and also by Steven Kivelson, Daniel Rokhsar, and James Sethna. The idea was that singlet valence bonds form between the spin-qubits on neighboring Cu ions, and these resonate between different possible pairings of nearest-neighbor qubits, as shown in Figure 9.14. It was argued by Nicholas Read and the author that in many cases the consequences of these resonances on the square lattice are very different from those in benzene. Whereas in benzene the final ground state has full hexagonal symmetry, here the resonances actually lead to *bond order* with structures like those in Figure 9.12, which break the symmetry of the square lattice. The reason for this is illustrated in Figure 9.14: regular bond-ordered configurations have more possibilities of plaquettes for resonance, and this lowers the energy of states with higher symmetry. Of course, each bond-ordered state has partners related to it by the symmetry of the square lattice (e.g. the state in Figure 9.12 has three partners obtained by rotating it by multiples of 90° about any lattice point), but the amplitude of the tunneling between these macroscopically distinct states is completely negligible.

black and white squares on a chess board. The second state is a paramagnet in which the spins are paired in valence bonds described by the quantum state (9.15).

With this understanding in hand, we can now address the physics of the electron charge, and the insulator–superconductor quantum transition with increasing $\delta$ in

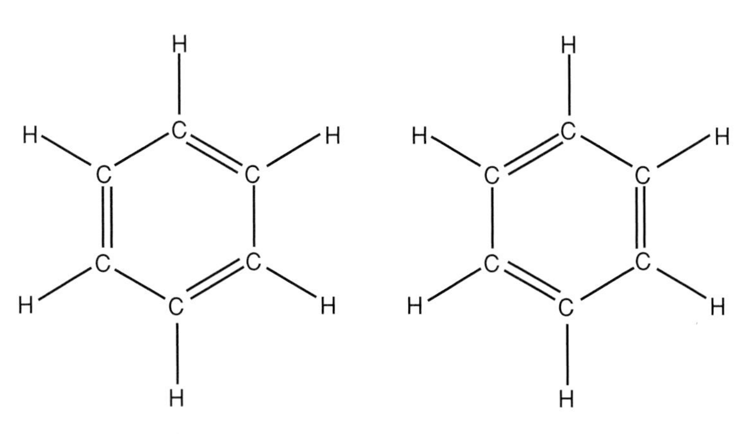


**Figure 9.13.** The two structures of benzene proposed by Kekulé. The second of the double lines represent valence bonds between $\pi$ orbitals, which are similar to the valence bonds in (9.12). The stability of benzene is enhanced by a quantum-mechanical resonance between these two states.

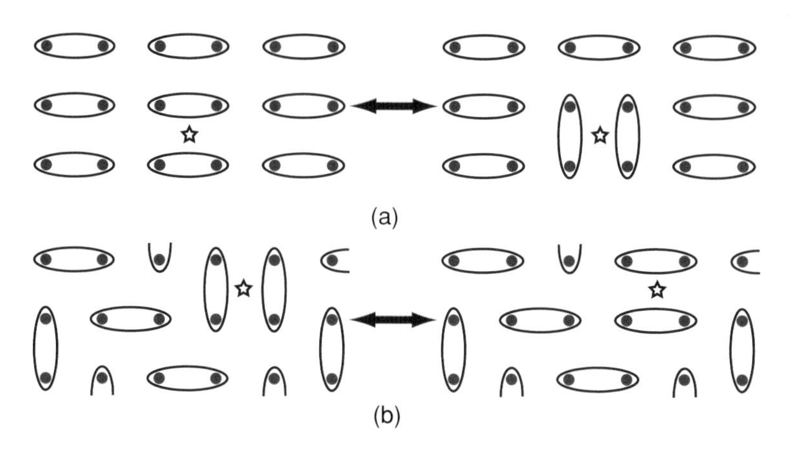


**Figure 9.14.** Resonance of valence bonds around the plaquettes of a square lattice. Shown are resonances for two different valence-bond arrangements where the plaquettes are marked with a star. Note that there are many more possibilities of resonance in (a) than in (b), and this effect selects ground states with bond order.

$La_{2-\delta}Sr_{\delta}CuO_4$. A picture of the insulating state was shown in Figure 9.10. We show an analogous picture of the superconductor in Figure 9.15. Note first that there are Cu sites with no electrons: these are the holes, and, as we have discussed earlier, they are able to hop rapidly across the entire crystal without paying the price associated with occupancy of two electrons on the same Cu site. The motion of these holes clearly perturbs the regular arrangement of magnetic moments at $\delta = 0$ in Figure 9.10, and this enhances their quantum fluctuations and reduces the average magnetic moment on every site. Eventually, the magnetic moment vanishes, and we obtain a paramagnet in which the electrons are all paired in singlet bonds, as shown in Figure 9.15. Unlike the valence bonds in Figure 9.12, the valence bonds in Figure 9.15 are mobile and can transfer charge across the system – this is illustrated in Figure 9.16, where the motion of the holes has rearranged the valence bonds and resulted in a net transfer of electronic charge in the direction opposite to the motion of the holes. This suggests an alternative interpretation of the valence bonds: they are the *Cooper pairs* of electrons, which can move freely, as a pair, across the system with holes present. Their facile motion suggests that these composite bosons will eventually undergo Bose–Einstein condensation, and the resulting state will be a superconductor. We have therefore presented here a simple picture of the crossover from the insulator to a superconductor with increasing $\delta$.

The essence of our description of the insulator–superconductor transition was the "dual life" of a valence-bond pair. In the insulator, these valence bonds usually prefer to crystallize in regular bond-ordered states, as shown in Figure 9.12, and in the results

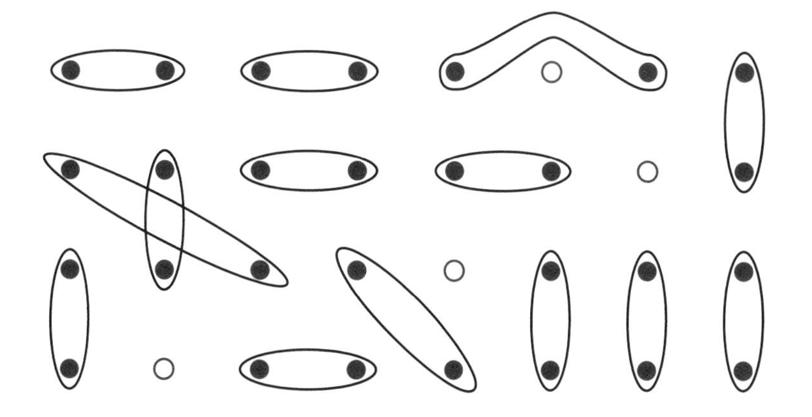

**Figure 9.15.** A snapshot of the superconducting state in $La_{2-\delta}Sr_{\delta}CuO_4$ for large $\delta$. The open circles represent the holes, or Cu sites with no electrons. The spins on the remaining sites are paired with each other in the valence bonds of Figure 9.12. We argue in the text that the valence bonds in the superconductor possess vestiges of the bond order of Figure 9.12, and this bond order is enhanced and experimentally detectable near vortices in the superflow of Cooper pairs.

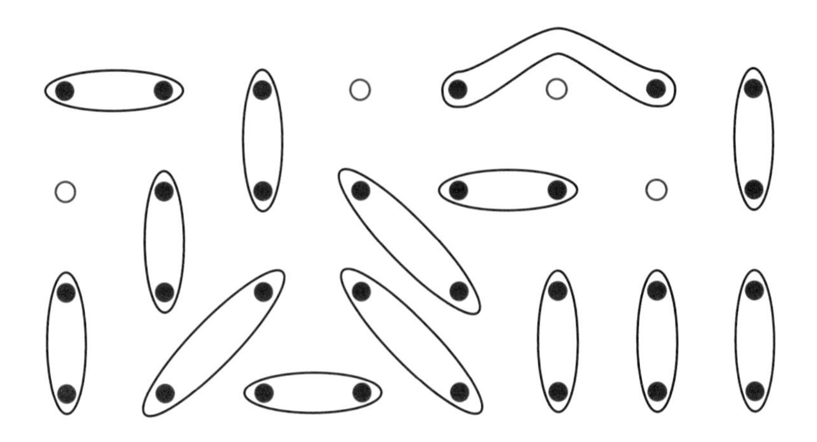

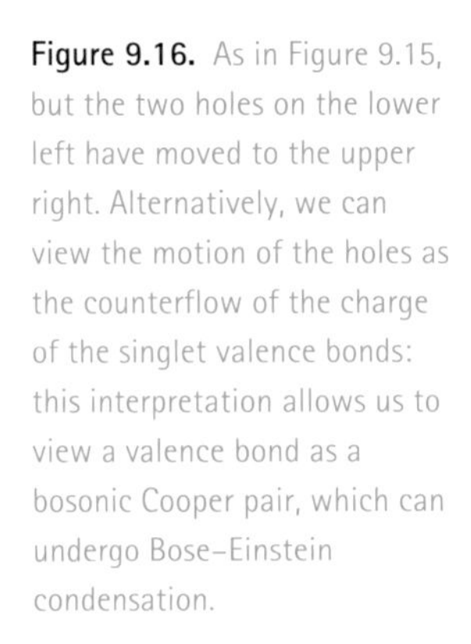

**Figure 9.16.** As in Figure 9.15, but the two holes on the lower left have moved to the upper right. Alternatively, we can view the motion of the holes as the counterflow of the charge of the singlet valence bonds: this interpretation allows us to view a valence bond as a bosonic Cooper pair, which can undergo Bose–Einstein condensation.

of Sandvik and collaborators. The valence bonds do resonate among themselves, and thus change their orientation, but this resonance does not lead to motion of any charge. In the superconductor, the valence bonds become mobile and rather longer-ranged, and transmute into the Cooper pairs of the theory of Bardeen, Cooper, and Schrieffer. The Bose–Einstein condensation of these pairs then leads to high-temperature superconductivity.

While this picture is appealing, does it have any practical experimental consequences? Can the dual interpretation of a valence bond be detected experimentally? One theoretical proposal was made in 2001 by Eugene Demler, Kwon Park, Ying Zhang, and the author, reasoning as follows. Imagine an experiment in which we are locally able to impede the motion of the Cooper pairs. In the vicinity of such a region, the valence bonds will become less mobile, and so should behave more like the valence bonds of the paramagnetic insulator. As we have seen in Figure 9.12, such "stationary" valence bonds are likely to crystallize in a regular bond-ordered arrangement. An experimental probe that can investigate the electronic states on the scale of the Cu lattice spacing (which is a few ångström units; an ångström unit is $10^{-10}$ m) should be able to observe a regular modulation associated with the bond order. We also proposed a specific experimental mechanism for impeding the motion of the Cooper pairs: apply a strong magnetic field transverse to the sample – this induces vortices around which there is a supercurrent of Cooper pairs, but the cores of the vortices are regions in which the superconductivity is suppressed. These cores should nucleate a halo of bond order. Matthias Vojta had studied the relative energies of different types of bond order in the superconductor, and found that, in addition to the modulation with a period of two lattice spacings shown in Figure 9.12, states with a period of four lattice spacings were stable over an especially large range of doping.

On the experimental side, Jennifer Hoffman, Seamus Davis, and collaborators carried out experiments at the University of California in Berkeley examining the electronic structure in and around the vortex cores on the surface of the high-temperature superconductor $Bi_2Sr_2CaCu_2O_{8+\delta}$ (closely related experiments have also been carried out by Craig Howald, Aharon Kapitulnik, and collaborators at Stanford University). In a remarkable experimental tour-de-force, they were able to obtain sub-ångström-scale resolution of electronic states at each energy by moving a "scanning tunneling microscope" across a clean surface of $Bi_2Sr_2CaCu_2O_{8+\delta}$. A picture of the observed structure of the electronic states around a single vortex is shown in Figure 9.17. Notice the clear periodic modulation, with a period that happens to be exactly four lattice spacings. Is this an experimental signal of the bond order we have discussed above? While this is certainly a reasonable possibility, the issue has not yet been settled conclusively. There continues to be much debate on the proper interpretation of these experiments

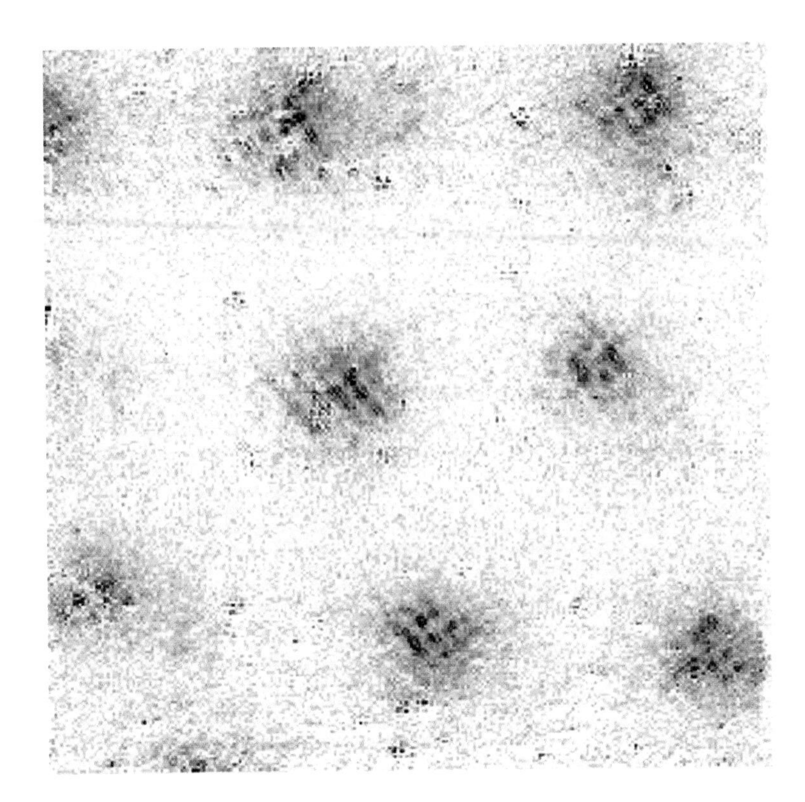

**Figure 9.17.** Measurement of Hoffmann, Davis, and collaborators of the local density of states on the surface of the high-temperature superconductor $Bi_2Sr_2CaCu_2O_{8+\delta}$ by STM measurement. A magnetic field has been applied perpendicular to the plane of the paper, and the dark regions are centered on the vortices in the superflow of the Cooper pairs; the diameter of each dark region is about 100 Å. There is a checkerboard modulation in each dark region with a period of four Cu-lattice spacings, and this is possibly a vestige of bond order, as discussed in the text. (Figure from J. E. Hoffman *et al. Science* 295, 466 (2002)).

and on the relative role of single quasiparticle and valence-bond excitations. Whatever the final answer turns out to be, it is exciting to be involved in a debate that involves fundamental theoretical issues concerning the physics of quantum phase transitions coupled with remarkable experiments at the cutting edge of technology.

## 9.5 Conclusions

This chapter has given a simple overview of our efforts to understand "*quantum matter.*" Quantum mechanics was originally invented as a theory of small numbers of electrons interacting with each other, with many features that jarred with our classical intuition. Here, we hope that we have convinced the reader that many of these counter-intuitive features of quantum mechanics are also apparent in the bulk properties of macroscopic matter. We have discussed macroscopic phases whose characterizations are deep consequences of the quantum-mechanical principle of superposition, and which can undergo phase transitions driven entirely by Heisenberg's uncertainty principle. Theoretical work on the classification of distinct phases of quantum matter continues today: many examples that make intricate use of quantum superpositions to produce new states of matter have been found. Some of this work is coupled with experimental efforts on new materials, and this bodes well for much exciting new physics in the century ahead.

## FURTHER READING

Technical information and original references to the literature can be found in the article by S. L. Sondhi, S. M. Girvin, J. P. Carini, and D. Shahar, *Rev. Mod. Phys.* **69** (1997) 315 and in the book *Quantum Phase Transitions* by S. Sachdev, Cambridge, Cambridge University Press, 1999.

A more detailed discussion of the application of the theory of quantum phase transition to the cuprate superconductors appears in the article "Order and quantum phase transitions in the cuprate superconductors," S. Sachdev, *Rev. Mod. Phys.*, July 2003, http://arxiv.org/abs/cond-mat/0211005. A related perspective on "fluctuating order" near quantum critical points is given in the article "How to detect fluctuating order in the high-temperature superconductors," by S. A. Kivelson, E. Fradkin *et al.*, http://arxiv.org/abs/cond-mat/0210683. For a review of quantum phase transitions in metals, see "Quantum phase transitions" by Thomas Vojta, http://arxiv.org/abs/cond-mat/0010285.

## Subir Sachdev

Subir Sachdev is Professor of Physics at Harvard University. His research interests span many areas of quantum many-body theory and statistical physics: he is especially interested in the physical properties of materials in the vicinity of a quantum phase transition – a transformation in the state of a system induced by Heisenberg's uncertainty principle. He has shown how these theoretical ideas can help in the understanding of the cuprate superconductors, intermetallic compounds with metallic moments, trapped atomic gases, and other physical systems. He is the author of numerous scientific articles and of the book *Quantum Phase Transitions*, published by Cambridge University Press in 1999.

He began his undergraduate studies at the Indian Institute of Technology in New Delhi in 1978, and then transferred to the Massachusetts Institute of Technology, obtaining his undergraduate degree in physics in 1982. He obtained a Ph.D. in theoretical physics from Harvard University in 1985. After a brief stay at AT&T Bell Laboratories, he joined the faculty of Yale University in 1987, and has been there ever since. He was an Alfred P. Sloan Foundation fellow, winner of the American Physical Society's Apker Award, a Presidential Young Investigator, Matsen and Ehrenfest lecturer, and is a fellow of the American Physical Society. He has also held visiting faculty appointments at the Universities of Paris, Grenoble, and Fribourg, and at the Institut Henri Poincaré.

For more information, visit his website, http://sachdev.physics.harvard.edu.

# III Quanta in action

# 10 Essential quantum entanglement

Anton Zeilinger

## 10.1 How it all began

In 1935, a decade after the invention of quantum theory by Heisenberg (1925) and Schrödinger (1926), three papers appeared,[1] setting the stage for a new concept, quantum entanglement. The previous decade had already seen momentous discussions on the meaning of the new theory, for example by Niels Bohr (see his contribution to *Albert Einstein: Philosopher-Scientist* in Further reading); but these new papers opened up the gates for a new, very deep philosophical debate about the nature of reality, about the role of knowledge, and about their relation. Since the 1970s, it has become possible for experimenters to observe entanglement directly in the laboratory and thus test the counter-intuitive predictions of quantum mechanics whose confirmation today is beyond reasonable doubt. Since the 1990s, the conceptual and experimental developments have led to new concepts in information technology, including such topics as quantum cryptography, quantum teleportation, and quantum computation, in which entanglement plays a key role.

In the first of the three papers, Einstein, Podolsky, and Rosen (EPR) realized that, if two particles had interacted in their past, their properties will remain connected in the future in a novel way, namely, observation on one determines the quantum state of the other, no matter how far away. Their conclusion was that quantum mechanics is incomplete and they expressed their belief that a more complete theory may be found. Niels Bohr replied that the two systems may never be considered independent of each other, and observation on one changes the possible predictions on the other. Finally, Erwin Schrödinger, in the paper which also proposed his famous cat paradox, coined the term "entanglement" (in German *Verschränkung*) for the new situation and he called it "not one, but *the* essential trait of the new theory, the one which forces a complete departure from all classical concepts."

The discussion on these issues was for a long time considered to be "merely philosophical." This changed when John Bell discovered in 1964 that such entangled systems

[1] A. Einstein, B. Podolsky, and N. Rosen, Can quantum-mechanical description of physical reality be considered complete?, *Phys. Rev.* **47** (1935) 777.
N. Bohr, Can quantum-mechanical description of physical reality be considered complete?, *Phys. Rev.* **48** (1935) 696.
E. Schrödinger, Die gegenwärtige Situation in der Quantenmechanik, *Naturwissenschaften* 23 (1935) 807, 823, and 844. English translation in *Proc. Am. Phil. Soc.* **124** (1980). Reprinted in J. A. Wheeler and W. H. Zurek (eds.), *Quantum Theory and Measurement*, Princeton, Princeton University Press, 1983.

*The New Physics for the Twenty-First Century*, ed. Gordon Fraser.
Published by Cambridge University Press. 

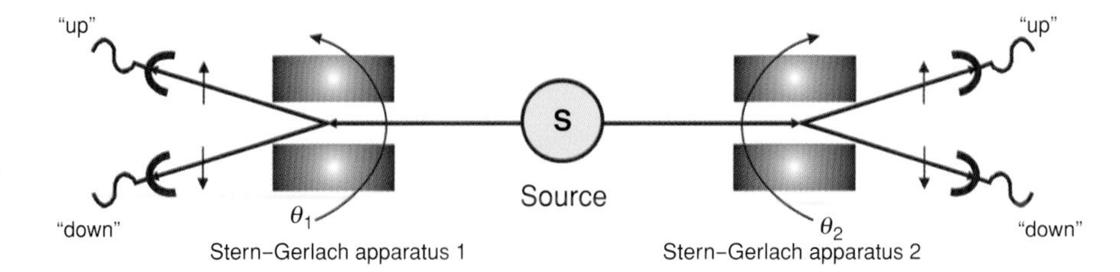


**Figure 10.1.** The classic *Gedankenexperiment* to demonstrate quantum entanglement between two particles emitted from a single source.

allow experimental tests of a very important philosophical question – that of whether the world can be understood on the basis of local realism. Local realism is the worldview that observations can be explained on the basis of each of two assumptions: firstly, that measurement results, at least in those cases in which one can make predictions with certainty, correspond to some element of reality; and secondly, that they are independent of whatever action might be performed at a distant location at the same time. By now, a number of experiments have confirmed the quantum predictions to such an extent that a local-realistic worldview can no longer be maintained.

While the work on entanglement was initially motivated by purely philosophical curiosity, a surprising development happened more recently. Methods were discovered whereby entanglement plays a crucial role in novel ways of encoding, transmitting, and processing information. These discoveries benefited greatly from the observation that multi-particle entanglement has even more striking features than two-particle entanglement, and from the enormous experimental progress, making manipulation of entanglement of individual quantum states standard routine in the laboratory.

## 10.2 The physics of entanglement

To discuss the essential features of entanglement, consider a source S (Figure 10.1) that emits pairs of particles, particle 1 to the left and particle 2 to the right. Let us assume that these are spin-$\frac{1}{2}$ particles and that each particle is subjected to a spin measurement using a Stern–Gerlach apparatus. Each of the Stern–Gerlach apparatuses can be rotated around the beam axis to be oriented along the angle $\theta_1(\theta_2)$. Let us further assume that the source consists of spin-0 particles which decay into two spin-$\frac{1}{2}$ particles. Thus, by conservation of angular momentum, the spins of the two particles must be anti-parallel. If one particle has its spin up ($\uparrow$) along the arbitrary direction $z$, then the other particle 2 must have its spin down ($\downarrow$) along that direction. Therefore, there are two possibilities for both particles, either the first one is up ($\uparrow_1$) and the second is down ($\downarrow_2$) or the first one is down ($\downarrow_1$) and the second one is up ($\uparrow_2$). Since the $z$ direction can be chosen arbitrarily, the spins of the two particles must be anti-parallel along any direction. If we perform the experiments of Figure 10.1 with many pairs of particles, we therefore observe the following behavior: for any orientation of the Stern–Gerlach magnets, each of the detectors fires completely randomly, that is, a specific particle has the same probability of 50% of triggering either its "up" detector or its "down" detector. Yet, for the two systems oriented in parallel ($\theta_1 = \theta_2$), once the, say, "up" detector of particle 1 has fired, we know with certainty that the "down" detector of particle 2 will register on the other side and vice versa.

One might be tempted to assume that these results can be understood on the basis of the assumption that the source emits particles in a statistical mixture, half $\uparrow_1\downarrow_2$ and half $\downarrow_1\uparrow_2$. While such an assumption can explain the perfect correlations if the Stern–Gerlach systems are oriented along one specific direction, it cannot explain the results along arbitrary directions. We are forced to assume that, rather than being in a statistical mixture of $\uparrow_1\downarrow_2$ and $\downarrow_1\uparrow_2$, the two particles are in a coherent superposition of both

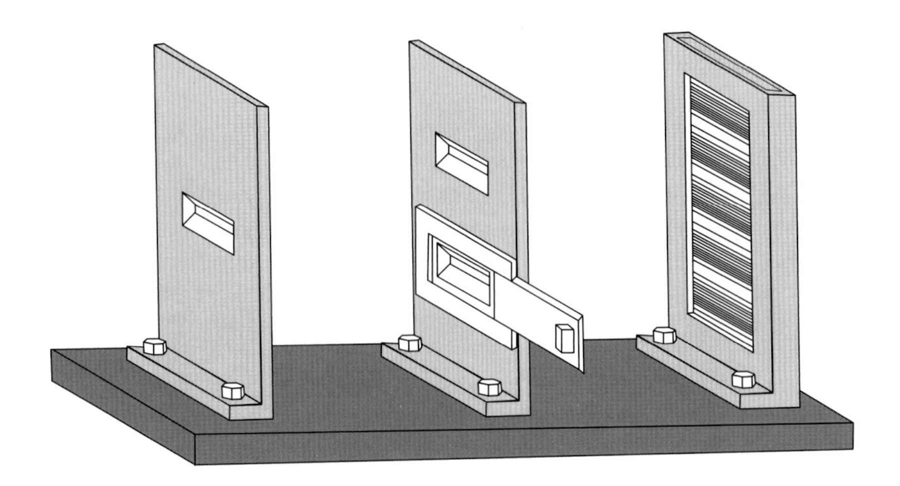

**Figure 10.2.** The enigmatic double-slit experiment. It is wrong to assume that a single particle has to pass through one or other slit. Instead, the slit passage is completely unknown and even a single particle is in a superposition of passing through one slit and of passing through the other.

possibilities. This is a coherent superposition very much in the same way as for a particle passing through the famous double-slit assembly (Figure 10.2). As is well known there, it is wrong to assume that the particle passes through either slit, but rather that there is complete ignorance of slit passage and therefore the particle is in a superposition of passing through one slit and through the other. In the same way, in the situation we are discussing (Figure 10.1), there is complete ignorance as to whether the two particles are in $\uparrow_1\downarrow_2$ or in $\downarrow_1\uparrow_2$ and the two particles are a superposition of both.

From a conceptual point of view, this means that neither of the two particles is in a well-defined spin state, i.e. neither has a well-defined spin before it is measured. Once it is measured, it gives randomly one of the two possible results, $\uparrow$ or $\downarrow$. But then the other particle is immediately projected into the corresponding opposite state, independently of how large the separation between the two particles at the moment of measurement may be. Einstein criticized exactly this point, saying in his Autobiographical Notes in the volume *Albert Einstein: Philosopher-Scientist* that the "real factual situation" of the second particle must be independent of whatever measurement is performed on its distant sibling. He therefore concluded that quantum mechanics, prescribing different quantum states for the second system depending on which specific measurement is performed on the first, cannot be a complete description of physical reality.

We have just seen that, from the measurement on one particle, we can predict with certainty the measurement result on the other particle should the Stern–Gerlach systems be oriented in parallel. Thus, EPR suggested that one can introduce an element of physical reality corresponding to that spin-measurement result. They further argued that, since the two measurements can be space-like separated, i.e. separated in such a way that no signal traveling at the speed of light can communicate between the two measurements, an element of reality can be assigned independently of the measurement on the other particle.

In other words, the specific argument may be understood in a very simple way. If we perform a measurement on the first particle, we can predict with certainty what the measurement result on the second one would be, should it be measured. However, since it is very far away, it is reasonable to assume that its features are independent of whether one cares to observe the first one and of which kind of measurement is performed on the first one. This is not to say that the two results are independent of each other. Indeed, consider, in contrast, the case of identical twins, Charlie and John. Most of the features are perfectly correlated, for instance, they may both have black hair. But the fact that Charlie's hair is black is independent of whether we care to make an observation of the color of the hair of his twin brother John.

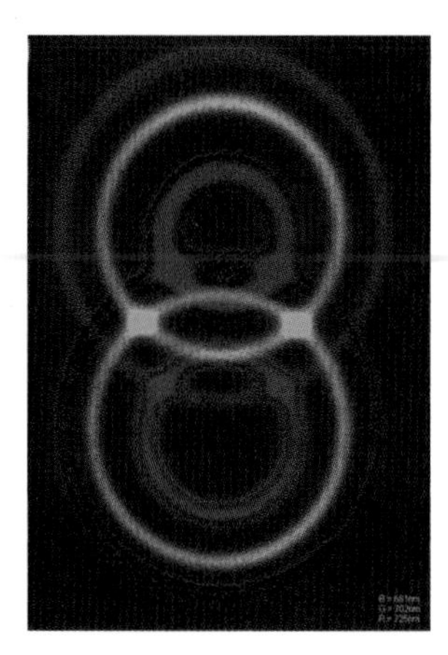

**Figure 10.3.** Entanglement. A photograph of the light emitted in type-II parametric down-conversion (false colors). The polarization-entangled photons emerge along the directions of the intersection between the green rings. They may be selected for an experiment by placing small holes there. (Photo by Paul Kwiat and Michael Reck. Courtesy of the Institute of Quantum Optics and Quantum Information, Vienna.)

Furthermore, since there is complete freedom of which measurement direction is chosen for the first particle, and since perfect correlations exist for all measurement directions, each particle must contain an element of reality following EPR for every possible measurement direction. For Charlie and John, this means that they must already carry instructions, the genes, with determine all physical features that are identical, independently of what observation we care to perform on either one. It has been shown by John Bell that such a model is unable to explain the correlations between measurement results when the two Stern–Gerlach apparatuses are not oriented parallel or anti-parallel, but along oblique directions. There are many mathematical formulations of the resulting Bell inequality. The simplest was first formulated by Eugene Wigner. For three arbitrary angles, $\alpha$, $\beta$, and $\gamma$, the following inequality must be satisfied: $N(\uparrow_\alpha, \uparrow_\beta) \leq N(\uparrow_\alpha, \uparrow_\gamma) + N(\uparrow_\beta, \downarrow_\gamma)$. Here, for example, $N(\uparrow_\alpha, \uparrow_\beta)$ means the number of cases when the "up" detector in the Stern–Gerlach apparatus of particle 1 oriented along direction $\alpha$ registers simultaneously with the "up" detector of particle 2 with the Stern–Gerlach apparatus oriented along direction $\beta$. Quantum mechanics predicts a sinusoidal variation of these correlations: $N(\uparrow_\alpha, \uparrow_\beta) = \frac{1}{2}N_0 \sin^2(\alpha - \beta)/2$ and $N(\uparrow_\alpha, \downarrow_\beta) = \frac{1}{2}N_0 \cos^2(\alpha - \beta)/2$. It is easy to see that Bell's inequality is violated for many possible combinations of measurement directions, for example for angles such that $\alpha - \beta = \beta - \gamma = 60^\circ$. Therefore at least one of the assumptions entering into the derivation of Bell's inequality must be incorrect. This, besides the assumptions of reality and locality discussed above, is also the assumption of counterfactual definiteness, assuming that the system can be assigned properties independently of actual observation.

Interestingly, at the time of Bell's discovery there existed no experimental evidence allowing one to discriminate between quantum physics and local realism as defined in Bell's derivation. After the realization, independently by John Clauser and Michael Horne in the late 1960s, that the polarization-entangled state of photons emitted in atomic cascades can be used to test Bell's inequalities, the first experiment was performed by Stuart J. Freedman and John F. Clauser in 1972. There have now been many such experiments. The earliest to reveal large violation of a Bell-type inequality were the experiments by Alain Aspect, Philippe Grangier, and Gérard Roger in 1981 and 1982. The most widely used source for polarization-entangled photons today utilizes the process of spontaneous parametric down-conversion in nonlinear optical crystals. A typical picture of the emerging radiation is shown in Figure 10.3.

Aside from one very early experiment, all results agreed with quantum mechanics, violating inequalities derived from Bell's original version using certain additional assumptions. If Charlie and John were entangled quantum systems, we are forced to conclude that, while they exhibit perfect correlations of their features, for example, both having black hair, the assumption that they had black hair before observation is wrong, and it is even wrong to assume that some hidden property determines which color of hair they should have, should they be observed. In contrast, the specific color of hair they will show upon observation is completely undetermined, but, once it is observed, one exhibits randomly some color, in our case black, and then the other, no matter how far away he is, will also exhibit black hair, should the color of his hair be observed.

While, for two-particle entanglement, any contradiction between quantum mechanics and local realism occurs for the statistical predictions of quantum mechanics only, the case in which one can make predictions with certainty of the features of the second particle from a measurement on the first one can indeed be explained by a local-realistic model. This might seem to be natural since, after all, predictions with certainty are the realm of classical physics. Surprisingly, as discovered in 1987 by Daniel Greenberger, Michael Horne, and Anton Zeilinger (GHZ), this is not the case for entanglement of three or more systems. There, indeed, situations arise in which, on the basis of measurements

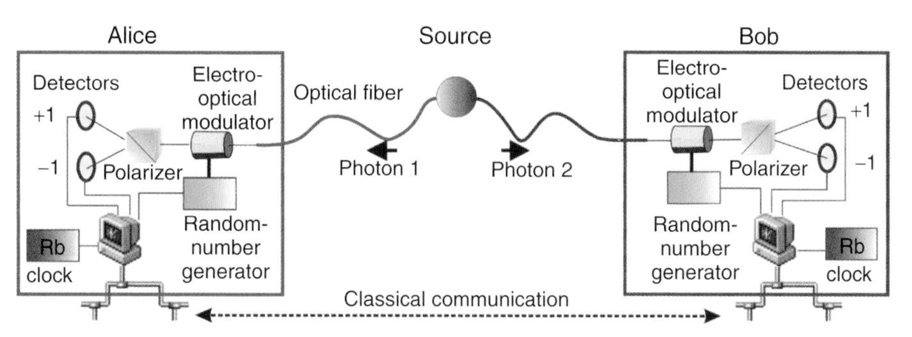


Figure 10.4. Investigating quantum mechanics. A long-distance experiment with independent observers to probe the Bell inequality. Considerable experimental effort is required to eliminate possible objections and make the conclusions universally valid.

of $N-1$ particles, a local realist predicts the last particle to be, say, $\uparrow$, whereas quantum mechanics predicts it with certainty to be $\downarrow$. Such violently nonclassical behavior has been confirmed by experiment.

For a long time, one problem was that, due to limitations in detector efficiency, only a small subset of the entangled pairs emitted by the source could be registered. It was therefore suggested that, while the subset registered violates a Bell inequality and appears to agree with quantum mechanics, the complete set of all pairs emitted would not. This so-called "detection loophole" was closed in an experiment done in 2000 by the group of David Wineland on entanglement between atomic states of ions in a magneto-optical trap. In that experiment, one could still save local realism by assuming that the atoms sitting very close to each other exchange information about the measurements performed. This loophole can be closed only by separating the two measurement stations by very large distances and performing the measurement so fast that no communication at the speed of light is possible. It was closed in an experiment by Wheis in 1998 in which two entangled photons were coupled into long glass fibers and the polarization correlation over a distance of the order of 400 m was measured. The decision regarding which polarization to measure on each photon was made shortly before they arrived at their respective measurement stations (Figure 10.4).

By now, entanglement has been distributed over larger and larger distances – 10.7 km having been obtained by the Geneva group of Nicolas Gisin using glass fibers. More recently, it was also possible to distribute entanglement over large distances in free space (Figure 10.5). In both experiments, a violation of a Bell inequality was observed, confirming the predictions of quantum mechanics.

While all loopholes for local realism have now been closed by experiments, it might still be argued that Nature behaves in a local-realistic way, assuming that, in the experiments closing either loophole, Nature makes use of the other loophole. While such a position is in principle logically possible, it is extremely unlikely that Nature is

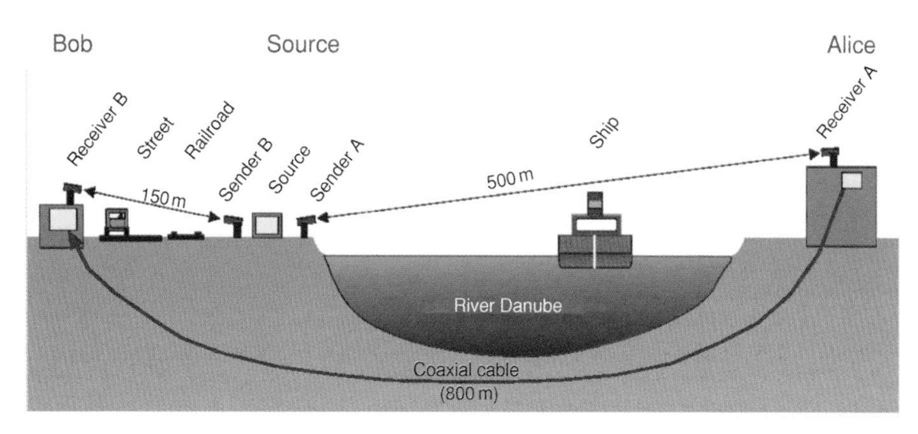


Figure 10.5. Long-distance entanglement in free space across the River Danube in Vienna. Experiments demonstrating entanglement across larger and larger distances confirm the predictions of quantum mechanics.

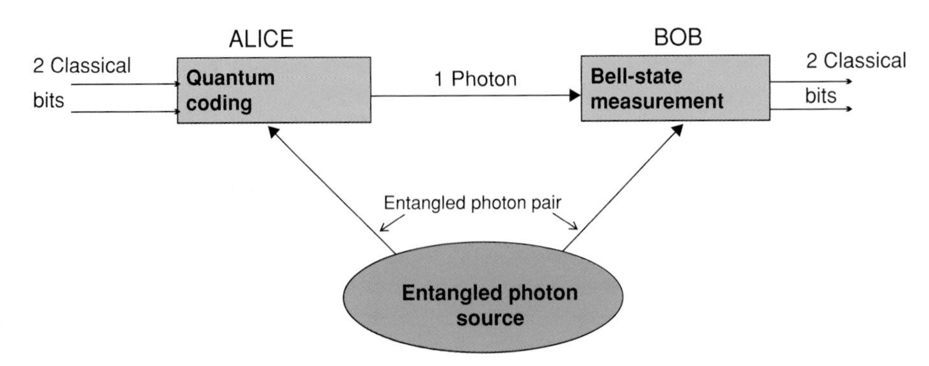


**Figure 10.6.** The emerging field of quantum information technology provides completely new methods for the secure encoding and processing of information. The quantum dense-coding technique shown here is an entanglement-based method to encode more than one bit per photon manipulated.

so vicious. Nevertheless, an experiment some day closing all loopholes simultaneously would finally close all debate. It appears that such an experiment, for didactic reasons preferably even performed with three-particle entangled GHZ states, will most likely use entanglement of atoms, since atomic states can be detected with very high efficiency. Besides the "conservative" approach of having these atoms created in a common source and propagated to distant measurement locations, a more recent idea is to have these atoms sitting already in individual atomic traps at distant locations and becoming entangled by individual photons propagating between them. It is evident that there is a lot of room for exciting experimental development.

## 10.3 Entanglement and information

As mentioned above, research into the field of entanglement was originally motivated by philosophical considerations. Interestingly, the new field of quantum information technology utilizes the fact that quantum systems allow encoding and processing of information in completely novel ways, which are impossible in classical physics.

In any classical coding scheme, be it a knot in a handkerchief or the states of the most advanced high-speed computers, two physically distinct classical states are used to encode individual bits of information. In an electronic circuit, for example, the bit value "0" may correspond to a certain voltage level, and the bit value "1" to another voltage level.

In quantum coding schemes, one assigns two different orthogonal quantum states to the bit values. The novel nonclassical feature here is that these quantum states can be in a superposition of the two states resulting in a superposition of "0" and "1." In 1993, Ben Schumacher of Kenyon College coined the name qubit for systems representing such quantum superpositions of bit values. Furthermore, if one has more than one qubit, these can even be in entangled states. Intuitively, it is immediately clear that such entangled states provide a novel communication link, since the quantum states of the entangled particles are intimately connected with each other, and measurement on one immediately changes the state of all others. While the first-ever concept of utilizing quantum information, the unforgeable quantum money proposed by Stephen Wiesner (*c.* 1970), used single qubits only, entanglement has become a key feature of more advanced communication schemes such as quantum teleportation, quantum dense coding, and an entanglement-based version of quantum cryptography.

In quantum dense coding (Figure 10.6), Alice can send two bits of information to Bob with a single photon if they share a pair of entangled photons. Alice has access to only one of the two entangled photons while Bob obtains the other one directly. Alice then can perform four different operations on his photon, changing the total entangled state which then can be analyzed by Bob performing joint Bell-state measurements on the two photons. These Bell states are the four maximally entangled states of two photons. As with two qubits, there are four different such Bell states, so four different kinds of information can be encoded by Bob and sent to Alice, corresponding to two classical bits of information.

## 10.4 Quantum teleportation

Quantum dense coding was the first experimental demonstration of the basic concepts of quantum communication utilizing entanglement. An even more interesting example is quantum teleportation.

Suppose that Alice has an object she wants Bob to have. In a situation in which, as we assume, it is impossible to send the object itself, she could, at least in classical physics, scan all the information contained and transmit that information to Bob, who, with suitable technology, would then be able to reconstruct the object. Unfortunately, such a strategy is not possible, because quantum mechanics prohibits complete determination of the state of any individual object.

Interestingly, quantum mechanics itself provides a strategy that will work. In 1993, Charles Bennett, Gilles Brassard, Claude Crépeau, Richard Josza, Asher Peres, and William Wooters – in a Canada–UK–France–Israel–USA collaboration – showed that quantum entanglement provides an elegant solution for the problem. Suppose (Figure 10.7) that Alice wants to teleport the unknown quantum state $\psi$ to Bob. They both agree to share in advance an entangled pair of qubits, known as the ancillary pair. Alice then performs a joint Bell-state measurement on the teleportee (the photon she wants to teleport) and on her ancillary photon. She randomly obtains one of the four possible Bell-state results. This measurement projects Bob's other ancillary photon into a quantum state uniquely related to the original. Alice then transmits the result of her measurement to Bob (two classical bits of information corresponding to the four possible Bell-state results). Bob finally performs one of four operations on his ancillary photon, thus obtaining the original state and completing the teleportation procedure.

It is essential to understand that the Bell-state measurement performed by Alice projects the teleportee qubit and her ancillary qubit into a state that does not contain any information about the initial state of the teleportee. Furthermore, by that Bell-state measurement, the teleportee becomes entangled with Alice's ancillary. This means that the teleportee loses all its individual information, since, in the resulting maximally entangled state, the individual entangled particles do not enjoy their own quantum states. Therefore, more precisely, no information whatsoever about the original state is retained at Alice's location. It is all directly transferred over to Bob. In perfect teleportation, the final teleported photon is indistinguishable from the original. It has all its features, and the original has lost them. Thus, since the notion of being the original has no sense besides that of carrying all properties of the original, Bob's final photon is not a copy, but it is the original photon which has been teleported.

The experiments are extremely challenging, since they require performance of a Bell-state measurement jointly on the teleportee photon and on one of the independently produced entangled ancillaries. Since the notion of entanglement implies indistinguishability of the entangled systems, one has to guarantee in the Bell-state measurement that the photons lose all their individuality. That is, it must not be possible on the basis of the measurement results to conclude which photon was which.

In the experiment, this was achieved using photons created by a femtosecond pulsed laser in such a way that they arrive at exactly the same time within their very short coherence length at the Bell-state analyzer.

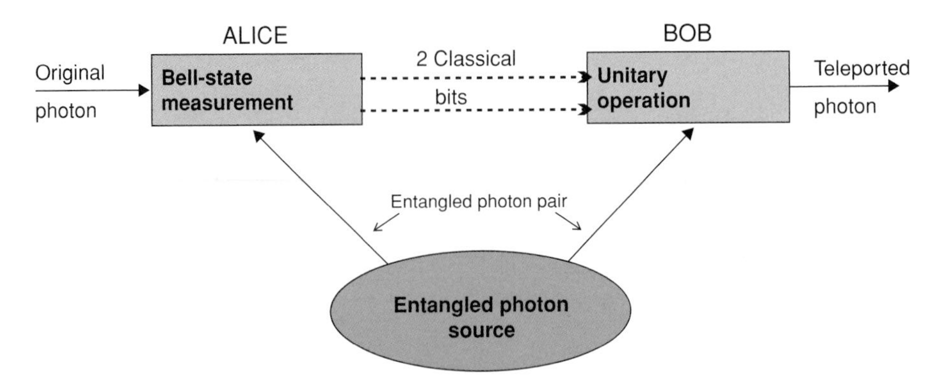


Figure 10.7. The principle of quantum teleportation. If Alice has an object she wants Bob to have, but which is impossible to send, she could, at least in classical physics, scan all the information contained and transmit it to Bob, who, with suitable technology, could then reconstruct the object. Unfortunately, this is not possible, because quantum mechanics prohibits complete determination of the state of any object. However, quantum mechanics, via quantum entanglement, provides an elegant way to teleport, if both participants agree to share in advance an entangled pair of qubits.

## 10.5 Entanglement-based quantum cryptography

The history of cryptography can be seen as the history of a continuous battle between the code-makers and the code-breakers. It appears that, with the advent of quantum cryptography, the code-makers can claim final victory. There are two different approaches to quantum cryptography, one based on individual photons, individual qubits, which was first proposed by Bennett and Brassard in 1984, and one based on entangled states proposed by Artur Ekert in 1991. They have significant similarities. Here, we will briefly discuss the latter one.

A common feature of both approaches is that they realize, using individual quantum systems, the discovery by Gilbert S. Vernam (see Chapter 11, Section 11.7) of an encryption system that is secure against eavesdropping. The basic idea there is to use a key to encrypt a message that is completely random and is used only once, a so-called one-time pad. The reason for its security is that then the transmitted, encoded message has no structure and no redundancy at all and therefore cannot be deciphered by attacks with high-speed computers looking for systematic features of the message. The second feature in common is that both approaches exploit the fact that any eavesdropper invariably has to do some measurement on the individual quanta propagating between Alice and Bob and any measurement modifies the quantum state unless the observer knows already which measurements will be performed on the quanta.

A typical realization of entanglement-based quantum cryptography utilizes the set up of Figure 10.4. The essential point here is that, because of the perfect correlations between the observations on the two photons, Alice, who may operate the left-hand measurement apparatus, and Bob, who may operate the right-hand one, obtain for each individual pair obtained from the source the same result randomly, either "0" or "1," as we may name them. Therefore, if they measure many successive pairs, they obtain the same random sequence on both sides. This random sequence can then be used to encode the information (Figure 10.8). The encoding of the information is very simple. The original message is represented in binary form, and the random key is added bit by bit, modulo 2. Since the key is completely random, the transmitted encrypted message is also random. Bob then, since he knows the key, can easily undo the encoding.

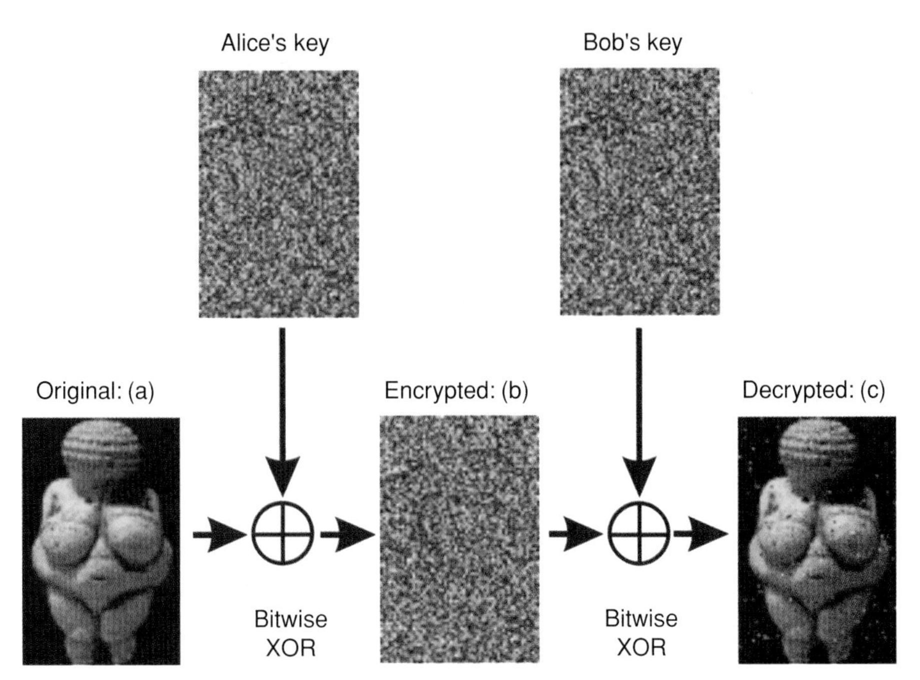


**Figure 10.8.** Transmission of a secret message by Jennewein *et al.* using entanglement-based cryptography using the set up shown in Figure 10.4. Because of the perfect correlations between the observations on the two photons, Alice and Bob obtain for each individual pair obtained from the source the same random result. If they measure many successive pairs, they obtain the same random sequence, which can then be used to encrypt and decrypt the information. (XOR is the "exclusive or" operation which is equivalent to binary addition modulo two.)

## 10.6 Outlook

Besides its application in quantum cryptography and quantum teleportation, entanglement is also an essential feature of quantum computation. In a quantum computer, the information is represented in qubits instead of classical bits. Therefore, also here, one may obtain superposition of entanglement and information. A quantum computer is then a machine that processes in a coherent way the information represented in superposed and entangled qubits. There are various schemes being investigated at present for a possible realization of the processors in quantum computers. The information may be represented as quantum states of photons, atoms, and ions, either in free space or inside a solid, as currents in superconductors, and more. It will be interesting to see in the future which of the procedures best manages to overcome the problem of decoherence, i.e. the loss of information to the environment. An essential question is scalability, that is, realization of multi-qubit quantum processors, once quantum processors with a few qubits have been made to operate successfully. There will be specific applications for which quantum computers will be of enormous advantage. As has been argued by Richard Feynman, a quantum computer would be the ideal machine to simulate the behavior of complex quantum systems. Following the discovery by David Deutsch in 1984 that quantum computers can solve certain problems much faster than classical ones by utilizing superposition, in 1994 Peter Shor proposed an algorithm that would allow one to factor large numbers into their prime factors significantly faster than can be done with any classical computer. This would allow essential present classical encryption codes to be broken – an important step for the code-breakers. Yet, with the discovery of quantum cryptography, the code-makers appear to have the upper hand.

The extraordinary experimental progress during the past decade on producing, handling, and applying entangled quantum states has also contributed to opening up again the fundamental debate on the conceptual foundations of quantum mechanics. Since a local-realistic view has now become unreasonable, the question is the following: what does quantum mechanics tell us about the nature of reality and the role of our knowledge and of information in the world? It appears that, certainly at least for entangled quantum systems, it is wrong to assume that the features of the world which we observe, the measurement results, exist prior to and independently of our observation. An important clue may be seen in the fact that it is never possible to make any statement about observations without referring to information obtained in some way. Therefore, the notion that a reality exists independently of information is a senseless concept. One possible position to take is that what we do in our daily endeavors, both as scientists and as ordinary human beings, is to collect information, order it in some way, construct a picture of the world that we hope corresponds to some reality, and act accordingly. It is safe to assume that information plays a much more central role than admitted hitherto. Indeed, one may arrive at a natural understanding of entanglement if one realizes (a) that the information carried by two qubits can be used either to encode their properties individually or to encode joint properties, and (b) that localization in space and time – the property of being able to assign features to systems here and now – is secondary.

## FURTHER READING


1. H. C. von Baeyer, In the Beginning was the Bit, *New Scientist*, February 17, 2001.
2. J. S. Bell, *Speakables and Unspeakables in Quantum Mechanics*, Cambridge, Cambridge University Press, 1987.
3. R. A. Bertlmann and A. Zeilinger (eds.), *Quantum [Un]speakables, From Bell to Quantum Information*, Berlin, Springer-Verlag, 2002.
4. D. Bouwmeester, A. Ekert, and A. Zeilinger, *The Physics of Quantum Information. Quantum Cryptography, Quantum Teleportation, Quantum Computation*, Berlin, Springer-Verlag, 2000.
5. C. Macchiavello, G. M. Palma, and A. Zeilinger (eds.), *Quantum Computation and Quantum Information Theory. A Reprint Volume*, Singapore, World Scientific, 2000.
6. P. A. Schilpp (ed.), *Albert Einstein: Philosopher-Scientist*, Evanston, The Library of Living Philosophers, 1949.
7. A. Zeilinger, Quantum teleportation, *Scient. Am.*, April 2000, p. 50; an updated version appeared in the *Scientific American* collection *The Edge of Physics*, spring 2003.
8. A. Zeilinger, *Einsteins Schleier*, Munich, C. H. Beck, 2003.


## Anton Zeilinger

Professor Anton Zeilinger and his group have demonstrated many fundamental predictions of quantum theory and their extraordinary implications for our view of the world. His group's achievements include quantum teleportation (1997), the demonstration of entanglement of more than two particles (Greenberger–Horne–Zeilinger states, 1998), and quantum cryptography based on quantum entanglement (1999). The group's quantum interference experiments with "buckyball" molecules in 1999 set a record for the largest objects to have demonstrated quantum behavior so far. The investigation of quantum properties with objects of increasing complexity explores the transition between the quantum world and the classical world.

Anton Zeilinger was born in 1945 in Austria, and has held positions at universities in Munich, Vienna, Innsbruck, and Melbourne, at MIT, in Oxford, and at the Collège de France. Among his many awards and prizes are an Honorary Professorship at the

University of Science and Technology of China, the Senior Humboldt Fellow Prize of the Alexander von Humboldt-Stiftung, and the King Feisal Prize. He is Fellow of the American Physical Society and Member of the German Order pour le Mérite and various academies.

For further information: http://www.quantum.univie.ac.at.

# 11 Quanta, ciphers, and computers

Artur Ekert

## 11.1 Introduction

Computation is an operation on symbols. We tend to perceive symbols as abstract entities, such as numbers or letters from a given alphabet. However, symbols are always represented by selected properties of physical objects. The binary string

10011011010011010110101101

may represent an abstract concept, such as the number 40 711 597, but the binary symbols 0 and 1 have also a physical existence of their own. It could be ink on paper (this is most likely to be how you see them when you are reading these words), glowing pixels on a computer screen (this is how I see them now when I am writing these words), or different charges or voltages (this is how my word processor sees them). If symbols are physical objects and if computation is an operation on symbols then computation is a physical process. Thus any computation can be viewed in terms of physical experiments, which produce outputs that depend on initial preparations called inputs. This sentence may sound very innocuous but its consequences are anything but trivial!

On the atomic scale matter obeys the rules of quantum mechanics, which are quite different from the classical rules that determine the properties of conventional computers. Today's advanced lithographic techniques can etch logic gates and wires less than a micrometer across onto the surfaces of silicon chips. Soon they will yield even smaller parts and inevitably reach a point at which logic gates are so small that they are made out of only a handful of atoms. So, if computers are to become smaller in the future, new, quantum technology must replace or supplement what we have now. The point is, however, that quantum technology can offer much more than cramming more and more bits onto silicon and multiplying the clock-speed of microprocessors. It can support an entirely new kind of computation, known as quantum computation, with qualitatively new algorithms based on quantum principles.

The potential power of quantum phenomena to perform computations was first adumbrated in a talk given by Richard Feynman at the First Conference on the Physics of Computation, held at MIT in 1981. He observed that it appeared to be impossible, in general, to simulate the evolution of a quantum system on a classical computer in an efficient way. The computer simulation of quantum evolution typically involves an exponential slowdown in time, compared with the natural

*The New Physics for the Twenty-First Century*, ed. Gordon Fraser.
Published by Cambridge University Press. 

evolution, essentially because the amount of classical information required to describe the evolving quantum state is exponentially larger than that required to describe the corresponding classical system with a similar accuracy. However, instead of viewing this intractability as an obstacle, Feynman regarded it as an opportunity. He pointed out that, if it requires that much computation to work out what will happen in a quantum multi-particle interference experiment, then the very act of setting up such an experiment and measuring the outcome is equivalent to performing a complex computation.

The foundations of the quantum theory of computation were laid down in 1985 when David Deutsch, of the University of Oxford, published a crucial theoretical paper in which he described a universal quantum computer and a simple quantum algorithm. Since then, the hunt has been on for interesting things for quantum computers to do, and at the same time, for the scientific and technological advances that could allow us to build quantum computers.

A sequence of steadily improving quantum algorithms led to a major breakthrough in 1994, when Peter Shor, of AT&T's Bell Laboratories in New Jersey, discovered a quantum algorithm that could perform efficient factorization. Since the intractability of factorization underpins the security of many methods of encryption, including the most popular public-key cryptosystem, RSA (named after its inventors, Rivest, Shamir, and Adelman), Shor's algorithm was soon hailed as the first "killer application" for quantum computation – something very useful that only a quantum computer could do. We mention in passing that in December 1997 the British Government officially confirmed that public-key cryptography was originally invented at the Government Communications Headquarters (GCHQ) in Cheltenham. By 1975, James Ellis, Clifford Cocks, and Malcolm Williamson from GCHQ had discovered what was later rediscovered in academia and became known as RSA and Diffie–Hellman key exchange.

By some strange coincidence, several of the superior features of quantum computers have applications in cryptanalysis. Once a quantum computer is built many popular ciphers will become insecure. Indeed, in one sense they are already insecure. For example, any RSA-encrypted message that is recorded today will become readable moments after the first quantum factorization engine is switched on, and therefore RSA cannot be used for securely transmitting any information that will still need to be secret on that happy day. Admittedly, that day is probably decades away, but can anyone prove, or give any reliable assurance, that it is? Confidence in the slowness of technological progress is all that the security of the RSA system now rests on.

What quantum computation takes away with one hand, it returns, at least partially, with the other. Quantum cryptography offers new methods of secure communication that are not threatened by the power of quantum computers. Unlike all classical cryptography, it relies on the laws of physics rather than on ensuring that successful eavesdropping would require excessive computational effort.

The promise of quantum cryptography was first glimpsed by Stephen Wiesner, then at Columbia University in New York, who, in the early 1970s, introduced the concept of quantum conjugate coding. He showed how to store or transmit two messages by encoding them in two "conjugate observables," such as linear and circular polarization of light, so that either, but not both, of the observables may be received and decoded. He illustrated his idea with a design of unforgeable bank notes. Unfortunately, Wiesner's paper was rejected by the journal to which he sent it! A decade later, Charles H. Bennett, of the IBM T. J. Watson Research Center, and Gilles Brassard, of the Université de Montréal, who knew of Wiesner's ideas, began thinking of how to use them for secure communication. However, at the time their work was perceived as mere curiosity and remained largely unknown. In 1990, independently and initially unaware of the earlier work, the current author, then a Ph.D. student at the University of Oxford, developed

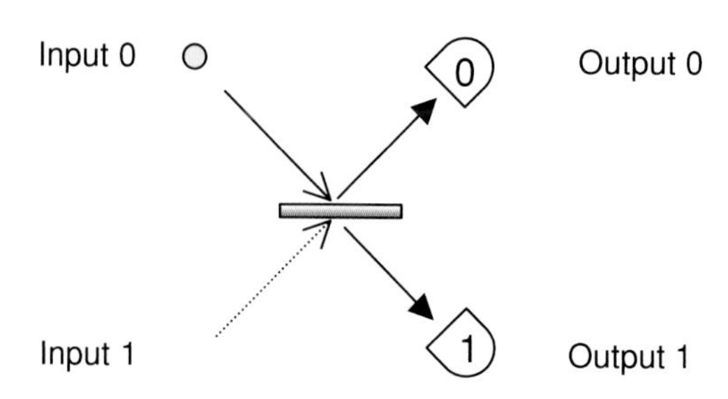


**Figure 11.1.** A half-silvered mirror, or a beam-splitter, as a simple quantum logic gate.

a different approach to quantum cryptography, which was based on peculiar quantum correlations known as quantum entanglement. This new approach was more familiar to experimental physicists working on the foundations of quantum mechanics and was noticed very quickly. Since then quantum cryptography has evolved into a thriving experimental area and is quickly becoming a commercial proposition.

This popular account is aimed to provide some insights into the fascinating world of quantum phenomena and the way we exploit them for computation and secure communication.

## 11.2 From bits to qubits

To explain what makes quantum computers so different from their classical counterparts, we begin with a basic chunk of information, namely one bit. From a physicist's point of view a bit is a physical system that can be prepared in one of two different states, representing two logical values: no or yes, false or true, or simply 0 or 1. For example, in digital computers, the voltage between the plates in a capacitor represents a bit of information: a charged capacitor denotes bit value 1 and an uncharged capacitor bit value 0. One bit of information can also be encoded using two different polarizations of light or two different electronic states of an atom. However, if we choose a quantum system, such as an atom, as a physical bit then quantum mechanics tells us that, apart from the two distinct electronic states, the atom can be also prepared in a "coherent superposition" of the two states. This means that the atom is both in state 0 and in state 1. Such a physical object is called a quantum bit or a qubit.

To become familiar with the idea that a qubit can represent "two bit values at once" it is helpful to consider the following experiment. Let us try to reflect a single photon off a half-silvered mirror, i.e. a mirror that reflects exactly half of the light which impinges upon it, while the remaining half is transmitted directly through it (Figure 11.1). Such a mirror is also known as a beam-splitter.

Let the photon in the reflected beam represent logical 0 and the photon in the transmitted beam represent the logical 1. Where is the photon after its encounter with the beam-splitter, in the reflected or in the transmitted beam? Does the photon at the output represent logical 0 or logical 1?

One thing we know is that the photon doesn't split in two; thus it seems sensible to say that the photon is *either in the transmitted or in the reflected beam* with the same probability of each situation. That is, one might expect the photon to take one of the two paths, choosing randomly which way to go. Indeed, if we place two photodetectors behind the half-silvered mirror directly in the lines of the two beams, the photon will be registered with the same probability either in the detector "0" or in the detector "1." Does it really mean that after the half-silvered mirror the photon travels in either reflected or transmitted beam with the same probability of 50%? No, it does not! In fact the photon takes "two paths at once."

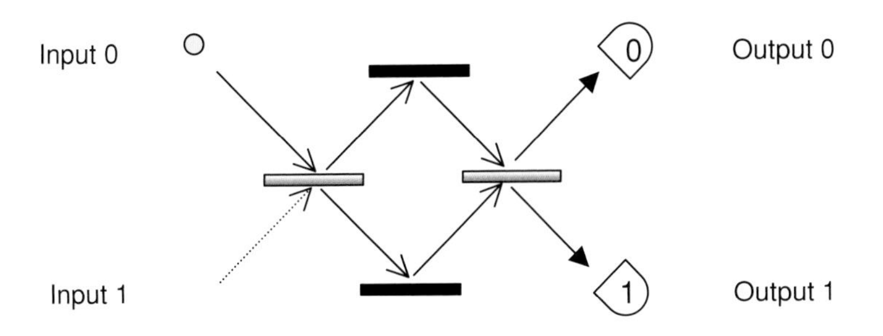


**Figure 11.2.** Two concatenated beam-splitters affect a logical NOT gate. Thus each beam-splitter separately represents an inherently quantum operation called the square root of NOT.

This can be demonstrated by recombining the two beams with the help of two fully silvered mirrors and placing another half-silvered mirror at their meeting point, with two photodetectors in the direct lines of the two beams (Figure 11.2).

If it were merely the case that there was a 50% chance that the photon followed one path and a 50% chance that it followed the other, then we should find a 50% probability that one of the detectors registers the photon and a 50% probability that the other one does. However, that is not what happens. If the two possible paths are exactly equal in length, then it turns out that there is a 100% probability that the photon reaches the detector "1" and 0% probability that it reaches the other detector "0." Thus the photon is certain to strike the detector "1."

The inescapable conclusion is that the photon must, in some sense, have traveled both routes at once, for if either of the two paths were blocked by an absorbing screen, it would immediately become equally probable that "0" or "1" is struck. In other words, blocking off either of the paths illuminates "0"; with both paths open, the photon somehow is prevented from reaching "0."

Furthermore, if we insert slivers of glass of different thicknesses into each path (see Figure 11.3) then we can observe a truly amazing quantum interference phenomenon. We can choose the thickness of the glass, and hence the effective optical length of each path, in such a way that the photon can be directed to either of the two detectors with any prescribed probability.

In particular, when the *difference* between the thicknesses of the two slivers is chosen appropriately the photon will certainly emerge at detector "0" instead of detector "1." The photon reacts only to the *difference* between the thicknesses of the slivers located in the two different paths – more evidence that the photon must have traveled both paths at once – and each path contributes to the final outcome. Thus the output of the beam-splitter does not represent *either* "0" *or* "1" but rather a truly quantum superposition of the two bit values.

From a computational point of view a beam-splitter is an elementary quantum logic gate, operating on a single qubit. It is quite a remarkable logic gate, which can be called the square root of NOT ($\sqrt{\text{NOT}}$) because the logical operation NOT is obtained as the result of two consecutive applications of beam-splitters (see Figure 11.2). This

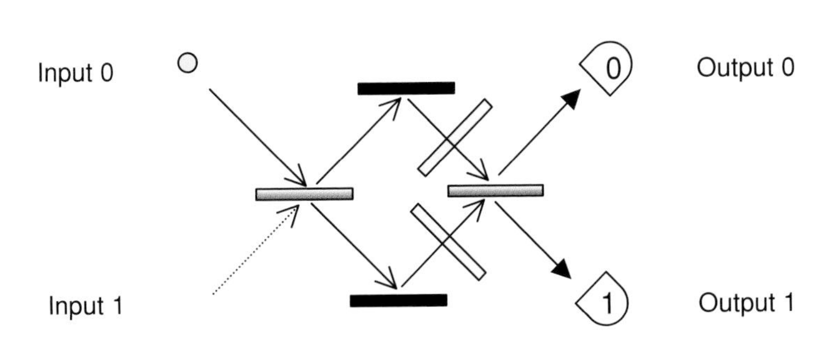


**Figure 11.3.** Quantum interference.

purely quantum operation has no counterpart in classical logic and forms one of the elementary building blocks of a quantum computer.

## 11.3 Entanglement

The idea of superposition of numbers can be pushed further. Consider a register composed of three physical bits. Any classical register of that type can store, in a given moment of time, only one out of eight different numbers, i.e. the register can be in only one out of eight possible configurations such as 000, 001, 010, . . .,111. A quantum register composed of three qubits can store, in a given moment of time, all eight numbers in a quantum superposition. It is quite remarkable that all eight numbers are physically present in the register, but it should be no more surprising than a qubit being both in state 0 and 1 at the output of a beam-splitter. If we keep adding qubits to the register we increase its storage capacity exponentially, i.e. three qubits can store 8 different numbers at once, four qubits can store 16 different numbers at once, and so on; in general $L$ qubits can store $2^L$ numbers at once.

Some superpositions of numbers can easily be obtained by operating on individual qubits, whereas some require joint operations on two qubits. For example, a superposition of 00 and 10 can be generated by starting with two qubits in state 00 and applying the square root of NOT to the first qubit. If we subsequently apply the same operation to the second qubit we will turn the 00 component into a superposition of 00 and 01, and the 10 component into a superposition of 10 and 11, which all together gives an equally weighted superposition of the four numbers: 00, 10, 01, and 11. Each of the two qubits is in a superposition of 0 and 1, and the register made out of the two qubits is in a superposition of all possible binary strings of length 2. In general, a register of $L$ qubits can be prepared in a superposition of all binary strings of length $L$ by applying the square root of NOT to each qubit in the register. However, operations on individual qubits will never turn a quantum register in state 00 into, say, an equally weighted superposition of 00 and 11, or a superposition of 01 and 10. Such superpositions are special and their generation requires special quantum logic gates operating on two qubits at a time. Qubits in such superpositions are said to be entangled. They cannot be described by specifying the state of individual qubits and they may together share information in a form that cannot be accessed in any experiment performed on either of them alone. Erwin Schrödinger, who was probably the first to be baffled by this quantum phenomenon (Figure 11.4), writing for the Cambridge Philosophical Society in 1935, summarized it as follows:

*When two systems, of which we know the states by their respective representatives, enter into temporary physical interaction due to known forces between them, and when after a time of mutual influence the systems separate again, then they can no longer be described in the same way as before,* viz. *by endowing each of them with a representative of its own. I would not call that one but rather the characteristic trait of quantum mechanics, the one that enforces its entire departure from classical lines of thought. By the interaction the two representatives [the quantum states] have become entangled.*

There are many entangling operations on two qubits. They require some form of interaction between the qubits. One of them, which is analogous to the square root of NOT, is the square root of SWAP. The classical SWAP operation interchanges the bit values of two bits, e.g. 00 $\rightarrow$ 00, 01 $\rightarrow$ 10, 10 $\rightarrow$ 01, 11 $\rightarrow$ 11. However, there is no classical two-bit logic gate such that its two consecutive applications result in the logical operation SWAP. Still, the square root of SWAP does exist. It is an inherently

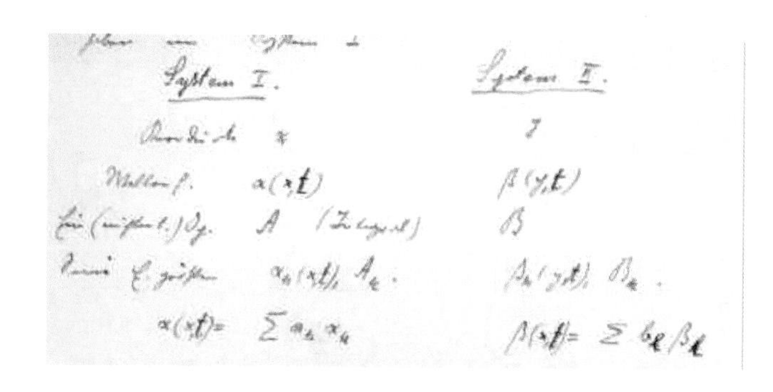


**Figure 11.4.** Schrödinger's note on quantum states of interacting subsystems dating back to 1932–33. This seems to be the first known reference to the concept of quantum entanglement. It was discovered recently by Matthias Christandl and Lawrence Ioannou, of Cambridge University, in the Schrödinger archive in Vienna.

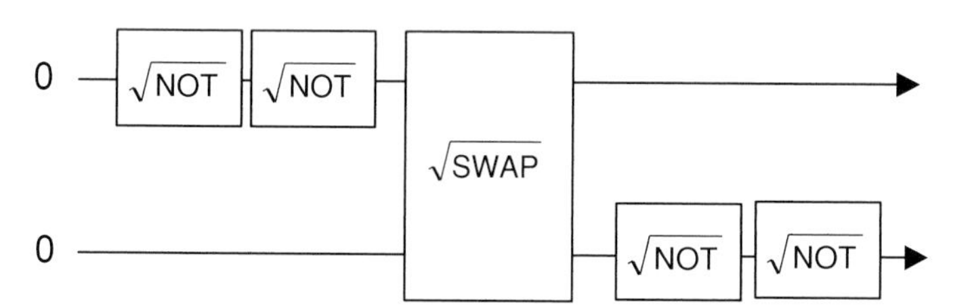


**Figure 11.5.** An example of a quantum Boolean network operating on two qubits. The qubits are represented by the horizontal lines and quantum logic gates by rectangular icons. The operations are performed from the left to the right. The input consists of two qubits in state 00. The output is an entangled state of the two qubits – an equally weighted superposition of 00 and 11.

quantum logic gate and can be obtained by switching on for a prescribed period of time the "exchange interaction" between qubits. The resulting quantum dynamics takes the input state 01 half way toward the swapped state 10 and effectively generates a superposition of 01 and 10.

## 11.4 Quantum Boolean networks and their complexity

The square root of SWAP together with the square root of NOT and phase shifters (operations equivalent to those induced by the slivers of glass in Figure 11.3) allow the construction of arbitrary superpositions of binary strings of any length. They form an adequate (universal) set of quantum gates. Once we can implement these operations with sufficient precision we can perform any quantum computation. This is very reminiscent of constructing classical computation out of simple primitives such as logical NOT, AND, OR, etc.

For example, if you need a superposition of 00 and 11, you can construct it using only the square root of NOT and the square root of SWAP. Start with two qubits in state 00, apply the square root of NOT twice to the first qubit; this gives state 10. (Of course, this is just applying logical NOT, but the point is to use only the prescribed adequate gates.) Then apply the square root of SWAP to obtain the superposition of 10 and 01. Now comes an interesting part; once the register has been prepared in an initial superposition of different numbers, quantum gates perform operations that affect all numbers in the superposition. Thus, if we apply the square root of NOT twice to the second qubit, it will turn 10 into 11 and 01 into 00, which gives the superposition of 00 and 11. It is convenient to illustrate this sequence of operations in a diagram, shown in Figure 11.5.

Such graphical representations are called quantum Boolean networks or quantum circuits. Quantum computers, for all practical purposes, can be viewed as quantum Boolean networks operating on many qubits.

In order to solve a particular problem, computers, be they classical or quantum, follow a precise set of instructions that can be mechanically applied to yield the solution to any given instance of the problem. A specification of this set of instructions is called an algorithm. Examples of algorithms are the procedures taught in elementary schools for adding and multiplying whole numbers; when these procedures are mechanically applied, they always yield the correct result for any pair of whole numbers. Any algorithm can be represented by a family of Boolean networks $N_1$, $N_2$, $N_3$, . . ., where the network $N_n$ acts on all possible input instances of size $n$ bits. Any useful algorithm should have such a family specified by an example network, $N_n$, and a simple rule explaining how to construct the network $N_{n+1}$ from the network $N_n$. These are called uniform families of networks.

The big issue in designing algorithms or their corresponding families of networks is the optimal use of physical resources required to solve a problem. Complexity theory is concerned with the inherent cost of computation in terms of some designated elementary operations such as the number of elementary gates in the network (the size of the network). An algorithm is said to be fast or efficient if the number of elementary operations taken to execute it increases no faster than a polynomial function of the size of the input. We generally take the input size to be the total number of bits needed to specify the input (for example, a number $N$ requires $\log_2 N$ bits of binary storage in a computer). In the language of network complexity, an algorithm is said to be efficient if it has a uniform and polynomial-size network family. Problems for which we do not have efficient algorithms are known as hard problems.

From the point of view of computational complexity, quantum networks are more powerful than their classical counterparts. This is because individual quantum gates operate not just on one number but on superpositions of many numbers. During such evolution each number in the superposition is affected and as a result we generate a massive parallel computation, albeit in one piece of quantum hardware. This means that a quantum gate, or a quantum network, can in *only one* computational step perform the same mathematical operation on different input numbers encoded in coherent superpositions of $L$ qubits. In order to accomplish the same task any classical device has to repeat the same computation $2^L$ times or one has to use $2^L$ processors working in parallel. In other words, a quantum computer offers an enormous gain in the use of computational resources such as time and memory.

Here we should add that this gain is more subtle than the description above might suggest. If we simply prepare a quantum register of $L$ qubits in a superposition of $2^L$ numbers and then try to read a number out of it then we obtain only one, randomly chosen, number. This is exactly like in our experiment in Figure 11.1, in which a half-silvered mirror prepares a photon in a superposition of the two paths but, when the two photodetectors are introduced, we see the photon in only one of the two paths. This kind of situation is of no use to us, for although the register now holds all the $2^L$ numbers, the laws of physics allow us to see only one of them. However, recall the experiments from Figures 11.2 and 11.3 in which just the single answer "0" or "1" depends on each of the two paths. In general, quantum interference allows us to obtain a single, final result that depends logically on all $2^L$ of the intermediate results. One can imagine that each of the $2^L$ computational paths is affected by a process that has an effect similar to that of the sliver of glass in Figure 11.3. Thus each computational path contributes to the final outcome. It is in this way, by virtue of the quantum interference of many computational paths, that a quantum computer offers an enormous gain in the use of computational resources – though only in certain types of computation.

## 11.5 Quantum algorithms

What types of computation benefit from quantum computing? As we have said, ordinary information storage is not one of them, for although the computer now holds all the outcomes of $2^L$ computations, the laws of physics allow us to see only one of them. However, just as the single answer in the experiments of Figures 11.2 and 11.3 depends on information that traveled along each of two paths, quantum interference now allows us to obtain a single, final result that depends logically on all $2^L$ of the intermediate results. This is how Shor's algorithm achieves the mind-boggling feat of efficient factorization of large integers.

### 11.5.1 Shor's algorithm

As is well known, a naive way to factor an integer number $N$ is based on checking the remainder of the division of $N$ by some number $p$ smaller than $\sqrt{N}$. If the remainder is 0, we conclude that $p$ is a factor. This method is in fact very inefficient: with a computer that can test for $10^{18}$ different $p$'s per second (this is faster than any computer ever built), the average time needed to find the factor of an 80-digit-long number would exceed the age of the Universe.

Rather than this naive division method, Shor's algorithm relies on a slightly different technique to perform efficient factorization. The factorization problem can be related to evaluating the period of a certain function $f$ that takes $N$ as a parameter. Classical computers cannot make much of this new method: finding the period of $f$ requires evaluating the function $f$ many times. In fact mathematicians tell us that the average number of evaluations required to find the period is of the same order as the number of divisions needed with the naive method we outlined first. With a quantum computer, the situation is completely different – quantum interference of many qubits can effectively compute the period of $f$ in such a way that we learn about the period without learning about any particular value $f(0), f(1), f(2), \ldots$ The algorithm mirrors our simple interference experiment shown in Figure 11.3. We start by setting a quantum register in a superposition of states representing 0, 1, 2, 3, 4, . . . This operation is analogous to the action of the first beam-splitter in Figure 11.3 which prepares a superposition of 0 and 1. The next step is the function evaluation. The values $f(0), f(1), f(2), \ldots$ are computed in such a way that each of them modifies one computational path by introducing a phase shift. This operation corresponds to the action of the slivers of glass in Figure 11.3. Retrieving the period from a superposition of $f(0), f(1), f(2), \ldots$ requires bringing the computational paths together. In Figure 11.3 this role is played by the second beam-splitter. In Shor's algorithm the analogous operation is known as a "quantum Fourier transform." Subsequent bit-by-bit measurement of the register gives a number, in the binary notation, which allows the period of $f$ to be estimated with a low probability of error. The most remarkable fact is that all this can be accomplished with a uniform quantum network family of polynomial size!

Mathematicians believe (firmly, though they have not actually proved it) that, in order to factorize a number with $L$ binary digits, any classical computer needs a number of steps that grows exponentially with $L$: that is to say, adding one extra digit to the number to be factorized generally multiplies the time required by a fixed factor. Thus, as we increase the number of digits, the task rapidly becomes intractable. No one can even conceive of how one might factorize, say, thousand-digit numbers by classical means; the computation would take many times as long as the estimated age of the Universe. In contrast, quantum computers could factor thousand-digit numbers in a fraction of a second – and the execution time would grow at most as the cube of the number of digits.

### 11.5.2 Grover's algorithm

In Shor's algorithm all computational paths are affected by a single act of quantum function evaluation. This generates an interesting quantum interference that gives observable effects at the output. If a computational step affects just one computational path it has to be repeated several times. This is how another popular quantum algorithm, discovered in 1996 by Lov Grover of AT&T's Bell Laboratories in New Jersey, searches an unsorted list of $N$ items in only $\sqrt{N}$ or so steps.

Consider, for example, searching for a specific telephone number in a directory containing a million entries, stored in the computer's memory in alphabetical order of names. It is easily proved (and obvious) that no classical algorithm can improve on the brute-force method of simply scanning the entries one by one until the given number is found, which will, on average, require 500 000 memory accesses. A quantum computer can examine all the entries simultaneously, in the time of a single access. However, if it is merely programmed to print out the result at that point, there is no improvement over the classical algorithm: only one of the million computational paths would have checked the entry we are looking for, so there would be a probability of only one in a million that we would obtain that information if we measured the computer's state. But if we leave that quantum information in the computer, unmeasured, a further quantum operation can cause that information to affect other paths, just as in the simple interference experiment described above. It turns out that, if this interference-generating operation is repeated about 1000 times (in general $\sqrt{N}$ times), the information about which entry contains the desired number will be accessible to measurement with probability 50% – i.e. it will have spread to more than half the terms in the superposition. Therefore repeating the entire algorithm a few more times will find the desired entry with a probability overwhelmingly close to unity.

One important application of Grover's algorithm might be, again, in cryptanalysis, to attack classical cryptographic schemes such as the Advanced Encryption Standard (AES). Cracking the AES essentially requires a search among $2^{128} \approx 3 \times 10^{38}$ possible keys. If these can be checked at a rate of, say, a billion billion ($10^{18}$) keys per second, a classical computer would need more than the age of the Universe to discover the correct key, whereas a quantum computer using Grover's algorithm would do it in less than 20 seconds.

## 11.6 Building quantum computers

In principle we know how to build a quantum computer: we can start with simple quantum logic gates, such as the square root of NOT, the square root of SWAP, and phase gates, and try to integrate them together into quantum networks (circuits).

However, subsequent logical operations usually involve more complicated physical operations on more than one qubit. If we keep on putting quantum gates together into circuits we will quickly run into some serious practical problems. The more interacting qubits are involved the harder it tends to be to engineer the interaction that would display the quantum interference. Apart from the technical difficulties of working at single-atom and single-photon scales, one of the most important problems is that of preventing the surrounding environment from being affected by the interactions that generate quantum superpositions. The more components the more likely it is that quantum computation will spread outside the computational unit and will irreversibly dissipate useful information to the environment. This process is called decoherence.

Without any additional stabilization mechanism, building a quantum computer is like building a house of cards: we can build a layer or two, but when one contemplates building a really tall house, the task seems hopeless. Not quite!

In principle we know how to handle errors due to decoherence, provided that they satisfy some assumptions, e.g. that errors occur independently on each of the qubits, that the performance of gate operations on some qubits does not cause decoherence in other qubits, that reliable quantum measurements can be made so that error detection can take place, and that systematic errors in the operations associated with quantum gates can be made very small. If all these assumptions are satisfied, then fault-tolerant quantum computation is possible. That is, efficient, reliable quantum-coherent quantum computation of arbitrarily long duration is possible, even with faulty and decohering components. Thus, errors can be corrected faster than they occur, even if the error-correction machinery is faulty.

The first models of quantum computers, proposed in the 1980s, were, understandably, very abstract and in many ways totally impractical. In 1993 progress toward a practical model took an encouraging turn when Seth Lloyd, then at Los Alamos National Laboratory, and subsequently Adriano Barenco, David Deutsch, and Artur Ekert, of the University of Oxford, proposed a scheme for a potentially realizable quantum computer. Lloyd considered a one-dimensional heteropolymer with three types of weakly coupled qubits, which were subjected to a sequence of electromagnetic pulses of well-defined frequency and length. Barenco, Deutsch, and Ekert showed how to implement quantum computation in an array of single-electron quantum dots (see Figure 11.6).

Although these early schemes showed how to translate mathematical prescriptions into physical reality, they were difficult to implement and to scale up, due to decoherence. In 1994 Ignacio Cirac and Peter Zoller, from the University of Innsbruck, came up with a model that offered the possibility of fault-tolerant quantum computation. It was quickly recognized as a conceptual breakthrough in experimental quantum computation. They considered a trap holding a number of ions in a straight line. The ions would be laser cooled and then each ion could be selectively excited by a pulse of light from a laser beam directed specifically at that ion. The trap would therefore act as a quantum register with the internal state of each ion playing the role of a qubit. Pulses of light can also induce vibrations of ions. The vibrations are shared by all of the ions in the trap and they enable the transfer of quantum information in between distant ions and the implementation of quantum logic gates (Figure 11.7). In mid 1995 Dave Wineland and his colleagues at the National Institute of Standards and Technology in Boulder, Colorado, used the idea of Cirac and Zoller to build the world's first quantum gate operating on two qubits.

Today there are many interesting approaches to experimental quantum computation and related quantum technologies. The requirements for quantum hardware are simply stated but very demanding in practice. Firstly, a quantum register of multiple qubits must be prepared in an addressable form, and isolated from environmental influences, which cause the delicate quantum states to decohere. Secondly, although weakly coupled to the outside world, the qubits must nevertheless be strongly

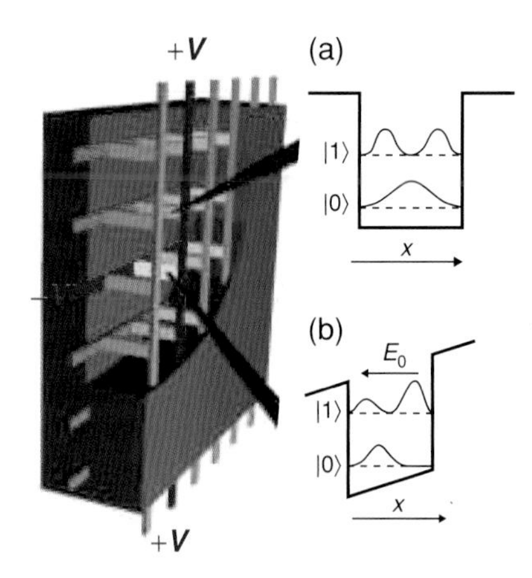


**Figure 11.6.** One of the first experimental schemes for quantum computation used an array of single-electron quantum dots. A dot is "activated" by applying a suitable voltage to the two metal wires that cross at that dot. Similarly, several dots may be activated at the same time. Because the states of an active dot are asymmetrically charged, two adjacent active dots are each exposed to an additional electric field that depends on the other's state. In this way the resonance frequency of one dot depends on the state of the other. Light at carefully selected frequencies will selectively excite only adjacent, activated dots that are in certain states. Such conditional quantum dynamics are needed in order to implement quantum logic gates.

coupled together through an external control mechanism, in order to perform logic-gate operations. Thirdly, there must be a readout method to determine the state of each qubit at the end of the computation. There are many beautiful experiments showing that these requirements, at least in principle, can be met. They involve technologies such as linear optics, nuclear magnetic resonance, trapped ions, cavity quantum electrodynamics, neutral atoms in optical lattices, interacting quantum dots, and superconducting devices, among many others. They are but a sample of the emerging field of quantum information technology. At the moment it is not clear which particular technology will be the ultimate winner.

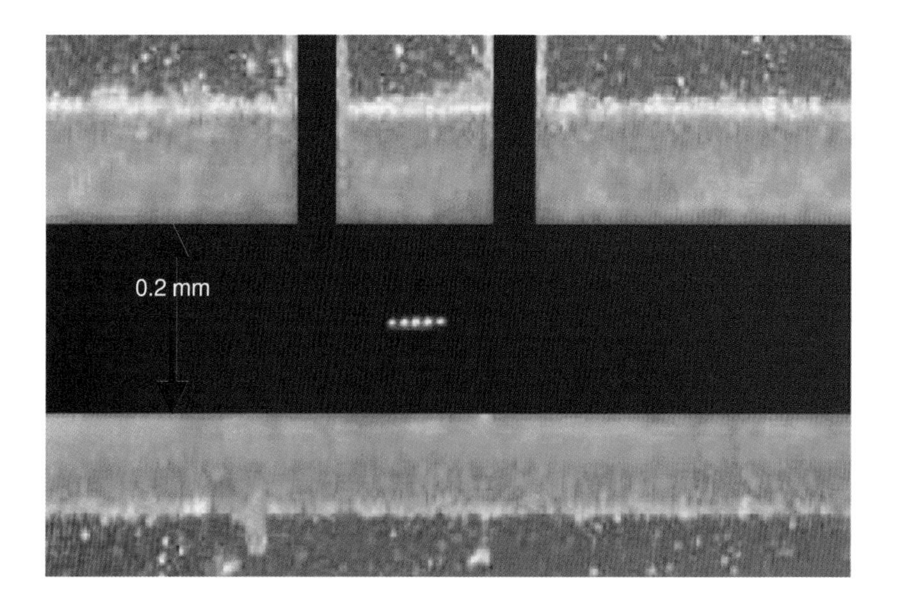


**Figure 11.7.** A photograph of five beryllium ions in a linear ion trap. The separation between ions is approximately 10 μm. Pulses of light can selectively excite individual ions and induce vibrations of the whole line of ions. The vibrations "inform" other ions in the trap that a particular ion has been excited; they play the same role as a data bus in conventional computers. This picture was taken in Dave Wineland's laboratory at the National Institute of Standards and Technology in Boulder, USA.

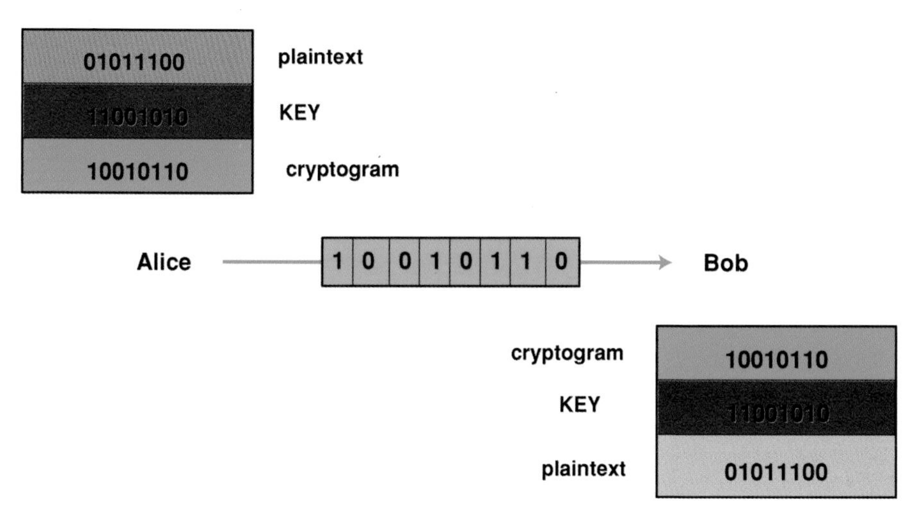


Figure 11.8. A one-time pad. The modern version of the "one-time pad" is based on binary representation of messages and keys. The message is converted into a sequence of 0's and 1's and the key is another sequence of 0's and 1's of the same length. Each bit of the message is then combined with the respective bit of the key using addition in base 2, which has the rules $0 + 0 = 0, 0 + 1 = 1 + 0 = 1, 1 + 1 = 0$. Because the key is a random string of 0's and 1's the resulting cryptogram – the plaintext plus the key – is also random and therefore completely scrambled unless one knows the key. The message is recovered by adding (in base 2 again) the key to the cryptogram.

## 11.7 The art of secure communication

We have mentioned that several of the superior features of quantum computers have applications in cryptanalysis, i.e. in the art of breaking ciphers. The quantum answer to quantum cryptanalysis is quantum cryptography.

Despite a long and colorful history, cryptography became part of mathematics and information theory only in the late 1940s, mainly as a result of the work of Claude Shannon of Bell Laboratories in New Jersey. Shannon showed that truly unbreakable ciphers do exist and, in fact, they had been known for over 30 years. The one-time pad (see Figure 11.8), devised in about 1918 by an American Telephone and Telegraph engineer named Gilbert Vernam, and Major Joseph Mauborgne of the US Army Signal Corps, is one of the simplest and most secure encryption schemes. The message, also known as a plaintext, is converted into a sequence of numbers using a publicly known digital alphabet (e.g. ASCII code) and then combined with another sequence of random numbers called *a key* to produce a cryptogram. Both sender and receiver must have two exact copies of the key beforehand; the sender needs the key to encrypt the plaintext, the receiver needs the exact copy of the key to recover the plaintext from the cryptogram. The randomness of the key wipes out various frequency patterns in the cryptogram that are used by code-breakers to crack ciphers. Without the key the cryptogram looks like a random sequence of numbers.

There is a snag, however. All one-time pads suffer from a serious practical drawback, known as the key-distribution problem. Potential users have to agree secretly, and in advance, on the key – a long, random sequence of 0's and 1's. Once they have done this, they can use the key for enciphering and deciphering and the resulting cryptograms can be transmitted publicly such as by radio or in a newspaper without compromising the security of messages. But the key itself must be established between the sender and the receiver by means of a very secure channel – for example, a very secure telephone line, a private meeting, or hand-delivery by a trusted courier. Such a secure channel is usually available only at certain times and under certain circumstances.

So users who are far apart, in order to guarantee perfect security of subsequent crypto-communication, have to carry around with them an enormous amount of secret and meaningless (as such) information (cryptographic keys), equal in volume to all the messages they might later wish to send. For perfect security the key must be as long as the message; in practice, however, in order to avoid having to distribute keys too often, much shorter keys are used. For example, the most widely used commercial cipher, the DES, depends on a 56-bit secret key, which is reused for many encryptions over a period of time. This has been replaced by the Advanced Encryption Standard (AES), with at least a 128-bit key. This simplifies the problem of secure key distribution and storage, but does not eliminate it.

Mathematicians tried very hard to eliminate the problem. The 1970s brought a clever mathematical discovery of the so-called public-key cryptosystems. They avoid the key-distribution problem but unfortunately their security depends on unproved mathematical assumptions, such as the difficulty of factoring large integers.

## 11.8 Quantum key distribution

Physicists view the key-distribution problem as a physical process associated with sending information from one place to another. From this perspective eavesdropping is a set of measurements performed on carriers of information. Until now, such eavesdropping has depended on the eavesdropper having the best possible technology. Suppose that an eavesdropper is tapping a telephone line. Any measurement on the signal in the line may disturb it and so leave traces. Legitimate users can try to guard against this by making their own measurements on the line to detect the effect of tapping. However, the tappers will escape detection provided that the disturbances they cause are smaller than the disturbances that the users can detect. So, given the right equipment, eavesdropping can go undetected. Even if legitimate users do detect an eavesdropper, what do they conclude if one day they find no traces of interception? Has the eavesdropping stopped? Or has the eavesdropper acquired better technology? The way round this problem of key distribution may lie in quantum physics.

We have already mentioned that when two qubits representing logical 0 and 1 enter the square root of a SWAP gate they interact and emerge in an entangled state, which is a superposition of 01 and 10. They remain entangled even when they are separated and transported, without any disturbance, to distant locations. Moreover, results of measurements performed on the individual qubits at the two distant locations are usually highly correlated.

For example, in a process called parametric down-conversion we can entangle two photons in such a way that their polarizations are anti-correlated. If we choose to test linear polarization then one photon will be polarized vertically, $\updownarrow$, and the other one horizontally, $\leftrightarrow$. The same is true if we choose to test circular polarization; one photon will carry left-handed, $\circ$, and the other right-handed, $\bullet$, polarization. In general, the perfect anti-correlation appears if we carry out *the same* test on the two photons. The individual results are completely random, e.g. it is impossible to predict in advance whether we will obtain $\updownarrow$ or $\leftrightarrow$ on a single photon. Moreover, the laws of quantum physics forbid a simultaneous test of both linear and circular polarization on the same photon. Once the photon is tested for linear polarization and we find out that it is, say, vertical, $\updownarrow$, then all information about its circular polarization is lost. Any subsequent test for circular polarization will reveal either left-handed, $\circ$, or right-handed, $\bullet$, with the same probability.

Both linear and circular polarization can be used for encoding the bits of information. For example, we can agree that for linearly polarized photons ↕ stands for "0" and ↔ for "1," and for circularly polarized photons "0" is represented by ○ and "1" by •. However, for the decoding of these bits to be meaningful the receiver must know in advance which type of test, or measurement, to carry out for each incoming photon. Testing linear (circular) polarization on a photon that carries one bit of information encoded in circular (linear) polarization will reveal nothing.

The quantum key distribution which we are going to discuss here is based on distribution of photons with such anti-correlated polarizations. Imagine a source that emits pairs of photons in an entangled state, which can be viewed both as a superposition of ↕↔ and ↔↕, and as a superposition of ○• and •○. The photons fly apart toward the two legitimate users, called Alice and Bob, who, for each incoming photon, decide randomly and independently of each other whether to test linear or circular polarization. A single run of the experiment may look like this:

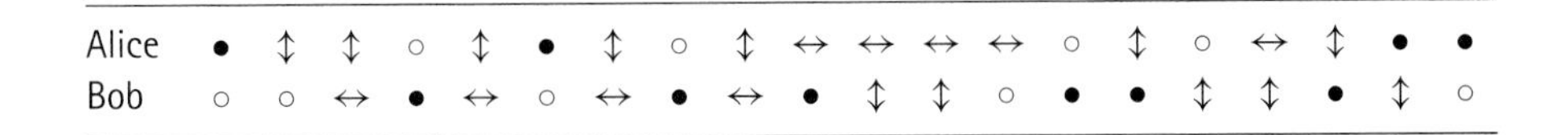

| Alice | • | ↕ | ↕ | ○ | ↕ | • | ↕ | ○ | ↕ | ↔ | ↔ | ↔ | ↔ | ○ | ↕ | ○ | ↔ | ↕ | • | • |
|---|---|---|---|---|---|---|---|---|---|---|---|---|---|---|---|---|---|---|---|---|
| Bob | ○ | ○ | ↔ | • | ↔ | ○ | ↔ | • | ↔ | • | ↕ | ↕ | ○ | • | • | ↕ | ↕ | • | ↕ | ○ |

For the first pair both Alice and Bob decided to test circular polarization and their results are perfectly anti-correlated. For the second pair Alice measured linear polarization whilst Bob measured circular. In this case their results are not correlated at all. In the third instant they both measured linear polarization and obtained perfectly anti-correlated results, etc.

After completing all the measurements, Alice and Bob discuss their data in public so that anybody can listen including their adversary, an eavesdropper called Eve, but nobody can alter or suppress such public messages. Alice and Bob tell each other which type of polarization they measured for each incoming photon but they do not disclose the actual outcomes of the measurements. For example, for the first pair Alice may say "I measured circular polarization" and Bob may confirm "So did I." At this point they know that the results in the first measurement are anti-correlated. Alice knows that Bob registered ○ because she registered •, and vice versa. However, although Eve learns that the results are anti-correlated she does not know whether it is ○ for Alice and • for Bob, or • for Alice and ○ for Bob. The two outcomes are equally likely, so the actual values of bits associated with different results are still secret.

Alice and Bob then discard instances in which they made measurements of different types (shaded columns in the table below).

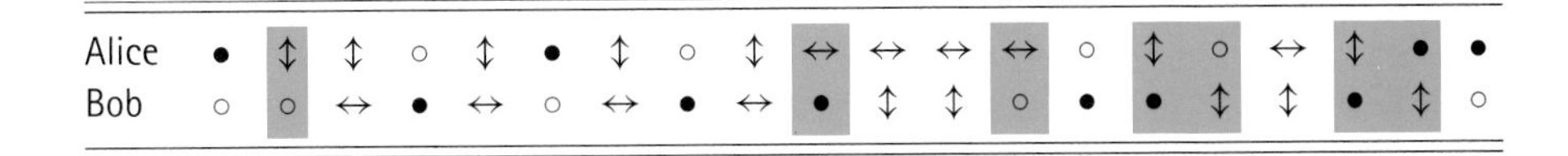

| Alice | • | ↕ | ↕ | ○ | ↕ | • | ↕ | ○ | ↕ | ↔ | ↔ | ↔ | ↔ | ○ | ↕ | ○ | ↔ | ↕ | • | • |
|---|---|---|---|---|---|---|---|---|---|---|---|---|---|---|---|---|---|---|---|---|
| Bob | ○ | ○ | ↔ | • | ↔ | ○ | ↔ | • | ↔ | • | ↕ | ↕ | ○ | • | • | ↕ | ↕ | • | ↕ | ○ |

They end up with shorter strings, which should now contain perfectly anti-correlated entries. They check whether the two strings are indeed anti-correlated by comparing, in public, randomly selected entries (shaded columns in the table below).

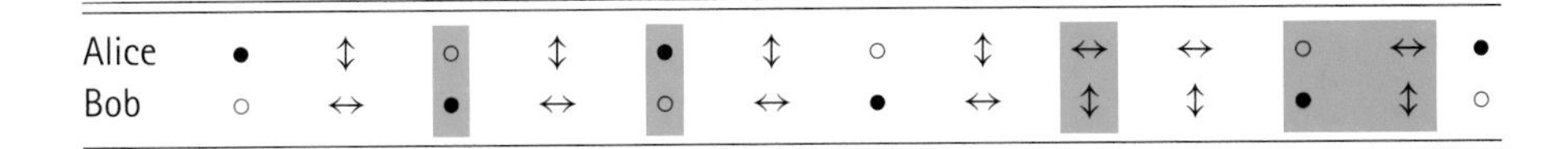

| Alice | • | ↕ | ○ | ↕ | • | ↕ | ○ | ↕ | ↔ | ↔ | ○ | ↔ | • |
|---|---|---|---|---|---|---|---|---|---|---|---|---|---|
| Bob | ○ | ↔ | • | ↔ | ○ | ↔ | • | ↔ | ↕ | ↕ | • | ↕ | ○ |

The publicly revealed entries are discarded and the remaining results can be translated into a binary string, following the agreed-upon encoding, e.g. as in the table below.

| Alice | • | ↕ | ↕ | ↕ | ○ | ↕ | ↔ | • |
|---|---|---|---|---|---|---|---|---|
| Bob | ○ | ↔ | ↔ | ↔ | • | ↔ | ↕ | ○ |
| KEY | 1 | 0 | 0 | 0 | 0 | 0 | 1 | 1 |

In order to analyze the security of the key distribution let us adopt the scenario that is most favorable for eavesdropping, namely we will allow Eve to prepare all the photons and send them to Alice and Bob!

Eve's objective is to prepare the pairs in such a way that she can predict Alice's and Bob's results and that the pairs pass the anti-correlation test. This is impossible. Suppose that Eve prepares a pair of photons choosing randomly one of the four states: ↕↔, ↔↕, ○•, or •○. She then sends one photon to Alice and one to Bob. Let us assume that Alice and Bob measure the same type of polarization on their respective photons. If Alice and Bob choose to measure the right type of polarization then they obtain anti-correlated results but Eve knows the outcomes; if they choose to measure the wrong type of polarization then, although the outcomes are random, they can still obtain anti-correlated results with probability 50%. This will result in 25% of errors in the anti-correlation test and in Eve knowing, on average, every second bit of the key. Eve may want to reduce the error rate by entangling the photons. She can prepare them in a state that is a superposition of ↕↔ and ↔↕ and also of ○• and •○. In this case the pairs will pass the test without any errors but Eve has no clue about the results. Once ↕↔ and ↔↕ (or ○• and •○) are "locked" in a superposition there is no way of knowing which component of the superposition will be detected in a subsequent measurement. More technical eavesdropping analysis shows that all, however sophisticated, eavesdropping strategies are doomed to fail, even if Eve has access to superior technology, including quantum computers. The more information Eve has about the key, the more disturbance she creates.

The key-distribution procedure described above is somewhat idealized. The problem is that there is, in principle, no way of distinguishing errors due to eavesdropping from errors due to spurious interaction with the environment, which is presumably always present. This implies that all quantum key-distribution protocols that do not address this problem are, strictly speaking, inoperable in the presence of noise, since they require the transmission of messages to be suspended whenever an eavesdropper (or, therefore, noise) is detected. Conversely, if we want a protocol that is secure in the presence of noise, we must find one that allows secure transmission to continue even in the presence of eavesdroppers. Several such protocols were designed. They are based on two approaches, namely on purification of quantum entanglement, proposed in this context by Deutsch, Ekert, Jozsa, Macchiavello, Popescu, and Sanpera, and on classical error correction, which was pioneered by Dominic Mayers. Subsequently the two approaches have been unified and simplified by Peter Shor and John Preskill.

Experimental quantum cryptography has rapidly evolved from early demonstrations at the IBM T. J. Watson Research Laboratory in Yorktown Heights in the USA and the Defence Research Agency in Malvern in the UK to several beautiful experiments that demonstrated full quantum key distribution both in optical fibers and in free space. Quantum cryptography today is a commercial alternative to more conventional, classical cryptography.

## 11.9 Concluding remarks

When the physics of computation was first investigated systematically in the 1970s, the main fear was that quantum-mechanical effects might place fundamental bounds on the accuracy with which physical objects could realize the properties of bits, logic gates, the composition of operations, and so on, which appear in the abstract and mathematically sophisticated theory of computation. Thus it was feared that the power and elegance of that theory, its deep concepts such as computational universality, its deep results such as Turing's halting theorem, and the more modern theory of complexity, might all be mere figments of pure mathematics, not really relevant to anything in Nature.

Those fears have not only been proved groundless by the research we have been describing, but also, in each case, the underlying aspiration has been wonderfully vindicated to an extent that no one even dreamed of just twenty years ago. As we have explained, quantum mechanics, far from placing limits on what classical computations can be performed in Nature, permits them all, and in addition provides entirely new modes of computation. As far as the elegance of the theory goes, researchers in the field have now become accustomed to the fact that the real theory of computation hangs together better, and fits in far more naturally with fundamental theories in other fields, than its classical approximation could ever have been expected to.

Experimental and theoretical research in quantum computation and quantum cryptography is now attracting increasing attention both from academic researchers and from industry worldwide. The idea that Nature can be controlled and manipulated at the quantum level is a powerful stimulus to the imagination of physicists and engineers. There is almost daily progress in developing ever more promising technologies for realizing quantum information processing. There is potential here for truly revolutionary innovations.

### FURTHER READING

For a lucid exposition of some of the deeper implications of quantum computing see *The Fabric of Reality* by David Deutsch (London, Allen Lane, The Penguin Press, 1997). For a popular introduction to quantum information technology see *Schrödinger's Machines* by Gerard Milburn (New York, W. H. Freeman & Company, 1977). Julian Brown in his *Minds, Machines, and the Multiverse* (New York, Simon & Schuster, 1999), gives a very readable account of quantum information science. The Centre for Quantum Computation (http://cam.qubit.org) has several Web pages and links devoted to quantum computation and cryptography.

## Artur Ekert

Artur Konrad Ekert is the Leigh Trapnell Professor of Quantum Physics at the Department of Applied Mathematics and Theoretical Physics (DAMTP), University of Cambridge and a Professorial Fellow of King's College, Cambridge. He is also a Temasek Professor at the National University of Singapore. He is one of the pioneers of quantum cryptography. In his doctoral thesis (Oxford, 1991) he showed how quantum entanglement and nonlocality can be used to distribute cryptographic keys with perfect security. He has worked with and advised several companies and government agencies and has made a number of contributions to quantum information science.

# 12 Small-scale structures and "nanoscience"

Yoseph Imry

## 12.1 Introduction

The atomic nature of matter is well documented and appreciated. Solids are made from regular (in the case of crystals) or irregular (for the example of glasses) arrangements of atoms. Nature finds a way to produce various materials by combining constituent atoms into a macroscopic substance.

Imagine the following hypothetical experiment. Suppose that one builds up, for example, a crystal of silicon by starting with a single atom and adding atoms to make a small number of unit cells, repeating the process until a large enough crystallite has formed. A fundamental question is the following: at what stage in this process, i.e. at what size of the crystallite, will it approximately acquire the "bulk" properties? The precise answer to this question may well depend on what exactly one intends to do with the crystal.

While this may sound like a purely thought experiment, remarkable modern fabrication methods are close to being able to achieve this, using at least two independent approaches. Molecular-beam epitaxy (Figure 12.1) is able to grow high-quality crystalline layers one by one with remarkable control. Atomic-resolution methods are already able to deposit atoms selectively on some surfaces (Figure 12.2).

Some of the answers, but probably not the deep conceptual ones, may be provided by large-scale computer simulations. Their rapidly increasing power already enables rather precise calculations on tens or even hundreds of atomic units. This increasing power is due in a large measure to the computer units themselves becoming more miniaturized, so that next year's computer will be significantly smaller and more powerful than the present one. Linear sizes of hundreds of atoms for a computer element and linewidths of tens of atoms are imminent. One of Moore's laws states that the size of computer elements shrinks by roughly a factor of two every 18 months and is a valid description of the current progress in miniaturization. It follows that these elements should reach atomic dimensions in less than 15 years. That will mark the limit of miniaturization. However, the question posed above becomes very relevant even before the ultimate atomic limit is reached. It is very likely that, on scales that are small but still much larger than atomic, the operation of advanced computer elements will eventually become hampered due to their being so small that they no longer obey the usual rules.

There is no universal answer to the question "when does the system no longer obey all the relevant macroscopic laws?" For many applications, being very small is not

*The New Physics for the Twenty-First Century*, ed. Gordon Fraser.

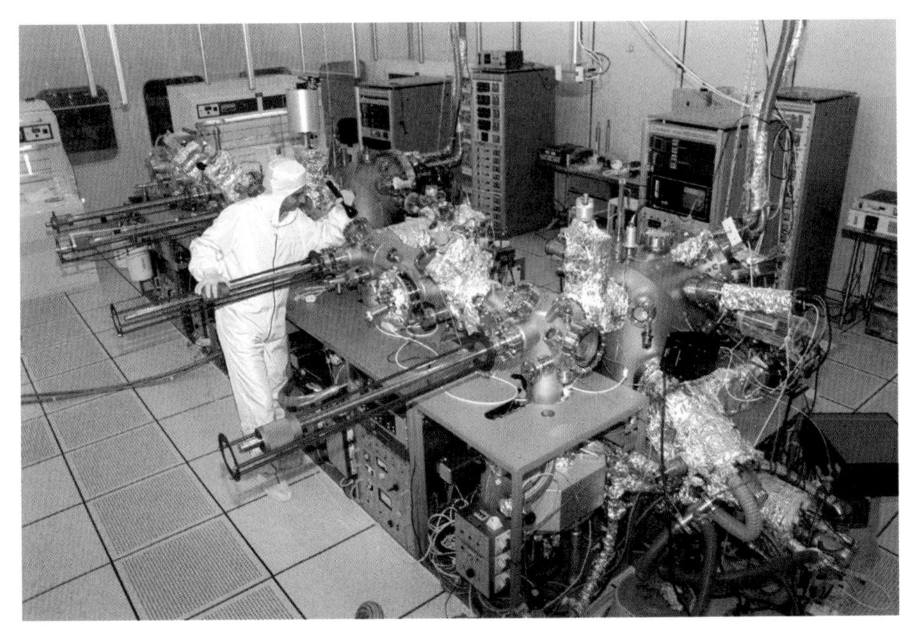

Figure 12.1. A molecular-beam-epitaxy (MBE) growth/analysis system at the Braun Submicron Center at the Weizmann Institute. Inside an ultra-high-vacuum chamber, this method is able to grow epitaxial layers (such as gallium arsenide and aluminum–gallium arsenide) one by one with excellent control over materials and purity. Such layers (produced using MBE, or often less exacting methods) form the basis of many semiconductor devices.

necessarily detrimental. As a very simple example, small diamond crystallites may well be good enough or even better than large ones for specific delicate industrial tasks. One might even hope that the new rules introduced by very small sizes will allow further new and powerful ideas for real-life applications. Very interestingly, by combining small units, still much larger than a single atom, in an ingenious way, novel materials with improved qualities become possible.

These applications are not just in the realm of space exploration and the like. Many people have used tough carbon-fiber-based materials in their tennis rackets or ski equipment. The science of cleverly designed man-made novel materials made from

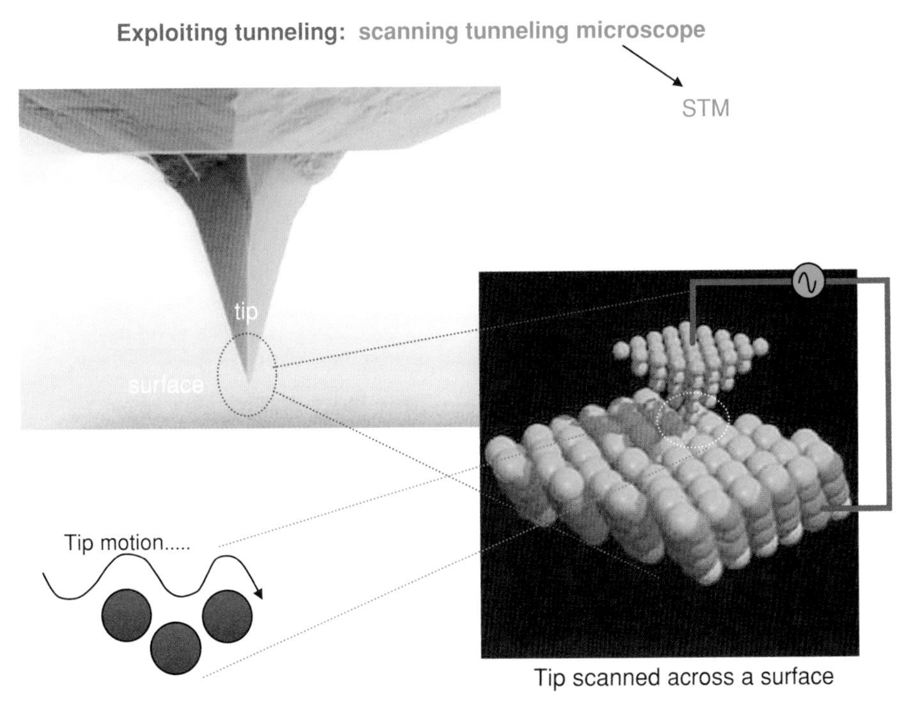


Figure 12.2. The atomic force microscope (AFM) and the scanning tunneling microscope (STM) have powerful capabilities for treating and analyzing surfaces. The figure shows a schematic view of a STM, with the inset showing a STM in action. Below is the resulting atomic arrangement.

"nano-units" is very dynamic and has many conceivable applications. Such ideas are used in some theories suggested to explain high-transition-temperature superconductivity (see Chapter 8). The idea is that these rather complicated systems spontaneously form some type of intrinsic inhomogeneity, with regions in space having varying charge-carrier density, either compact or having the form of, for example, electronic chain-type structures. Novel lubricants based on inorganic molecular structures have also been suggested, as well as selective chemical catalysis through solvated nanocrystals and porous nanomaterials. Nevertheless, the concept of achieving molecular-scale electronics (and photonics) is one of the most exciting major goals of nanoscience.

Pioneering ideas for molecular-scale electronics were proposed in 1959 when Richard Feynman imagined the possibility of manipulating and controlling things on a small scale. He described the levels of miniaturization then accessible as "the most primitive, halting step," and anticipated "a staggeringly small world that is below." "In the year 2000, when they look back at this age," he continued, "they will wonder why it was not until the year 1960 that anybody began seriously to move in this direction. Why cannot we write the entire 24 volumes of the *Encyclopedia Britannica* on the head of a pin?"

These developments highlight the question proposed above: suppose that our small-scale system has not yet reached the range where it is similar to the "bulk" and its properties are still appreciably different. This can be disadvantageous for some applications, but the properties due to the small size may open up novel powerful applications that would not otherwise be possible. The increased strength of whiskers and fibers has been mentioned. Possibly even more exciting is that there may be electronic properties that could be exploited as well.

We begin with a few examples based on the electromagnetic properties of small systems. The modified semiconductor band gap of small crystallites with respect to that of the bulk can change the spectral absorption or emission edge of such crystallites and hence their color as well. Likewise, the color of a small metal particle should change because of modifications in the electronic mode ("plasmon") frequencies. The ability of such a system to radiate can also be greatly modified. Related ideas are already being used for more efficient, lower-threshold lasers.

Much of this chapter looks at the implications for fundamental physics. We know that the electrons in atoms must be described by quantum mechanics. However, macroscopic systems behave classically in most situations. How exactly does the behavior change on going from one limit to another? Could we use this knowledge to answer basic questions about quantum mechanics? Perhaps these quantum properties can be harnessed for novel applications, such as quantum computation (see Chapter 11).

Two major technological revolutions of the second half of the twentieth century that greatly influenced our lifestyle were semiconductors and lasers. The principles of both are based on quantum mechanics, but once that is understood, much can be achieved by designs based on a semiclassical description. However, here we have in mind even more exotic possibilities. The strange and counter-intuitive possibilities offered by quantum mechanics, such as tunneling, interference, and superposition, suggest a number of unprecedented applications. One example that already exists is the type of extremely sensitive measurement devices called superconducting quantum interference devices (SQUIDs) – see Chapter 8 – which are already being manufactured commercially. The delicate experiments shown later (see Figure 12.11) on "mesoscopic persistent currents" were done using a tiny on-chip SQUID.

The most sensitive and powerful device for treating and analyzing the surface of a metal is the "scanning tunneling microscope" (STM), whose name implies its reliance on quantum tunneling. Figure 12.2 shows an example of the atomic-scale analysis power of this method. There are several extremely useful variations on this technique.

As well as applications in chemistry, electronics, magnetism, and photonics, there is also interest in small-scale mechanics. A burgeoning field of mechanical engineering is the use of technologies developed for microelectronics and nanoelectronics to make very small-scale mechanical components. Elements such as beams, wheels, gears, shifters, screws, cantilevers, and cutters, among many others, have been investigated, with varying degrees of success. Optical switches rely on moving a silicon mirror by a fraction of a wavelength. There are also important applications for miniature sensors of all kinds. Micromachined acceleration sensors are already used as airbag actuators in cars. A micron-size engine is another challenging goal. Nature has endowed some micro-organisms with their own motors and navigational facilities. Medical devices – such as small cutters or applicators – carried and micro-manipulated inside arteries do not seem impossible. Nanoscience opens up not only exciting fundamental research issues, but also a large range of real-life applications.

## 12.2 Mesophysics and nanophysics, the tools and capabilities

Traditionally, systems of interest to physicists have been divided into the macroscopic and the microscopic realms, where the latter implies atomic and molecular sizes or smaller. Recently, research in the intermediate – mesoscopic – regime has achieved significant scientific successes. This field is characterized by the need to use the microscopic laws of quantum mechanics, while, on the other hand, the samples can be made and operated by essentially ordinary macroscopic methods. This involves linear size scales from a few to thousands of atoms (see Figure 12.4), and reliable fabrication and analysis methods exist down to the scale of about fifty atoms. The term "nano" characterizes the low end of this range.

Methods based on extremely powerful scanning electron microscopy (STM, see Figure 12.2) and atomic-force microscopy (AFM) that allow impressive control of both fabrication and analysis on the scale of single atoms are being developed. Controlling man-made molecular structures on the level of individual atoms (including position and orientation) is possible in principle, and impressive physical measurements such as that of the local (atomic-scale) density of electronic states near an appropriate defect in a superconductor have been made. A variety of surface probes can measure, for example, the local electrostatic potential on a sample's surface with a resolution approaching 10 nm, which is comparable to the electronic Fermi wavelength in semiconducting systems (see Figure 12.3). Besides these powerful techniques that rely on surface-probe microscopy, efforts to control, for example, tiny metallic clusters of precise and reproducible numbers of atoms (for example gold clusters of precisely 55 atoms) could lead to new possibilities.

There are two distinct approaches to manufacturing small-scale devices. Microelectronics starts with a large system and then divides it, reducing its dimensions by a variety of well-controlled methods. Most of these are lithographic, being based on altering selected parts of the sample (while protecting the other parts by suitable "masks") using various forms of radiation and removing the unwanted parts by chemical methods. Impressive production rates have been achieved for standardized structures. For specialized samples used in research and development, very high resolutions have been achieved, but appear to reach their smallest feature capability around the 10-nm scale (1 nm = $10^{-7}$ cm, about the size of three atoms).

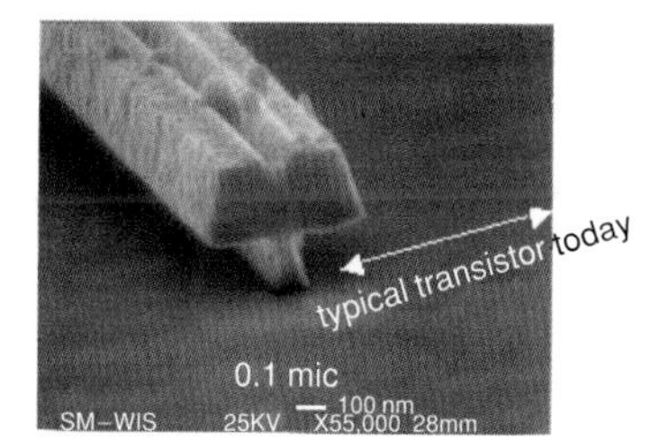


1 mm = 1000 microns
1 micron = 1000 nanometers
1 nanometer = 10 ångström units (A typical atomic spacing in a solid is a few ångström units)

**Figure 12.3.** Physics on the small scale involves linear size scales from a few to thousands of atoms, and there are reliable micro-fabrication and analysis methods down to the scale of about 50 atoms. Nanophysics deals with the low end of this range down to the atomic scale. The images (D. Mahalu and M. Heiblum) show (left) scales of 10 $\mu$m, with a circuit element compared with a human hair, and (right) of 0.1 $\mu$m, with a small state-of-the-art transistor, compared with one commercially available in 2001.

**Figure 12.4.** Local compressibility of a two-dimensional electron gas (S. Ilani *et al.* 2001). Measurement of discrete charging events in a low-density two-dimensional gas of holes, which fragments on the nanoscale into microscopic charge configurations that coexist with the surrounding metallic compressible phase. The black lines in the figure are the individual charging events of these fragments. The mutual electrostatic interactions between them are manifested by the vertices formed at their crossings. The color map represents inverse electronic compressibility. Inset: the measurement circuit utilizes a single-electron transistor as a local detector of compressibility.

The other manufacturing approach is "from the ground up," namely combining microscopic units. Chemistry has always used this principle, controlling chemical composition via the ingredients and often the catalysts as well. Traditionally this process was carried out "in a beaker" rather than on the level of a single molecule. However, new atomic-scale methods may be used to achieve controlled synthesis of individual chemical units.

An example of such large-scale preparation of high-quality materials is provided by the MBE method, shown in Figure 12.1. Extremely high-quality semiconductor systems, including the two-dimensional (2D) electron-gas case (see Figure 12.4), are prepared using this method by growing individual lattice layers in a high vacuum with impressive control of parameters. Similar methods are also used to grow extra-thin metallic layers. Such methods can be used to grow two different semiconductors on top of each other, especially if their lattice parameters are matched. The classic examples are GaAs and AlAs and their mixtures, but other combinations are possible. The different band structures (mainly the energy gap) and work functions usually cause some transfer of charge between the two adjacent materials in order to reach equilibrium conditions. Electrons are attracted to the remaining holes and the well-known dipole layer is formed at the interface. This leads to a space-dependent potential energy or "band-bending" near the interface. Thus, potential wells, barriers, and doped thin layers (including strictly 2D situations, etc.) can be formed. The physical properties of these synthetic materials can be engineered to a large extent. By using sandwiches of one type of semiconductor between two layers of another, further types of wells and barriers can be made. Such structures can be repeated to form, for example, a periodic superlattice in the growth direction.

There are also combinations of the bottom-up and top-down approaches. The high-quality 2D electron gas mentioned above can be laterally manipulated by electrostatic gates deposited above the uppermost semiconducting layer. By biasing these gates, desired patterns can be created in the 2D electron gas. Some of these techniques were developed in the microelectronics industry whereas others evolved from them. This leads to impressive control over the design of man-made materials. This control is continually improving, but may be limited by fundamental factors. The STM-type methods and ingenious chemistry will undoubtedly push the limits further.

A great deal of research goes into finding new materials with complex structures and properties that can be well controlled by manipulations at the molecular level. A goal of chemistry research is to synthesize and control molecules, allowing great versatility in the construction of new materials. Nanoscience relies on the use of small building blocks to generate functional nanostructures in a controlled way. An example is supramolecular chemistry, where the relevant concept is "self-assembly," which is closer to traditional chemistry. Serious attempts are being made to organize matter on the nanometer scale using a variety of methods, including colloidal systems, surface chemistry, and biological molecules passivating nanoparticles by forming a layer on their surfaces. Small particles can also be grown in onion-like layers to control their quantum properties. Arrays of such particles, consisting of semiconductor compounds (CdS) or metals, have already been used to demonstrate mesoscopic behavior at room temperature. Magnetic molecules and clusters that will allow new studies of the quantum magnetism of nanostructures are also being prepared. Nanocrystals can be fabricated by a variety of chemical methods, including the use of wet-solution sono-generators and electrochemical deposition. Chemically sensitive probes are of particular interest in scanning microscopy.

Special molecular structures such as the fullerenes and carbon nanotubes (see Chapter 18) have been analyzed and experimented on. Some of these will be reviewed briefly

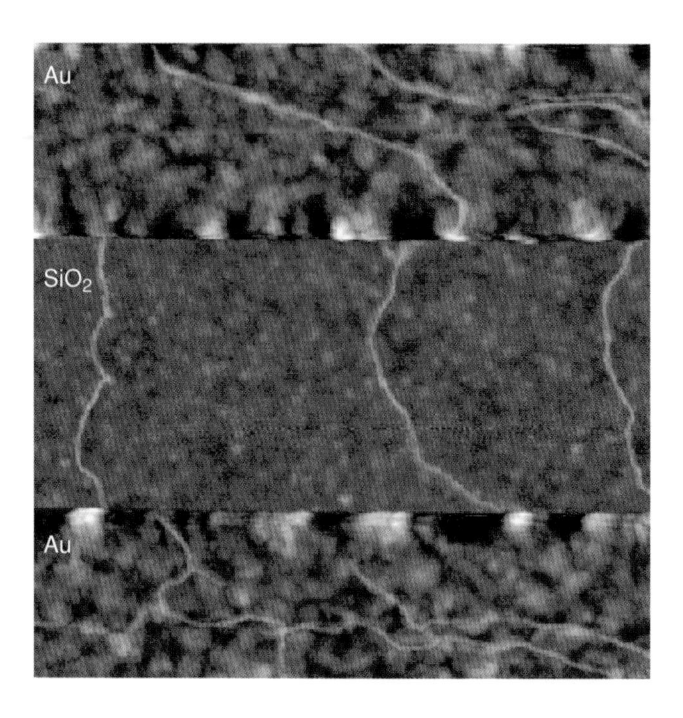


**Figure 12.5.** DNA molecules bridging two metallic electrodes. (From C. Dekker *et al.*, Delft, http://www.mb.tn. tudelft.nl/ SingleDNAmolexp. html.)

later. In biologically relevant large molecules and polymers, such as DNA, the extremely rich enzymatic mechanisms allow various groups and molecules to be cut and "pasted" at specific sites. Interesting ideas for using biomolecular technology to create a reproducible substrate for extra-small electronic circuitry have been suggested and partially demonstrated. Such molecules can be made to connect two metallic electrodes and their electrical conductivity addressed (see Figure 12.5 – however, an even more advanced situation, with nanotubes, will be mentioned later, see Figure 12.13). There is a real hope that, with selective doping, a technology resembling that of ordinary semiconductors (but based on carbon!) may be achieved. It is likely that tunneling and electrical transport measurements on individual biologically relevant molecules will soon be feasible and the electrical transport mechanisms clarified. Questions such as whether that transport is electronic or ionic in the "undoped" case and how to dope such systems (as is routinely done in semiconductors) might soon have definitive answers.

There is a tremendous interest in the chemistry of small clusters and grains with surface-to-volume ratio of order unity. One of the main reasons is the extreme relevance to catalysis, a basic understanding of which is still one of the major unsolved problems of chemistry. Before outlining some specific examples of interesting nanosystems, we first review some of the physics involved.

Going down in scale to the molecular level may lead to more reproducible devices with more robust features than those that can be achieved with mesoscopic devices produced via the "top-down" approach. It is now envisaged that well-known quantum mesoscopic phenomena (such as conductance quantization – see Section 12.3) will operate at room temperature for nanoscale samples. Thus, quantum interference phenomena should be ubiquitous on these scales.

Since mesoscopic physics deals with the intermediate realm between the microscopic scale and the macroscopic scale, it can give us very fundamental information on the transition between these two regimes. Macroscopic-type electrical measurements can be sensitive to quantum phenomena in mesoscopic samples at low temperatures. The latter are needed in order to preserve the phase coherence of the electrons, which will be discussed later.

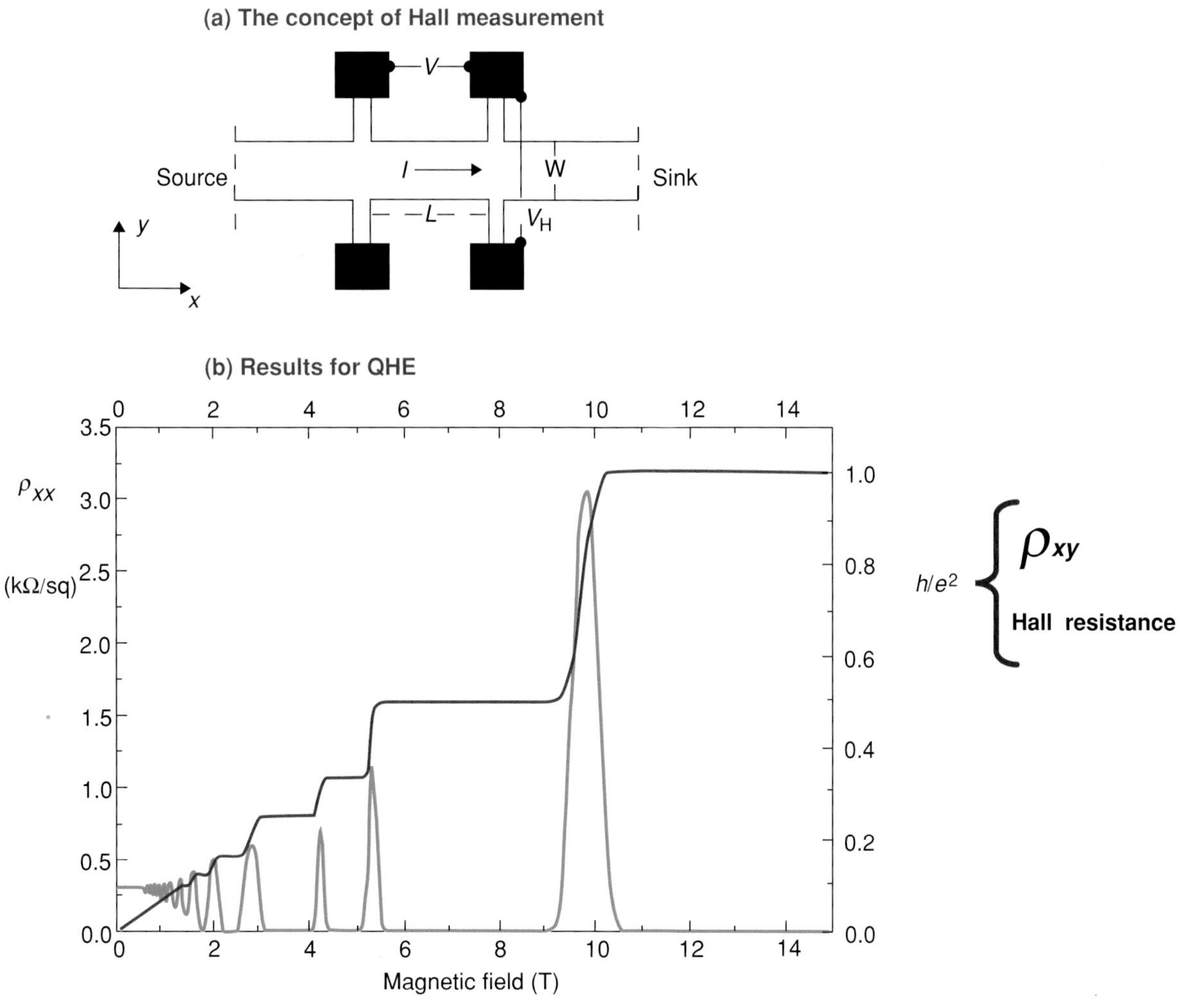

**Figure 12.6.** (a) A schematic diagram of an experiment to study the quantum Hall effect in which the transverse (Hall) conductance of a two-dimensional electron gas shows surprising behavior – see part (b). (From von Klitzing *et al.* (1982).) The Hall conductance is the ratio between the current flowing in a certain direction and the voltage applied perpendicular to that direction. In classical physics, this simply yields the carrier density in the material. With reasonably low temperatures and a strong magnetic field perpendicular to the electron gas, the dependence of the Hall conductance on parameters such as the carrier density and the magnetic field becomes very different, giving plateaus (red line). The green line shows the longitudinal resistance of the device. The magnetic field is applied in the $z$ direction.

## 12.3 Quantization of conductances

In 1982, it was discovered experimentally by von Klitzing, Pepper, and Dorda that a quantity called "the Hall conductance" of a good-quality 2D electron gas can have a surprising quantum-mechanical behavior. That conductance is the ratio between the current flowing in a certain direction and the voltage applied perpendicular to that direction, and when a magnetic field is applied perpendicularly to both (see Figure 12.6(a)). In classical physics, this conductance simply yields the carrier density in the material (this has become a standard tool in semiconductor science). With reasonably low temperatures and a strong magnetic field perpendicular to the 2D electron gas, the dependence of the Hall conductance on parameters such as the carrier density and the magnetic field itself becomes very different (Figure 12.6(b)). It exhibits plateaus on which its value is given to a very good approximation by integer multiples of a novel "quantum conductance unit":

$$G_Q = e^2/h, \tag{12.1}$$

where $e$ is the charge of the electron and $h$ is Planck's constant ($\approx 6.6 \times 10^{-27}$ cgs units), whose appearance implies that we are dealing with a fundamentally quantum phenomenon.

It is instructive to understand how this quantum conductance unit comes about. In classical electromagnetism, there is a natural conductance unit, the inverse of the impedance of free space. In cgs units (which are convenient to interpret basic physics results), that impedance is the inverse of the velocity of light, $c$. In engineering units, it is approximately given by 30 ohms ($\Omega$). The ability of an antenna to radiate depends on the magnitude of its impedance, compared with the above unit. In quantum physics, there is an interesting, dimensionless, pure number, the "fine-structure constant" $\alpha = 2\pi e^2/(hc) \approx 1/137.04$. It represents the strength of the coupling of the charged particle to the electromagnetic field. On dividing the impedance $1/c$ by $\alpha$, one obtains $G_Q$ and finds that it is roughly equal to 25 k$\Omega$. This determination is so accurate that it is used for a resistance standard and for determining the fine-structure constant $\alpha$. This discovery, which started a new chapter in solid-state physics and led to many further surprising discoveries, provided an important additional impetus for mesoscopic physics.

Following the discovery of the quantum Hall effect, it became clear that electrical quantities such as conductances can display discrete "quantized" values, the basic quantum unit being just $G_Q$. There is a deep and very useful relationship, due to Landauer, between ordinary conductance in the quantum regime and electron transmission. According to this relation, in the simplest case, the conductance of a small obstacle inserted in a one-dimensional line is basically the above quantum conductance unit multiplied by the quantum-mechanical transmission coefficient of the obstacle.

A quantum particle has a finite probability of entering classically inaccessible regions, but, on the other hand, may be reflected from small obstacles that are classically fully penetrable. From this, one can predict the conductance of very small orifices connecting large electron reservoirs. This conductance, in units of $G_Q$, is limited by the "number of quantum channels" (transverse states below the Fermi energy, which gives in a sense the number of parallel lines equivalent to an ideal finite-cross-section wire). In special cases, such as a ballistic (pure) orifice with a tapered shape, the integer values can be approached, because the relevant transmission coefficients can approach unity.

This has been found experimentally (Figure 12.7(a)), for carefully prepared "quantum point contacts" (QPCs) in the 2D electron gas at low temperatures. For a 2D electron gas, the QPC is created by depositing, using lithographic techniques, two rather narrow metallic "gates" that almost touch each other at their tips. A negative bias depletes the electron gas beneath the gates, leaving a small opening that serves as the orifice and is controlled by the gate voltage. Figure 12.7(a) shows the original results presented by Van Wees *et al.* (which were obtained concurrently by Wharam *et al.*) in 1988. This figure shows the conductance of such a QPC as a function of the gate voltage. When the latter becomes less negative, the orifice opens up and its conductance exhibits a tendency to have plateaus at integer multiples of $e^2/h$, as shown in Figure 12.7(a). More recently, the conductance-quantization effect has been discovered to be of relevance in various naturally forming atomic-scale contacts at room temperatures. Figure 12.7(b)

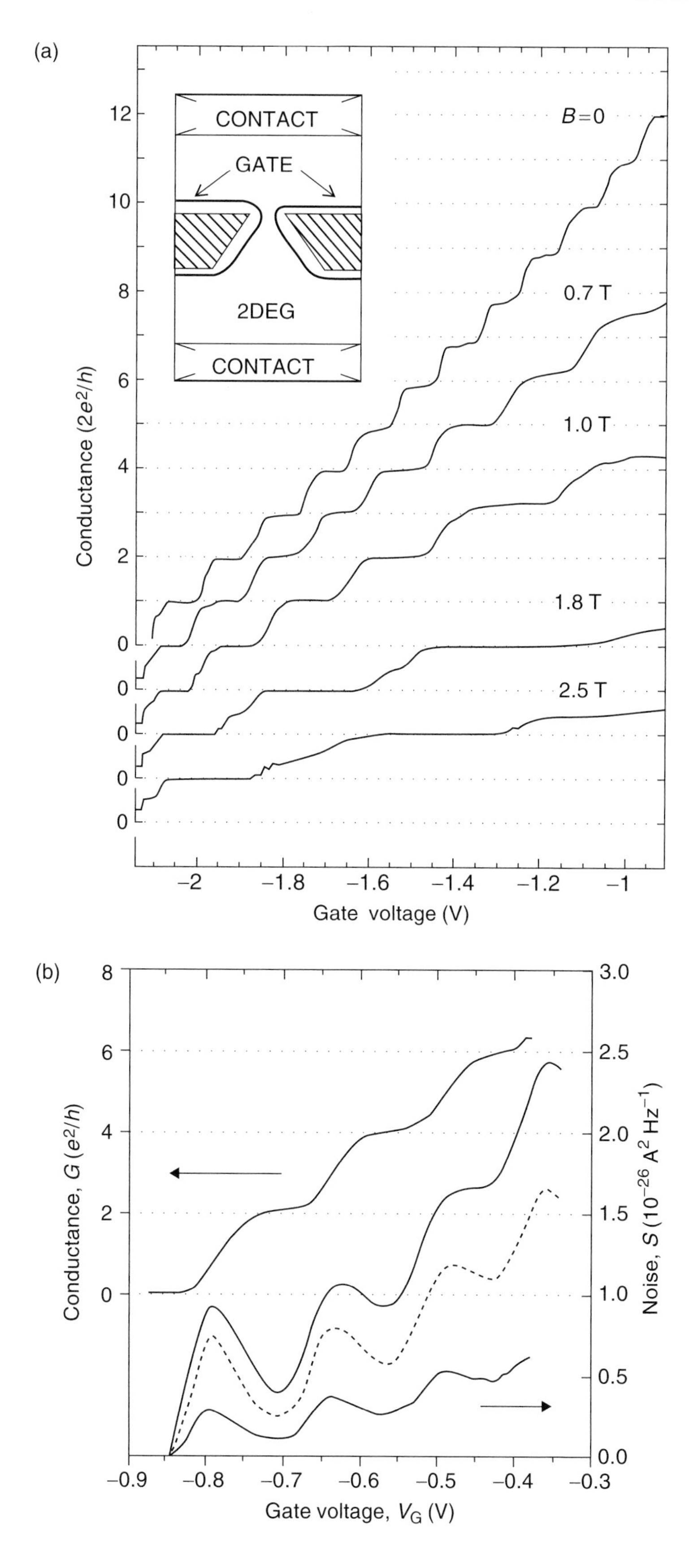


**Figure 12.7.** (a) In special cases, integer values of the ordinary (longitudinal) conductance can be approached. This has been found experimentally for carefully prepared quantum point contacts in a two-dimensional electron gas at low temperatures (0.6 K). This diagram (the original results obtained by Van Wees *et al.* in 1988) shows the quantization of the quantum point-contact conductance versus the gate voltage, for several magnetic fields at 0.6 K. (b) These results obtained in an experiment by Reznikov *et al.* show the conductance of a quantum point contact in a two-dimensional electron gas as a function of the gate voltage, and the spectral density of the noise power due to the current through the device. This non-equilibrium "shot noise" due to the discreteness of the electronic charge was predicted to peak around the transitions among the plateaus, as demonstrated by the results. The scale on the right applies to the lower curves, giving the shot-noise power at various transport voltages.

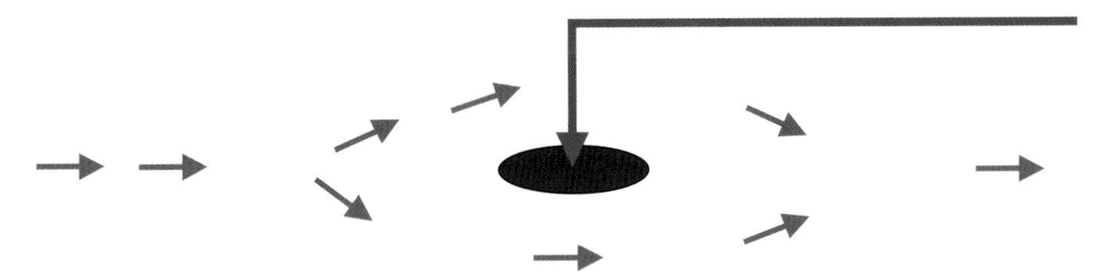


**Figure 12.8.** The principle of a mesoscopic Aharonov–Bohm interferometer. The wave-particle enters from the left and splits into two paths, which encircle the magnetic field (thick blue line) and subsequently combine. For small voltages, the current is proportional to the voltage and depends periodically on the flux. This happens even if the magnetic field vanishes, or its effect is negligible, in the wires confining the electrons.

shows an experiment by Reznikov *et al.*, which was performed using the QPC as a tool. This study also obtained the spectral density of the noise power due to the current through the device. This non-equilibrium "shot noise," which is due to the discreteness of the electronic charge, was predicted to peak around the transitions between consecutive plateaus, as demonstrated by the results. The shot noise is an interesting manifestation of the discrete-particle nature of the electron, although it obviously also plays a role when the electron's wavelike properties are dominant, such as in tunneling through a barrier. The sensitivity of the shot noise to the statistics and correlations of the particles makes it an interesting subject for further study.

## 12.4 Aharonov–Bohm conductance oscillations

The Landauer formulation enables one to consider the conductance of a mesoscopic ring as a function of an Aharonov–Bohm (AB) flux, $\Phi$, through its opening (Figure 12.8). The conductance is predicted to depend periodically on the flux, $\Phi$, with a period

$$\Phi_0 = hc/e \tag{12.2}$$

in agreement with experiment (Figure 12.9 shows the first experiment, performed in 1985 in a gold ring – some of the many subsequent experiments will be mentioned later) and with general theorems stating that $\Phi$ is equivalent to a phase change of

$$2\pi\Phi/\Phi_0. \tag{12.3}$$

This equivalence is the basis of the AB effect in quantum mechanics. Magnetic fields can even influence physics far from where they are applied, due to the description of the magnetic field in terms of a "vector potential," which can be nonzero even if the field itself vanishes. It is this vector potential which appears in the quantum-mechanical equations of motion. The AB effect, although perhaps counter-intuitive (like many other aspects of quantum physics), is now well established.

The conductance of macroscopically long cylinders with mesoscopic radii was found experimentally to have a period of $\Phi_0/2$, one-half that of Equation (12.2). This factor of two was found to be due to the difference between sample-specific and impurity/ensemble-averaged behavior. The latter implies averaging over many

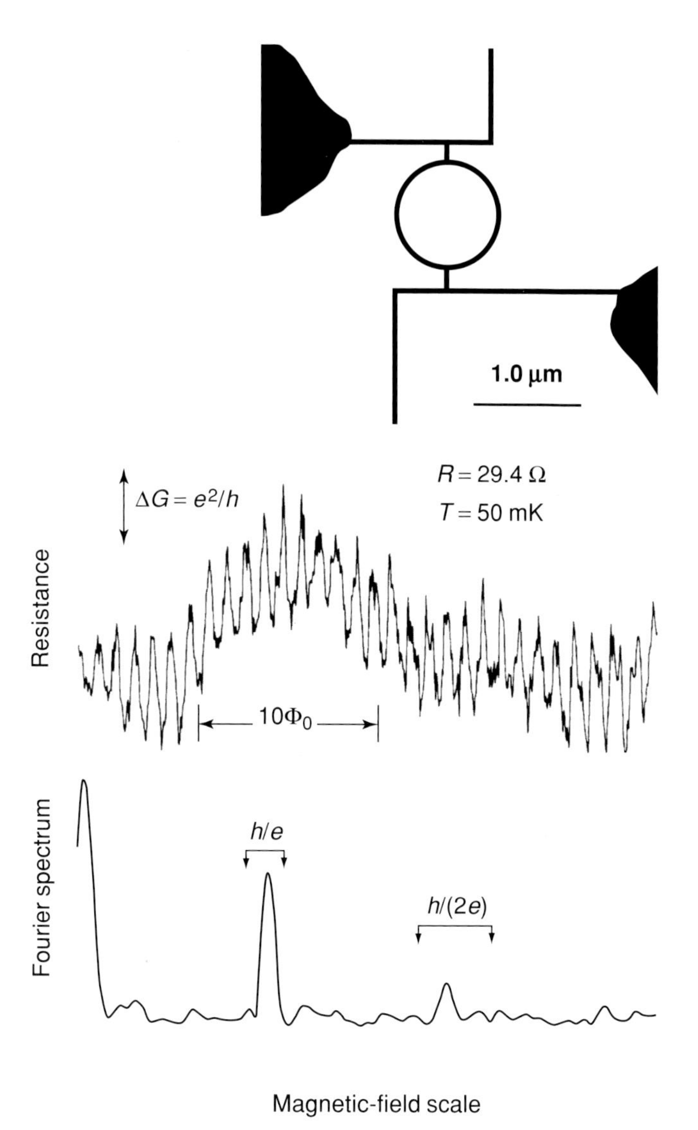


**Figure 12.9.** A schematic diagram of an experiment to demonstrate the Aharonov–Bohm effect in metals and the results obtained. The top curve shows the resistance of the device as a function of the magnetic field. The oscillation period corresponds to a flux $h/e$ inside the ring.

samples, all having the same macroscopic parameters, but differing in the detailed placement of impurities and defects. The long cylinder consists of many such samples, whose resistances are added incoherently. Hence the total resistance displays the ensemble-averaged behavior, in which the fundamental $\Phi_0$ period is averaged out. This insight opened the way to a consideration of "mesoscopic" (sample-dependent) fluctuations in the conductance (and possibly in other physical properties as well).

## 12.5 Persistent currents in rings and cylinders

When the mesoscopic ring is isolated from its leads (Figure 12.10), the AB flux through it continues to play a role. For free electrons, it is straightforward to see that the energy levels will depend on the flux, as will the total energy (and the free energy, which is the relevant quantity at finite temperatures). Since the thermodynamic equilibrium current is the derivative of the free energy with respect to the flux, it follows that an equilibrium circulating current can flow in the ring and will be periodic in the flux, with a period $\Phi_0$. For this to happen, the electrons must remain coherent (see below) while going at least

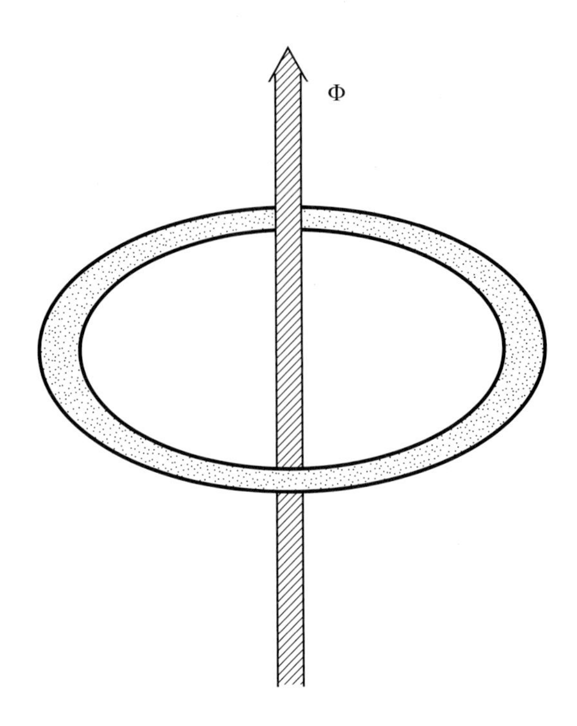


**Figure 12.10.** When a mesoscopic ring is isolated from its leads, the Aharonov–Bohm flux through it continues to play a role, and persistent equilibrium currents are possible. Since the thermodynamic equilibrium current is the derivative of the free energy with respect to the flux, it follows that an equilibrium circulating current can flow in the ring and will be periodic in the flux.

once around the ring. (In the classical case, when there are no coherence effects, the free energy does not depend on the flux.) It was thought that scattering of the electrons by defects, imperfect edges, etc. would cause this current to dissipate and make the equilibrium current vanish in real systems. However, one of the fundamental quantum-mechanical notions that mesoscopic physics helped to clarify is that elastic scattering by defects does **not** eliminate such coherence effects. The impurities present (random) static potentials to the electron. Well-defined coherent wave functions prevail, and the energy levels are sharp and flux-dependent, as long as the disorder is not strong enough to cause "localization."[1] The resulting "persistent currents" never decay. This initially surprising fact was also demonstrated experimentally (Figure 12.11). These currents can even be larger than those of non-interacting electrons. Many-body theory is necessary to explain their magnitudes and some questions remain unsolved.

The fact that scattering from static impurity does not destroy phase coherence is obviously what enables the AB oscillations in the conductance to happen. Elastic scattering does not cause phase incoherence – only inelastic scattering can,[2] as will be discussed in some detail later. Since the strength of the inelastic scattering decreases at low temperatures, one may reach the regime where electrons stay coherent over the whole sample. This is an important condition for flux sensitivity and for persistent currents to exist. For many experimental configurations, this means that sizes in the micron range are useful at temperatures of a few Kelvin. In the nanometer size range, quantum effects should be observable at higher temperatures, even including room temperature. Later we will consider the loss of coherence of an electron wave due to the interaction with other degrees of freedom.

[1] This means that the disorder is strong enough for the electrons to become trapped in finite regions of space. Such trapping is actually easier to achieve in quantum mechanics than in the classical case, due to interference effects, as discovered in 1958 by P. W. Anderson. This concept is often relevant for the physics of real, impure, conductors.

[2] As will become clear, here the term "inelastic" implies simply changing the quantum state of the environment. It is irrelevant how much energy is transferred. This also includes zero energy transfer – flipping the environment to a state degenerate with the initial one.

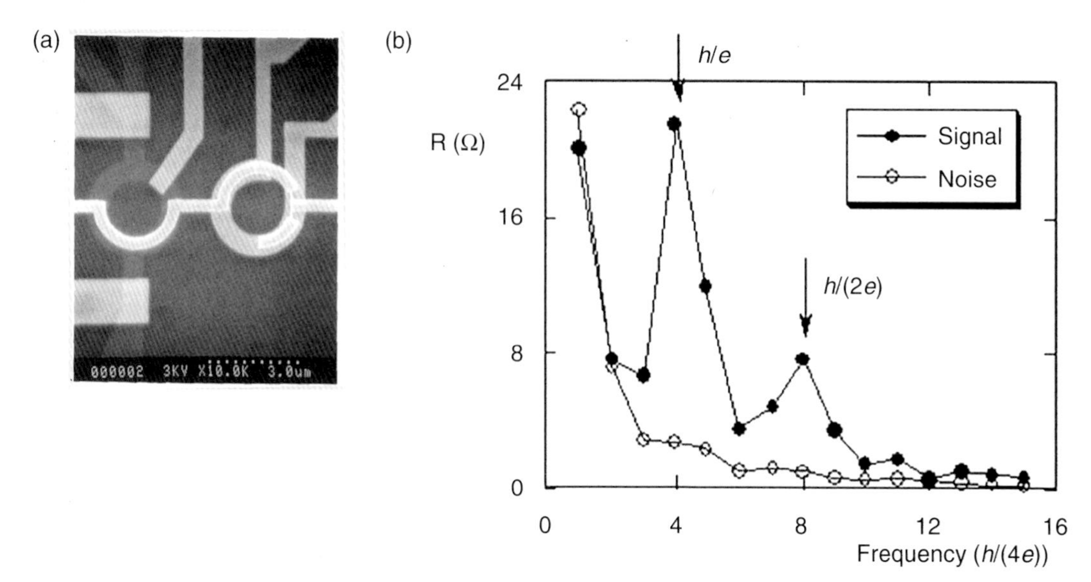


**Figure 12.11.** (a) Aharonov–Bohm effects can produce persistent currents in tiny rings, despite the presence of defects and impurities. This shows ring structures built with high-quality semiconductor microtechnology, using gates to connect and disconnect the ring to electrical contacts. An on-chip superconducting quantum interference device (SQUID) was used to measure the magnetic moment produced by the current. (b) The graph shows the Fourier components of the current versus flux, peaking at $h/e$ and its harmonics, demonstrating the existence of persistent currents.

## 12.6 Coulomb blockade

The so-called "Coulomb blockade" is an interesting and potentially useful phenomenon based on straightforward electrostatic considerations. Suppose that an isolated small piece of matter (a "quantum dot") is stripped of its conduction electrons. It has then a positive charge $Ne$, where $N$ is an integer – the total charge of the positive ions. Obviously this large positive super-ion will usually lower its total electrostatic energy by binding exactly $N$ electrons – thus achieving charge neutrality. If one insisted on adding (or removing) $n$ extra electrons to (or from) this neutral quantum dot, the energy cost would be $(ne)^2/(2C)$, where $C$ is the total capacitance of the dot, including its self-capacitance and (half) its mutual capacitance with the environment. These are determined by the geometry and the effective dielectric constant $\varepsilon$. For a dot size in the nanometer range, this energy is typically in the electron-volt range (roughly $10^4$ degrees), much larger than all typical energies in an experiment when one tries to pass current through the dot (which implies charging it temporarily). Therefore, when the dot bonds two wires, the conductance between them will effectively vanish, a situation described as "Coulomb blockade."

If one now changes the electrostatic potential of the dot, by connecting it to a (continuous) voltage source, or to an electrostatic "gate," sometimes dubbed a "plunger" (see Figures 12.12 and 12.14), then, when this gate voltage is varied continuously, it reaches special values where the optimal $N$ is half integer, and the lowest energy states of the dot, neglecting changes in the single-particle energy differences, are the integers $N - \frac{1}{2}$ and $N + \frac{1}{2}$. The conductance of the dot is now finite, since the vanishing of the above energy differences makes it easy to add or subtract an electron to or from the dot, which is a necessary step for the conduction process. The latter situation can be described as suppressing the Coulomb blockade by bringing the "dressed" resonance through the dot level to the Fermi level of the leads. The linear conductance of the dot can approach $G_Q$ (for symmetric couplings of the dot to the two leads). As a function of the gate voltage, the dot's linear conductance will exhibit peaks on a very small background. The width of these peaks is given by the width of the dot states, which are induced by the coupling to the leads. This is shown in Figure 12.12. Also shown are the total charge on the dot and its electrostatic potential as functions of the gate voltage. The former exhibits plateaus with integer values of the charge. These plateaus,

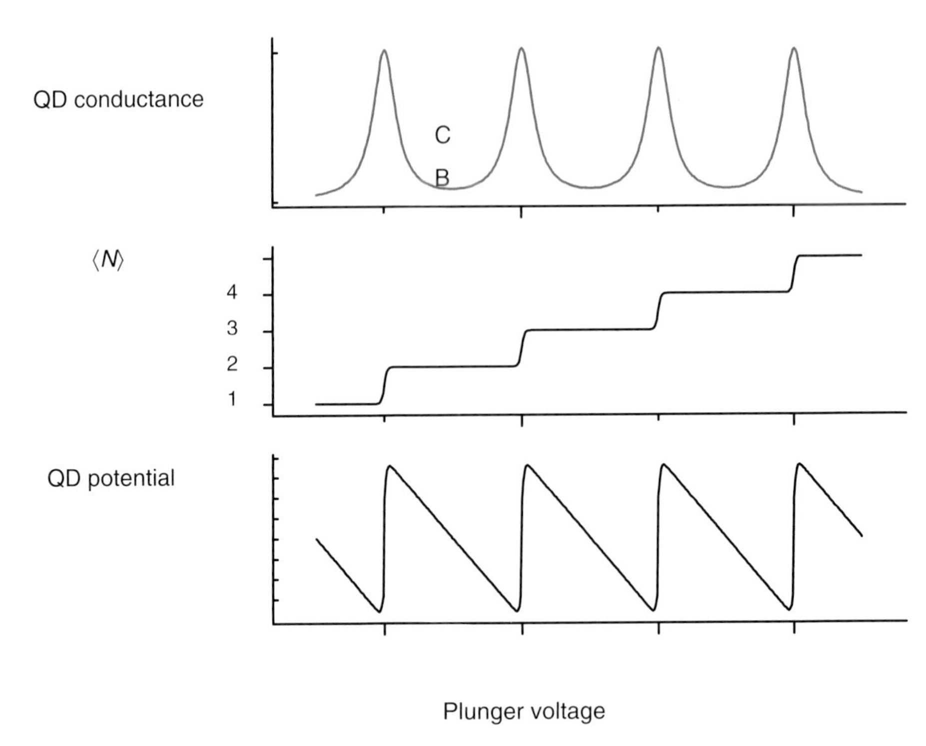


**Figure 12.12.** The phenomenon of "Coulomb blockade" as observed by Buks *et al.* in the context of controlled dephasing (Figures 12.15 and 12.16), for a "quantum dot" (QD) bridging two conducting lines. The "plunger voltage" changes the potential energy of the dot. When its discrete quantum levels are brought one by one into resonance with the Fermi levels of the lines, a peak of the differential conductance with height approaching the quantum conductance $G_Q$ follows (top graph) and the total number of electrons on the QD increases by one (middle graph). The QD electrostatic potential (lower graph) is steadily and linearly decreased, but jumps by $(e^2/2C)$ across each charging event (where $C$ is the effective capacitance of the quantum dot). The term "Coulomb staircase" (middle graph) describes the total charge on the dot plotted against the source–drain (transport) voltage (here shown versus the plunger voltage).

and similar ones for the total current as a function of the source–drain voltage for a given gate voltage, are termed the "Coulomb staircase."

The sharp change with gate voltage of the current near the peak resembles that seen in a transistor, and it can indeed be used to construct a field-effect-transistor (FET) amplifier. Figure 12.13 shows a FET constructed using a nanotube, employing a similar principle. The FET amplifiers or switches extensively used in many everyday applications (including computer elements) are microelectronic devices using similar approaches. The novelty here is in employing single electrons, hence the term "SET" – single-electron transistor.

## 12.7 Decoherence – dephasing of quantum interference

The dephasing of quantum interference ("decoherence") is well understood as being due to the coupling of the measured, interfering, wave-particle with other degrees of freedom ("the environment"). The interference is due to there being at least two paths that traverse the system and interact with the environment, recombining at the

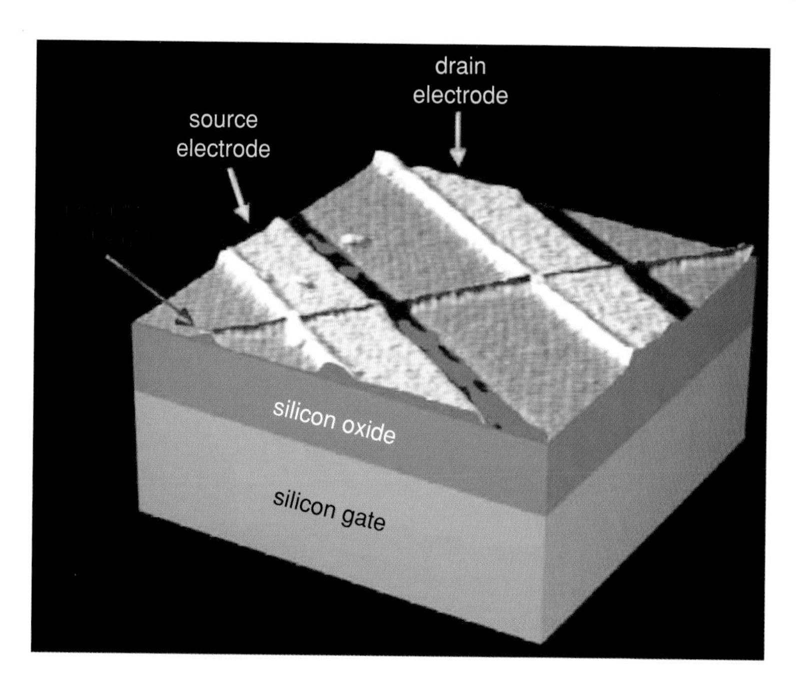


**Figure 12.13.** A one-dimensional transistor based on a nanotube bridging two electrodes and gated by the doped silicon below by C. Dekker *et al.*

measurement point after the interaction with the environment has been switched off. Dephasing occurs when one can tell which of the two paths the wave followed. This is possible if and only if by following one of them the wave leaves a so-called "which-path" trace in the environment by flipping it to a quantum state distinct from that left by the other path.

An alternative description is that the dynamical fluctuations of the environment can cause an uncertainty in the relative phase of the two waves, which, once it becomes large enough, will eliminate the interference. These two seemingly different descriptions can be shown to be precisely equivalent. This is due to the interactions between the wave and the environment, and between the environment and the wave, being identical (as in Newton's third law) and to a fundamental relationship between the fluctuations of any system and its excitability. Clearly the vibrations of a soft spring are easier to excite than those of a stiff one, and the former has larger fluctuations – a floppy system is both noisier and easier to excite than a stiff one. These broad statements can be formulated into a general relationship, "the fluctuation–dissipation theorem" (FDT), which can be proved for equilibrium systems, and also holds (as has been found via research in mesoscopic physics) for a large class of non-equilibrium, steady-state ones. It guarantees that, just when a partial wave has left a trace in the environment (compared with another partial wave), the fluctuations of the environment induce a large enough uncertainty in the relative phase of the two waves to wash away the interference.

Useful theoretical relationships based on the above predict how the rate of loss of phase coherence for an electron in a real piece of metal depends, for example, on the temperature. These predictions are fully confirmed by experiment. Except for very special situations, this rate vanishes at the zero-temperature limit, because under these conditions no excitation can be exchanged between the particle and the environment. This follows from the principles of thermodynamics and is consistent with the general statement that purely "elastic" scattering of an electron by lattice defects retains phase coherence. This is one of the important physical principles that mesoscopic physics has helped to reinforce. A corollary is that the "decoherence rate" increases with temperature. Therefore, when a sample becomes smaller, the electrons will stay coherent over its whole length at higher temperatures. The numbers are such that in most conductors the sample size over which the electrons will be coherent at ambient "room" temperature is of the order of tens of ångström units. Thus, nanoscale systems are very special in that electrons moving in them should fully obey quantum rules. This point, already made earlier, emphasizes the importance of such systems.

This also leads to a number of general observations.

- The phase uncertainty remains constant when the interfering wave does not interact with the environment. Thus, if a trace is left by a partial wave in its environment, this trace cannot be wiped out after the interaction is over. The proof follows from a quantum-mechanical property called "unitarity," which is related to the total probability of finding the given particle anywhere always being 100%.
- If the *same* environment interacts with the two interfering waves and both waves emit *the same* excitation of the medium, each of the partial waves' phases becomes uncertain, but the relative phase is unchanged. This is due to the fact that the two emerging states of the environment are identical. A well-known example is "coherent inelastic neutron scattering" in crystals, which is now a routine method for measuring the dispersion of many collective excitations (lattice vibrations, spin waves, . . .). This is due to the coherent addition of the quantum amplitudes for the neutron exchanging *the same* excitation with *all* scatterers in the crystal.
- Excitations (phonons, photons, etc.) with wavelengths larger than the spatial separation of the interfering paths do not dephase the interference. It might be thought that this is because of their energy being too low to cause dephasing. However, rather than the amount of energy transferred, the ability of the excitation produced to identify the path is the key. This is physically related to the fact that the excitation must influence the relative phase of the paths. The situation is similar to that of the Heisenberg microscope, whose resolving power is wavelength-limited: the radiation with wavelength $\lambda$ cannot resolve the two paths well if their separation is smaller than or on the order of $\lambda$.
- Dephasing may occur by coupling to a discrete (finite) or to a continuous (effectively infinite) environment. In the latter case, the excitation may become infinitely remote and the loss of phase can be regarded in practice as irreversible. However, in special cases it is possible, even in the continuum case, to have a finite probability of reabsorbing the created excitation and thus retaining coherence. This happens in some special situations, for example for the Hall effect in insulators.

One extremely instructive experiment can be carried out in a sophisticated mesoscopic system (Figure 12.14). In this experiment, the "which-path" information gradually accumulated with time. Dephasing should then have occurred once that information became reliable, as will be explained. This experiment also demonstrated how the dephasing rate (defined as $1/\tau_\phi$, where $\tau_\phi$ is the dephasing time) can be controlled by adjusting experimental parameters. The QPC, alluded to earlier, can be a sensitive indicator when biased to be on the transition between two quantized conductance plateaus. There, it is rather sensitive to small changes in parameters, for example in the electrostatic field nearby. This sensitivity may be used to detect the presence of an electron in one of the arms of an interferometer, provided that the two arms are placed asymmetrically with respect to the QPC. Measurements confirming "which-path" detection by the QPC were performed by Buks *et al.* at the Weizmann Institute. The AB oscillations were measured in a ring interferometer. On one of its arms the transmission was limited by a "quantum dot" where the electron wave would resonate

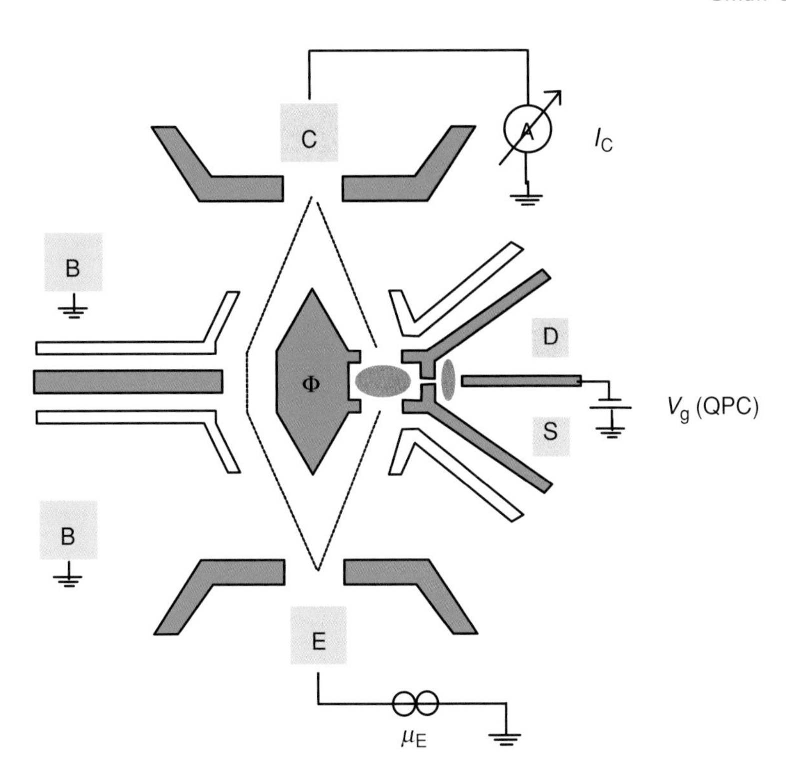


**Figure 12.14.** A schematic diagram of the apparatus used to demonstrate controlled dephasing of quantum interference.

for a relatively long and controllable (to a degree) "dwell time" $\tau_D$. A QPC was placed near that arm and the degree of dephasing due to it (determined by the ratio $\tau_\phi/\tau_D$) could be inferred from the strength of the AB conductance oscillations. Figure 12.15 shows the results.

Clearly, a necessary and sufficient condition for dephasing is $\tau_\phi < \tau_D$. The experimental results and their dependences on several experimental parameters were in good agreement with a qualitative theory developed by the authors (Figure 12.15). The underlying idea was that $\tau_\phi$ is the time over which the "which-path" information becomes reliable. The condition for that was that the signal accumulated in the measurement has to be larger than the corresponding noise. The former is the difference in the QPC current due to the electron to be detected, accumulated over the time $\tau_\phi$. The latter is basically the effect of the shot noise (Figure 12.7(b)) due to the same measurement current, over the same time. Subsequently several more sophisticated systematic theoretical calculations have reproduced the same results.

The new interesting feature of this non-equilibrium dephasing is that a finite current is flowing in the detector and, with increasing time, each electron transmitted adds its contribution to the decrease of the overlap of the two states of the environment produced by the two partial waves. The dephasing time $\tau_\phi$ is the time for which these two states become close enough to being orthogonal (no overlap). As discussed above, the reduction in overlap is conserved when further thermalization of the transferred electrons in the downstream contact occurs. This is ensured by unitarity, as long as no further interaction with the interfering electron takes place. The alternative picture is that non-equilibrium (shot-noise) fluctuations of the current in the QPC create a phase uncertainty for the electron in the quantum dot. While the equivalence of these two pictures is guaranteed by the discussion above, it is interesting and nontrivial to see how it emerges in detail. This is all the more interesting since an older argument according to which the current fluctuations in the QPC cause dephasing seems superficially to contradict the idea that dephasing needs to overcome the

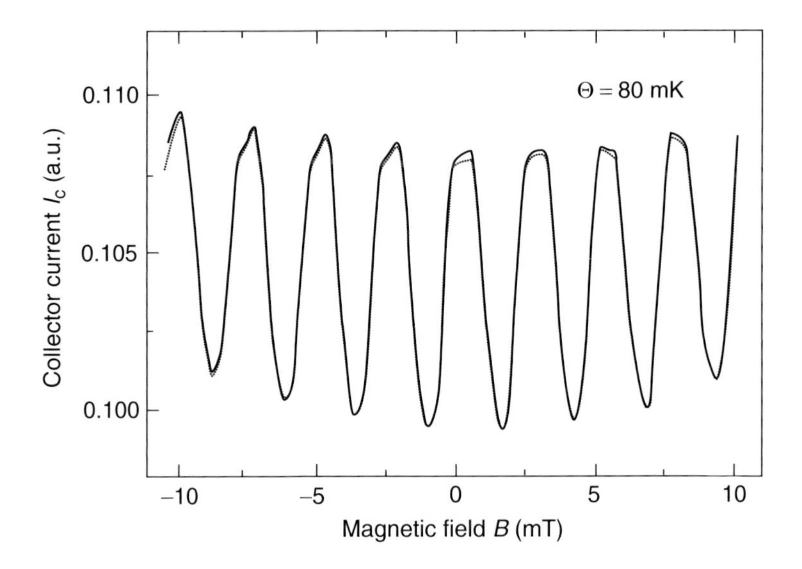


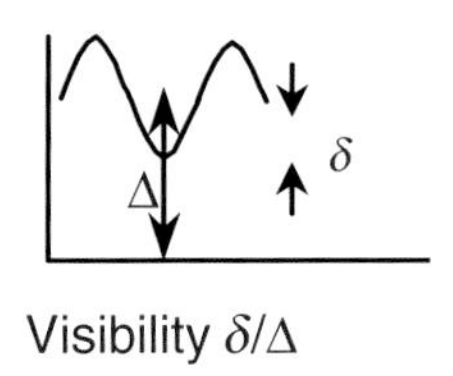


**Figure 12.15.** Controlled dephasing of Aharonov–Bohm oscillations (Buks *et al.*). The oscillations are seen to decrease when the current through the adjacent quantum point contact increases. Below – the relative magnitude of the oscillations ($\delta/\Delta$) is reduced when the quantum point-contact current increases.

shot-noise fluctuations, which therefore could be thought to oppose dephasing. However, all these seemingly conflicting pictures produce, as they must, identical results.

## 12.8 Examples of nanosystems

An important trend in mesoscopic physics has been and will continue to be toward smaller and smaller sizes, the nanoscales. For smaller systems, both the quantum and the Coulomb effects will occur at higher temperatures, as has already been mentioned. In addition, unique new possibilities for creating various types of smaller systems begin to open up. A nanoscale transistor-type structure is shown in Figure 12.13.

A very pertinent example is provided by small quantum dots. Besides the fact that the Coulomb charging energy becomes larger with decreasing size, enabling easier observation, for example, of the "Coulomb staircase" (Figure 12.12) and single-electron transistor action, qualitatively new phenomena become possible. An example is a beautiful demonstration of the Kondo effect in small semiconducting quantum dots containing on the order of tens of electrons. In bulk systems, the Kondo effect follows from the tendency of an unpaired impurity spin to form a singlet state with the conduction electrons near the Fermi level. As a result, a resonance is formed at or very near the Fermi energy. This creates additional electron scattering and resistance at very low temperatures, where the scale is the "Kondo temperature," $T_K$.

For a quantum dot (which plays the role of the "impurity") connected to leads, the same Kondo resonance yields an extra conductance, which approaches the quantum limit $2G_Q$ ("the unitarity limit" – the factor 2 is due to the two spin directions of the electron). Since this resonance is always at the Fermi energy, it leads to a much

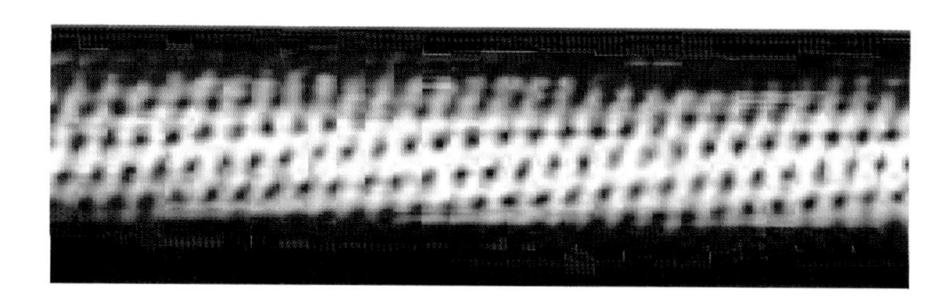

**Figure 12.16.** A carbon nanotube. (Courtesy of C. Dekker *et al.*).

enhanced conductance in the whole of the relevant Coulomb valley. This dramatically increased conductance with decreasing temperature between the Coulomb blockade peaks has been observed experimentally. It in fact approached the unitarity limit. In subsequent experiments $T_K$ was confirmed to increase markedly with decreasing dot size, and smaller dots are needed to bring it to a measurable regime. The influences of various control parameters that can decrease the effect, other than the temperature, such as a smaller coupling to the leads, a larger transport voltage, and a magnetic field, were examined and found to be in agreement with expectations. Another interesting development is due to the fact that in mesoscopic systems it is possible to measure the phase of the electronic transmission amplitude via, for example, AB-type interference.

Another example of a strongly correlated electronic system, with nontrivial properties, occurs in single-channel one-dimensional conductors. This can be achieved in semiconductors by cleaved-edge overgrowth with MBE. Another interesting example is provided by certain carbon nanotubes, discussed later.

Some recently discovered organic systems hold the promise of nanometer-scale molecular conductors or two-dimensional electronic systems of interest to both physics and technology. One example is provided by various structures based on the graphite phase of carbon, such as fullerene molecules and carbon nanotubes (Figures 12.13, 12.16, and 12.17). Despite these initial difficulties, such systems are still very interesting. For example, in layers of coupled fullerene molecules, appropriate doping is possible and leads to large 2D electron (or hole) densities. This can make them conducting, or even superconducting, at relatively high temperatures.

Of special fundamental interest are single-wall nanotubes obtained, in principle, by rolling a layer of two-dimensional graphite, called graphene, into a hollow cylinder parallel to the $z$ axis and having a nanometer-scale diameter (Figure 12.16). Graphene is a poor conductor, having a very small effective density of states at the Fermi energy. Theoretical predictions of when these nanotubes become metallic or semiconducting depend on the details of how exactly the graphene sheet is wound and connected to itself. The prediction is that all "armchair nanotubes" (Figure 12.17 – in which the $z$ direction is parallel to one of the hexagon sides) and one-third of the "zigzag nanotubes" (in which the $z$ direction is perpendicular to one of the hexagon sides) should be metallic. These ideas have received ample experimental support both from transport and from scanning-tunneling-probe measurements. Conducting nanotubes offer a unique possibility of realizing a purely one-dimensional conductor as well as a superconductor. Transistor-type structures employing nanotubes bridging two electrodes have been demonstrated at Delft and IBM – Figure 12.13 displays a gated nanotube transistor from IBM.

Such nanoscale systems offer great promise both for fundamental research and for applications. Ideas to employ DNA molecules as templates for electronic circuits are also of great interest. Such networks have been created and the challenge is how to endow them with (easily controllable) conductivity.

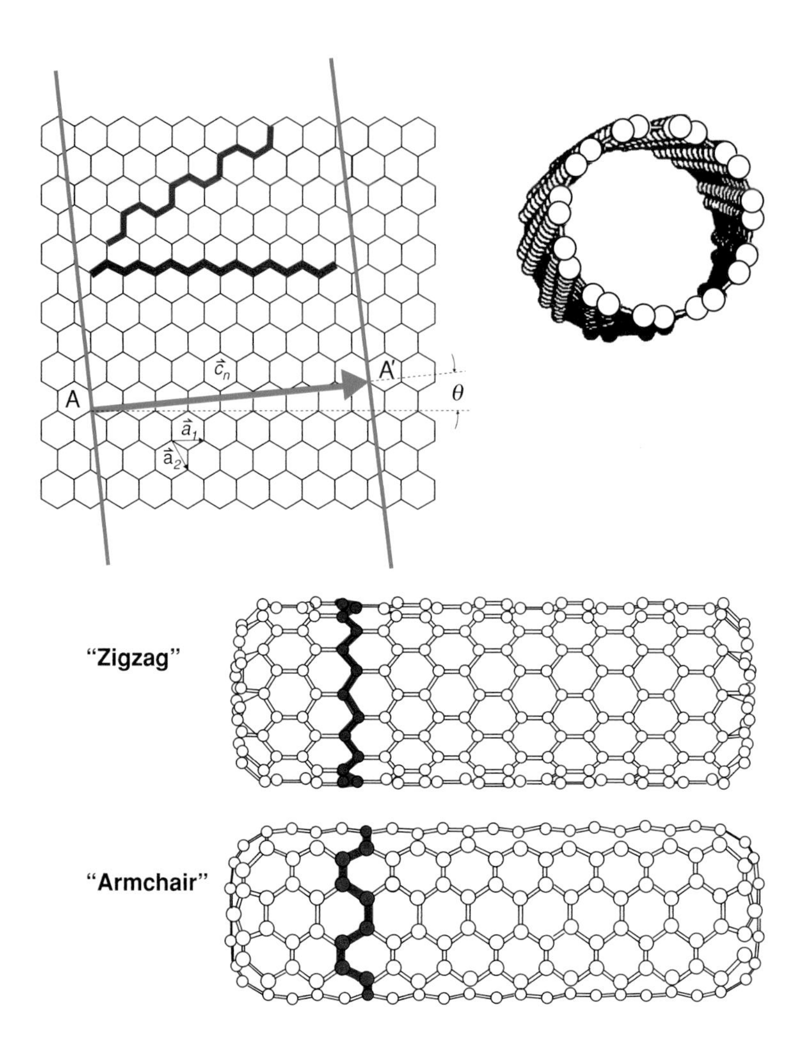


**Figure 12.17.** Various types of carbon nanotubes can be obtained by rolling and gluing a graphene sheet.

## 12.9 Conclusions

The uniqueness of the science of mesoscale and nanoscale systems lies in combining the possibility of addressing fundamental physics issues with advances in material science leading to practical applications. An additional example for the power of the experimental methods is shown in Figure 12.18 – the intriguing observation by STM techniques of a shadow, or "mirage," produced by a Kondo impurity in an appropriate man-made elliptical atomic geometry.

One class of nanosystems not discussed in this chapter is that which uses atomic traps and modern optical cooling techniques of quantum optics. Combinations of a small number of photons and possibly of atoms, for example in an optical cavity, are also possible. This is covered in Chapter 7.

The Coulomb blockade discussed earlier is only a simple approximate way to incorporate the all-important electron–electron interactions (which can typically be of the same order of magnitude as, or even larger than, the kinetic energy). More subtle electron-correlation effects are now at the forefront of current research in this field. The Kondo effect alluded to above, superconductivity, and the anomalies of lower-dimension systems all come into this category.

**Figure 12.18.** Some unusual phenomena are due to the wave nature of electrons and their correlations around impurities. This shows images of man-made elliptical arrangements of atoms on a metallic surface, both prepared and visualized using a scanning tunneling microscope, Figure 12.2. In addition to the boundaries of the atomic "corral" a specific magnetic impurity was deposited at desired points inside the latter. Experiments reveal some unusual phenomena due to the wave nature of the electrons and their correlations around the impurity. Placing the impurity just at a focal point of the ellipse created a shadow (a "Kondo mirage" – D. Eigler *et al.*, IBM Almaden) in the other focus. These examples and others can be found at http://www.almaden.ibm.com/almaden/media/image_mirage.html.

However, many fundamental specific unresolved problems remain. Despite tremendous advances, the problem of the very narrow (one-dimensional, molecular-thickness) conductor is not yet fully understood, due to the interplay of disorder and interactions. This is also true of the purely one-dimensional superconductor, which should be sensitive to quantum fluctuations. A small enough grain or cluster should not be a superconductor, but opinions differ as to how exactly to define experimentally whether a small particle is superconducting.

Opening up a smaller length scale makes new physical phenomena possible, and nanoscales should continue this tradition. The frontiers between the microscopic and macroscopic realms and between natural and man-made molecules are already being explored. An ambitious long-range goal would be to make the connection to biological systems, which in a sense are functional combinations of mesoscale units with the ability to replicate.

## Acknowledgements

A. Aharony, H. Bouchiat, E. Braun, O. Cheshnovsky, C. Dekker, D. Eigler, O. Entin-Wohlman, U. Gavish, M. Heiblum, D. Mahalu, M. Roukes, U. Sivan, A. Stern, and A. Yacoby are thanked for help and discussions. This contribution was submitted in 2002.

## FURTHER READING

1. Y. Imry, *Introduction to Mesoscopic Physics*, Oxford, Oxford University Press, 1997. A review of some aspects of nanoscience.
2. G. Binnig, H. Rohrer, Ch. Gerber, and E. Weibel, Surface studies by scanning tunneling microscopy, *Phys. Rev. Lett.* **49** (1997) 57.
3. J. M. Lehn, *Supramolecular Chemistry. Concepts and Perspectives*, Weinheim, VCH, 1995.
4. L. Esaki, in *Proceedings of the 17th International Conference on the Physics of Semiconductors*, San Francisco, August 1984, eds. J. D. Chadi and H. A. Harrison, New York, Springer, 1985.
5. A. C. Gossard, *IEEE J. Quantum Electron.* **QE-22** (1986) 1649.
6. L. J. Geerligs, C. J. P. M. Harmans, and L. P. Kouwenhoven (eds.), *The Physics of Few-Electron Nanostructures*, Amsterdam, North-Holland, 1993.
7. C. Dekker, Carbon nanotubes as molecular quantum wires, *Phys. Today*, May 1999, p. 22.
8. L. Salem, *The Molecular Orbital Theory of Conjugated Systems*, 3rd edn., Reading, MA, W. A. Benjamin, 1972.
9. M. S. Dresselhaus, G. Dresselhaus, and P. C. Eklund, *Science of Fullerenes and Carbon Nanotubes*, New York, Academic Press, 1996.
10. R. Saito, G. Dresselhaus, and M. S. Dresselhaus, *Physical Properties of Carbon Nanotubes*, London, Imperial College Press, 1998.
11. E. Braun, Y. Eichen, U. Sivan, and G. Ben-Yoseph, *Nature* **391** (1998) 775.

# Yoseph Imry

Yoseph Imry is the Max Planck Professor of Quantum Physics in the Department of Condensed Matter Physics, the Weizmann Institute of Science, Rehovot, Israel. After experimental M.Sc. work at the Hebrew University, Jerusalem, his Ph.D. work, concluded in 1966 and mostly theoretical, was performed at the Soreq Research Center, while studying at the Weizmann Institute. It addressed proton dynamics in hydrogen-bonded ferroelectrics and in water, employing what can be learned from neutron scattering. Later, and in postdoctoral work at Cornell, he concentrated on the theory of phase transitions and superfluidity. At an early stage, he began to study the special effects due to finite size on systems undergoing phase transitions. This was one of the precursors for mesoscopic physics, on which he embarked about ten years later.

After moving to Tel Aviv University in 1969, his work on phase transitions concentrated on scaling ideas and then on random disordered systems. In parallel, he worked on superconductivity and on Josephson physics, with its many nonlinear phenomena. In the late 1970s his interests shifted to disordered electronic systems, employing scaling ideas and continuing into mesoscopic and nanoscopic physics. He was one of the pioneers of these fields, being deeply involved in predicting some novel phenomena. These include conductance quantization, Aharonov–Bohm conductance oscillations, normal-metal persistent currents, and the relevance of sample-specificity. He studied electron decoherence in some depth and among his more recent interests are the theory of quantum and non-equilibrium noise, novel effects in quantum interferometry and electron transport, superconductivity in nanoparticles, and the connection to molecular-scale systems. Work on these subjects continued when he moved to the Weizmann Institute in 1986, where he initiated the Braun Center of Submicron Research. His *Introduction to Mesoscopic Physics* (Oxford University Press) is a standard textbook in the field. He has visited and worked in many research centers, including the University of California (Santa Barbara and San Diego), Brookhaven National Laboratory, the IBM Research Center, Yorktown Heights, New York, for which he was a Distinguished Science Advisor, Yale University, the University of Karlsruhe with a

Humboldt award, the University of Leiden, where he held the Lorentz Chair, and the Ecole Normale Supérieure in Paris, where he held the Pascal Chair.

He served for six years as the director of the Albert Einstein Minerva Center for Theoretical Physics and for three years as the director of the Maurice and Gabriella Goldschleger Center for Nanophysics, both at the Weizmann Institute. He is a member of the European Academy of Arts and Sciences and of the Israeli Academy of Sciences, and has received several awards, including the Weizmann, Rothschild, and Israel prizes. He is a special advisor on nanoscience to the President of Ben-Gurion University of the Negev and a member of Israel's National Committee on Nanotechnology.

# Calculation and computation

# 13 Physics of chaotic systems

Henry D. I. Abarbanel

## 13.1 Introduction

Depending on one's point of view, the realization that solutions to even simple deterministic dynamical systems could produce highly irregular – chaotic – behavior happened 40 years ago with the publication of Edward Lorenz' seminal paper "Deterministic nonperiodic flow" or probably more than 100 years ago with Poincaré's study of complicated orbits in three-body problems of classical Hamiltonian mechanics. Each study indicated the prevalence of complex orbits in classical state space when only a few degrees of freedom were involved. Each study was an unpleasant surprise to physical scientists, and Poincaré's work was roundly ignored for more than half a century, while Lorenz' results were reported in a geosciences journal read by a relatively small group of atmospheric scientists.

Each result, one on the celestial mechanics of Hamiltonian systems and the other on a severe approximation to the dissipative fluid dynamics of convection, had no place in the mainstream pursuits of the day. This was in remarkable contrast to the development of the wave equation for nonrelativistic quantum theory, or the crystal structure of DNA. Both of these were at the core of widely identified important problems and were developments for which a huge body of scientists was prepared. Scientists were not even looking in the right direction when chaotic behavior in deterministic systems was found.

A meta-theme of physical science in 1963, and a worldview still held in many engineering circles, was that linearization of nonlinear problems near an "operating" point, or equivalently perturbation theory near a "good approximate solution" would usually carry the day, and, if not, approximate summation of perturbation series would address any remaining issues. Seeking analytical solutions near "operating points" or for "good approximate solutions" was the clear goal of scientists. Lorenz' results, if attended to, were often thought to be errors due to approximations in the numerical solutions of the three ordinary differential equations he solved, and were not seen as a fundamental change in paradigm.

Writing an article such as this about 40 years after Schrödinger's equation was first written down, say in 1970, one could happily point out whole bodies of mathematical results accompanied by new physical discoveries, including explanations for low-temperature superconductivity, calculations to high accuracy of the magnetic moment of the electron, and the invention of the transistor and the laser. The problems in

*The New Physics for the Twenty-First Century*, ed. Gordon Fraser.
Published by Cambridge University Press. 

physics waiting to be solved by quantum mechanics resulted in a flood after 1927. Not so after 1963 and Lorenz.

We will discuss many things here but touch on none as dramatic as a laser. The realization of chaos in classical systems was much more of a solution looking for problems rather than problems, identified and agreed on, looking for a framework for solutions. Thus we will not be reporting on new technological devices flowing from new insight into nonlinear problems, nor will we dwell on longstanding scientific issues finally resolved by the appearance of deterministic chaos.

Instead we might characterize the realization of chaos as providing a need to re-examine how we view "stochastic" processes, how we think of "statistical" mechanics, what we view as "noise," and how we balance the value of numerical solutions to scientific problems versus that of analytical, closed-form solutions. The recognition of chaos provided an opportunity to realize that the global behavior of nonlinear systems, truly every system of interest in the natural world, is at the essence of the description of most phenomena we observe, and that the success of linearization and perturbation theory is a rare event to be admired rather than followed.

This chapter will, indeed, talk about physics as traditionally seen to be the study of the physical, inanimate world but not exclusively. Both the success of twentieth-century physics and its limitations have led to an increasing blurring of its traditional boundaries, so discussions of biological systems have taken their place in "physics" as well.

Chaos and nonlinear dynamics have produced a fresh view of the behavior of classical systems, living or inert, solid and liquid, which has emerged from a few decades of analysis, typically numerical, but analytical in important arenas, of many properties of nonlinear systems. There are few, if any, devices changing our lives that are based on studies of chaos and nonlinear dynamics, but it takes no special insight to say that there will be. Perhaps it is immodest to recall the story told of Michael Faraday when asked by Britain's Chancellor of the Exchequer what good his work on the arcane subject of electricity and magnetism could possibly be for practical people. Faraday is reported to have responded, "Sir, someday you will be able to tax it!" Well, perhaps so here too.

With taxation, and of course its scientific underpinnings in mind, we plan to discuss several topics related to chaos in small dissipative systems. We will concentrate on descriptive matters connected with instability in nonlinear systems, characterization of the attractors (see Box 13.1) of such systems, analysis of data produced by experiments on such systems, prediction and model building resulting from data analysis for nonlinear systems, as well as synchronization of chaotic systems (perhaps with application to communications) and some ideas of how one could control chaotic oscillations, if desired.

Fermi is reported to have said that "physics is what physicists do late at night," and in that vein, by dint of the hard, late-night work of many colleagues, who might not even possess formal degrees in "physics," I trust that the material we discuss will be seen as "new physics" and as new developments in methods for analyzing and understanding scientific problems.

## 13.2 Instability; strange attractors

The hallmark of many systems, mechanical or electrical, classical or quantum, physical or biological, is their exhibition of stable behavior in the sense of accomplishing a task or as a description of their motion in time. We learn in university mechanics courses that small displacements of bound Keplerian orbits, motion of bodies in our own Solar

BOX 13.1 ATTRACTORS

Solutions to differential equations describing physical systems with dissipation exhibit motion, long after initial transients have died out, in a space with dimension smaller than that of the dynamical system itself. The geometric figure on which the solution trajectory lies is called an *attractor* because trajectories from a set of initial conditions are drawn, attracted, to this time-asymptotic set. Some attractors are geometrically regular, but, when chaotic motion occurs in the dynamics of the system, the attractor can be a set with fractal dimension and has been termed a *strange attractor.*

The simplest example of an attractor may be given by the one-dimensional equation for the dynamical variable $x(t)$:

$$\frac{dx(t)}{dt} = -ax(t), \qquad a > 0.$$

The solutions $x(t) = x(0)e^{-at}$ are attracted to the point $x = 0$, regardless of the initial condition $x(0)$. The space of dynamical variables is one-dimensional, and the attractor has dimension zero.

Another example that is easy to visualize is the two-dimensional system $(r(t), \theta((t))$ satisfying

$$\frac{dr(t)}{dt} = -ar(t) + br(t)^2, \qquad a > 0,$$
$$\frac{d\theta(t)}{dt} = 1.$$

If $b < 0$, then every point in the plane is attracted to the origin, regardless of the initial condition $(r(0), \theta(0))$. Sets of points in dimension two, the number of dynamical variables, are attracted to a set of dimension 0. If $b > 0$, then all points $(r(0), \theta(0))$ in the plane are attracted to the circle at $r = a/b$; sets of points in dimension two are attracted to a set of dimension one.

Both these examples have regular geometric forms for attractors. The Lorenz equation as described in the main text has a different kind of attractor. It comprises a set of dimension 2.06, and is properly called strange. We can show that there is an attracting region for the Lorenz equations, but cannot give a nice form for it. The argument is this. Recall the Lorenz equations

$$\frac{dx(t)}{dt} = \sigma(y(t) - x(t)),$$
$$\frac{dy(t)}{dt} = -y(t) + rx(t) - x(t)z(t),$$
$$\frac{dz(t)}{dt} = -bz(t) + y(t)x(t).$$

Using these equations one can show that

$$\frac{1}{2}\frac{d(x(t)^2 + y(t)^2 + (z(t) - r - \sigma)^2)}{dt}$$
$$= -\left[\sigma x(t)^2 + y(t)^2 + \left(z(t) - \frac{\sigma + r}{2}\right)^2\right] + b\left(\frac{\sigma + r}{2}\right)^2$$

and, for $(x, y, z)$ large enough, the right-hand side is always negative. This means that the "distance" $x(t)^2 + y(t)^2 + (z(t) - r - \sigma)^2$ decreases whenever $(x, y, z)$ is large enough. Points on the orbit that are outside a certain compact region of $(x, y, z)$ space are always returned into that compact region; that is, all points on all orbits are within the region or are attracted to the region. This argument tells us that for the Lorenz system there is an attracting region, though it tells us nothing about the attractor within that region.

The set of initial conditions in the state space of the dynamical system that are attracted to the same attractor is called the basin of attraction. Dissipative differential equations may have many basins of attraction for the same values of parameters in the differential equations, and hence many different attractors for the same parameters. The attractor to which an orbit is attracted will then depend on the initial conditions.

An orbit of the system starting at some initial point in the state space will move toward the attractor, which has dimension less than that of the state space, and asymptotically in time move on the attractor itself. In a typical case the movement to the attractor is exponentially rapid, governed by the *negative* Lyapunov exponents of the system. If there are positive Lyapunov exponents, namely chaos in the orbits, then the way in which the points of the attractor are visited is different for every initial condition, but the set of points visited, namely the attractor, is the same. All orbits look different as a function of time, but, since the attractor is the same, we can characterize the underlying dynamical system using properties of the attractor.

System, lead to other bound Keplerian orbits with different energies. Small changes lead to small alterations in behavior. Slightly perturbing a hyperbolic Keplerian motion leads again to hyperbolic motion, and, while both lead to escape from neighborhoods, both motions are "the same."

Nonlinear systems in high enough dimensions in general do not have this kind of stability. Even classical Keplerian motion in three space dimensions would be unstable were $1/r$ classical potentials not possessed of an unusually high symmetry. Keplerian physics describes motion in three coordinates and three canonical momenta. Because both angular momentum (for all central potentials) and the SO(4) symmetry associated with angular momentum and the conservation of the Runge–Lenz vector hold for $1/r$ potentials, the reduced problem, motion in six-dimensional space labeled by the two constants of such motion, is both integrable and effectively one-dimensional, thus not possessed of enough degrees of freedom to exhibit chaos. From the point of view of the sources of modern science, this must be regarded as truly essential. Imagine ancient astronomers developing tools for the description of chaotic planetary motions and interpreting celestial phenomena with limited temporal predictability! The notions of the development of agriculture-based society and seasons for producing crops would surely be remarkably different if planetary motions were highly chaotic.

Motions of systems with few degrees of freedom, one and two in particular, are regular because of geometric constraints. Autonomous one-dimensional oscillators with a scalar dynamical variable $x(t)$ satisfy

$$\frac{dx(t)}{dt} = f(x)$$

and always have a solution

$$t - t_0 = \int_{x(t_0)}^{x(t)} \frac{dy}{f(y)},$$

when the integral is finite starting at some initial time $t_0$ at which $x(t)$ is $x(t_0)$. This system is called *one-dimensional* not because $x(t)$ is a scalar but because only the first derivative of $x(t)$ enters the equation, and thus only $x(t_0)$ needs specification to determine $x(t)$ for all $t > t_0$.

Systems in dimension two are also regular. Such a system requires a dynamical variable $x(t)$ and typically its first time derivative $dx(t)/dt_0 = \dot{x}(t_0)$ to be specified so $x(t)$ for all $t > t_0$ can be determined. A typical example of this in physics is a one-dimensional Hamiltonian system with a coordinate $x(t)$ and a conjugate momentum $p(t)$. We must give both $x(t_0)$ and $p(t_0)$ to determine the motion of this. The Poincaré–Bendixson theorem tells us that, for bounded motion in dimension two, regardless of the dynamics, linear or nonlinear, only fixed points $(x(t), p(t)) =$ constant or periodic motion $(x(t+T), p(t+T)) = (x(t), p(t))$, for some $T$, is permitted. The proof is essentially geometric. Since motion cannot leave the plane of $(x, p)$ and orbits of an autonomous system cannot cross one another in a transverse manner (the solution is unique), only fixed points or motions for which orbits are tangential, namely repeat the same orbit periodically, are allowed. Both linear motions $H(x, p) = p^2/(2m) + m\omega^2 x^2/2$, namely the simple harmonic oscillator, and nonlinear motions $H(x, p) = p^2/2 + k(1 - \cos x)$, namely the simple pendulum, satisfy this result, as do dissipative examples in dimension two. As noted, because of symmetries and their attendant conserved quantities, Keplerian motion fits into this category.

Amazingly, even though the extension from two degrees of freedom (two initial conditions required to specify the state space) to three changes everything, except for Poincaré at the turn of the last century, no serious attention was paid to the implications that a lot of interesting physics lives in dimensions three and larger.

What happens in dimension three (and larger)? Essentially the geometric constraints are lifted, and orbits that might have been forced to cross transversally in dimension two now simply pass by each other and periodicity, while possible, appears an unlikely outcome. It was realized by the early 1970s (though a decade earlier by Ruelle and Takens in work disconnected from Lorenz' efforts), that periodic motions, perhaps quasi-periodic where incommensurate frequencies are involved, are "structurally unstable." This means, really, that while periodic motions may occur for some values of the parameters specifying the dynamics, small changes in those parameters can typically lead to nonperiodic motion – namely, back to Lorenz.

The original Lorenz system is actually an excellent example for demonstrating all of this. The three differential equations for that system are distilled from the partial differential equations of fluid dynamics describing the motion of a fluid with viscosity and thermal conductivity confined between flat two-dimensional surfaces at $x_3 = h$, held at temperature $T$, and at $x_3 = 0$, held at $T + \Delta T$. For $\Delta T$ small enough, there is no net macroscopic fluid motion and heat is conducted from bottom to top with a profile $T(x_3) = \mathrm{T} + (1 - x_3/h)\Delta T$. As $\Delta T$ is increased this conduction becomes unstable and the fluid moves against gravity and viscosity to carry thermal energy more efficiently than would conduction alone. Chandrasekhar describes the physics here in great detail, relating it to a minimization of a "free energy" as well. Lorenz enters with a set of three functions of time $(x(t), y(t), z(t))$, which are the time-dependent coefficients of functions of coordinate space $(x_1, x_2, x_3)$ that satisfy the physical boundary conditions. These dynamical variables satisfy

$$\frac{dx(t)}{dt} = \sigma(y(t) - x(t)),$$
$$\frac{dy(t)}{dt} = -y(t) + rx(t) - x(t)z(t),$$
$$\frac{dz(t)}{dt} = -bz(t) + y(t)x(t),$$

where $\sigma$ is the ratio of thermal conductivity to viscosity, $r$ is a dimensionless expression of the stress due to the temperature imbalance $\Delta T$ and the forces of gravity and viscosity restraining fluid motion, and $b$ is a dimensionless rescaling of the height of the fluid.

We anticipate that, if $r$ is small enough, the thermal conduction

$$T(x_3) = T + (1 - x_3/h)\Delta T$$

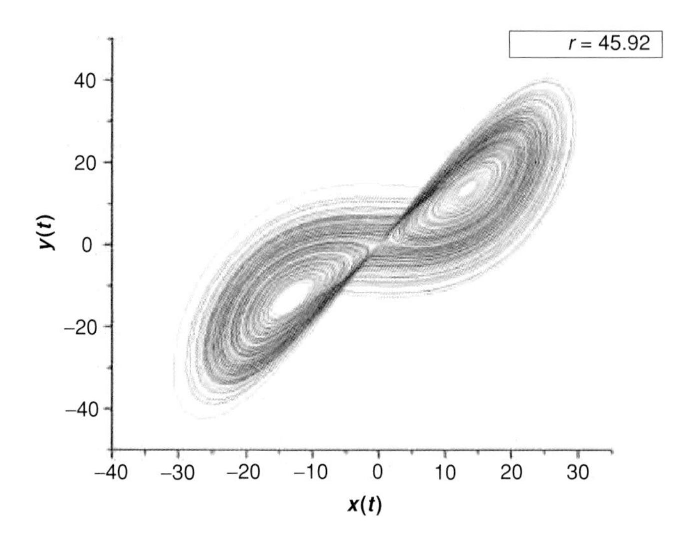


**Figure 13.1.** Chaotic motion of a fluid between two flat two-dimensional plates, as investigated by Edward Lorenz in his pioneering work. There are no fixed-point solutions of the dynamical equations in this regime of $r$. Instead, the solutions follow a complicated orbit to an attractor that is not a regular geometric figure: a "strange attractor."

profile, which has been subtracted from the complete motion, should be a solution; this means $(x_0, y_0, z_0) = (0, 0, 0)$. This is a fixed point of the Lorenz equations; the others are $(x_0, y_0, z_0) = (\pm\sqrt{b(r-1)}, \pm\sqrt{b(r-1)}, r-1)$ and are the simultaneous zeros of the left-hand side of the equations (the vector field) when all time derivatives are zero.

If we perturb these equations by small time-dependent changes from the fixed point to $x(t) = x_0 + X(t)$, $y(t) = y_0 + Y(t)$, and $z(t) = z_0 + Z(t)$, and then linearize the equations in $(X(t), Y(t), Z(t))$, we find

$$\begin{pmatrix} \frac{dX(t)}{dt} \\ \frac{dY(t)}{dt} \\ \frac{dZ(t)}{dt} \end{pmatrix} = \begin{pmatrix} -\sigma & \sigma & 0 \\ r - x_0 & -1 & -x_0 \\ x_0 & x_0 & -b \end{pmatrix} \begin{pmatrix} X(t) \\ Y(t) \\ Z(t) \end{pmatrix}.$$

This description of the motion is accurate as long as $(X,Y,Z)$ all remain small. If any of the eigenvalues of the constant matrix on the right-hand side are positive, the variables grow in an unbounded way, so the solution $(x_0, y_0, z_0)$ becomes linearly unstable, and the motion of the system departs from $(x_0, y_0, z_0)$. For $(x_0, y_0, z_0) = (0, 0, 0)$ the eigenvalues are $\lambda = -b$, and the solutions of $\lambda^2 + \lambda(\sigma + 1) + \sigma(1 - r) = 0$. If $r < 1$, then all eigenvalues are negative. When $r > 1$, one eigenvalue is positive and $(x_0, y_0, z_0) = (0, 0, 0)$ is no longer a solution of the equations.

The fixed point $(x_0, y_0, z_0) = (\pm\sqrt{b(r-1)}, \pm\sqrt{b(r-1)}, r-1)$ is not possible for $r < 1$, but, for $r > 1$, becomes the stable solution. When

$$r > \frac{\sigma(\sigma + b + 3)}{\sigma + b + 1},$$

these two fixed points become unstable. There are no more fixed points, and a simple numerical integration of the equations shows that the solution is no longer periodic but follows an orbit as shown in Figure 13.1.

This orbit is not periodic. One can tell this from its Fourier spectrum, shown in Figure 13.2. All frequencies bounded by the length of the time series and the integration time step are present.

The geometric figure seen in Figure 13.1 is reached exponentially rapidly by any initial condition for the Lorenz equations, and it is called the attractor of the system. The fixed points which have become unstable now for $r > 1$ were also attractors. This attractor is not a regular geometric object and is called a *strange attractor*.

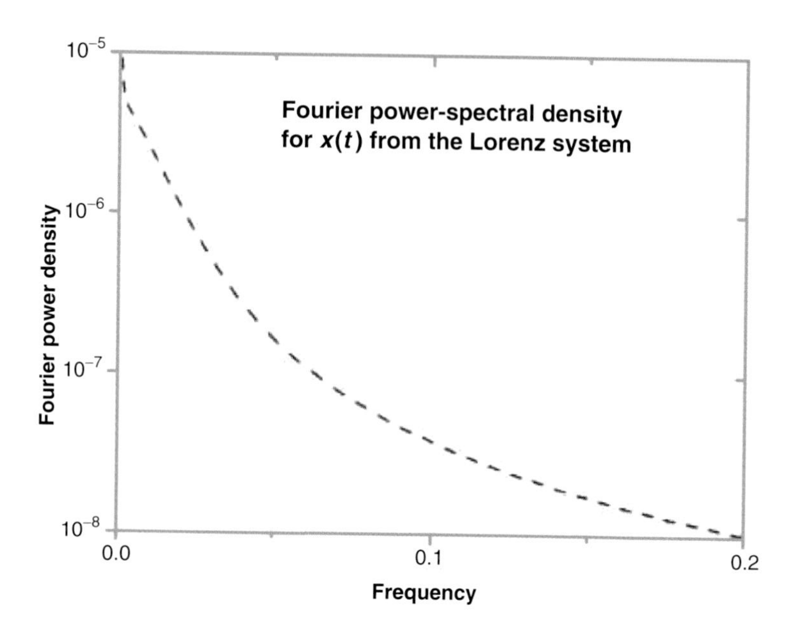


**Figure 13.2.** The Fourier spectrum of the orbit shown in Figure 13.1 covers a wide range of frequencies, showing that the attractor, moreover, is not periodic.

Though Lorenz' example was probably the earliest uncovered, there are myriads of low-dimensional dynamical systems known 40 years later that exhibit nonperiodic or chaotic motions. All of them that are rooted in physical or biological systems manifest some form of "route to chaos" through the development of instabilities as one solution gives way through changes in parameters, such as $r$. In this example, we have a sense that increasing $r$ means increasing the stress on the fluid leading to the excitation of complex motions. What was surprising and remains significant is that such complex behavior is apparent in such a simple system. The nonlinearities in this fluid system are quadratic only and arise from the advective terms $\boldsymbol{v} \cdot \nabla \boldsymbol{v}$ in the Navier–Stokes equations.

This example is not only classical, but also exposes the key idea in understanding the qualitative behavior of many nonlinear systems: they are driven by energy or momentum or vorticity flow from one scale in space or time to another. In the simple Lorenz system, the Rayleigh number $r$ characterizes the dimensionless *stress* on the system, and as this *control parameter* is varied, the state reached by the system after transients die out may change. These bifurcations are characteristic of each system.

## 13.3 Characterizing strange attractors

We are familiar with the characterization of linear systems using the values of the eigenmodes of their oscillation. Identification of an oscillating drumhead can be achieved by knowing all the eigenfrequencies, and these can be identified by Fourier transforming the acoustic signal from the drum. The Fourier transform contains three pieces of information: the frequencies at which spectral peaks appear, the magnitudes of the peaks, and the phase of the signal. The last two are not characteristic of the drum but represent the strength with which we beat the drum and the time at which we began beating it. Only the eigenfrequencies identify the drum with precision.

Since we do not have spectral peaks to use in identifying chaotic systems, indeed working in Fourier space in analyzing nonlinear systems typically makes life harder, we have to turn to other quantities. The basic idea is that all initial conditions for the system produce orbits leading to motion on the attractor, so we should be able to characterize some things about the nonlinear behavior by characterizing the attractor. This cannot be a complete description of the system, as thinking about an attractor that is a fixed point will tell us readily. There are many systems with fixed-point attractors for some parameter values, and noting that the attractor is a fixed point does not distinguish among them. Nonetheless, there are things that can be learned even if we do not have a complete description.

We know that the attractor has a dimension that is less than that of the full dynamical system, call it $D$. The full system is determined by orbits in $D$-dimensional space characterized by the vectors $\mathbf{x}(t) = [x_1(t), x_2(t), x_3(t), \ldots, x_D(t)]$ which satisfy the differential equation $d\mathbf{x}(t)/dt = F(\mathbf{x}(t))$. A volume in this space evolves as

$$\frac{dV(t)}{dt} = \int_{\text{surface around the volume}} \nabla \cdot F(x(t)) d^D x$$

and the divergence of the vector field, on average over the space, should be negative. For the three-dimensional Lorenz vector field $\nabla \cdot F(\mathbf{x}(t)) = -(\sigma + b + 1)$, which means that $V(t) = V(0)e^{-(\sigma+b+1)t}$, and thus goes to zero for whatever initial condition or value of the parameters. Zero volume does not mean a point, it simply means that the dimension of the attractor $d_{\mathrm{A}} < D$. This suggests that evaluating the dimension of the attractor would be one way to characterize it, though we would need many more characteristics to make the identification complete.

As it happens, there is an infinite number of dimensions associated with a set of points in a $D$-dimensional space. Perhaps the most intuitive is associated with the idea that, in a sphere of radius $r$ in this space, the number of points on the geometric object that lie within the sphere should behave more or less as $r^{d_{\mathrm{A}}}$ for "small enough" $r$. So, *locally*, counting the number of points in the sphere $N(r)$ and taking the limit of smaller and smaller $r$ should lead to

$$d_{\mathrm{A}} = \lim_{r \to 0} \frac{\log N(r)}{r}$$

as the dimension of the cloud of points on the attractor. This is called the *box-counting dimension* of the attractor. As mentioned, it is just one dimension characterizing the "volume" occupied by the points traced out by the dynamical system. One can evaluate this quantity for the Lorenz system, and one finds $d_{\mathrm{A}} \approx 2.06$. The fact that this is not an integer might be surprising, but if the dimension were an integer and the motion were on a regular geometric point, line, or surface, it would be unlikely to be nonperiodic, as it is. The interpretation is that all the periodic possibilities in the motion are now *unstable*, so the orbit moves from the region of an unstable periodic object to the region of another, continuing forever and never settling down to motion on a regular object. This provides an intuitive connection between the fractal dimension of the attractor and the nonperiodicity of the motion. Often the other dimensions turn out to be very near this one, so, in a practical sense, the characterization of the attractor by this means is probably achieved.

This calculation has assumed that we know all $D$ dynamical variables for the system and are interested, as we already know a lot about the system, in knowing in what subspace we will find the attractor. A more likely scenario is that we have *not* observed all $D$ components of the state vector, but perhaps only one. It would seem that the chance of learning anything about the nonlinear source of our signal is rather minimal. However, a remarkable set of observations comes to our rescue.

Suppose that we have observed just one component of the state of the system, call it $s(t) = g(x(t))$, where $g(\cdot)$ is some scalar function in $D$-dimensional space. We presume that there is some set of differential equations underlying the dynamics we are observing, so that if we were able to construct from $s(t)$ a sufficiently large number of derivatives we might well have reconstructed the relation among them which constitutes the original differential equation for the source of the signal. However, we have sampled the signal at times $t = t_0 + n\tau_s$, where $t_0$ is the time at which we begin observations and $\tau_s$ is the sampling time. To evaluate the derivative of $s(t)$ we would approximate it by

$$\frac{ds(t)}{dt} = \frac{s(t+\tau_s) - s(t)}{\tau_s},$$

which is a high-pass filter applied to our signal $s(t)$. This means in practice that any noise in the signal is emphasized over the signal itself by this kind of filtering. Of course, if we could achieve $\tau_s \to 0$, this would solve the problem, but we cannot. The attempt to evaluate higher derivatives of $s(t)$ makes things worse.

However, if we look at the expression for $ds(t)/dt$ we see that the only new piece of information beyond $s(t)$ is $s(t + \tau_s)$, and this suggests that forming a state description using vectors formed out of $s(t)$ and its time delays should allow a full description of the system without forming the high-pass filters required for derivatives. David Ruelle first made this observation around 1979 and suggested that a $d$-dimensional vector

$$y(t) = [s(t), s(t + T\tau_s), \ldots, s(t + (d-1)T\tau_s)]$$

would suffice. For appropriate integers $T$ and $d$, $d$ tells us how many components of the vector we need, while $T$ tells us what multiple of the chosen sampling time will describe the behavior of the system.

To select $T$ we need to make the components of the vector $y(t)$ somewhat independent of each other by assuring that each provides an independent view of the dynamics of the system. Choosing $d$ requires us to determine how many components are required to capture the observed dynamical behavior.

Selecting $T$ requires us to specify some kind of correlation between the measurements at $t$ and at $t + \tau_s$. A nonlinear correlation between measurements at this time is provided by the *average mutual information* between all measurements $s(t)$ and measurements $s(t + \tau_s)$. This quantity is the amount of information in *bits* determined about $s(t + \tau_s)$ when we know $s(t)$:

$$I(T) = \sum_{\{s(t), s(t+T\tau_s)\}} P(s(t), s(t + T\tau_s)) \log_2\left(\frac{P(s(t), s(t+T\tau_s))}{P(s(t))P(s(t+T\tau_s))}\right).$$

$P(s(t))$ is the distribution, normalized with respect to unity, of the values of $s$ taken over time, and $P(s(t), s(t + T\tau_s))$ is the joint distribution of such values at time $t$ and at time $t + \tau_s$. If the measurements at these times are independent, then $P(s(t), s(t + T\tau_s)) = P(s(t)), P(s(t + T\tau_s))$ and $I(T)$ is 0. In general $I(T) \geq 0$. A *prescription* for choosing $T$ is to look for the first minimum of $I(T)$, and this provides a compromise for values of $T$ that are too small and those which are too large.

To select a value for $d$, we imagine that our observed system is working in some high dimension and recognize that our observation is in one dimension. This means that points in the higher dimension can be expected to lie near each other even though they may be well separated in the higher dimension. Points near each other by projection are neighbors, but false neighbors. If we use vectors in dimension two $y(t) = [s(t), s(t + T\tau_s)]$, then fewer false neighbors will appear. Similarly, if we use $y(t) = [s(t), s(t + T\tau_s), s(t + 2T\tau_s)]$ fewer still will be present. Finally, if we use a large enough value for $d$, we will have zero false neighbors.

Using these ideas on the Lorenz system, we find a value for $T$ that depends on $\tau_s$ and consistently find $d = 3$. This permits us, from scalar observations that are typical of data, to reconstruct the state of the system, more precisely a proxy vector for the state of the system $y(t) = [s(t), s(t + T\tau_s), s(t + 2T\tau_s)]$. The points in this space are equivalent

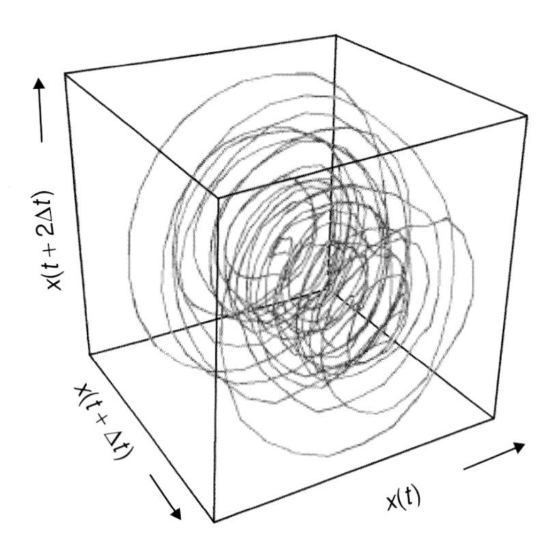


**Figure 13.3.** A reconstruction of the attractor for a chaotic system (in this case a nonlinear electronic circuit) from actual data.

to the points on the attractor of the solution of the underlying differential equation. An example of this procedure is shown in Figure 13.3 for data from a nonlinear electronic circuit.

In this space we may evaluate the various fractal dimensions or any other quantity characteristic of the system. One particularly important characteristic focuses on the instabilities of a nonlinear system that exhibits chaos. We presume that we have determined the (proxy) state vector of the system and through one means or another have established a dynamical rule that connects the state at time $t$ to the state at time $t+1$, where 1 is some unit time:

$$y(t+1) = F(y(t)).$$

Now we ask about the time development of a nearby state $y(t) + \Delta(t)$. For small $\Delta(t)$, we have $\Delta(t+1) = DF(y(t)) \cdot \Delta(t)$, where $DF(y(t))$ is the Jacobian of partial derivatives $DF_{ab}(y) = \partial F_a(y)/\partial y_b,\ a, b = 1, 2, \ldots, d$. If we iterate this rule for $L$ steps along the orbits in proxy space, form the orthogonal matrix $[\{DF^L(y)\}^{\mathrm{T}} \cdot DF^L(y)]^{1/(2L)}$, where $DF^L(y) = DF(y(t+L)) \cdot DF(y(t+L-1)) \ldots DF(y(t))$, then in the limit $L \to \infty$, Oseledec proved in 1968 that this is finite and its eigenvalues $e^{\lambda_1}, e^{\lambda_2}, e^{\lambda_3}, \ldots, e^{\lambda_d}$ are finite. Also the *Lyapunov exponents* $\lambda_1 \geq \lambda_2 \geq \lambda_3 \geq \ldots \geq \lambda_d$ are independent of the starting point $y(t)$. The spectrum of Lyapunov exponents is unchanged under a smooth change of coordinates, so evaluating them in the proxy space is the same as evaluating them in the original (unknown) dynamical space. Thus these exponents are invariant characteristics of the underlying dynamics.

If any of the $\lambda_a$, $a = 1, 2, \ldots, d$ are positive, then the system is locally unstable. However, if $\lambda_1 + \lambda_2 + \cdots + \lambda_d < 0$, then the system is globally stable. Furthermore, if one of the exponents is zero, this means that the underlying system is described by a differential equation since this means that we can displace the orbit along the flow in time and the perturbation simply resets time. For the Lorenz system with $r > r_c$, one exponent is positive, meaning that the system is chaotic, one is zero, meaning that the system is described by a differential equation, and $\lambda_1 + \lambda_2 + \lambda_3 = -(\sigma + b + 1) < 0$.

The bottom line in all this is that one can characterize an observed time series by forming proxy vectors $y(n) = [s(n), s(n-T), \ldots, s(n-(d-1)T)]$ using the sampling time as a unit of time. In this reconstructed or proxy space one can calculate all the quantities characteristic of the original system producing the signal $s(t)$ while not knowing the system itself.

### 13.3.1 Prediction using observed time series

An interesting example of the use of a proxy space comes when we wish to *predict* new values of the measured variable, here $s(t)$. Once we have unfolded the attractor in dimension $d$, and determined how many degrees of freedom are used to describe motion on the attractor using a local form of false nearest neighbors, we can look at the neighbors of any point $y(n)$ in the proxy space and ask where these neighbors go in one time step. Call the nearest neighbors $y^{(r)}(n)$, $r = 1, 2, \ldots, N_B$ ordered by their distance from $y(n)$ We know the time label associated with each point, and we know where that point goes in $K$ time steps: $y^{(r)}(n) \rightarrow y(n + K; r)$. The nearest neighbors are not necessarily near in time label, since the orbit can go quite far from $y(n)$ before returning to its neighborhood. Similarly $y(n + K; r)$ need not be nearest neighbors of $y(n + K)$. Since we know where a proxy space cluster of points goes in time $K$, we can use this information to provide a method for interpolating where points near this cluster might go. To do this we select a set of interpolating functions $\phi_m(x)$ defined on $d$-dimensional vectors $x$. Using these, we seek a description of where $y^{(r)}(n)$ arrives in time $K$:

$$y(n + K; r) = F(y^{(r)}(n)) = \sum_{m=1}^{M} c(m, n)\varphi_m\left(y^{(r)}(n)\right).$$

Using the information we have on $y^{(r)}(n)$ and $y(n + K; r)$, the coefficients $c(m, n)$ are determined locally in proxy space by a linear least-squares method.

Now we are presented with a new measurement $S(i)$, and we want to predict $S(i + K)$. To do this we form the new proxy state vector $Y(i) = [S(i), S(i - T), \ldots, S(i - (d - 1)T)]$ and identify that proxy vector in the original set $y(q)$ closest to $Y(i)$. Using the interpolating formula whose coefficients we have just determined, we predict that the proxy state vector following $Y(i)$ would be

$$Y(i + K) = F(Y(i)) = \sum_{m=1}^{M} c(m, q)\varphi_m(y(q)).$$

Next we look for the proxy state vector closest to $Y(i + K)$, and from the interpolating formula for that location we predict $Y(i + 2K)$, and so forth. The limit on the time forward we can predict is determined by the largest Lyapunov exponent as determined earlier.

An example illustrates this. We begin with a time series of voltage from a circuit with a hysteretic element as the nonlinearity. The voltage is sampled at 2 kHz and part of the data set is shown in Figure 13.4. Using the tools for time-series analysis above, we learn that $d = 3$ both globally and locally. This provides a proxy state-space picture of the attractor, which we show in Figure 13.5.

Using local polynomials in proxy space as interpolating functions, we can create a predictor that gives the results shown in Figure 13.6. Here $K = 15$ and a local linear predictor is used. The results are quite accurate, as one can see.

On evaluating the average prediction error using this method for various forward prediction times $K$, one finds the results shown in Figure 13.7, where we see clearly that prediction is limited to about 30–40 steps before the error is about 0.5, so prediction by these methods is close to just chance.

This method of predicting works well when one has a lot of data, namely the system attractor is well visited by the data points seen as proxy state vectors. However, we learn nothing really about the dynamics of the system, we can only predict the future of the observed quantity $s(t)$, and we have no idea what happens when the control parameters of the underlying system are changed.

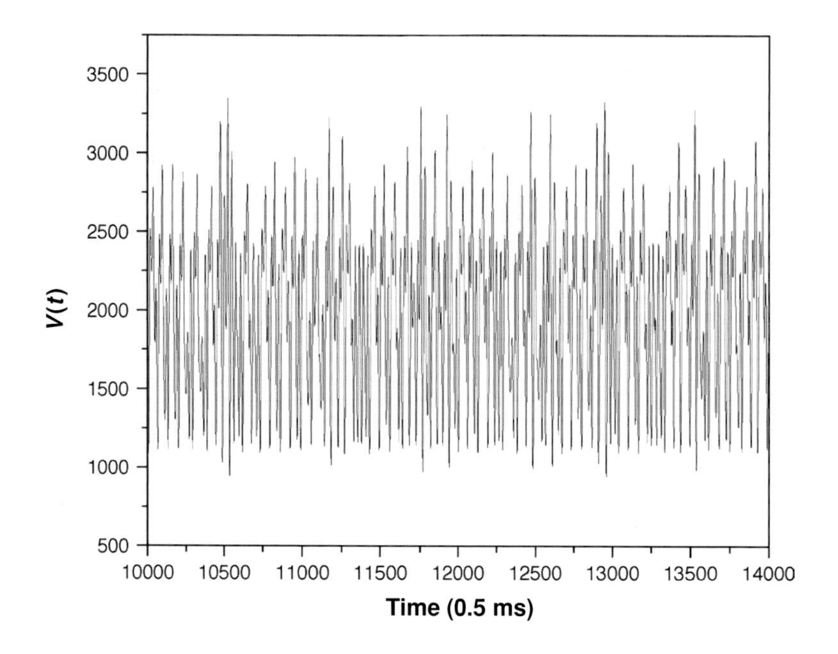


**Figure 13.4.** Observed chaotic voltage variations in an electronic circuit where hysteresis introduces nonlinearity.

To go beyond this, one has to make a global model of the *vector field* $F(\,)$ associated with a map $y(n+K) = F(y(n))$ or differential equation $dy(t)/dt = F(y(t))$. This is, from a purely algorithmic point of view, not possible. Instead, one must select a form for the vector field from considerations about the physics of the system, and that form will contain parameters that must be determined from the data. The methods for determining those parameters are not universal, and, while much has been written about the methods, none has emerged as preferred in all cases.

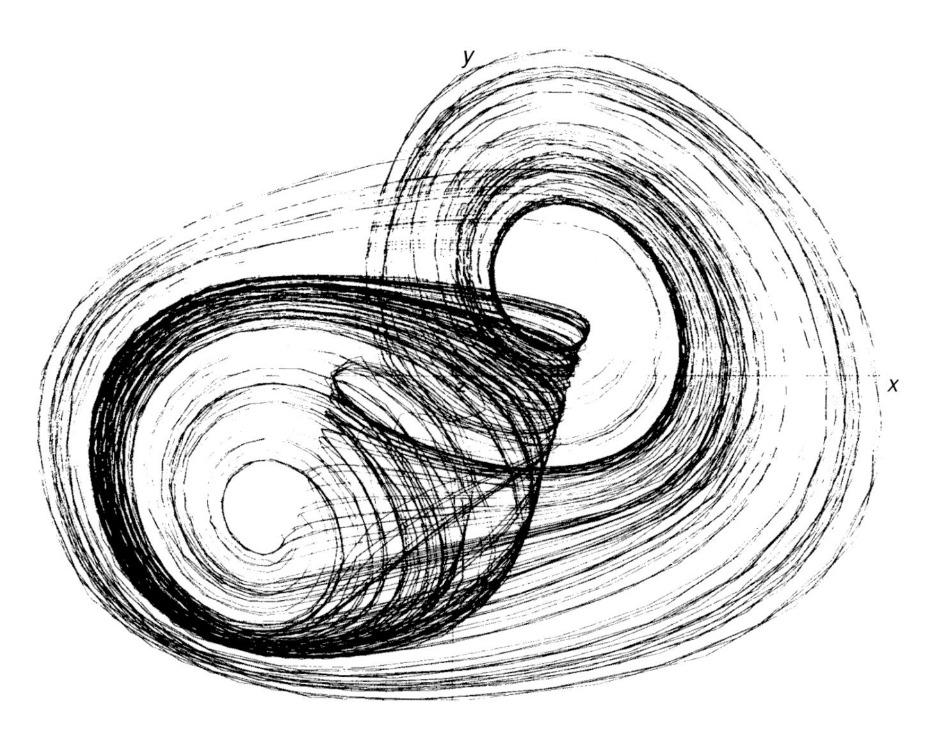


**Figure 13.5.** The reconstructed attractor of the behavior shown in Figure 13.4. The coordinates $(x, y, z)$ are $(V(t), V(t-\tau), V(t-2\tau))$, where the time lag is 4.5 ms. Knowing the form of the attractor, prediction becomes possible within the limits set by the largest Lyapunov exponent.

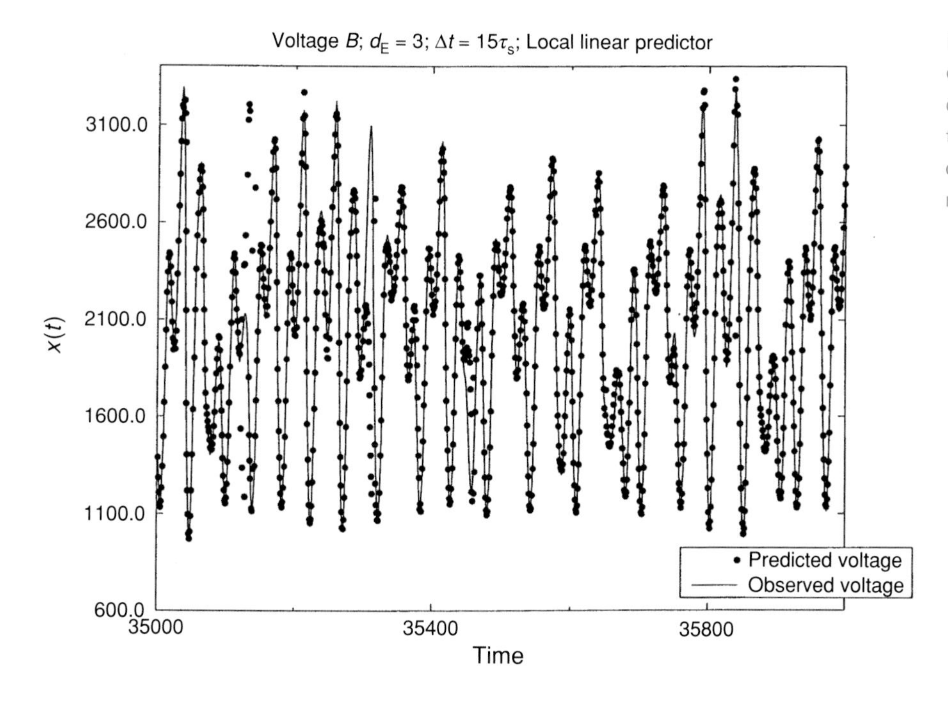


**Figure 13.6.** Predicting chaos – the predicted behavior of voltage against time using the attractor of Figure 13.4, compared with actual observed results.

## 13.4 Synchronization of chaotic oscillators; chaotic communication

An important aspect of the nonlinear dynamics of physical and biological systems is their ability to synchronize with each other even when each system is chaotic. This was studied in the early and mid 1980s by groups in Japan and the USSR, but particularly

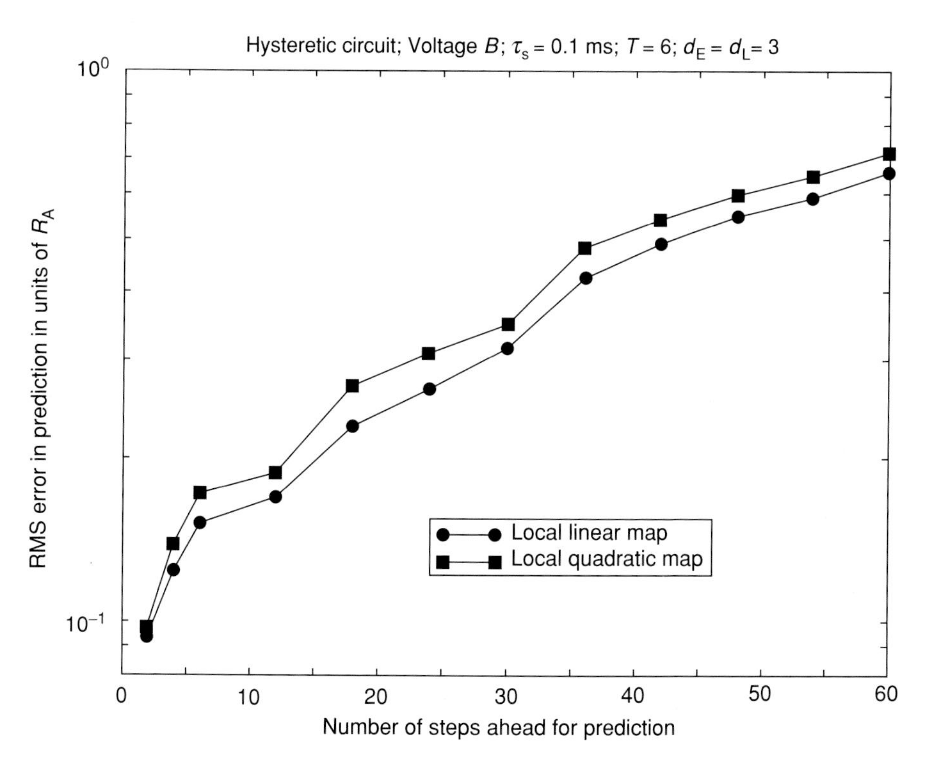


**Figure 13.7.** However, such predictions are restricted – the further ahead the prediction, the worse the result, which after about 40 iterations becomes equivalent to mere chance.

in the work of Pecora and Carroll in the early 1990s. The idea of Pecora and Carroll was to consider two identical systems and transmit one of the dynamical variables from one (the driver or transmitter) to the other (the receiver). For the driver we split the $d$-dimensional state variables $x_D(t)$ into one transmitted variable $u_D(t)$ and the rest $x_D(t) = [u_D(t), W_D(t)]$. The driver equations are

$$\frac{du_D(t)}{dt} = F_u(u_D(t), W_D(t)),$$

$$\frac{dW_D(t)}{dt} = F_W(u_D(t), W_D(t)),$$

and this driving system is autonomous. Now the variable $u_D(t)$ is taken to the receiver system and replaces the equivalent receiver variable $u_R(t)$:

$$\frac{dW_R(t)}{dt} = F_W(u_D(t), W_R(t)).$$

Synchronization occurs when the $d - 1$ variables not transmitted are equal $W_R(t) = W_D(t)$.

A necessary condition for this to be a stable state of the coupled systems is that small deviations from the synchronization manifold $x_D(t) = x_R(t)$ decay back to equality. Only the variables $W(t)$ need be checked in the Pecora and Carroll idea, and this requires that the deviation $\xi(t) = W_D(t) - W_R(t)$ satisfying

$$\frac{d\xi(t)}{dt} = DF_W(u(t), W(t)) \cdot \xi(t)$$

have solutions that satisfy $\xi(t) \rightarrow 0$ for long times. This means that the eigenvalues of the Jacobian $DF_W(u(t), W(t))$, which is conditional on the driving forces $u(t)$, must be negative. These are called the *conditional Lyapunov exponents*. For the Lorenz system, for example, the conditional exponents are negative for either $x(t)$ or $y(t)$ as the transmitted variable, but for $z(t)$ transmitted, one is positive. Synchronization occurs rapidly in the first cases, but not in the second.

There is a vast body of literature on synchronization, and one can hardly cover all of the aspects here. It is clear that the ability of nonlinear systems to synchronize allows large networks to be built of elements that can synchronize to perform tasks requiring coherent signals arising from the collective action of large numbers of individual elements, and can act in an unsynchronized fashion at other times.

One arena in which synchronization has been investigated concerns the ability to communicate using chaotic transmitters and receivers. The chaotic signal is used as a "carrier" of the information to be communicated. One must modulate the message onto the chaotic oscillation in some manner – this is usually quite easy. Then one must demodulate the information at the transmitter, and this is not always so easy.

While one might think that there is something special about chaotic systems that would make communication using them somehow "secure," it is clear that security is not one of the special features of nonlinear communication strategies. One can achieve secure communications by encrypting the messages to be sent before transmitting them using a chaotic transmitter, and then upon receipt, decrypting them as one would any such signal. The advantage of chaotic communications appears to lie in the possibility of using lower-power, less-expensive components for communications since one need not worry about maintaining linearity in all elements as in conventional systems. Indeed, in conventional systems one uses the linearity of the transmitter, the communications channel, and the receiver as key features of the system. Nonlinearities result in the mixing of Fourier components of signals and produce interference among the transmitted symbols of a message. This is not an issue with the use of nonlinear methods, but, if they depend on synchronization, then the methods are often

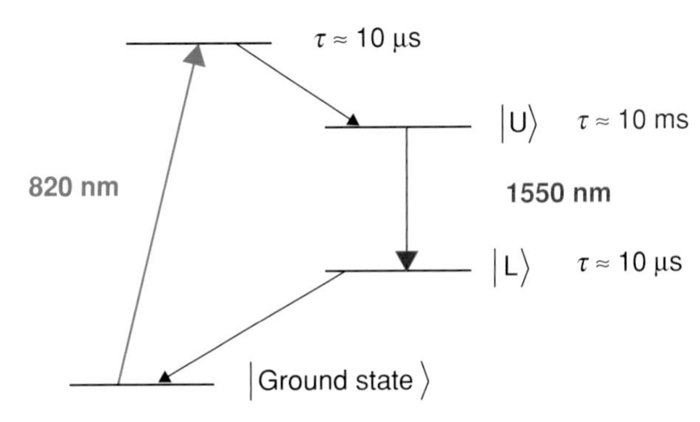


**Figure 13.8.** Energy levels of an erbium-doped laser system used by Roy in 1998 to demonstrate synchronization of and communication with chaotic optical systems.

quite sensitive to noise and attenuation in the communications channel between the receiver and transmitter. The issue is that synchronization typically requires that the transmitted signal be replicated precisely at the receiver, and noise and attenuation make that unlikely.

There is a communications system quite familiar to all of us that does not have this feature: the use of pulses or action potentials by nervous systems to communicate relies on timing rather than waveform, so such a system is robust against noise and attenuation. The time of arrival of a signal can be noted even if the signal is attenuated and the precise shape of the transmitted waveform is altered.

A communications scheme taking advantage of this can be stated in a general form that is easy to synchronize. In this scheme one divides the transmitted signal into two parts as before:

$$\frac{du_T(t)}{dt} = F(u_T(t), W_T(t)),$$
$$\frac{dW_T(t)}{dt} = G(u_T(t), W_T(t)),$$
$$\frac{du_R(t)}{dt} = F(u_T(t), W_R(t)),$$
$$\frac{dW_R(t)}{dt} = G(u_T(t), W_R(t)).$$

As in the Pecora–Carroll method, the receiver system is being run open loop, but now the driving through $u_T(t)$ is used in the dynamics of the receiver too. This permits modulation of the driving signal with a response by the receiving signal.

An optical example of this which was studied experimentally by Raj Roy and his students in the late 1990s is composed of two lasers, each having an erbium-doped fiber as the active element. The energy diagram of the laser set up is shown in Figure 13.8. The active medium is pumped by a laser diode at about 820 nm, and this brings the atomic levels to a state that decays rapidly to a very-long-lived "upper" level, lasting about 10 ms. This makes the transition to the "lower" level by emitting a photon of wavelength about 1550 nm. Of course, the actual picture is more complicated, but this suffices to explain what is seen. The output of the active medium is reinjected into the erbium-doped fiber after traversing a few tens of meters of normal fiber. The advantage of the 1550-nm light is that it is at a minimum of absorption for off-the-shelf optical fiber. The propagation of the light around the "external" fiber is described by a nearly linear solution of Maxwell's equations; the only nonlinearity is associated with the intensity-dependent index of refraction. The dynamical equations are given by

$$E_T(t+\tau) = M_E(w_T(t), E_T(t)),$$
$$\frac{dw_T(t)}{dt} = Q - \frac{1}{T_1}\left(w_T(t) + 1 + \frac{A}{G}\left|E_T(t)\right|^2(e^{Gw_T(t)} - 1)\right)$$

for the transmitter laser and

$$E_R(t+\tau) = M_E(w_R(t), E_T(t)),$$
$$\frac{dw_R(t)}{dt} = Q - \frac{1}{T_1}\left(w_R(t) + 1 + \frac{A}{G}|E_T(t)|^2(e^{Gw_R(t)} - 1)\right)$$

for the receiver laser. Here we see that the transmitter electric field $E_T(t)$ is injected into the receiver. The receiver has its internal laser dynamics given by the population inversion $w_R(t)$. $M_E(w(t), E(t))$ represents the propagation around the undoped fiber.

To see whether synchronization is possible in this laser system, we subtract one of the equations for the population inversion from the other, finding

$$\frac{d(w_T(t) - w_R(t))}{dt} = -\frac{1}{T_1}\left((w_T(t) - w_R(t)) + \frac{A}{G}|E_T(t)|^2\, e^{Gw_R(t)}(e^{G(w_T(t)-w_R(t))} - 1)\right).$$

Using $e^x - 1 \geq x$, we have

$$\frac{d(w_T(t) - w_R(t))}{dt} \leq -\frac{(w_T(t) - w_R(t))}{T_1}\left(1 + \frac{A}{G}|E_T(t)|^2 e^{Gw_R(t)}\right),$$

so

$$w_T(t) - w_R(t) \to 0, \qquad \text{as } t \to \infty,$$

showing that synchronization is possible. Numerical simulations verify this stability.

One may modulate a message by any invertible operation onto the electric field $E_T(t)$. Represent this by $m(t) \otimes E_T(t)$ and reinject this into the active medium in the transmitter and into the active medium of the receiver. The output at the receiver is $E_R(t) = E_T(t)$, so, by inverting the rule which modulated the message onto $E_T(t)$, one recovers $m(t)$.

The use of erbium-doped fiber lasers for this purpose is not very attractive since the very long lifetime of the upper lasing state, about 10.2 ms, means that the population inversion is effectively frozen out of the dynamics, and no chaos is possible expect via the very weak nonlinear index of refraction of the undoped fiber. Matching this from transmitter to receiver is difficult.

However, on using these ideas in the dynamics of a semiconductor laser system, the timescales associated with the lasing-state lifetimes and the decay rates of photons in the lasing cavity change from milliseconds to nanoseconds. One cannot modulate the phase of the light, however, so the intensity $|E_T(t)|^2$ is modulated and chaos is introduced by an electro-optical delay from a diode output proportional to the intensity and feedback with a delay into the bias current of the semiconductor laser. This scheme has been demonstrated in the laboratory of J.-M. Liu at UCLA. Communication bandwidths of nearly 1 GHz can be achieved.

As noted, all of these methods suffer when attenuation in the communications channel (an optical fiber in our examples) or noise is present. One can emulate the way nervous systems communicate by developing a transmitter that produces identical pulses such that the interpulse interval is chaotic. This can be achieved by the use of simple iterated maps in the transmitter, which govern the time at which a pulse generator produces a pulse of voltage. This pulsed voltage can then drive an antenna and wireless transmission of the pulse time can be set to a receiver. The receiver has an identical iterated map and, upon receiving a pair of pulses, can "predict" the sequence of interpulse intervals to follow. If the signal is interrupted in transmission, the receiver will automatically resynchronize after a small number of pulses during which it relearns the interpulse interval. The receiver knows only about the timing of the pulses, and it does not need knowledge of the pulse shape. As long as the pulses carry enough energy to be detectable, attenuation is not an issue because the energy, while attenuated, arrives within a precise time window. Modulation is achieved at the

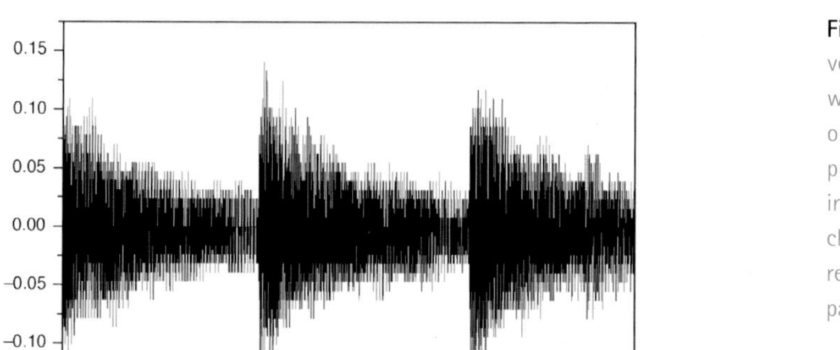


**Figure 13.9.** The amplitude versus time of a speech waveform. This is modulated onto the output of a chaotic pulse generator. Although the interval between pulses is chaotic, a suitably tuned receiver can "predict" the pattern of the received signals.

transmitter by delaying a pulse to represent a "1" and leaving untouched all pulses representing "0"s.

Figure 13.9 shows a speech signal that was digitized and modulated onto a chaotic pulse generator working as described. The pulse intervals in the transmitted signal are now chaotic and a sample of this is shown in Figure 13.10. At the receiver, the pulses were detected and the interpulse interval compared with the one expected from synchronization with the transmitter. Figure 13.11 shows the reconstructed "message."

This method of chaotic pulse modulation for communication has quite robust behavior in the presence of noise, and favorably competes with the best linear methods available in terms of bit error rates as a function of signal-to-noise ratio. It will *never* be as good as the best linear method because, to achieve synchronization of chaotic systems, one must use a part of the available channel capacity to communicate the state of the transmitter to the receiver so that the full state of the transmitter can be reproduced (the definition of synchronization) at the receiver. In systems using simple iterated maps to produce chaotic signals, this may actually be a small penalty.

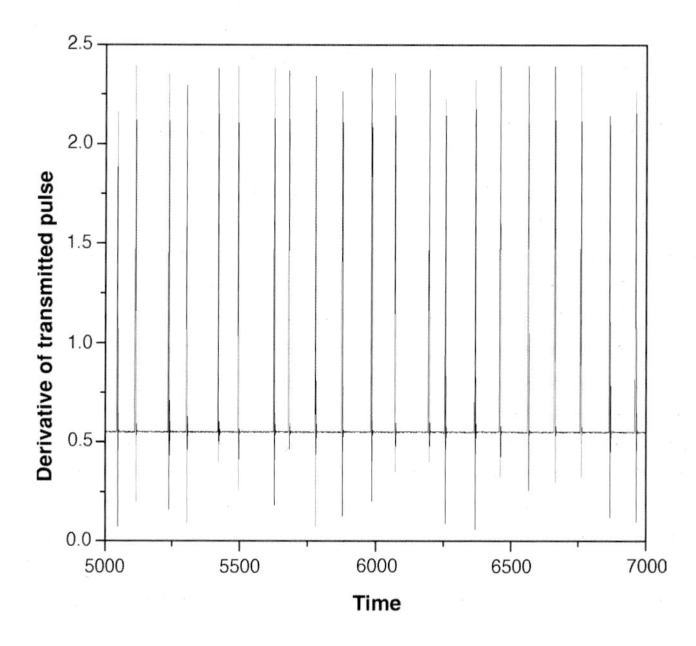


**Figure 13.10.** The chaotic transmitted signals from the modulation shown in Figure 13.9.

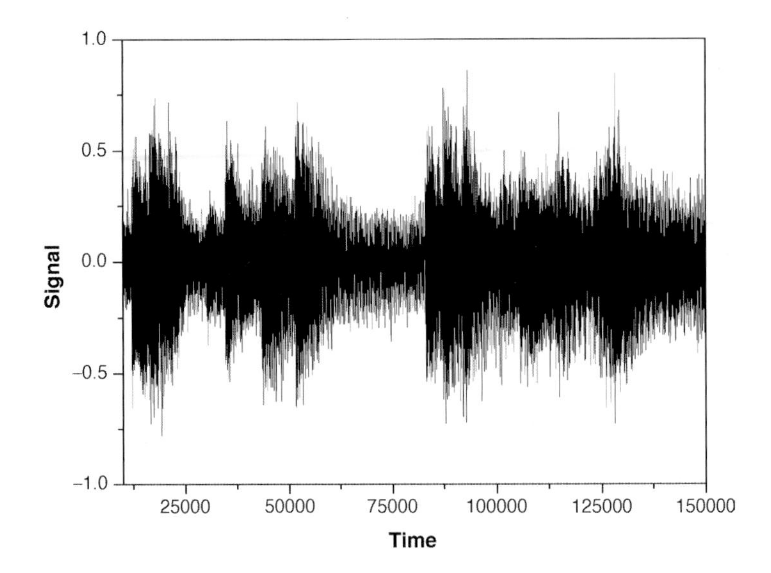


**Figure 13.11.** Despite having chaotic carrier signals, the original signal can be recovered by comparing interpulse intervals (compare with Figure 13.9).

## 13.4.1 Control of chaos

Chaotic orbits may be described as composed of an infinite number of *unstable* periodic orbits. One may think of the path to chaos induced by implementing changes in the stress on a physical system by varying some control parameter as destabilizing the periodic orbits which appear when that stress is small. A fixed point or DC solution to the nonlinear behavior is to be interpreted as an orbit of period infinity.

Ott, Grebogi, and Yorke suggested around 1990 that by altering the chaotic orbit of a system using small external forces one might be able to force selected unstable periodic orbits to become stable again. The idea is easily stated in dimension two, where a periodic orbit (unstable) is characterized by a stable and an unstable direction. These are denoted as $\hat{s}$ and $\hat{u}$, respectively. As an orbit of this system approaches this periodic orbit, it is drawn into the periodic orbit along the stable direction $\hat{s}$ and repelled along the unstable direction $\hat{u}$. If we could perturb the orbit so that it would lie on the stable direction with a time-dependent perturbation $\delta p(n)$ (time is discrete), then we could effectively alter the original system so the selected period orbit is stabilized. This has acquired the name *control of chaos*. In maps in dimension two a nice formula for $\delta p(n)$ is found by considering the map of the state of the system $w(n) \rightarrow F(w(n), p) = w(n+1)$. Suppose that we are near a fixed point at $w_0$ for some value of the parameter $p$. We observe that the orbit is not at $w_0$, but is located at $w_0 + \Delta(n)$. On using the map and taking the parameter $p$ to be time-dependent, $p \rightarrow p + \delta p(n)$, we find that $\Delta(n)$ will become

$$\Delta(n+1) = g\,\delta p(n) + DF(w_0) \cdot [w(n) - g\delta p(n)],$$

where $g = \partial w_0(p)/\partial p$ is the derivative of the fixed-point value with the parameter $p$. If we write the Jacobian $DF(w_0)$ as $DF(w_0) = \lambda_u \hat{u} f_u + \lambda_s \hat{s} f_s$ using the stable and unstable directions of the Jacobian, and then require that $w(n+1)$ lie along the stable direction, this means taking

$$\delta p(n) = \frac{\lambda_u}{\lambda_u - 1} \frac{w(n) \cdot f_u}{g \cdot f_u}.$$

This rule works rather well when one can determine all the required quantities, and it has been demonstrated in numerous examples. Using more conventional

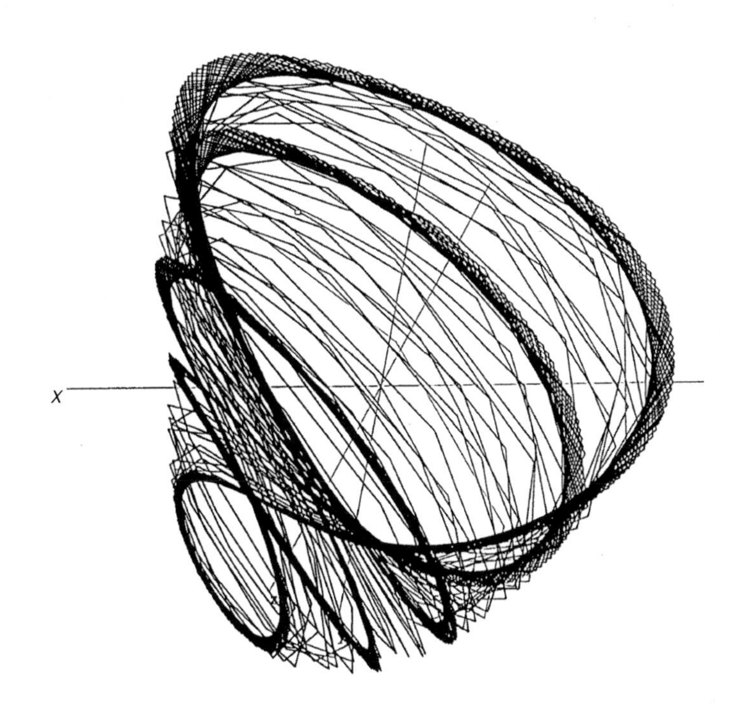


**Figure 13.12.** The strange attractor reconstructed from electronic circuit data. The coordinates $(x, y, z)$ are $(V(t), V(t-\tau), V(t-2\tau))$, where the time lag is 6.4 ms.

control-theory language, the instability of the periodic orbit is associated with an eigenvalue of the Jacobian matrix lying outside the unit circle in the eigenvalue plane. The rule suggested here takes that unstable eigenvalue and moves it to the maximally stable position: the origin of the plane. The stable eigenvalue is already within the unit circle and remains untouched. This is a quite severe demand on the eigenvalues, for, if one moved the unstable eigenvalue into the unit circle, stability would also be achieved.

An optimal control formulation of the problem lends itself to this less restrictive demand. Working in proxy state space, one formulates a "black-box" version of the dynamics $y(t+1) = F(y(t))$ and adds to this an external control force $u(t)$. Locate an unstable periodic orbit using one of many methods; basically, one looks for near recurrences of values of the orbit in proxy space. Find a sequence of points $s_1, s_2, \ldots, s_k, \ldots$ on the unstable periodic orbit, and minimize

$$\frac{1}{2}\left|s_k - F(y(t)) - u(t)g(y(t))\right|^2 + \frac{Bu(t)^2}{2},$$

where $B$ is a constant, and $g(y)$ is a smooth function. This determines the control force $u(t)g(y(t))$ to be applied as

$$u(t) = \frac{-g^{\mathrm{T}}(y(t)) \cdot [F(y(t)) - s_k]}{B + g^{\mathrm{T}}(y(t)) \cdot g(y(t))}.$$

Everything in this formula is known, so for a given unstable periodic orbit and a known constant $B$ and function $g(y)$, it can be preprogrammed and applied in real time.

For data from an electronic circuit the reconstructed state-space attractor is seen in Figure 13.12. This formulation was used to control the chaotic oscillation to several periodic orbits.

The timing of the system is shown in Figure 13.13. The sampling time was 3.2 ms, and the period of the unstable periodic orbit was 14 ms. When the orbit passed through a plane at $t = 0$, the system was allowed to continue for 5 ms with no control. Then the control force with $g$ constant was applied for 5 ms.

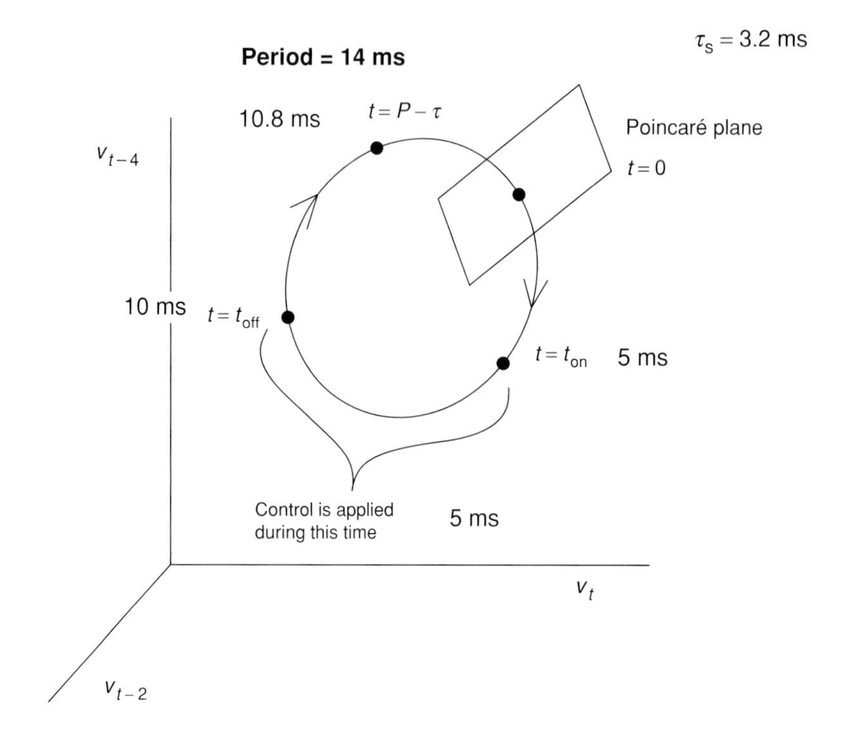


**Figure 13.13.** Controlling chaos – by applying appropriately timed control signals, the behavior of a circuit can be pushed to another orbit (periodic and previously unstable) within the attractor of Figure 13.12.

In Figure 13.14 we can see how the chaotic orbit under this control approaches the selected (formerly) unstable periodic orbit, and in Figure 13.15 we see how the voltage – the observed quantity in the circuit – moves toward a periodic signal while the control force moves toward zero as the orbit lies closer and closer to the selected periodic sequence.

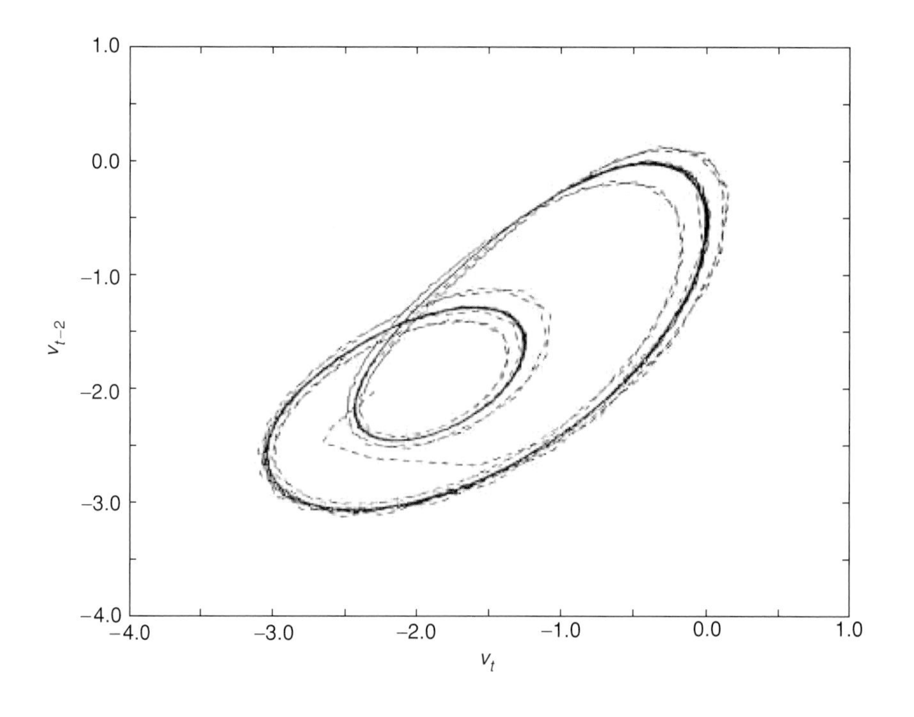


**Figure 13.14.** Under a suitable driving force, the attractor of Figure 13.12 can be driven to a highly simplified orbit. Here, the orbit converges to a selected periodic orbit.

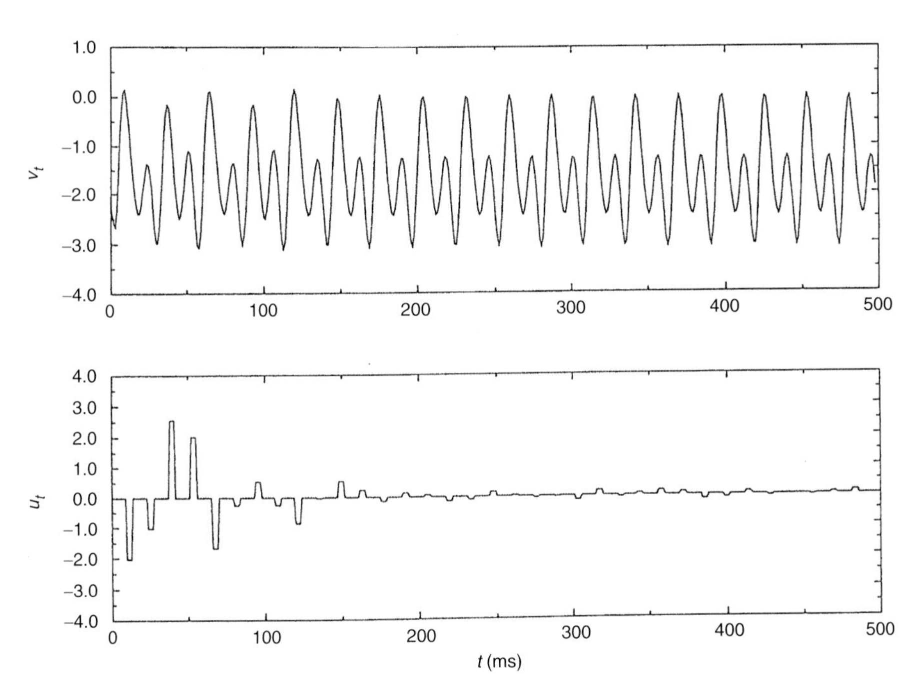


**Figure 13.15.** Under the conditions of Figure 13.14, a periodic voltage signal emerges where once there was chaos.

## 13.5 Chaos in simple physical systems: perspective

The key ingredient in the analysis and use of chaotic oscillations in the study of physical systems is the ability to reconstruct an equivalent or proxy state space from one or a limited number of measurements. From this reconstruction one can address the issue of classifying the source of the observed signals as well as the possibility of manipulating the system to achieve goals such as synchronization or control. There are many other tasks that can be accomplished in this proxy space, and the references include pointers to many of them. It is important to note that the methods described in this chapter have found application for those circumstances under which the dimension of the reconstructed space is "small." "Small" is a moving target since it is often determined in terms of computing capability, and that is increasing apace. Except when subsystems of an extended dynamical system, a system with spatial variation as well as temporal dynamics, synchronize to reduce the effective degrees of freedom, the methods described here might not be terribly successful, and the reader is referred to Chapter 14.

The proxy state-space methods described here are a starting point for the study of physical or biological systems that one would like to model on the basis of observed data. They define the terms of the discussion: the dimension of the active degrees of freedom, invariant characteristics of the attractor, . . . but they do not address the very important issue of modeling the observed dynamics so one may make predictions of behavior when physical parameters take the system outside the realm of observation or provide dynamical equations that may be used to gain insight into the physical properties of the material (fluids, optical systems, acoustics, plasmas, . . .) generating the signal. This is the traditional and powerful methodology used by physicists with dramatic success over several centuries.

One of the techniques we have described here is that of making "black-box" models of the observed dynamics that may have interesting applications, but this approach does not provide physical models of a traditional sort. One can, of course, make such models and the methods described indicate how many degrees of freedom one must use in such models. A likely outcome of such modeling efforts would be a set of differential equations, typically ordinary differential equations or possibly ones involving a time delay representing a coarse graining of partial differential equations and the propagation of some field. These approximate equations will differ from those one might write down using Maxwell's equations or other fundamental physical rules, and they will contain phenomenological parameters that one would like to determine from observations.

There is a serious challenge in this and, at the time of writing, it is an unsolved problem how one may determine parameters in a "reduced" model of dynamics from time-series observations alone when the orbits of the dynamics are chaotic. Part of the problem is that one must determine both the parameters' values and the values of the *initial conditions* of the orbit, for the observed orbit depends sensitively on these initial conditions when the system has positive Lyapunov exponents. The other interesting aspect of the problem is that understanding how the phenomenological parameters in a "reduced" model describing orbits on an attractor vary when physical parameters in a "large" physically based model are changed means that one must work on the common attractor of two systems, reduced and large physical, which may have quite different dimensions. One possibility is to synchronize the small and large systems on their common attractor and extract parameters' values as well as initial conditions in that setting. Efficient and practical ways to do this are not yet known, and addressing this will add substantial value for physicists interested in the physical properties of chaotic systems.

## FURTHER READING


1. H. D. I. Abarbanel, M. I. Rabinovich, and M. M. Sushchik, *Introduction to Nonlinear Dynamics for Physicists*, Singapore, World Scientific, 1993.
2. H. D. I. Abarbanel, *Analysis of Observed Chaotic Data*, New York, Springer-Verlag, 1996.
3. R. Badii and A. Politi, *Complexity: Hierarchical Structures & Scaling in Physics*, Cambridge, Cambridge University Press, 1999.
4. H. Kantz and T. Schreiber, *Nonlinear Time Series Analysis*, Cambridge, Cambridge University Press, 1997.
5. D. Kaplan and L. Glass, *Understanding Nonlinear Dynamics*, New York, Springer-Verlag, 1995.
6. F. C. Moon, *Chaotic and Fractal Dynamics*, New York, John Wiley and Sons, 1992.
7. A. Pikovsky, M. Rosenblum, and J. Kurths, *Synchronization: A Universal Concept in Nonlinear Science*, Cambridge, Cambridge University Press, 2001.
8. D. Ruelle, *Phys. Today* **47** (1994) 24.
9. H.-G. Schuster, *Deterministic Chaos*, 2nd edn., New York, VCH, 1988.
10. J. M. T. Thompson and B. Stewart, *Nonlinear Dynamics and Chaos*, 2nd edn., Chichester, John Wiley and Sons, 2001.
11. E. Ott, *Chaos in Dynamical Systems*, 2nd edn., Cambridge, Cambridge University Press, 2002.


## Henry D. I. Abarbanel

Henry D. I. Abarbanel received the B.S. in Physics from Caltech and his Ph.D. in Physics from Princeton University. He worked on the quantum theory of fields at Princeton, the Stanford Linear Accelerator Center, Fermilab, and the University of California, Berkeley. He moved to the University of California, San Diego in 1983 to pursue his interest in nonlinear dynamical systems at the Department of Physics and at the Scripps Institution of Oceanography at UCSD. In the Fall of 1983, in conjunction with others throughout the University of California, he helped establish the Institute for Nonlinear Science (INLS), which was approved by the UC Regents in January 1986. He has served as Director of the INLS since then.

At UCSD/INLS Abarbanel has investigated various aspects of nonlinear dynamics, ranging from fundamental algorithms for identifying nonlinear systems to the use of chaotic electrical and optical systems as communications devices. In the past few years he has established a neurophysiology laboratory at the INLS where he and his colleagues are investigating the role of nonlinear dynamics in the understanding of biological neural systems and developing numerical and electronic models for individual neurons and neural systems that have known functionalities. The overall focus of his research has been to utilize the power of nonlinear dynamics in the understanding of physical and biological systems while developing the fundamental algorithms for employing that power when they are required.

# 14 Complex systems

Antonio Politi

## 14.1 Introduction

Ever since physics came into existence as a scientific discipline, it has been mainly concerned with identifying a minimal set of fundamental laws connecting a minimal number of different constituents, under the more or less implicit assumption that knowledge of these rules were a sufficient condition for explaining the world we live in. Such a program has had remarkable success, to the extent that processes occurring on scales that range from that of elementary particles up to those of stellar evolution can nowadays be satisfactorily described. Nevertheless, it has meanwhile become clear that the inverse approach is a source of unexpected richness and can hardly be included with the original microscopic equations. The existence of different phases of matter (gas, liquid, solid, and plasma, together with perhaps glasses, granular materials, and Bose–Einstein condensates) provides the most striking evidence that the adjustment of a parameter (e.g. temperature) can dramatically change the organization of the system.

Moreover, not only can the same set of microscopic laws give rise to different structures, but also the converse is true: different systems can, for example, converge toward the same crystalline configuration. One of the basic ingredients that makes the connection between different levels of description complex, though very intriguing, is the presence of nonlinearities. As long as each atom deviates slightly from its equilibrium position, the equations of motion can be linearized and the dynamics thereby decomposed into the sum of independent evolutions of so-called normal modes. In some cases, especially when disorder is present, such a decomposition cannot be easily performed (and very interesting physics can indeed arise, e.g. Anderson localization) but is, at least in principle, feasible. The presence of nonlinear interactions greatly changes this scenario: apart from a few specific cases, the system dynamics is no longer integrable, and it is no longer possible to break the original high-dimensional problem into many independent simple ones. At equilibrium, this seeming difficulty becomes a great simplification, as a generic trajectory (see Chapter 13) extends over the whole of phase space, justifying the equiprobability of microscopic configurations. This is the cornerstone of equilibrium statistical mechanics, which is built on the assumption that the probability of each state depends only on its energy, through the well known Boltzmann–Gibbs factor $\exp[-H/(k_B T)]$, where $T$ is the temperature and $k_B$ the Boltzmann constant.

*The New Physics for the Twenty-First Century*, ed. Gordon Fraser.
Published by Cambridge University Press. 

**Figure 14.1.** A phase-space portrait of the Chirikov–Taylor map for $K = 0.96$. The colored regions correspond to different ergodic components. Black lines correspond to quasi-periodic trajectories – the remnants of harmonic behavior.

Thus the observation by Fermi, Pasta, and Ulam in 1953, who, studying a chain of nonlinear oscillators, found classes of trajectories that repeatedly returned close to the initial condition, without extending over the available phase space, was a surprise. Even more unexpectedly, a few years later Kolmogorov, Arnold, and Moser rigorously proved that quasi-periodic motion (i.e. the superposition of independent oscillations) survives when a small amount of nonlinearity is added to an integrable system. A qualitative idea of the overall scenario can be appreciated from Figure 14.1, where the phase portrait of a simple dynamical system, the Chirikov–Taylor map, is presented,

$$\begin{aligned} I_{n+1} &= I_n + K \sin\theta_n, \\ \theta_{n+1} &= \theta_n + I_n. \end{aligned} \tag{14.1}$$

The colored regions correspond to separate ergodic components, while the black lines correspond to quasi-periodic trajectories. However, it has become progressively clear that, in systems with many degrees of freedom, the existence of two types of behavior may at most give rise to a slow convergence toward equilibrium, rather than presenting a real challenge to thermodynamics.

Besides this obstacle, of purely dynamical origin, to phase-space filling, a further major source of slow and complex dynamics has been identified in the existence of different macroscopic phases. The study of spin and structural glasses has indeed shown that phase space may be organized into many valleys separated by high (free-energy) barriers. The connection between the structuring of such valleys and the microscopic laws is still unclear, although much progress has been made in the past two decades (as will be discussed later).

Although composite systems still offer challenging questions today, the class of phenomena occurring in systems steadily maintained out of thermodynamic equilibrium is far richer and more complex, with the generation and functioning of living beings presenting the ultimate challenge. The very first difficulty encountered when dealing with non-equilibrium systems is the absence of a simple prescription like the Gibbs–Boltzmann factor with which to attribute a probability to each microscopic configuration. A second major difficulty (but also the source of a rich variety of effects) is that evolution need not consist of a simple convergence toward a minimum of the free energy.

In spite of such differences, one is, in a very broad sense, faced with the same problem as in equilibrium statistical mechanics: the emergence of nontrivial behavior from the repeated application (either in space or in time) of an apparently simple rule. Next, the question of whether sufficiently general mechanisms can be identified to classify a fairly large number of different non-equilibrium phenomena arises. I will go on to illustrate some prominent examples and indicate their possible common links.

## 14.2 Changing levels of description

Hydrodynamics is a valuable way to explore the connections between the microscopic and macroscopic world. The most important equation describing the dynamics of an incompressible fluid was first derived in 1823 by Navier, slightly later also by Stokes, namely the Navier–Stokes equation (NSE),

$$\frac{\partial \boldsymbol{v}}{\partial t} + \boldsymbol{v}\nabla\boldsymbol{v} = \nu\,\nabla^2\boldsymbol{v} - \frac{\nabla p}{\rho} + \frac{F}{\rho}, \tag{14.2}$$

where $\boldsymbol{v}$ is the fluid velocity, $\rho$ the density, $\nu$ the kinematic viscosity, $p$ the pressure, and $F$ an external force. The NSE is a mesoscopic equation, i.e. it arises from a spatial and temporal coarse-graining under the assumption that in each, not well-defined, small volume the fluid is locally at equilibrium. Within this framework, the only relevant macroscopic variables are those corresponding to conservation laws (such as conservation of momentum, mass, and energy), since their evolution is controlled by boundary fluxes, while the others can rapidly converge to equilibrium without constraint. Insofar as the NSE is a dissipative equation (contracting phase-space volumes), nontrivial behavior can be sustained only if energy is continuously pumped in (typically by stirring the fluid on a macroscopic scale – see the force $F$ in Equation (14.2)).

In the case of weak forcing, the asymptotic regime is stationary and basically homogeneous in space; at large forcing, turbulent behavior sets in: this regime, although familiar, is not yet fully understood and remains mysterious. Since the NSE is important in a number of applications (e.g. the design of aerodynamical shapes), many efforts to find effective algorithms for integrating it have been made. This was quite clear to von Neumann (see Box 14.5), who in the early 1950s had already envisaged hydrodynamics as one of the major fields of application for the new computing machines.

Lattice-gas automata have been introduced precisely with the goal of performing massively parallel simulations. A lattice-gas automaton is basically a set of particles characterized by a fixed (small) number of velocities and moving on a discrete lattice. An example is illustrated in Figure 14.2, where red and green arrows correspond to positions and velocities at two consecutive time steps (time is here assumed to be a discrete variable). Whenever two particles happen to reach the same site, they collide and change velocity according to rules that ensure conservation of momentum and

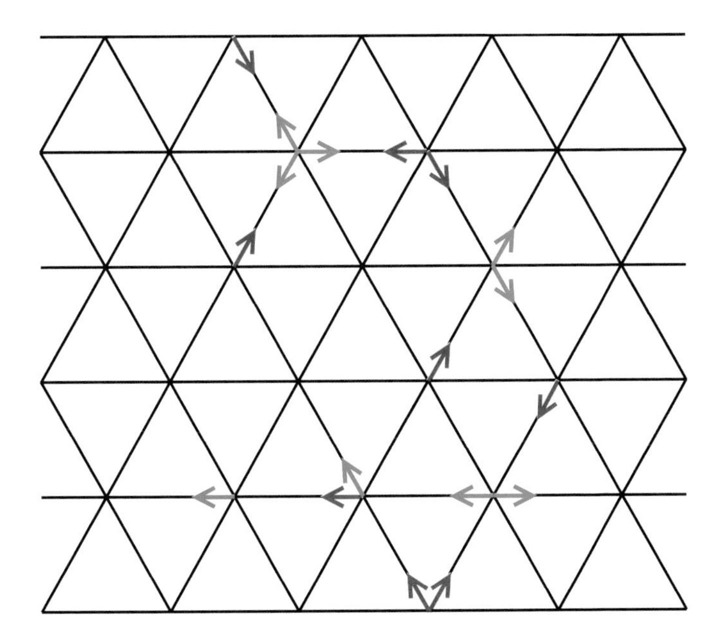

**Figure 14.2.** An example of lattice-gas dynamics on a triangular lattice. Each arrow represents a single particle and its trajectory. Red and green colors correspond to two consecutive time steps.

energy. Typically, the lattice anisotropy is reflected in the anisotropy of the coarse-grained behavior; however, in 1986, Frisch, Hasslacher, and Pomeau discovered that the two-dimensional NSE is satisfied also by the mesoscopic dynamics of particles moving on a hexagonal lattice (and characterized by only one velocity). It is thus interesting to note that NSE-type dynamics can also arise from a non-Hamiltonian microscopic evolution: a clear indication that describing evolution at a higher level does not necessarily require detailed knowledge of what is happening at a lower level.

Thermodynamic relationships were established before realizing that they could be derived from a Hamiltonian, and still today, in many cases, knowledge of the microscopic interactions is not sufficiently accurate to allow macroscopic observables to be determined. In fact, efforts to develop a non-equilibrium statistical mechanics follow two complementary routes. On the one hand are those who, working along the lines of nineteenth-century physics, aim to prove that free energy and entropy are also fruitful concepts away from equilibrium. For instance, Oono and Paniconi conjectured in 1998 that steady states can be assimilated to standard equilibrium ones if the heat produced by dissipation $Q_{\rm hk}$ (and called by them "housekeeping" heat) is subtracted from the total heat exchanged by the system $Q_{\rm tot}$. More precisely, they argue that it is only the excess heat $Q_{\rm ex} = Q_{\rm tot} - Q_{\rm hk}$ that plays the role of an entropy-like quantity away from equilibrium.

On the other hand, other researchers aim at determining general expressions from microscopic laws. Unfortunately, this is hindered by the absence of a general prescription (like the Boltzmann–Gibbs factor) with which to attribute a probability to each configuration. However, Jona-Lasinio and collaborators showed in 2002 that an entropy can be effectively introduced in a class of stochastic systems of interacting particles. At variance with standard equilibrium systems, the entropy turns out to be a nonlocal functional: the implications of this have yet to be clarified.

## 14.3 The onset of macroscopic structures

Dynamical phase transitions leading to the onset of patterned structures perhaps constitute the most significant challenge for the development of non-equilibrium statistical mechanics. Morphogenesis appears in many different contexts, ranging from fluid convection to sand ripples on the beach and geological formations. It has even been argued that pattern formation played a role in the early stages of the expansion of the Universe.

Most theoretical work on the onset of spatial structures has been concerned with the study of suitable partial differential equations (PDEs). It is remarkable that, even though diffusion smears out any inhomogeneity, translationally invariant equations may give rise to patterned structures. Turing was the first to predict, in his pioneering work in 1952, that stationary spatial inhomogeneities can arise in a context of two interacting chemical species when one of the two substances diffuses much faster than the other. In order to illustrate how this can arise, consider a simplified model of an activator $a$ interacting with a substrate:

$$\begin{aligned}\frac{\partial a}{\partial t} &= D_a \nabla^2 a + \rho_a(a^2 s - a),\\ \frac{\partial s}{\partial t} &= D_s \nabla^2 s + \rho_s(1 - a^2 s).\end{aligned} \tag{14.3}$$

In this model, an initial local growth of $a$ (for $s$ sufficiently large) is accompanied by consumption of the substrate $s$, whose fast diffusion in turn inhibits propagation of the high-density activator state. More technically, the loss of stability of the homogeneous state can be investigated by performing a linear stability analysis: translational invariance implies that the eigendirections are plane wave, but not all of them are equally stable: those whose wavevectors lie in a suitable region are amplified and, as a result, inhomogeneous spatial structures naturally arise. Depending on whether translational or rotational invariance is broken, different patterns may appear. As an example, Figure 14.3 plots the transverse electric field in an experiment in which a laser beam modulates the refractive index of a liquid-crystal light valve, which in turn modulates the phase of the input light.

As a result of nonlinear dynamical laws, different types of patterns that are linearly stable may exist for the same parameter values: this multi-stable scenario is reminiscent of first-order phase transitions. The asymptotic stability can be inferred from the dynamics of the interfaces separating the different regions, but, at variance with the equilibrium scenario, detailed balance is not obeyed. This is another way of saying that, in general, there is no shortcut to solving the dynamical problem. Furthermore, the pattern may also lose its linear stability and give rise, through a so-called secondary bifurcation, to even less ordered and possibly time-dependent regimes.

## 14.4 A paradigm: the complex Ginzburg–Landau equation

Many aspects of the detailed off-equilibrium phenomenology are captured by another "universal" PDE, the complex Ginzburg–Landau equation (CGLE),

$$\frac{\partial A}{\partial t} = \varepsilon A + (1 + ic_1)\nabla^2 A - (1 - ic_3)|A|^2 A. \tag{14.4}$$

In contrast to the NSE, the CGLE is perturbative in that it is valid in the vicinity of a transition point where the coefficient $\varepsilon$ of the linear term changes sign, denoting a change of stability for the state characterized by a zero value of the "order parameter" $A$. The real Ginzburg–Landau equation ($c_1 = c_3 = 0$) was introduced to describe the behavior of free energy in a second-order equilibrium phase transition, such as the onset of superconductivity. From a dynamical point of view, it is not particularly

(a)

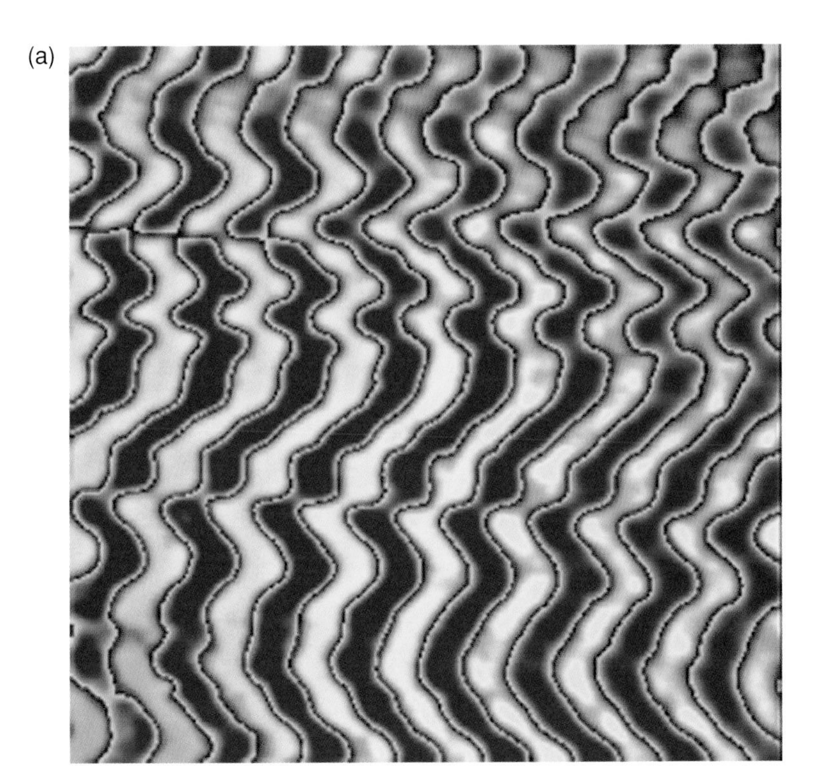

(b)

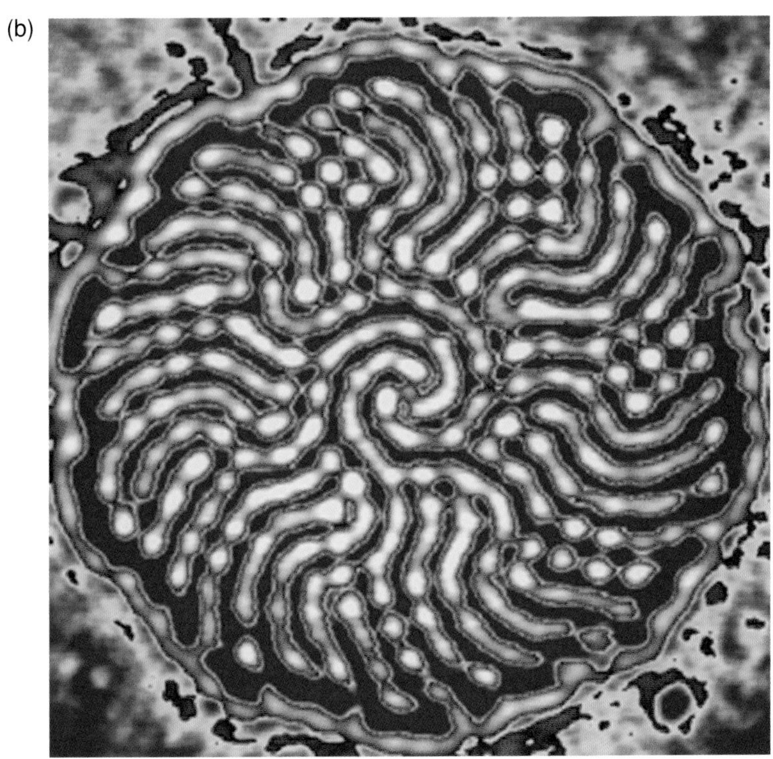

**Figure 14.3.** Two examples of patterns spontaneously emerging in the transverse field of a laser beam, as a result of the nonlinear feedback due to the interaction with a liquid crystal. (Courtesy of P. L. Ramazza.)

interesting, since it describes purely relaxational dynamics (it can be easily seen that the free energy $\mathcal{F} = -\varepsilon|A|^2 + \frac{1}{2}|A|^4 + |\nabla A|^2$ is indeed a Lyapunov functional, decreasing until it reaches a minimum). The full CGLE was phenomenologically introduced in 1971 by Newell and Whitehead and later recognized to be one example of amplitude equations describing the slow behavior (in space and time) of a suitable field in spatially distributed systems. This typically happens in the vicinity of bifurcations (see Box 14.1); in the case of the CGLE, it arises whenever all wavelike perturbations are damped except for a small set of modes with wavelengths close to a critical value $q_c$, which are instead marginally amplified. Additionally, the dynamics has to be invariant under $A \to Ae^{i\phi}$, which is the case if $A$ is the amplitude of a wave in a translationally invariant system. This remark about gauge invariance is not simply a technical aside about the character of the equation, but instead exemplifies the crucial role played by symmetries in determining the character of a transition and providing a natural framework in terms of which to classify them. The importance of symmetries can hardly be underestimated: a condition for the validity of the CGLE is the existence of a localized maximum of the instability in the wavevector space (see Box 14.1, Figure A): this is very natural in one dimension, but much less so in two and three. Indeed, if the system is isotropic, an entire ring of eigenmodes becomes simultaneously unstable (see Box 14.1, Figure B) and accounting for such degeneracy requires a different model.

The CGLE is basically a normal form, meaning that any bifurcation satisfying the above criteria can always be reduced to such a minimal form, provided that the proper variables are introduced. Indeed, the CGLE has successfully been applied to a wide variety of physical problems ranging from Rayleigh–Bénard convection to laser instabilities, chemical reactions, etc. Although, rigorously speaking, the CGLE is the result of a perturbative approach and, as such, applies only close to threshold, the classes of phenomena it gives rise to are rather general and persist away from criticality; for this reason it is still being studied extensively.

A remarkable feature of the CGLE is that, when $c_1$ and $c_3$ are different from zero, not only can time-dependent dynamics asymptotically persist, but also significantly different classes of behavior can be found. Formally speaking, a partial differential equation like the CGLE involves infinitely many degrees of freedom, since it describes a "field" rather than a single variable. Nevertheless, only those degrees that remain active after a long time are truly important. Near a bifurcation, the wavenumber $q_c$ of the most unstable mode represents a natural self-generated scale: if the inverse of the system's size is on the order of the width of the unstable region (see Figure A), only a few degrees of freedom are excited and can contribute to the dynamical complexity. Beautiful experiments on low-dimensional chaotic dynamics have been carried out precisely under such conditions (see Chapter 13). In virtually infinite systems, many degrees of freedom are simultaneously active and one speaks of spacetime chaos. In the phase-turbulence regime depicted in Figure 14.4(a), the dynamics is determined by the phase of $A(t)$, the amplitude being a "slaved" variable that exhibits small oscillations; amplitude turbulence (see Figure 14.4(b)) is instead characterized by the appearance of defects, spacetime points where the amplitude is equal to 0 and the phase is undefined. In two dimensions, zero-amplitude points can be generically found at any time: they are the core of spiral solutions.

## 14.5 Connecting microscopic and macroscopic worlds

The deterministic PDE description of non-equilbrium dynamics should not let us forget the statistical nature of the underlying physical system and, in particular, the presence

## BOX 14.1 BIFURCATIONS

A phase transition is a qualitative change in the macroscopic behavior of a statistical system. Most of the fascination of phase transitions is due to the fact that they arise in systems with infinitely many components (the so-called thermodynamic limit). Whenever the proper order parameter(s) are identified, a phase transition can be schematically described as a bifurcation – a qualitative change occurring in a dynamical system with a finite number of degrees of freedom. In the latter context, the qualitative change arises when, after tuning a suitable control parameter, a given solution becomes unstable and is replaced by a different asymptotic dynamics. This is exemplified by the Hopf bifurcation, described by the model dynamics $\dot{u} = (\alpha + i\beta)u - |u|^2 u$, where $u$ is a complex variable. $u = 0$ is a solution for all parameter values, but it is stable only for $\alpha < 0$. For $\alpha > 0$, the evolution converges toward the periodic behavior $u(t) = \sqrt{\alpha}\, e^{i\beta t}$.

Whenever spatial degrees of freedom are included, the evolution is described by a partial differential equation (if no long-range interactions are present). In this case, stability analysis of a solution $u_0(\boldsymbol{x})$ amounts to linearizing the equation around $u_0(\boldsymbol{x})$ and thereby finding eigenmodes and eigenvalues of the resulting linear operator. The solution $u_0(\boldsymbol{x})$ is stable if the real part $\lambda$ of all eigenvalues is negative. The scenario of Figure A refers to a two-dimensional context, where $q_x$ and $q_y$ are the components of the wavevectors labeling the various eigenmodes. In the case depicted, all modes falling within the circular region are unstable. Their number depends on the size of the system, since the linear spacing of modes supported by a system of size $L$ is approximately equal to $1/L$.

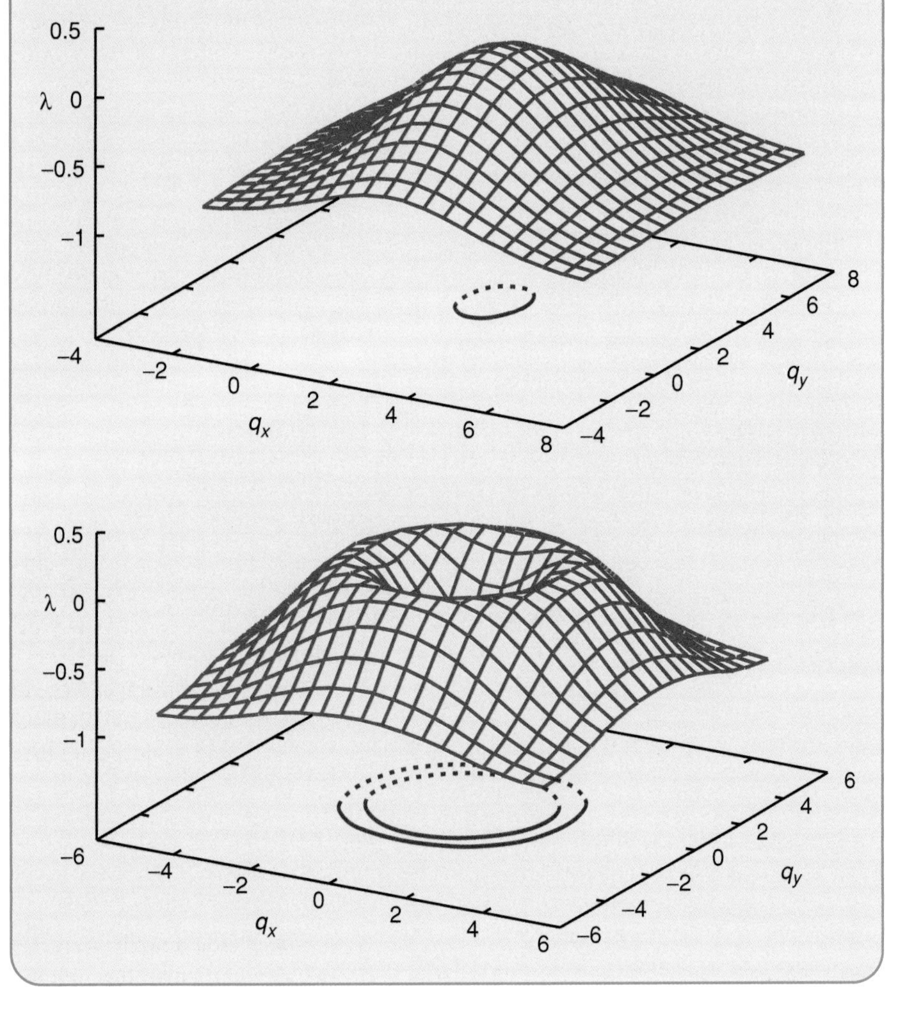


**Figure A.** The stability spectrum of a generic two-dimensional system: $\lambda$ is the real part of the eigenvalue associated with the eigenmode of wavenumbers $q_x$ and $q_y$. Unstable modes are confined within the region surrounded by the blue line that corresponds to $\lambda = 0$.

**Figure B.** Unlike the spectrum plotted in Figure A, here isotropy implies the existence of a full ring of unstable modes.

**Figure 14.4.** Two spacetime patterns (time points downwards) from the one-dimensional complex Ginzburg–Landau equation: (a), obtained for $c_1 = 3.5$ and $c_3 = 0.5$, corresponds to the so-called phase-turbulence regime; and (b), obtained for $c_1 = 3.5$ and $c_3 = 1.0$, corresponds to amplitude turbulence. (Courtesy of L. Pastur.)

of thermal fluctuations. Even the onset of a stationary pattern is indeed accompanied by nontrivial behavior of the fluctuations at the transition point. The first theory going beyond linear analysis was developed by Swift and Hohenberg in 1977 in the context of Rayleigh–Bénard convection. They predicted that a full account of fluctuation dynamics would transform the continuous transition into a first-order one. Because of the typical smallness of fluctuations, it was not until 1991 that the first quantitative experimental studies were performed by Rehberg and collaborators, who investigated electro-convection in a nematic liquid crystal. Even leaving aside the question of a convincing understanding of the critical behavior, it is important to realize the subtle character of the connection between microscopic disorder and macroscopic order.

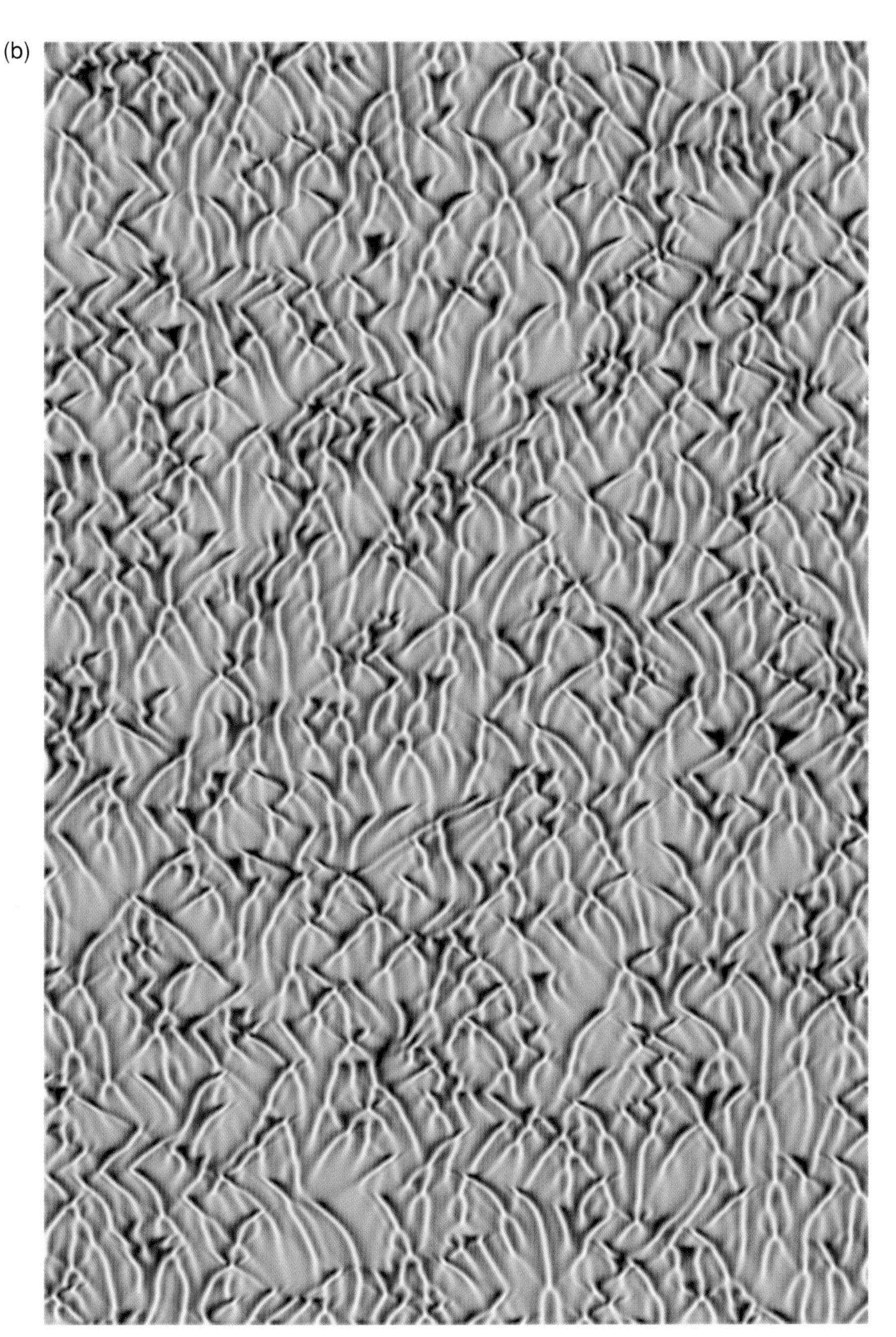


**Figure 14.4.** *(cont.)*

The NSE was derived under the assumption that a fluid can be seen as a sequence of contiguous subsystems, each in a different state of local equilibrium. This appears to contrast with the existence of self-generated long-range correlations in, for example, a periodic pattern.

It is therefore crucial to investigate directly how collective motion can emerge from disordered microscopic evolution, when the only difference with respect to equilibrium regimes is, for example, in the boundary conditions that are modified to maintain a steady flux of energy or matter. At equilibrium, detailed balance drives the evolution toward a minimum of the free energy; how, away from equilibrium, can oscillations arise and be self-sustaining?

## BOX 14.2 CHEMICAL CHAOS

A simple model to illustrate how collective motion can arise and sustain itself away from equilibrium is the reaction scheme introduced by Willamowski and Rössler in 1980. It involves three chemical species X, Y, and Z, plus the further elements $A_1$–$A_4$ whose concentrations are kept constant in time,

$$\begin{aligned} &A_1 + X \Longleftrightarrow 2X, \quad X + Y \Longleftrightarrow 2Y, \quad A_5 + Y \Longleftrightarrow A_2, \\ &X + Z \Longleftrightarrow A_3, \quad A_4 + Z \Longleftrightarrow 2Z. \end{aligned} \tag{14.5}$$

Wu and Kapral showed in 1993 that this scheme can be implemented as a lattice-gas model by representing each of the three interacting species as an ensemble of different particles moving on a discrete lattice. Simultaneously, they assumed the lattice to be populated by a stationary probabilistic distribution of $A_1$–$A_4$ and inert particles. Each time two reacting particles happen to occupy the same site, the corresponding reaction is assumed to occur with a probability proportional to the reaction rate. On the other hand, only elastic collisions can occur with the inert particles.

If the reaction rates are sufficiently small in comparison with the diffusive rate, one expects the dynamics to follow from the mass-action law, i.e.

$$\begin{aligned} \dot{x} &= \kappa_1 x - \kappa_{-1} x^2 - \kappa_2 xy + \kappa_{-2} y^2 - \kappa_4 xz + \kappa_{-4}, \\ \dot{y} &= \kappa_2 xy - \kappa_{-2} y^2 - \kappa_3 y + \kappa_{-3}, \\ \dot{z} &= \kappa_4 xz + \kappa_{-4} + \kappa_5 z - \kappa_{-5} z^2. \end{aligned}$$

For a suitable choice of the transition rates, the evolution is chaotic, as can be appreciated by looking at a typical trajectory plotted in Figure C.

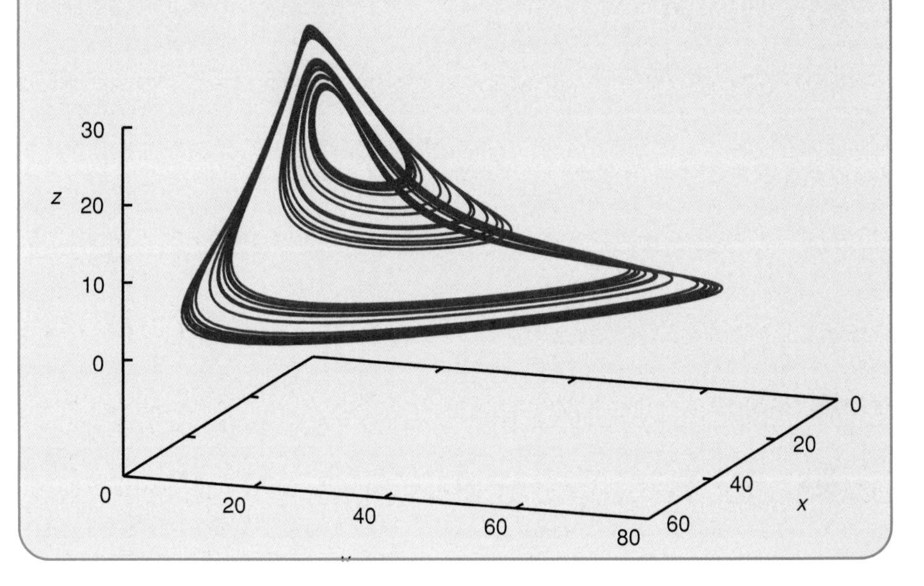


**Figure C.** A three-dimensional representation of chaotic dynamics arising in the Willamowski–Rössler model.

Meanwhile, in the mid 1980s it became clear that lattice-gas automata of the type discussed before can effectively simulate fluid dynamics, and this tool is exploited to unveil the onset of nontrivial macroscopic dynamics. In this context, the simplest configuration is perhaps interacting chemical species, since a lattice-gas model can be easily implemented by introducing different kinds of particles.

The numerical studies performed by Wu and Kapral in 1993 with the Willamowski–Rössler model (see Box 14.2) have confirmed that, in the limit of well-stirred reactants, all concentrations are spatially homogeneous and their evolution accordingly follows

### BOX 14.3 COUPLED MAPS

One of the basic objectives in the investigation of complex phenomena is to capture their essential properties by means of simple "microscopic" models. This is the philosophy behind coupled-map lattices (CMLs), a tool introduced independently in the 1980s by Kaneko and Kapral. Analogous to a lattice gas, a CML is a system in which both space and time variables are discrete, but, unlike lattice gases, the local variable is continuous. The most widely adopted scheme is

$$x^i_{n+1} = f(x^i_n) - \varepsilon \sum_{j \in \mathcal{N}(i)} f(x^j_n), \tag{14.6}$$

where the index $i$ labels the position in the lattice, $\varepsilon$ gauges the coupling strength with all sites within the neighborhood $\mathcal{N}(i)$, and the function $f(x)$ is a transformation of the unit interval onto itself. A typical choice for $f(x)$ is the logistic function $1 - ax^2$ first studied by the ecologist May in the 1970s as a prototype model for population dynamics.

The rationale is that a continuous-time evolution like that of a chain of oscillators can be viewed as a CML if the differential equations are integrated from time $n\,\Delta t$ to time $(n+1)\Delta t$. However, rather than actually performing the integration (a time-consuming task), one guesses the functional dependence of the variables at later times from the initial condition. This approach has proved very effective in the investigation of low-dimensional chaos (see Chapter 13).

the mean-field dynamics. This is not a trivial result, since it can be argued that, in the presence of macroscopic chaos, thermal fluctuations are amplified, thus leading to a breakdown of the deterministic description. Instead, one learns that the unavoidable internal noise acts in the same way as the external noise, blurring the shape of the attractor. Yet, if the system is large enough, a coherent chaotic motion cannot be stably sustained, as conjectured by T. Bohr and collaborators in 1987, who pointed out that, in the presence of noise, a droplet corresponding to a different phase of macroscopic dynamics can always be created and grow. We illustrate this phenomenon by a period-3 dynamics. In this case, the three phases A, B, and C can always be labeled in such a way that A invades B, B invades C, and C invades A. Therefore, if a droplet of A spontaneously arises inside B, it will grow if it is large enough to overcome surface tension, and the same happens for C inside A and so on (see Figure 14.5), where A, B, and C correspond to different colors in a self-evident way, so the generic macroscopic behavior must necessarily be that of an incoherent mixture of the three phases.

Despite this apparently perfect argument, Chaté and Manneville showed convincingly in the early 1990s that macroscopic collective behavior can arise in coupled-map lattices (see Box 14.3) of more than two dimensions. It is interesting to note that the coherent behavior arises despite local chaoticity and in the absence of any conservation laws. Since no analytical explanation has been put forward, it is still unknown whether this macroscopic behavior is simply a metastable state. However, it is also possible either that the droplet argument does not apply, since not all equilibrium-statistical-mechanics concepts extend to off-equilibrium regimes, or that synchronicity of the updating rule is the crucial ingredient ensuring stability of the collective behavior.

A class of systems in which the connection between the two levels of description has fruitfully been investigated is that of highly connected networks, where each element is directly coupled with many other elements. Examples of such systems are laser arrays and networks of Josephson junctions, but the most prominent can be found in biological systems: brain and metabolic-cell networks.

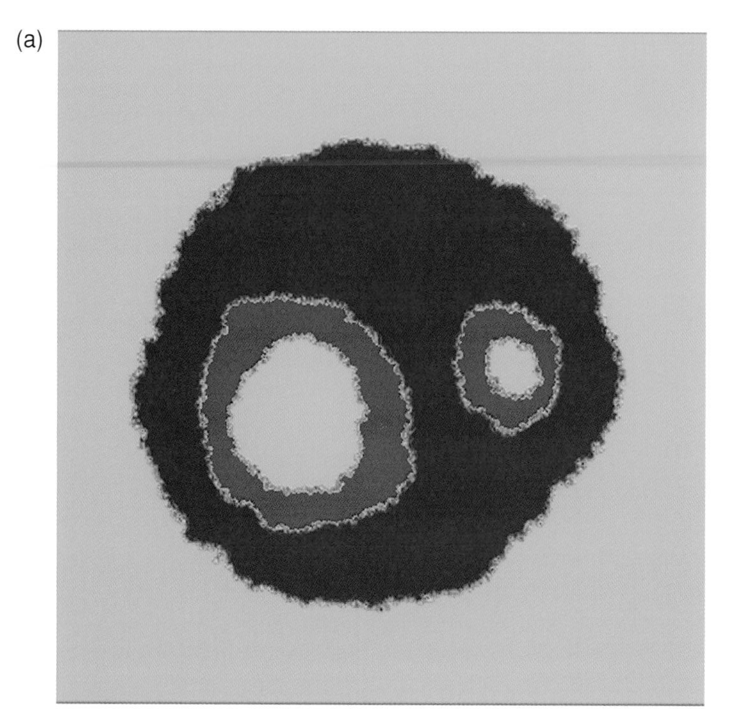


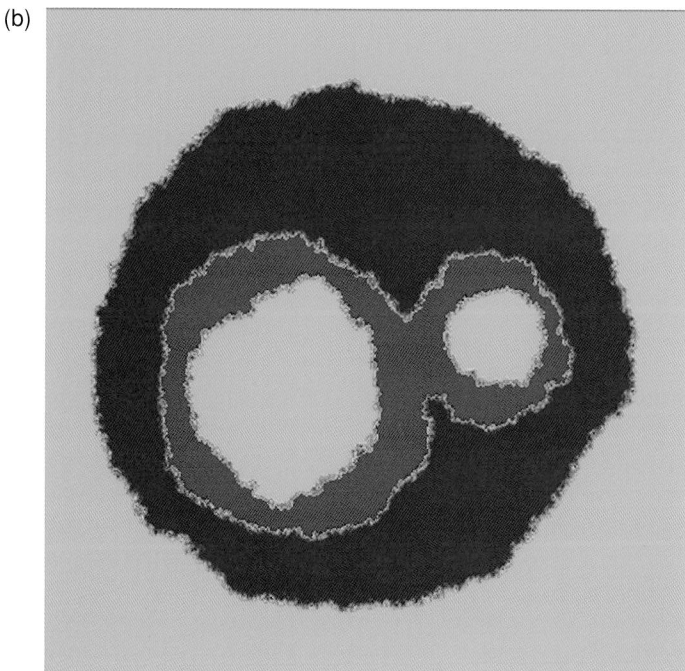


**Figure 14.5.** An example of droplet growth in which three different phases eat each other (green to red to blue to green), so none of them is thermodynamically stable; (b) shows the evolution of (a).

The simplest model is an ensemble of coupled maps, with the neighborhood $\mathcal{N}(i)$ extending to the whole set of elements. Its collective dynamics can be studied by solving the equation for the distribution $W_n(x)$ of elements in the state $x$ at time $n$. It is called the Perron–Frobenius equation, but is practically the equivalent of the Liouville equation for Hamiltonian systems,

$$W_{n+1}(x) = \int dy\,\delta(x - f(y) + \varepsilon h_n) W_n(y). \tag{14.7}$$

It owes its nonlinear character to the global coupling contained in the definition of $h_n$,

$$h_n = \int dy\, f(y) W_n(y). \tag{14.8}$$

In suitable parameter regions $h_n$ exhibits a nontrivial dynamical evolution, as shown in Figure 14.6(a). This is quite a surprising result, since it implies that hidden coherence among the network elements must exist, despite any expectation that local instabilities would destroy correlations. Kaneko and collaborators in 1999 further clarified the nature of this collective motion by adding local noise. They found that its dimension (the number of active degrees of freedom) decreases on increasing the noise variance $\sigma^2$. This can be qualitatively appreciated by comparing the two panels in Figure 14.6, while the result of a quantitative analysis, based on the computation of the dimension from the Kaplan–Yorke formula, is presented in Figure 14.7. Although it is seemingly counter-intuitive, one can understand this dependence by noticing that the noise destroys the coherence responsible for the macroscopic motion.

Given the simplicity of this model, one can hope to gain some insight about the link between the two levels of description. Vulpiani and collaborators evaluated in 2000 the instantaneous growth rate $\lambda(\delta)$ for various values of the perturbation amplitude $\delta$ applied to the macroscopic observable $h_n$. For small enough $\delta$, $\lambda$ reduces to the usual maximum Lyapunov exponent (i.e. the exponential growth rate of a generic infinitesimal perturbation).

This can be seen in Figure 14.8, where all data sets converge to the value that is a measure of the microscopic chaos. Furthermore, Figure 14.8 shows also another plateau for $\delta > 0.2/\sqrt{N}$, which corresponds to macroscopic chaos. As a result, one is led to conjecture that microscopic and macroscopic evolutions are mutually separated, even though one would expect the latter to be contained in the former. On the other hand, results of studies performed in two-dimensional fluids by Posch and collaborators indicate that some properties of the macroscopic evolution can be identified from the growth rates of all perturbations (the so-called Lyapunov spectrum).

## 14.6 Turbulence

So far we have looked at phenomena in which the macroscopic and microscopic scales are well separated. This occurs whenever a specific scale can be unambiguously identified because, for example, correlations decay exponentially so that all relevant properties are confined to a finite and possibly small range of times or lengths: milliseconds if we speak of $CO_2$-laser dynamics or millimeters in Rayleigh–Bénard convection in a 1-cm cell. However, in many natural phenomena many scales are present simultaneously.

Turbulence is certainly one of the most prominent such examples. In spite of the interest it has attracted over many centuries (starting with Leonardo da Vinci) and the fact that a meaningful mathematical model became available in the nineteenth century (the NSE equation), turbulence with vortices and eddies over many different scales (see Figure 14.9) has not yet been solved completely. Turbulence is of fundamental importance in many applied problems, ranging from the diffusion of pollutants in the atmosphere or the oceans to the design of optimal aerodynamical shapes. The dynamical properties of the ruling equation depend basically only on the Reynolds number $Re = vL/\nu$, where $L$ and $v$ are the characteristic length and velocity of the fluid. It is precisely this property that makes it possible to test small models in wind tunnels, where smaller spatial scales can be compensated by larger velocities. In the limit of

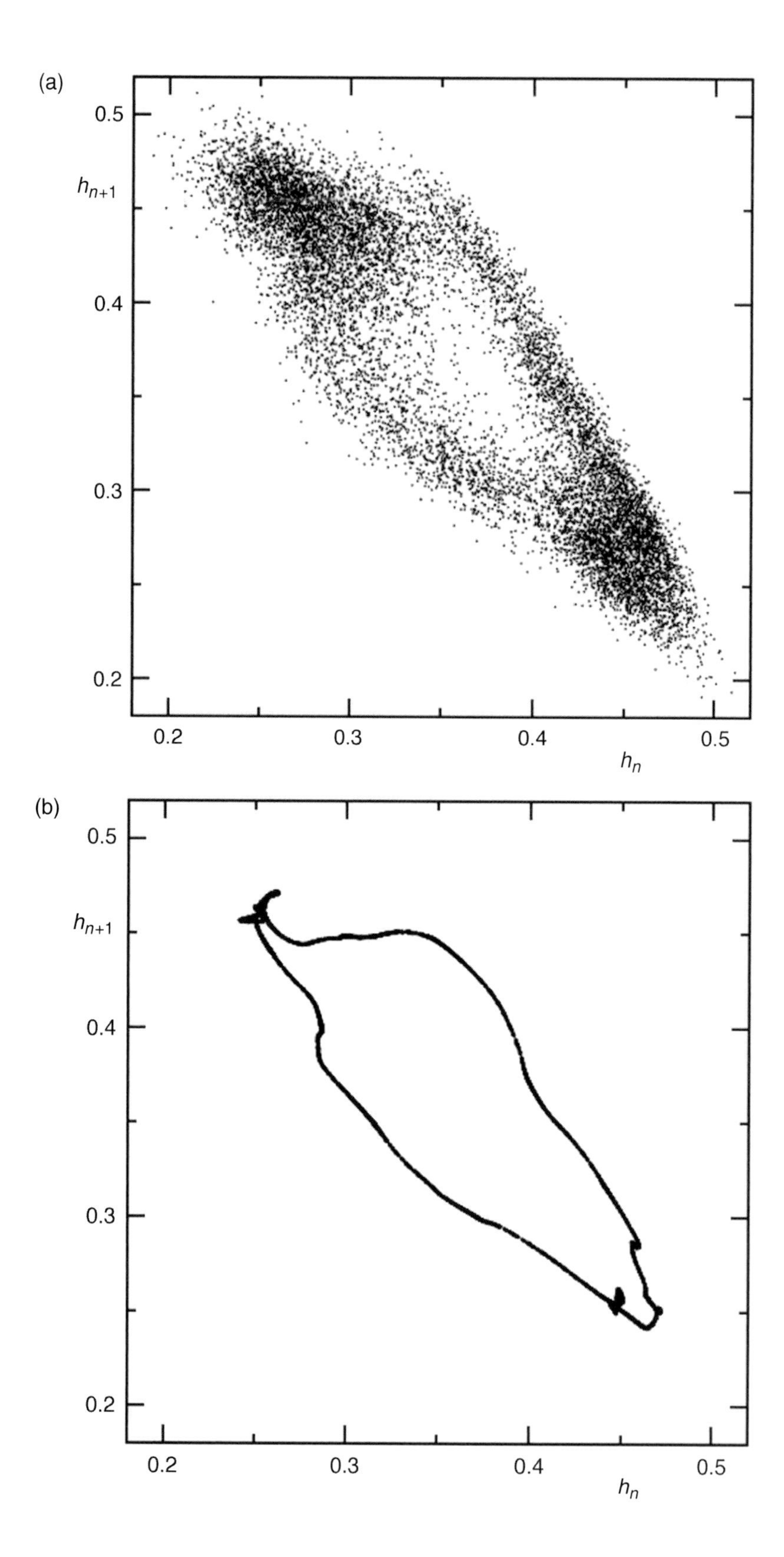


**Figure 14.6.** Collective behavior in globally coupled logistic maps ($f(x) = 1 - ax^2$) for $a = 1.86$ and coupling strength $\varepsilon = 0.1$; $h_n$ is the average value of the state variable at time $n$: (a) without external noise and (b) with noise amplitude $\sigma^2 = 2.7 \times 10^{-6}$.

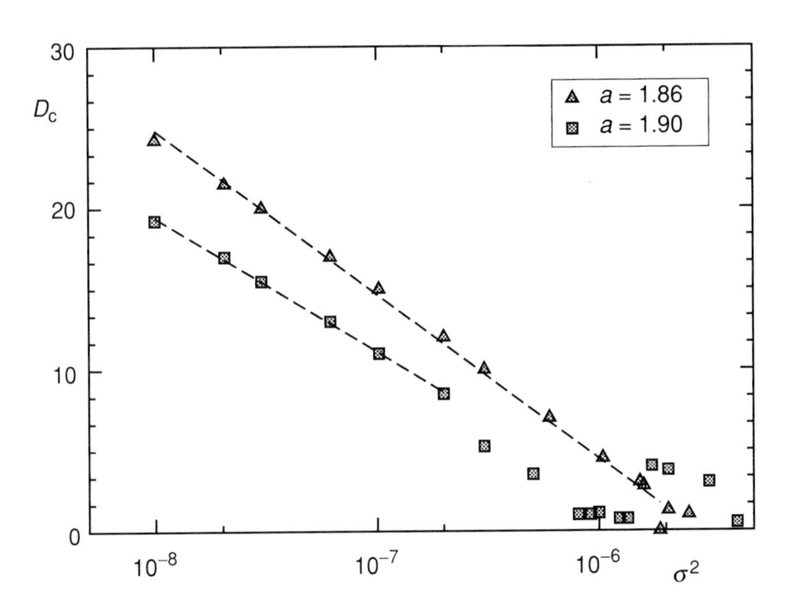


**Figure 14.7.** The dimension $D_c$ of the collective motion of globally coupled logistic maps versus the noise amplitude for two values of $a$ (see Figure 14.6). (Courtesy of K. Kaneko and T. Shibata.)

large Reynolds numbers, viscosity can be neglected except for the small spatial scales on which it becomes effective, transforming kinetic energy into heat. The difficulty of understanding turbulence arises from the singularity of such a limit; indeed, one might naively think that the NSE could be well approximated by its inviscid (and conservative) limit, the Euler equation augmented with a suitable small-scale cutoff (to prevent unphysical ultraviolet divergences). However, this is not the case, since even in the limit $Re \to \infty$, the energy flow from the large scales where it is supposedly injected (e.g. by stirring the fluid) down to the dissipative ones is finite. As a result, the fluid remains far from equilibrium and no natural smallness parameter can be used to develop a perturbative approach.

One of the few rigorous results concerns the third moment $S_3 = \langle \delta v^3 \rangle$ of the longitudinal velocity difference $\delta v = |\boldsymbol{v}(\boldsymbol{x}+\boldsymbol{r}) - \boldsymbol{v}(\boldsymbol{x})| \cdot \boldsymbol{r}/|\boldsymbol{r}|$. In 1941, under the assumption of a homogeneous and isotropic velocity field, Kolmogorov showed that $S_3$ is proportional to distance and to the energy-dissipation rate. As a result, since the energy flow

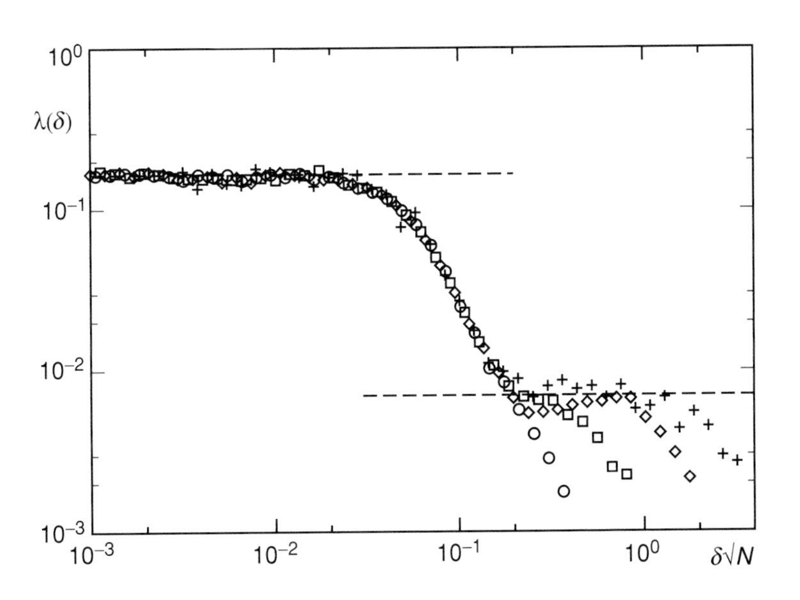


**Figure 14.8.** The finite-amplitude Lyapunov exponent $\lambda(\delta)$ versus the perturbation amplitude in an ensemble of globally coupled maps for various numbers of elements (crosses, diamonds, squares, and circles correspond to $N = 100$, 300, 1000, and 3000, respectively), showing how data sets switch from microscopic to macroscopic chaos. (Courtesy of M. Cencini and A. Vulpiani.)

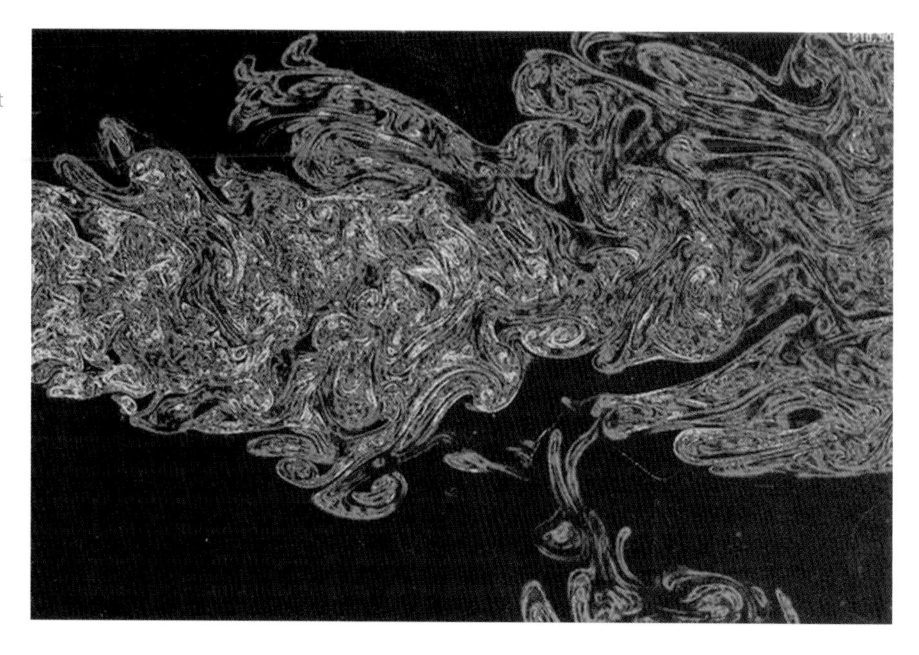

**Figure 14.9.** A two-dimensional image of an axisymmetric water jet. Different colors represent, in some nonlinear scaling, various magnitudes of the "dissipation" rate; magnitudes increase from deep blue to red and, finally, to white. (Courtesy of K. Sreenivasan.)

is constant, $\delta v \approx |\boldsymbol{r}|^{1/3}$ and the energy contained in the modes of wavenumber $k$ scales as $k^{-5/3}$. On the one hand, this result is universal: the scaling rate does not depend on the scale at which the energy is pumped. On the other hand, the result follows, rigorously speaking, from an equation for the third moment of velocity differences. There is no reason why the generalized structure function $S_n = \langle \delta v^n \rangle$ scales as $|\boldsymbol{r}|^{n/3}$ for all values of $n$. In fact, after the first experimental confirmation of the Kolmogorov prediction by Grant in 1962, many experiments demonstrated the existence of deviations from a purely linear dependence on $n$: this is also considered as the signature of multifractal behavior. Such deviations are due to fluctuations in the scaling factor (referred to as intermittency): if, on passing from $r$ to $\mu r$, the velocity difference is not multiplied everywhere by the same factors, then the scaling behaviors of the various moments $S_n$ depend on their distribution. For those who tend to associate scaling with the existence of a fixed point of some renormalization transformation, the existence of persistent fluctuations over the entire range of resolutions comes as a surprise.

To go beyond Kolmogorov's scaling considerations, many different routes have been explored, but a definite answer is still sought. Here, we mention two approaches that have helped gain some insight. The first deals with so-called shell models, first introduced by Novikov, which consist in collectively describing all Fourier modes characterized by the same wavenumber (the "shell") with a single variable, under the assumption that the internal degrees of freedom are irrelevant at least for the scaling properties of the energy cascade. One of the peculiarities of the power-law behavior observed in the turbulent regime is its coexistence with (macroscpically) chaotic dynamics. Indeed, it is usually assumed that chaos must imply a finite correlation time: on the timescale corresponding to a few units of the inverse of the (maximum) Lyapunov exponent, two trajectories diverge and thus become uncorrelated. How is it then possible to have long-range properties? As we have seen in the previous section, in high-dimensional systems nonlinearities may drastically change the evolution of small (but not infinitesimal) perturbations. When and how a coarse-grained description gives rise to a qualitatively different dynamics cannot be easily ascertained. It is nevertheless gratifying to notice that the mesoscopic description provided by shell models is characterized by many nearly vanishing Lyapunov exponents, as observed by Yamada and Ohkitani in 1988 (see Figure 14.10): the existence of a very weakly

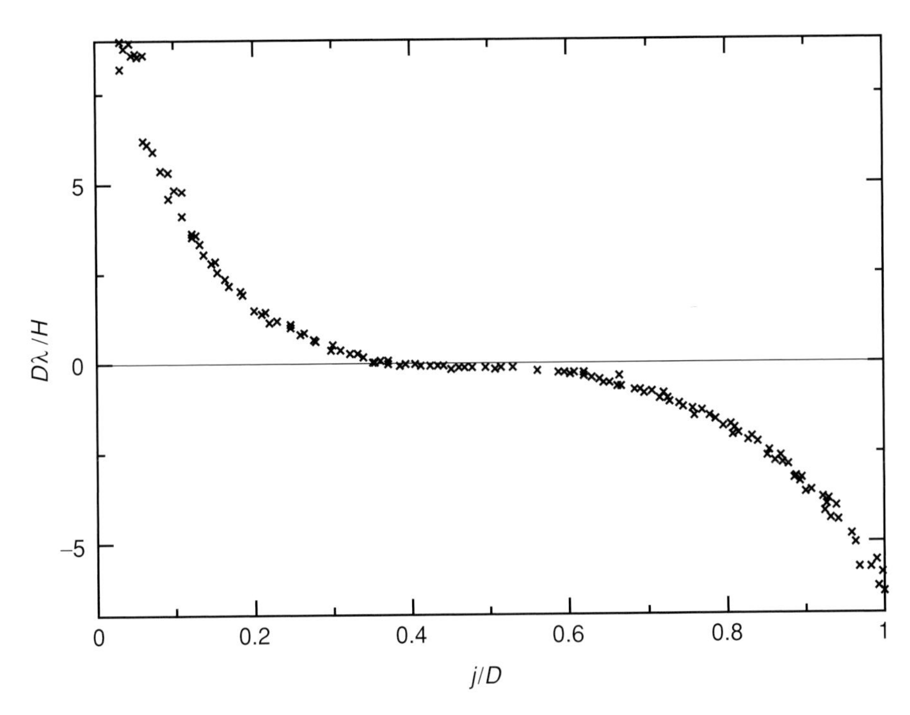


**Figure 14.10.** The Lyapunov spectrum in a shell model of fully developed turbulence. The label $j$ is rescaled to the dimension $D$ of the attractor, while the amplitude $\lambda$ is scaled by $D/H$, where $H$ is the Kolmogorov–Sinai entropy (i.e. the sum of the positive exponents). With these scales, the area of the positive part of the curve is equal to unity and the integral of $j/D$ is zero. Data points for various dissipation rates are superimposed. (Courtesy of M. Yamada and K. Ohkitani.)

unstable direction seems to link with the idea that long-range correlations can exist only in the presence of some weak chaos (at some level of description).

Another route that has successfully been explored is the simplified dynamics of a passive scalar. This approach involves studying the dynamical properties of the flow induced by an externally assigned velocity field, without having to deal with the difficulty of accounting for the feedback. Within this context, Kraichnan discovered in 1994 that the scalar field can display intermittent behavior even in the absence of any intermittency in the velocity field. In this context, the problem consists of investigating the scaling behavior of the structure function $S_n = \langle(\theta(x+r) - \theta(x))^n\rangle$, where $\theta$ is the advected field. A very instructive and clear way of looking at this problem is the Lagrangian approach, namely following the trajectories of particles that transport the scalar field. In practice, $S_n$, involving products of up to $n$ different terms, can be viewed and determined as the correlations among $n$ particles that move according to the same velocity field. The discovery in the mid 1990s of statistically conserved quantities in the dynamics of clusters of such particles has been shown to be the origin for a nonlinear scaling behavior of the various moments of the structure function.

To illustrate the existence of such conserved quantities, consider the much simplified context of a purely random field: in this case, given any three particles $\mathbf{r}_i(t)$ ($i = 1, 2, 3$), it follows straightforwardly that $\langle(\mathbf{r}_1 - \mathbf{r}_2)^2\rangle - \langle(\mathbf{r}_2 - \mathbf{r}_3)^2\rangle$ is constant in time because all the particles diffuse in the same way. More generally, the resulting conservation laws can be seen as collective effects regulating the behavior of the $n$ particles forced by the same field: it is a sort of "average synchronization." The importance of such conservation laws eventually lies in the implications for the anomalous scaling behavior of $S_n$.

## 14.7 Self-organized criticality

A completely different class of (quasistatic) phenomena in which power laws have unambiguously been observed is that of earthquakes, described by the

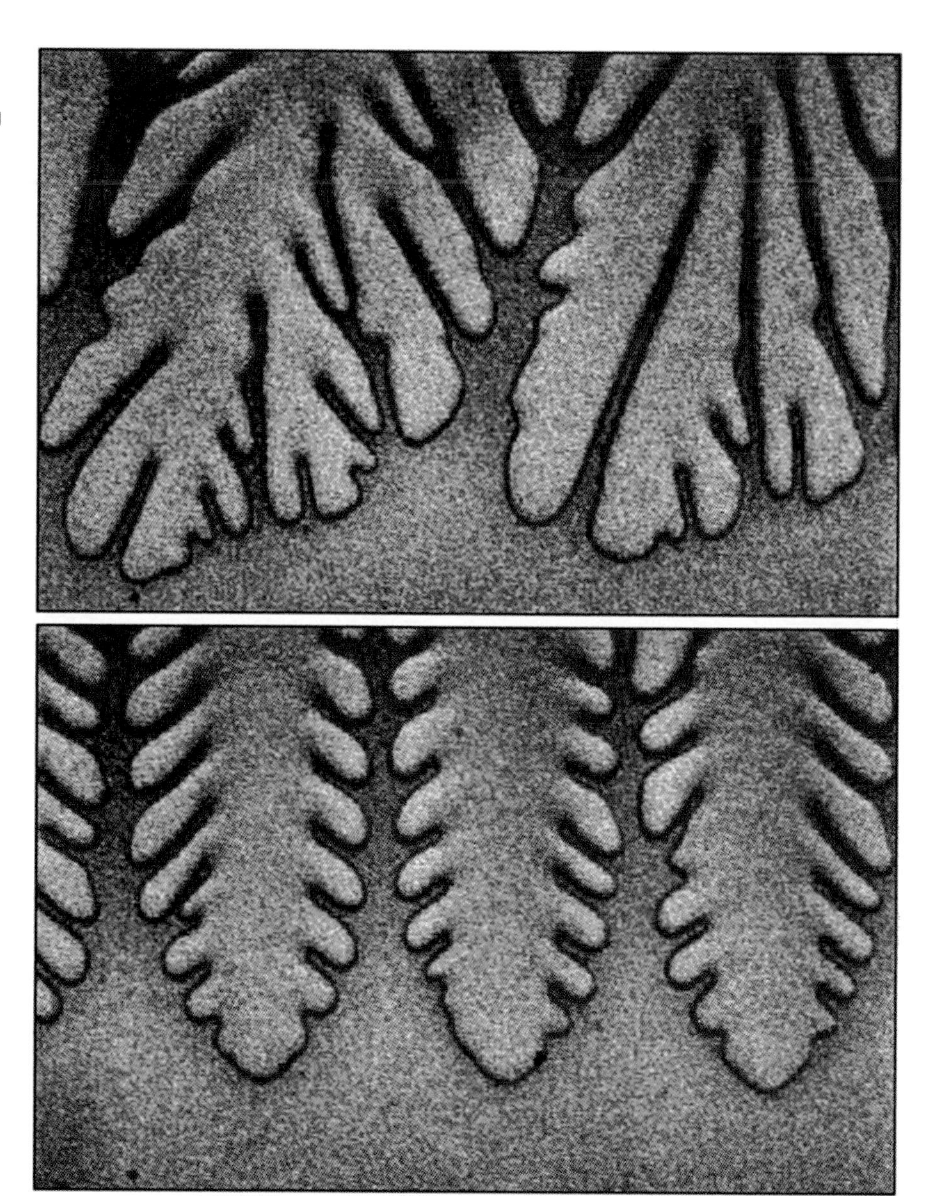

**Figure 14.11.** Solidification of an organic alloy between two glass slides. The solid is growing downward into the liquid. (Courtesy of E. Bodenschatz.)

Gutenberg–Richter law, in which the number of seismic movements involving an energy larger than $E$ scales as $N = E^{-b}$, with $b$ around 0.6 for large quakes and 0.8 for smaller ones. Although it is clear that large phenomena are rarer than small ones, they are not so exceptional as to be unobservable: on the contrary, how to predict them is an important problem. Other examples are the ubiquitous $1/f$ noise and the fractality of coasts as pointed out by Mandelbrot. All these examples cry out for a general and robust explanation of how power laws can spontaneously emerge. From equilibrium statistical mechanics, we have learned that correlations generically decay exponentially; it is only in the vicinity of a continuous phase transition that power-law dependences may arise. Indeed, in the early 1970s Wilson showed that, upon repeatedly rescaling the space variables (i.e. by applying the so-called renormalization-group transformation – RGT), the Hamiltonian may converge toward a fixed point. Since the fixed point is unstable, it can be reached only if the Hamiltonian has no component

**Figure 14.12.** A pattern formed by pumping air into an oil layer contained between two 300-mm-diameter glass plates spaced only 0.13 mm apart. (Courtesy of M. G. Moore, E. Sharon, and H. L. Swinney, University of Texas at Austin.)

along the expanding direction(s). This implies that the observation of scale invariance requires the fine-tuning of a parameter, until a phase-transition point is exactly met.

A problem that has attracted much interest and in which power laws spontaneously arise is that of dendritic growth: a dendrite is a crystal with a tree-like structure formed, e.g., during solidification (such as frost on glass) and in ion-deposition processes (see Figure 14.11). Similar phenomena occur also in a porous medium filled with a liquid, when a second immiscible fluid characterized by a much lower viscosity is injected (see Figure 14.12). This Hele–Shaw flow is important for the petroleum industry, which exploits this phenomenon to extract oil from rocks by injecting water.

Some insight comes from simple models introduced in the hope of capturing the relevant mechanisms. The most successful is certainly diffusion-limited aggregation, which was introduced by Witten and Sander in 1981 for irreversible colloidal aggregation. It consists of a recursive process: to a connected cluster of particles, a further particle is added at a large distance (see Figure 14.13). The new particle is assumed to perform a random walk until it touches the cluster and irreversibly sticks to it. By repeatedly adding particles, a highly branched, fractal cluster is progressively generated. The reason for the spontaneous onset of the branched structure is the increasingly small probability of the newly added particles penetrating the narrow fjords. The structure grows with a local velocity proportional to the probability density of random walkers $\rho$ which, in turn, satisfies the Laplace equation ($\nabla^2 \rho = 0$). A similar interpretation holds in the context of Hele–Shaw fluid flow if the probability density is replaced with the pressure field, but there are subtle differences between the two

**Figure 14.13.** A diffusion-limited-aggregation (DLA) cluster. Various colors correspond to different arrival times of particles. The black line shows the random motion of a newly arriving particle.

cases. The evolving cluster is fractal with a dimension that is about 1.71 in planar systems and, as for turbulence, the probability of attaching to a given site is multifractal. As in turbulence, the difficulty of obtaining an accurate description stems from the absence of a suitable smallness parameter for a perturbative approach. Despite such difficulties, much progress has been made in two dimensions by exploiting techniques from conformal field theory; and quite surprising is a recently discovered analogy with two-dimensional quantum gravity.

The ubiquitous presence of power laws led Bak to conjecture that in Nature robust mechanisms can maintain out-of-equilibrium systems in a critical state without the need to fine-tune some control parameter. Bak viewed sand dynamics as a good example of a scenario that he called "self-organized criticality" (SOC). Dry sand can be heaped only until a suitable slope is reached, after which larger slopes induce a flow that re-establishes the maximal sustainable slope. The essence of SOC is that this self-adjusting state is critical in the sense that small deviations may induce avalanches of all sizes. A simplified model of sandpile dynamics used to identify the basic mechanisms and to perform a quantitative analysis consists of a variable number $Z$ of grains assigned to each node of a square lattice. Grains are then randomly added until $Z$ becomes larger than or equal to 4, in which case four grains are displaced to the neighboring sites. Such a displacement can, in turn, trigger toppling of other sites – an avalanche (see Figure 14.14). At the boundaries, grains are lost to the outside world. New grains are dropped only after a previous avalanche is complete. After an initial transient, the system spontaneously reaches a regime in which avalanches of all sizes may appear, i.e. the probability of generating an avalanche of size $N$ is $N^a$.

The basic ingredients considered responsible for the onset of SOC are (i) slow driving; (ii) the existence of many marginally stable states – each new grain is added only after the previous avalanche has stopped (since avalanches can last a very long time, it is clear that this can only be an approximation to the real world, but it does not diminish

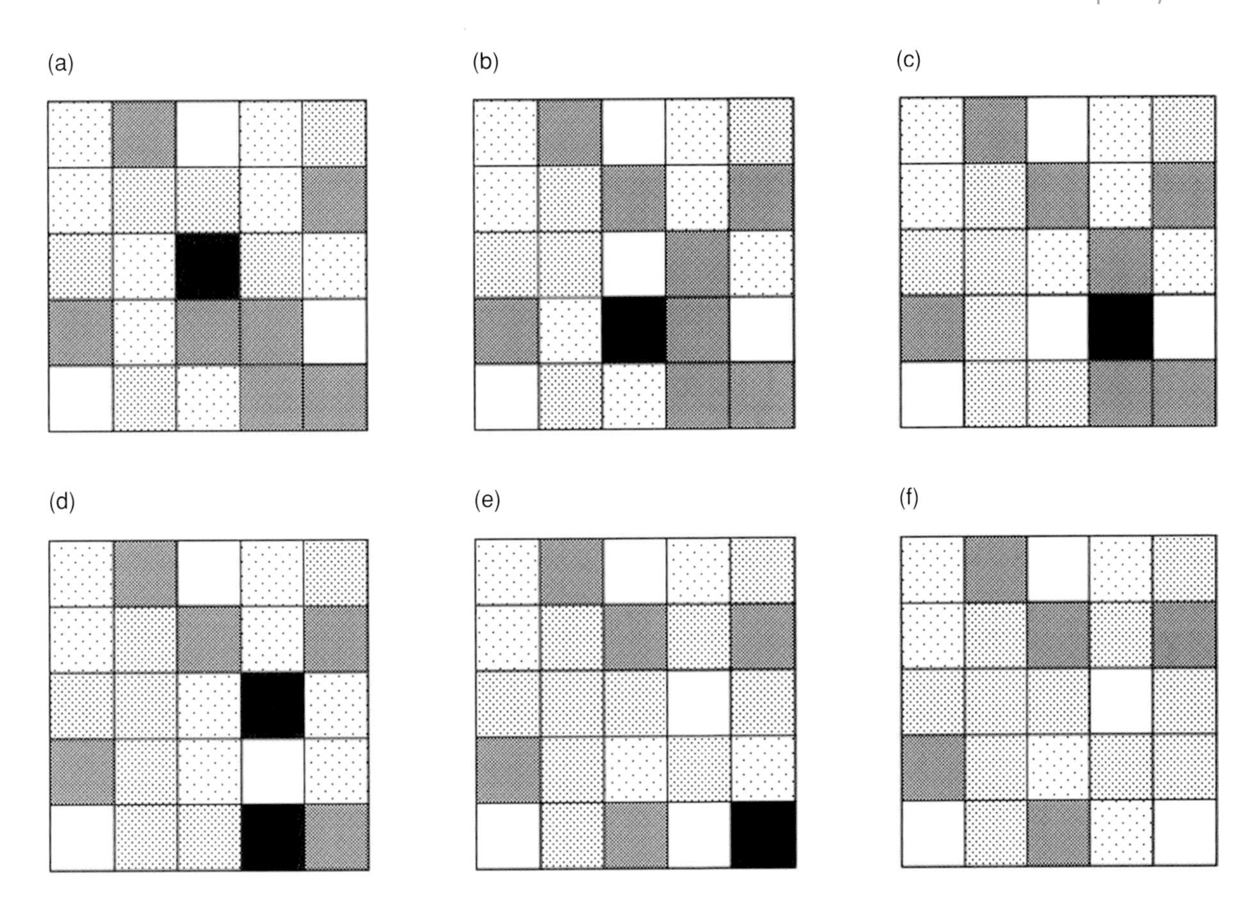


Figure 14.14. An example of the Bak–Tang–Weisenfeld model for sandpile dynamics in a 5 × 5 system. Gray levels from white to black correspond to occupation numbers from 0 to 4. The sequence (a)–(f) shows the avalanche following the initial toppling of the central site.

the relevance of the argument); and (iii) many metastable states, which are needed in order to ensure the existence of a distributed memory of the accumulated stress.

Additionally, this model is conservative (the number of grains is conserved during the avalanche except when it touches the boundary). Although, as shown by Kardar, this helps to ensure long-range correlations, it does not appear to be truly necessary. In fact, no conserved quantity exists in the more realistic model introduced by Carlson and Langer in 1989 to simulate stick–slip dynamics. They considered a chain of blocks and springs as in Figure 14.15. The crucial ingredient is the dependence of the friction force $F$ on the velocity (see Figure 14.16 for a qualitative plot). As a result, as long as the force applied is smaller than $F_0$, a block "sticks," otherwise it slips. The simultaneous motion of several blocks is the equivalent of the avalanche in the previous context. If the pulling process is sufficiently slow, a power-law distribution is observed.

More important than the observation of SOC in simplified models is its observation in real experiments. Although some have revealed only avalanches of specific sizes, evidence of power laws has been found as well. An experiment performed at IBM in 1990 closely paralleled the simplified model: 0.6-mg particles of aluminum oxide were dropped one at a time onto the middle of a circular disk. The resulting distribution of avalanches is scale-invariant over a large range for smaller disks, while relaxation oscillations were observed in the case of the largest disk (of diameter 3 inches). In another experiment performed in Norway rice grains were dropped between two narrowly separated plexiglas plates. The experimenters found instead that power-law distributions arise with long-grained rice, whereas avalanches have a typical size with more rounded rice. The difference between the two experiments has been attributed to the non-negligible role of inertia in the latter case. Thus, even though SOC

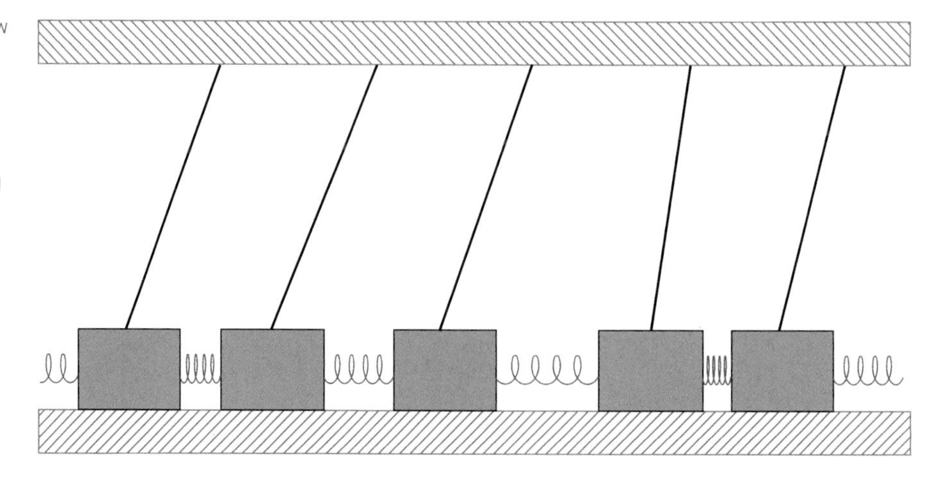

Figure 14.15. A schematic view of the model introduced by Carlson and Langer to describe stick–slip dynamics. Each block is harmonically coupled to its neighbors and to a common rail that moves slowly, pulling all blocks simultaneously. Finally, movement on the underlying surface is affected by the nonlinear friction described in Figure 4.16.

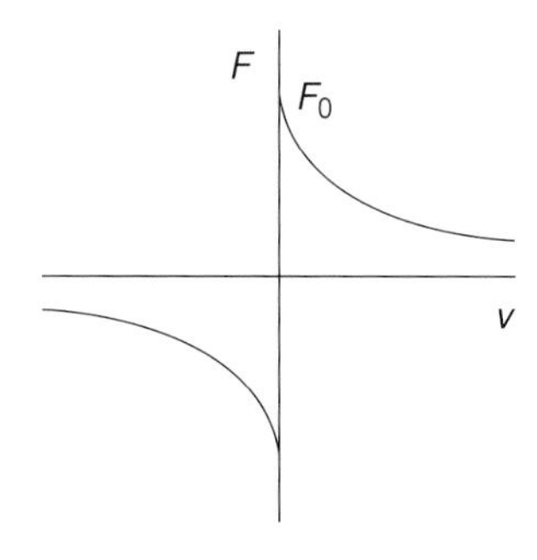


Figure 14.16. The dependence of the friction force felt by each block of Figure 4.15 on its velocity. The discontinuity schematizes the difference between static and dynamic friction.

plausibly exists in finite regions of parameter values, it remains to be understood what the necessary and sufficient conditions for it to occur are.

The main theoretical problem in SOC is understanding how scaling behavior can be robustly generated. Setting up a proper RGT amounts to looking at the process with various spatial resolutions (e.g. by interpreting all sites in a box of size $2^k$ as a single site) and thereby determining a recursive expression for the dynamical law $L$ at two consecutive levels of the coarse-graining process (e.g. $k$ and $k + 1$). If the dynamical law is described with a sufficient accuracy by the set of parameters $P$, the latter step amounts to writing $P_{k+1} = L(P_k)$. A fixed point $P_0 = L(P_0)$ represents a scale-invariant solution. However, unlike equilibrium problems, for which knowledge of the Hamiltonian is sufficient to solve the problem (at least in principle), here the iteration process requires incorporating some "dynamical ingredients": in particular, it is necessary to determine the probability density of critical sites (those that topple on receiving a "quantum" of energy) at each level of description. In 1991 Pietronero and collaborators proposed solving the problem by imposing a stationarity requirement – that the amount of "energy" falling on each cell compensates, on average, for the flow of "energy" toward neighboring cells. This constraint is not simply a theoretical trick and is an essential element for SOC. Indeed, it is generally believed that the dynamics spontaneously provides the necessary feedback to stabilize the otherwise unstable fixed point of RGT.

## 14.8 Granular media

The above-mentioned examples of SOC deal with granular media. Granular media are employed in many industrial processes: pharmaceutical products often derive from powders; detergents, cement, and grain are further examples of granular media that need be transported and stocked. Understanding grain dynamics would therefore be extremely useful, and for instance would prevent the explosion of silos induced by electrostatic forces.

Technically speaking, a granular medium is a conglomerate of particles of sizes above 1 μm: above such a size, thermal fluctuations (on the order of $k_B T$) are negligible, while mutual friction is not. Although grains are "macroscopic" objects, one can consider them as "microscopic" elements characterized by a few degrees of freedom (position, size, and so on) to construct thereby a macroscopic description. In this sense, one can loosely (but not always) view a granular medium as a fluid interacting with a heat bath at zero temperature. It is precisely the presence of friction, together with the

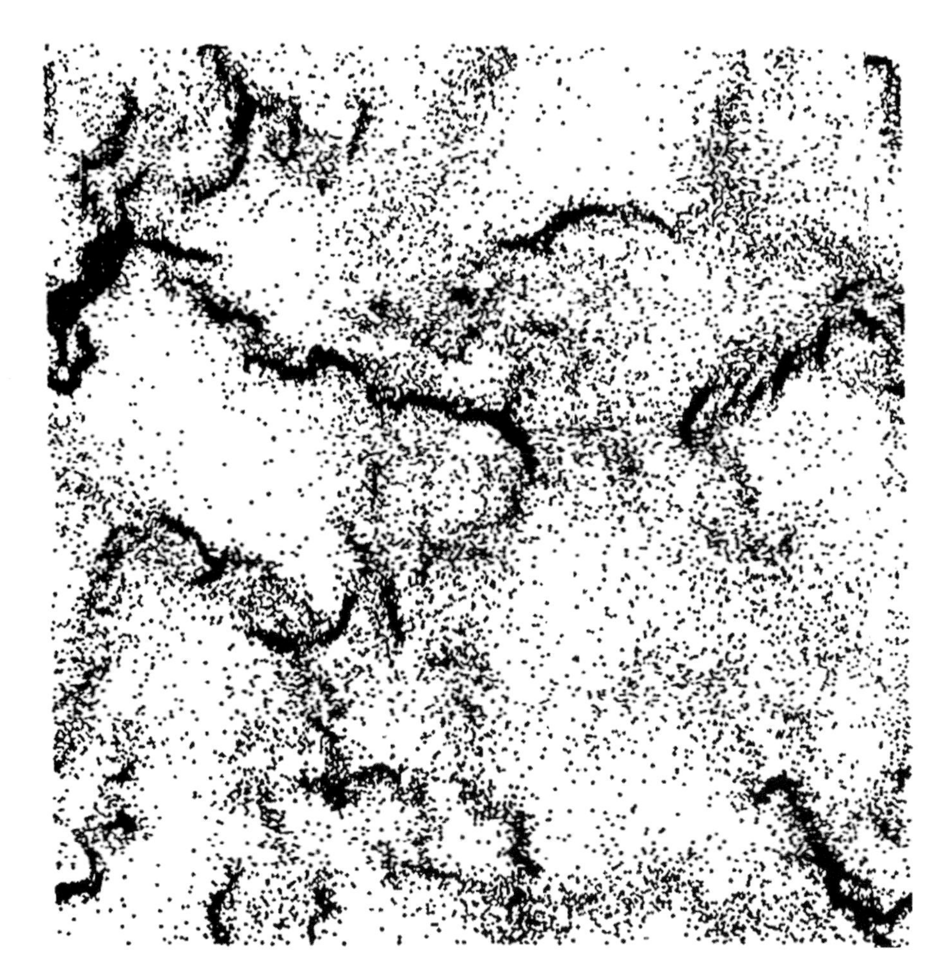

Figure 14.17. A typical configuration of an ensemble of 40 000 particles after approximately 500 collisions per particle (the coefficient of restitution is 0.6). (Courtesy of I. Goldhirsch and G. Zanetti, *Phys. Rev. Lett.* 70 (1993) 1611.)

absence of thermal fluctuations, that induces the existence of many metastable states. Everybody knows that, when sand is poured into a box, the final state depends on how this was done.

The large number of metastable states is not only responsible for peculiar dynamical properties, but is also an obstacle to carrying out experiments under well-controlled conditions and drawing reliable conclusions. Edwards and Grinev argued in 2002 that the real question is that of whether "there is a way of depositing sand into a container in such a fashion that one can make physical predictions, and find what are the circumstances when the smallest amount of information is required about the sample packing."

Anyhow, at least for lower densities, it is natural to ask whether one can average out fast local motion of the single particles and thereby obtain a Navier–Stokes-type equation. In general, a slow (in space and time) dynamics is ensured by the existence of conserved quantities, since they can vary only because of their flow across neighboring regions. Unfortunately, the inelastic collisions in granular media contribute to dissipation of energy. It is, therefore, not entirely a surprise to see strong clustering in numerical simulations of, e.g., two-dimensional hard-disk systems (see Figure 14.17). Perhaps more unexpected are the experimental results of Nagel and collaborators, who in 1995 found quite a rapid motion close to the boundaries of a vibrating container filled with seeds. No longitudinal motion is expected close to a boundary in a normal fluid!

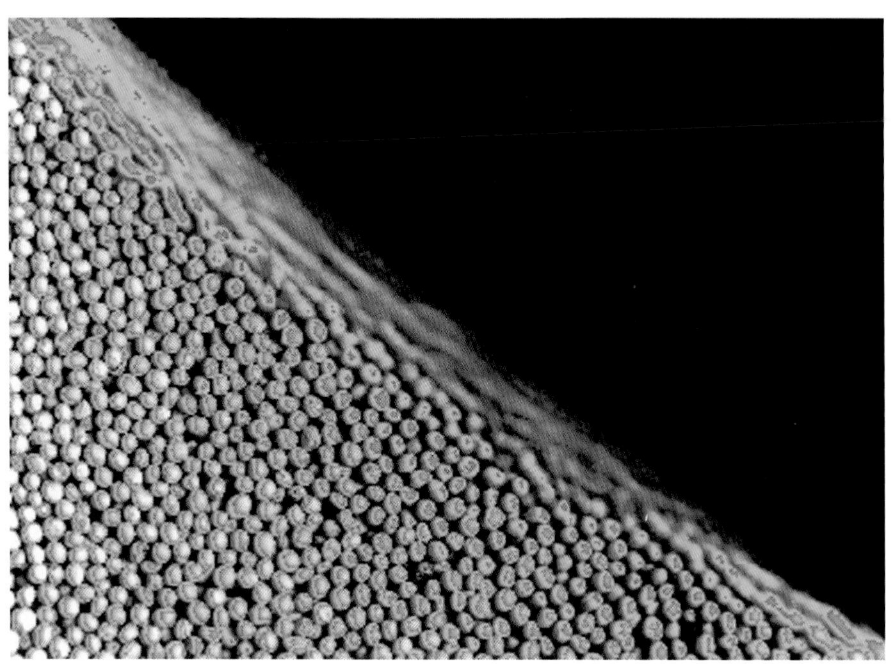

Figure 14.18. A snapshot of an avalanche in a pile of mustard seeds. Motion is confined to a thin layer. (Courtesy of S. Nagel.)

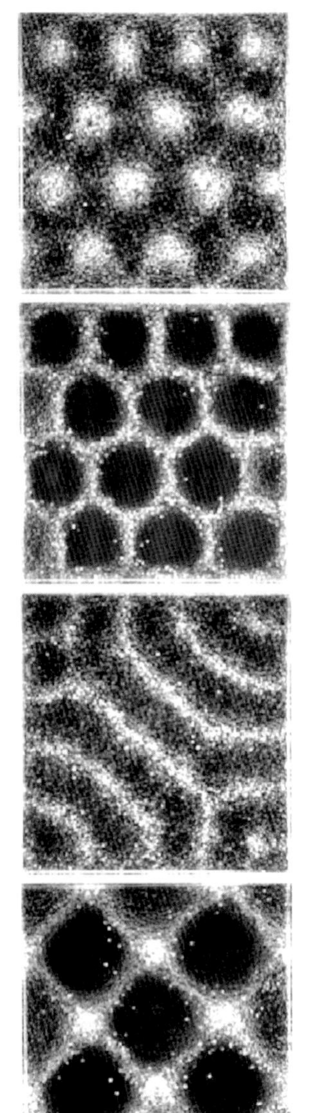

Figure 14.19. Examples of patterns arising in sand deposited on a vertically vibrating surface, for various frequencies and amplitudes of the modulation. (Courtesy of H. Swinney *et al.*, *Phys. Rev. Lett.* 80 (1998) 57–60.)

The peculiarity of boundary dynamics in granular media is confirmed by another experiment, which was designed to study the generation of avalanches. Figure 14.18 shows how the motion of the grains (here mustard seeds) is restricted to the top layers, so that different parts of the system behave simultaneously as a fluid and as a solid. As a result of all these considerations, it is doubtful whether a general hydrodynamic description of granular media can be constructed. Nonetheless, there are instances where standard out-of-equilibrium patterns appear in granular media, suggesting that their dynamics is controlled by a few macroscopic variables. Figure 14.19 shows a series of four examples for sand on a vibrating plate.

At larger densities, further peculiar features can be observed in granular media, like the extremely slow evolution typical of glassy regimes. This was observed in an experiment by Nagel and collaborators in 1995, with spherical glass particles. After loosely packing beads inside a tube, they repeatedly let the tube vibrate and observed a logarithmic growth of the density (determined from the height of the powder in the tube). The essence of this behavior can be captured by a simple parking-lot model (see Box 14.4), where slowness arises from the fragmented nature of the excluded volume.

Besides giving evidence of "jamming" phenomena, the experiments by this group have the further merit of opening an entirely new perspective. On progressively increasing the tapping force, it was observed that the density increases until it flattens when the close-packing regime is reached (full circles in Figure 14.20). On subsequently decreasing the tapping force, it was found that the density keeps increasing until it reaches the maximum close-packing value (diamond-shaped points). In contrast to the former branch, the latter is reversible: increasing or decreasing the tapping force makes it possible to move back and forth along the same sequence of macrostates. It is therefore tempting to look for a thermodynamic description. In particular, Edwards argues that the reversibility of the process is an indication that all microstates compatible with a given density are accessible and it should therefore be possible to define an entropy as in ordinary statistical mechanics. The difference is that one has to count all microscopic configurations corresponding to the same global volume rather than to the same energy. Accordingly, one can replace the well-known thermodynamic definition

BOX 14.4 THE PARKING-LOT MODEL

The parking-lot model is one of the simplest systems exhibiting glassy behavior. Identical particles (cars) of unit size can either leave the parking lot with a rate $k_-$ or make an attempt to park with a rate $k_+$. Not all attempts are successful, because parked cars are randomly positioned (there are no delimiting lines) and the attempts themselves are made at random positions (see for instance the unsuccessful attempt depicted in Figure D). For $k_- \to 0$, the equilibrium density of occupied sites approaches unity (no empty space), but the temporal convergence is logarithmically slow: when the density is high, many empty spaces are typically available but none of them is large enough for a car to fit in, and an exponentially large number of random moves must be made in order to create locally a "quantum" of space.

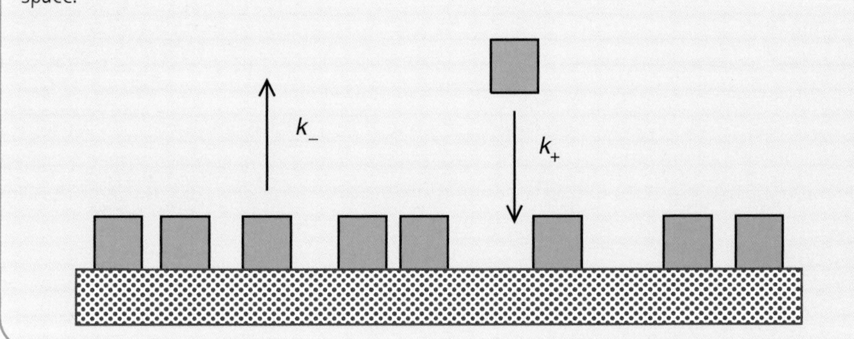


**Figure D.** The parking-lot model. "Automobiles" (cyan boxes) can either leave or enter the parking lot at rates $k_-$ and $k_+$, respectively. In the latter case, the move is accepted only if a large enough space is available to host the "automobile."

of temperature $T = dE/dS$ with an analogous relation $X = dV/dS$, where $X$ is a suitable temperature. Edwards goes on to suggest the same experiment on a horizontal tube filled with two different granular media A and B separated by a movable membrane. In the reversible branch of the tapping process, the position of the membrane will always

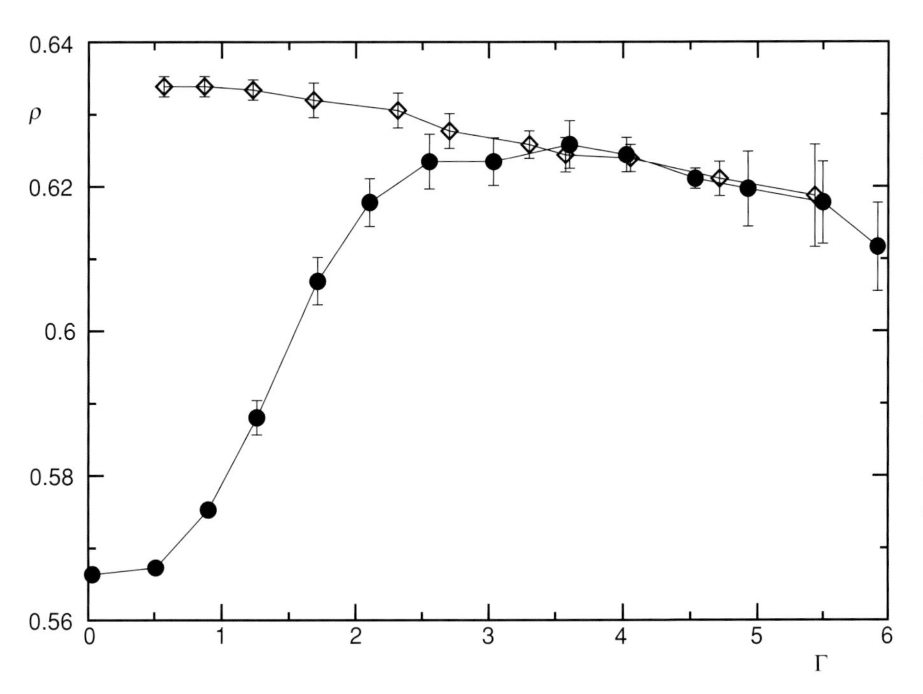


**Figure 14.20.** The concentration (quantified by the packing fraction $\rho$) of a pile of monodisperse beads in a tube as a function of the tapping force $\Gamma$ (measured in units of the acceleration due to gravity, $g = 9.8\ \mathrm{m\,s^{-2}}$). Full circles correspond to the irreversible branch constructed by increasing the tapping force from zero. Diamonds correspond to the reversible branch along which the system can move once the close-packing regime has been reached. (Courtesy of S. Nagel.)

be the same when the same tapping force is applied. The mysterious quantity $X$ should then express how the two media are mutually in "equilibrium."

## 14.9 Glassy behavior

So far we have mainly considered systems kept away from equilibrium by continuous injection of energy; however, as we have partly seen in granular media, there exists another regime in which the convergence to equilibrium is so slow that, in practice, the asymptotic state is never reached. This is the case of supercooled liquids, known also as structural glasses, which behave both as highly viscous liquids and as amorphous solids.

The slow dynamics of a glass is often depicted as random motion in a highly complex energy landscape characterized by several minima. After a rapid and deep quenching, a liquid does not appear to converge toward a crystal structure, even though this state is an absolute minimum of the energy, since many intervening barriers have to be overcome. Since the complexity of the dynamics is a direct consequence of the structure of the energy landscape, for some time physicists directed their interest toward the simpler problem of understanding glassy behavior in spin systems, where the relevant microscopic variable is the magnetic moment (examples are dilute solutions of magnetic impurities in noble metals such as CuMn, as well as magnetic insulators and some amorphous alloys). The advantage of these systems is that their energy landscape is assigned a priori, rather than being the result of a self-organized process as in structural glasses.

However, even in the "simplified" spin-system context, the link between the type of interaction and the structure of the energy landscape is far from obvious. Furthermore, one must carefully distinguish energy wells found in phase space from free-energy valleys. Neighboring minima in phase space are typically separated by barriers whose height is of order unity (independently of the system's size) and can thus be overcome by thermal fluctuations at any temperature. Only macroscopic barriers, involving sizable rearrangements of an extensive number of variables, may contribute to the birth of truly metastable states (separated by barriers whose height is proportional to the system's size). Whenever this is the case, different "replicas" of the same system may be found in macroscopically different states, and one speaks of replica-symmetry breaking.

For many years it was believed that two ingredients were required for glassy behavior: disorder and "frustration" (conflicting constraints arising from the attempt to optimize distinct energy contributions independently). A quite general class of spin models is defined by the Hamiltonian

$$\mathcal{H} = -\sum_{\mathcal{N}} J_{i_1,i_2,\ldots,i_p} s_{i_1} s_{i_2} \ldots s_{i_p}, \tag{14.9}$$

where $s_i$ is a classical variable assuming values $\pm 1$ and the interactions are restricted to suitable sets of $p$-tuples. For $p = 2$ and random exchange interactions $J_{ij}$ extending to all pairs of particles, the above Hamiltonian reduces to the well-known Sherrington–Kirkpatrick (SK) model, which has contributed much to establishing the proper analytical tools for describing spin glasses. In the SK model, frustration is induced by the simultaneous presence of ferromagnetic and antiferromagnetic interactions (the coupling constants are symmetrically distributed around 0), which give conflicting contributions to the alignment of each spin. One of the major difficulties in the construction of a macroscopic theory is the identification of the proper order parameter. In most cases this corresponds to a directly accessible observable like the magnetization. The solution of the SK model found by Parisi in 1980 indicates that in spin glasses the order parameter is a much more complex and indirect object: the distribution $P(q)$ of mutual differences between generic pairs $(\sigma\,\tau)$ of equilibrium states, where the difference is quantified by the so-called overlap parameter $q = \langle \sigma_i \tau_i \rangle$.

In the 1990s it was a surprise to find that glassy behavior may appear even in the absence of disorder. In particular this is the case of the model defined by Equation (14.9) for $p = 3$ and with purely ferromagnetic interactions ($J = 1$). In this model, the homogeneous solution $s_i = 1$ is certainly the minimal-energy state: however, if a spin happens to flip because of thermal fluctuation, this automatically introduces antiferromagnetic interactions, which, in turn, favor the opposite alignment of all pairs of spins, i.e. disorder emerges spontaneously!

Although much progress has been made in understanding the glassy regime, many fundamental questions remain. Results of experimental and numerical studies suggest that there is a true transition toward an amorphous phase. The evidence comes from two independent observations. On the one hand, experiments using various glasses at various temperatures reveal that the relaxation time $t_r$ (determined e.g. from viscosity measurements) diverges according to the heuristic Vogel–Fulcher law, $t_r = \exp[A/(T - T_{VF})]$. In the so-called strong glasses, $T_{VF} = 0$ and, accordingly, the above reduces to a standard Arrhenius law, whereas for fragile glasses (such as $SiO_2$), $T_{VF}$ is finite. On the other hand, measuring the specific heat in the (supercooled) liquid and in the crystal phase at the same temperature can determine the configurational entropy $S(T) = S_{liq} - S_{cry}$. Kauzmann in 1948 was the first to notice that $S(T)$ decreases with $T$ and becomes negative below a temperature $T_K$. Since it makes no sense to attribute a negative entropy to a thermodynamic phase, one is led to conjecture that $T_K$ represents a real transition point. The consistency of this scenario is strengthened by the remarkable observation of Richert and Angell, who in 1999 noticed that $T_K$ is very close to $T_{VF}$. Their statement is, however, somehow weakened by the absence, in either case, of experimental results close enough to the hypothetical critical temperature.

In mean-field-type models, where all particles interact with each other, a second important critical temperature $T_c > T_K$ has been identified as the one below which the phase space is split into an exponentially large number of disconnected ergodic components. In finite-dimensional systems, $T_c$ is no longer a well-defined quantity (leaving aside some preliminary evidence of a topological signature) since the free-energy valleys communicate with each other and, as a result, one "simply" observes increasingly slow (upon decreasing the temperature) relaxation phenomena. In fact, experimentally, the glass transition is identified via an anthropomorphic concept: the temperature $T_g$ below which the relaxation time is larger than $10^3$ s.

Below $T_g$ (at least until the relaxation time remains finite), one might imagine that glassy behavior is nothing but a slow convergence toward equilibrium and that the physics has been understood once the proper timescale is identified. Experiments performed with colloidal, orientational, and spin glasses and also with polymers indicate that aging – as the phenomenon is called – is far richer and even possesses universal features. Aging experiments require that (i) the system is rapidly quenched to a temperature $T_0 < T_g$ and is allowed to relax for some time $t_w$; after which (ii) a perturbation is introduced (by, e.g., an external field), whose effect is finally measured $t$ units later. Over a wide range of timescales, the response to the perturbation depends on the two times $t_w$ and $t$ through their ratio $t/t_w$, as can be appreciated in Figure 14.21, which plots the results of an experiment by Vincent and collaborators. The data collapse implies that the response time depends on the (arbitrary) waiting time: a beautiful indication of the presence of infinitely many timescales.

An even more intriguing aspect of aging was discovered in 1958 by Kovacs, who observed that the glassy state of a polymer can "remember" the cooling process. The now standard protocol for investigating memory effects was introduced by Nordblad

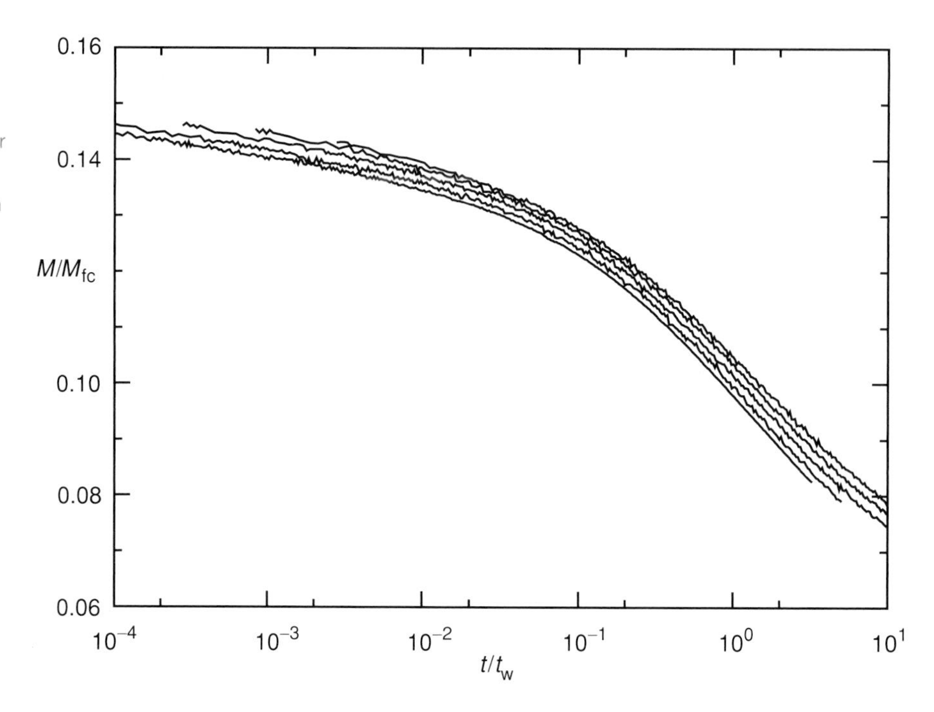


**Figure 14.21.** Suitably normalized thermo-remanent magnetization versus time $t$ scaled to the waiting time $t_w$ for five values of $t_w$ (300, 1000, 3000, 10 000, and 30 000 s from right to left) in an experiment with Ag : Mn (2.6% Mn) spin glass at $T = 9\text{ K} = 0.87T_g$. (Courtesy of E. Vincent.)

and consists of various cooling/heating steps during which a suitable observable (e.g. susceptibility or the dielectric constant) is recorded. In the first step, the system is cooled down from above $T_g$ to some minimum temperature $T_m$ and heated to determine a reference behavior of some observable $\mathcal{O}$. In the second step, the cooling process is interrupted for several hours at some intermediate temperature $T_i$, before being resumed. Figure 14.22 shows the results of an experiment by Ciliberto and collaborators on plexiglass. Both curves give the difference with respect to the reference curve. One sees that (i) the susceptibility, after dropping down at $T = T_i$ (as a result of a natural

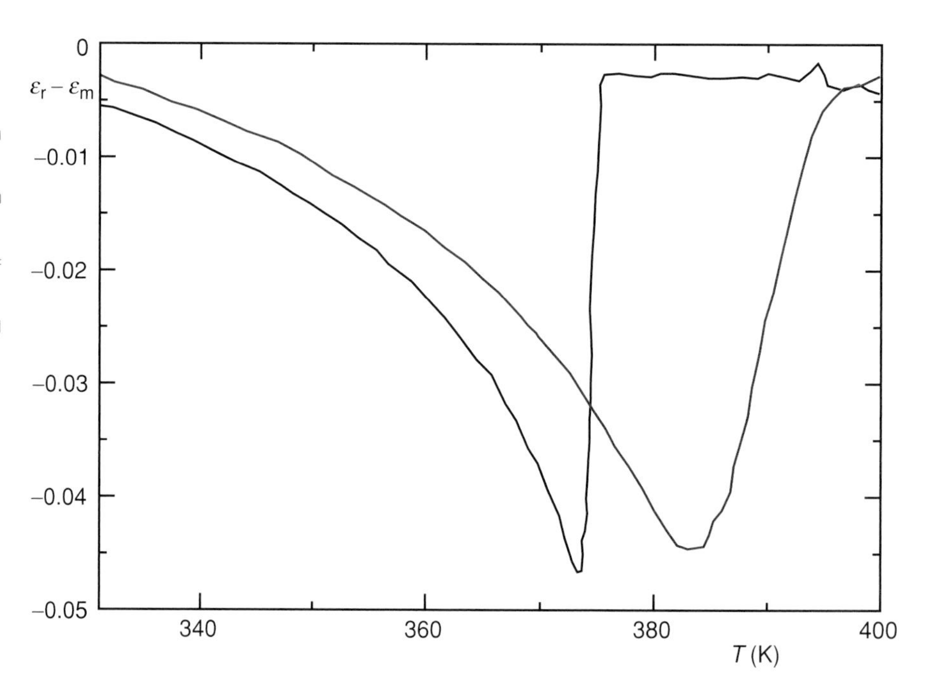


**Figure 14.22.** The real part of the dielectric constant $\varepsilon_r$ versus temperature in an experiment with plexiglass. The black branch shows what happens on decreasing the temperature from 415 K at a cooling rate of 20 K h$^{-1}$ with a 10-h stop at $T =$ 374 K. The red branch shows the result of subsequently increasing the temperature at the same rate. For clarity, the value of $\varepsilon_m$ of the dielectric constant in a similar experiment performed without intermediate stops has been subtracted. (Courtesy of S. Ciliberto.)

slow convergence toward equilibrium), tends to return to the reference curve; and (ii) during the heating branch, the susceptibility exhibits again (almost) the same drop around $T = T_{\rm i}$ even without a long interruption. This scenario suggests a hierarchical organization of the free-energy valleys, but a convincing theoretical explanation is still lacking.

## 14.10 The computational approach

So far, we have looked at many significantly different phenomena at the macroscopic level. Is there some common description to make a fruitful comparison with the final goal of accommodating all these phenomena into possibly just a few classes? Many of the models appear as (deterministic/stochastic) rules processing a finite number of symbols: this is the case for lattice-gas models introduced to describe fluid behavior, the sandpile model, and spin systems; but it is also known that chaotic dynamics can be transformed (without loss of information) into a sequence of symbols by suitably partitioning the phase space and sampling a trajectory at fixed time intervals. It is therefore natural to interpret different physical systems in terms of suitable automaton rules and analyze them with the tools used in the mathematical theory of computation. This approach is particularly appropriate for formulating problems such as the following: given a certain pattern $S = b_1 b_2 \ldots$ (for simplicity we refer to one-dimensional strings), find the minimal model capable of producing $S$. In fact, identification of the minimal model would allow different processes to be compared and possibly reveal unexpected degrees of universality.

Formalization of this problem requires a precise and general definition of a "model" as a tool to manipulate symbols. Various equivalent approaches were introduced in the past century, but the universal Turing machine (UTM) is definitely the most transparent, since it is the formalization of a class of machines that are now familiar to everybody: general-purpose digital computers with infinite memory. Although Turing machines are believed to embody the most powerful class of computing automata, this statement has not been proved rigorously (more powerful machines could arise from quantum computation, given its strongly "parallel" character). Since not all problems can be transformed into manipulations of integer numbers, Blum and collaborators in the 1990s extended the approach by including real-number manipulations; although conceptually important, this generalization does not appear to lead to striking new phenomena.

Because of the strong analogy between UTMs and digital computers, it is easy to convince oneself that each model has a size (the size of the program) and the identification of the minimal model is thus a well-known problem. The length of the minimal description is called the Kolmogorov–Chaitin algorithmic information. This has an objective meaning, since its variation with the UTM is bounded and independent of the sequence under investigation. It is considered akin to information, because of analogies with Shannon information $H$, but $H$ is a very different quantity since it follows from probabilistic concepts applied to ensembles of sequences. For instance, given a formally infinite and stationary binary sequence $S$, one can determine the probability $p_i$ of each subsequence of length $n$ and thereby compute $H = -\sum_i p_i \ln p_i$ (where ln denotes a base-2 logarithm). The information is maximal for random sequences of independent bits, in which case $H$ is equal to the number $n$ of bits. In the algorithmic theory of computation, a code of length $N$ is attributed a probability $2^{-N}$ as if the code were the result of a random assembly of independent symbols. With this interpretation in mind, it is clear that the role of the minimal code length $K$ is similar to that of the information $H$. Thus, $K$ is not only a measure of randomness, but,

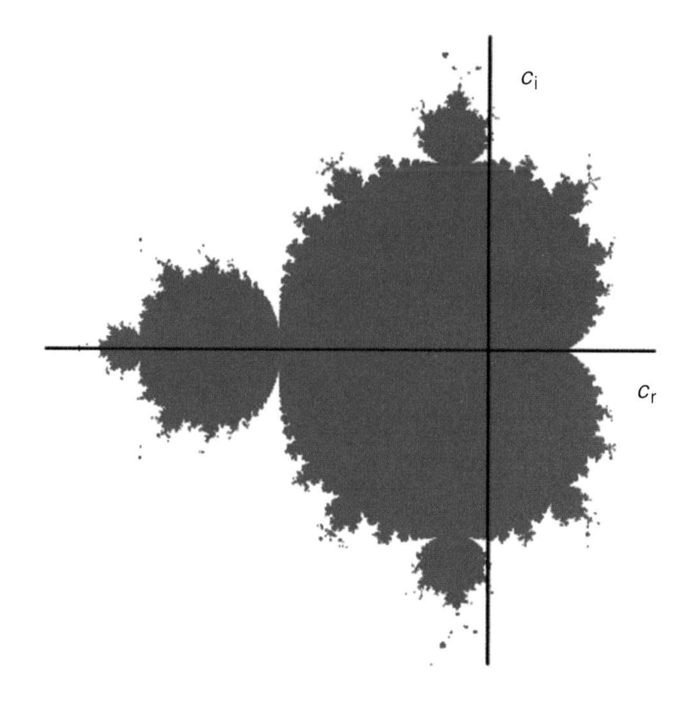


**Figure 14.23.** The Mandelbrot set. $c_r$ and $c_i$ are the real and imaginary parts of the parameter $C$ in the mapping $z_{n+1} = z_n^2 + c$. Set points are those which do not lead to a divergence on starting from the initial condition $z_0 = 0$.

moreover, the most refined one, since it takes into account the possibility that a sequence can be compressed by a suitable algorithm. Thus it is instructive to note that all random-number generators employed in numerical simulations (such as transformations of the type $u_{n+1} = (au_n + b) \mathrm{mod} A$) are not truly random, since they are obtained precisely by iterating a finite simple algorithm!

The importance of the algorithmic approach does not lie in the possibility of introducing an objective indicator, but in the underlying idea of looking for the most compact description of the phenomenon of interest. However, this ultimate goal is like the Holy Grail: it cannot be reached, since there is no algorithmic procedure to identify the minimal model. In fact, algorithmic information is uncomputable in generic sequences (a consequence of the famous undecidability of Gödel's theorem of 1931). What are the consequences of such a negative result? They are perhaps not as dramatic as might appear at first glance.

An illuminating example is the uncomputability of the Mandelbrot set, defined as the ensemble of complex numbers $c = c_r + ic_i$ such that iterations of the mapping $z_{n+1} = z_n^2 + c$ do not lead to a divergence when starting from the initial condition $z_0 = 0$. Roger Penrose was the first to suggest that this set might be uncomputable and the statement was later rigorously proved by Blum and collaborators. The implications for physics are modest, because numerical tools allow reconstruction of the Mandelbrot set to any desired degree of accuracy, as demonstrated by Figure 14.23, which required a few minutes of programming and seconds to generate the points! Insofar as physics is an experimental science and models are approximate descriptions of reality, the possibility of determining only those trajectories that remain bounded for a finite number of iterations does not represent a serious limitation.

More serious would be the consequences of undecidability, if it turned out, for example, that it is conceptually impossible to predict the scaling behavior of some observable. For instance, accurate numerical simulations by Grassberger on some SOC models show that the divergence is accompanied by anomalously large oscillations. If this is incipient evidence of computational irreducibility, it might be a serious obstacle to constructing a theory of out-of-equilibrium systems. Whatever the answer to this question, a crucial point that must be clarified in many-component systems is understanding how information can be processed effectively. This question is connected both to the functioning of living beings and to our ability to construct artificial systems able to manipulate information. The study of nonlinear systems has revealed various mechanisms of information transport.

Deterministic chaos is often viewed as a flow of information from the least to the most significant digits: this flow is quantified by the Kolmogorov–Sinai entropy, a measure of the amount of information needed to describe a generic trajectory over a certain time. However, such a flow is more relevant conceptually than practically, since exponential amplification of perturbations is also responsible for rapid loss of memory. In fact, removing local instability is a necessary condition for effective information processing. One possibility could be to stabilize some of the infinitely many periodic orbits embedded in chaotic systems, since multi-stable systems, switching between different stable states, provide a meaningful framework for a reliable computation. Much interest has been given to the control of chaos over the past ten years (also motivated for other reasons – see Chapter 13) and various stabilization schemes have been introduced.

Besides suitably engineered chaotic systems, it is known that many stable orbits also naturally arise in Hamiltonian systems when a small dissipation is combined with external forcing. This occurs whenever many elliptical orbits exist in the Hamiltonian system, since many of them are turned into stable orbits by the small perturbation. An example of this can be found by introducing dissipation into the Chirikov–Taylor map by modifying the first of the two relations in Equation (14.1) as $I_{n+1} = (1 - \nu)I_n + K \sin \theta_n$, where $\nu$ is a small number. Yorke and collaborators have been able to identify more than 100 low-period attractors.

Even though chaos opens up a wide variety of configurations, local instability must be eliminated if information is to be processed meaningfully. One exception is the class of systems in which the phase space is partitioned into different chaotic components, with more than one ergodic component. Sinha and Ditto have exploited this idea, showing in 2002 that spacetime chaos can exhibit computing capability. On the other hand, the study of systems with many degrees of freedom has shown the existence of yet another mechanism that could be responsible for generating a large variety of stable dynamics: in the presence of strong nonlinearities in the evolution rule (with reference to coupled-map lattices, an example is the quasi-continuous map depicted in Figure 14.24), although the Lyapunov exponents may be negative, finite-amplitude perturbations may nevertheless propagate, giving rise to seemingly erratic behavior. This has been called "stable chaos," to stress that, while it is irregular, it is nevertheless stable against small perturbations. Because of linear stability, this type of dynamics is akin to computation, a statement indirectly confirmed by a relationship found by me and my collaborators with cellular automata.

## 14.11 Information processing in the brain

Brain function is increasingly being investigated by physicists under the assumption that it is possible to understand how information is processed without necessarily having a detailed understanding of the underlying biochemistry. Single-neuron dynamics has therefore become an important object of investigation and several different levels of abstraction are currently being employed for modeling neuron networks.

It is well known that various types of neurons exist, but most of the differentiation is presumably a consequence of the need to provide specific functions for sensory input and motor output. Therefore, as long as one is concerned with information

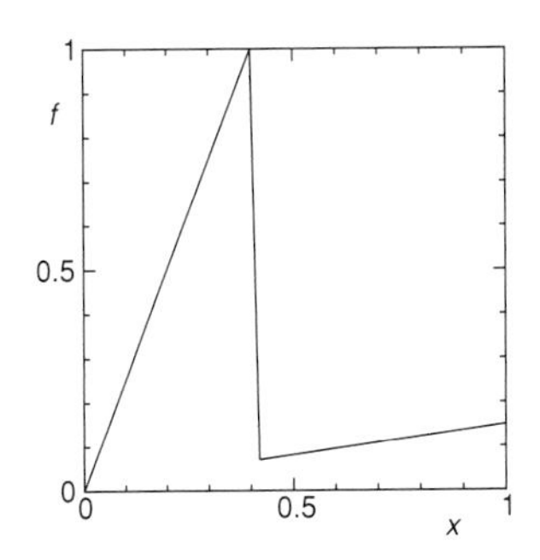


**Figure 14.24.** An example of a discrete mapping $x$ to $f(x)$ that may give rise to "stable chaos" if combined with diffusive spatial coupling.

processing, disregarding such neuron variation should not be a severe limitation. Neurons work by emitting spike-like signals every few milliseconds, followed by a so-called refractory period during which it is much harder to stimulate a further spike, since the neuron is recovering energy. This is described by the celebrated Hodgkin–Huxley (HH) model, which was derived in the early 1950s to describe the dynamics of the electric voltage $V$ in the squid axon. Besides $V$, the model involves three auxiliary variables connected with ionic (sodium and potassium) concentration gradients across the neural membrane.

Neuron–neuron coupling is ensured by spikes traveling from the axon of the emitting neuron to the dendrites of the receptors. This deserves comment, since it has been demonstrated that propagation of fixed-shape spikes can occur only in the presence of nonlinear mechanisms compensating for broadening due to unavoidable dispersion phenomena. It is remarkable that nonlinear mechanisms have been naturally selected to propagate information, while man-made devices are still mostly based on linear propagation, such as within optical fibers: this is because the low level of noise does not significantly corrupt the marginally stable linear waves. Neural coupling can be either excitatory or inhibitory, depending on whether the incoming pulse encourages or suppresses emission in the receiving neuron.

Since the dynamics of coupled HH neurons is difficult to analyze, a further level of simplification is often introduced with the additional hypothesis that what really matters is that the response rate of the single neuron is considered to be either large or small. Accordingly, biological neurons are often treated as spin-like (or bit-like) variables. In this philosophy, the attractor neural network (ANN) is a brain metaphor that has proved to be useful in the context of storage and retrieval of patterns. In 1982, Hopfield introduced a much simplified model that nevertheless allows quite remarkable conclusions to be drawn. Aware of the high degree of connectivity of the human brain (each of the $10^{11}$ neurons is typically connected to $10^4$ other neurons), he introduced a mean-field model, in which all neurons influence each other. In this model, in the absence of noise, the state of the $i$th neuron is updated according to the weighted sum of the incoming signals, $U_i = \sum_{j \le N} J_{ij} s_j$ (where $N$ is the number of neurons). If $U_i$, roughly corresponding to the membrane potential, is larger than some threshold $U_{\rm th}$ the neuron fires ($s_i = 1$); otherwise it remains inactive ($s_i = 0$). In the presence of noise, $U_i$ fluctuates and the new state is chosen by probability.

The ANN is said to recognize a given pattern $S(k)$, defined as a suitable sequence of bits $S(k) = b_1(k), \ldots, b_N(k)$, if it converges to a final state such that $s_i = b_i(k)$ for all values of $i$, having started from an initial condition where equality holds only for a few neurons (meaning that only a hint is given). How many different patterns can be simultaneously stored and confidently recognized in the presence of noise?

A rather detailed and almost analytical solution has been found under the further simplifying hypothesis that $J_{ij} = J_{ji}$, since the model then becomes formally equivalent to the Sherrington–Kirkpatrick spin-glass model, with ferromagnetic and antiferromagnetic interactions replaced by excitatory and inhibitory ones, while the noise amplitude and the threshold $U_{\rm th}$ play an analogous role to temperature and an external magnetic field, respectively. One should admit that the assumption of symmetric interactions has no justification in biological neurons, and was introduced merely to guarantee convergence to fixed points and allow an analogy with the spin-glass problem.

The only relevant difference from the Sherrington–Kirkpatrick problem is that the exchange energies $J_{ij}$ are not directly defined as random variables, but must be fixed in order to ensure convergence toward pre-assigned patterns $S(k)$. A simple prescription, though not the optimal one, is $J_{ij} = \sum_k b_i(k) b_j(k)$. Within this framework, Amit, Gutfreund, and Sompolinsky proved in 1985 that the ANN can store up to $0.14N$

different patterns, all with the same degree of stability. From the very definition of $J_{ij}$ it is clear that the information about the patterns is encoded in the strength of the connection: in a true biological context, it has to be explained how the system parameters can spontaneously reach such values. To address this problem, networks with time-dependent exchange interactions are increasingly being investigated. The beauty of the ANN is that the storage capacity is independent of any assumption regarding the structure of the patterns to be memorized. The importance of this ANN lies in its robustness against noise. In a noisy environment, large fluctuations can always occur. However, as long as the configurations corresponding to the patterns to be recognized are separated by macroscopic free-energy barriers, noise cannot induce jumps between neighboring valleys, and misrecognitions are highly unlikely. Such robustness is welcome, since it also implies that the consequence of sudden failure of neurons is negligible, and we know that many of our neurons die daily! For completeness, it should also be pointed out that, besides the prescribed stable states, the ANN turns out to exhibit further "metastable" configurations that in principle could be used to store additional patterns. Unfortunately, as Amit points out in his book (see Further reading), this is not a free lunch, because such minima necessarily correspond to superpositions of the other patterns.

As for a comparison with a real brain, if we imagine our neuronal system as the collection of $10^7$ globally coupled networks, each containing $N = 10^4$ neurons, we can conclude that one can store the fairly large number of $10^7 \times 0.14N^2 = 10^{14}$ bits. On the other hand, some 50 years ago von Neumann estimated that a human is presumably able to memorize $10^{20}$ bits of information. Part of the difference between these two numbers can be covered by noticing that much of the information we store is redundant (many of the images we remember are similar), but further new mechanisms need to be discovered.

Another possibility is that single neurons process more information than that assumed by the simple ANN. As already pointed out by Mackay and McCulloch in 1952, information may be encoded not just in the contrast of high versus low spiking rates, but also in the separation in time between consecutive spikes: on introducing a time resolution $\delta t$, one can translate a spike train into a sequence of 0's and 1's, depending on whether a spike is contained in each given interval (because of the refractory period, it makes no sense to choose $\delta t$ smaller than 3 ms). Could it be that meaningful information is effectively encoded in such a symbolic representation? A beautiful experiment performed in 1996 on the sensory neuron H1 of the fly's visual system (H1 codes the horizontal motion of a given pattern) produced results suggesting that the answer is probably yes. Indeed, in 1997 Bialek and collaborators first measured the information content in the fly's response to a time-dependent stimulus consisting of computer-generated moving stripes. Then, they measured the (conditional) information associated with fluctuations of the response to the same stimulus. Clearly, this second contribution corresponds to meaningless information produced by the system, possibly due to noise or internal chaos. By subtracting the latter information from the former, they found that the response signal contains information down to the minimal acceptable temporal resolution (2–3 ms). The route to understanding how such information can propagate and be processed is still quite long, but the experiment suggests that it is worth continuing in this direction.

## 14.12 Perspectives

Understanding the behavior of complex systems is one of the main challenges for the physics of the twenty-first century. The study of many different out-of-equilibrium

### BOX 14.5 JOHN VON NEUMANN

*The sciences do not try to explain, they hardly even try to interpret, they mainly make models. By a model is meant a mathematical construct which, with the addition of certain verbal interpretations, describes observed phenomena. The justification of such a mathematical construct is solely and precisely that it is expected to work.*

János (John) Louis von Neumann (1903–1957) was certainly one of the preeminent scientists of the twentieth century. Even though in his time there was little perception that complex systems could become the object of intensive research activity, in retrospect, von Neumann contributed de facto more than anybody else to the advancement of this field. Without forgetting his relevant papers on the formalization of quantum mechanics, in 1927 he set the basis for the future development of game theory and recognized the conceptual relevance of questions about completeness, consistency, and decidability of mathematical theories: problems that were formalized a year after by his teacher D. Hilbert, and afterwards solved by K. Gödel.

Besides being attracted by abstract issues, he was challenged by the development of mathematical methods for the solution of specific problems like the determination of global weather patterns and the evolution of shock waves. As a result, he perceived much earlier than his colleagues the importance of automatic computation as a means for gaining insight about highly nonlinear regimes. Being involved at the outset as a consultant in the project for the construction of the Electronic Numerical Integrator and Calculator (ENIAC), his deep knowledge of formal logic allowed him to make substantial progress in setting up the proper architecture: he was so successful that actual computers still conform to the ideas he outlined in a draft report.

Computers represented, however, for von Neumann an intermediate step toward the synthesis of a general theory encompassing logic and large-scale computation, with the perspective of eventually understanding brain functioning and living matter. This project led him to introduce cellular automata and he was even able to "construct" one such model capable of both simulating a Turing machine and self-replicating. His premature death prevented him from making further progress.

systems indicates that, even though a general theory cannot yet be constructed, several general classes of behavior do exist. Such universality classes are possible building blocks for higher-level theories to bridge the gap between the physics of microscopic objects and the laws which govern living beings. To proceed along this route, it will certainly be helpful to understand how far concepts of equilibrium statistical mechanics can be extended away from equilibrium, entropy being the most important. We shall presumably learn a lot also by discovering when and why entropy cannot be extended to strongly non-equilibrium regimes: should it be complemented by other macroscopic variables?

In trying to discern the emerging picture, the availability of fast computing devices is inducing a change in the meaning of "explanation." This term is increasingly used to indicate that a simple simulation or model captures the essence of a phenomenon: this chapter is full of such examples. Is this a temporary situation, awaiting the development of a new general theory, or does it in itself herald a new working method? The latter hypothesis would be very much in the spirit of von Neumann (see the quotation in Box 14.5).

## FURTHER READING


1. D. Amit, *Modeling Brain Function*, Cambridge, Cambridge University Press, 1989.
2. R. Badii and A. Politi, *Complexity: Hierarchical Structures and Scaling in Physics*, Cambridge, Cambridge University Press, 1997.
3. P. Bak, *How Nature Works*, New York, Springer-Verlag, 1996.
4. T. Bohr, M. H. Jensen, G. Paladin, and A. Vulpiani, *Dynamical Approach to Turbulence*, Cambridge, Cambridge University Press, 1999.
5. P. Grassberger and J. P. Nadal, *From Statistical Physics to Statistical Inference and Back*, Dordrecht, Kluwer, 1994.
6. M. Mezard, G. Parisi, and M. A. Virasoro (eds.), *Spin Glass Theory and Beyond*, Singapore, World Scientific, 1987.
7. P. Manneville, *Dissipative Structures and Weak Turbulence*, San Diego, Academic Press, 1990.


# Antonio Politi

Antonio Politi was born in Florence in 1955. He studied at the University of Florence, where he obtained his Laurea in Physics in 1978. Since 1981 he has been working at the Istituto Nazionale di Ottica where, from 1994, he has managed the quantum-optics research unit. In 2004 he became responsible for the Florence section of the newly founded Istituto dei Sistemi Complessi of the CNR (Consiglio Nazionale delle Ricerche). He has contributed to our understanding of the behavior of generic nonlinear dynamical systems, initially focusing his attention on low-dimensional chaos, and later investigating systems with many degrees of freedom. The latter class of problems provides a natural bridge to problems of statistical mechanics out of thermodynamic equilibrium, including, for example, anomalous transport and several phase transitions.

He is a member of the American Physical Society and an associate editor for *Physical Review* E and *The European Physical Journal* D. In 2004, he was awarded the Gutzwiller Fellowship by the Max Planck Institute for the Physics of Complex Systems in Dresden.

# Collaborative physics, e-Science, and the Grid – realizing Licklider's dream

Tony Hey and Anne Trefethen

## 15.1 Introduction

### 15.1.1 e-Science and Licklider

It is no coincidence that it was at CERN, the particle-physics accelerator laboratory in Geneva, that Tim Berners-Lee invented the World Wide Web. Given the distributed nature of the multi-institute collaborations required for modern particle-physics experiments, the particle-physics community desperately needed a tool for exchanging information. After a slow start, their community enthusiastically adopted the Web for information exchange within their experimental collaborations – the first Web site in the USA was at the Stanford Linear Accelerator Center. Since its beginnings in the early 1990s, the Web has taken by storm not only the entire scientific world but also the worlds of business and recreation. Now, just a decade later, scientists need to develop capabilities for collaboration that go far beyond those of the Web. Besides being able to access information from different sites they want to be able to use remote computing resources, to integrate, federate, and analyze information from many disparate and distributed data resources, and to access and control remote experimental equipment. The ability to access, move, manipulate, and mine data is the central requirement of these new collaborative-science applications – be they data held in a file or database repositories, data generated by accelerators or telescopes, or data gathered from mobile sensor networks.

At the end of the 1990s, John Taylor became Director General of Research Councils at the Office of Science and Technology (OST) in the UK – roughly equivalent to Director of the National Science Foundation (NSF) in the USA. Before his appointment to the OST, Taylor had been Director of HP Laboratories in Europe and HP as a company have long had a vision of computing and IT resources as a "utility." Rather than purchase expensive IT infrastructure outright, users in the future would be able to pay for IT services as they require them, in the same way as we use the conventional utilities such as electricity, gas, and water. In putting together a bid to government for an increase

*The New Physics for the Twenty-First Century*, ed. Gordon Fraser.

in science funding, Taylor realized that many areas of science could benefit from a common IT infrastructure to support multidisciplinary and distributed collaborations. He therefore articulated a vision for this type of collaborative science and introduced the term "e-Science":

e-Science is about global collaboration in key areas of science, and the next generation of infrastructure that will enable it.

As a result of Taylor's initiative, the UK has invested over £250M in the UK e-Science program, the funding spanning the five years from 2001 to 2006.

It is important to emphasize that e-Science is not a new scientific discipline in its own right – rather, the e-Science infrastructure developed by the program is intended to allow scientists to do faster, better, different research. This is best illustrated by an example. The UK program recently funded a major e-Science project on "integrative biology." This is a £2.3M project led by Oxford University to develop a new methodology for research on heart disease and cancer. The project involves four other UK universities and the University of Auckland in New Zealand. Denis Noble's group at Oxford is world-renowned for its research into models of the electrical behavior of heart cells. In the Bioengineering Department at the University of Auckland in New Zealand, Peter Hunter and his team are doing pioneering research into mechanical models of the beating heart. Both groups are therefore currently doing world-class research in their own specialist areas. However, the project intends to connect researchers in these two groups into a scientific "virtual organization." This virtual organization is an environment that allows researchers in the project – and only researchers in the project – routine access to the models and data developed both in Oxford and in Auckland, as well as allowing them access to the UK High Performance Computing (HPC) systems. Building such a virtual organization will not be a trivial achievement. Of course, researchers have long been able to access resources at a remote site: here the intent is to put in place a comprehensive infrastructure that can authenticate each user and authorize them to access specific resources at each site, avoiding problems with firewalls and multiple administrative authorities. By providing a powerful and usable e-Science research environment in which these two groups can combine their research activities, it will be possible to follow a path from specific gene defects to heartbeat irregularities that neither group could do independently. It is in this sense that "e-Science" can enable new science.

What e-Science infrastructure is required to support these new ways of working? For e-Science applications, scientists need to be able to access many different sets of resources in a uniform, secure, and transparent way. The collaboration – or virtual organization (VO) – must be able to set and enforce policies and rules that define who can use what resource or service, when, and at what cost. The VO may be for a long-term collaboration across many institutions or it may be for a specific project or transaction taking anything from a few minutes to several weeks and involving only a few sites. The infrastructure technologies, on which e-Science depends, will need to be able to identify the resources, apply the policies, and monitor the usage dynamically and automatically. The various institutions are likely to use different user-authentication procedures, possess computing and storage systems purchased from multiple vendors, operate a variety of firewalls and security systems, and have different user policies. Furthermore, the various institutional resources are likely to be running different operating systems and adhering to different software standards. With present mechanisms, users need to know about each resource, both its existence and how to use it, and need to log into each system individually.

Of course, these problems are not new – the computer-science community has been grappling with the challenges of distributed computing for decades. Such an e-Science infrastructure was in fact very close to the vision that J. C. R. Licklider ("Lick") took with him to ARPA in the early 1960s when he initiated research projects that led to the ARPANET. Larry Roberts, one of his successors at ARPA and principal architect of the ARPANET, described this vision as follows:

> Lick had this concept of the intergalactic network which he believed was everybody could use computers anywhere and get at data anywhere in the world. He didn't envision the number of computers we have today by any means, but he had the same concept – all of the stuff linked together throughout the world, that you can use a remote computer, get data from a remote computer, or use lots of computers in your job. The vision was really Lick's originally.

The ARPANET of course led to the present day Internet – but the killer applications have so far been email and the Web rather than the distributed-computing vision described above. In the early 1960s, Licklider only envisaged needing to connect a small number of scarce, expensive computers at relatively few sites. Over the past thirty years Moore's law – Gordon Moore's prediction that the number of transistors on a chip would double about every 18 months so that the price–performance ratio is halved at the same time – has led to an explosion in the numbers of supercomputers, mainframes, workstations, personal computers, and personal data assistants that are now connected to the Internet. Although the operation of Moore's law – at least if the IT industry continues to be based on silicon technology – must come to an end when feature sizes of transistors begin to approach atomic dimensions, the IT revolution still has perhaps a decade or more to run. Already we are beginning to see intelligent sensors and radio-frequency tagging devices (RFIDs) being connected to the network. In addition, new high-throughput experimental devices are now being deployed in fields as diverse as astronomy and biology and this will lead to a veritable deluge of scientific data over the next five years or so.

### 15.1.2 Cyberinfrastructure, e-infrastructure, and the Grid

Scientific collaborations have long been connected by the high-speed national research networks that constitute the underlying fabric of the academic Internet. Under the banner of "e-Science," scientists and computer scientists around the world are now collaborating to construct a set of software tools and services to be deployed on top of these physical networks. Since this software layer sits between the network and the application-level software, it is often referred to as "middleware." A core set of middleware services for scientists will provide them with the capability to set up secure, controlled environments for collaborative sharing of distributed resources for their research. Collectively, these middleware services and the global high-speed research networks comprise the new "e-infrastructure" (in Europe) or "cyberinfrastructure" (in the USA) for collaborative scientific research. This middleware is often called Grid middleware.

The term "Grid" was first used in the mid 1990s to denote a proposed distributed-computing infrastructure for advanced science and engineering. At that time, the idea was driven by a desire to use distributed-computing resources as a meta-computer, and the name was taken from the electricity power grid – with the analogy that computing power would be made available for anyone, anywhere to use. The Grid was a product of developing technologies in high-performance computing (HPC) and networking, together with the 1980s "Grand Challenges" research program in the USA. These scientific grand challenges required geographically disparate teams to work together on large-scale problems whose computational demands often exceeded the computational power that any one HPC system could provide. In order to create meta-computers for use in these challenges several toolkits that had suitable protocols and tools to allow users to log on securely to remote computers were developed, including the well-known Legion and Globus systems. Other projects focused on issues such as scheduling across a number of distributed computers and the mass storage of scientific data. The Condor project of Miron Livny's group at Wisconsin is a particularly successful example of a system that allows users to harvest unused cycles on an organization's computers. Condor is now used routinely by many universities around the world and by a number of companies. Reagan Moore's Data Intensive Computing Environment (DICE) group at the San Diego Supercomputer Center produced the Storage Resource Broker (SRB) system to handle large amounts of scientific data. They particularly emphasized the need for high-quality metadata to annotate distributed collections of scientific data.

In 2001, Ian Foster, Carl Kesselman, and Steve Tuecke recognized the broader relevance of Grid middleware. They redefined the Grid in terms of infrastructure to enable collaboration:

> The Grid is a software infrastructure that enables flexible, secure, coordinated resource sharing among dynamic collections of individuals, institutions and resources.

This vision of the Grid in terms of the middleware necessary to establish the virtual organizations needed by scientists is surprisingly close to Licklider's original vision. Unfortunately, although the vision is there, present-day versions of "Grid" middleware – such as Globus, Condor, and Unicore – support only a small part of the functionality required for a viable e-Science infrastructure. Nevertheless, a vision of the Grid middleware that will allow scientists to set up Grids tailored for their specific e-Science collaborations has proved to have universal appeal. This is Taylor's challenge for e-Science and for the NSF's Cyberinfrastructure Initiative. Many other countries around the world – both in Europe and in the Asia–Pacific region – have developed similar national programs. The European Union sees the development of robust Grid middleware as a key component of its necessary research infrastructure to create a true European Research Area (ERA). The formidable challenge both for computer scientists and for the IT industry is to develop a set of Internet middleware services that are not only secure, robust, and dependable but also able to be deployed routinely by all kinds of scientists to create their specialist Grids.

### 15.1.3 An outline of this chapter

The rest of the chapter will provide a more detailed consideration of application requirements and the elements of Grid technology. Section 15.2 describes some physics applications that exemplify the requirements of the e-Science infrastructure. In Section 15.3 we outline the fundamental Grid services and technologies under development, with some discussion of the unsolved problems and issues. Section 15.4 speculates on applications of Grid technologies in industry and commerce. The final section offers some conclusions.

## 15.2 Collaborative-physics projects

In this section we present a brief description of some of the global Grid activities in the physics domain. These projects have specific scientific goals but at the same time provide an understanding of the functionality required by the underpinning e-Science infrastructure. The projects are generally collaborations between physicists at different institutions – with the motivation of doing new science – and computer scientists interested in understanding the challenges posed by these applications and adding to the present level of functionality, performance, or robustness of the Grid middleware. There are many more physics-related Grid projects than we describe here; this selection is merely intended to illustrate some of the potential for Grid-empowered collaborative physics. We are aware that we have an emphasis on projects funded by the UK e-Science program, but reassure the reader that they are only intended to be representative, not an exhaustive survey of worldwide Grid projects. Pride of place has also been given to particle-physics applications and in particular to efforts to build a persistent global infrastructure with capabilities far beyond those of the Web that the particle-physics community invented.

## 15.2.1 Particle physics

### The Standard Model

Particle physicists around the world strive to answer the fundamental questions concerning particles of matter and the forces acting between them. A key part of their present understanding of nature is the so-called "Standard Model," which was developed twenty years or so ago (see Chapter 4). This is based on relativistic quantum-field theories of electrodynamics and the strong and weak interactions. In quantum electrodynamics (QED), the familiar electromagnetic forces between charged particles such as electrons are mediated by massless vector bosons – photons. Similarly, the weak and strong forces are carried by massive – in the case of the weak force – and massless – for the strong force – vector bosons. The candidate theory of the strong interactions is quantum chromodynamics (QCD). In QCD, the strong forces between the fundamental building blocks of matter – the quarks – are mediated by a family of massless vector bosons called gluons. A particular challenge for particle theorists is to understand how the gluons conspire so that the quarks – with their non-integer charges – are "confined" and never seen as free particles. This is believed to be directly related to the fact that, unlike the electrically neutral photons, the gluons themselves carry a QCD charge. Standard approaches, such as perturbation theory, can never reveal confinement and theoretical particle physicists are exploring nonperturbative aspects of QCD by performing detailed simulations of QCD on some of the most powerful supercomputers ever built. We shall return to these theorists' needs for a QCD Grid.

Experimental particle physicists are currently more concerned with exploring another aspect of the Standard Model, the theory of the weak interactions. In the 1970s, after important work by Gerard 't Hooft and Martinus Veltman, Sheldon Glashow, Abdus Salam, and Steven Weinberg were awarded the Nobel Prize for their "GSW" unified model of the electromagnetic and weak interactions. The key feature that any successful model of the weak and electromagnetic interactions must explain is how the weak vector bosons can acquire mass rather than being massless like the photon. A crucial insight came from Peter Higgs in Edinburgh, who had been able to show how the introduction of a new scalar particle could result in a situation in which the natural symmetry of a "gauge theory" such as the GSW theory – with all vector bosons massless – was effectively "hidden." The situation is somewhat similar to the ground state of a ferromagnet – we believe that the forces between the atoms of the ferromagnet are rotationally invariant but the ground state of the magnet clearly violates rotational invariance, since it is magnetized along one particular direction.

(In fact, a better analogy is the case of a superconductor: here the Cooper pairs play a role analogous to that of the Higgs boson. A more detailed discussion of "spontaneous symmetry breaking" in gauge theories would be a diversion here – the interested reader is referred to the recommended reading at the end of this chapter.) The GSW model therefore introduces a scalar Higgs particle to generate the masses of the W and Z weak vector bosons. However, the Higgs boson also has another role in the GSW model: it is also responsible for generating all the fermion masses – such as those of the electron and muon.

Despite the fact that the Higgs boson apparently plays such a key role in the GSW model, experimentalists have so far found no direct sign of it in their experiments to date. However, there is a clue within the model itself as to what energy region might reveal effects due to the Higgs boson. The argument is as follows. The perturbative predictions of the model have been verified spectacularly by detailed experiments carried out over the last decade or more. However, if the mass of the Higgs boson is greater than the order of 1 TeV or so, the theory predicts that there should be a breakdown of these perturbative predictions. It is for this reason that particle physicists feel confident that there will be some sign of the Higgs boson in the TeV energy range that will be accessible at the new accelerator under construction at CERN on the French–Swiss border near Geneva.

The search for the Higgs boson has been going on for over ten years, both at CERN's Large Electron Positron Collider (LEP) in Europe and at Fermilab in the USA. At the CERN laboratory in Geneva, the LEP machine has been dismantled and the LEP tunnel is being re-engineered to house what will be the world's most powerful particle accelerator – the Large Hadron Collider (LHC). The machine is scheduled to be operational by 2007 and the new generation of LHC experiments will be able to observe and record the results of head-on collisions of protons at energies of up to 14 TeV. This will allow the experimentalists to explore the GSW model of the weak and electromagnetic interactions in the all important TeV energy range. Although the primary goal is to find signs of the Higgs boson, particle physicists are also hoping for indications of other new types of matter. In particular, "string" theories (see Chapter 5) that attempt to unify gravity with the strong, weak, and electromagnetic forces postulate the existence of a new type of symmetry called "supersymmetry." This requires all the known particles to have as-yet-undiscovered supersymmetric partners with a different spin: scalar electrons, scalar quarks, and scalar neutrinos – light-heartedly dubbed "selectrons," "squarks," and "sneutrinos" – as well as fermionic partners of the gauge bosons – called "gluinos," "winos," "zinos," and "photinos." Such particles, if found, may both shed light on the "dark-matter" problem of cosmology and provide some tangible evidence that string theories are not just a beautiful mathematical diversion.

## 15.2.2 Experimental particle physics: the LHC computing Grid

Finding experimental evidence for the existence of the Higgs particle is a major technological challenge. Although the LHC will provide an almost ideal laboratory for this search, the characteristic signals for the Higgs are very rare and subtle. The challenge is analogous to looking for a needle in several thousand haystacks. Experiments at the LHC will generate petabytes of data per year and will also be on a scale greater than any other previous physics experiments. The major experiments involve a collaboration of over 100 institutions and over 1000 physicists from all over the world. The petabytes of experimental data, although initially generated at CERN, are far too much to be stored and analyzed centrally in Geneva. Thus a vast amount of data will need to be distributed to sites all over the world for subsequent analysis by the hundreds of teams of physicists at their collaborating institutions. Furthermore, besides the large volumes of experimental data, in order to perform their analysis, particle physicists must also create large samples of simulated data for each experiment in order to understand the detailed behavior of their experimental detectors. Thus the e-Science infrastructure required for these LHC experiments goes far beyond the capability to access data on static Web sites. The experimental particle physicists need to put in place LHC Grid infrastructure that will permit the transport and data mining of such huge distributed data sets. The Grid middleware must also provide the capability for physicists to routinely set up appropriate data-sharing/replicating/management services and facilitate the computational needs of the simulation and analysis in a distributed manner.

There are two Grid projects with a significant particle-physics component funded by the European Union – the DataGrid and DataTag projects. There is also significant national funding for the LHC experimental infrastructure from the CERN member states, such as the GridPP project originating from the UK. In the USA there are three major Grid projects focusing on the particle-physics LHC middleware – the NSF GriPhyN and iVDGL projects and the DOE PPDataGrid project. We shall restrict our discussion to the UK GridPP project, which is the largest single Grid project funded in the UK and has been awarded over £30M funding in two phases. This project represents just the UK portion of the global effort by physicists preparing for the LHC. GridPP is necessarily integrated with the international activities of the European Union and the USA.

The goal of the GridPP project is to provide a collaborative environment and infrastructure for the particle physicist in the UK to take part in the LHC experiments that include teams across the globe. Hence interoperability is of key importance, as shown on the project plan. This interoperability will also allow the project to leverage Grid middleware developed in other projects – a key desire of international efforts to develop a global Grid infrastructure.

The second column of the project plan shown in Figure 15.1 represents the EU DataGrid (EDG) project, in which GridPP, representing the UK's Particle Physics and Astronomy Research Council (PPARC), is one of six principal partners. The goals of the EDG project are to develop and test a global particle-physics infrastructure for disseminating, managing, and analyzing the LHC data. Although particle physics was undoubtedly the dominant application area in EDG and particle physicists played a key role in driving the project forward to its successful conclusion, there are two other application areas. These are biology and medical image processing, and Earth observations. The EDG work packages (WPn) attempted to cover the set of requirements for creating and implementing an e-infrastructure of Grid technologies over the multi-tiered, hierarchical resource centers indicated in column 4. The eight EDG work packages were Resource Management, Data Management, Information and Monitoring Services, Fabric Management, Mass Storage Management, Deployment and Testing, Network Services, and Applications.

The LHC will provide facilities for a number of experiments, as shown under "Applications" on the project plan. These projects all have similar but different requirements for the Grid middleware infrastructure since they each have their own particular simulation and analysis software, and different computational and storage needs. The GridPP project has therefore focused on the development of standard interfaces to leverage the activities of one experiment across to others. Atlas and LHCb are two of the LHC experiments and their software infrastructure is called GAUDI/ATHENA. This software is designed to support the whole suite of application software, including simulation, reconstruction, and analysis. To provide a flexible Grid interface, the projects have come together to develop the GANGA (GAUDI/ATHENA and Grid Alliance) front-end. This user interface facilitates configuration of experiments, submission of computational

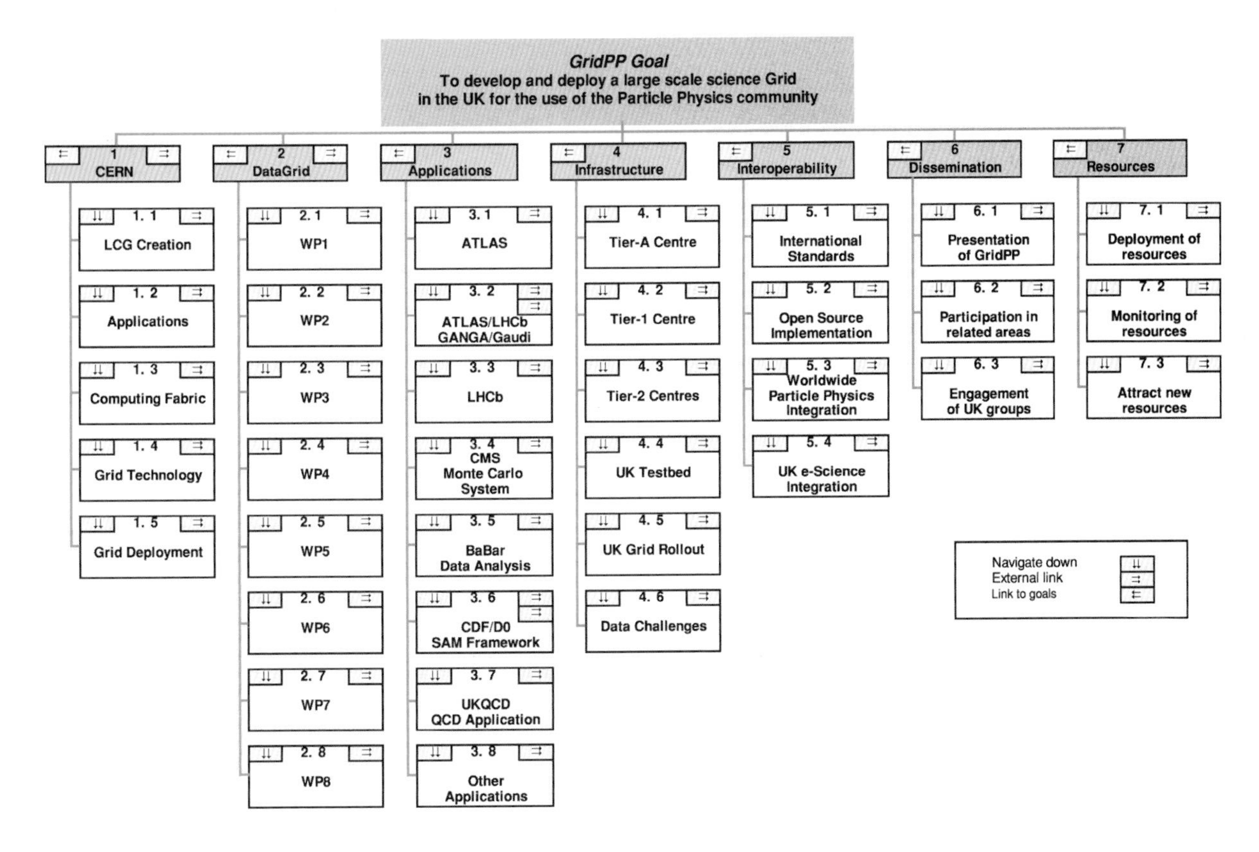


Figure 15.1. The project plan of GridPP for UK particle physics. The goal is to provide a computing environment and infrastructure for particle physicists working in worldwide collaborations. (Courtesy of David Britton, GridPP, Imperial College.)

jobs to the Grid resources, file transfer between Grid sites, monitoring of Grid activity, accounting on the resources used, collection of output and results, and reporting on Grid jobs. The GANGA front-end relies on middleware from other projects, including the Globus project, led by Ian Foster and Carl Kesselman, and the EU EDG project, to perform the low-level Grid operations, and attempts to provide a transparent environment for the user. The GridPP team have also been contributing to the ATCOM (Atlas Commander) Monte Carlo production tool for generating simulations for particular data sets, which will also be integrated into the GANGA interface. Creating open community interfaces to Grid technologies, leveraging ongoing activities, and incorporating both legacy and new applications are thus crucial to any of these large-scale efforts.

The first phase of the GridPP project came to an end in September 2004. The second phase of GridPP received approval for a further £16M funding. The goal of this second phase is to take the prototype infrastructure that has been created in collaboration with the EDG and Globus projects up to the quality required of a full production Grid. This will entail not only scaling up the infrastructure to the required resource level for the LHC experiments but also ensuring that the services provided are robust and secure and able to guarantee appropriate levels of quality of service. Moving from a prototype Grid to a production Grid is a big challenge for all the Grid technology projects described here. As for the first phase of GridPP, the project will be working closely with an EU project. This is a new EU research infrastructure project called EGEE – Enabling Grids for E-Science in Europe – which began in April 2004. The goal of EGEE will be to move from a prototype trans-national particle-physics-based Grid to a full production-level pan-European "Grid of Grids" supporting a wide range of scientific endeavor. The embryonic LHC Computing Grid and the particle physicists are the dominant, initial application but the EGEE project plans to develop a more generic middleware infrastructure than one just for use by particle physicists.

**Figure 15.2.** The nationwide UK Particle Physics Grid. This is a major component of the EU's Research Infrastructure project EGEE – Enabling Grids for E-Science in Europe. The goal is to move from a prototype trans-national particle-physics Grid to a full production-level pan-European "Grid of Grids" supporting a wide range of science.

### 15.2.3 Theoretical particle physics: the QCDGrid

As discussed above, the QCD gauge field theory is believed to describe the interactions between quarks and gluons. In order to predict quantities such as the masses of familiar particles such as the proton and the pion, it is necessary to solve QCD in a regime in which the usual tools of perturbation are not valid. Since the quarks are themselves confined – do not appear as free particles – QCD forces must become very large as the interquark separation is increased. It was for this reason that, in the 1970s, Ken Wilson (Nobel Prize 1982) formulated a version of QCD on a discrete four-dimensional spacetime "lattice." In this form, "lattice QCD" is amenable to numerical solution based on a Monte Carlo evaluation of the relevant path integral. Despite the considerable ingenuity of the lattice-gauge theorists in developing new and more efficient algorithms, numerical simulation of QCD is extremely computer-intensive. In addition, in order to reduce the errors caused by approximating the theory on a lattice and obtain results relevant to the continuum limit, the theorists need to use the largest lattice that they can simulate on the fastest computers available. This explains why lattice QCD theorists have been major users of supercomputers for the past decade or more.

The UK QCD project is collaborating with Columbia University and IBM to build a multi-teraflop machine designed specifically for lattice QCD computations. The

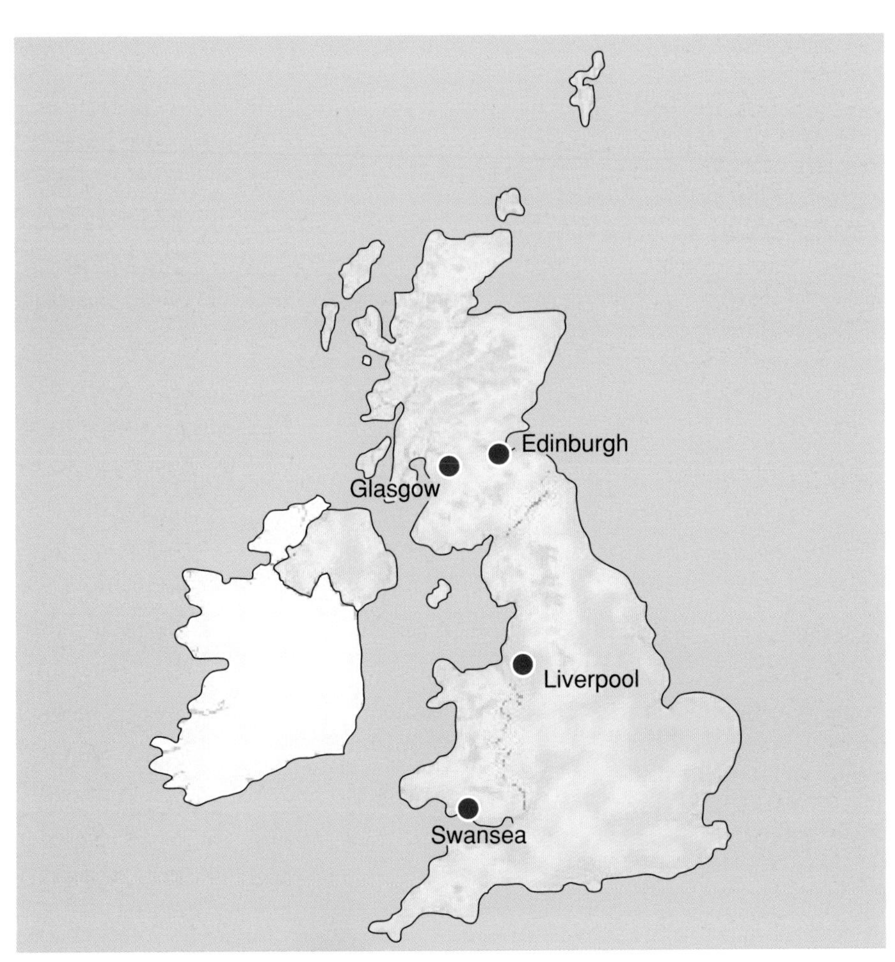


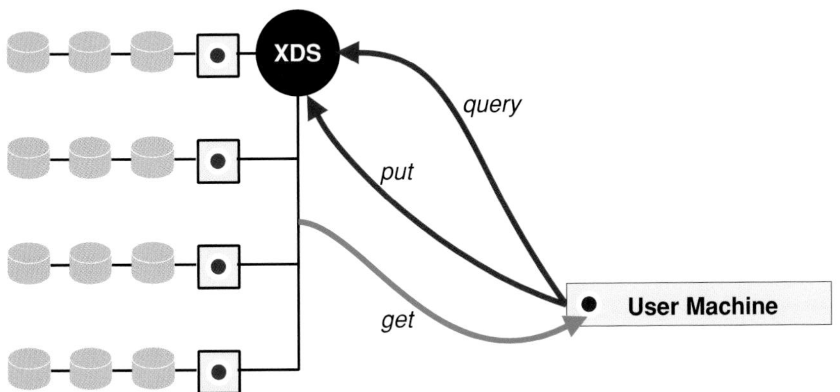


**Figure 15.3.** The UK QCDGrid system has data nodes at four sites to provide multi-terabyte storage for quantum-chromodynamics calculations of particle properties. To safeguard data, no one node is independent of the others. The data are replicated across the system so that if a node were to fail the data could be recovered. (Courtesy of Tony Doyle, University of Glasgow.)

"QCDOC" machine came on line in 2004 and as a 5-teraflop supercomputer is capable of creating vast amounts of data. The physicists who were responsible for creating the data naturally make the initial physics analysis of the data but, since the terabytes of simulated QCD data have been produced at great cost, it is sensible to archive the raw data for possible later analysis by others. This is the problem that the QCDGrid project is tackling. The project is therefore creating a number of data nodes at four sites across the UK (see Figure 15.3) that will provide multi-terabyte storage on farms of RAID arrays.

A total of 2.3 terabytes of data storage is available at a node in Edinburgh and 3.0 terabytes in Liverpool. Data security and recovery is very important and hence no one node is independent of the others. The data are replicated across the system in such a way that, if a node were to fail, then the data can be recovered. To retain their usefulness the data must be annotated with fairly complex metadata that capture all the relevant physical parameters and the algorithmic details used in their production. Metadata schema are under development and a metadata catalog will facilitate the mapping of metadata onto data and provide a high-level mechanism for researchers to query and retrieve data easily. The project has proposed an international standard for QCD data. The software used by QCDGrid is based on that developed in the EDG and Globus projects. Since the data, once stored, are not subject to change or updating, data-coherency problems are limited for the QCDGrid distributed archive.

## 15.2.4 Astronomy: an international virtual observatory

The International Virtual Observatory Alliance (IVOA), or as it is sometimes known, the World Wide Telescope, is an ambitious example of an attempt to enable a new generation of collaborative science. The goal is to provide uniform access to a federated, distributed repository of astronomical data spanning all wavelengths from radio waves to high-energy gamma rays. At present, astronomical data using different wavelengths are taken with different telescopes and stored in a wide variety of formats at many different locations. Virtual-observatory projects in the USA, the UK, and elsewhere around the world are collaborating to create a "data warehouse" for astronomical data that will enable new types of astronomy to be explored. The ability to combine data from different wavelengths, spatial scales, and time intervals holds the promise of revealing as-yet-unrecognized features that cannot be identified with a single data set. In addition, for the first time astronomers will also be able to test results of large-scale simulations against large volumes of observational data. A further benefit is that the data would be accessible from anywhere with a sufficiently high-bandwidth connection to the Internet instead of being held in repositories in sometimes remote and poorly networked sites.

There are many significant challenges to creating such an international virtual observatory. The first hurdle is partly sociological – the data sets are owned and curated by different groups – and making data more widely and publicly available in a form that other scientists could use will be a major cultural shift for some groups. The data are also completely heterogeneous and span the entire electromagnetic spectrum from gamma rays, X-rays, ultraviolet, optical, and infrared, right through to radio wavelengths. A further complication on the horizon is the development of the next-generation high-volume sky surveys that will produce many terabytes of data each year. For example, starting in 2006, the Visible and Infrared Survey Telescope for Astronomy (VISTA) will be generating around 500 Gbytes per night for the next 12 years.

In the past two to three years, twelve virtual-observatory projects have been funded around the world and they came together to form the International Virtual Observatory Alliance. Table 15.1 shows the partners in the Alliance.

The only way to achieve the International Virtual Observatory goal is through collaboration. The international astronomy community, ahead of many other scientific communities, agreed a common data format in 1980, the Flexible Image Transport System (FITS). However, this data-format standard is concerned with syntax rather than semantics and the community is now working to create further appropriate metadata and semantic data standards. The first standard to be agreed is an XML standard for astronomical tables, VOTable.

Table 15.1. *International Virtual Observatory Alliance partners*

| | |
|---|---|
| AstroGrid (UK) | http://www.astrogrid.org |
| Australian Virtual Observatory | http://avo.atnf.csiro.au |
| Astrophysical Virtual Observatory (EU) | http://www.euro-vo.org |
| Virtual Observatory of China | http://www.china-vo.org |
| Canadian Virtual Observatory | http://services.cadc-ccda.hia-iha.nrccnrc. gc.ca/cvo/ |
| German Astrophysical Virtual Observatory | http://www.g-vo.org/ |
| Italian Data Grid for Astronomical Research | http://wwwas.oat.ts.astro.it/idgar/IDGAR-home.htm |
| Japanese Virtual Observatory | http://jvo.nao.ac.jp/ |
| Korean Virtual Observatory | http://kvo.kao.re.kr/ |
| National Virtual Observatory (USA) | http://us-vo.org/ |
| Russian Virtual Observatory | http://www.inasan.rssi.ru/eng/rvo/ |
| Virtual Observatory of India | http://vo.iucaa.ernet.in/~voi/ |

As well as the data model, the community must provide the services that allow access to data, seamless querying of data, exchange methods, and registers of the services available. The registers will allow astronomers, or indeed others, to find appropriate resources easily. In the USA, the National Virtual Observatory (NVO) project is pioneering the development of tools built from Web-services technology (see below) and exploiting the power of relational databases to perform relevant astronomical queries. In the UK, the AstroGrid project is building a prototype Grid of resources, databases, computational resources, data-mining services, and visualization services that is intended to be interoperable with the IVOA. The AstroGrid project has been working closely both with the GridPP project and with the myGrid bioinformatics project in the UK since all of these projects have similar requirements. They all need to access and analyze heterogeneous data from multiple sources and provide appropriate metadata, provenance, and ontology services. Members of the project team are also creating tools that provide the user with their own virtual workspace, which they call MySpace. A prototype International Virtual Observatory infrastructure involving the Australian, UK, and US teams has now been demonstrated.

An interesting aspect of the AstroGrid project has been the mechanisms that have been put in place to allow the teams to collaborate successfully. Their tools are all interactive and Web-based. They allow team members to post information, to deposit documents, code, meeting minutes, and the like, with the system keeping a record of the events, making it simpler to collate information for distribution and reports. The AstroGrid environment is based on the Wiki technology and provides a valuable infrastructure for their development environment. Another valuable contribution of the virtual-observatory community will be to engage the interest of schoolchildren in science and astronomy. The e-Star project in the UK is building a Grid infrastructure for the remote control of a network of telescopes. By combining remote use of instruments with the virtual-astronomy data warehouse, schoolchildren will be able to engage in genuine astronomy projects of real relevance.

### 15.2.5 Astrophysics and cosmology

The Cosmos consortium, led by Stephen Hawking of Cambridge University, is concerned with the origin and structure of the Universe. Their goal is to model the history

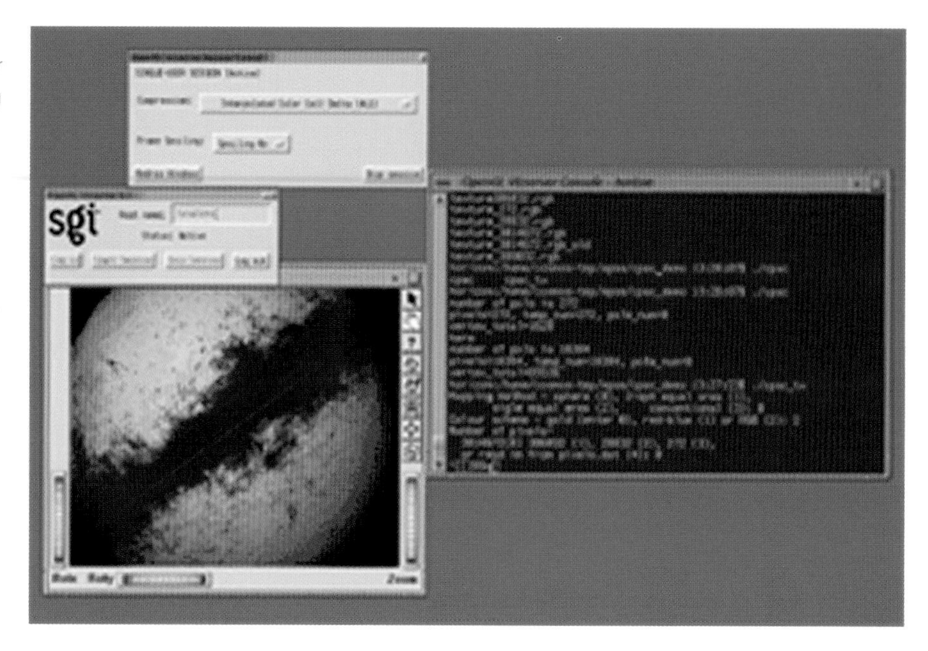


**Figure 15.4.** The CosmoGrid project provides technologies for the Cosmos consortium studying the origin and structure of the Universe. The project has two models for remote visualization, one based on a visual-server approach, allowing multi-user collaboration and visualization of centralized data. (Courtesy of Paul Shellard, DAMTP, Cambridge University.)

of the Universe from the first fractions of a second after the Big Bang, right through to the present, some 13.7 billion years later. The models developed lead to nonlinear simulations that require large-scale supercomputing and create large amounts of data. The consortium has members at Imperial College, Sussex, Portsmouth, and Cambridge, with a supercomputer at Cambridge. The CosmoGrid project (see Figure 15.4) is developing Grid technologies to allow the consortium members to collaborate remotely. The emphasis is on being able to control the facilities at Cambridge, share data, and collaboratively visualize and analyze the data.

Remote visualization of data is a generic concern for many communities and this project is building a number of technologies and services that have been developed. However, there are unresolved issues regarding appropriate adaptive algorithms to provide a uniform quality of service across an inhomogeneous network. At present the project has two models for providing remote visualization, one based on a visual server and the other on data sharing. The visual-server approach uses the technology based on the SGI Vizserver product that allows multi-user collaboration and visualization of centralized data. The project team is extending these visualization capabilities through their CSKY software.

The second model, data sharing, is being used when the network link is poor. This model is dependent upon replicated data sets, closer to the remote site, and again provides the capability to remotely share control of the visualization applications. The tension is, as ever, over whether or not to move the data or the computation, and how best to respond to network-capacity constraints, in order to deliver the visual detail and quality of service required.

## 15.2.6 Fusion physics

Security within a virtual organization is crucial to the implementation of a Grid application, and one consortium that is dealing with this aspect of collaboration is the FusionGrid project. FusionGrid is a project run by the National Fusion Collaboratory, with over 1000 researchers in over 40 institutions in the USA. This community is focused on advances in magnetic fusion, both in terms of new scientific models and

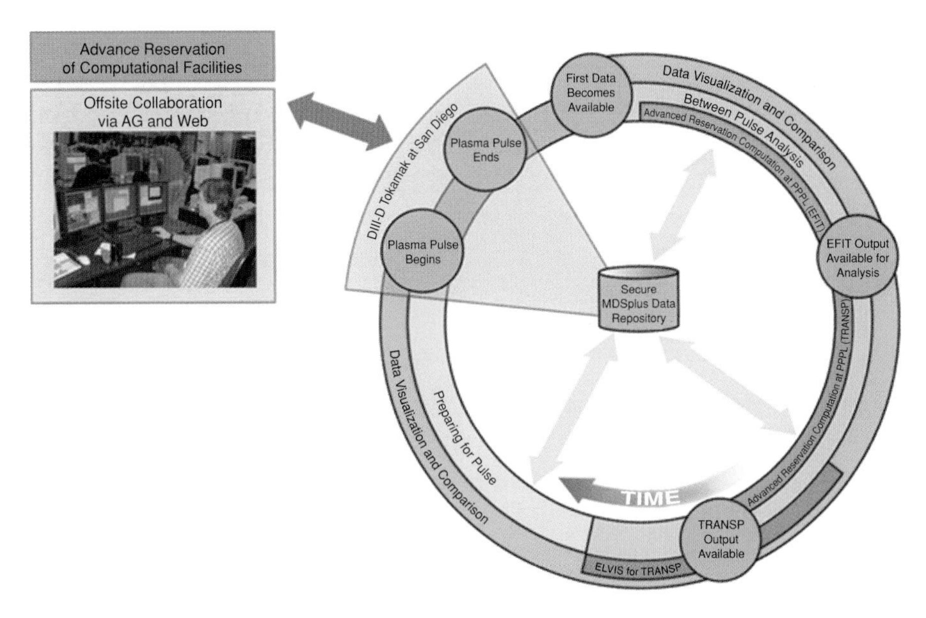


Figure 15.5. The FusionGrid project run by the US National Fusion Collaboration includes over 1000 researchers in over 40 institutions. The coordination of experiments at three large facilities requires large-scale computation and visualization with extensive involvement of other experts. Top left is the virtual control room.

also in terms of the incorporation of new technology to enable more efficient use of experimental facilities by creating an infrastructure capable of integrating experiment, theory, and modeling. The Collaboratory is focused on the design and implementation of experiments at three large facilities – the Alcator tokamak at MIT, the DIII-D National Fusion Facility in San Diego, and NSTX (the National Spherical Torus experiment) at Princeton. Magnetic-fusion experiments at these facilities take the form of a day of pulses – one pulse every 15 min. The electrical currents that create the huge fields of the tokamak heat the magnets and restrict the duration of each pulse to only a few seconds. It then takes 15 min of cooling time before the next pulse – thus giving a crucial 15 min of analysis time! During these 15 min the scientists need to analyse the data created in the pulse – up to 10 000 separate measurements, 250 megabytes of data – in order to "design" the next pulse. This analysis requires large-scale computation, visualization, and, of course, consultation with other experts in the field.

In bringing the disparate facilities, simulations, and teams together, the Collaboratory has adopted a common data-acquisition and -management system (MDSplus) and legacy data analysis and simulation codes are converted to a common data format. The project is developing metadata standards and new data models separating metadata from the experimental data. As noted above, an area that has been of great importance is security of data and of access and authorization of collaborators. Levels of trust between institutions have been formally defined and policies enforced. The project is using authentication mechanisms from the Globus toolkit and Akenti software to implement an appropriate authorization model. The Akenti authorization services provide the functionality to specify and enforce convenient access policies for use of resources by multiple remote collaborators. These policies can be based on individual-user identification, or can be defined by groups or roles within the collaboration. Use of the system in this collaboration is likely to lead to extensions of the services provided within this virtual organization.

In November 2003, the National Fusion Collaboratory demonstrated their virtual control room at SC2003 in Phoenix (shown in Figure 15.5). This demonstration showed off-site scientists collaborating with scientists at the DIII-D facility as though they were in a single control room. The off-site scientists had access to status information, data, and computational resources just as though they were located in the control room.

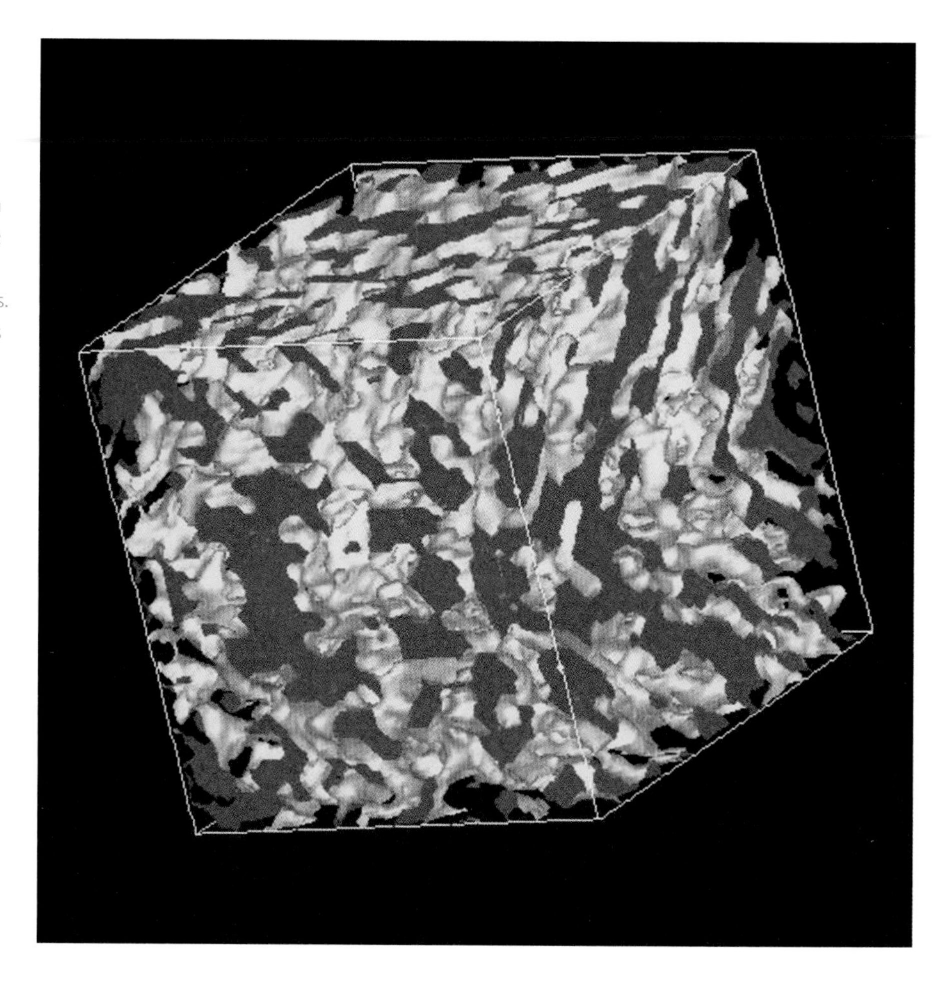

**Figure 15.6.** The RealityGrid project combines high-performance computing, real-time visualization, and computational steering for a range of applications. Although computer simulations may take days, the analysis of the resulting data may take months. The project aims to remove this bottleneck. This simulation, by the TeraGyroid collaboration studying defect dynamics in liquid crystals, used over 6000 processors. (Courtesy of Peter Coveney, RealityGrid, UCL.)

They were able to share applications with concurrent control between the multiple sites. AccessGrid technology was used to allow detailed discussion between the teams of scientists.

### 15.2.7 Condensed-matter physics

The RealityGrid project in the UK (see Figure 15.6) is a project that combines high-performance computing, real-time visualization, and computational steering in a number of application domains. The project is concerned with the issue that, for many in-silico experiments, the simulation may take days but the analysis of the resulting data may take months. The project's mission is to "move the bottleneck out of the hardware and back into the human mind."

The project includes two applications where experimentation and computation are brought together. The first is in X-ray microtomography. Microtomography is a technique similar to a medical CAT scan, whereby X-ray images are taken at many angles around the object and the projection computed. In microtomography, however, the spatial resolution is unprecedented – 2 μm – facilitating the obtaining of images of the detailed elemental composition of the structures inside materials. RealityGrid combines the resources required to carry out the data collection – expensive synchrotron beam time – together with the simulation and visualization. The testbed on which they are piloting the application is providing a model for the European Synchrotron Facility.

The second such example is the London University Search Instrument (LUSI), which is perhaps the first instrument to apply combinatorial techniques to materials-science problems, in this case for the study of ceramic materials. The LUSI provides high volumes of experimental data and with the RealityGrid infrastructure this can be fed directly into the high-performance analysis tools. The infrastructure will also allow researchers from other institutions to query, analyze, and add to the LUSI database.

One of the key technologies being developed within the RealityGrid project is a suite of tools for enabling application steering. These are being used within the project to allow scientists to steer lattice-Boltzmann fluid simulations running on a variety of HPC systems. In order to steer the application, the scientist needs to be able to interface directly with the simulation in real time, as it is being computed. The RealityGrid project goes beyond simply allowing parameter changes to be made to steer the computation, by also providing services to visualize the complex data sets – facilitating a more intuitive interaction with the simulation. This level of visualization requires significant computational resources and thus might also be running a resource remote from the scientist or the real-time simulation. These problems therefore involve quite complex scheduling requirements and issues concerning co-allocation of resources at different sites and quality-of-service guarantees between both those sites and the scientist. The RealityGrid consortium has demonstrated prototype versions of such capabilities at a number of scientific and visualization conferences.

The RealityGrid team has worked with the NSF TeraGrid consortium to realize the TeraGyroid project – Grid-based Lattice-Boltzmann Simulations of Defect Dynamics in Amphiphilic Liquid Crystals. The successful conclusion of this collaborative experiment represented an achievement both in the use of an international Grid infrastructure and in deriving new scientific results from simulation of soft condensed matter. The computational experiment used the supercomputing facilities in the UK – CSAR and HPCx – together with the TeraGrid machines at the US National Center for Supercomputing Applications (NCSA) and the San Diego Supercomputer Center, to calculate and steer a lattice-Boltzmann computation involving in excess of one billion lattice sites. In total, the TeraGyroid experiment used more than 6000 processors and consumed 17 teraflops of computation across the two continents, together with using large visualization systems at Manchester for the interactive visualization component. A transatlantic network of 10 Gbytes $s^{-1}$ between Europe and the USA was required in order to make the project feasible. This project is at the HPC end of the spectrum of Grid applications but the present prototype infrastructure is far from production quality. However, the experiment has provided some insights into the future for large-scale simulation and visualization.

One important lesson that was learnt from this experiment is that collaborative use of international facilities is at present far from routine. The success of this project depended significantly on "heroic" efforts made by a large team of technical experts. The true challenge for e-Science is to make such collaborative-science experiments easy for scientists to set up and run without such a large amount of technical system support.

## 15.2.8 Atomic physics

Electronic collisions drive many of the key chemical and physical processes and are equally important in the fields of laser physics, plasma physics, atmospheric physics, and astrophysics. Accurate simulation of these phenomena is difficult since an infinite number of continuum states of the target atoms must be taken into account. A group at Queen's University Belfast has developed techniques that allow the accurate study of collisions of electrons with hydrogen-like atoms. The techniques are based on a generalized R-matrix method, which they have implemented in a software package, 2DRM. This package is used worldwide to generate atomic-collision data for the applications listed above. The researchers have now adapted their application to run on the UK e-Science Grid and this constitutes a fine example of a computationally focused legacy application migrating to a Grid environment.

The algorithm is such that it can be broken down into a set of independent components, each of which has different computational needs. This makes it very suitable for computation on a Grid of heterogeneous resources such as the UK e-Science Grid. The original software package was developed using the MPI message-passing standard. To run effectively on the Grid, the software was re-engineered by removing the MPI communications and creating a "workflow processor" to manage the computation and a resource-allocation service to match tasks to resources. Each component has been implemented so that it can be run on any of the resources on the UK e-Science Grid, allowing the application to take up dynamically any resource as it becomes available. The application has run on the UK e-Science Grid for more than 6 months, demonstrating that, in spite of some fragility, the UK e-Science Grid, using middleware based on the Globus Toolkit, can generate useful scientific data. The team also developed a number of other middleware services, including the workflow processor and additional software to support resource discovery, resource allocation, and monitoring tools. Such utilities will become the norm for any future Grid environment.

In migrating this legacy application, a number of generic issues surfaced. These included the requirements of particular compilers and libraries across the heterogeneous set of resources, the need for some type of notification mechanism for system and software upgrades that have been implemented on the resources, the requirement of a platform-independent data format (rather than binary format), and the capability to adapt to the floating-point accuracy of particular machines. Present management tools to handle such issues are relatively immature, but are progressing through projects such as the UK e-Science Grid, the DOE Science Grid, the TeraGrid, and the EU EGEE project.

### 15.2.9 Applications of idle-cycle harvesting

To some people the Grid is synonymous with high-performance computing, to others with "idle-cycle harvesting." As we have emphasized, the scope of e-Science applications is much broader than just these two types of computer-centric application areas. Nevertheless, using the idle processing capacity on desktop or personal computers is a type of Grid application that can yield very tangible benefits both for science and for industry. The types of applications that are able to take advantage of such a framework generally require many similar or identical, relatively small computations across a parameter sweep or set of data. The most famous of these is probably the SETI@home project, launched on May 13, 1999, which provides a screen-saver program to download for anyone willing to offer idle cycles. The screen-saver program analyzes data from sophisticated radio telescopes and receivers searching for signals that might indicate the existence of extraterrestrial intelligence. Over 3.8 million people from 226 countries have downloaded and run the SETI@home screen saver, downloading chunks of data to be analyzed and providing, at times, an average throughput of over 25 teraflops.

Note that, although such a large computational throughput as 25 teraflops often induces the popular press to refer to this as "supercomputing," it is important to recognize that this confuses "throughput" with "capacity" computing. Thus 25 000 PCs each running at 1 gigaflop with identical copies of a program but on different data sets gives 25 teraflops throughput. However, this is entirely different from one program on one very large data set running at 25 teraflops on the Earth Simulator. Grid computing does not replace very-high-end computing but can satisfy different needs.

Another such application is the Great Internet Mersenne Prime Search – GIMPS. The GIMPS project allocates different prime-number candidates to contributing machines. On November 13, 2003, after 2 years of computation on more than 200 000 computers, Michael Shafer, a chemical engineering student at Michigan State University, found the 40th known Mersenne prime, $2^{20\,996\,011} - 1$. Entropia, a commercial "Grid" company, provide the middleware for this project. Another company that has developed similar "Grid" technology is United Devices, which is also involved in a number of scientific applications. Working with a team from Oxford University, they have initiated projects for screening billions of molecules for suitability as candidate drugs for the treatment of anthrax and also for some types of cancer. The company established the PatriotGrid as a mechanism for the general public to offer their computers to help combat bio-terrorism.

In the UK, the Climateprediction.net project exploits this type of Grid computing. The project is using this type of idle-cycle model to compute probabilistic climate forecasts by running a vast parameter sweep of the same climate model and using a weighted ensemble of forecasts to make conjectures regarding future risks and the sensitivity of such models to various input parameters. This project goes beyond simply farming out tasks. Each of the climate-model computations results in several megabytes of data. In general, the data remain on the computational node that generated them, so federated data services and remote visualization are required in order to query and analyze the very distributed data. The project currently has more than 45 000 PCs signed up, including some schools.

A research group collaborating with the Cambridge e-Science Centre has used the UK e-Science Grid (see Figure 15.7) to calculate the diffusion of ions through radiation-damaged crystal structures. In this case, the Grid is built using the Globus Toolkit, so the researchers have to have the necessary digital authentication certificates to be authenticated before they are allowed to use the Grid – rather a different model from the use of independent PCs in the examples above. The computations are based on Monte Carlo simulations and a sweep across a parameter space, requiring very little input data and resulting in very little output data. The Condor-G client tool was used to manage the farming out of jobs to resources. This type of simulation can always use up any available resource, since, in general, the results will have more statistical significance, the more computations have been completed. In this application case, the larger the number of CPUs involved in the simulation, the higher the temperatures that can be modeled and the better the statistical representation that can be obtained.

## 15.3 e-Infrastructure and the Grid

### 15.3.1 User requirements

From the examples described above we can informally deduce some generic requirements for the e-Science infrastructure. All of the projects feature distributed collaborations with a mix of computing and data demands. The particle physicists have a truly global community and have a multi-petabyte source of data from the LHC in Geneva. They require the infrastructure to be able to move data efficiently and securely and utilize "throughput" computing resources throughout their collaboration. The virtual

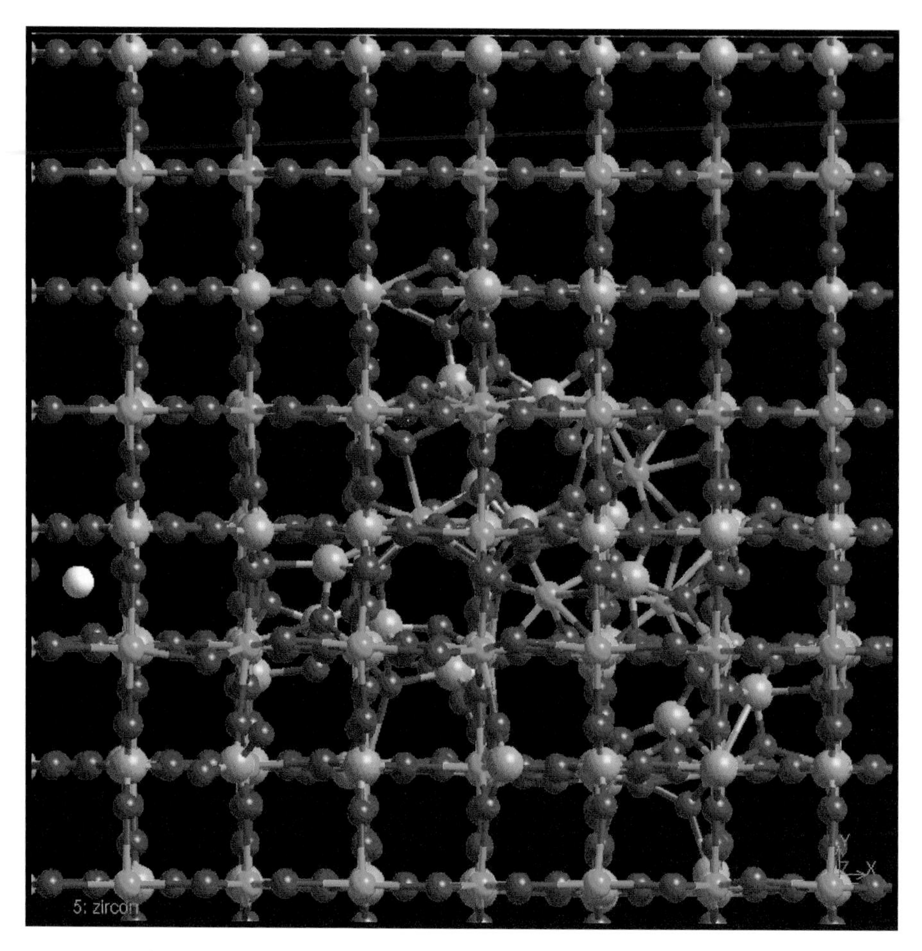


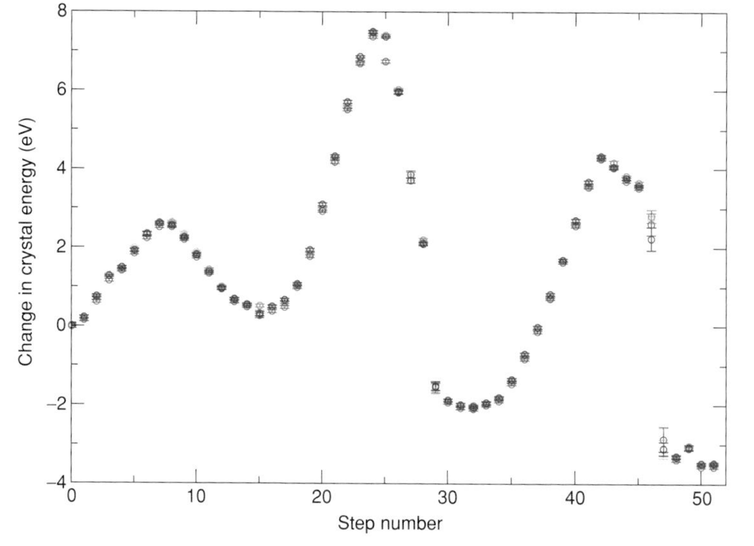


**Figure 15.7.** A group collaborating with the Cambridge e-Science Centre used the UK e-Science Grid for Monte Carlo simulations of ion diffusion through radiation-damaged zircon ($ZrSiO_4$, zirconium silicate) crystal structures. Zircon is a naturally occurring mineral that appears to be highly resistant to radiation damage, making it a candidate for encasing radioactive waste. In the simulation, the migrating ions are marched in small steps toward the required end point, with the entire crystal being relaxed at every such step. The picture shows a small section of the starting configuration; the blue balls represent zirconium ions, the brown silicon, the red oxygen, and the white ball is a migrating sodium ion. Note that the damaged region is due to a very-low-energy event (~ keV). The simulations were run at a temperature of 10 K, with the sodium ion marching toward the center of the damaged region in small steps of ~0.2 Å. The graph shows the change in crystal energy as the ion penetrates the damaged region. The core of the damaged region provides a lower-energy state for the sodium ion, protected by a shell of disordered ions due to the decay. (Courtesy of Mark Calleja, Earth Sciences, Cambridge University.)

organizations for the LHC experiments will be fairly long-lived since the institutions involved will be engaged in a project lasting a decade or more. The astronomers have a very different data-centric focus, more akin to commercial data warehouses and data-mining scenarios. They require the infrastructure to support a distributed-data warehouse that allows federation and queries across multiple data sets. Most of the other projects are more computer-centric, with some just needing access to supercomputers and others satisfied with Condor-style idle-cycle harvesting. Distributed visualization, computational steering, and remote control of equipment and facilities are also featured.

There are also some generic challenges for the e-Science infrastructure that need to be addressed for all these applications – these include the obvious properties such as security and dependability as well as scalability and performance. In addition, scientists should be able to identify the resources that they need, negotiate some sort of "price" if necessary, be authenticated and authorized as someone allowed access to those resources, and integrate them into their own collaborative Grid application. To make this possible there need to be directories of Grid resources and a simple mechanism for querying them: once the type of resource or set of resources has been located, a resource broker can then instantiate the job or workflow – a set of predefined operations – for the user. Last but not least, robust mechanisms for billing and payment are needed and the whole area of business models for dynamic virtual organizations needs further consideration.

### 15.3.2 History

Probably the most popular Grid infrastructures of the 1990s were the C-based versions of the Globus Toolkit, Condor, and the Storage Resource Broker (SRB), although other international offerings such as Legion, Ninf, Nimrod, and Unicore are also widely known, along with commercial systems such as Entropia and Avaki.

The Globus Toolkit, developed by Ian Foster, Carl Kesselman, and their teams, sets out to provide three elements necessary for computing in a Grid environment – resource management, information services, and data management. The resource-management "pillar" involves a set of components concerned with the allocation and management of Grid resources. Information services provide information about Grid resources while the data-management pillar provides the ability to access and manage data stored in files. This includes the well-known GridFTP component that uses multiple channels to move files at high speed between Grid-enabled storage systems. Underpinning these three pillars, the Globus Toolkit uses the Grid Security Infrastructure (GSI) to enable secure authentication and communication over an open network like the Internet. The GSI uses public-key encryption, X.509 digital certificates, and the Secure Sockets Layer (SSL) communication protocol. Some such authentication process is clearly necessary if users registered at one site are to be authorized to use resources at another.

The Condor system, developed by Miron Livny's group at Wisconsin, grew out of Livny's Ph.D. thesis on "Load Balancing Algorithms for Decentralized Distributed Systems" from 1983. Condor is a resource-management system for computer-intensive jobs. It provides a job-management mechanism, scheduling policy, resource monitoring, and resource management. Users submit jobs to Condor and the system chooses where and when to run them, monitors their progress, and informs the user on completion. Condor is not only able to manage dedicated compute clusters but also can scavenge and manage otherwise wasted CPU power from idle desktop workstations across an entire organization. Condor-G represents the marriage of technologies from the Globus Toolkit and Condor, in particular using the Globus protocols for secure interdomain communication. The Condor system is widely used in academia worldwide, and is also used by several companies.

The SRB system is produced by Reagan Moore and his Data Intensive Computing Environment (DICE) group at the San Diego Supercomputer Center. Their SRB middleware organizes distributed digital objects as logical "collections" distinct from the particular form of physical storage or the particular storage representation. A vital component of the SRB system is the metadata catalog (MCAT) that manages the attributes of the digital objects in a collection. Provision of high-quality metadata is vital not only for access but also for storing and preserving scientific data. Moore and his colleagues distinguish four types of metadata for collection attributes in their system: SRB metadata for storage and access operations; provenance metadata based on the Dublin Core; resource metadata specifying user-access arrangements; and discipline metadata defined by the particular user community. The SRB provides some facilities to integrate and federate data stored in different formats.

As can be seen from the above descriptions, these systems certainly provide some of the functionality needed by scientists for their e-infrastructure. However, in the main these systems focus on meeting the requirements of relatively simple computer-intensive applications with data held in files. To move on from these still rather fragile systems we must turn to developments in distributed-computing technology.

### 15.3.3 Distributed computing and Web services

Internet-scale distributed computing is a difficult problem that has resisted a fully robust and satisfactory solution up to now. With the proliferation of network-connected devices, the scalability of distributed middleware is now a key problem for the IT industry. In addition, the IT industry has learnt lessons from experience with object-oriented systems such as CORBA, DCOM, and Java. Systems like CORBA and DCOM worked reasonably over local-area networks (LANs) but proved to be rather "brittle" over wide-area networks (WANs). One of the principal causes for such fragility was that the object-oriented style tended to encourage building links to objects rather tightly. In Internet-scale distributed systems, this leads to problems with clients holding pointers to objects held behind firewalls and being managed within different administrative domains or organizations.

To overcome these problems, a more loosely coupled style of building middleware for distribution over WANs is required, with, ideally, some guarantees of interoperability of their respective systems by the major middleware vendors. The major players in the IT industry have therefore come together to focus on Web services as an appropriate technology for Internet-scale distributed computing. At the time of writing, only the lower-level Web services had been agreed upon and there was still debate on the higher-level services, with different vendor groups proposing specifications for a variety of services. Despite this routine jockeying for advantage among the major IT companies, the importance of the Web-services movement cannot be over-emphasized. For the first time, Microsoft and IBM have stood side by side and pledged that their Web-service offerings will be interoperable. This offers the promise of some real stability in this complex space, with the interoperability of Web services hiding differences in operating systems and other components. Furthermore, the Web-service movement has attracted major investment in application-development environments – such as .NET and WebSphere – as well as good support tools. Given this investment by the IT industry, it is clear that any "future-proof," Internet-scale, e-Science middleware must be based on Web services.

Web services are an example of a service-oriented architecture in which resources or operations are presented as "services," to be "consumed" by users. A service is an entity that provides some capability to clients that is effected by exchanging messages. Describing services in terms of exchanging messages leads naturally to distributed implementations. The basic model for a Web service is the publish–subscribe triangle shown in Figure 15.8: a service is published by a service provider and discovered

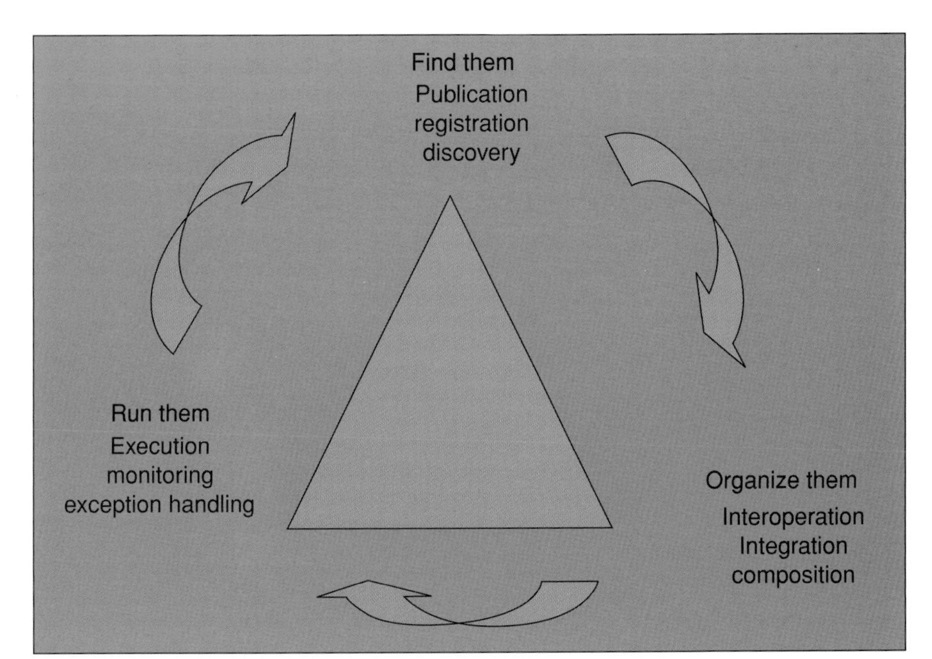


**Figure 15.8.** The basic model for a Web service is the publish–subscribe triangle: a service provider is discovered by the would-be user, who can then deal directly with the provider.

by the would-be user, who can then enter into a contract directly with the service provider. This provides the much looser coupling required for distribution of services over a WAN. The user may (or might not) need to see the publish–subscribe model. Something as commonplace as sending an email message to an individual seems not to fit easily into such a model until we remember that in fact email messages are usually sent to a server, on which your computer is listed in a directory, and delivered when the recipient is logged on. This is, in effect, a publish–subscribe model, although it does not appear so to the user. Until recently the Web-service community was focused on relatively simple transactions with more or less static collections of resources. Typical Grid operations are long-running jobs that can involve several messages and require knowledge of the state of the operation or resource. For this reason, a Grid service is sometimes described as a "stateful" Web service. There is a lively debate in the Global Grid Forum about the best way to handle both "internal" and "interaction" states in Web services. Almost all Web services encapsulate an internal state – such as the balance in your bank account. What is at issue is how best to access this internal state and how the state describing the interaction should be best handled.

In 2003, the Global Grid Forum agreed to base the new Grid infrastructure on a "service-oriented architecture" similar to Web services. The proposed Grid middleware architecture is called the Open Grid Services Architecture (OGSA) and working groups at the Global Grid Forum continue to debate the technical specifications of the core services required. The architecture allows the development of layers of services that provide the software stack from the scientific application down to the resources that it uses. Figure 15.9 is illustrative of this architecture and shows the structure of the services that have been developed. The lower-level infrastructure layer is referred to as the Open Grid Services Infrastructure (OGSI). The services in this layer provide the basic building blocks on which the higher-level services can operate.

In this approach to Grid middleware, Grid services are just a specific collection of Web services. However, the first specification and implementation of OGSI brought to the surface a number of issues. One of these was the fact that the specification of the Grid service was sufficiently distinct from standard Web-service specifications that the numerous Web-service tools that had been created for commercial activities were not

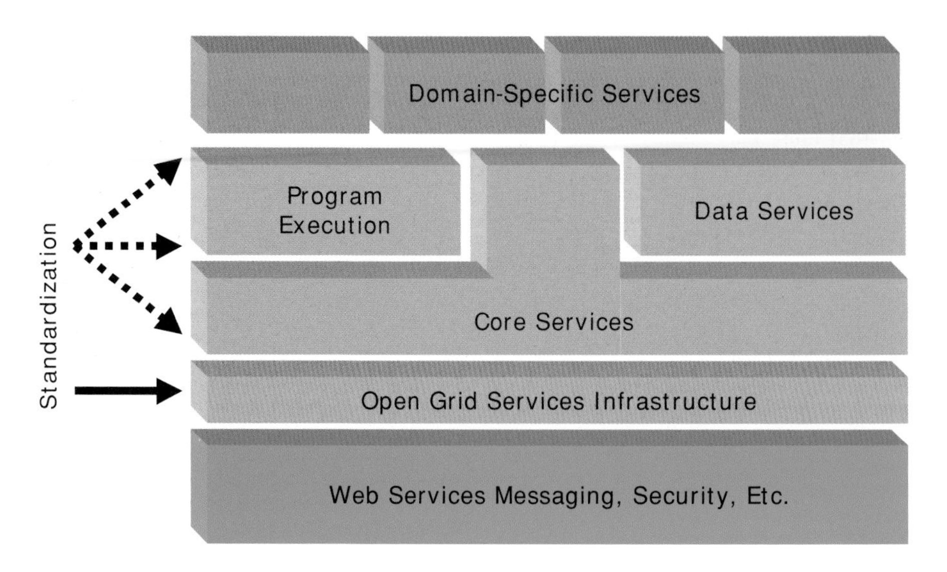


**Figure 15.9.** The lower-level Open Grid Services Infrastructure (OGSI), agreed at the Global Grid Forum in 2003, provides a foundation for higher-level services. (Courtesy of Ian Foster, ANL.)

usable. Furthermore, the OGSI architecture also had an object-oriented flavor that did not fit comfortably with Web-service philosophy. It was in this context that Parastatidis *et al.* at Newcastle provided an analysis of the OGSI in which they proposed that the OGSI functionality could be implemented with a set of emerging Web-service specifications, in a way that was sympathetic to the Web-service architecture. After a period of discussion, in January 2004 a new set of Web-service specifications collectively known as the WS-Resource Framework (WSRF) was proposed by a consortium consisting of Akamai, the Globus Alliance, Hewlett Packard, IBM, SAP, Sonic Software, and TIBCO. This includes an event-notification service called WS-Notification. With this proposed framework, the new layered architecture will simply have the functionality of the OGSI, but expressed using the WSRF as shown in Figure 15.10.

This proposal – that a Grid service conforms to a set of Web-service standards – represents great progress and will eventually lead to a stable basis on which to build the e-Science middleware infrastructure that we require. However, these proposed new Web-service specifications are now open to discussion. In particular, these new

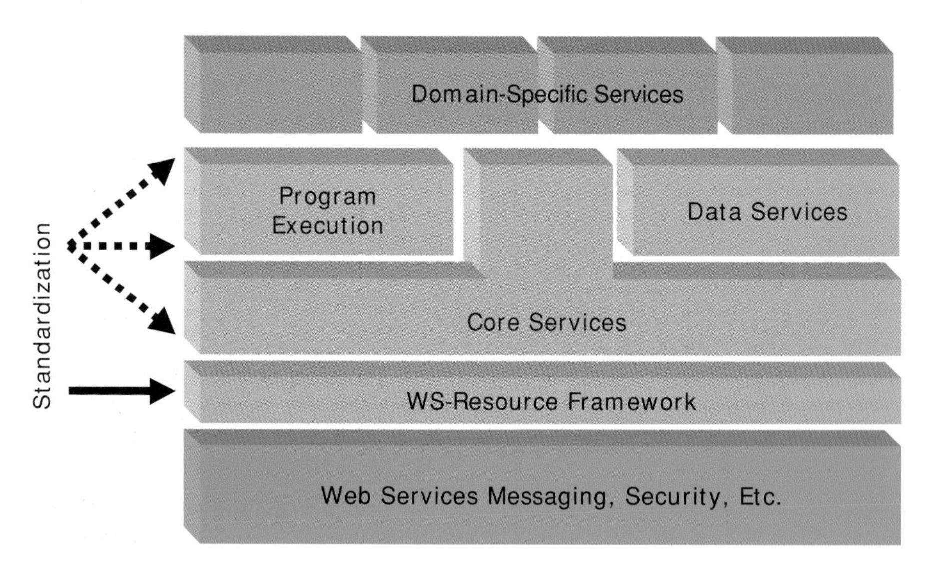


**Figure 15.10.** The Grid Stack with the Web Service Resource Framework and the functionality of the Open Grid Services Infrastructure (OGSI – see Figure 15.9). (Courtesy of Ian Foster, ANL.)

specifications are not yet being supported by companies such as Microsoft and Oracle, which, in some cases, have rival proposals.

It is therefore to be expected that there will be some inter-company cut and thrust and that there may well be some changes to the proposed WSRF standard before a stable "WS-Grid" standard emerges. Given the large number of Grid projects nearing completion around the world at present, it is important to build a low-level Grid middleware infrastructure on stable foundations – perhaps before a stable WS-Grid standard appears. In this context, many projects are implementing Grids based on Web-service specifications that have been agreed or are close to agreement – as well as exploring the specific WSRF proposal. A possible set of such Web services could comprise WS-I, WS-Security, WS-Addressing, and BPEL for example. For messaging services, Geoffrey Fox and colleagues have analyzed the differences between WS-Reliable Messaging (from IBM and Microsoft) and WS-Reliability (from Oracle and Sun) and found only minor differences. Thus, it is probably valid to include one of these proposed Web-service messaging standards in the safe subset.

### 15.3.4 Further user issues

Security is a key requirement for any application or individual user wishing to exploit the Grid. Users will want to know that their transactions or analyses are not being tampered with, that the results that they receive are trustworthy, and that their identity cannot be "stolen." They need assurance regarding the services and resources that they are using and they are likely to want a single sign-on system that allows access to a set of resources with a single authentication. Some of these issues have been solved or partially solved, and there are Web-service standards being developed for security such as WS-Security, WS-Trust, and WS-Policy. However, there are many other issues to be resolved – policy implementation, scalability of authentication, authorization technology, and federated security, as well as understanding trust models and policy implementation across domains. There is also further work required in order to understand how to implement trust relationships between virtual organizations and how they evolve and are implemented across domains and resources. Legal issues such as non-repudiation and the validity of electronic contracts will also be important issues for commercial virtual organizations.

Reliability and dependability of Grid services is a related area that needs consideration. It is important that, in the case of a failure of a service or a component part, there is an appropriate failure mechanism and the infrastructure is sufficiently robust that data are not lost or corrupted. As indicated above, Web and Grid services operate by sending messages between directories and services. If a system error causes a failure then the infrastructure needs to be sufficiently robust that all services involved are able to recover effectively with a high level of transparency for the user. There has been a large amount of research around the issue of distributed transactions but more research is needed in the context of a dynamic set of resources and organizations.

The vision of the Grid is that a user will be able to find automatically the services or resources that they require (as indicated in Figure 15.8 above). This vision relies heavily on there being definitions of the services and resources available for automatic perusal – which in turn requires that a semantic definition (or ontology) is available. Whilst ontologies are being developed in some areas such as design engineering, chemistry, bioinformatics, and health, they will probably be needed across all application areas. Further work will be required to ensure that these separate worlds defined by their own ontologies are interoperable. The requirement of semantics goes beyond the application areas and down through the stack of Grid services. An important strength of this architecture is the capability to compose, federate, and create complex services – this will be possible only with suitable semantic definitions. Figure 15.9 illustrates that the user services described are the higher-level services dependent upon the stack of services beneath – all of which will be affected by heterogeneity and all of which need to be driven by user requirements. It is also at this level that one can imagine personalization services, which have a semantic representation of the user and are capable of presenting resources and capabilities in that context.

In the OGSA framework there will certainly need to be services to assist in data access and integration from distributed databases. The UK OGSA-DAI project is developing such middleware to allow users access to data from multiple, distributed sources in addition to traditional centralized storage facilities. However, as has been emphasized by Reagan Moore and his group in San Diego, there is much to be done regarding data management in the broadest meaning of the term. In order for data collections to be useful, they need to be curated effectively. This requires high-quality metadata describing the data. For the next generation of high-throughput scientific experiments, annotation with metadata will need to be performed automatically. There is also the interesting question of provenance. In this world of shared data and collaboratories where any authorized individual can access and possibly change data, there is the question of tracking the provenance of the data – roughly speaking, who invented what, when. Many projects are considering these issues in a variety of disciplines and contexts. One of the major such projects in the UK is the myGrid project which is developing the technology for bioinformatics. Similar issues arise in other areas such as astronomy and chemistry.

As well as curating the data appropriately, the scientist will need to analyze the data. The most effective analysis of large amounts of data normally involves visualization. In the Grid environment, there may be multiple sites that need to collaborate in analyzing the data and therefore share the visualization. The sites might not have equal network capacity, in which case the visualization service needs to be sufficiently sophisticated to adapt to the techniques and resources used for the visualization according to the capacity – providing an equal level of service to all collaborators. These technologies need also to operate on widely heterogeneous platforms from palmtops to workstations and beyond.

For there to be a sustainable e-Science infrastructure, there are many other issues that need to be resolved. Fixing a value to a service is a complex issue, depending upon the availability of the service, the service provider, the possible availability of competing services, and the requirements of the user. Moreover, the value will undoubtedly be dynamic, changing from any one instant to the next. Given the services model, models for licensing of software and applications need to be re-thought. These issues are not entirely technical but they will all require technical implementations of the policy solutions and are certain to have an impact on scientific research involving simulations and databases.

### 15.3.5 Grid management

In considering the management of a Grid, many of the issues outlined above are relevant, but from a different perspective. For example, security remains a prominent issue, but now is an issue of protecting the resource, service, and fabric, and providing assurance domains. Non-repudiation, confidentiality, and auditing will also be necessary for the delivery of commercial integrated services.

Heterogeneity of resources will be an ever-challenging issue for Grids. Computers come in many shapes and sizes and the present push into small and mobile devices like hand-held, portable, and even wearable computers will accelerate the onset of this ubiquitous information-processing world. This pervasive computing model is very closely related to Grid technology and e-Science, with a combination of decentralized and distributed computing involving an array of resources allowing dynamic connectivity via the Internet. This brings with it many challenges regarding software development and deployment, but also further security issues related to using wireless communications and with all kinds of new user interfaces, creating risks of loss of confidentiality, integrity, and availability.

The scale of heterogeneity of the systems now being deployed is beyond anything that has been dealt with before. There are many questions concerning the quality of service that can be expected or demanded in such environments. Before any issues regarding quality of service can be addressed, there must be a capability for robust monitoring of services. Monitoring across ownership domains as well as different technology bases is a problem area that is not well understood – it will require new technological advances and is likely to require communication and agreement across suppliers.

Developing software within this services architecture also creates new challenges. The distributed nature of the services requires complex test harnesses, in particular, with federated services that might have multiple dependences. Indeed, there will be a requirement for new software-management tools – as services are updated, the updates may need to be propagated through related services and onto resources that may have different operating systems and different mechanisms for installation and deployment of software. These tools will need to be scalable and automatic – "autonomic" methods will need to be developed.

One of the main threats to the success of the Grid is the sheer complexity of the multilevel infrastructure. As indicated above, this is a consideration both for users and for the developer. At present we rely on human intervention and administration to manage these complex systems, to recognize faults, and to repair them. Consider the situation of a large virtual organization with many thousands of computers, hundreds of distributed databases, scientific facilities or instruments, and a middleware infrastructure built on layers of services with software agents allowing many millions of transactions and computations every minute. Human management of such a system is not realistic. The solution must again lie in the development of autonomic technologies that are capable of self-configuration, self-optimization, and self-management.

An autonomic system is a system analogous to our human nervous system – self-healing and adaptive with some sense of itself – keeping track of the devices, networks, etc., which at any point in time define it. Ideally, it needs to be able to recover from damage by some means, be capable of improving its own performance, adapt to the dynamic changes allowed by Grid services, defend itself against attackers, and to some extent anticipate users' actions. IBM has a very active program considering these issues within their family of IT products. The capabilities of autonomic systems are in their infancy and likely to develop incrementally through systems and operations across the board.

## 15.4 Grids for industry and commerce

From the (non-exhaustive) set of examples described above, it is clear that Grid technology will be important to the sciences. Will the Grid, like the Web before it, cross over from academic applications to industry and commerce? Although our discussion so far has been in the context of scientific research, it is clear that much of the Grid middleware technology being developed will cross over to business and industrial applications.

### 15.4.1 Engineering applications

The ability to share resources in a secure, dynamic fashion is becoming as important to industry as it is to science. In the intensely competitive marketplaces of the automotive, aerospace, energy, electronics, and pharmaceuticals industries, it is increasingly important to achieve a faster time to market by decreasing the time for the full design life cycle. This often requires an improvement in design collaboration across departments and partners that host the individual specialists who form the design team. In addition, many companies are facing pressure to reduce IT investment and increase return on investment and are looking to optimize their own computing and data resources. It is often the case that companies will need to complete large simulations as part of a design phase, hence requiring high-performance computing, or will need to curate large amounts of data, or to visualize the results of the simulations. Within this framework, a company would have the opportunity to use the resources of a partner to complete these important but infrequent tasks.

For aircraft-engine maintenance, the company needs to ensure critical safety levels but wants to increase efficiency in terms of engine maintenance and thereby increase the number of hours an engine could be in use. In the UK e-Science DAME project, a consortium comprising the Universities of York, Oxford, Sheffield, and Leeds, together with Rolls Royce and its information-system partner Data Systems and Solutions, is building the capability to monitor engines in real time, in flight, and analyze the data collected to identify any anomalies in performance before the plane lands. At any one time there are thousands of Rolls Royce engines in flight and the amount of data being collected quickly runs into gigabytes. These data need to be analyzed alongside historical data that are stored at two data warehouses across the globe – currently in the UK and in the USA. There is a need for high-performance computing capability in terms of the amount of data to be managed, the analysis to be completed, and the integrated nature of the resources being used. Distributed diagnosis as exemplified in this application is a generic problem that is fundamental to many fields such as medicine, transport, and manufacturing. The DAME project should therefore also produce valuable middleware infrastructure and experience that can benefit many other important application areas.

In the AEC (architecture/engineering/construction) industry, large projects are tackled by consortia of companies and individuals, who work collaboratively for the duration of the project. In any given project, the consortium may involve design teams, product suppliers, contractors, and inspection teams who will inevitably be geographically distributed. This is a good example of the type of virtual organization we are discussing here – an organization that comes together for the life of a project, during which time they need to be able to share resources and information across the domains of their own companies. At the end of the project, the virtual organization is dissolved since the companies will generally have no interest in continuing to share information and resources. The Welsh e-Science Centre, the Civil Engineering Division at Cardiff University, and BIWTech are assessing a Grid-based system that will (1) allow interactive, collaborative planning and management; (2) provide availability, delivery, and costing information for products and supplies; and (3) support the integration of three-dimensional geometric product models into a computer-aided-design (CAD) environment for architects and design teams. The system will bring together items of software that have been designed to tackle these issues individually, together with a product and supplies database, and integrate them into a shareable, secure environment. In this case, the emphasis is on allowing the companies to adapt their legacy products in order to be able to provide them as a service on the Grid in a dynamic, secure fashion.

### 15.4.2 Disaster response

As a further example of a dynamic virtual organization, consider the case of a natural disaster, say a major fire: not only is it necessary to have all the emergency services collaborating but it may also be the case that the fire involves toxic fumes being emitted into the environment. A futuristic vision of the effective use of Grid technologies would be an environment that allowed the emergency services to engage immediately with experts in toxicity and pollution, to run real-time simulations of the incident that incorporated geospatial data and meteorological forecast data, and to calculate and visualize the likely spread of the pollution and thereby assist in recommending appropriate levels of evacuation. This scenario requires engagement with specific individuals who may be in different parts of the country, easy access to large amounts of computer power, access to topographical data, and collaboration with meteorologists for secure access to their weather models and data. The results of these simulations and the resulting visualization need to be shared with the disparate members of the team. Clearly, there are also military analogs of the scenario outlined above. Since these military scenarios will increasingly involve multinational forces, the imperative will be for Grid middleware that guarantees interoperability between different IT systems.

### 15.4.3 Health Grids

The implications of Grid technologies for health and medical research are potentially enormous. For example, the UK National Health Service (NHS) can be thought of as a set of overlapping virtual organizations made up of patients, doctors, specialists, and nurses, all of whom need access to particular resources, but only within particular timeframes and with specific authorizations. Healthcare is also a clear example where distributed resources can be more effectively utilized – even when those resources might be individual specialists! This ability to set up a team to collaborate from different locations is one of the key motivators for e-Science virtual organizations. A project that illustrates this very well is the Cambridge e-Science Centre telemedicine project. The partners in the project are the West Anglia Cancer Network (WACN), the Department of Radiology (University of Cambridge and Addenbrookes Hospital), the Cambridge e-Science Centre, Macmillan Cancer Relief, and Siemens Medical Solutions. Cancer services across the NHS are developing to meet the challenges set out in the recent NHS Cancer Plan. To do this, clinical networks require effective and timely communication of information, including diagnostic images from radiological and pathological investigations. In the WACN, clinicians are currently traveling large distances to provide remote clinical services and to meet other specialists to discuss diagnosis and treatment. In this project, the team is investigating the use of AccessGrid and commercial video-conferencing technologies in distributed case conferences, providing access to remote microscopes and patient data, thus reducing the travel times of scarce medical experts.

### 15.4.4 Virtual organizations

In a sense, virtual organizations have existed as long as collaboration. However, the new technology, together with this age of information-based commerce, provides a new opportunity for creating a universal approach to the integration of business assets. Industries around the globe are recognizing that they must be familiar with these technologies and early to adopt them, in order to remain internationally competitive. Virtual organizations are becoming increasingly important as multinational companies acquire globally managed resources, as smaller organizations work to reduce IT investment and increase return on investment, and as increasingly many activities and developments require collaborations of disparate teams and resources. It is also clear that the utility computing model based on on-demand cycles, data storage, and network bandwidth is beginning to become a reality and we believe that over the next five years this activity will expand with provision of many kinds of services (building on the outsourcing of computing and data and leading to application and specific service provision across the Grid), which will lead to the creation of new business models.

It is likely that the new service infrastructure will create new opportunities both for user industries and for suppliers. It allows a new methodology for providing services – what was once packaged as an application and installed on a company computer or database can now be offered as a service, which may but need not run on the company machine. It offers the business opportunity of being able to offer federated services to the end user, that is, a service that is made up of component services. In general, it will force suppliers to change their mode of operation and it will open up new markets to which they will need to supply products. It also opens technological innovation to mid-size companies, whereas the applications, integration, and infrastructure solutions to date have addressed the issues of large organizations with large budgets.

The user industries will benefit since this service technology should allow relatively easy integration of legacy applications into this new middleware infrastructure to allow use within a virtual organization. There may be some initial barriers to this, due to the skills base of the company. This is always an issue for any technological innovation, but it is compounded in this case by the fact that both Web services and Grid services are still evolving and standards have not yet been agreed upon – this means that at this point in time substantial investment is required. In fact it is clear that the main cost of ownership is people – the actual IT costs continue to decline but the expertise level of individuals needed to build and operate these systems is at present increasing. Grid technology is attempting to change this equation and reduce the total cost of ownership of corporate IT resources. Although Grid middleware has a long way to go before it can really be said to be autonomic, it is clear that Grid services are likely to be as important to industry and commerce as they will be to scientific research.

Given the parallel requirements of science, industry, and commerce, it is clear that the research community is likely to reap significant benefits from the alignment of Web services and Grid services.

## 15.5 Conclusions

The evolving e-infrastructure based on Grid technologies will constitute the basis for a new methodology for scientific research. Increasingly, computational-physics simulations and physics experiments involve many collaborators in universities and laboratories around the world. This is not to say that the day of a small team focusing together has passed – merely that Grid technologies offer the potential for new ways of doing science.

The range of applications described in this chapter illustrates the breadth of this potential and yet constitutes only a small subset of ongoing e-Science projects around the world. These examples also illustrate that there are many generic issues among applications – both in physics and in other disciplines – that are best tackled by multi-disciplinary teams. Building such teams or communities is a difficult task and requires much energy and encouragement to bring projects and domain experts together and to achieve engagement across the disciplines. The examples demonstrate some of the benefits of cross-discipline collaboration. There is also a moral – computational physicists should look beyond their own domains before investing in "new" solutions to what may be "old" problems in another discipline!

Consider the case of data curation. In general, librarians (digital and traditional) and social scientists are much more aware of the issues regarding curation of digital data. Scientists often have little knowledge of the technologies and tools available or of the processes that need to be considered when creating data archives to be used – and kept in a usable form for the future. In the UK, a national Digital Data Curation Centre has been created at Edinburgh University to act as a source of reference for research and development both in the data-management community and in the e-Science community. Data are valuable assets that need to be maintained for future scientists and, in some cases, made available for wider community use.

As in the case of the NHS example discussed above, scientists are often part of other communities, such as teachers, doctors, and policy makers, and it is essential that the e-infrastructure that is being created can also serve these communities. This is beginning to happen as the health services are beginning to become involved in e-Science projects, and government departments such as transportation, environment, and crime prevention begin to take these technologies on board.

E-learning has perhaps the most overlap in terms of technology requirements and scientific digital resources. An example of this is in the schools-telescope project being developed by the e-Star project. This is a project that illustrates the use of Grid technologies to bring together remote telescopes and astronomy databases. There are several efforts in the UK that are trying to allow schoolchildren access to major telescopes from the classroom, but this is not an easy challenge. For example, it may be difficult for a teacher to book a slot time that coincides with the class time and, of course, the weather at the telescope site may be bad. With the e-Star Grid technology, the teacher will be provided with much more flexibility and have the capability of bringing in images from databases as well as accessing a remote telescope.

Our experience with the e-Science program in the UK has also shown us that the most difficult issues are often not technical but social, and that much effort is required for "social engineering." In order for Licklider's vision to be realized, a very broad community must be educated about Grid technology – not only those using the technology but anyone in the chain of resource owner to service provider. In universities this means vice chancellors, computer-service directors, librarians, researchers, students, firewall managers, and system administrators. Engaging such a broad community means that the barriers to entry must be made as low as possible – which in turn means that issues such as human factors and user-interface design will be just as important as engineering the underlying Grid middleware.

We are now in the stage of development at which many of the tools and services required for a global e-infrastructure are beginning to mature. There is still a long way to go before such infrastructure can offer a secure, production Grid service to scientists. However, we have already glimpsed some of the benefits that will be delivered by this new generation of scientific endeavor. In the next few years, we believe that the

Web-service standards will settle down and we will see steady improvements in the reliability and usability of Grid technology. There is the real possibility that, in the not too distant future, we may accept the presence of such e-infrastructure just as routinely as we now accept the Internet and the Web.

## FURTHER READING

Over the past year there have been many publications on Grid applications and technologies. Fortunately, two edited collections of papers on Grids that provide a comprehensive source of information and references have appeared recently. These are *Grid Computing: Making the Global Infrastructure a Reality*, edited by Berman, Fox, and Hey (New York, Wiley, 2003), and *The Grid 2: Blueprint for a New Computing Infrastructure*, 2nd edition, edited by Foster and Kesselman (Morgan, Kaufmann). Both these collections contain articles that cover the range of new technologies and applications and provide both a snapshot of the field at this time and a view of the way ahead. The reader interested in knowing more about the particle-physics Standard Model is referred to *Gauge Theories in Particle Physics*, 3rd edition, in two volumes, by Aitchison and Hey (Bristol, Institute of Physics Publishing, 2004). A comprehensive overview of the Semantic Grid together with developments in ontologies and knowledge management is contained in the January/February 2004 issue of *IEEE Intelligent Systems*. More details of the Reality Grid project are contained in a paper published in the September/October 2003 edition of *Contemporary Physics*.

## Acknowledgements

The authors would like to thank the members of the consortia and projects discussed in this chapter for their assistance and for the provision of images. They also thank Vijay Dialani, Wolfgang Emmerich, Neil Geddes, and Paul Watson for helpful comments on early drafts of the paper.

## Tony Hey

In April 2001 Tony Hey was seconded from his post as Dean of Engineering in the University of Southampton to the UK Engineering and Physical Sciences Research Council as Director of the UK e-Science Core Programme. The £250M UK e-Science initiative is now well known around the world and the Core Programme has played a key part in building a new, multidisciplinary community. The UK e-Science program has considerable engagement with industry.

Tony Hey has been Professor of Computation in the Department of Electronics and Computer Science at the University of Southampton since 1987 and was Head of Department for five years from 1994. During his research career he has spent sabbaticals at several world-class research centers – Caltech, MIT, and IBM Research, Yorktown Heights. He co-edits the international computer-science journal *Concurrency and Computation: Practice and Experience* and is on the editorial board of several other international journals. His primary research expertise is in the field of parallel computing and he has published over 100 refereed papers and written and edited several books. These include *The Feynman Lectures on Computation* and two popular science books – *The New Quantum Universe* and *Einstein's Mirror*. Tony Hey is a fellow of both the British Computer Society and the Institute of Electrical Engineers, a chartered engineer, and a fellow of the Royal Academy of Engineering. In 2005 he became Microsoft's Corporate Vice-President for Technical Computing.

## Anne Trefethen

Anne Trefethen is Deputy Director of the UK e-Science Core Programme. The Core Programme is multifaceted and is tasked with the essential role of creating a Grid infrastructure to support e-Science applications through the development of appropriate middleware and infrastructure and collaborating with UK industry. Before joining the e-Science Core Programme, she was the Vice President for Research and Development at NAG Ltd, responsible for leading technical development in the range of scientific, statistical, and high-performance libraries produced by NAG.

From 1995 she was the Associate Director for Scientific Computational Support at the Cornell Theory Center, one of four national supercomputer centers in the USA at that time. She had worked for seven years on high-performance computing in the area of parallel linear algebra and parallel scientific applications, both at the Cornell Theory Center and as a research scientist at Thinking Machines Corporation. In 2003 she was a visiting professor in the Advanced Computational Modelling Centre at the University of Queensland, Australia.

# Science in action

# 16 Biophysics and biomolecular materials

Cyrus R. Safinya

## 16.1 Introduction

The application of basic physics ideas to the study of biological molecules is one of the major growth areas of modern physics, and emphasizes well how physics principles ultimately underpin the whole of Nature. This chapter focuses on the collective properties of biological molecules showing examples of hierarchical structures, and, in some instances, how structure and dynamics enable biological function. In addition, supramolecular biophysics casts a wide web with contributions to a broad range of fields. Among them are, in medicine and genetics, the design of carriers of large pieces of DNA containing genes for gene therapy and for characterizing chromosome structure and function; in molecular neurosciences, elucidating the structure and dynamics of the nerve-cell cytoskeleton; and in molecular cell biology, characterizing the forces responsible for condensation of DNA in vivo, to name a few. Concepts and new materials emerging from research in the field continue to have a large impact in industries as diverse as cosmetics and optoelectronics. A separate branch of biophysics dealing with the properties of single molecules is not described here due to space limitations and the availability of excellent reviews published in the past few years.

If one looks at research in biophysics over the last few decades one finds that a large part has been dedicated to studies of the structure and phase behavior of biological membranes. Membranes of living organisms are astoundingly complex structures, with the lipid bilayer containing membrane-protein inclusions and carbohydrate-chain decorations as shown in a cartoon of a section of the plasma membrane of a eukaryotic cell, which separates the interior contents of the cell from the region outside of the cell (Figure 16.1). The common lipids in membranes are amphiphilic molecules, meaning that the molecules contain both hydrophilic ("water-liking") polar head groups and hydrophobic ("water-avoiding") double tail hydrocarbon chains. Plasma membranes contain a large number of distinct membrane-associated proteins, which may traverse the lipid bilayer, be partially inserted into the bilayer, or interact with the membrane but not penetrate the bilayer.

Membrane proteins are involved in a range of functions: for example, from energy transduction to hormone receptors initiating cell-signaling cascades, and to the coordinated generation of the action potential along axons in nerve-cell communication.

*The New Physics for the Twenty-First Century*, ed. Gordon Fraser.

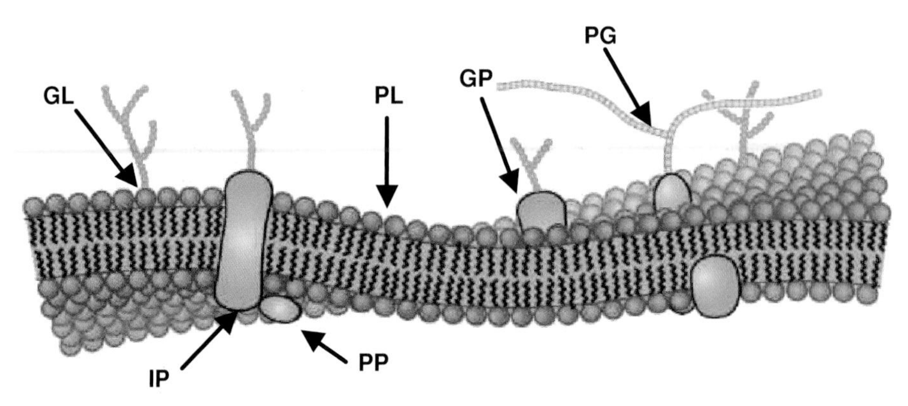


Figure 16.1. A cartoon of a plasma membrane based on the fluid mosaic membrane model of S. J. Singer and G. L. Nicolson (Salk Institute for Biological Studies) described in their seminal paper in the journal *Science* published in 1972. The asymmetric membrane contains a variety of phospholipids (PL), glycolipids (GL), glycoproteins (GP), and cell-surface proteoglycans (PG). The glycosaminoglycan chains of PGs protruding from the membrane surface are 95% carbohydrate. The membrane-associated proteins may traverse and protrude from both sides of the bilayer (IP, integral protein), penetrate only partially, or remain in the aqueous environment next to the membrane (PP, peripheral protein). The important lipid cholesterol is not shown.

Carbohydrate groups linked to proteins (glycoproteins) or lipids (glycolipids) surround the outer leaflet of the plasma membrane, which, together with cell-surface proteoglycans (PG, Figure 16.1), forms the glycocalyx, a carbohydrate-rich region surrounding the cell, with important roles in cell–cell recognition and adhesion of the cell to the extracellular matrix.

Much of our current understanding of the biophysics of membranes can be traced to the work of several early pioneers. These include Richard Bear, Ken Palmer, and Francis Schmitt on account of their X-ray-diffraction studies, during the 1940s at Washington University in St. Louis, elucidating the multi-lamellar nature of the nerve-cell myelin sheath, and Vitorrio Luzzati on account of his daring studies nearly 20 years later, at the molecular-genetics laboratory in Gif-sur-Yvette, of various structures and shapes adapted by lipids extracted from various tissues. In more recent times during the late 1970s and early 1980s important advances were made by Graham Shipley and co-workers at Boston University's School of Medicine in understanding the phase behavior of well-defined model membranes of synthetic lipids from which complicating factors, such as the membrane-associated proteins, had been removed (Figure 16.2(a)). The assumption was that the approximations nevertheless allowed one to retain the essential biophysical properties of natural membranes, for example, the nature of chain ordering of lipid tails in different phases.

In model membrane systems, proteins are typically treated as impurities, with their presence leading to a renormalization of the bare parameters, such as the elastic moduli of out-of-plane bending and in-plane compression, characterizing the membrane. It is only in the very recent past that the importance of lipids not just to the biophysics of cells but also to the biology of cells finally became better appreciated. This is most evident in the realization of key lipid–protein interactions in lipid-rafts – nanometer-sized lipid domains – where lipids play a critical role in the proper maintenance of receptor proteins involved in cell signaling events.

Phospholipid molecules tend to form closed bilayer shells known as liposomes or vesicles when dispersed in an aqueous medium because of the amphiphilic nature of

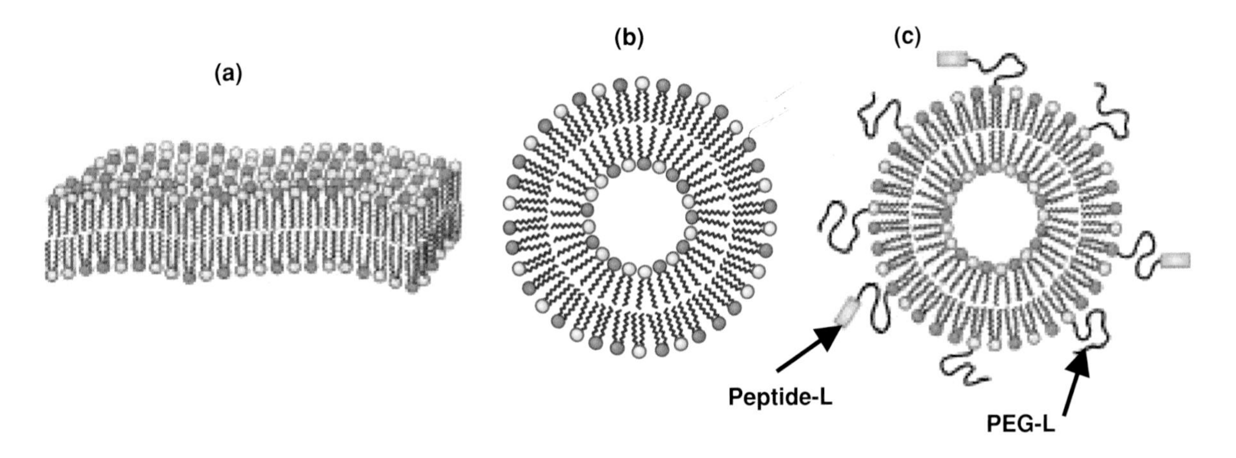


Figure 16.2. (a) A model biological membrane containing two types of lipids. (b) A uni-lamellar vesicle (also referred to as a liposome). (c) A vesicle coated with polymer lipids and peptide lipids used in drug-delivery applications. The peptide enables the vesicle to bind specifically to certain cell-surface receptors for targeted delivery of molecules. The polymer (poly(ethylene glycol), PEG) coat shields the vesicle from immune-system molecules and cells.

the molecules (Figure 16.2(b)). The hydrophobic interaction, which prevents contact between the tails and the nearby aqueous environment, favors these structures with the vesicles consisting of either a single lipid bilayer or a few concentric multi-bilayers. From the moment of their discovery by A. D. Bangham and co-workers in Cambridge during the early 1970s, uni- and multi-lamellar vesicles have received much attention because of their similarities to living cells and their potential for encapsulating and segregating water-soluble materials from bulk solution. The latter property is expected to have a dramatic impact on the medical field via their use as drug and gene carriers and we will come back to this point in more detail later.

Results of recent studies show that attaching poly(ethylene glycol) (PEG) to a biological macromolecule can dramatically increase blood-circulation times in delivery applications. Both peptides and proteins can be protected by covalently attached PEG for in-vivo administration of therapeutic enzymes, and so-called "stealth" liposomes consisting of vesicles covered with PEG lipids hydrophobically anchored to the membrane (Figure 16.2(c)) are currently in use as drug carriers in therapeutic applications. The inhibition of the body's immune response to these PEG-coated liposomes has been attributed to a polymer-brush-type steric repulsion between PEG-coated membranes and molecules of the immune system. Vesicles are also used in the cosmetics industry as controlled-chemical-release agents in formulations of lotions, gels, and creams.

The variety of forms and shapes assembled in Nature is simply staggering, and as we describe later in the section on protein assembly, the eukaryotic cell itself provides an unlimited source of inspiration. This has led to a large increase in biophysical research aimed at elucidating collective interactions between different proteins leading to supramolecular structures, spanning lengths from the ångström-unit to the micrometer scales, with the holy grail of elucidating the relation to cell function. For example, filamentous actin, a major component of the cell cytoskeleton, which is also a key building block of muscle, is thought to assemble to form either a tightly packed bundle phase, or loosely packed two- and three-dimensional gel network structures in vivo. The distinct functions resulting from these highly regulated structures interacting with other biomolecules and motors include cell shape and mechanical stability, cell adhesion and motility, and cell cytokinesis, the physical splitting of a cell into two daughter cells.

A closely related field to biophysics is the rapidly emerging field of biomolecular materials. The term "biomolecular materials" was coined by Hollis Wickman of the US National Science Foundation in the late 1980s and refers to materials that result from research at the interface of physics, biology, and materials science. This highly interdisciplinary research enterprise claims practitioners from diverse disciplines, including biophysics and soft condensed matter, chemical synthesis and biochemistry, and molecular cell biology and genetic engineering. Biomolecular materials may be comprised of self-assembled and functionalized interfaces where the functionality is derived from biomolecules, which may be manipulated at the molecular level. For example, two-dimensional solid phases of the membrane protein bacteriorhodopsin and bacterial surface layers are of current interest for the development of materials in technological areas as diverse as molecular electronics and optical-switch applications, use as molecular sieves, and the lithographic fabrication of nanometer-scale patterns. Functionalized biomolecular interfaces, which include receptor proteins, are currently being developed as chemical and biological sensors.

Over the past decade, Erich Sackmann and co-workers in Munich have developed methods of functionalizing surfaces by the sequential layering of soft polymeric ultra-thin films followed by deposition of self-assembled monolayer or bilayer lipid molecules. Sackmann notes that such biomolecular materials provide ideal model substrates for cells for in-vitro studies of tissue/organ development and engineering, which involve cell proliferation, locomotion, and adhesion.

In many cases the biomolecular materials have uses in biomedical applications, for example, in developing self-assembled drug- and gene-delivery vehicles. As we describe later, researchers have been working feverishly on unraveling the structure and function properties of complexes of cationic (positively charged) membranes and DNA, which are currently being used in clinical nonviral gene-therapy trials worldwide. In particular, what is the precise nature of the structures of the cationic-membrane–DNA complexes? What is the relation between the structure and the ensuing interactions between complexes and cellular membranes which allow DNA to be transported across the membrane?

At present the common mechanisms leading to large-scale hierarchical structures in biological systems include self-assembly at thermal equilibrium, such as for lipids, and out-of-equilibrium assembly in energy-dissipative systems. The latter include materials under flow and biomolecule assembly and disassembly involving hydrolysis of the high-bond-energy molecules adenosine triphosphate (ATP) and guanosine triphosphate (GTP). The GTP-dependent assembly and disassembly of microtubules, which are involved in cell functions such as cellular transport and chromosome separation in a dividing cell, is a well-known example. In the latter systems, the out-of-equilibrium structures may reach a steady state. A third, less common, paradigm in biological assembly is irreversible aggregation, which happens when two molecules or aggregates collide and stick irreversibly because the binding energy is far greater than the thermal energy. The structure is now out of equilibrium even though dissipation is absent. In many biologically relevant cases irreversible assembly occurs in the aging processes, for example, in aggregation of proteins with altered structure in neurodegenerative diseases.

## 16.2 Membranes

Our current understanding of structures and phase behavior in isolated biological membranes has to a large degree been a direct result of the knowledge acquired using multilayers, which exhibit weak coupling between the fluid membranes from layer to layer (labeled $L_\alpha$, Figures 16.3(a) and (b)). Studies of the lamellar $L_\alpha$ phase have also been motivated by the desire to understand how lipid–lipid and lipid–protein interactions relate to their biological activity. Furthermore, researchers worldwide have always considered multi-lamellar fluid membranes as nearly ideal models for elucidating the fundamental interactions between membranes in living systems.

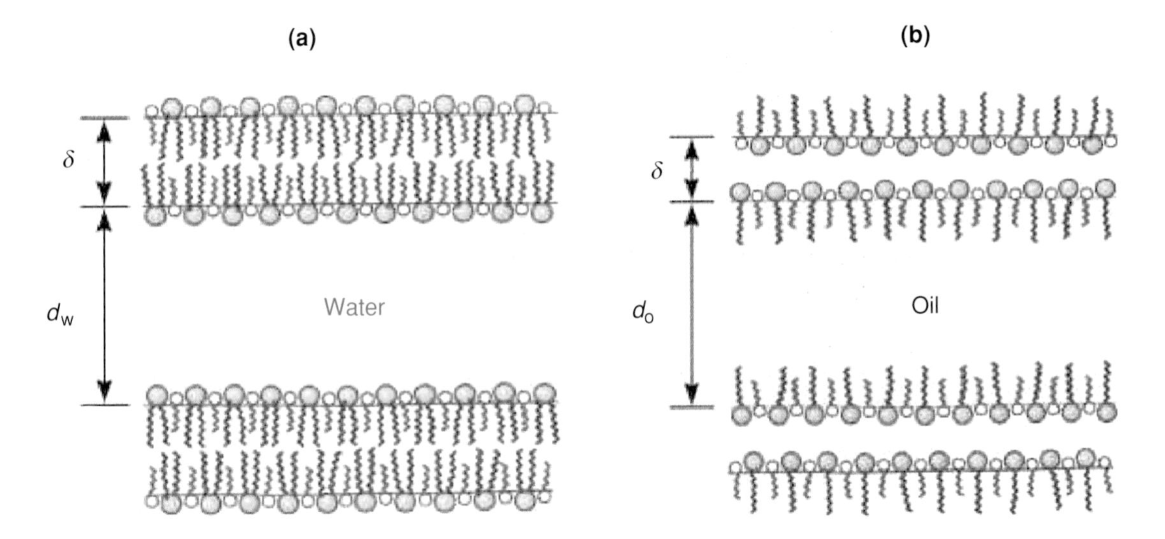


**Figure 16.3.** The multi-lamellar $L_\alpha$ phase. (a) The biological membranes are separated by water. (b) The "inverted" membranes (two surfactant monolayers separated by a thin layer of water) are separated by oil (hydrophobic molecules). In (a) the surfactant is a double-tailed lipid. Co-surfactant (short-single-tail) molecules are shown mixed in both membranes. The membrane thickness is denoted by $\delta$. The thicknesses of water and oil regions are denoted by $d_w$ in (a) and $d_o$ in (b). The center-to-center distance between the membranes is given by $d = \delta + d_W$ for (a) and $d = \delta + d_o$ for (b).

For physicists, membranes are also considered ideal prototypes for elucidating the statistical-mechanical properties of both ordered and fluid two-dimensional interfaces embedded in three dimensions. In contrast to strictly two-dimensional systems, the height-fluctuation degree of freedom of membranes is expected to result in fundamentally new behavior both with regard to the physics of the problem and also regarding predictions of properties of new materials. As we describe, the advent of high-brightness synchrotron-radiation sources has made possible the high-resolution X-ray-scattering and -diffraction experiments needed to elucidate the structural and statistical-mechanical nature of biological membranes.

An important realization regarding biological membranes is that often they have negligible surface tension; thus, geometric shapes and fluctuations govern the free energy. The observation of the flickering of red blood cells with the optical microscope was the first instance in which the importance of thermal height fluctuations in single membranes was realized. One may ask about the relative importance of these fluctuations for the interactions between membranes, for example, between the plasma membranes of approaching cells and in the membranes of the myelin sheath that wrap around the nerve cell. Two fluctuating or rippled interfaces are expected to repel each other. These simple ideas were in fact what led Wolfgang Helfrich of the Free University in Berlin to the description of the undulation repulsive forces between membranes in 1978. The undulation interaction arises from the difference in entropy between a fluctuating "free" membrane (Figure 16.4(a)) and a "bound" membrane between neighboring bilayers (Figure 16.4(b)). From a biophysical viewpoint the nature of a fluid membrane surface may in some cases have a profound influence on the precise mechanism of membrane–membrane interactions, which influence adhesion properties in cell–cell interactions.

For fluid membranes, the bending-rigidity modulus $\kappa_c$ associated with the restoring force to layer bending is the important modulus, which in many cases will determine the physical state of the membrane. One would expect thermal fluctuations to be important only if $\kappa_c$ is of order $k_B T$. It is in the multilayer $L_\alpha$ phase of membranes where the bending rigidity was purposefully softened that the Helfrich undulation forces were conclusively demonstrated in quantitative high-resolution synchrotron X-ray-scattering experiments beginning in the mid 1980s and extending to the mid 1990s. The strength of this interaction scales as $\kappa_c^{-1}$, and, for flexible membranes with $\kappa_c \sim k_B T$, it completely overwhelms the attractive van der Waals interaction

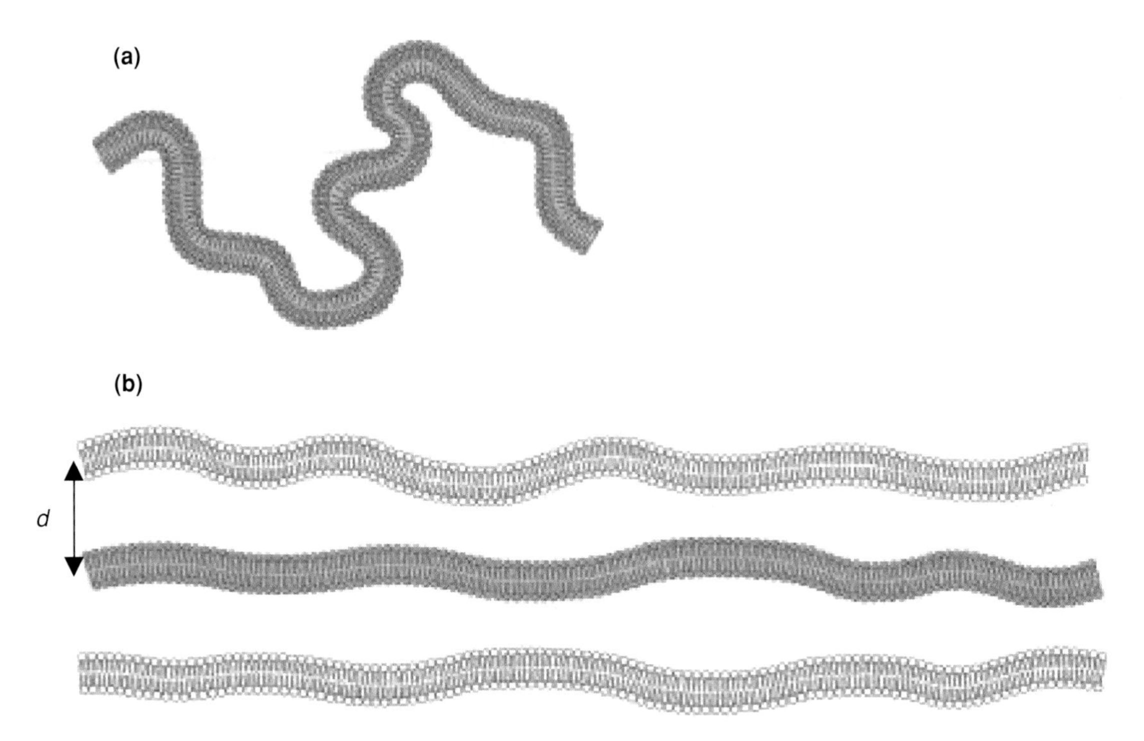


**Figure 16.4.** Schematic drawings of a free membrane (a) compared with a membrane (red) confined between neighbors in a multi-lamellar $L_\alpha$ phase (b). The free membrane has more space to explore and is in a higher entropic state.

and stabilizes the membranes at large separations of several hundred or even a few thousand ångström units.

This new regime is distinct from the classical regime in which largely separated charged membranes are stabilized because of their mutual electrostatic repulsion. Furthermore, because the undulation forces are entropically driven, the interaction is universal, with its dependence determined entirely by geometric and elastic parameters such as the layer spacing and the layer rigidity. Synchrotron experiments have further shown that this universal regime consisting of fluctuating membranes can be accessed by thinning and thus dramatically lowering $\kappa_c$ of an initially rigid biological membrane. In the early 1980s, Reinhart Lipowsky (then at the University of Munich) and Stan Leibler (then at CEA Saclay) produced a theory showing that, in the very strongly fluctuating regime, this effective long-range repulsive force may lead to a complete unbinding of membranes. On a more general level this type of interaction is found in other condensed-matter systems. For example, the dominant contribution to the free energy of a polymer confined between walls and that associated with wandering walls of incommensurate phases are similarly entropic in origin.

In the late 1980s and early 1990s work at Harvard University by David Nelson and Luca Peliti and separately at the University of Pennsylvania by Tom Lubensky and Joseph Aronovitz delineated the properties of tethered solid membranes. By taking into account the coupling between the in-plane phonons and the out-of-plane height fluctuations, Nelson and Peliti predicted that the bending rigidity of solid membranes would diverge with the size of the membrane and stabilize a new flat phase. In contrast, fluid membranes are always crumpled on scales larger than the persistence length of the membrane, which remains finite because the bending rigidity softens with increasing membrane size. Lubensky and Aronovitz showed that the elastic constants of the solid membranes are also renormalized, in particular, the shear modulus of such solid membranes was found to vanish as the size of the membrane increased to infinity.

These are indeed extremely novel results. For example, they imply that a flat two-dimensional solid membrane behaves as a special type of liquid phase (in the sense that the renormalized shear modulus softens to zero at large length scales due to height fluctuations) even though each molecule is part of the tethered membrane with known position and not free to diffuse as in a true liquid phase. Experimentalists worldwide are still busy synthesizing and studying possible experimental realizations of these novel phases of matter which can be expected to have many important technological applications similar to commercial applications of polymers. In 1995, Leo Radzihovsky of the University of Colorado and John Toner of the University of Oregon produced a theory predicting that anisotropic solid membranes may form tubular solids. Mark Bowick and collaborators at Syracuse University verified the existence of the tubular phase in numerical simulations in 1997.

### 16.2.1 Interactions between fluid membranes with large bending rigidity

In typical model biological membranes comprised of a single lipid with thickness ranging from 35 to 70 Å, $\kappa_c$ is significantly larger than the thermal energy. Stan Leibler has shown that in real biological membranes local bending rigidities may be significantly lower due to the presence of proteins (Figure 16.1). In studies designed to measure the interactions between two membranes, the free energy has been found to be consistent with that between two flat membranes interacting via van der Waals and hydration forces. The van der Waals interaction per unit area between two membranes with separation $d$ and thickness $\delta$ is well approximated by the expression $F_{\mathrm{vdW}} = -[H/(12\pi)][1(d-\delta)^2 + 1(d+\delta)^2 - 2/d^2]$, where the Hamaker constant $H = [(\varepsilon_{\mathrm{w}} - \varepsilon_{\mathrm{m}})/(\varepsilon_{\mathrm{w}} + \varepsilon_{\mathrm{m}})]^2 k_{\mathrm{B}}T$ and $\varepsilon_{\mathrm{w}}$ and $\varepsilon_{\mathrm{m}}$ are the zero-frequency dielectric constants of water and the nonpolar oily membrane regions (Figure 16.3(a)). For such multilayer systems the interaction is attractive and varies as $-1/d^2$ for small $d\delta$, with a crossover to $-1/d^4$ for $d \gg \delta$.

In a series of landmark studies over the last 20 years, Adrian Parsegian of the National Institutes of Health in Maryland and Peter Rand from Brock University in Ontario and co-workers have shown that, for small separations, the dominant force governing the interactions between phospholipid membranes is the hydration force. This strong repulsive potential acts to prevent the approach of phospholipid bilayers embedded in water. The hydration energy per unit area has been shown to have the form $F_{\mathrm{hydration}} = A_{\mathrm{H}} \exp[-(d-\delta)/\lambda_{\mathrm{H}}]$, where $A_{\mathrm{H}} \approx 4k_{\mathrm{B}}T$ Å$^{-2}$ at room temperature, and $\lambda_{\mathrm{H}}$ is a microscopic length of order 2 Å. This interaction dominates for distances less than 10 Å but is negligible for distances larger than 30 Å. For charge-neutral membranes with $\kappa_c > k_{\mathrm{B}}T$, the inter-membrane potential has a minimum for $d - \delta =$ 30 Å resulting from the competition between the long-range van der Waals attraction and the short-range repulsive hydration interaction.

### 16.2.2 The Helfrich undulation interaction between flexible fluid membranes

Wolfgang Helfrich first realized the importance of membrane-height fluctuations and its subsequent effect on the interaction between two fluid membranes. The phenomenological approach to describing curvature fluctuations in a fluid membrane begins by considering the elastic free energy per unit area: $F = 0.5\kappa_c(C_1 + C_2 - C_o)^2 + \kappa_G C_1 C_2$. Here, $C_1 = 1/R_1$ and $C_2 = 1/R_2$ are the principal curvatures, with the principal radii of curvature of points on the membrane being given by $R_1$ and $R_2$. The spontaneous radius of curvature $R_0 = 1/C_0$ describes the tendency of the membrane to bend toward the "inside" or "outside" aqueous media and is nonzero for membranes with asymmetric bilayers resulting, for example, from different lipid components in each monolayer. The first term in $F$ describes the elastic energy of bending a membrane away from its spontaneous radius of curvature, with $\kappa_c$ the energy cost. The modulus $\kappa_G$ controls the Gaussian curvature $C_1 C_2$ of the membrane, and, because the integral of the second term in $F$ over a closed surface is a topological constant due to the Gauss–Bonnet theorem, we ignore it in our consideration of large closed membrane surfaces.

As was first clarified in a landmark paper that appeared in 1961 by Schulman and Montagne of Columbia University, the central assumption in the description of the membrane is that it consists of an incompressible film of lipid molecules with a well-defined area per molecule undergoing curvature fluctuations while keeping its total area constant. Thus, the usual surface-tension term of the interface is neglected. If one considers small displacements of a liquid membrane of area $A$ from the $x$–$y$ plane characterized by a height function $h(x, y)$, then the elastic free energy $F$ leads, at equilibrium, to height fluctuations that grow with the membrane's size as $\langle h^2 \rangle \propto A(k_B T/\kappa_c)$. This violent growth in height fluctuations (Figure 16.4(a)), which scales with the linear dimension of the membrane, results from the inherent statistical-mechanical properties of two-dimensional objects with negligible surface tension. The amplitude of the thermally driven undulations is controlled by the bending rigidity $\kappa_c$.

For a stack of membranes, the large height fluctuations of any single sheet will be cut off because of excluded-volume effects resulting from its neighbors (Figure 16.4(b)). When $h = \langle h^2 \rangle^{1/2}$ equals the interlayer spacing $d$, collisions between undulating membranes will occur. From statistical considerations, patch sizes of order $A_p \propto d^2[\kappa_c/(k_B T)]$ will experience a collision. Because of the steric constraint that membranes cannot cross each other, a collision will result in a loss of configuration for the membrane, decreasing its entropy and raising the free energy by an amount of order $k_B T$. Figure 16.4 shows that a free membrane (a) has more phase space to explore than does a membrane bound between its neighbors (b). Thus, the effective (entropic) increase in free energy per unit area resulting from collisions of fluctuating membranes is $F_{\text{undulation}} \propto k_B T/A_p = d^{-2}[(k_B T)^2/\kappa_c]$. This is the physical origin of the undulation interaction, which compares the free energy of a free membrane with that of one bound between its neighbors. This long-range repulsive undulation interaction was originally described in a seminal theoretical paper by Wolfgang Helfrich in 1979 but received very little attention because most biological membranes were thought to have large bending rigidities so that undulation forces would be negligible. It is important to realize that this repulsive undulation interaction is long range, falling off as $1/d^2$, and competes effectively with the long-range van der Waals attraction only if the membrane has low rigidity $\kappa_c = k_B T$.

### 16.2.3 Synchrotron X-ray scattering from flexible fluid lamellar membranes reveals the presence of the Helfrich undulation interactions

As it turns out, in contrast to typical model biological membranes comprised of a single lipid type with $\kappa_c \gg k_B T$, solutions of a mixture of surfactant and co-surfactant molecules that result in the microemulsion phase, a bicontinuous structure consisting of floppy interfaces that separate water from oil regions (Figure 16.5), have an interfacial bending rigidity $\kappa_c$ of order $k_B T$. In a landmark paper on microemulsions, P. G. De Gennes (Nobel Prize 1991) and C. Taupin of the Collège de France had in

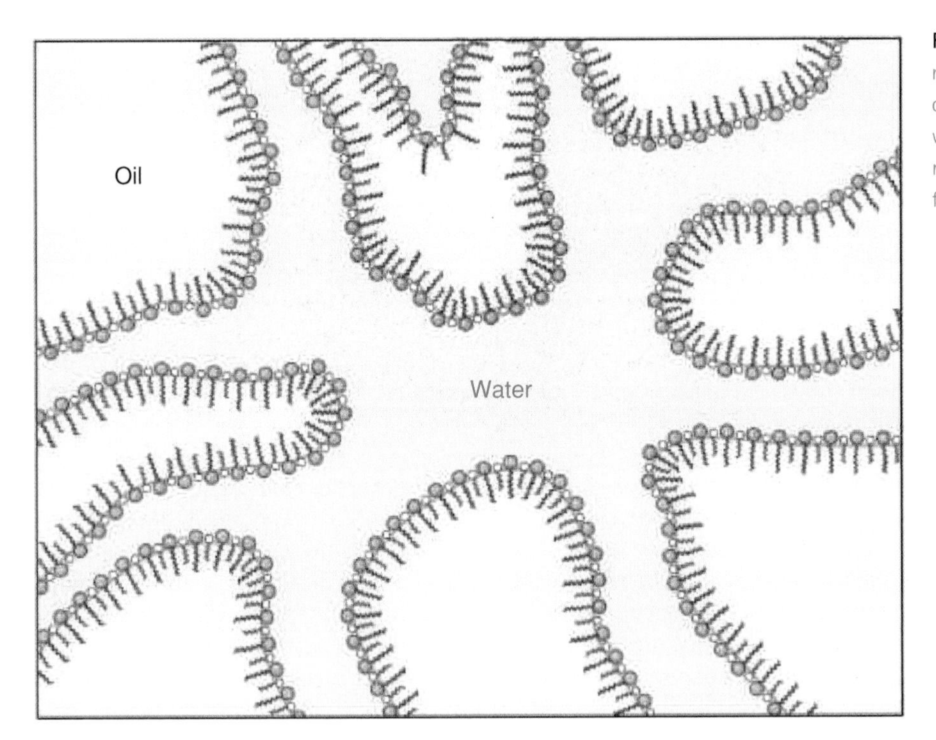


**Figure 16.5.** A cartoon of the microemulsion phase consisting of a bicontinuous structure with surfactant/co-surfactant monolayers separating oil-rich from water-rich regions.

fact reasoned in 1982 that it was the low rigidity of these interfaces that resulted in the stability of the bicontinuous phase. Indeed, it is in the lamellar phases of similar surfactant–co-surfactant systems with floppy interfaces that the existence of the Helfrich undulation forces was unambiguously demonstrated in a series of synchrotron X-ray-diffraction experiments by Didier Roux, the author, and co-workers at the Exxon Research and Engineering Laboratories in Clinton, New Jersey, and, independently, in neutron-diffraction experiments by Grégoire Porte and Patricia Bassereau at Montpellier during the mid to late 1980s.

In the X-ray study, the lamellar ($L_\alpha$) phase was comprised of the quaternary mixture of sodium dodecyl sulfate (SDS, the surfactant), pentanol (a co-surfactant), water, and dodecane as a function of dodecane dilution shown schematically in Figure 16.3(b). This system has distinct advantages for studies of interactions between membranes, including the absence of electrostatic interactions between the membranes and the ability to increase the inter-membrane separations over a large range between $d \sim 3.0$ and larger than 50.0 nm. The unique properties of this lamellar phase with low $\kappa_c$ allowed one to probe the long-range van der Waals and the postulated Helfrich undulation interactions.

The difference between $L_\alpha$ phases with high and low bending rigidities lies in the presence of co-surfactant molecules for soft membranes. Although co-surfactant molecules, which typically consist of long-chain alcohols such as pentanol or hexanol, are not able to stabilize an interface separating hydrophobic and hydrophilic regions, when mixed in with longer-chain "true" surfactants, they are known to lead to dramatic changes in interface elasticities. Compression models of surfactant chains developed in the 1980s by Igal Szleifer and Avinom Ben-Shaul at the Hebrew University, and William Gelbart at the University of California at Los Angeles (UCLA), showed that the bending rigidity of membranes $\kappa_c$ scales with chain length $l_n \propto \delta_m$, the membrane thickness, where $n$ is the number of carbons per chain, and with the area per lipid chain $A_L$ as $\kappa_c \propto l_n^3/A_L^5$. Thus, because the mixing of co-surfactants with lipids is expected to lead to a thinner membrane, this will result in a strong suppression of $\kappa_c$ making the membrane highly flexible.

Synchrotron X-ray-diffraction experiments by Eric Sirota, Didier Roux, and the author at the Exxon Research and Engineering Laboratories have shown that the addition of the co-surfactant pentanol to dimyristoyl phosphatidyl choline (DMPC) membranes of lamellar phases with a molar ratio of 2–4 leads to a significant decrease of $\kappa_c$ from about $20k_BT$ to about $2k_BT$. In the absence of co-surfactant, the $L_\alpha$ phase of DMPC is stable up to a maximum dilution of 40% water, which corresponds to an interlayer separation $d = 60$ Å. This corresponds to the spacing for which the van der Waals attraction balances the hydration repulsion. In the presence of co-surfactant (Figure 16.3(a)) one is able to separate the membranes to very large spacings with $d > 200$ Å. Because the significantly thinner DMPC–pentanol membrane is charge neutral, the dilution cannot be a result of repulsive electrostatic forces and the repulsion arises from long-range repulsive undulation forces.

Lev Landau and Rudolf Peierls first realized in the 1930s that three-dimensional materials whose densities are periodic in only one direction, such as the smectic-A phase of thermotropic liquid crystals consisting of a stack of layers of oriented rod-shaped molecules and the lamellar $L_\alpha$ phase of membranes, are marginally stable against thermal fluctuations that destroy the long-range crystalline order. The result is that the $\delta$-function Bragg diffraction peaks of the structure factor at (0, 0, $q_m = mq_o = m2\pi/d$) ($d$ is the interlayer spacing and $m = 1, 2, \ldots$ is the harmonic number) are replaced by weaker algebraic singularities. Working with De Gennes at the Collège de France, A. Caillé first derived the structure factor for the smectic-A phase, which has the same elastic free energy as that for the $L_\alpha$ phase. The calculation is based on the De Gennes–Landau elastic free energy: $F_{\text{smectic-A}} = 0.5\{B(\partial u/\partial z)^2 + K[(\partial^2 u/\partial x^2) + (\partial^2 u/\partial y^2)]^2\}$. Here, $u(r)$ describes the locally varying membrane-height displacement in the $z$ direction normal to the layers, away from the flat reference-layer positions, and $B$ and $K = \kappa_c/d$ are the bulk moduli for layer compression and layer curvature.

The asymptotic form of the structure factor for the $L_\alpha$ and smectic-A phases is described by the power laws $S(0, 0, q_z) \propto |q_z - q_m|^{-2+\eta_m}$ and $I(q_\perp, 0, q_m) \propto q_\perp^{-4+2\eta_m}$, where $q_z$ and $q_\perp$ are the components of the wavevector normal and parallel to the layers, and the power-law exponent $\eta_m = m^2 q_o^2 k_B T/[8\pi(BK)^{0.5}]$ describes the algebraic decay of layer correlations. The orientationally averaged structure factor behaves asymptotically as $S(q) \propto |q - q_m|^{-1+\eta_m}$, where $q^2 = q_\perp^2 + q_z^2$. This form is used to analyze X-ray-diffraction data from unoriented samples of the $L_\alpha$ phase. The observation of power-law behavior describing the algebraic decay of correlations in smectic-A liquid crystals was first described in a landmark paper – a "tour-de-force" of experimental ingenuity – by Jens Als-Nielsen (Copenhagen University), Robert Birgeneau (UC Berkeley and MIT), and David Litster (MIT) in the early 1980s, nearly half a century after this remarkable state of matter had first been imagined by Landau and Peierls.

X-ray measurements normally give information on the structure of a system studied. However, in low-dimensional systems like the $L_\alpha$ and smectic-A phases, in which large thermal fluctuations prevent long-range crystalline order and lead to enhanced thermal diffuse scattering in the structure factor, thermodynamic properties, such as the elastic moduli, can be obtained reliably from a line-shape analysis yielding the exponent $\eta_m$. It is highly unlikely that Landau and Peierls would ever have expected their ideas to lead to a powerful probe of interactions between biological membranes in diffraction experiments.

Figure 16.6 (top) shows plots of the X-ray-diffraction data through the first harmonic of the structure factor versus $q - G$ ($G = 2\pi/d$, where $d$ is the interlayer spacing), for three $L_\alpha$-phase samples of the DMPC–pentanol–water system as the interlayer

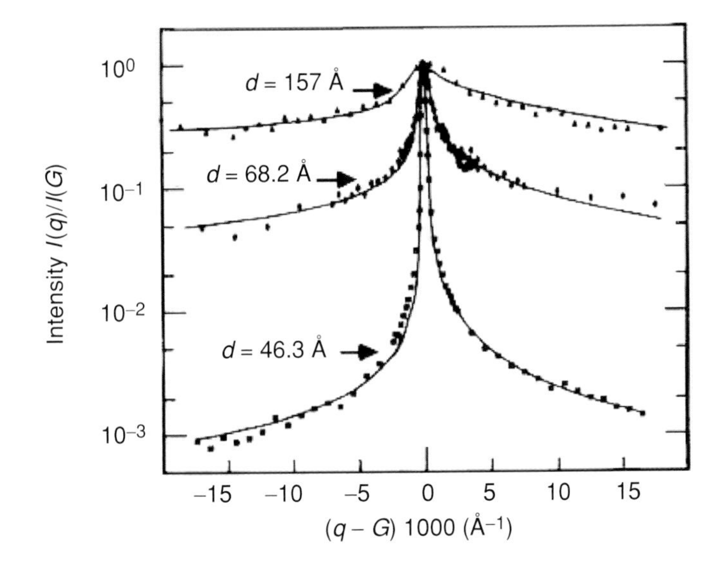


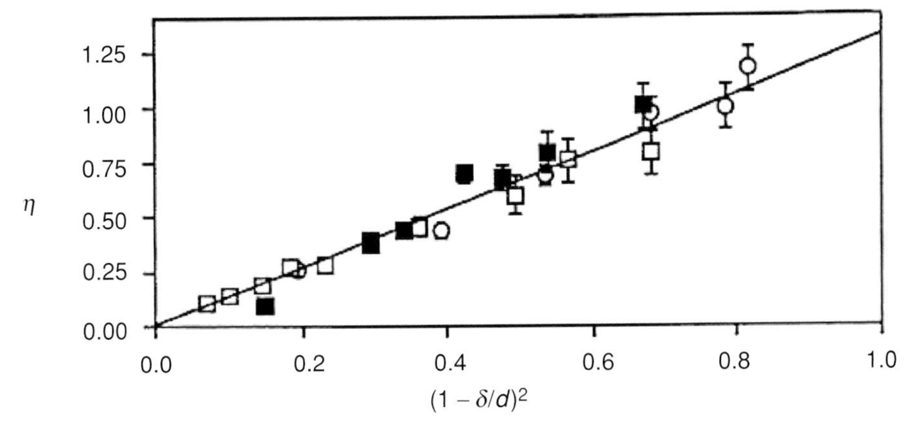


**Figure 16.6.** Top: the normalized X-ray-scattering profile through the first harmonic plotted on a semi-logarithmic scale for three lamellar $L_\alpha$ samples as a function of increasing interlayer spacing $d$ (see Figure 16.3(a)). The increase in tail scattering as $d$ increases is due to thermal fluctuations of the membranes, which prevents long-range order of the stack of membranes as was predicted by Landau and Peierls in the 1930s. Bottom: the exponent $\eta_1$ as a function of $(1 - \delta/d)^2$ describing the divergence of the X-ray structure factor for a lamellar $L_\alpha$ system ($\delta$ is the membrane thickness). The solid line is the prediction of Helfrich's theory of undulation interactions between flexible fluid membranes. The solid squares are for dimyristoyl phosphatidyl choline membranes separated by water, and the open circles and squares are for sodium dodecyl sulfate membranes separated by brine and oil, respectively. The membranes were rendered floppy (low bending modulus) by the addition of pentanol as the co-surfactant. Reprinted with permission from C. R. Safinya, E. B. Sirota, D. Roux, and G. S. Smith, *Physical Review Letters* **62**, 1134 (1989). Copyright 1989 by the American Physical Society. See also C. R. Safinya *et al.*, *Phys. Rev. Lett.* **57**, 2718 (1986) and D. Roux and C. R. Safinya, *J. Physique* **49**, 307 (1988).

separation increases. An obvious feature of the profiles is the tail scattering, which becomes dramatically more pronounced (indicative of a large increase in the exponent $\eta_1$) as $d$ increases between $d = 46.3$ Å and $d = 157$ Å. The solid lines through the data are results of fits of the structure factor to the data, from which it is found that the exponent $\eta_1$ increases by about a factor of ten from 0.1 to 1. For $L_\alpha$ phases with

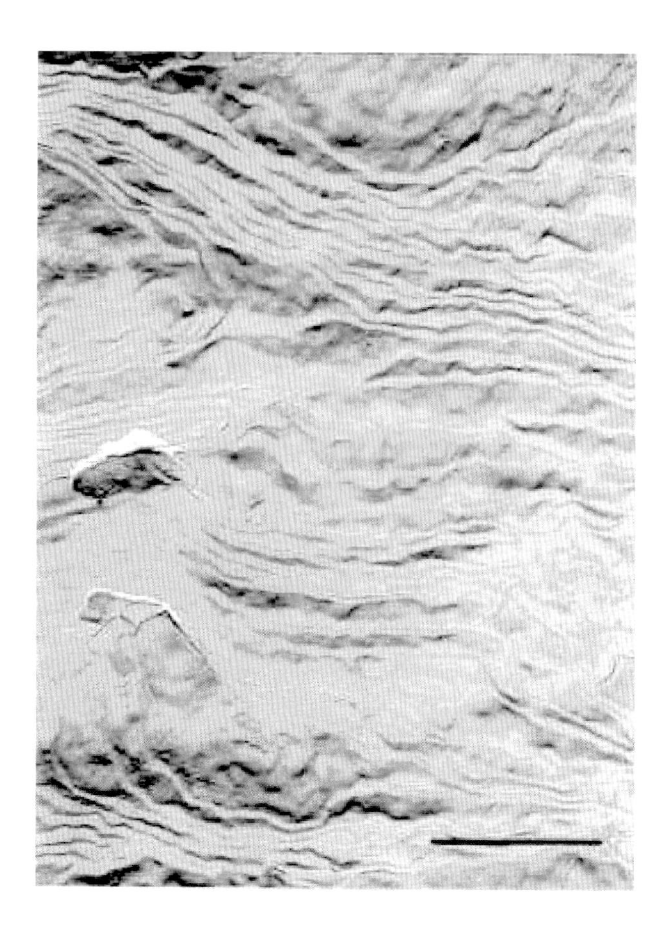

**Figure 16.7.** A freeze-fracture electron micrograph of a multi-lamellar $L_\alpha$ phase shows the highly flexible membranes which give rise to Helfrich's repulsive undulation forces between neighboring membranes. The membranes (a mixture of cetylpyridinium bromide and n-hexanol) are separated by brine. The distance between membranes measured by neutron diffraction is 360 Å. The scale bar shows 1 μm. (Courtesy of Reinhard Strey. Reprinted with permission from R. Strey *et al.*, *Langmuir* 6, 1635 (1990). © 1990 American Chemical Society.)

$\kappa_c = k_B T$, $F_{\text{undulation}} \propto d^{-2}[(k_B T)^2/\kappa_c] \approx 0.2 k_B T/d^2$ should dominate over the van der Waals interaction $-0.03 k_B T/d^2$. Thus, because $B$ is proportional to the second derivative of $F_{\text{undulation}}$, the Helfrich theory predicts that the structure-factor exponent describing the algebraic decay of layer correlations should have an elegant universal form: $\eta_1 = 1.33(1 - \delta/d)^2$.

Figure 16.6 (bottom) plots $\eta_1$ versus $(1 - \delta/d)^2$ for three systems: DMPC–pentanol–water (solid squares) and SDS–pentanol membranes diluted with dodecane (open squares) and brine (open circles). The theoretical prediction for $\eta_1$ is drawn as a solid line. The universal behavior is now clear and is in remarkable agreement with the prediction of the Helfrich theory for multi-lamellar fluid membranes where undulation forces dominate. The power-law exponent $\eta_1$ has a simple universal functional form dependent only on geometric factors because it has its origins in entropic considerations. This is in stark contrast to situations in which the interactions are dominated by electrostatic or van der Waals interactions where details such as the solution salt conditions and the dielectric constants enter into the equations describing the interactions. Figure 16.7 shows a beautiful freeze–fracture electron micrograph obtained by Reinhard Strey at the Max Planck Institute (MPI) for Biophysics in Göttingen, of a stack of membranes in the $L_\alpha$ phase of a mixture of surfactant and co-surfactant molecules, for which neutron diffraction by Porte and Bassereau had shown that the Helfrich undulation forces dominate. The highly wrinkled nature of the individual membranes leading to the undulation interactions is clearly visible.

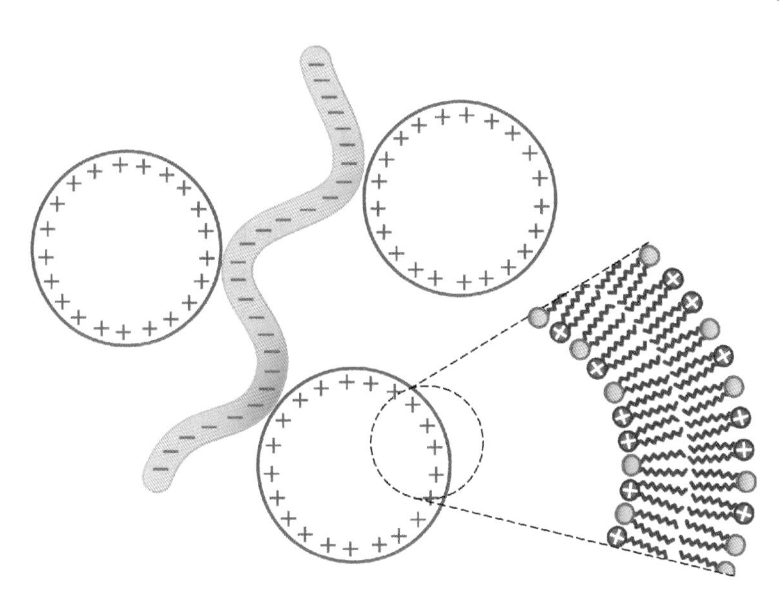

Figure 16.8. A cartoon of the bead-on-string model of cationic liposomes (beads) complexed with DNA (string).

## 16.3 DNA–membrane complexes

Since the early 1990s and continuing to the present, very many groups worldwide have increasingly been studying biological materials consisting of deoxyribose nucleic acid (DNA) adsorbed onto oppositely charged cationic (positively charged) lipid membranes for a variety of purposes. From a fundamental biophysics perspective, DNA adsorbed onto membranes is a model for studies of the phases and statistical-mechanical properties of DNA in a two-dimensional world with implications for the behavior of nucleic acids in vivo. Such studies, for example, may shed light on the packing nature of the RNA genome found in the rod-shaped tobacco mosaic virus where nucleic acid is adsorbed on the inner curved surface of the positively charged capsid shell (a protein container of the genome). Scientists are also hopeful that atomic-force-microscope images of DNA chains stretched onto cationic surfaces will one day be of sufficient resolution to distinguish among the four bases adenine (A), thymine (T), cytosine (C), and guanine (G) of DNA, thus enabling a method for rapid sequencing of DNA. Although gene sequencing is done in a completely different manner, nonetheless, DNA–membrane studies by others have opened up entirely new research directions. Helmuth Möhwald and his group at the MPI for Colloid and Surface Science in Golm and others are currently developing strategies to produce highly oriented DNA chains on surfaces, which may be suitable for the development of nanoscale masks in lithography and in diagnostics applications to detect sequence mutations related to specific diseases.

In the biomedical field researchers are complexing cationic liposomes to stretches of foreign DNA, containing genes and regulatory sequences, in order to carry them into cells. In a landmark paper on applications in nonviral gene therapy, published in 1987 in the *Proceedings of the National Academy of Sciences USA*, Phillip Felgner (then at Syntex Research in Palo Alto) and co-workers described experiments in which incubation of mammalian cells with complexes of cationic liposomes (CLs) and DNA led to the expression of the foreign genes contained within the DNA. They reasoned that this was because CL–DNA complexes are adsorbed, through the electrostatic interactions, onto the outer membranes of mammalian cells, which contain negatively charged receptor molecules. The authors hypothesized that such complexes would consist of long DNA chains decorated by distinct spherical liposomes (Figure 16.8).

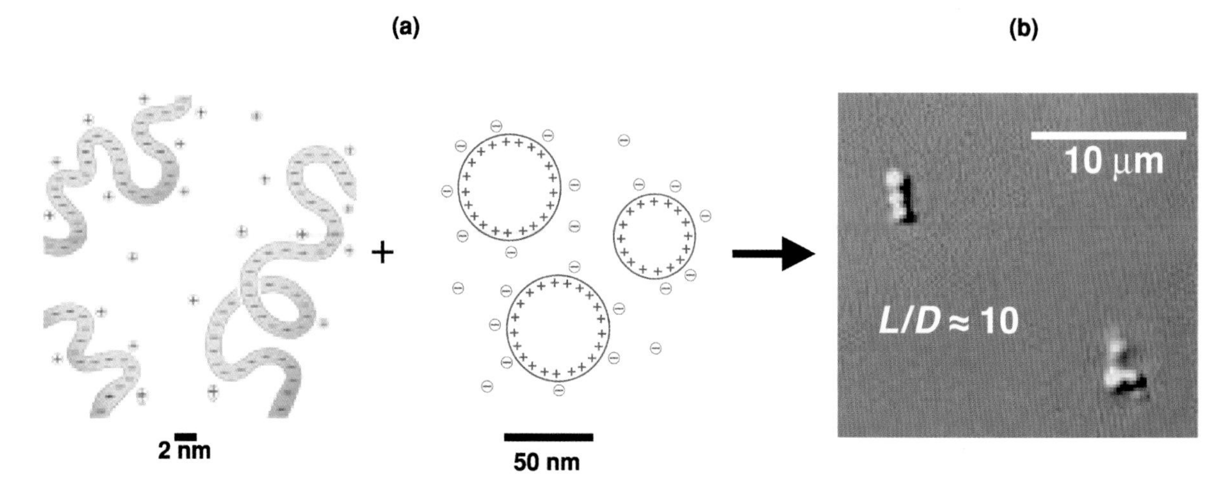


Figure 16.9. (a) A schematic diagram of DNA (with negative phosphate groups) mixed with cationic liposomes. (b) High-resolution differential-interference-contrast optical-microscopy images of cationic liposome–DNA complexes, showing distinct globules in a mixture of lipid-to-DNA weight ratio $L/D = 10$. Dynamic light scattering shows that the average globule diameter is 200 nm. Reprinted in part with permission from J. O. Rädler *et al.*, *Science* 275, 810 (1997).

Although the hypothesized structure turned out not to be correct, the empirical discovery that CL–DNA complexes are capable of delivering their DNA cargo across the outer cell membrane followed by the expression of the exogenous gene (DNA transcription followed by translation of messenger RNA into protein) was nevertheless a truly remarkable achievement in the field. The work of Felgner *et al.* was followed over the next few years by a series of seminal papers by Leaf Huang and co-workers from the University of Pittsburgh, which, taken collectively, has spawned the field of nonviral gene delivery. At present, the field consists of a very large contingent of researchers ranging from pharmacologists, biophysicists and chemists to chemical engineers and biomedical scientists, who are focusing on elucidating the mysteries of DNA delivery by cationic liposomes and polymers.

As we describe, synchrotron X-ray-diffraction techniques revealed the spontaneously assembled hierarchical structure of CL–DNA complexes. In the most common form a sandwich structure with DNA layered between the cationic membranes is observed. On rare occasions the DNA chains are encapsulated within lipid tubules, which on larger length scales are arranged in a hexagonal super-structure. The solution of the structure of DNA–lipid complexes is important to the current worldwide attempts at clarifying the function of the CL–DNA complex; namely, the mechanisms at the molecular and self-assembled levels that enable the CL–DNA complex to cross the cellular-membrane barrier and deliver the exogenous DNA for the purpose of protein production. Recent laser-scanning confocal microscopy and cell-transfection experiments have led to the first hints of the relation between the distinctly structured assemblies and the mechanisms by which DNA is transported into cells.

### 16.3.1 The lamellar $L_\alpha^C$ phase of cationic liposome–DNA complexes

When DNA is mixed with small cationic liposomes with diameter of order 100 nm (Figure 16.9(a)) a transition may be observed with an optical microscope, in the differential-interference-contrast mode, whereby the strongly fluctuating loose mixture rapidly collapses into condensed CL–DNA complexes of diameter on the order of 1 μm (Figure 16.9(b)). By labeling the DNA and lipid molecules separately, one can observe the co-localization of DNA and lipid within the globules in fluorescence microscopy. When viewed with a polarizing microscope, the globules are found to be optically birefringent, which suggests that the internal structure is liquid crystalline.

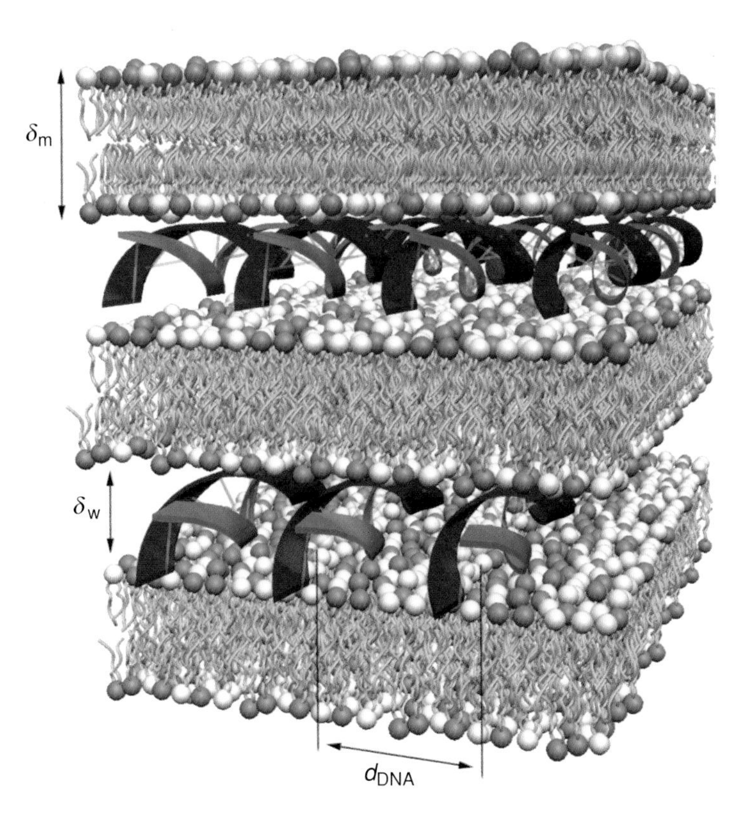


**Figure 16.10.** A model of the lamellar $L_\alpha^C$ phase of cationic liposome–DNA complexes with alternating lipid layers and DNA monolayers as deduced from synchrotron X-ray-diffraction data. The interlayer spacing is $d = \delta_w + \delta_m$. Reprinted with permission from J. O. Rädler *et al.*, *Science*, 275, 810 (1997) and I. Koltover *et al.*, *Science* 281, 78 (1998).

The first work on CL–DNA complexes which led to the elucidation of the structure on the ångström-unit scale was a study by Joachim Raedler, Ilya Koltover, Tim Salditt, and the author performed at the University of California at Santa Barbara (UCSB) during the period 1995–1997 using the quantitative technique of synchrotron X-ray diffraction (XRD). These initial experiments used linear λ-phage DNA containing 48 502 base pairs, giving a contour length of 16.5 μm. The liposomes consisted of mixtures of the cationic lipid dioleoyl trimethyl ammonium propane (DOTAP) and the charge-neutral lipid dioleoyl phosphatidyl choline (DOPC). The synchrotron XRD data were indicative of a complete topological rearrangement of liposomes and DNA into a multilayer structure with DNA intercalated between the lipid bilayers (Figure 16.10). Figure 16.11 plots XRD data of the CL–DNA complexes as a function of increasing $\Phi_{DOPC}$, the weight fraction of DOPC in the DOPC–DOTAP cationic liposome mixtures. At $\Phi_{DOPC} = 0$ (bottom profile) the two sharp peaks at $q = 0.11$ and $0.22\,Å^{-1}$ correspond to the $(00L)$ peaks of a layered structure with an interlayer spacing $d = \delta_m + \delta_w = 57\,Å$. The thickness of a DOTAP bilayer is $\delta_m = 33 \pm 1\,Å$, which leaves a thickness of $\delta_w = d - \delta_m = 24\,Å$ for the water gap. This is sufficient to accommodate a monolayer of hydrated B-DNA. (B-DNA is the normal form of DNA found in biological systems.) The broad peak at $q_{DNA} = 0.256\,Å^{-1}$ arises from DNA–DNA correlations and corresponds to a DNA inter-axial spacing of $d_{DNA} = 2\pi/q_{DNA} = 24.55\,Å$.

The packing nature of DNA chains adsorbed on the lipid bilayers can be elucidated in experiments in which the total lipid area is increased by addition of neutral lipid to the DOTAP bilayers with the overall charge of the CL–DNA complex being maintained at zero by fixing the ratio DOTAP/DNA = 2.20 (wt./wt.) at the isoelectric point of the complex. The XRD scans in Figure 16.11, where the arrows point to the DNA peak, show that $d_{DNA} = 2\pi/q_{DNA}$ increases from 24.54 to 57.1 Å, as $\Phi_{DOPC}$ increases from 0 to 0.75 upon addition of neutral lipid (with $L/D$ = (DOPC + DOTAP)/DNA increasing from 2.2 to 8.8). The most compressed DNA inter-axial spacing of ≈24.55 Å at

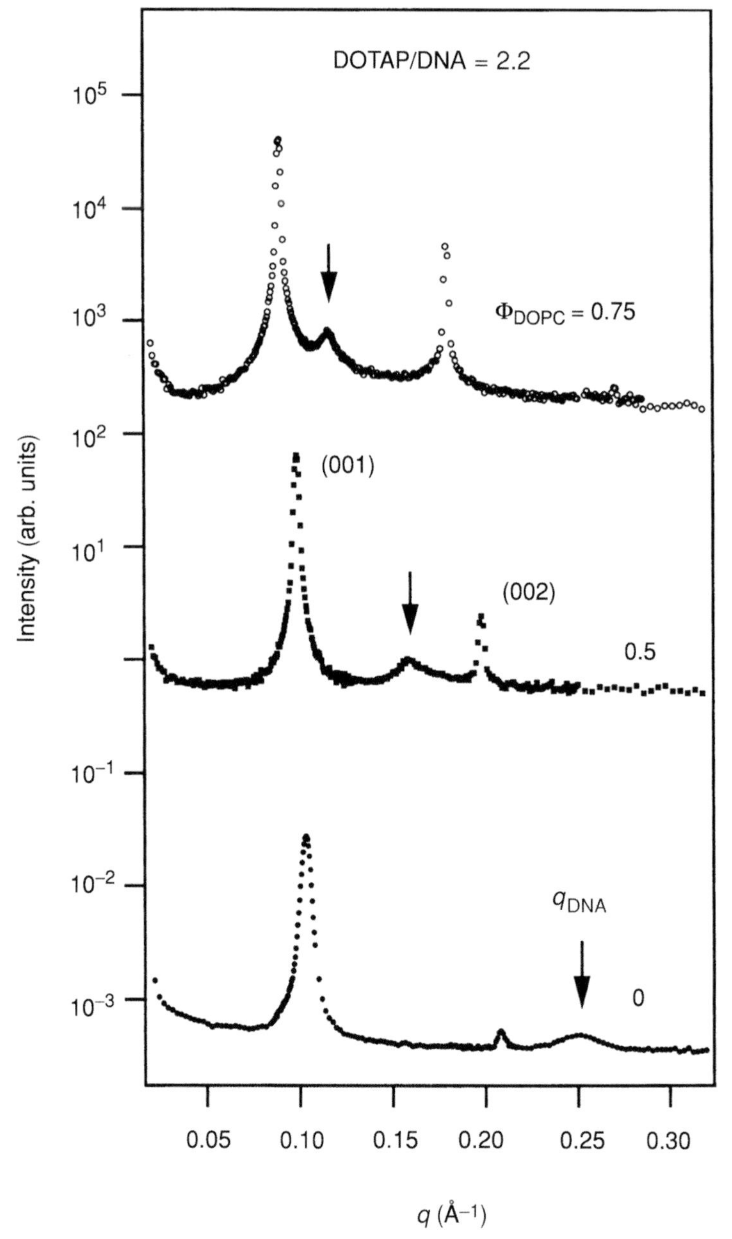


**Figure 16.11.** X-ray-diffraction data of CL–DNA complexes at constant DOTAP/DNA = 2.2 with increasing DOPC/DOTAP ratio, showing the DNA peak (arrow) moving toward smaller $q$, indicating that $d_{DNA}$ increases as the volume fraction of DOPC $\Phi_{DOPC}$ increases. DOPC is the zwitterionic lipid dioleoyl phosphatidyl choline. DOTAP is the positively charged lipid dioleoyl trimethylammonium propane. Reprinted with permission from I. Koltover *et al.*, *Biophys. J.* 77, 915 (1999). ©1999 Biophysical Society. See also J. O. Rädler *et al.*, *Science*, 275, 810 (1997).

$\Phi_{DOPC} = 0$ approaches the short-range repulsive hard-core interaction of the B-DNA rods.

At the isoelectric point of the complex all of the DNA chains are adsorbed on the available cationic membrane with no excess lipid or DNA coexisting with the complex. Under these conditions the DNA inter-axial spacing can be calculated from simple geometry if one assumes that the DNA chains adsorbed between the lipid bilayers form a one-dimensional lattice, which expands to fill the increasing lipid area as $L/D$ increases: $d_{DNA} = (A_D/\delta_m)(\rho_D/\rho_L)(L/D)$. Here, $\rho_D = 1.7\,\mathrm{g\,cm^{-3}}$ and $\rho_L = 1.07\ \mathrm{g\ cm^{-3}}$ denote the densities of DNA and lipid, respectively, $\delta_m$ the membrane thickness, and $A_D$ the DNA area. $A_D = W(\lambda)/(\rho_D L(\lambda)) = 186\,\text{Å}^2$, where $W(\lambda)$ = weight of $\lambda$-DNA = $31.5 \times 10^6/(6.022 \times 10^{23})$ g and $L(\lambda)$ = contour length of $\lambda$-DNA = $48\,502 \times 3.4$ Å.

Figure 16.12 plots $d$ and $d_{DNA}$ as functions of $L/D$. The solid line in Figure 16.12 obtained from the equation for $d_{DNA}$ shows the agreement between the model and data

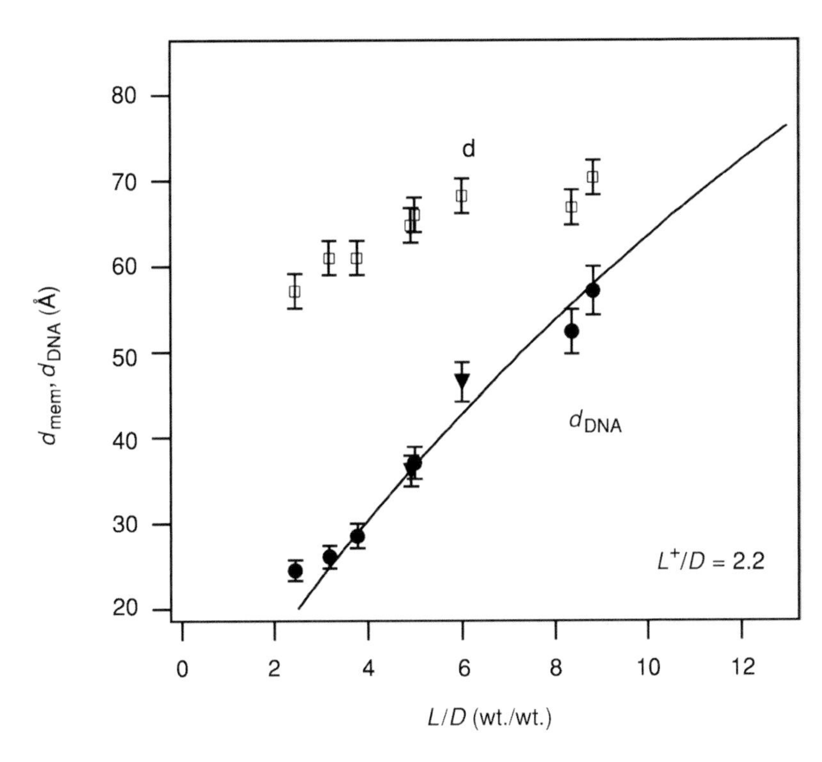


**Figure 16.12.** The DNA inter-axial distance $d_{DNA}$ and the interlayer distance $d$ in the $L_\alpha^C$ phase plotted as a function of lipid/DNA ($L/D$) (wt./wt.) ratio at the isoelectric point of the complex DOTAP/DNA = 2.2. $d_{DNA}$ is seen to expand from 24.5 Å to 57.1 Å. The solid line through the data is the prediction of a packing calculation in which the DNA chains form a space-filling one-dimensional lattice. Reprinted in part with permission from J. O. Rädler *et al.*, *Science*, **275**, 810 (1997).

over the range of measured inter-axial spacing from 24.55 to 57.1 Å. The observed behavior is illustrated schematically in Figure 16.13 with the DNA inter-axial spacing increasing as the lipid area increases. The observation of a variation in the DNA inter-axial distance as a function of the lipid-to-DNA ($L/D$) ratio in multilayers unambiguously demonstrates that XRD directly probes the behavior of DNA in CL–DNA complexes. An analysis of the line widths of the DNA peaks by Tim Salditt has led to the conclusion that chains have positional correlations extending to nearly ten neighbors (Figure 16.14). Thus, the DNA chains form a finite-sized one-dimensional array adsorbed between membranes; that is, a finite-sized two-dimensional smectic phase of matter. Tom Lubensky has shown that on larger length scales the lattice should melt into a two-dimensional nematic phase of DNA chains due to dislocations.

The experimental finding of an increase in $d_{DNA}$ indicates that the DNA inter-chain interactions are long-range and repulsive, otherwise a two-phase regime with high and low densities of DNA adsorbed on membranes would have resulted upon expansion

**Figure 16.13.** A top-view schematic diagram of the one-dimensional lattice of DNA adsorbed on the membrane, showing the increase in distance between DNA chains as the membrane charge density is decreased with the addition of neutral lipid (which increases the area of the membrane) at the isoelectric point.

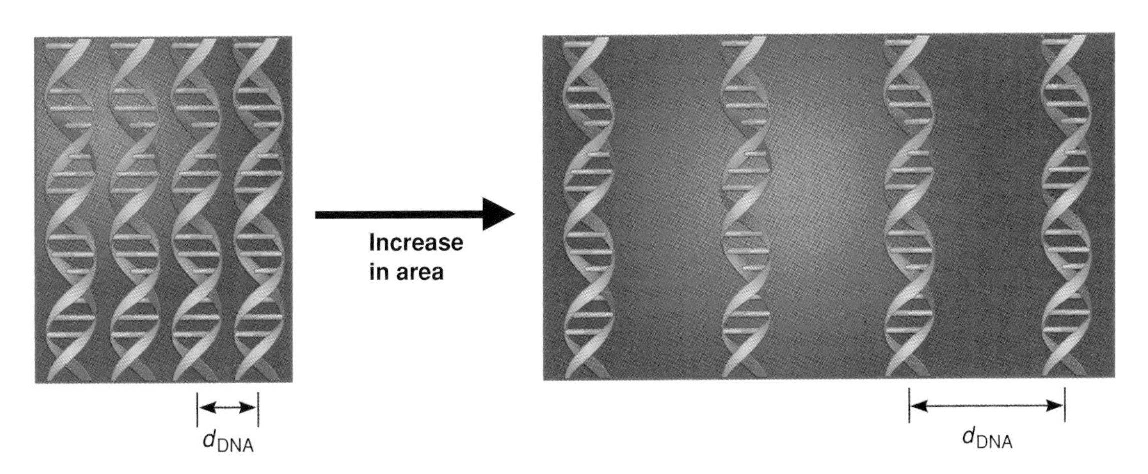

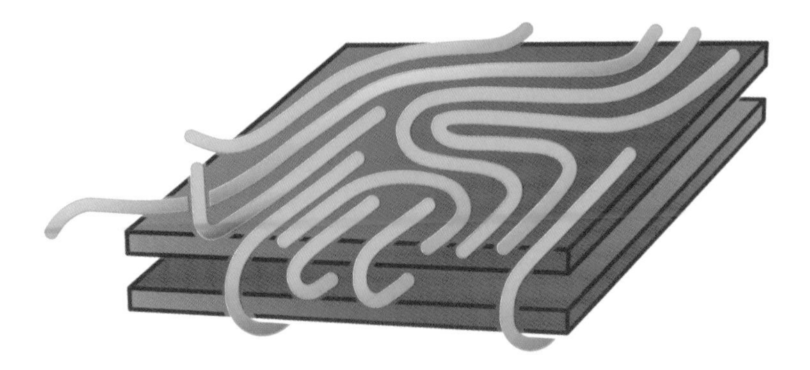

**Figure 16.14.** A cartoon of cationic liposome–DNA complexes in the lamellar $L_\alpha^C$ phase, showing two bilayers and a top monolayer of DNA chains. X-ray diffraction from the DNA–DNA correlation peak shows that the chains are correlated over distances of order ten neighbors as shown here.

of the total lipid area. Indeed, a 1998 analytical model of the electrostatics of DNA–cationic-membrane complexes developed by Robijn Bruinsma and Jay Mashl at UCLA predicts a repulsive power-law fall-off of the DNA inter-chain interactions.

Motivated by structural studies of DNA–lipid complexes of interest to gene therapy, Tom Lubensky and C. O'Hern of the University of Pennsylvania and independently Leo Golubovic and M. Golubovic of the University of Virginia described in 1998 their theoretical finding of the possibility of entirely new "sliding phases" of matter in layered structures such as three-dimensional stacks of two-dimensional crystals. The ground state of these systems is considered to be a mixed columnar–lamellar phase in which the lipid forms a regular smectic phase and the DNA forms a periodic aligned columnar lattice with a single periodic row sandwiched between pairs of lipid lamellae. This phase would have the symmetry of an anisotropic columnar phase. At higher temperatures positional coherence between DNA columns in adjacent layers could be lost but without destroying orientational coherence of DNA columns between layers. Or both positional and orientational coherence between layers could be lost. The theory indicates that the orientationally ordered but positionally disordered "sliding" phase can exist and that there can be a phase transition between it and the columnar phase. This would be a remarkable new phase of matter. Orientational coherence between layers can also be lost at sufficiently long length scales where nonlinearities and dislocations and disclinations in the two-dimensional smectic lattice of DNA become important.

## 16.3.2 Counter-ion release drives the self-assembly of DNA–membrane complexes

The spontaneous self-assembly of DNA with cationic membranes is a result of the gain in entropy due to the release of the counter-ions, bound to the vicinity of both DNA and membrane, back into solution. To see this we first consider a DNA chain (Figure 16.15, top left) which has a bare length $l_o = 1.7$ Å between the negative phosphate groups. This is less than the value of the Bjerrum length in water $l_B \equiv e^2/(\varepsilon k_B T) = 7.1$ Å, which defines a distance at which the Coulomb energy between two unit charges is equal to the thermal energy $k_B T$. In a landmark paper on polyelectrolyte physics, which appeared in 1969, Gerald Manning of Rutgers University showed that a mean-field analysis of the nonlinear Poisson–Boltzmann equations predicts that counter-ions will condense on the DNA backbone if $l_B > l_O$ and that condensation will proceed until the Manning parameter $\xi \equiv l_B/l_0^*$ approaches unity. Here, $l_0^*$ is the renormalized distance between negative charges after counter-ion condensation has occurred, as shown schematically in Figure 16.15 (top left).

A similar analysis shows that, close to a membrane with charge density $\sigma$, nearly half of the counter-ions are contained within the Gouy–Chapman layer $l_{G\text{-}C} \equiv e/(2\pi l_B \sigma)$

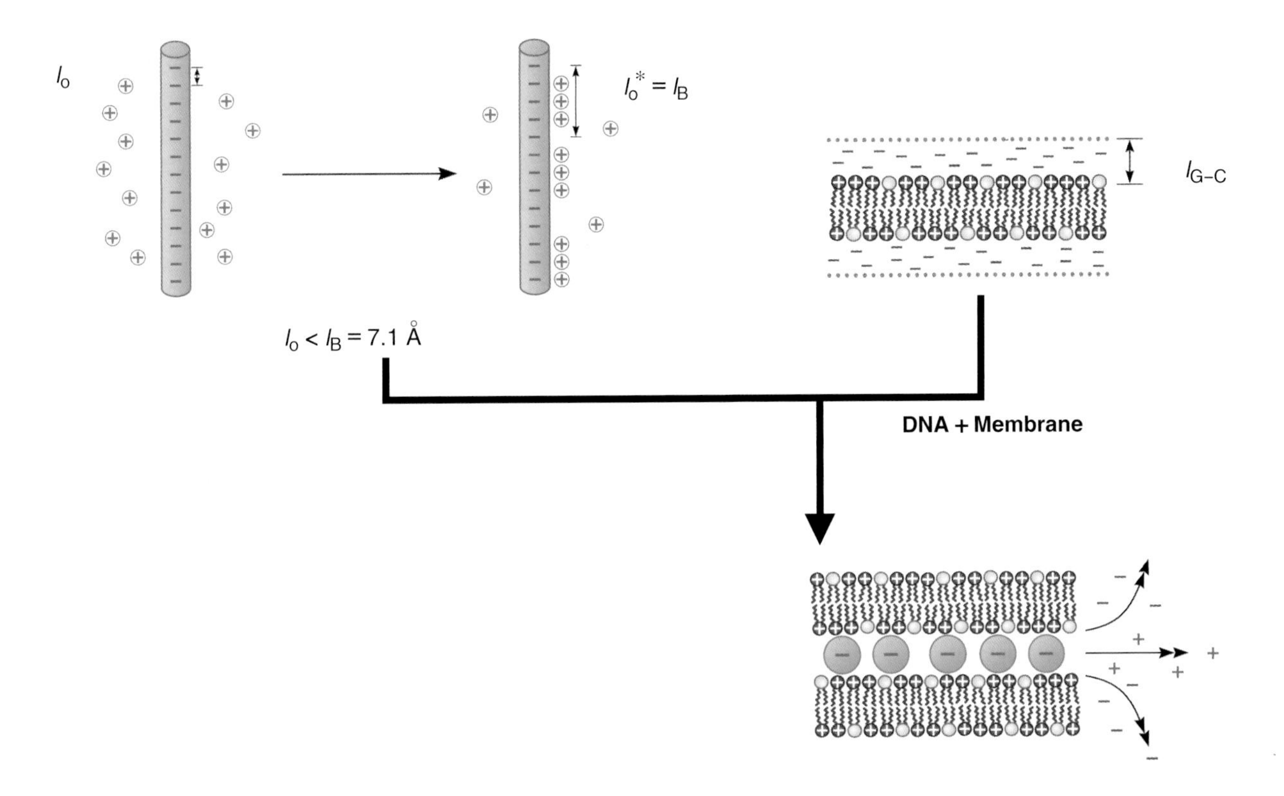


Figure 16.15. Top left: DNA chains have a condensed Manning layer if the bare length is less than the Bjerrum length ($l_o < l_B$). Top right: cationic membranes attract negative counter-ions within the Gouy–Chapman layer $l_{G-C}$. Bottom right: the complexation of DNA and cationic membranes leads to the release of bound counter-ions, which increases their entropy and lowers the free energy.

(Figure 16.15, top right). The condensation of DNA and cationic membranes allows the cationic lipids to neutralize the phosphate groups on the DNA, in effect replacing and releasing the originally condensed counter-ions into solution (Figure 16.15, bottom right). Therefore, the driving force for self-assembly is the release of counter-ions, which were bound to DNA and cationic membranes, into solution.

### 16.3.3 Overcharging of CL–DNA complexes; analogies to DNA–histone complexes in vivo

If one ponders the question of the overall charge of CL-DNA complexes, one may naively expect that, on either side of the isoelectric point, CL-DNA complexes would be charge neutral, coexisting with either excess DNA or excess liposomes. The true answer has turned out to be most interesting and unexpected, namely, in the presence of excess liposomes, CL-DNA complexes are positively overcharged and upon addition of DNA undergo charge reversal across the isoelectric point and become negatively overcharged. As we describe below, the cationic histone proteins, which condense DNA into nucleosome particles in cells, exhibit a remarkably similar behavior.

The optical micrographs in Figure 16.16 show this behavior for CL-DNA complexes for three lipid ($L$) to DNA ($D$) ratios (the lipid is DOTAP plus DOPC in 1:1 ratio). At low DNA concentrations, 1-μm CL–DNA globules are observed (Figure 16.16, $L/D \approx 50$). As the isoelectric point ($L/D = 4.4$) is approached as a result of the addition of DNA, the globular condensates flocculate into large aggregates of distinct globules (Figure 16.16, $L/D \approx 5$). Upon further addition of DNA, the size of the complex becomes smaller and is stable in time again (Figure 16.16, $L/D \approx 2$). Electrophoretic mobility measurements confirm that for $L/D \approx 50$ the complexes are positively charged, whereas for $L/D = 2$ the complexes are negatively charged. The positively and negatively charged globules

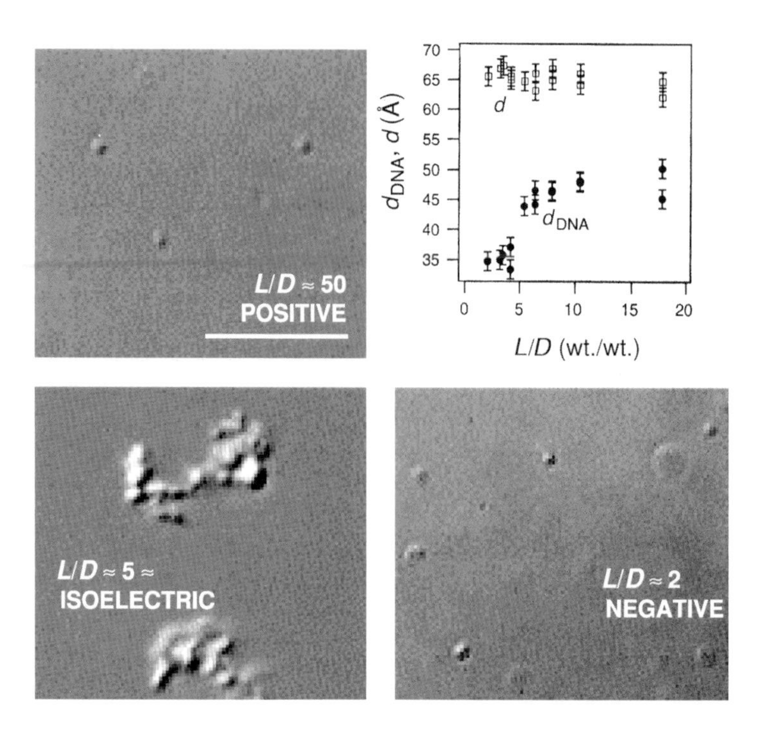


**Figure 16.16.** Top left and bottom: high-resolution differential-interference-contrast optical images of CL–DNA complexes forming distinct condensed globules in mixtures of various lipid-to-DNA weight ratios ($L/D$). $L/D = 4.4$ is the isoelectric point. CL–DNA complexes are positively charged for $L/D \approx 50$ and negatively charged for $L/D \approx 2$. The bar is 10 μm. Top right: the spacings $d$ and $d_{DNA}$ (see Figure 16.10 for definitions) as functions of $L/D$, below and above the isoelectric point of the complex. Reprinted with permission from J. O. Rädler *et al.*, *Science*, 275, 810 (1997).

at $L/D \approx 50$ and $L/D \approx 2$, respectively, repel each other and remain separate, whereas, as $L/D$ approaches 5, the nearly neutral complexes collide and tend to stick due to van der Waals attraction. Thus, the charge reversal and overcharging properties of complexes are clearly evident.

The degrees of packing of DNA chains adsorbed between membranes below and above the isoelectric point differ. As $L/D$ decreases with increasing amounts of DNA from 18 to 2 (Figure 16.16, top right), $d_{DNA}$ abruptly decreases from ≈44 Å in the positively charged regime ($L/D > 4.4$), to ≈37 Å in the negatively charged regime ($L/D < 4.4$) with more chains packed between the lipid bilayers.

On the theoretical side, Robijn Bruinsma explained the physics of overcharging of CL–DNA complexes in an analytical model in 1998 for the limit of a low concentration of cationic lipid. In the negatively charged regime in which CL–DNA complexes coexist with excess DNA, the system lowers its free energy if additional DNA chains enter complexes (leading to negative overcharging) and release their bound counter-ions within the volume of the complex. Similarly, in the positively charged regime excess lipids enter the complex and release their bound counter-ions. Most remarkably, a numerical treatment of the structure and thermodynamics of CL–DNA complexes, *valid at any lipid concentration*, which was published in 1998 by Avinoam Ben-Shaul, William Gelbart, Sylvio May, and Daniel Harries (all from the Hebrew University), is able to describe most of the experimental findings.

Let us now turn to the phenomenon of condensation and de-condensation of DNA in biological systems. In eukaryotic cells undergoing division, DNA condenses into chromosomes to ensure that daughter cells receive one full copy of the cell's genome. De-condensation is necessary during much of the life of cells because proteins have to access the DNA template, for example, during transcription when the molecular machine RNA polymerase attaches to the template and transcribes DNA into RNA, and in gene regulation when various transcription factors attach to specific DNA sequences and turn genes on or off. At the smallest length scales the process of DNA condensation begins with the polymer wrapped twice around a histone core particle (an octameric assembly of four different histone particles, H2A, H2B, H3, and H4), forming a

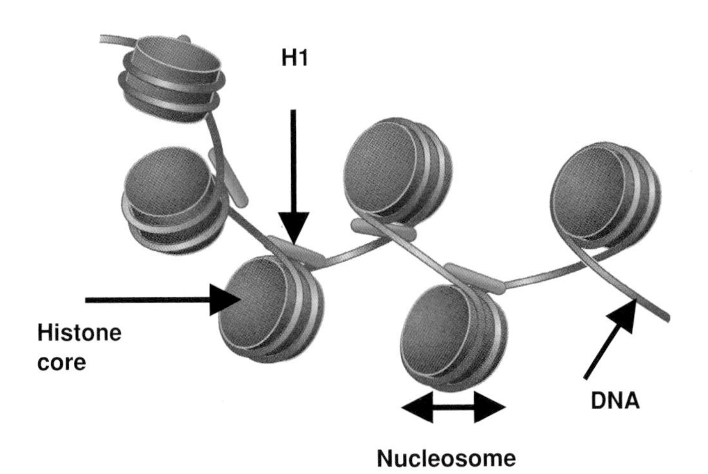


Figure 16.17. A cartoon of DNA wrapped twice around each histone core to form the bead-on-string structure representing the first level of DNA packing as described in the text. Redrawn from M. Grunstein, *Scient. Am.* 267:68 (1992).

nucleosome unit, which has a bead-on-string-like structure (Figure 16.17). The nucleosome particle has a size of 10 nm and its structure has recently been elucidated by synchrotron X-ray crystallography.

In the presence of a fifth linker histone particle (H1) and under physiological salt conditions the nucleosomes further assemble into a helical pattern forming what is referred to as the 30-nm chromatin fiber. On its pathway to forming ultra-condensed chromosomes with length scales readily visible in the light microscope, chromatin further assembles into a higher-ordered structure. However, the mechanisms involved and the structures at these length scales are currently not understood.

Unexpectedly, the nucleosome particle is believed to be negatively overcharged in vivo, with excess negative phosphate groups from DNA compared with cationic charge from histones. The condensation of DNA by cationic lipid and the physics of the overcharging properties of the CL–DNA particles and nucleosomes are most likely closely related. Theoretical papers that appeared in 1999, by William Gelbart, Robijn Bruinsma, and Stella Park (UCLA), and independently by Phillip Pincus, Edward Mateescu, and Claus Jeppesen at UCSB, on the packing nature of DNA–histone complexes suspended either in solution (where the entropy of released counter-ions is important), or in vacuum (where electrostatics dominate), have indeed revealed that the nucleosome particle should be overcharged.

### 16.3.4 The inverted hexagonal $H_{II}^{C}$ phase of CL–DNA complexes

So far we have described CL–DNA complexes containing the neutral lipid DOPC mixed in with the cationic lipid, which leads to a lipid–DNA sandwich structure (Figure 16.10). A second commonly used neutral lipid in CL–DNA mixtures is dioleoyl-phosphadtidylethanolamine (DOPE). Synchrotron X-ray-diffraction work by I. Koltover, T. Salditt, J. Radler, and the author has shown that DOPE-containing CL–DNA complexes may give rise to a completely different structure in which each DNA chain is contained within a lipid tubule (i.e. a cylindrical inverse micelle) with the tubules arranged on a hexagonal lattice as shown schematically in Figure 16.18. The CL–DNA structure (labeled $H_{II}^{C}$) is reminiscent of the inverted hexagonal $H_{II}$ phase of DOPE in water, which has been elucidated by John Seddon of Imperial College, London, and Sol Gruner of Cornell University during the last 20 years, with the water space inside the lipid micelle filled by DNA. How do we understand the different spontaneously produced

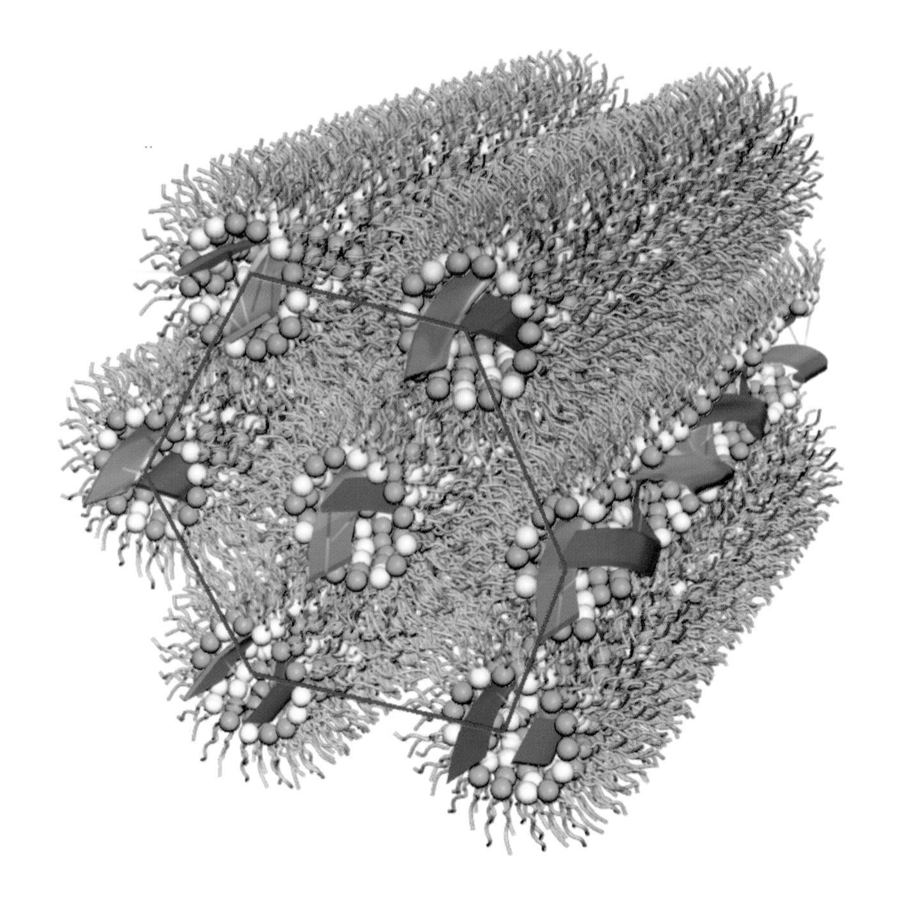

**Figure 16.18.** A model of the inverted hexagonal $H_{II}^{C}$ phase of cationic liposome–DNA complexes with DNA contained within lipid tubules arranged on a hexagonal lattice as deduced from synchrotron X-ray-diffraction data. Reprinted with permission from I. Koltover *et al.*, *Science* 281, 78 (1998).

structures that CL–DNA complexes may adopt? We expect that the interplay between the electrostatic and membrane elastic interactions in CL–DNA complexes determines their structure. Indeed, theoretical work in 1998, by Avinoam Ben-Shaul and Sylvio May, predicted that electrostatic interactions are expected to favor the inverted hexagonal $H_{II}^{C}$ phase, which minimizes the charge separation between the anionic groups on the DNA chain and the cationic lipids. However, the electrostatic interaction may be resisted by the membrane elastic cost of forming a cylindrical monolayer membrane around DNA. For example, many lipids have a cylindrical shape, with the head-group area nearly matched to the hydrophobic tail area, and tend to self-assemble into lamellar structures with a spontaneous curvature $C_0 = 0$. Other lipids have a cone shape, with a smaller head-group area than tail area, and give rise to a negative spontaneous curvature, $C_0 < 0$. Alternatively, lipids with a larger head group than tail area have $C_0 > 0$.

In landmark work during the 1970s, J. Israelachvili, D. Mitchell, and B. Ninham of the University of Canberra presented convincing evidence that, in many lipid systems, the "shape" of the molecule, which influences the spontaneous curvature of the membrane $C_0$, will also determine the actual curvature $C$ which describes the structure of the lipid self-assembly. This is clearly true if the bending rigidity of the membrane is large ($\kappa_c/(k_B T) \gg 1$), because then a significant deviation of $C$ from $C_0$ would cost too much elastic energy. However, we see that, if the bending cost is low with $\kappa_c \approx k_B T$, then $C$ may deviate from $C_0$ without costing much elastic energy, especially if another energy is lowered in the process. Thus, for large bending rigidity the structure of the CL–DNA complex is determined by the curvature favored by the lipid mixture, whereas a lower $\kappa_c$ should lead to a structure favored by electrostatics. The observation of the $H_{II}^{C}$ phase as DOPC is replaced by DOPE is consistent with the fact that the spontaneous curvature of the mixture of DOTAP and DOPE is driven negative with $C_0$ = volume fraction of DOPE × $C_0^{\mathrm{DOPE}}$.

### 16.3.5 The relation between the structure of CL–DNA complexes and gene delivery

As we described earlier, a key reason why the properties of CL–DNA complexes have been the subject of intense scientific scrutiny worldwide stems from their known ability to carry DNA (and thus genes of interest) into cells for therapeutic applications. One aspect of the research is focused on elucidating the structures of CL–DNA complexes and the structure–function relation in transfection studies, which measure the amount of protein produced in cells encoded by the foreign gene.

The lamellar $L_\alpha^C$ CL–DNA complexes, which typically exhibit low transfection behavior compared with hexagonal $H_{II}^C$ complexes, have been redesigned in the last few years, by Nelle Slack, Alison Lin, Ayesha Ahmad, Heather Evans, and Kai Ewert in the author's group at UCSB, to achieve transfection comparable to that achieved with $H_{II}^C$ complexes. The transfection studies were quantified by measuring the expression of the firefly luciferase gene by counting the photons emitted by the gene. The results of the study on the highly complex $L_\alpha^C$-cell system revealed an unexpected simplicity whereby the charge density of the membrane ($\sigma_M$) was found to be a key universal parameter for transfection.

At the same time as the transfection studies, three-dimensional imaging with a laser scanning confocal microscope by Alison Lin and Heather Evans revealed distinct interactions between CL–DNA complexes, comprised of either the $L_\alpha^C$ phase or the $H_{II}^C$ phase and mouse fibroblast cells. The combined confocal imaging and transfection data led to a model of cellular entry via $L_\alpha^C$ CL–DNA complexes shown schematically in Figure 16.19. The initial electrostatic attraction between cationic CL–DNA complexes and mammalian cells is mediated by negatively charged cell surface sulfated proteoglycans (Figure 16.19(a)). Confocal images revealed no evidence of fusion with the plasma membrane, implying that $L_\alpha^C$ complexes entered via the endocytic pathway (Figures 16.19(b) and (c)). At low $\sigma_M$, mostly intact $L_\alpha^C$ complexes were observed inside cells, implying that DNA was trapped by the lipid, whereas at high $\sigma_M$, confocal images revealed lipid-free DNA inside cells. Because the observed DNA is in a condensed state, it must reside in the cytoplasm (Figure 16.19(e)) because of the lack of DNA-condensing molecules in the endosome. The simplest mechanism of escape from the endosome is through fusion with the endosomal membrane (Figure 16.19(d)).

In comparison with lamellar complexes, DOPE-containing $H_{II}^C$ complexes exhibit almost no dependence on $\sigma_M$. However, $H_{II}^C$ complexes do exhibit evidence of fusion with the cell plasma membrane, which does not occur for $L_\alpha^C$ complexes. The mechanism of transfection by DOPE-containing $H_{II}^C$ complexes appears to be dominated by other effects; for example, possibly the known fusogenic properties of inverted hexagonal phases as depicted in Figure 16.20. A CL–DNA $H_{II}^C$ complex is shown in Figure 16.20(a) approaching either the plasma or the endosomal membrane, which would occur if the complex entered through endocytosis. The outermost lipid monolayer, which must cover the $H_{II}^C$ complex due to the hydrophobic effect, is at an opposite curvature to that of the preferred negative curvature of the lipids coating DNA inside the complex. This elastically frustrated state of the outer monolayer (which is independent of $\sigma_M$) may drive the rapid fusion with the plasma or endosomal membrane, leading to release of a layer of DNA and a smaller $H_{II}^C$ complex as shown in Figure 16.20(b).

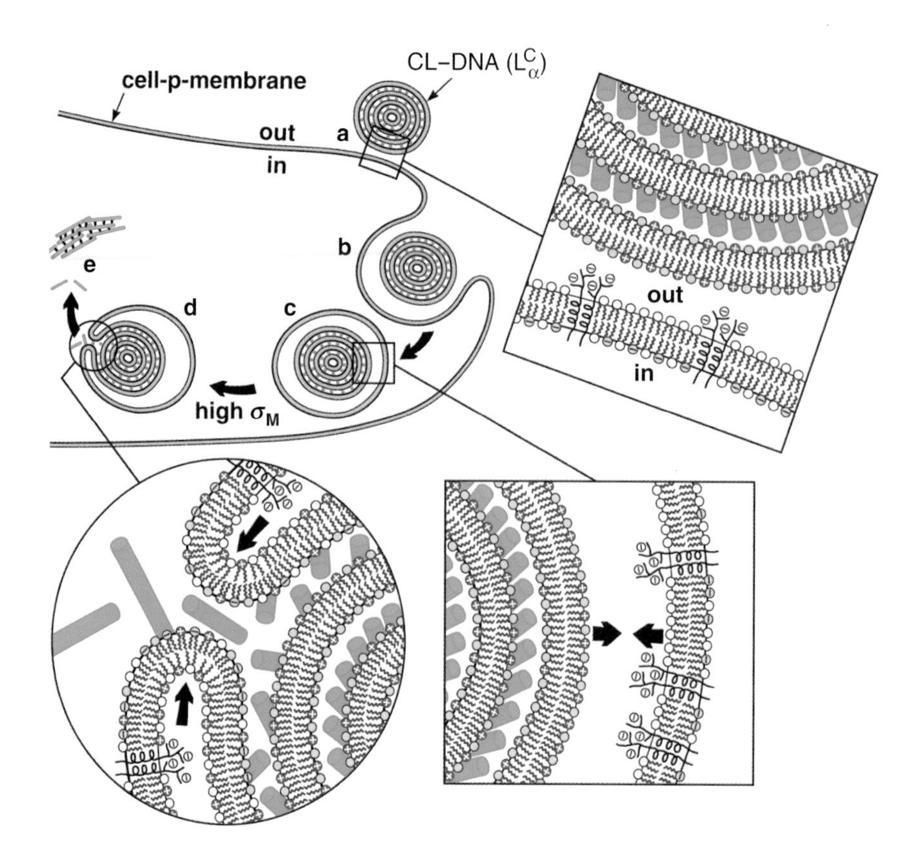


**Figure 16.19.** A model of the cellular uptake of $L_\alpha^C$ CL–DNA complexes. Complexes adhere to cells due to electrostatic interactions (a) and enter through endocytosis (b and c). Primarily complexes with high membrane charge density ($\sigma_M$) escape the endosome through activated fusion (which is consistent with transfection efficiency data) (d), with released DNA forming aggregates (observed in microscopy) with oppositely charged cellular biomolecules (e). Expanded views: (c) arrows indicate attractive electrostatic interactions; and (d) arrows point to bending modes of membranes. Reprinted with permission from A. Lin *et al.*, *Biophys. J.* 84 (5), 1 (2003). © 2003 Biophysical Society.

Fluorescent labeling of both lipid and DNA is now beginning to allow researchers worldwide to probe the spatial and temporal distribution of complexes inside the cell to look for fusion events and other destabilizing processes leading to release of DNA from the complex and localization near and across the cell nucleus. A key goal of these ongoing biophysical studies is to use the knowledge gleaned in the design of optimal carriers of genetic material for gene-delivery purposes. The field of nonviral gene therapy is at a very early stage at which most of the work is empirical and real progress at the clinical level will only follow careful biophysical and biochemical work at the cellular level.

## 16.3.6 DNA in two dimensions

Researchers have been working for more than a decade on developing methods to align DNA on surfaces in general and in particular on ultra-smooth mica surfaces with ångström-unit-level roughness. The initial reason was for rapid sequencing of DNA molecules associated, for example, with the Human Genome Project. In this post-genomic "proteomics era" one still aspires to develop techniques by which sections of DNA can be sequenced rapidly and at minimal cost, for example, to look for DNA-sequence mutations indicative of specific disease states during a routine visit to the doctor. Groups of researchers have been pursuing a similar line of research by using the atomic force microscope (AFM) to image DNA adsorbed on surfaces with the ultimate goal of distinguishing among the four letters used in our genome. A beautiful AFM image of circular DNA molecules adsorbed on positively charged lipid bilayers supported on cleaved mica can be seen in Figure 16.21. This and similar images have been produced in the laboratory of Zhifeng Shao at the University of Virginia Medical

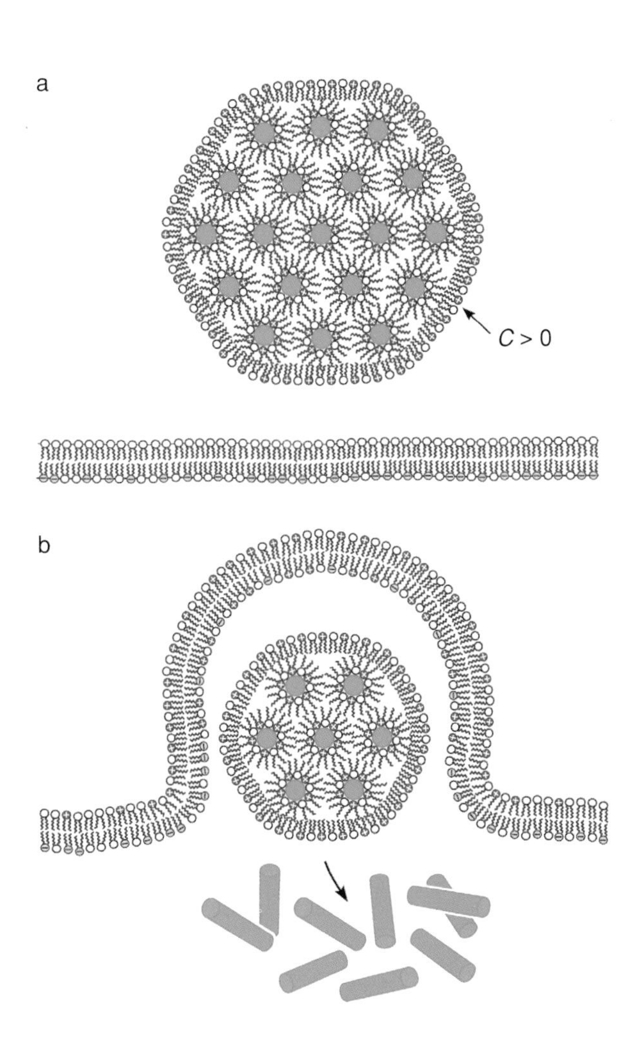


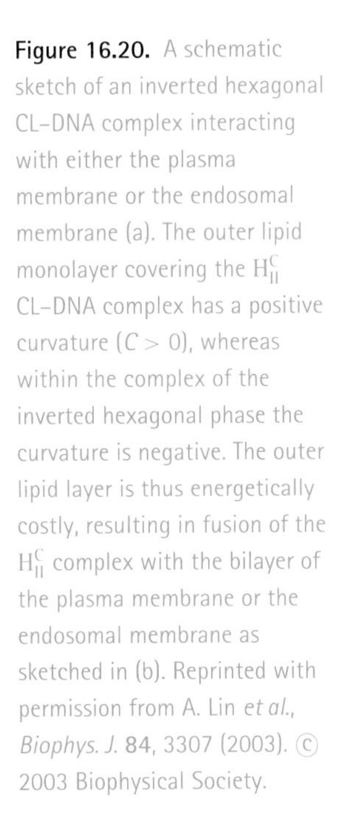

**Figure 16.20.** A schematic sketch of an inverted hexagonal CL–DNA complex interacting with either the plasma membrane or the endosomal membrane (a). The outer lipid monolayer covering the $H_{II}^{C}$ CL–DNA complex has a positive curvature ($C > 0$), whereas within the complex of the inverted hexagonal phase the curvature is negative. The outer lipid layer is thus energetically costly, resulting in fusion of the $H_{II}^{C}$ complex with the bilayer of the plasma membrane or the endosomal membrane as sketched in (b). Reprinted with permission from A. Lin *et al.*, *Biophys. J.* 84, 3307 (2003). © 2003 Biophysical Society.

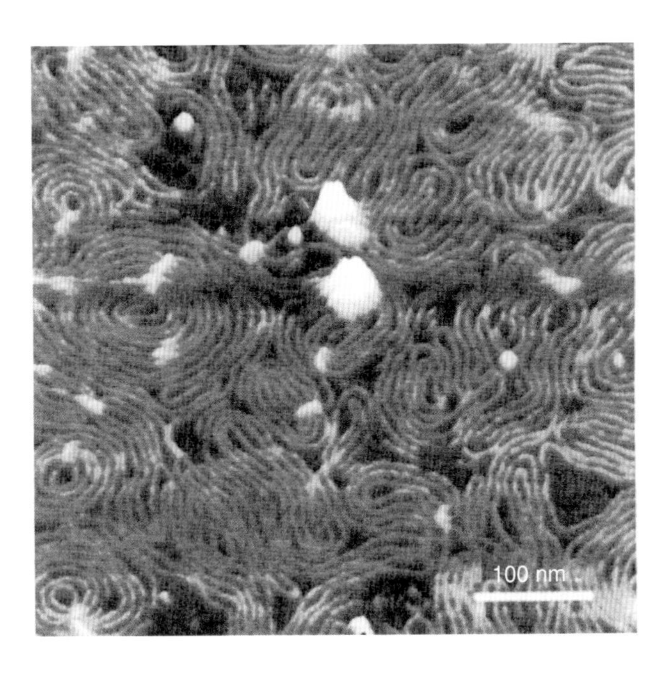


**Figure 16.21.** A high-resolution atomic-force-microscopy image of plasmid DNA adsorbed on a cationic bilayer coating a freshly cleaved mica surface. The highly packed DNA chains are clearly visible. The measured width of DNA is 2 nm, which is similar to the diameter of B-DNA. Reprinted by permission of the Federation of the European Biochemical Societies and Z. Shao from "High-resolution atomic-force microscopy of DNA: the pitch of the double helix," by J. Mou *et al.*, *FEBS Letters* 371, 279–282, 1995. (Courtesy of Zhifeng Shao.)

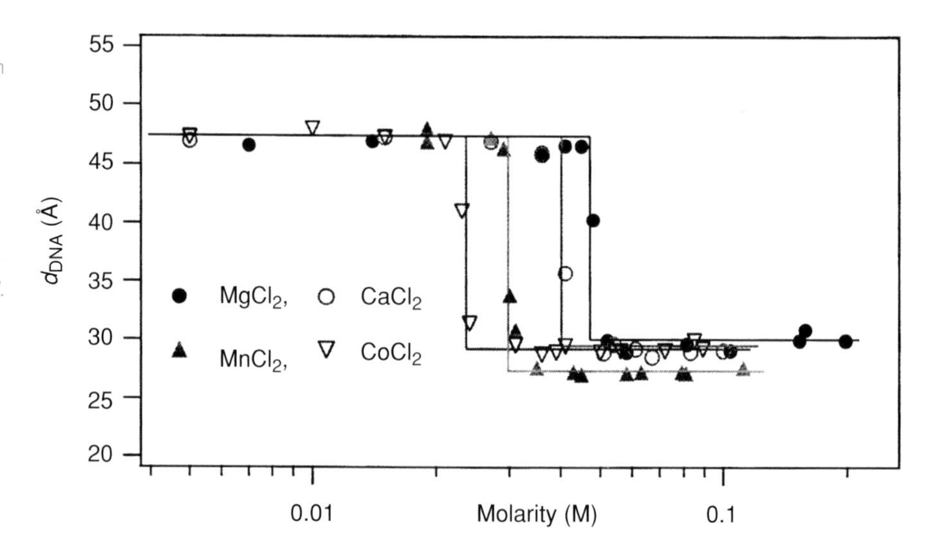


**Figure 16.22.** Variation of the DNA inter-axial spacing ($d_{DNA}$) in lamellar $L_\alpha^C$ CL–DNA complexes with the concentrations of four different divalent salts, showing the rapid collapse of $d_{DNA}$ into the condensed state of DNA. Reprinted with permission from I. Koltover *et al.*, *Proc. Natl. Acad. Sci. U.S.A.* 97, 14 046 (2000). © 2000 National Academy of Sciences, U.S.A.

School during the mid 1990s. The group of Jie Yang at the University of Vermont has also reported similar work more recently.

These preparations, which represent the current state of the art, have enabled the direct observation of the periodic helical modulation of the double-stranded DNA molecules, which was measured to be 3.4 nm and is thus consistent with the pitch of the hydrated B-DNA state found in vivo. The handedness of the DNA molecules is also visible. In the near future, we should expect to see AFM studies of DNA aimed at sequencing and imaging of DNA with associated proteins, such as gene-regulating transcription factors, and molecular machines, such as those involved in transcription and replication of DNA.

Aside from the structural elucidation of DNA–lipid complexes, the DNA–lipid multilayers described above (Figure 16.10) constitute ideal experimental models for understanding the nature of DNA packing in two-dimensions. One example will be described here, but in fact this is an area clearly worthy of future studies. In 2000, Ilya Koltover, Katherine Wagner, and the author studied the forces between DNA chains bound to cationic membranes in $L_\alpha^C$ CL–DNA complexes. The experiments were carried out in the presence of electrolyte counter-ions common in biological cells ($Ca^{2+}$, $Mg^{2+}$, and $Mn^{2+}$). Prior to the work, it was known that electrostatic forces between DNA chains in bulk aqueous solution containing divalent counter-ions remain purely repulsive and that condensation of DNA required a counter-ion valence $Z \geq 3$. As we see in Figure 16.22, as the electrolyte concentration is increased, the DNA inter-axial spacing $d_{DNA}$, measured by X-ray diffraction, remains constant until a sharp transition into a condensed phase of DNA chains with $d_{DNA} = 28.9 \pm 0.5$ Å is observed just above a critical concentration of divalent ions ($M^*$). In this highly condensed state the surfaces of neighboring DNA helices are separated by a distance of 4 Å, namely the ionic diameter of the hydrated divalent ions.

Thus, in striking contrast to bulk behavior, in which forces between DNA chains remain repulsive, the synchrotron X-ray-diffraction experiments reveal that the electrostatic forces between DNA chains adsorbed on surfaces *reverse from repulsive to attractive* above a critical concentration of divalent counter-ions and lead to a DNA-chain-collapse phase transition (Figure 16.23). This demonstrates the importance of spatial dimensionality to interactions between DNA molecules, in which nonspecific counter-ion-induced electrostatic attractions between the like-charged polyelectrolytes overwhelm the electrostatic repulsions on a surface. This new phase is the most compact

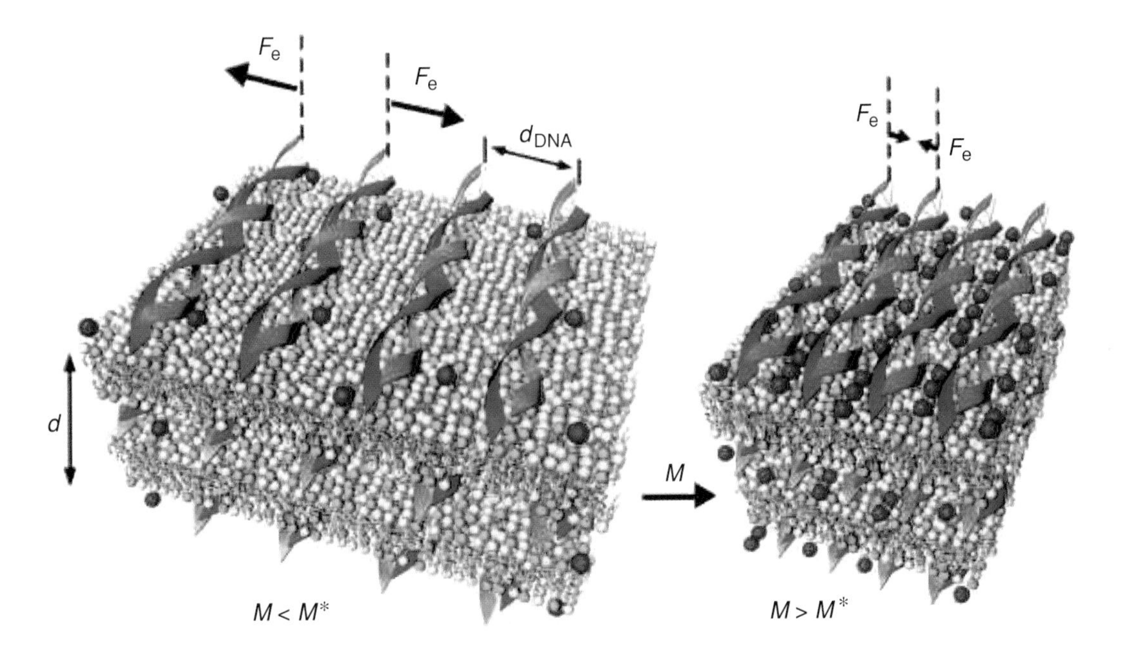


**Figure 16.23.** A schematic illustration of the data of Figure 16.22, showing the force reversal between DNA chains adsorbed on cationic membrane surfaces within the lamellar $L_\alpha^C$ phase. For divalent counter-ion concentrations $M < M^*$ the electrostatic forces ($F_e$) are repulsive. For $M > M^*$ the forces become attractive, leading to the DNA-condensation transition on the membrane. Reprinted with permission from I. Koltover *et al.*, *Proc. Natl. Acad. Sci. U.S.A.* 97, 14 046 (2000). © 2000 National Academy of Sciences, U.S.A.

state of DNA on a surface in vitro and may have potential applications in processing of organometallic materials and in high-density storage of genetic information.

## 16.4 Multi-lamellar onion phases

Within the last decade a group of researchers at Bordeaux led by Didier Roux have developed a novel non-equilibrium process leading to the formation of multi-lamellar vesicles comprised of surfactant molecules with controlled size. These so-called "onion phases" shown schematically in Figure 16.24(a) are a result of a hydrodynamic instability when lamellar $L_\alpha$ phases are subjected to shear flow. A suspension of onions may be seen in the electron micrographs of Figure 16.24(b). The structure obtained is made of concentric spherical stacked bilayers (Figure 16.24(c)). The onion phases are currently used in cosmetics and chemical-release applications and may have potentially important biomedical applications. As we described previously, there is currently a worldwide effort at developing synthetic carriers of DNA for gene-delivery applications. Although most laboratories are using cationic liposomes or polymers as carriers, there is a growing interest in producing carriers of DNA with neutral lipids. The main reason is that, whereas cationic lipids may be toxic when used in large quantities in vivo, neutral synthetic lipids are known to be more biocompatible.

In the case of onions made with neutral phospholipids, the three functions resulting from the presence of positively charged molecules, namely condensation of DNA, the making of colloidal particles, and the adhesion to mammalian-cell surfaces containing negatively charged groups (Figure 16.19(a), blow up), cannot be achieved in a single step but need to be executed in several steps. The first step has been the incorporation of DNA into a neutral lamellar phase in the absence of electrostatic interactions. The difficulty, at this stage, is to keep the DNA inside the spherulites, taking into account the osmotic pressure of the DNA. The Bordeaux team (D. Roux, T. Pott, O. Diat, P. Chenevier, L. Navailles, O. Regev, and O. Mondain Monval) has discovered a pathway in the relevant phase diagram that leads to a novel new phase of matter in which DNA, a highly charged negative polyelectrolyte, is trapped within neutral membranes.

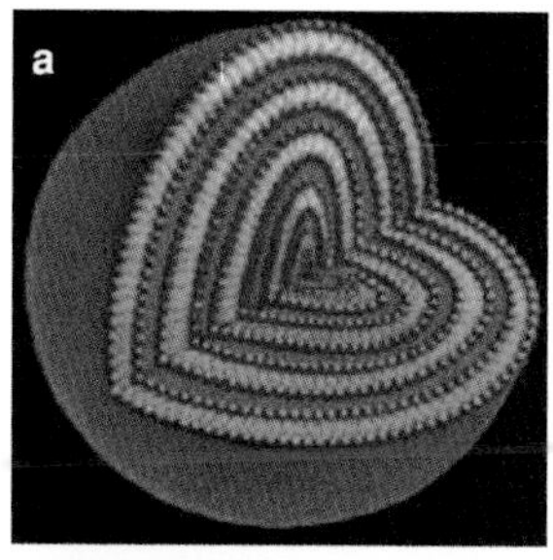


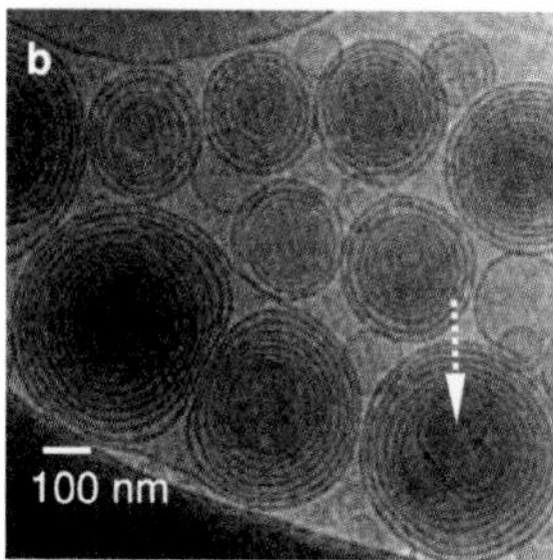


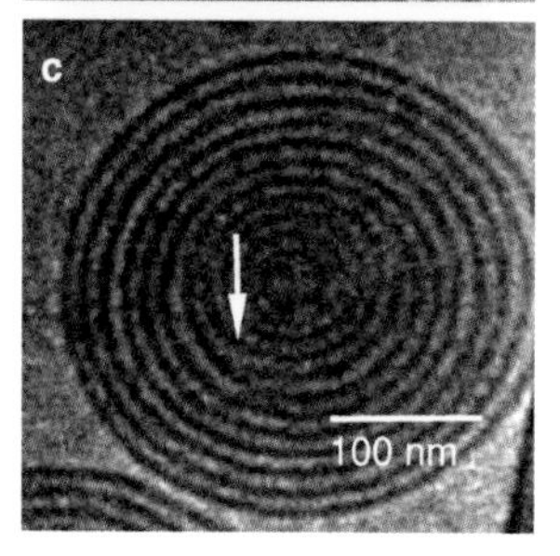


**Figure 16.24.** (a) An artistic representation of the spherulite onion phase; (b) and (c) electron microscopy (EM) of multi-lamellar onions. The technique used is cryogenic freezing followed by transmission EM of samples obtained after shearing a lamellar phase dispersed in an aqueous solution (phosphatidylcholine/water). (b) An ensemble of spherulites dispersed in water. (c) An enlargement of a spherulite, showing that the multi-lamellar structure extends to the center of the onion. The full arrow corresponds to dislocations and the dashed arrow to a reorganization of the core of the spherulites. (Courtesy of Didier Roux.)

The Bordeaux group's work to date has used short DNA strands containing 150 base pairs as a model system, in a charge-neutral lamellar phase of phosphatidylcholine. The X-ray structure studies by Roux and co-workers showed unambiguously that DNA is incorporated between the neutral lipid bilayers. They have shown that this system can be sheared and dispersed in an osmotically controlled aqueous solution by keeping the DNA molecules inside the spherulite structure. The team is now actively working on developing such neutral onions containing longer pieces of DNA of order 4000 base pairs, long enough to contain real genes for applications. As a final note one should appreciate that the observation of confinement of DNA by neutral membranes is a most remarkable scientific finding. What are the interactions that lead to this confinement not only of short pieces of DNA but also of long chains? Theoretical physicists will be pondering the interactions for many years to come.

## 16.5 Vesosomes – liposomes inside vesicles

Liposomal drug delivery is the oldest lipid-based method with proven benefits for delivery of small-molecule drugs for treating inflammation, chemotherapy, and pain relief. Liposomes used in such applications are typically 50–500 nm in diameter, and are synthetic versions of the basic components of natural human cell membranes (Figure 16.2(b)). By enclosing a drug in a liposome, drugs can be released throughout the body in various ways and the amount of time the drug is at its clinically useful concentration can be extended. The main difficulty of using uni-lamellar liposomes in drug delivery has been the difficulty in keeping a wide range of drugs inside the liposome while in the bloodstream. Since lipids are natural products, the immune system has a wide array of enzymes and receptors that target lipids for removal, destruction, or recycling. In the process, the drugs inside the liposomes are released prematurely, decreasing the benefits of encapsulation. Although a great deal of research has gone into altering the lipid composition of uni-lamellar liposomes, and protecting liposomes from enzymes by coating the liposome with a sugar or polymer coating (Figure 16.2(c)), this has not been enough to frustrate the body's immune system.

To overcome some of the difficulties of delivery by uni-lamellar liposomes, Joe Zasadzinski and his group members Scott Walker, Michael Kennedy, Ed Kisak, and Brett Coldren of the University of California at Santa Barbara have achieved an ingenious bioengineering feat in the past few years by inventing a process leading to a structure they refer to as the "vesosome," which consists of "double-bagging" both the liposomes and the drug inside of another bilayer membrane. The vesosome enables them to retain all of the benefits of the liposome and significantly extend release time (Figure 16.25). The outer bilayer of the vesosome can be stabilized by polymers, just like the uni-lamellar liposomes (Figure 16.2(c)), while both liposomes and other colloidal particles can be trapped inside. The colloids can be used as imaging agents – magnetic particles for MRI, for example. The drug is encapsulated using the same basic technology as that developed for liposomes.

The key to the vesosome self-assembly process is the "interdigitated" bilayer phase formed by adding ethanol to aqueous solutions of dipalmitoylphosphatidylcholine (DPPC) and other saturated lipids at room temperature. The added ethanol causes the DPPC bilayers to fuse and form large open sheets. The ethanol intercalates within the DPPC head groups; the larger area occupied by each head group causes the alkyl chains of DPPC to interdigitate into a crystalline packing. The bilayer is too rigid to form closed spheres, and hence forms extended sheets in solution. However, when the sheets are heated past the gel-to-liquid-crystalline, or melting, temperature, of the lipid bilayer, the alkyl chains of DPPC melt and no longer interdigitate. The ethanol is lost to solution, and the resulting sheets, which are now much less rigid, roll up to form spherical structures. In the process of rolling up to form spheres, the interdigitated sheets encapsulate whatever happens to be in the surrounding solution, including other vesicles to self-assemble into vesosomes (Figure 16.26).

The main benefit of the multicompartment structure has been that two membranes must be degraded in series before the drug is released. Recent results obtained by the Zasadzinski team show that the release of two important drugs, ciprofloxacin (an antibiotic) and vinblastince (for chemotherapy) is extended by factors that may be as large as an order of magnitude by using vesosomes in comparison with using unilamellar liposomes in a very similar way. We expect that work in the future will lead to important medical applications in vivo.

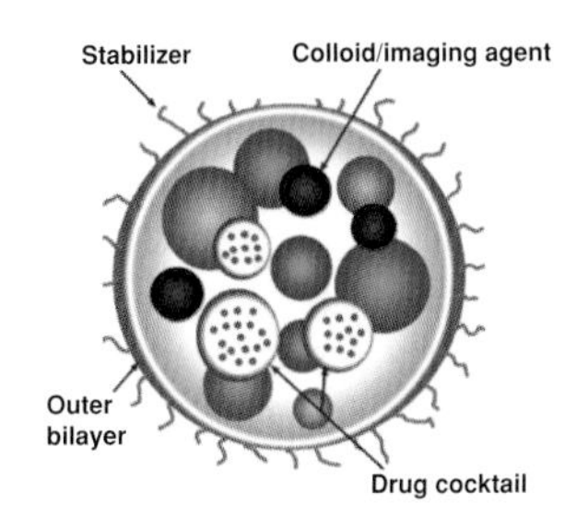


Figure 16.25. A schematic diagram of the vesosome – or liposomes inside a vesicle drug carrier. (Courtesy of Joe Zasadzinski.)

## 16.6 Protein assembly in vitro

An important goal in biophysics is the understanding of interactions leading to supramolecular structures of cytoskeletal proteins and associated biomolecules, and, most importantly, the elucidation of the roles that the structures play in cell functions. The cell cytoskeleton comprises three negatively charged filamentous proteins, which include filamentous actin of diameter 8.5 nm; the intermediate filaments, 10 nm in diameter; and microtubules, 25 nm in diameter. The cytoskeleton is involved in a range of cell functions including mechanical stability, cell locomotion, intracellular trafficking and signal transduction. Because of space limitations we will describe as a model system some recent work on interactions and structures in supramolecular assemblies of the actin cytoskeletal protein under in-vitro conditions.

We show in Figure 16.27 (right-hand side) laser scanning confocal microscope images of cultured mouse fibroblast cells spread on a glass cover plate where filamentous actin has been stained to view the actin cytoskeleton. Figure 16.27 presents a full three-dimensional stereo image created with confocal software, which can be seen by using color glasses. The actin cytoskeleton provides a structural framework for the mechanical stability of eukaryotic cells and a range of cell functions including adhesion, motility, and division. In cells, actin is found both in a monomeric globular (G) actin state and as polymerized filamentous (F) actin. Bundles, comprised of closely packed parallel arrangements of F-actin, and networks, containing F-actin chains criss-crossed at some large angle, form the most common known supramolecular structures in cells (Figure 16.27, left-hand side). Interactions between F-actin and distinct actin cross-linking proteins may lead to two-dimensional networks and bundles of F-actin

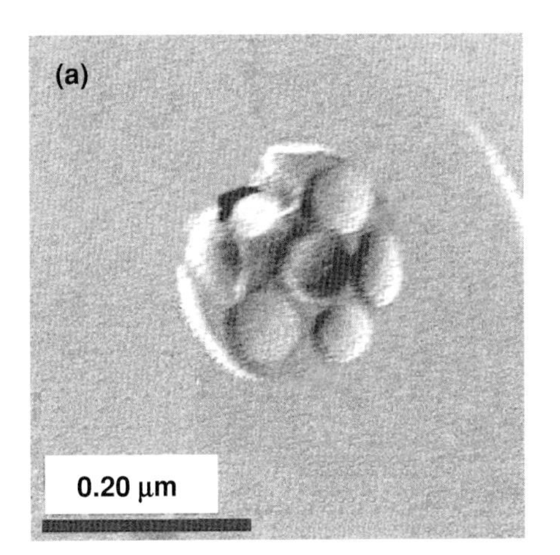


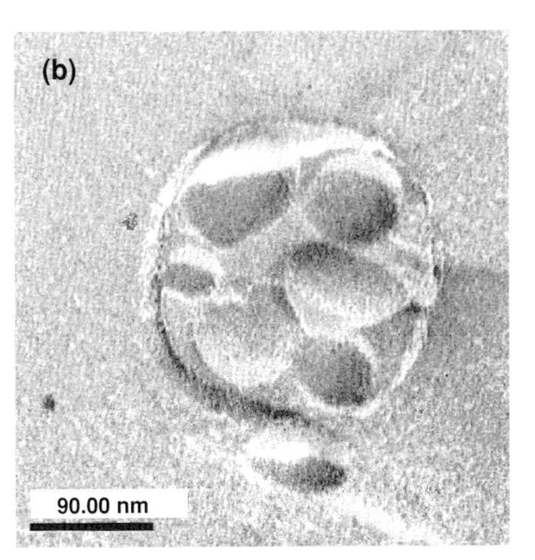


Figure 16.26. Freeze-fracture electron micrographs of two examples of vesosomes made from interdigitated DPPC sheets encapsulating DPPC interior vesicles. These structures can be made to be sufficiently small that they will circulate in the bloodstream without being removed by the immune system. DPPC is the zwitterionic lipid dipalmitoyl-phosphatidyl choline. (Courtesy of Joe Zasadzinski.)

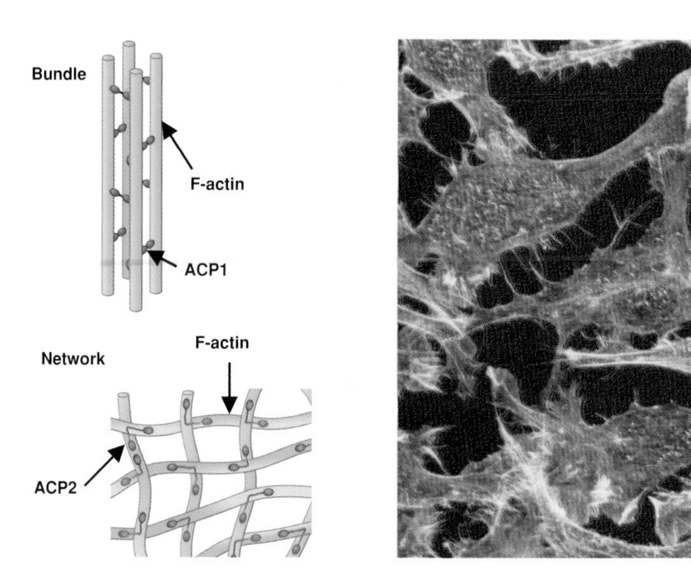


Figure 16.27. Left: a schematic drawing showing how different actin cross-linking proteins (ACPs) may cross-link actin filaments into a bundle or a network. Right: a laser scanning confocal-microscope three-dimensional image of fibroblast mouse cells fixed and stained (using Alexa-Fluor 488 Phalloidin) to visualize filamentous actin as described in the text. The figure represents a full three-dimensional stereo image created with confocal software, which can be seen by using standard color glasses for red/blue, red/green three-dimensional pictures (available from Rainbow Symphony, Inc., www.rainbowsymphony.com., or confocal-microscope companies like Zeiss and Leica). (Courtesy of Alison Lin.)

interacting with the plasma membrane to determine cell shape, or three-dimensional networks of F-actin imparting gel-like properties to the cytosol. F-actin filaments may associate with the plasma membrane, forming a network for regulation of the distribution of membrane proteins.

In Figure 16.27 (right-hand side) actin bundles can be seen traversing the length of two cells near the center of the image. These bundles are components of stress fibers, which end in focal adhesion spots responsible for cell adhesion, for example, to the extracellular matrix in the space between cells in vivo. Another cell can be seen in the top-right-hand quadrant of the image (a spherical doublet) during cytokinesis as cell division proceeds. A contractile ring of actin, comprised of F-actin bundles, myosin-II motors and associated biomolecules, begins to constrict the middle section of the cell (the depressed region) by using chemical energy, which eventually leads to the splitting of the cell in two, the fundamental process of cell growth. The stress fibers and contractile ring are comprised of F-actin as a key protein, yet, in the presence of other biomolecules, they are assembled into different supramolecular structures, each designed for the distinct functions of cell adhesion and cell splitting.

An interesting observation is that the dynamical behavior of F-actin can be seen in this static image: hours before this cell culture was fixed in preparation for imaging, the doublet cell undergoing division had been spread out similarly to the other cells, with stress fibers attaching the cell to the glass substrate. Upon entering the cell cycle, the stress fibers consisting of actin filaments were broken down with the help of other proteins. During the process of cytokinesis (the present image) long actin filaments were reformed as part of the contractile ring.

Much remains to be studied about the precise structural nature of these various F-actin supramolecular self-assemblies and the solution conditions which lead to the formation of various types of bundles, networks, and other new types of self-assemblies. Over the last two decades the groups of Paul Janmey (Harvard Medical School and the University of Pennsylvania) and Eric Sackmann (Technical University at Munich) have been responsible for a series of landmark experiments on the rheological properties of the actin cytoskeleton in vitro. Sackmann and co-workers have pioneered methods of making local rheological measurements in cells using magnetic beads. Although the viscoelastic properties of the F-actin cytoskeleton are crucial to cell survival and function, until recently very little was understood about them under physiological conditions in vitro. Fred MacKintosh of the Free University in Amsterdam has shown that classical rubber elasticity is not able to account for the unusually large shear moduli observed in experiments on actin gels by Janmey and Sackmann. Using a modern approach, MacKintosh developed a powerful model of semi-flexible actin gels, which is able to account for the unusual rheological behavior.

Because mechanical and rheological properties of cells are ultimately dependent on the precise structural state of the cytoskeleton, which often undergoes dynamical remodeling, the elucidation of the supramolecular structures of cytoskeletal proteins and associated biomolecules under physiological conditions constitutes a very important direction of continued biophysical research. Two examples will be described below to serve as a reminder that we are just in the early stages of uncovering new structures.

### 16.6.1 Examples of in-vitro structures of F-actin: stacked two-dimensional rafts and networks of bundles

In elegant light-scattering studies in the late 1990s, Paul Janmey and Jay Tang (Brown University) had shown that F-actin formed bundled phases in the presence of simple counter-ions such as $Ca^{2+}$, $Mg^{2+}$, and $Mn^{2+}$, which are electrolytes of biological relevance. Recent synchrotron X-ray-diffraction experiments carried out over the last few years by Gerard Wong (University of Illinois at Urbana Champaign), Youli Li (UCSB), Paul Janmey, Jay Tang, and the author have revealed that these simple counter-ions may modulate the interactions between F-actin molecules in a highly nontrivial manner, leading to phases consisting of distinct supramolecular assemblies. Figure 16.28(a) shows optical capillaries containing F-actin rods of length $\approx 1000$ Å in the presence of increasing electrolyte counter-ion concentration. From top to bottom the $Ca^{2+}$ concentrations are 2.5, 9.0, and 80 mM. The 9.0-mM sample has condensed into a translucent phase, whereas the 80-mM sample has condensed into an opaque phase. The same three samples observed in a light microscope under crossed polarizers are shown in Figures 16.28(b)–(d). The 2.5-mM uncondensed sample (b) is an isotropic phase, with no observable birefringence. In contrast, the observed birefringence with characteristic texture for the 9.0-mM (c) and 80-mM (d) samples reveals their liquid-crystalline nature.

At high concentrations of multivalent ions (80-mM), F-actin condenses into bundles of parallel, close-packed rods (see Figure 16.27, with ACP1 replaced by small ions). The synchrotron X-ray-diffraction studies of the intermediate-concentration phase (9 mM) by Wong *et al.* led to the surprising finding of a lamellar structure with a layer spacing nearly twice the F-actin hard-core diameter. The full structural analysis led to a model consisting of lamellar stacks of cross-linked two-dimensional F-actin rafts (labeled $L_{XR}$, Figure 16.29). The unit-cell structure of crossed sheets of F-actin in the $L_{XR}$ phase is

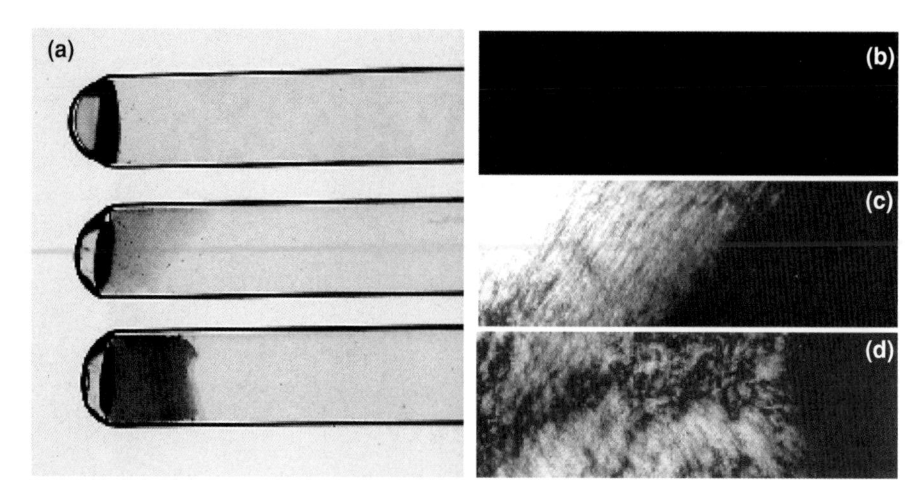


**Figure 16.28.** (a) Condensation of F-actin filaments (2 mg ml$^{-1}$, average length ~100 nm) with $CaCl_2$: the $CaCl_2$ concentrations are 2.5, 9.0, and 80 mM for the top, middle, and bottom capillaries, respectively. A condensed phase (the phase boundary can be observed) exists in the middle capillary at intermediate concentrations (a cartoon of this structure is shown in Figure 16.29). The bundled phase can be seen in the bottom capillary. (b)–(d) Polarized micrographs taken from the top, middle, and bottom capillaries, respectively. Reprinted with permission from G. C. L. Wong *et al.*, *Physical Review Letters* 91, 018103 (2003). © 2003 by the American Physical Society.

consistent with electrostatic considerations: Adrian Parsegian and co-workers showed in 1974 that repulsion between like-charged rods is maximal for parallel rods and minimal for rods crossed at 90°.

In a model published in 2001, Itamar Borukhov of UCLA and Robijn Bruinsma showed that, at low concentrations, below what is required for bundling, counter-ions will localize to actin crossings, in effect producing adhesion between two crossed filaments and potentially stabilizing the $L_{XR}$ phase. Significantly, the $L_{XR}$ structure may be relevant at physiological ionic strengths. The in-vitro data of Wong *et al.* show that about 3 mM $Mg^{2+}$ is required to condense 3000-Å F-actin into the $L_{XR}$ raft phase. For 1-μm and 10-μm filaments, 5 mM and 11 mM $Mg^{2+}$ are required. $Mg^{2+}$ levels within the cell can be as much as 10 mM.

Although more theoretical work is required before a full understanding of the stability of the lamellar raft ($L_{XR}$) emerges, the physics of attractive forces between like-charged polyelectrolytes leading to bundle formation in the presence of small counter-ions is better appreciated. The first step in understanding attractive forces between polyelectrolytes with the same charge is Manning condensation, whereby counter-ions are bound to the polyelectrolyte backbone when the Manning parameter $\xi = l_B/l_{actin} = 2.84$ is greater than unity. Here, $l_{actin} = 2.5$ Å for F-actin is the distance between the negative charges on the polyelectrolyte backbone (refer back to Figure 16.15). Counter-ion condensation strongly suppresses the repulsive electrostatic interactions between the polymers with increasing counter-ion valence $Z$ by reducing the effective charge per unit length to a fraction $1/(Z\xi)$ of its bare charge per unit length. In groundbreaking theoretical studies more than 30 years ago, Fumio Oosawa of the Institute of Molecular Biology at Nagoya University predicted that fluctuations in the bound counter-ions of neighboring chains are correlated over large distances, producing a long-range attractive force similar to an effective van der Waals attraction and falling off as $1/(\text{inter-axial distance})^2$.

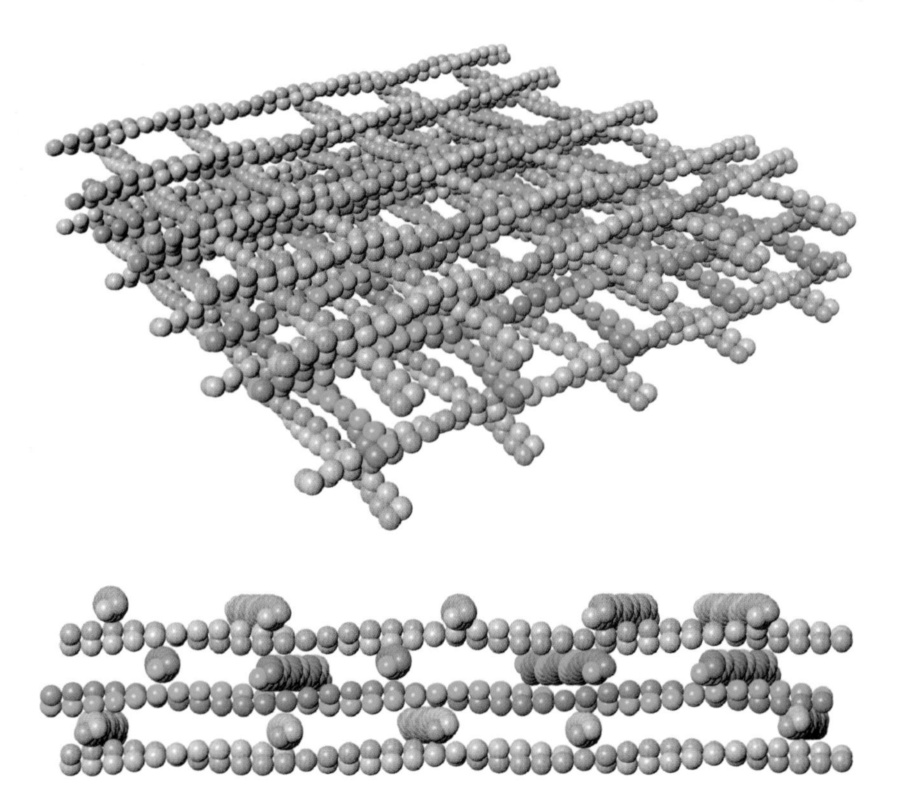

**Figure 16.29.** Two views of a model of the $L_{XR}$ phase in which rafts of F-actin rods (comprised of polymerized globular monomeric actin units) are cross-linked into a one-dimensional lamellar stack. One of the cross-linked bilayer rafts (i.e. a unit cell of the three-dimensional structure) has been highlighted in blue for clarity. Counter-ions (not shown for clarity), are localized to the crossing regions between actin rods. Reprinted with permission from G. C. L. Wong *et al.*, *Physical Review Letters* 91, 018103 (2003). © 2003 by the American Physical Society.

Recent numerical simulations and theories published in the late 1990s by Niels Gronbech-Jensen, Jay Mashl, Robijn Bruinsma, and William Gelbart at UCLA, and, independently, by the groups of Andrea Liu and Boris Shklovski at UCLA and the University of Minnesota, respectively, show that bound counter-ions on adjacent DNA chains develop positional correlations (like a one-dimensional Wigner crystal lattice at $T = 0$), resulting in a short-range exponentially decaying attractive force between chains, with strength and range increasing as $Z^2$ and $Z$, respectively. Liu's theory further predicts that kinetics limits the diameter of bundles, which are experimentally observed to be of finite size.

When filamentous actin is allowed to mix in-vitro with $\alpha$-actinin, an actin cross-linking protein purified from cells, a remarkable new type of biologically inspired polymer network with fundamentally new properties is spontaneously formed. Three-dimensional laser scanning confocal microscopy work by Olivier Pelletier, Linda Hirst, and Elena Pokidysheva in the author's laboratory has revealed a network of bundles on the micrometer scale (Figure 16.30). This is in contrast to the commonly observed network of single filaments observed in cells (Figure 16.27, bottom left). The branching (bifurcation) of the bundles on this mesoscale is evident and leads to a well-defined mesh size. X-ray-diffraction studies of the network allow one to look at the protein building blocks inside the bundles on the ångström-unit scale. The resulting structure shown schematically in Figure 16.31 reveals a distorted square lattice (with a unit-cell dimension of 30 nm) rather than the expected triangular lattice.

We expect future measurements of the mechanical properties of this network of bundles, using microrheology methods, to reveal length-scale-dependent elastic behavior. On the mesoscale, elastomer-like elasticity of the network should be observed. On the small scales within the bundle $\approx$250 nm we expect a nonzero shear and compression moduli resulting from the finite-sized quasi-two-dimensional lattice within the bundles. We point out that because the linkers form non-covalent bonds their position

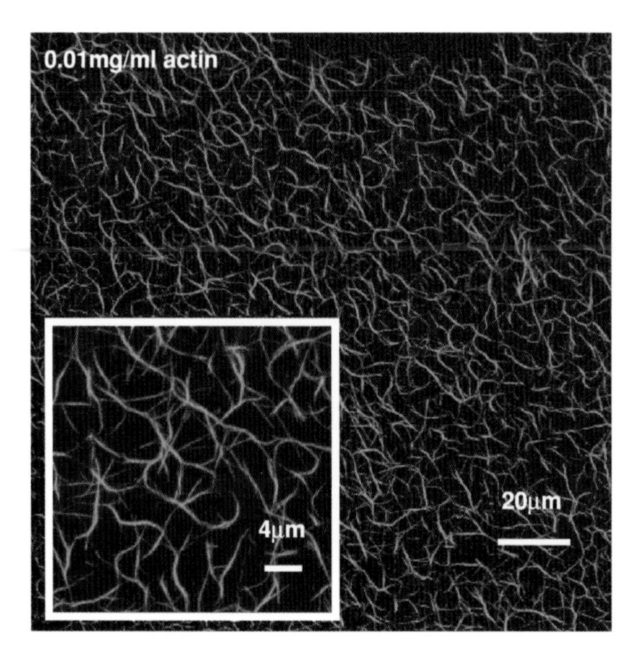


Figure 16.30. A laser scanning confocal-microscope image of a network of actin bundles induced by the actin cross-linking protein α-actinin. Reprinted with permission from O. Pelletier *et al.*, *Physical Review Letters* 91, 148102 (2003). © 2003 by the American Physical Society.

within the bundle is expected to be transient; thus, the dynamics of the branch points is expected to be quite interesting.

The two in-vitro structures described here (Figures 16.29 and 16.31) represent examples of entirely new supramolecular assemblies of F-actin since the bundled and network phases of single filaments (Figure 16.27, left-hand side) were described many years ago. These results suggest the rich variety of new supramolecular structures that await discovery by future researchers. We should expect many to be biologically

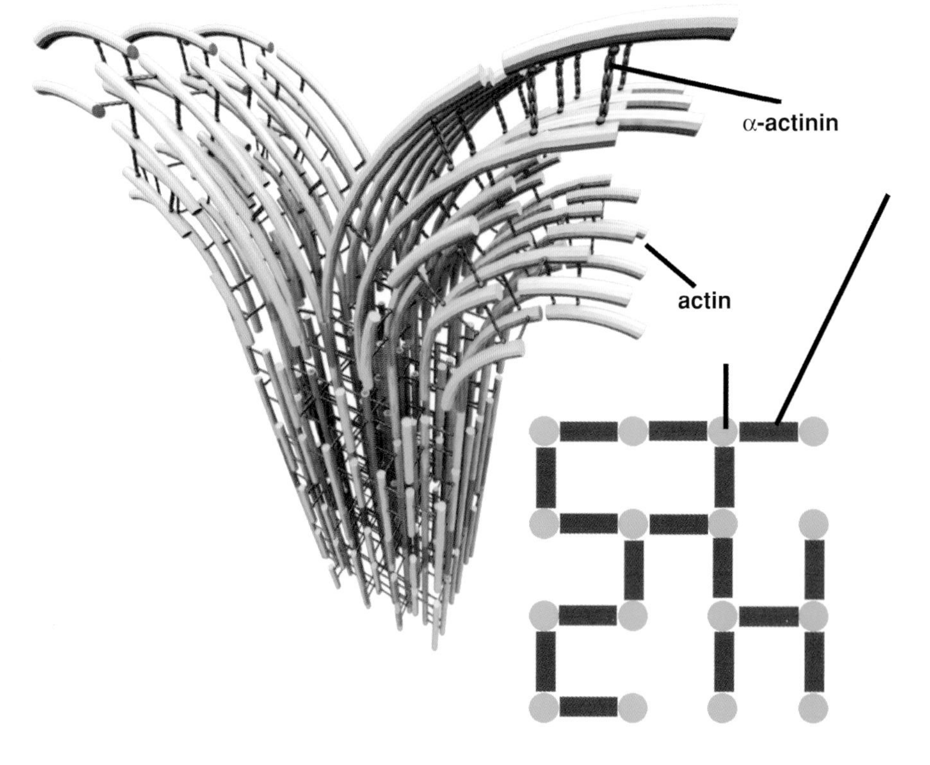


Figure 16.31. A model of the proposed structure of an F-actin bundle at a branching site, where F-actin is cross-linked with α-actinin. A cross-sectional projection of the bundle is also shown. X-ray diffraction for such bundles indicates a finite-sized disordered lattice with distorted square symmetry. Reprinted with permission from O. Pelletier *et al.*, *Physical Review Letters* 91, 148102 (2003). © 2003 by the American Physical Society.

relevant; that is, structures that are selected by the complex regulatory apparatus of the cell for particular functions.

## 16.7 Future work

With the unraveling of the human genome and the emerging proteomics era, the biophysics community is now challenged to elucidate the structures and functions of a large number of interacting proteins. Our understanding of modern biophysics has been increased significantly by our knowledge of the structures of single proteins determined by macromolecular crystallography, which normally provides clues relating structure to biological function. While genetic and biochemical studies will continue to reveal function, we believe that the elucidation of structure in self-assembling biological systems and protein networks will require interdisciplinary approaches. The simultaneous application of X-ray- and neutron-scattering and -diffraction techniques combined with electron microscopy and optical imaging, probing length scales from ångström units to many micrometers, should lead to the elucidation of the supramolecular structures. An example from the nerve cell serves to illustrate the point.

Two interacting nerve cells are shown schematically in the top part of Figure 16.32. We show in Figure 16.32 (middle) an electron micrograph of a longitudinal cut exposing a view of the cytoskeleton of a mouse nerve-cell axon taken by the neuroscientist Nobutaka Hirokawa of the School of Medicine at the University of Tokyo. One microtubule with attached vesicles is clearly visible near the middle. Microtubules act as tracks along which enzymes and precursor neurotransmitters are transported inside vesicles toward the ends of axons near an axon–dendrite or axon–cell-body synapse (see the top of Figure 16.32), for transmission of the electrochemical signal between nerve cells. In Hirokawa's marvelously clear image, molecular motors (the arrow points to one) may be seen transporting a large vesicle and a small vesicle. The microtubule is surrounded by neurofilaments (NFs), which are abundant in axons of vertebrate nerve cells and are responsible for structural stability of the long nerve-cell axons, for example, in the spinal chord. NFs are heteropolymers consisting of three distinct fractions, which assemble to form a core filament and long sidearm carboxy tails that extend away from the axis of the NF (Figure 16.32, bottom).

The image also shows extensive cross-bridging between neighboring NFs resulting from sidearm interactions and also interactions between neurofilaments and the microtubule. In-vivo expression levels of the three NF fractions are regulated over a narrow range of ratios. Over – or under – expression of NF fractions may be related to abnormal accumulation of NF fractions and NF-network disruptions, which are known hallmarks of motor neuron diseases such as amyotrophic lateral sclerosis (ALS). In many neurodegenerative diseases, microtubules and their assemblies have their structures disrupted, with detrimental consequences for the transport of neurotransmitters toward the synapse. Microtubules are stabilized by members of the class of microtubule-associated proteins (MAPs). In one well-known case it has been shown that overly phosphorylated MAP tau proteins unbind from microtubules and form paired helical filaments, which in turn aggregate to form neurofibrillary tangles implicated in Alzheimer's disease. Similar deposits are found in a number of other neurodegenerative diseases including ALS, Pick's disease, and Down's syndrome.

The careful work of neuroscientists, geneticists, and biochemists, over the past several decades, has led to the identification of a large group of key proteins and also the suspected changes in these proteins that lead to neurodegeneration and cell death. However, even though many key proteins and enzymes have been identified, we still have very little knowledge of the full three-dimensional supramolecular structures, which

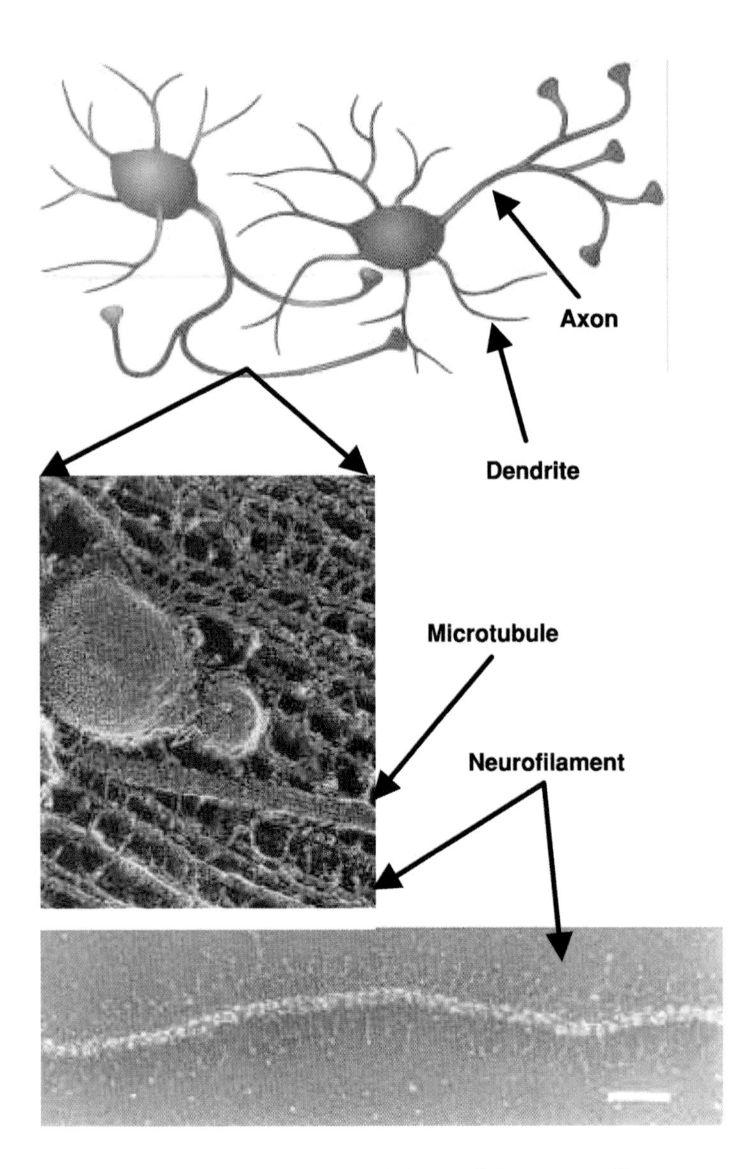


**Figure 16.32.** Top: two interacting nerve cells. The axon from the left-hand cell is forming synaptic junctions with the cell body and dendrite of the cell on the right. Middle: a rapid-freeze–deep-etch electron micrograph of a mouse nerve axon showing a single microtubule (MT, thick filament, lower part of image) with a transporting vesicle moving via molecular motors along the MT surrounded by neurofilaments (NFs). Extensive cross-bridging between NFs is readily seen. Also evident is the cross-linking between the NFs and the MT. Bottom: a rapid-freeze–deep-etch electron micrograph of a single NF. The sidearms are clearly visible. The bar is 50 nm. Reprinted in part from *Trends Cell Biol.* 6, N. Hirokawa, "Organelle transport along microtubules – the role of KIFs," 135–141, © 1996, with permission from Elsevier and Nobutaka Hirokawa.

are dynamically assembled when the cytoskeleton of the nerve cell is constructed and repaired throughout life. What do these protein structures look like and how do they fall apart during aging or under diseased conditions? What are the structures formed in neurofilament, G-actin, and tubulin mixtures in the presence of native MAPs and other associated proteins under in-vitro physiological conditions? The answers to these questions should allow us to determine the conditions which lead to destabilization and degeneration of the cytoskeleton, and to formulate strategies to prevent damage or repair damaged protein structures. By relating structure to function, the ultimate goal of molecular neuroscience is to reveal the inner workings of nerve cells at the molecular and supramolecular levels.

Future biophysical research will also undoubtedly lead to novel new biomolecular materials. Indeed, the lessons learned from Nature in in-vitro studies should lead to the application of similar concepts to the development of nanoscale miniaturized materials. An example involving F-actin combined with cationic lipids will be described. Hierarchical assembly, which is ubiquitous in Nature (e.g. in bone structure), is rapidly becoming the inspiration for inventing new materials. Using confocal microscopy and synchrotron X-ray diffraction, Gerard Wong, Alison Lin, Youli Li, Jay Tang, Paul

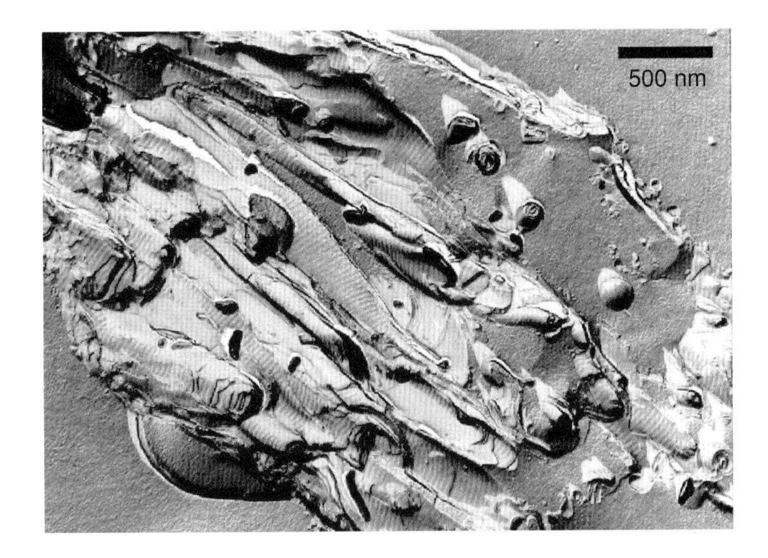


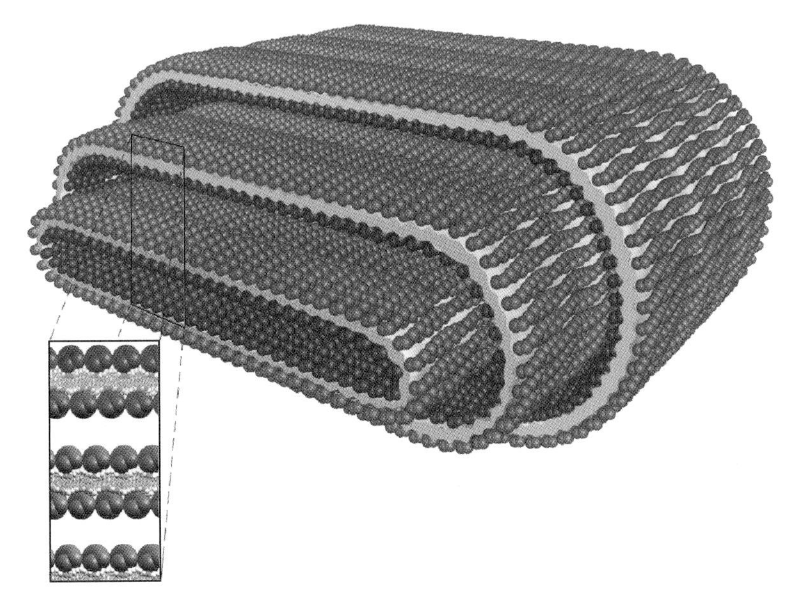

**Figure 16.33.** Top: a freeze-fracture electron micrograph of tubules of cationic liposome–F-actin complexes. The 36-nm corrugations visible on the fracture plane correspond to half the pitch of twisted F-actin. Bottom: a schematic model of cationic liposome–F-actin complexes showing the lipid bilayer (yellow) sandwiched between two layers of F-actin (blue, twisted strands), forming a three-layer composite membrane (inset). Reprinted with permission from G. C. L. Wong *et al.*, *Science*, **288**, 2035 (2000).

Janmey, and the author have found that cationic liposomes and F-actin form complexes consisting of a network of tubules, on the micrometer length scale, with the interior of the tubules consisting of a stack of three-layer membranes on the nanometer scale (Figure 16.33, bottom). The three layers consist of a lipid bilayer sandwiched between two monolayers of F-actin.

Freeze-fracture electron microscopy of the complexes shows a collection of the flattened multi-lamellar tubules with an average width of $\approx 0.25$ μm (Figure 16.33, top). A surface corrugation along the tube axis with a periodicity of $\approx 36$ nm can be seen on the fracture plane. This corresponds to the well-known (half) long-wavelength twist of F-actin. The persistence of these corrugations along the width of the tubules implies that the F-actin rods on the membrane are locked into a two-dimensional lattice, producing the model shown in Figure 16.33. We expect that the assembling forces leading to the formation of tubules are universal. Thus, for example, by replacing muscle actin and lipids with polymer analogs one may process robust polymer nanostructured tubules with applications in templating and chemical encapsulation and delivery.

A final example of an area of importance in future work is that of developing methods to make human artificial chromosomes (HACs) a readily available tool. The viability of HACs was demonstrated when, in 1997, H. F. Willard of Case Western Reserve University in Cleveland, Ohio and co-workers successfully transferred, although extremely inefficiently, sections of HACs of order a million base pairs into cells using cationic liposome-based carriers described earlier in this chapter. One of the most important advantages of nonviral over viral methods for gene delivery is the potential of transferring and expressing large pieces of DNA in mammalian cells. Current viral vectors have a maximum gene-carrying capacity of 40 000 base pairs. The future development of efficient HAC carriers will be extremely important for gene therapy. Because of their very large size capacity HACs would have the ability of delivering not only entire human genes, which in many cases are of sizes that exceed 100 000 base pairs (including exons and introus), but also their regulatory sequences, which are needed for the correct spatial and temporal regulation of gene expression. Aside from the medical and biotechnological ramifications in gene therapy, the development of efficient HAC vectors will also enable studies designed to characterize chromosome structure and function in complex organisms. Finally, HACs could be utilized in non-medical applications to produce designer proteins and peptides for applications in biomolecular materials.

The field of supramolecular biophysics and biomolecular materials offers the new generation of researchers the opportunity to work on a range of exciting problems. However, the likelihood of completely solving these problems is indeed very low due to the extreme complexity of biologically related research. Nevertheless, even small progress toward their solution tends to lead to large gains from a scientific and especially biomedical perspective. The interdisciplinary nature of the field means that significant progress requires a thorough understanding of the relevant biological issues in addition to the day-by-day collaboration and exchange of ideas among physical scientists, biologists, and medical researchers.

## Acknowledgements

The author would like to acknowledge support from the National Science Foundation and the National Institutes of Health of the USA. The author is deeply indebted to his wife Muriel Safinya for her continued support over many years.

## FURTHER READING

1. C. R. Safinya, Structures of lipid–DNA complexes: supramolecular assembly and gene delivery, *Current Opinion Struct. Biol.* **11** (4) (2001) 440. See also www.mrl.uscb.edu/safinyagroup.
2. A. Peters, S. L. Palay, and H. D. E. F. Webster, *The Fine Structure of the Nervous System*, 3rd edn., New York, Oxford University Press, 1991.
3. D. Bray, *Cell Movements: From Molecules to Motility*, 2nd edn., New York, Garland, 2001.
4. Lipid Biology (focus on lipids), *Science*, November 30, 2001.
5. R. I. Mahato and Sung Wan Kim (eds.), *Pharmaceutical Perspectives of Nucleic Acid-Based Therapeutics*, London, Taylor & Francis, 2002.
   This recent publication contains an up-to-date description of biophysical and biochemical aspects of DNA carriers for biomedical applications.
6. W. M. Gelbart, R. F. Bruinsma, P. A. Pincus, and V. A. Parsegian, DNA-inspired electrostatics, *Phys. Today*, September 2000, p. 38.

7. R. Lipowsky and E. Sackmann (eds.), *Structure and Dynamics of Membranes*, New York, North-Holland, 1995.
8. W. M. Gelbart, A. Ben-Shaul, and D. Roux (eds.), *Micelles, Membranes, Microemulsions, and Monolayers*, New York, Springer-Verlag, 1994.
9. S. A. Safran, *Statistical Thermodynamics of Surfaces, Interfaces, and Membranes*, Reading, MA, Addison-Wesley, 1994.

## Cyrus R. Safinya

Cyrus R. Safinya is a professor in the departments of Materials, Physics, and Molecular, Cellular, and Developmental Biology at the University of California at Santa Barbara. He obtained a B.S. in physics and mathematics from Bates College in 1975 and a Ph.D. in physics from the Massachusetts Institute of Technology in 1981 under the supervision of Robert J. Birgeneau. In 1981, he joined the Exxon Research and Engineering Company in New Jersey and conducted research on the structure of complex fluids and biological membranes. He joined the faculty of the University of California at Santa Barbara in 1992. His group's research is focused on elucidating structures and interactions of supramolecular assemblies of biological molecules. This includes understanding structures of DNA condensed by oppositely charged molecules in vitro and the relationship with condensed DNA in vivo, and developing a fundamental understanding of interactions between cell cytoskeletal proteins and their associated molecules that lead to their distinct structures on the ångström-unit scale to the many-micrometer scale. An important goal of the research is to relate the structures to cellular function. A major project of the group is the development of lipid carriers of DNA for gene delivery. He is the author or co-author of about 130 publications. He initiated the Gordon Research Conference on Complex Fluids in 1990 and the Materials Research Society Meeting on Complex Fluids in 1989. Dr. Safinya was awarded a Henri De Rothschild Foundation Fellowship by the Curie Institute in 1994. He was elected a Fellow of the American Physical Society in 1994 and of the American Association for the Advancement of Science in 1997.

# 17 Medical physics

Nikolaj Pavel

## 17.1 Introduction

Physics ultimately rules the processes in living organisms and thus is in that sense fundamental for understanding medicine. To cite some examples involving different branches of physics:

- the dynamics of forces in joints, e.g. the dependence of stress in an articulation → mechanics;
- the microstructure of bones and the role of compounds with high elasticity and high tensile strength → solid-state physics of articulations;
- control of blood circulation and the variable viscosity of blood and blood plasma → hydrodynamics;
- passive molecular transport through membranes via osmosis → thermodynamics;
- signal conduction in nerve cells → electrodynamics;
- image formation on the retina → optics; and
- hearing → acoustics and mechanics.

Many physical properties of tissues, substances, cells, and molecules and of their mechanisms of operation are exploited for diagnostics and therapy (e.g. ultrasonic imaging, electro- or magneto-encephalography, high-frequency electromagnetic-radiation therapy). Rather than these manifold relationships between physics and the phenomena of medicine and biology, medical physics today covers that part of physics in which new phenomema are exploited and new techniques developed explicitly for use in diagnostics and treatment. The largest area has to do with diagnostic methods, the most prominent being transmission radiography with X-rays, which was introduced soon after the discovery of X-rays by Röntgen in 1895. Many sections of this chapter cover the principles and recent developments of diagnostic tools, weighing their virtues as well as possible disadvantages (e.g. adverse side effects). The various techniques also provide complementary information. This is illustrated in the section on brain research, one of the most exciting areas of modern medicine and biology. The other important field of medical physics is therapy, where the focus is on those techniques with direct physical implications. As well as external or internal gamma-ray or particle

*The New Physics for the Twenty-First Century*, ed. Gordon Fraser.

> **BOX 17.1 TYPICAL ENERGIES FOR MEDICAL PHYSICS**
>
> Visible-light photons: 2–3 eV
> X-ray photons: 3–100 keV
> Gamma-ray photons and particles: 50 keV–5 MeV

radiation, therapies also involve treatment with acoustic shock waves and electromagnetic waves of relatively low frequencies. Covering all this in depth is not possible here, so this chapter focuses on the use of particle radiation (gamma rays, protons, ions) in therapy (see Box 17.1) and on the use of lasers.

In a somewhat broader sense, the structural analysis of macromolecules, which nowadays is done with synchrotron light, also belongs to medical research. Such information provides important input for drug development, where earlier trial-and-error or educated-guess approaches are being increasingly replaced or at least supplemented by systematic identification of promising candidates for active substances.

## 17.2 The future of radiology – low dose versus contrast

Transmission X-ray radiography, which has been used for over 100 years, is based on the partial absorption of X-rays in material, which depends on thickness ($x$) and the material-dependent absorption length ($\lambda$) through d'Alembert's law (Figure 17.1),

$$I(x) = I(0)\exp(-x/\lambda), \tag{17.1}$$

which describes the exponential decrease of beam intensity with thickness $x$.

Variations of integrated transmission length and $\lambda$ for different kinds of tissue – i.e. the variation of the absorption contrast – make for a variation of the integrated

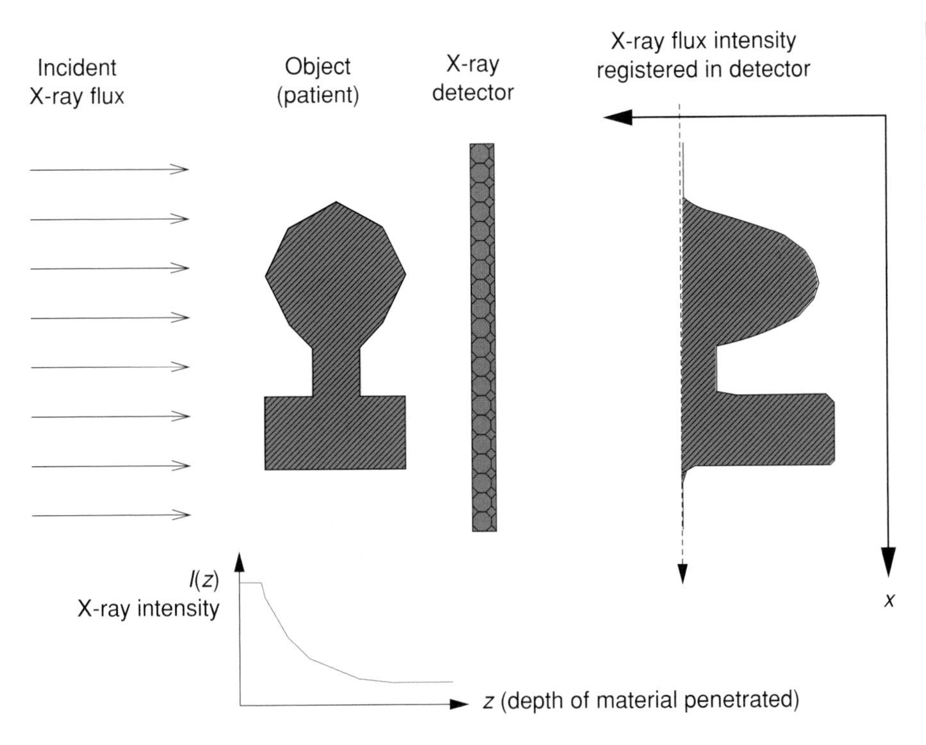


Figure 17.1. The principle of transmission radiography. The incident photon flux decreases exponentially when traversing material. The effect depends on the kind of material and its thickness. The recorded flux of photons behind the object is the result of the integrated attenuation of the incident flux and produces a projection image of the structure of the material (patient's body) in the X-ray beam.

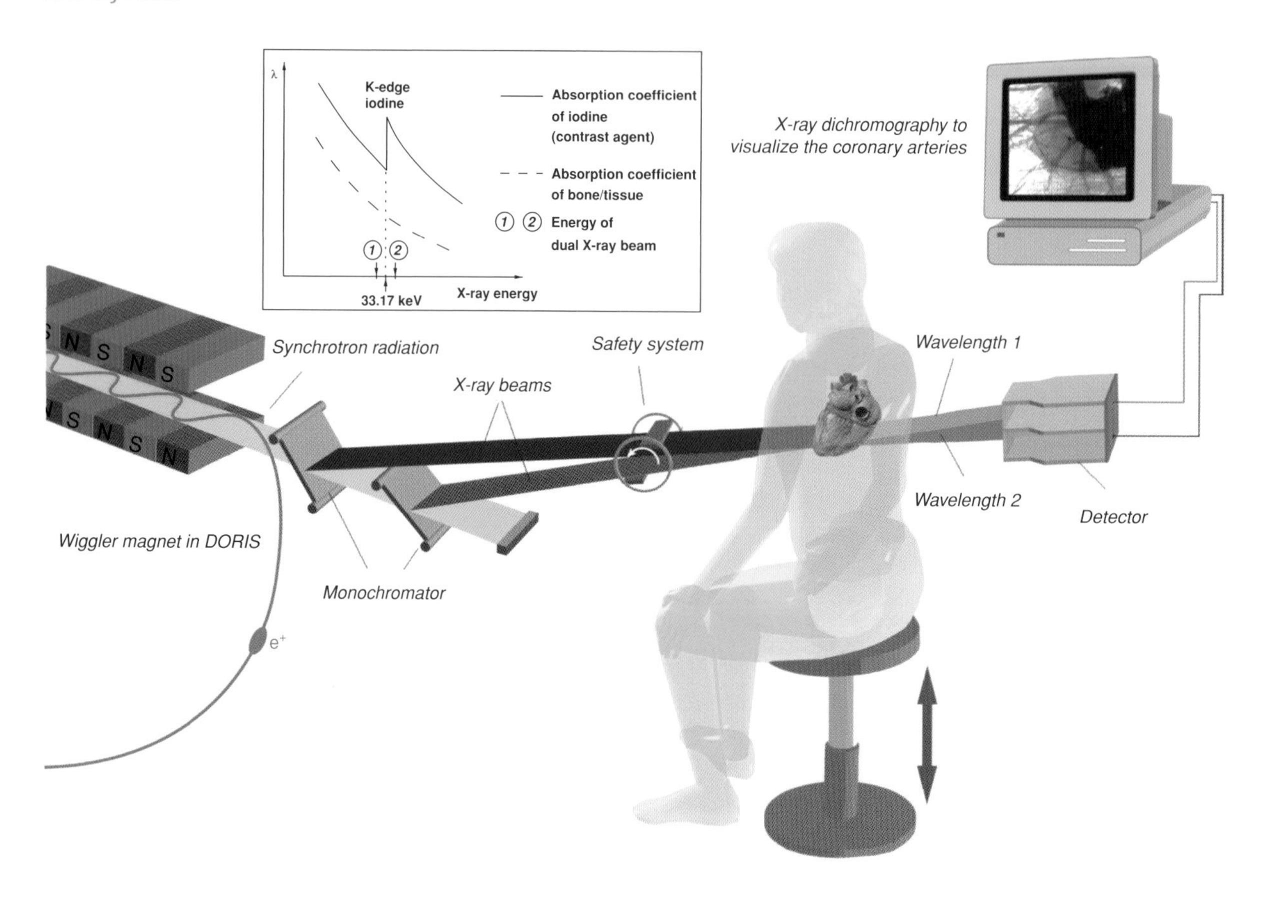


**Figure 17.2.** The principle of non-invasive coronary angiography with two monochromatic synchrotron-radiation beams with energies just below and above the K-edge in the X-ray-absorption coefficient of the contrast agent iodine. The beams, with a flat fan shape, cross a line segment of the heart. By moving the patient up and down, the whole heart is investigated. The energy dependence of the absorption coefficient of iodine as well as those of bone and soft tissue are shown in the energy range around the K-edge of iodine. (Courtesy of H. J. Besch and A. H. Walenta.)

photon flux recorded in an imaging detector, which can be expressed as the variation of the signal-to-noise ratio (SNR) of the imaging detector, $\Delta(\mathrm{SNR})$. Using computer tomography (CT), three-dimensional images can be reconstructed from a series of projected images. The challenge in radiography is that of how to detect small effects in a low-contrast environment while minimizing the dose. Because of the random statistics of the absorption process, $\Delta(\mathrm{SNR})$ decreases with the integrated number of photons in the image pixel – the radiation dose. Therefore high sensitivity for features with low absorption contrast requires high radiation doses:

$$\text{required dose} = \frac{1}{\mathrm{DQE}}(\Delta(\mathrm{SNR}))^2 \frac{1}{C_\mu^2}\frac{1}{d^4}, \tag{17.2}$$

where $C_\mu$ is the absorption contrast, $d$ the spatial size of the object to be identified, and DQE the detector's quantum efficiency. To decrease the required dose, meaning less risk for the patient, while maintaining high sensitivity, new techniques are being used, two of which will be described here because they involve interesting physics principles. The first is radiography with a contrast agent, which is used in coronary angiography and in gastro-intestinal investigations. The contrast agent with its much higher X-ray-absorption coefficient has to be injected with a sufficiently high concentration and the transmission images must be recorded within a short time. This is crucial for heart investigations, where the contrast agent has to be injected through a catheter (selective coronary angiography). Such invasive investigations have considerable disadvantages (stress, preparatory and subsequent treatment, complications).

Non-invasive methods, where the contrast agent is injected into the arm, require detection techniques more than 50–100 times more sensitive than those of

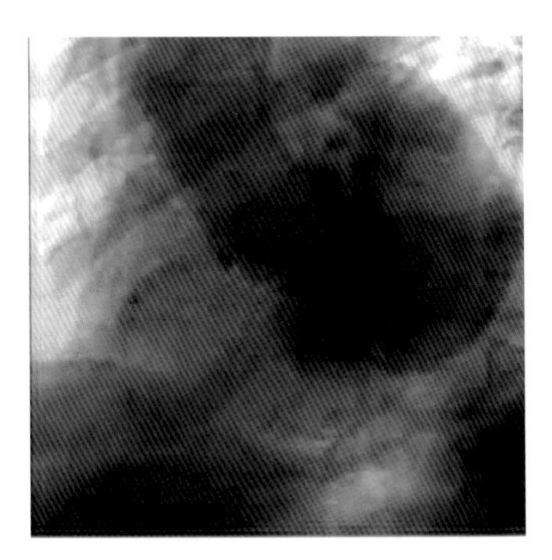
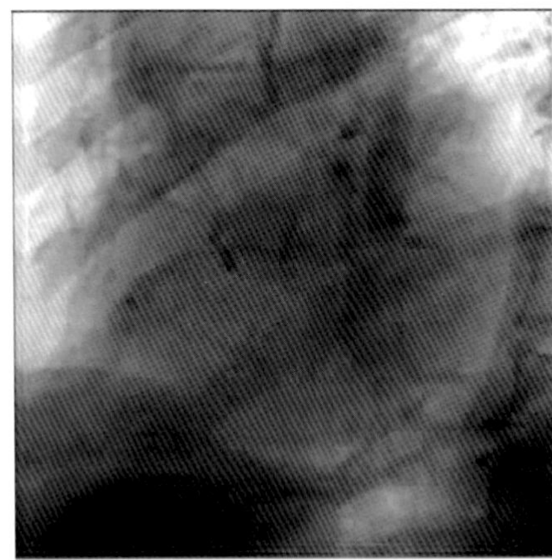
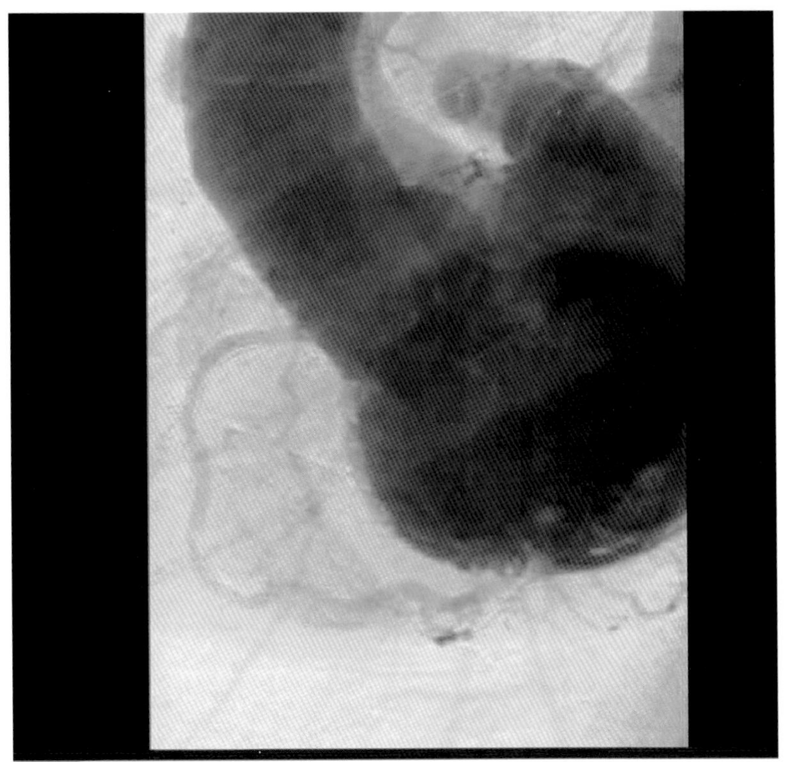

**Figure 17.3.** X-ray transmission images recorded at energies just below (left) and above (right) the K-edge of iodine. Below, the corresponding subtraction image is shown. (Courtesy of H. J. Besch and A. H. Walenta.)

conventional detectors. Non-invasive coronary angiography has been performed successfully on several hundred patients using a synchrotron beamline at HASYLAB (Hamburg).

The set up is shown schematically in Figure 17.2. Two monochromatic flat fan-like X-ray beams with energy just below and above the K-edge in the absorption spectrum of iodine are selected and made to cross at the patient's heart. The transmitted beam intensity is measured in two ultra-fast high-precision ionization chambers simultaneously for the two energies. While moving the patient through the beam, the heart is scanned at these two energies (see Figure 17.3). The scanning has to be performed in a fraction of a second because of the rapid movement of the heart. On subtracting one of the two images from the other, the structures filled with iodine become distinctively visible, despite the concentration of iodine being 20–50 times less than that

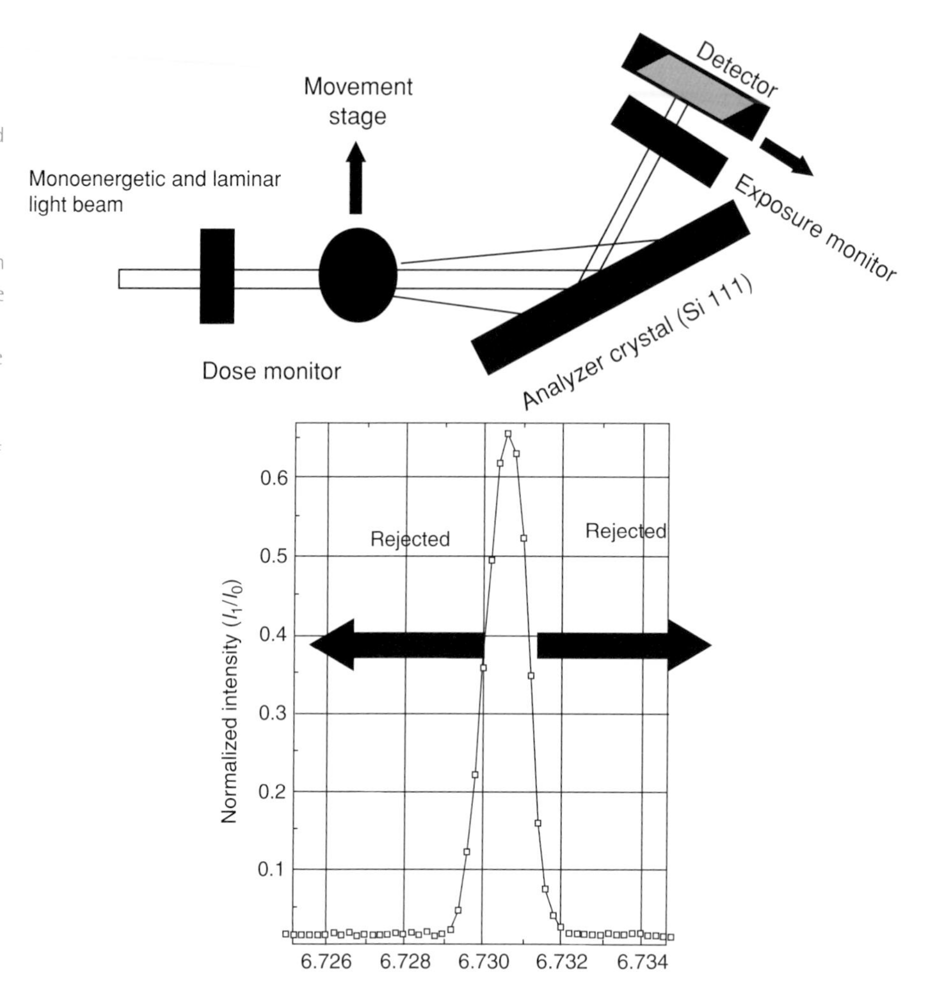


**Figure 17.4.** Diffractive enhanced imaging (DEI) with synchrotron light. The monochromatic beam is selected from the polychromatic original synchrotron light. The extremely small cross-section of the electron beam in the synchrotron and the distance from the source (19 m) yield an extremely low convergence and high coherence of the light. The "rocking curve" of the analyzer, i.e. the dependence of the reflectivity of the crystal for monochromatic X-rays ("light") on the reflection angle, has a very sharp peak around the Bragg angle.

used in conventional selective coronary angiography. The signals from surrounding tissue (including ribs) vanish from the subtracted image because the X-ray-absorption coefficient of tissue has similar values for the two energies used. Besides the speed of the measurement, a very high SNR and large dynamical range (19 bits) are essential in order to reveal and allow investigation of small features.

The second radiological technique uses the phase information of the scattered X-ray. Since the refractive index of X-rays deviates only marginally from unity and has only small material-dependent variations, a different technique from that employed for visible light has to be used. This is "diffractive enhanced imaging" (DEI). Again synchrotron radiation is used because a highly monochromatic and laminar X-ray beam is needed. In DEI the phase gradient across a wavefront corresponds to a change of the direction of propagation of the X-ray. This can be resolved using an analyzer crystal as illustrated in Figure 17.4.

The "rocking curve" of the analyzer, i.e. the dependence of the reflectivity of the crystal for monochromatic X-rays on the angle of reflection, exhibits a very sharp peak around the Bragg angle. When positioning the analyzer on the nominal Bragg angle for the incident X-ray, any change of the X-ray direction due to the object, i.e. the phase information, is transformed into information on X-ray intensity. An even better contrast can be achieved by positioning the analyzer on either side of the rocking curve where its slope is steepest. One obtains two intensity-modulated images from the

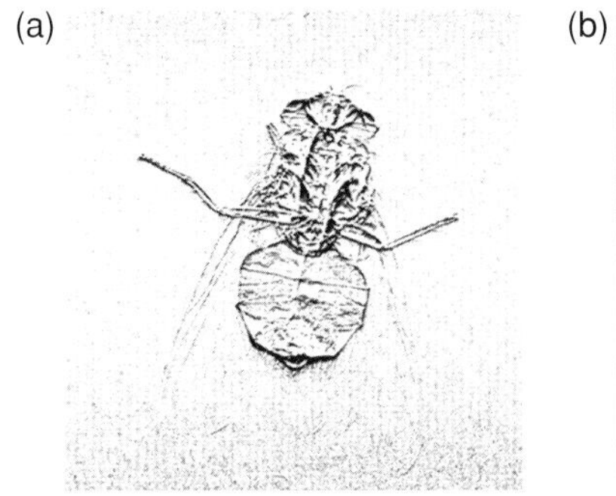


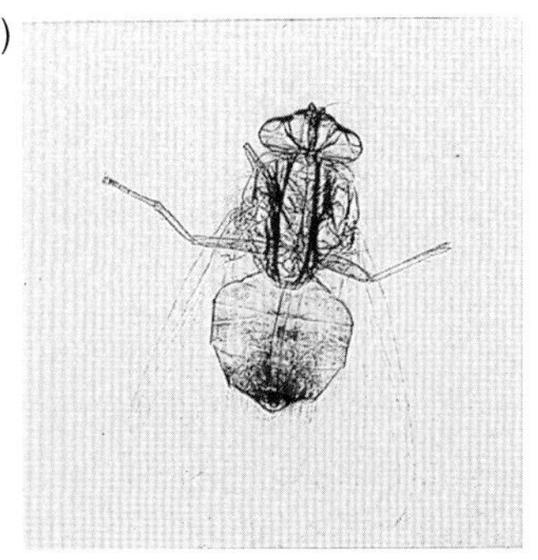


**Figure 17.5.** The apparent absorption image (a) and the refraction image (b) of a fly from DEI after image processing. (Courtesy of Rosa Sanmiguel, University of Siegen.)

left- and right-hand sides, which are described by

$$I_L = I_A\left(R(\theta_L) + \frac{\partial R}{\partial \theta}(\theta_L)\,\Delta\theta\right), \quad (17.3)$$

$$I_R = I_A\left(R(\theta_R) + \frac{\partial R}{\partial \theta}(\theta_R)\,\Delta\theta\right), \quad (17.4)$$

where $I_L$, $I_R$ is the intensity in a pixel of the image and $R(\theta)$ is the reflectivity of the analyzer, which is developed into a Taylor expansion around the positioning angles ($\theta_L$ and $\theta_R$) up to the linear term. These equations can be solved pixel by pixel for the quantity $I_A$ and $\Delta\theta$, resulting in an apparent absorption image ($I_A$), which is particularly sensitive to the variation of the refractive index in the object (refraction images). The effect is demonstrated in Figure 17.5, which shows the image of a small fly. In the refraction image, the contours of the object are very much enhanced, despite there being negligible conventional X-ray-absorption contrast. This shows that small lesions can be detected with hardly any absorption contrast. The increase of sensitivity for mammography with DEI is extraordinary. The benefit of early detection of small calcifications in the breast is widely acknowledged. These two examples show how specific physics phenomena (the K-edge in the absorption coefficient and phase contrast) are used to improve radiography. In addition, there are many developments of digital X-ray-imaging detectors providing higher spatial resolution, efficiency, and robustness. This will revolutionize radiography, but these developments belong more to engineering than to medical physics.

## 17.3 Radiation therapy – from cobalt bombs to heavy ions

The second field of medical physics with a long history is radiation therapy. Soon after their discovery, natural $\alpha$, $\beta$, and $\gamma$ forms of radioactivity were used for therapy, and X-rays were used for teletherapy.

Radiation hitting the body induces a series of very complex reactions, which modify cells and can be used to promote healing of healthy cells and destroy ill or malign cells (tumors). The understanding of these processes, which occur on different timescales, involves physics, chemistry, and biology. Any interaction of radiation with biological tissue results in deposition of energy, which in turn leads to ionization or excitation of atoms and molecules. All biological effects of radiation are induced by these processes, shown in Table 17.1.

The strength of the biological effect depends on the intensity and duration of the irradiation and the density of ionization per radiation particle, and varies with the type of cells and tissue. The reasons for and effects of the dependence on duration and

**Table 17.1.** *The processes induced by irradiation of the body (or part of the body) with their typical timescales; only the first two process steps are governed entirely by physics, the others by (bio)chemistry, biology, and medicine*

| Timescale (s) | Process step | |
|---|---|---|
| $\leq 10^{-18}$ | Deposition of energy by primary particle | |
| $10^{-15}$–$10^{-12}$ | Ionization/excitation on molecular level | |
| | ↓ | ↓ |
| | Formation of chemical radicals | Direct break-up of molecular bindings |
| $10^{-5}$ | Chemical reactions with radicals | |
| $1$–$10^{2}$ | New radiochemical products | |
| | ↓ | ↓ |
| Minutes–hours | Biochemical macromolecular alterations | |
| | ↓ | ↓ |
| ?? | Manifest direct cell death | Latent mutations |

## BOX 17.2 INTERACTIONS OF γ-QUANTA WITH MATTER

At low energy and in materials with high nuclear charge $Z$, gamma rays lose most of their energy through photoabsorption (see Figure 17.6). The incident photon is absorbed and all its energy transferred to one atomic electron, which interacts with the surrounding matter like a charged radiation particle (see Box 17.3). Since the range of such electrons is small (on the order of millimeters in tissue), photoabsorption leads to ionization within a small volume. Fluorescence photons may lead to some ionization clustering in the neighborhood, smearing energy deposition over a slightly wider range. At intermediate energies (≈0.5–1.5 MeV) and for lighter targets (like organic material), γ-quanta lose most of their energy via Compton scattering, in which the photon scatters inelastically from atomic electrons. The losses are not as large or as localized as in photoabsorption. Multiple Compton scattering leads to gradual energy attentuation to the range where photoabsorption dominates.

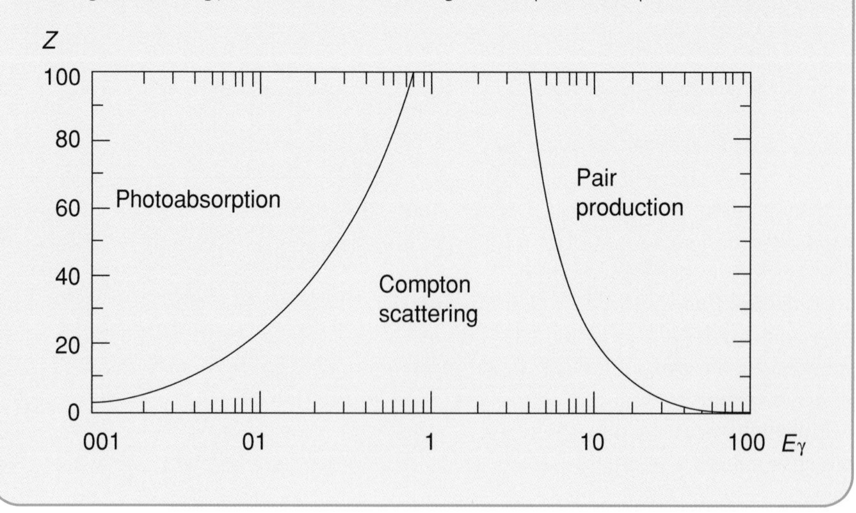


**Figure 17.6.** Gamma-ray interactions involve several mechanisms, depending on their energy ($E_\gamma$) and on target nuclear charge ($Z$). The regions in which each one of the three types of gamma-ray interaction dominates are indicated. (Courtesy of C. Grupen.)

intensity of the irradiation are well known. Recovery and repair mechanisms of the stricken cells, as well as the degree of toxicity and chemical reactivity of the products, play a key role. In all cases, radiation damage increases with the ionization density, which depends on how the energy of the radiation quantum is deposited. The energy deposition for $\gamma$ radiation differs from that of charged particles in two important ways (the case of neutrons will be treated later).

- The $\gamma$-quantum transfers energy to an atomic electron either by giving up its total energy in photoabsorption, or by losing a part of its energy by Compton scattering (see Box 17.2). In both cases, secondary electrons are produced. These lose their kinetic energy and stop after a certain distance, the range of the particle (see Box 17.3). The range of secondary electrons is relatively small (on the order of millimeters or less), so the energy of the incident $\gamma$ radiation is deposited within a small volume, provided that photoabsorption or Compton scattering has occurred. (The production of electron–positron pairs is not important in this energy regime.)
- Since photoabsorption and Compton scattering are random processes, the probability of such an interaction falls off exponentially in accord with d'Alembert's law, Equation (17.1). As a result, the ionization density decreases exponentially with the depth of penetration.

In contrast, charged particles such as electrons, protons, and ions constantly lose energy through Coulomb interactions with atomic electrons and eventually come to rest. On their way, the charged particles transfer small amounts of energy and momentum to many target electrons and ionize or excite the atoms. One can calculate only the average amount of energy loss per unit of material traversed ($dE/dx$), where $x$ is measured in units of centimeters or g cm$^{-2}$ (amount of mass traversed per area irradiated) (see Box 17.3). The strong increase of $dE/dx$ for low incident-particle velocities in Figure 17.7 is important, and makes heavy charged particles such as protons or ions useful for radiation therapy. As can be seen in Figure 17.8, heavy charged particles deposit the major fraction of their energy in a very small volume around their stopping point, while the energy loss in transit is comparatively small. Since the range, i.e. the stopping point, is also very sharply defined and can be precisely adjusted via the ion's energy, the location of energy deposition and thus of cell destruction can be very precisely controlled. The damage to surrounding tissue and the tissue along the beam's path is minimized by the strong localization and the large difference in ionization density seen in the Bragg curve (Figure 17.8).

Unfortunately, an enormous technical effort is needed for heavy-ion therapy, and this technique is therefore used only where all other methods would fail, or at least induce a high risk of collateral damage, e.g. in deep lesions in the brain. Mainly for economic reasons, most cases still use $\gamma$ sources in the form of "cobalt bombs" (strong, carefully collimated sources), while other $\gamma$ sources (linear electron accelerators with a target to produce bremsstrahlung) are used for teletherapy. Stereotactic irradiation with several beams is used to optimize the effect and minimize surface damage. Radiation therapy with $\beta$ particles (electrons) requires a beam with well-adjustable kinetic energies in the range 10–100 MeV. This can be achieved with the relatively small linear accelerators used in larger clinics. This $\beta$ radiation has a range of only a few centimeters, which is suitable for treating tumors in the skin or just below. This is quite effective since the surrounding tissue can be spared and the target region well defined, though not as precisely as with protons or ions, since electrons exhibit a significantly larger range straggling. This is due to the deviation of electrons by multiple Coulomb scattering (angular straggling) and bremsstrahlung, producing accompanying X-rays. For protons and heavier charged particles, bremsstrahlung is negligible since it is

## BOX 17.3 ENERGY LOSS OF CHARGED PARTICLES IN MATTER

The interaction of charged particles with the electrons of target atoms involves many individual inelastic scattering processes, each with only a very small energy transfer. The net result is an average energy loss of the projectile particle per amount of traversed material ($dE/dx$). Bethe and Bloch derived an analytical formula to calculate $dE/dx$, which is given here in a form suitable for medical applications:

$$-\frac{dE}{dx} \propto \frac{z^2_{\text{projectile}} Z^2_{\text{Target}} e^4}{v^2_{\text{projectile}}} \ln\left(\frac{m_e \beta^2 c^2 \gamma^2}{\langle I \rangle} - \beta^2\right) - C_{\text{capture}} - \delta_{\text{screening}}, \qquad (17.5)$$

where $z$ and $Z$ are the charges of the incident particle and the nuclei of the target material (the average $Z$ for biological tissue is $\approx 5$); $\beta$ is the velocity of the incident particle in units of the velocity of light ($v/c$), and $\gamma = 1/\sqrt{1-\beta^2}$. The average ionization potential $\langle I \rangle$ of the target material has a logarithmic dependence, giving a slow increase at large values of incident-particle velocities. The term $C_{\text{capture}}$ governs the decrease in energy loss due to electron capture at very low velocities so that the $1/v^2$ dependence becomes zero. The screening term damps the logarithmic rise of $dE/dx$ for relativistic particles so that $dE/dx$ approaches a plateau.

Plotting $dE/dx$ as a function of velocity $\beta$ (or the relativistic factor $\gamma$) gives a characteristic curve, which is independent of the mass of the incident particle and has a shallow minimum around $\beta \approx 0.8$ and a small (logarithmic) rise with increasing $\beta$ (Figure 17.7). On plotting instead $dE/dx$ against the kinetic energy of the particle, the shape of the curve remains the same, but different particle masses give curves that are shifted with respect to each other. This means that, at a given kinetic energy in the range of 0.1–10 MeV, heavier particles have a higher value of $dE/dx$. Most relevant for medical applications is the $1/\beta$ rise for low $\beta$ (there is a turnover at very low $\beta$ due to polarization effects, so the energy loss remains finite), showing that charged particles brought to a stop in tissue deposit most of their kinetic energy over the last few millimeters of their path.

The higher the particle mass, the larger this effect, as can be seen from the "Bragg curve," when $dE/dx$ is plotted versus the penetration depth $x$ (Figure 17.8). The range – the penetration depth at which charged particles have lost all their kinetic energy – depends linearly on the initial particle energy. The stronger this peak for heavy charged particles, the smaller the straggling of the range. This makes protons or ions very useful for radiation therapy. For multiply charged helium ions ($\alpha$ particles) and carbon ions, the ionization is stronger since $dE/dx$ increases quadratically with the charge $z$ of the incident particle (see Equation (17.5)).

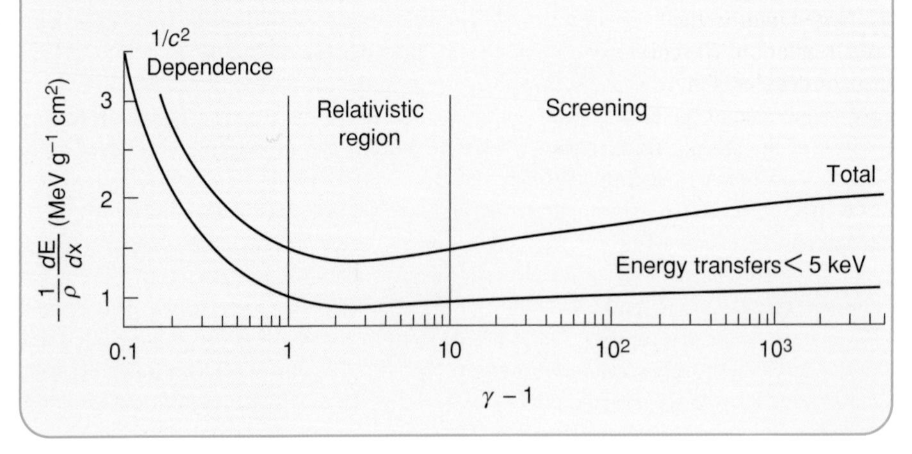


**Figure 17.7.** Energy loss $dE/dx$ of a charged particle in matter as a function of $\gamma = 1/\sqrt{1-\beta^2}$, $\beta = v/c$ where $v$ is the particle velocity and $c$ is the velocity of light. (Courtesy of A. Melissinos.)

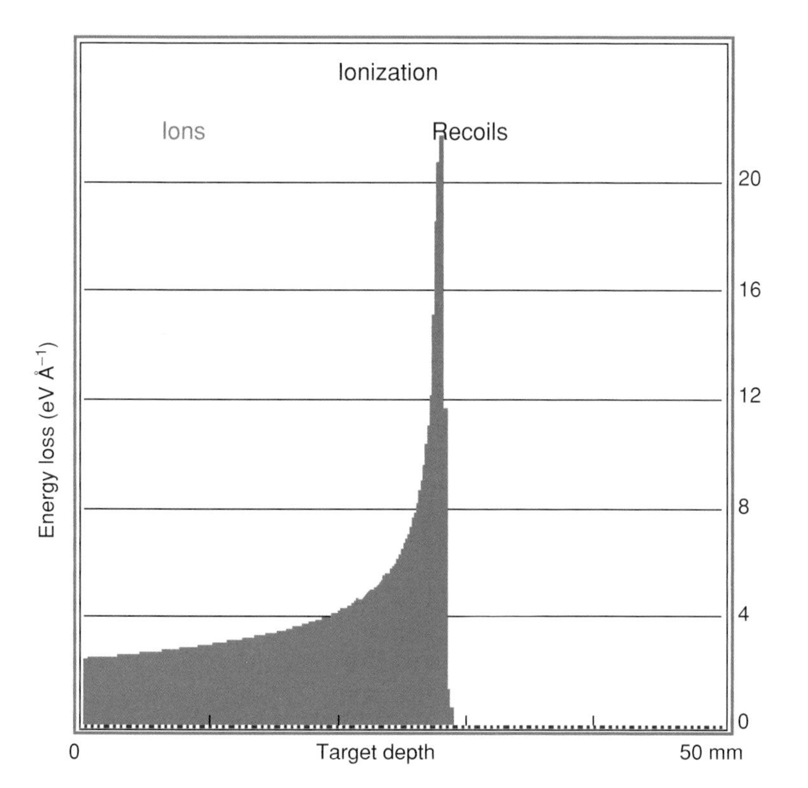


Figure 17.8. The energy loss in human tissue per unit depth of penetration, $dE/dx$, versus depth of penetration $x$ – the Bragg curve for carbon ions with a kinetic energy of 50 MeV per nucleon.

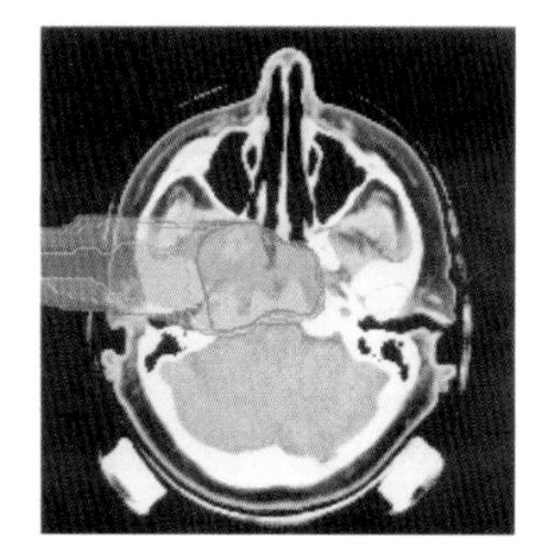

Figure 17.9. Ion radiation therapy of a large tumor close to a critical region (the brain stem). The colors indicate the calculated dose applied. Because of the steeply falling dose, the critical region can be spared almost completely. (Courtesy of G. Kraft.)

proportional to $1/m^2_{\text{particle}}$. In 1954 the first accelerator, a betatron, was built to produce a dedicated proton beam for radiation therapy and was used for more than 40 years to treat thousands of patients. Later proton-radiation-therapy centers were in some cases complemented with $\alpha$-particle beams. The most recent step is precise and effective treatment using carbon ions.

At the Institute for Heavy-Ion Research (GSI) in Darmstadt, Germany, a heavy-ion accelerator and storage ring are used to produce a beam of carbon ions with well-defined energy and controllable intensity for the treatment of very complicated tumors in the brain. Figure 17.9 shows the radiation plan for such a treatment. The target region is a large tumor in the base of the skull close to the brain stem, which has to be spared from irradiation. The color code indicates the dose during the therapy calculated on the basis of detailed simulations, which require knowledge not only of the beam characteristics but also of the anatomy of the patient's head. The critical region can be spared because of the steeply falling dose. Such calculations are validated by measurements with the ion beam on water-filled "phantoms" instrumented with position-sensitive dosimetric detectors (mostly precisely calibrated ionization chambers), before actual therapy.

As well as beam position and cross-section having to be precisely controlled, also the beam energy has to be fine-tuned by computer-controlled plastic wedges (Figure 17.10). Beam attenuation in the interleaved wedges is independent of beam position within the wedges, giving high precision. Figure 17.11 shows the ion-therapy center at the GSI. The patient's head is carefully positioned. In the background the end of the beam pipe is visible. Its intensity, cross-section, and position are continuously monitored by gas-filled high-rate ionization and wire chambers.

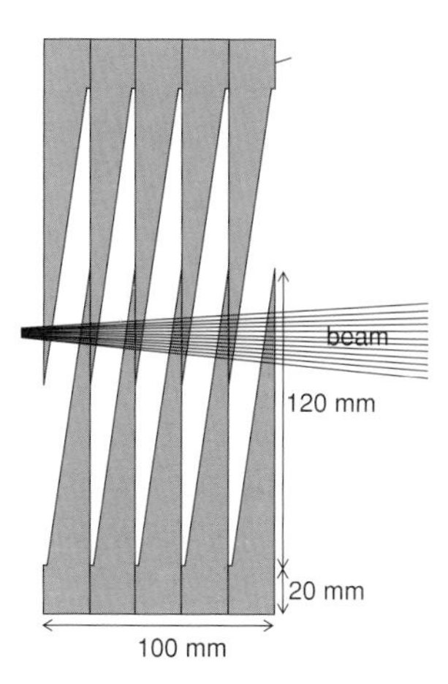


Figure 17.10. The kinetic energy of the ion beam is fine-tuned by two sets of interleaved plastic wedges. This allows precise control on a continuous scale, independently of the lateral beam position. This minimizes systematic errors of this critical parameter, which defines the depth of the dose.

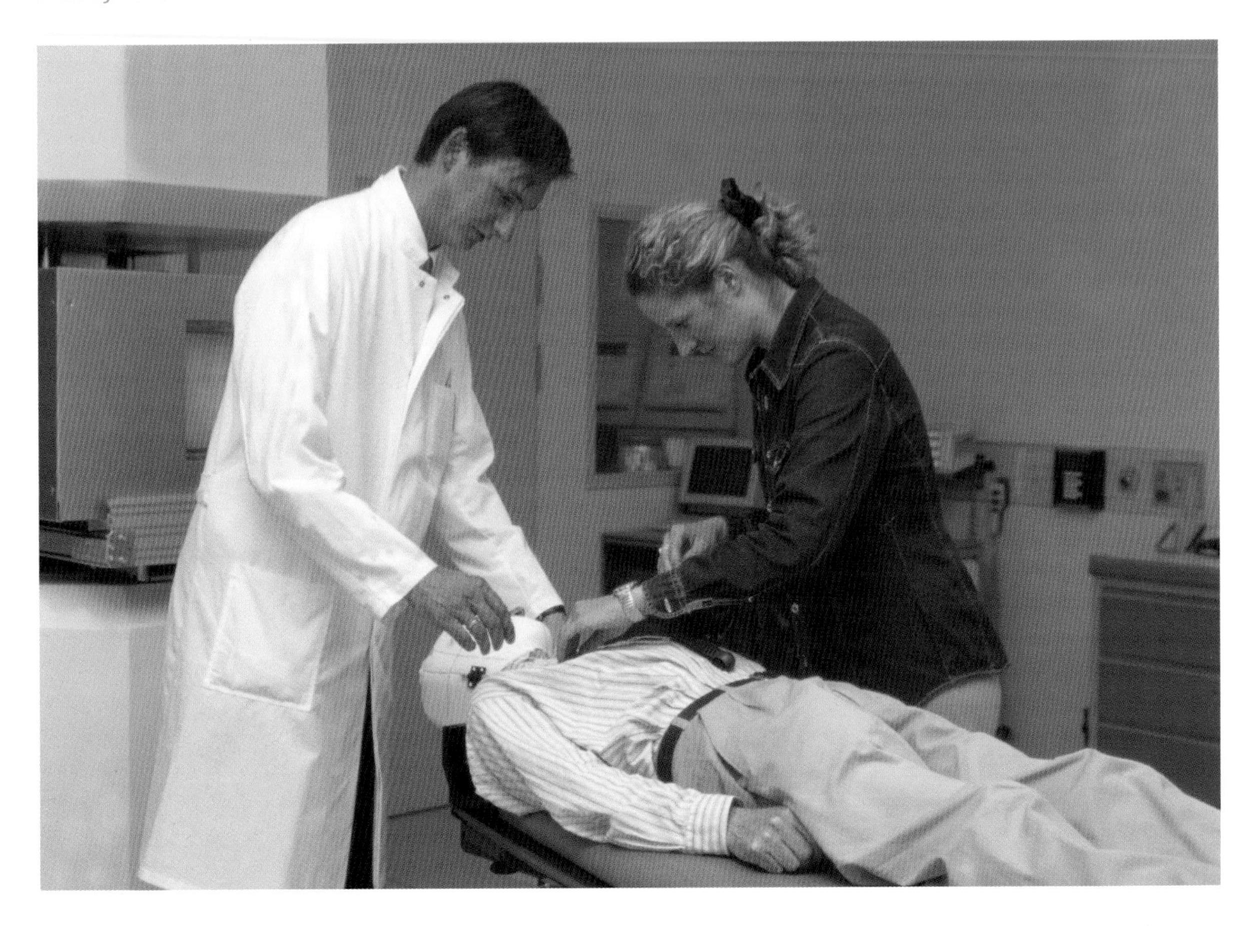

Figure 17.11. Ion radiation therapy at the GSI, Darmstadt, with a patient being prepared for irradiation. The head has to be fixed with an individually adapted mask for exact positioning of the target volume. The end of the ion-beam pipe with beam-monitor detectors (a multi-wire chamber and ionization detector) is visible on the left (Courtesy of G. Kraft.)

In addition to teletherapy with external beams or sources, some tumors have to be treated by brachytherapy. Here a suitable radioactive source ($\beta$- or $\gamma$-radioactive isotope) is brought close to the tumor. As well as careful calculations, this work also needs dosimeters with high spatial resolution and excellent precision. While in other fields of radiation therapy the ionization chamber is still the work-horse because of its reliable calibration and intensity measurements, dosimeters based on semiconductor detectors or scintillating fibers with position-sensitive photomultiplier readout are being developed. The required spatial resolution can be attained; however, there are still problems with calibration and time stability. Another technique for brachytherapy is the use of radiopharmaceuticals containing radiation-emitting isotopes. As in nuclear medicine (see Section 17.4), targeting is achieved by the optimal choice of carrier to maximize uptake only in the part of the body to be treated. In this field experts in radiochemistry, biochemistry, and medicine have to collaborate closely. This therapy is used for widespread tumors or metastases, provided that the affected parts can be targeted efficiently. Bone-marrow cancer and thyroid cancer are examples of where this technique is still considered valuable.

Radioisotope implants are also used for preventive or post-curative treatment, e.g. stents containing small quantities of suitably long-lived isotopes are used to avoid re-stenosis in heart arteries and veins. Neutrons play only a minor role for radiation therapy, and are used in some cases to induce a nuclear reaction with subsequent emission of protons or $\alpha$ particles. The target nuclei first have to be supplied by injection of suitable radiopharmaceuticals (see Section 17.4). With neutrons, there is only relatively little direct radiation damage, mainly by protons that acquire kinetic energy through elastic neutron scattering. This happens for neutron energies above $\approx$10 keV. The energies at which hadronic showers are created are far beyond those used in medicine. Neutrons were used to treat tumors hidden behind healthy organs and which could not be targeted by suitable radiopharmaceuticals. While today such cases are more likely to be treated with proton or ion beams, boron neutron-capture therapy (BNCT) is still used. Boron is introduced into the target region so that radiation with epithermal neutrons (in the energy range of a few keV) leads to neutron-capture reactions with subsequent $\alpha$-particle emission (e.g. ${}^{10}_{5}\mathrm{B} + \mathrm{n} \rightarrow {}^{7}_{3}\mathrm{Li} + \alpha(2.79\,\mathrm{MeV})$). The range of the recoil lithium nuclei and the low-energy $\alpha$ particles ($\approx 5-9\,\mu$m) is smaller than typical mammalian cell dimensions (10–20 $\mu$m), so all the energy is absorbed within a cell. Unfortunately, $\gamma$ radiation from other possible neutron reactions has a larger range and goes outside the target region, producing negative side effects. Most of the radiation-therapy techniques described here are by now well established, and development is mostly aimed at improving performance, with faster and more precise simulation for the planning of therapies, more precise monitoring of the radiation dose, and more accurate and better-collimated beams. This is mainly the task of medical engineering.

## 17.4 Nuclear medicine – trends and aspirations

For historical and practical reasons, nuclear medicine is somewhat artificially defined to be concerned only with those diagnostic methods whereby specific parts of the body (targets) are marked with radioactive tracers before being imaged and investigated using the emitted radiation, which in all cases is $\gamma$ radiation. Radiography and radiation therapy, which could also be called nuclear medicine from the physics point of view, are regarded as separate fields. Nuclear medicine began soon after the discovery of radioactivity as scintigraphy with natural radioactive isotopes using X-ray film to detect the emitted radiation. However, the natural isotopes available were usually quite inadequate for medical applications. Artificial radioactive isotopes were first produced in 1934, but only with the advent of particle accelerators (cyclotrons) for producing protons and $\alpha$ particles and nuclear reactors as neutron sources after World War II could artificial isotopes be produced in sufficient quantities and with the selectivity needed for broader medical applications. At the beginning of the 1970s tomographic images of three-dimensional objects were reconstructed from a series of projected images. While the mathematical proof of tomographic reconstruction had already been given by Radon in 1917, its numerical realization had to await the availability of suitable computer power. Computer tomography (CT) is applied in radiography (using transmitted $\gamma$ radiation) and in emission radiography. There are two types of the latter: single-photon-emission computer tomography (SPECT), using $\gamma$-quanta-emitting isotopes, and positron-emission tomography (PET). The principles of both these techniques are illustrated in Figure 17.12. In SPECT, many single-photon-emission events are used to obtain planar projective images. Since the projection angle has to be kept constant and known, a lead collimator is placed in front of the $\gamma$-ray detector to ensure that only those photons incident perpendicular to the entry window are accepted. Such a detector with a mechanical collimator is called an Anger camera (after its inventor, H. Anger) and today is still by far the most frequently used detector in nuclear medicine. Its principle and disadvantages will be described later.

The positrons ($e^+$) used for PET quickly annihilate, so that, almost at the positron-emission point, two annihilation $\gamma$-quanta of 511 keV are emitted. Because of momentum conservation, these two photons are emitted back to back. A single PET

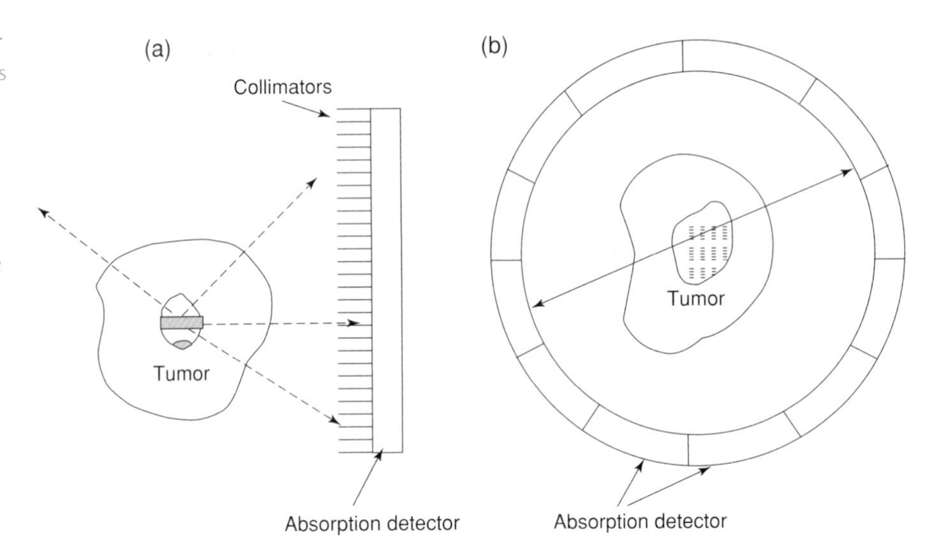


**Figure 17.12.** The principles for the nuclear-medicine techniques used today – SPECT (a) and PET (b). In SPECT, gamma-ray photons are picked up from a suitable introduced gamma-ray-active radiopharmaceutical. In PET, the organ is dosed with a suitable positron emitter. The positrons annihilate, producing characteristic back-to-back pairs of gamma photons. (Courtesy of H. Kamberaj.)

event is found by registering these two simultaneous photons using two independent $\gamma$-ray detectors. Placing the detectors in different positions around the object reveals the source. Computer tomography is used to reconstruct a full three-dimensional image from many such events. Nuclear-medical investigations require three essential criteria: suitable radioisotopes; suitable radiopharmaceuticals as carriers; and powerful imaging systems. For the choice of a radioisotope, several further criteria have to be considered.

- Biochemical compatibility: the element must not be toxic and should bind to biochemical molecules or macromolecules.
- There should be no parasitic radioactivity: $\beta$ or $\alpha$ emission would cause uncontrolled and unwanted radiation damage. The isotope should emit only $\gamma$-quanta of one specific energy. Additional $\gamma$-emission lines reduce image quality, since a small energy window for imaged $\gamma$-quanta is needed in order to minimize background.
- Suitable half-life $\tau$: clearly neither uranium $^{238}$U with $\tau = 4.51 \times 10^9$ years nor $^{24}$Na with $\tau = 2 \times 10^{-2}$ s would be a suitable isotope! The choice depends on the target to be investigated, the time needed for the preparation of the radiopharmaceutical, and the biological lifetime $\tau_{\text{biological}}$, which is determined by how the carrier is eliminated by the body. $\tau$ should be significantly less than $\tau_{\text{biological}}$, since any biological dysfunction would increase the effective lifetime and thus the effective dose administered to the patient. Typical isotope lifetimes are from several days down to about several hours.
- Quality and cost for the production and/or extraction of the isotope: the isotope has to be very pure to avoid adverse biochemical side effects or residual unwanted radioactivity. It is also desirable to produce the radiopharmaceutical as economically as possible, without necessarily using a cyclotron. For this, "generators" are very attractive – isotopes that undergo $\beta$ decay with a comfortably long lifetime into a metastable isotope, which is then used for the radiopharmaceutical. Because of this $\beta$ decay, the mother and metastable daughter isotopes can be separated chemically. In nuclear medicine, such long-lived $\beta$-active mother isotopes are called "cows" and the process of chemical extraction of a daughter isotope is termed "milking." Today, the most

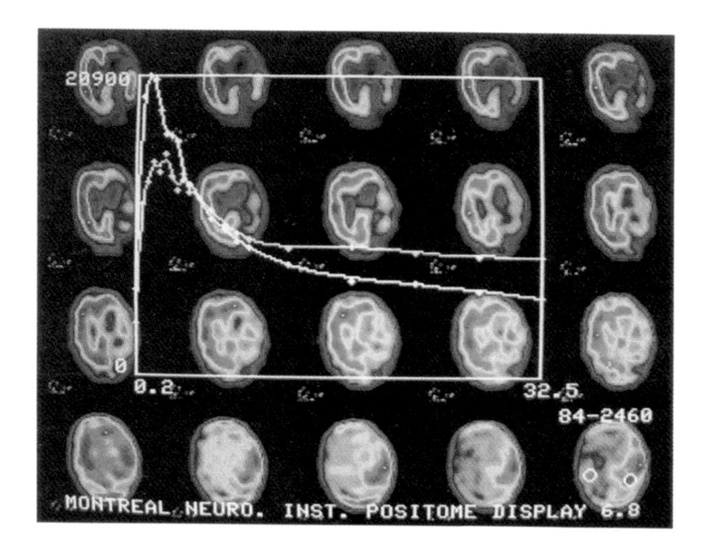


**Figure 17.13.** Investigation of brain tumors with PET using $^{18}$F as β-active isotope and $^{18}$FDG (fluor-dextro-glucose) as radiopharmaceutical to label specific metabolic activity. Owing to the difference between the glucose metabolism of malign cells and that of healthy brain cells, FDG is taken up with different timescales. After a certain time the tumor becomes particularly visible.

frequently used "cow" is $^{99}$Mo (molybdenum), which undergoes $\beta^-$ decay into $^{99m}$Tc (technetium). The metastable $^{99m}$Tc has a dominant $\gamma$ emission at 140 keV, which is favored for SPECT for reasons explained below. Earlier, $^{113m}$In (indium) with a $\gamma$-ray energy of 392 keV was used. It was extracted from the β-active mother isotope $^{113}$Sn (tin), which can easily be produced by neutron irradiation, like $^{99}$Mo. Unfortunately, almost all PET tracer isotopes cannot be produced in this way, but have instead to be directly produced using a proton or α-particle beam from a cyclotron. This is a significant drawback of the PET technique.

- The energy of the $\gamma$ radiation: this requires a compromise between two criteria that work in opposite directions. On the one hand, high $E_\gamma$ is desired in order to minimize angular straggling due to multiple Compton scattering. Such random deviation from the original photon direction leads to a blurring of the reconstructed image. The scattering probability varies only slightly with energy, but the higher the value of $E_\gamma$, the more the photons are scattered. On the other hand, higher $E_\gamma$ necessitates the use of a thicker $\gamma$ detector and a thicker lead collimator to avoid photons coming in at a slightly different angle punching through. The larger detector (NaI crystal) thickness reduces spatial resolution because of parallax, and the punch-through effect leads to a blurring of the image. However, thick collimators dramatically decrease efficiency. Multiple scattering is highest for inner organs. Nevertheless, today thinner lead collimators and NaI crystals, and consequently lower $E_\gamma$, like that of $^{99m}$Tc (140 keV), are most frequently used. A solution to this dilemma may be achieved by using the so-called Compton camera, a new technique that is still in the early stages of development. Its principle will be explained below.

The choice and production of suitable radiopharmaceuticals needs a combined effort of medicine, biochemistry, and radiochemistry. The objective is to allow the isotope to be injected into the organism without any toxic or other side effects and to be temporarily absorbed. The latter requires exact knowledge of the metabolism both of the target organ and of the whole body. The aim is to maximize the uptake in the target while minimizing that in the rest of the body. The time dependence of the uptake governs the data-taking for imaging. In some cases one can exploit the

**Table 17.2.** *Some of the radiopharmaceuticals with* $^{99m}$Tc *and the types of disease they are used to investigate*

| Carrier substance | Disease (organ) targeted |
|---|---|
| Sulfur colloid | Liver, bone marrow, spleen |
| Diphosphonate | Skeleton |
| Macroaggregated albumin | Pulmonary perfusion, arterial perfusion |
| Hexamethyl propyleneamineoxime (HMPAO) | Cerebral perfusion |
| Red blood cells | Ventricles, gastro-intestinal bleeding |
| Sestabimi (or teboroxime) | Myocardial perfusion |

difference between the time behaviors of the uptake in malign and surrounding tissue for optimizing contrast by choosing a suitable time window, as shown in Figure 17.13.

The other important biomedical process is the elimination of the radioactive substances by the natural metabolism, the time behavior of which can be described by an exponentially decreasing curve, often with more than one time constant, or "biological lifetime," if several metabolic processes are involved. As indicated above, the timescale for the metabolic removal of the radioactive isotope is one of the conditions controlling the choice of the isotope. A few examples for radiopharmaceuticals all derived from the same isotope, $^{99m}$Tc, are listed in Table 17.2. The standard imaging X-ray detector, which is nowadays used in the overwhelming number of cases, is the so-called Anger camera for SPECT (Figure 17.14).

The Anger camera consists of a large single inorganic scintillator crystal (NaI doped with thallium) equipped with photomultipliers (PMs), which transform visible light from scintillation into electrical signals. The photons are absorbed in this dense high-$Z$ material and the deposition of energy leads to the excitation of atoms. After a characteristic timescale, the atoms de-excite and emit light, mainly in the UV and blue range. A wavelength shifter, a material that absorbs in a higher energy range and emits at lower energies, can optimally match the wavelength of the produced light and the efficiency for photoelectron production by the PM. A single scintillation flash from an incident particle is seen by all the PMs, but the strength of the signal is proportional to its distance. The signals from the PMs are fed into a circuit to provide analog output proportional to the $x$ and $y$ coordinates of the flash. A third analog output is the sum of the PM signals, which is proportional to the energy of the incident $\gamma$-quantum. Thus the Anger camera is an energy- and position-sensitive detector. SPECT needs a lead pinhole collimator, as explained previously. The spatial resolution is a convolution of the intrinsic resolution of the camera without a collimator and the size of the collimator pinholes. Until about the early 1980s, Anger cameras worked only with analog electronics. Today the signals are digitized and the $x$ and $y$ coordinates determined using fast digital electronics. The rate capability is limited by the size of the single scintillator crystal and the duration of a single signal, including subsequent electronic processing, since time-overlapping hits cannot be resolved. Therefore modern "molecular Anger cameras" have crystals subdivided into sections, which are processed independently, so that the rate can be increased from the 50 kHz of analog electronics up to several hundred kilohertz. PET needs at least a double-headed detector system with two Anger-like detectors at 180° to capture the pair of annihilation photons simultaneously emitted back to back. Although the time information is in principle sufficient to define an annihilation event and its information

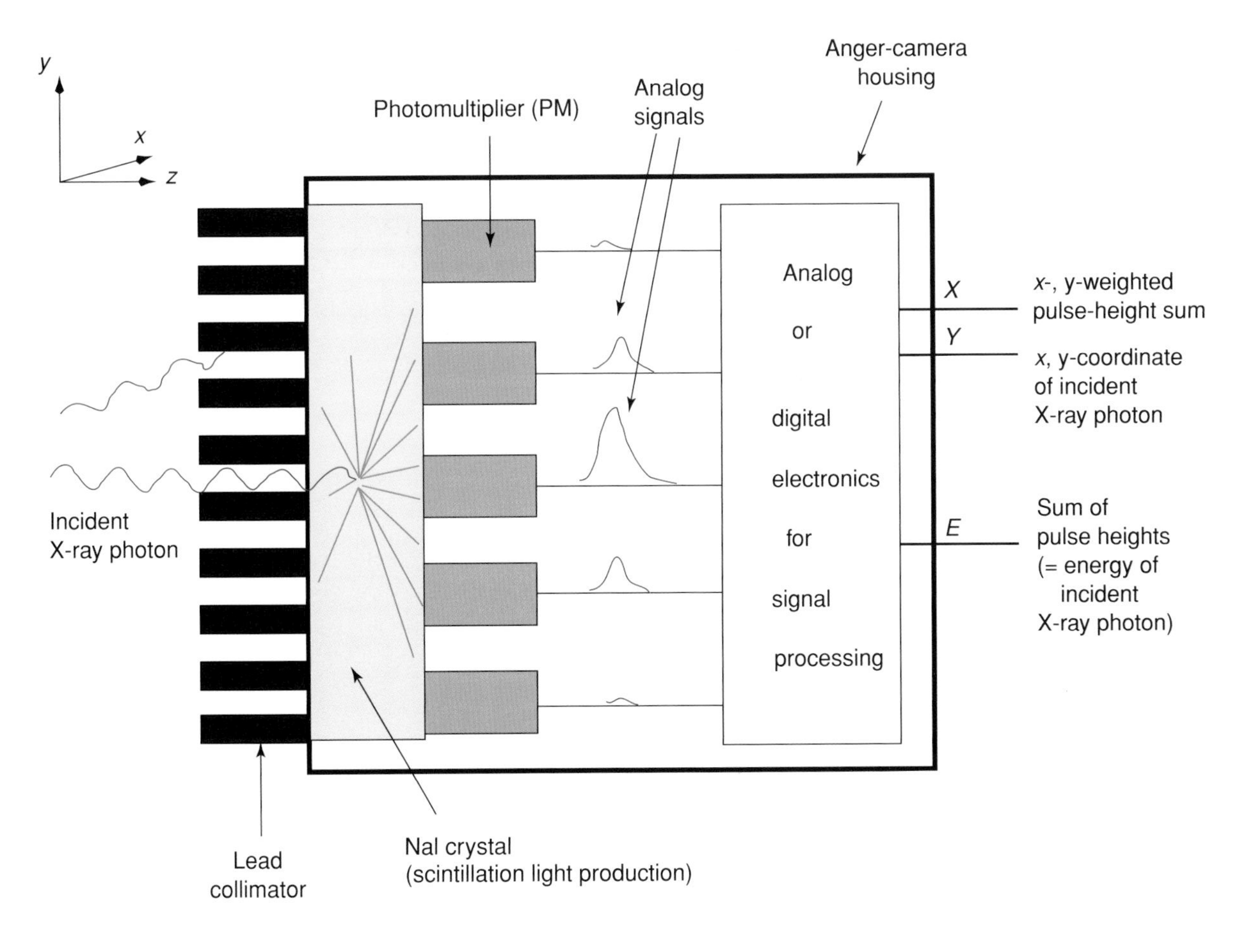


**Figure 17.14.** The principle of the Anger camera.

on the radioactive source (see Figure 17.12), collimators are used to reject other coincident events, which would otherwise blur the tomographic reconstruction. Since the energy of the photons is higher than that used in SPECT and the coincidence time window should be as small as possible, the scintillator material should have the following properties:

- high density and high nuclear charge $Z$ for good X-ray-detection efficiency;
- a short timescale for scintillation light production to give a fast signal with a steep rise for precise timing of the event; and
- high light yield per unit energy in the crystal to improve energy resolution (for further background rejection).

NaI doped with thallium has long been the most popular scintillator. However, a major effort, mainly in high-energy physics, aims to find better crystals. The properties of some of these are compared in Table 17.3. The material favored today for future development is lutetium oxyorthosilicate (LSO), mainly because of its more rapid scintillation, giving more precise time information, fewer false random coincidences (for PET), and a higher rate.

Advantages and disadvantages of these techniques are summarized in Table 17.4. Both SPECT and PET have disadvantages, some of them severe. The most severe disadvantage of SPECT is extremely low efficiency ($10^{-4}$), the fraction of the applied dose which can be used for diagnostics, since almost all the radiation is absorbed by the lead collimator. The collimator also limits spatial resolution, which in clinical

**Table 17.3.** *Properties for some of the modern scintillation materials which are relevant for medical physics*

| Property | NaI | LSO | BGO |
|---|---|---|---|
| Light yield w.r.t. NaI (%) | 100 | 15 | 75 |
| Time constant for scintillation ($10^{-9}$ s) | 230 | 40 | 300 |
| Density (g $cm^{-3}$) | 3.7 | 7.1 | 7.4 |
| Effective $Z$ | 51 | 75 | 66 |

practice is around 10 mm, because any increase would demand a collimator with finer pinholes and greater thickness. Both would lower efficiency dramatically. The other disadvantage comes from the dilemma of having to choose a suitable $\gamma$-ray energy, as outlined above. For diagnosis of deep lesions with relatively low $\gamma$-ray energies (typically 140 keV), the blurring effect due to multiple Compton scattering severely limits resolution and therefore diagnostic power. With PET some problems can be solved, e.g. the spatial resolution in clinical applications is much better and reaches 3–5 mm, depending on the depth of the target. The higher $\gamma$-ray energy (511 keV) is advantageous; however, thicker NaI scintillator crystals are needed for acceptable photon absorption in the crystal and this causes a large parallax error, since the depth of interaction is not measured in today's systems. The most severe drawbacks for PET are high costs and the expensive infrastructure (cyclotron) needed to prepare the isotopes. A new technique being developed for nuclear medicine involves the Compton camera. The principle is explained in Figure 17.15.

A Compton camera consists of two detector components. The key point is that the mechanical lead collimator is replaced by an "electronic collimator," a detector wherein a fraction of the emitted $\gamma$-quanta undergoes Compton scattering. In this scatter detector or "e-detector" the location of the Compton-scattering and the energy of the Compton-scattered electron $E_e'$ are measured as precisely as possible. This detector should be as close to the subject as possible. In the second detector (an absorption detector or "$\gamma$-detector"), which can be more distant, the impact point and energy of the Compton-scattered photon are measured. Since the kinematics of the inelastic scattering is well known, the Compton-scattering angle $\theta_{\text{Compton}}$ can be calculated from

**Table 17.4.** *Advantages and disadvantages of SPECT and PET with an indication of their importance (++, + , −)*

| Pro | Contra |
|---|---|
| SPECT | |
| Much clinical experience (++) | Very low efficiency (−) |
| Affordable price and running costs (+) | Low spatial resolution (−) |
| Low cost/ease of use (+) | Only low-$E_\gamma$ tracer usable (−) |
| Tracer nuclides available (+) | |
| PET | |
| Best spatial resolution so far (+) | High costs for apparatus (−) |
| Clinical practice (+) | High costs for production of radiopharmaceuticals (−) |

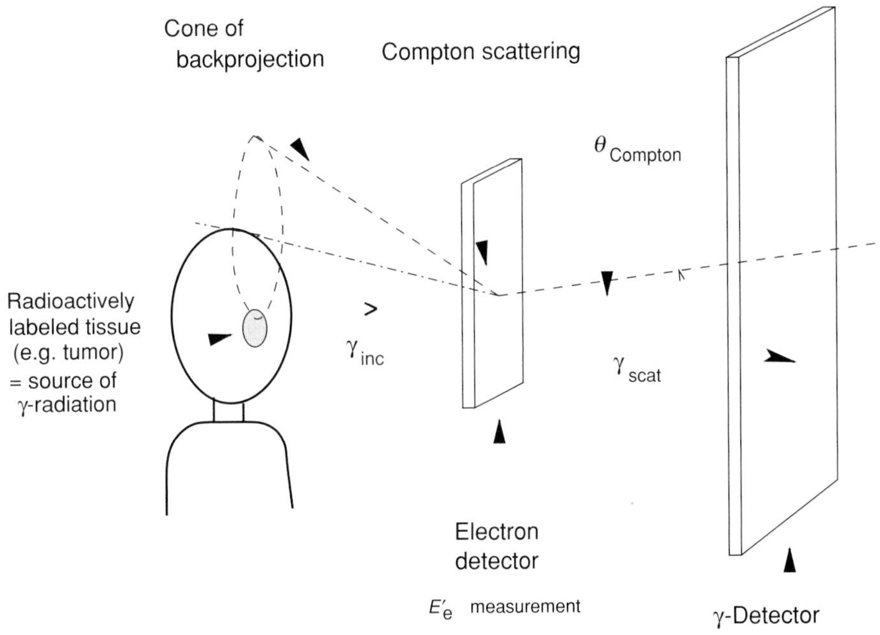

**Figure 17.15.** The principle of the Compton camera.

the measured value of $E'_e$ and the known incident-photon energy $E_0$:

$$\cos\theta_{\text{Compton}} = 1 - \frac{m_e c^2}{E_0}\frac{E'_e}{E_0 - E'_e} \tag{17.6}$$

(where $m_e$ is the electron mass and $c$ the velocity of light).

For each recorded Compton event (time-coincident event in both detectors) one knows that the source of photons is on a back-projected cone as illustrated in Figure 17.15. With the intersection of many cones (i.e. many Compton events), the spatial distribution of the source can be reconstructed. This is illustrated with a simulation of Compton-camera events for a line-shaped source in Figure 17.16.

The scattered-electron detector would use a special type of semiconducting silicon detector with high spectral resolution. To increase the fraction of Compton events which can be used in the reconstruction to 5%–10%, the scatter detector has to be built of many thin Si-detector layers that are individually read out. With this information, multiple Compton-scattering events can be identified and precise reconstruction gives the depth of interaction. For the absorption detector, a finely segmented scintillator detector with LSO crystals and avalanche diode readout on both sides (inner and outer) for additional measurement of the interaction depth in the absorption detector is envisaged. All this is needed in order to push the performance of such a camera to its limits, in principle offering new possibilities:

- increasing the efficiency by a large factor;
- using higher $\gamma$-ray energies to improve spatial resolution;
- attaining sub-millimeter resolution, since there is no limitation due to the finite distance of flight of the positron before its annihilation; and
- using small detectors for semi-endoscopic diagnostics.

This shows how fundamental research benefits other fields of science – both these types of detector were originally developed for experiments in fundamental physics (astrophysics and elementary-particle physics). New experience from high-energy physics experiments (microvertex detectors) will also be useful. Further such developments will surely open fresh fields in nuclear medicine.

**Figure 17.16.** The reconstruction principle for the Compton camera with simulated camera events emitted from a homogeneous line-shaped source (upper left). Upper right is the result after five events with the simple backprojection algorithm. Each of the cone sections corresponds to the possible locations of the source in the image plane, which has been fixed here for demonstration purposes. After 1500 backprojected events one can recognize the shape of the source (lower left). Iterative algorithms improve the quality of the image (lower right).

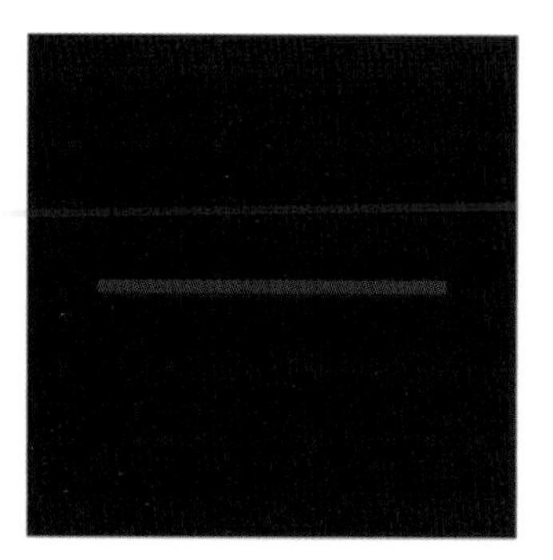

True (generated) emission distribution

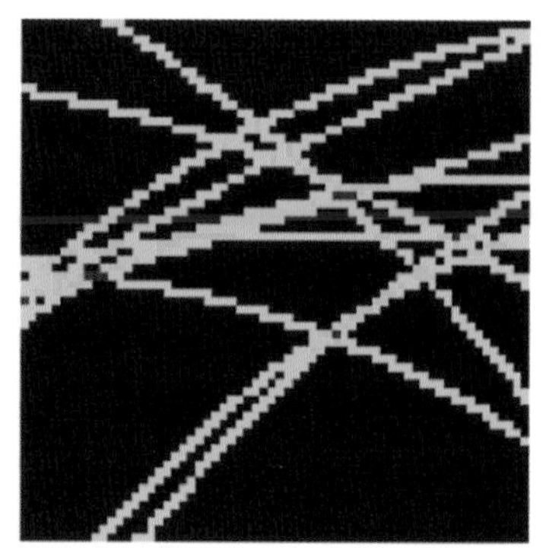

Reconstructed image with five Compton events

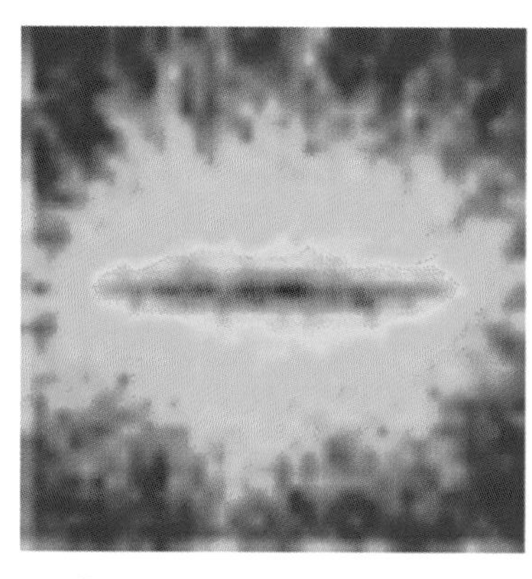

Reconstructed image with 1500 Compton events

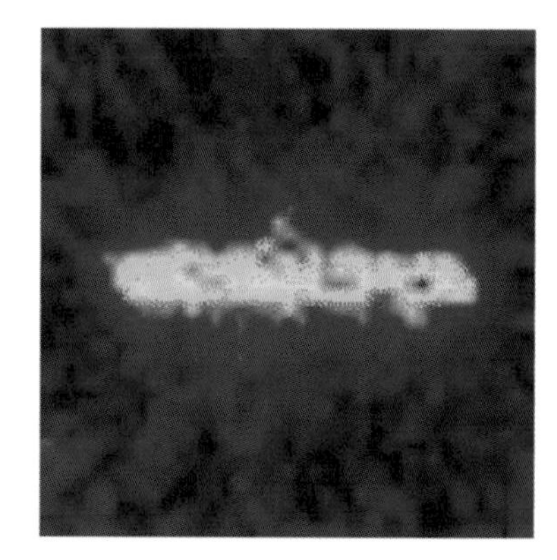

Reconstructed image with 1500 events using an iterative reconstruction algorithm

## 17.5 Nuclear-magnetic-resonance imaging – a view into the body

The aim of nuclear-magnetic-resonance imaging (NMRI or MRI) is to visualize the structure of body tissue with high spatial resolution and high contrast between different kinds of tissues. The basic idea is to measure the density distribution and orientation of nuclear spins, in this case of protons, which are exposed to strong external magnetic fields. The first direct measurement of nuclear spin was made by Rabi in 1938 using the resonant excitation of spin nuclear states, but it took until 1946 for the first NMR experiment to be performed (Bloch and Purcell), and another 27 years until the effect could be used for imaging diagostics. Since 1973 it has become one of the most powerful methods of non-destructive and non-invasive structure analysis, with many advantages. The most important are that NMRI is non-invasive, does not need contrast agents, and does not involve administration of any radioactive dose. Paul Lauterbur of Illinois and Peter Mansfield of Nottingham shared the 2003 Nobel Prize for Medicine for their development of NMRI. Today NMRI has reached an outstanding degree of sophistication, where many aspects of the interaction of nuclear spins with external magnetic fields and their immediate environment are exploited to increase the contrast, spatial resolution, and speed of the imaging process. The key for NMRI is the resonant excitation of nuclear spin states in an external magnetic field. For this the sample is

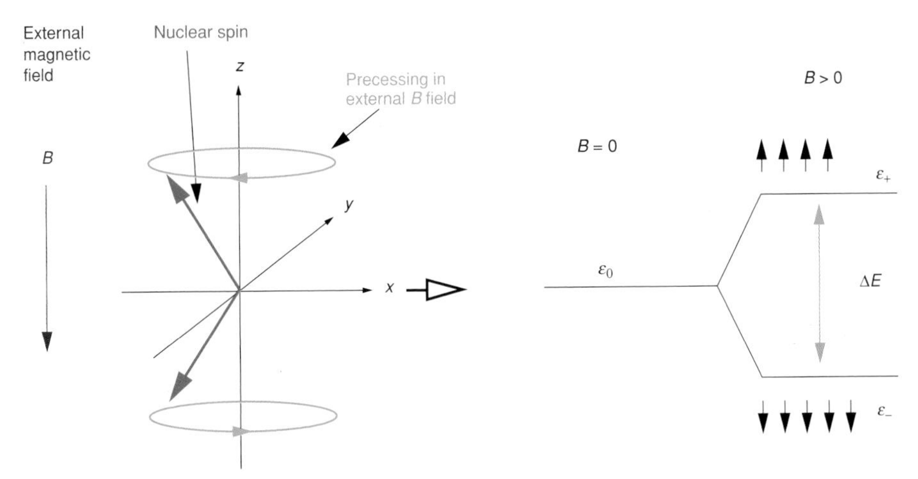


**Figure 17.17.** The possible orientation of the spin of a hydrogen atom (proton) in a static magnetic field $B_0$. This causes a splitting of the energy level of the nucleus into two levels separated by $\Delta E$. The slightly different population of these two states in thermal equilibrium is indicated symbolically.

exposed to a static magnetic field $B_0$. The magnetic moment $\boldsymbol{\mu}_I$ associated with the nuclear spin $I$ interacts with the field, which tends to align the spin. The angle between $\boldsymbol{\mu}_I$ and $B_0$ is quantized and characterized by the orientation quantum number $m_I$. The potential energy of the nuclear spin in the magnetic field depends on this angle and acquires a discrete energy $\epsilon_{\mathrm{m}}$. For hydrogen atoms the nuclear spin is the proton spin ($I = \hbar/2$, with the Planck constant $\hbar$), and there are two possible quantized spin orientations, up ($\uparrow$) and down ($\downarrow$), where the magnetic field is assumed to point upward as shown in Figure 17.17.

Correspondingly there are two energy levels, $\epsilon_{\mathrm{m}} = \epsilon_{\pm} = E(B = 0) \pm \Delta E$, where "+" is the spin-down state (anti-parallel to $B_0$) and "−" is the spin-up state. The energy difference $\Delta E$ between the levels increases linearly with the applied field. In thermal equilibrium the ratio of the numbers of nuclei with spin up and down is given by Boltzmann statistics:

$$n_{\downarrow}/n_{\uparrow} = \exp[-\Delta E/(k_{\mathrm{B}}T)], \tag{17.7}$$

where $k_{\mathrm{B}}$ is the Boltzmann constant and $T$ the temperature. For a proton-containing sample in a magnetic field of 1 Tesla at room temperature (300 K) this ratio is about 0.999 994, a difference of the order of one in a million. Nevertheless, this gives a measurable macroscopic magnetization of the sample ($M_0$), which in turn is proportional to the NMR signal as explained below. The larger $\Delta E$, the larger the difference in the equilibrium population and thus the NMR signature. Hydrogen nuclei are particularly well suited for NMR, with the highest ratio of magnetic moment associated with nuclear spin (gyromagnetic factor $g$), and high natural abundance. This is very suitable for medical applications, since the fraction of hydrogen in the body, in the form of water and proteins, is particularly high. NMR requires a magnetic field with a radio frequency (RF) $\omega_{\mathrm{rf}}$ in a direction perpendicular to $\vec{B}_0$, denoted by $B_{\perp}$ or $B_{\mathrm{rf}}$ in Figure 17.18.

The frequency $\omega_{\mathrm{rf}}$ is tuned such that the corresponding energy ($\hbar\omega_{\mathrm{rf}}$) is equal to the energy difference $\Delta E$ between spin states. In that case a resonant absorption of the RF-field quanta is possible and spins are flipped from the lower- to the higher-energy state. The macroscopic magnetization of the sample $M$, i.e. the $z$ component of $M$, $M_z$, thus changes direction. Depending on the duration and magnitude of the RF pulse, the $z$ component $M_z$ can be tuned to zero, so that $M$ is in the $x$–$y$ plane perpendicular to the $z$ axis, or it can be reversed to $-M_0$. Therefore such RF pulses are called "90°" and "180°" pulses, respectively. The resonance frequency for this excitation depends not only on the magnitude of the nuclear spin and its gyromagnetic factor ($g$), which

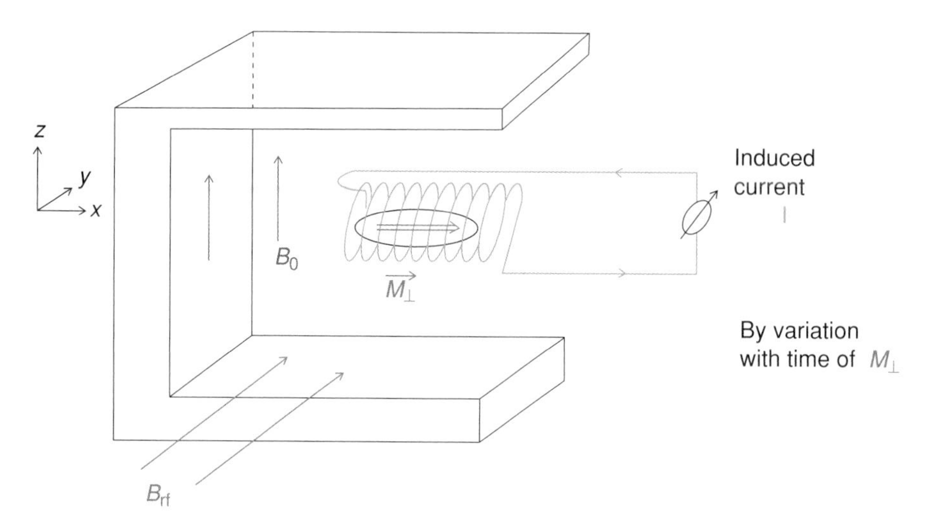


**Figure 17.18.** A schematic diagram of a NMR experiment. The static field $B_0$ is along the *z* axis; the high-frequency magnetic field for the radio-frequency (RF) pulses is along the *y* axis. The receiver coil to register the variation of the magnetization after a RF pulse is oriented with its axis in the *x* direction such that a resulting sample magnetization in the *x*–*y* plane induces a signal current. Since the vector of the resultant magnetization is rotating around the *z* axis with the Larmor frequency, the signal current in the coil also oscillates with that frequency.

are well known and fixed, but also on the effective magnetic-field strength at the location of the nucleus ($B_{\rm eff}$). This differs slightly from the external applied field $B_0$, depending on the molecular binding of the H atom and some physical properties of the environment. The external magnetic field is slightly screened by the atomic fields of other atoms. The NMR resonance frequency therefore depends in a characteristic way on chemical structure. This is extensively exploited for non-destructive analysis in chemistry. The screening also depends on the magnetic properties of neighboring atoms and crystals. This effect is now used to increase contrast using specific paramagnetic contrast agents. A NMR experiment is shown schematically in Figure 17.18. Besides an extremely homogeneous static magnetic field $B_0$ (here along the *z* axis), and the high-frequency magnetic field perpendicular to $B_0$ (here along the *y* axis) for the resonant spin excitation, a receiver coil to detect the variation of magnetization is needed. This induces a signal current in the coil. To understand the time dependence of the sample magnetization and of the NMR signal, one has to look at the motion of the nuclear spin vector in the presence of the magnetic fields $B_0$ and $B_\perp$. In the static magnetic field the nuclear spin $I$ precesses around the magnetic-field direction (here the *z* axis), where the angle between $B_0$ and $I$ is fixed by the orientation quantization (see Figure 17.19(a)). The precession (Larmor) frequency is given by $\omega_{\rm Larmor} = \Delta E/\hbar$ and depends on $B_0$.

It is helpful to view the spin motion in a reference frame that rotates with respect to the laboratory frame around the *z* axis with the Larmor frequency. Before applying a RF pulse, there is a net *z* component of the magnetization, but the resulting component of $M$ in the $x' - y'$ plane (perpendicular to the *z* axis), $M_\perp$, is zero, since the projections in this plane point randomly in all directions (see Figure 17.19(b)). The vector of the magnetic field of the RF pulse ($B_\perp$) rotates in the laboratory frame with the Larmor frequency and is fixed in the $x' - y'$ frame, say in the $y'$ direction. Under the influence of $B_\perp$ the spin moves toward the $x'$ direction, as a gyroscope would do if a torque were applied to it. After a "90°" pulse the spins are along the $x'$ axis, yielding a net

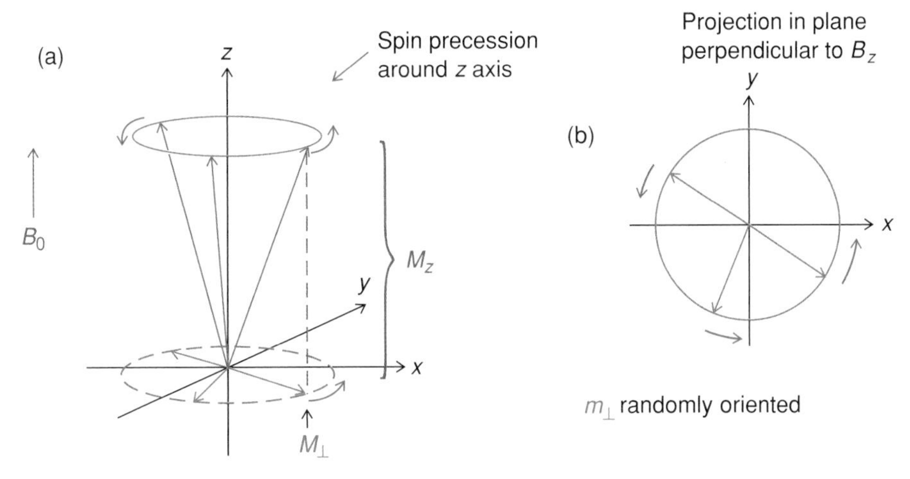


**Figure 17.19.** (a) The spin orientation in a static magnetic field $B_0$ in the z direction. The spin precesses around the z axis with the Larmor frequency and in the laboratory frame the projection of the magnetic moment onto the *z* axis yields a resultant magnetization in the *z* direction ($M_z$). $\theta$ is the angle between $B_0$ and the spin of the associated magnetic moment $\mu_I$. (b) Spin projections onto the *x*–*y* plane give a resultant magnetization perpendicular to *z* ($M_\perp$). Before a radio-frequency pulse, the spin directions are randomly distributed and $M_\perp = 0$.

magnetization of the sample in the plane perpendicular to the *z* axis, which rotates with $\omega_{\text{Larmor}}$ in the laboratory frame. Thus, in the stationary receiver coil, an oscillating current signal with the Larmor frequency is produced. If the excitation field is switched off, the original magnetization is restored because of two different relaxation effects. The first is the spin–lattice interaction which lets the spins flip back to the equilibrium state, so that $M_z$ changes from zero to the value before application of the RF pulse ($M_0$). This has an exponential time dependence with characteristic time constant $T_1$. At the same time the transverse magnetization $M_\perp$ decreases. However, another relaxation process lets $M_\perp$ decrease exponentially with a much shorter timescale $T_2$. This is due to interactions among the flipped spins, the spin–spin relaxation, which lets the spins which initially rotate in phase around the *z* axis become randomly oriented in the $x' - y'$ plane. These processes lead to an exponential decrease of the recorded NMR signal. Figure 17.20 shows schematically the time dependences of the magnetization along the static magnetic field and of that perpendicular to this direction. The timescales $T_1$ and $T_2$ are characterized by the type of tissue and differ considerably, as can be seen in Table 17.5. These effects are used to suppress or enhance the NMR signals of certain types of tissue in the region of interest by choosing an appropriate repetition time $T_R$ for the RF pulse. An example for the application of this technique, called $T_1$- and $T_2$-weighting, is given in Figure 17.21, where the result of an investigation of a baby's brain in pediatric MRI is shown.

NMR imaging requires measuring the spatial distribution of spin concentration. This is achieved by simultaneous application of different techniques: slicing or frequency encoding, gradient encoding, and phase encoding of position information. They are illustrated in Figure 17.22.

Frequency encoding is achieved by superimposing a *z*-dependent static magnetic field with a linear gradient (*z* dependence, $G_z = dB/dz$) to the static field $B_0$ so that the Larmor frequency of the spins depends on *z* ($\omega_{\text{Larmor}}(z)$). Choosing a frequency band for the RF pulse around $\omega_{\text{Larmor}}(z)$ means that only spins in the corresponding slice in the *z* direction resonate and can be measured. Thus, by changing $\omega_{\text{rf}}$, one can investigate the

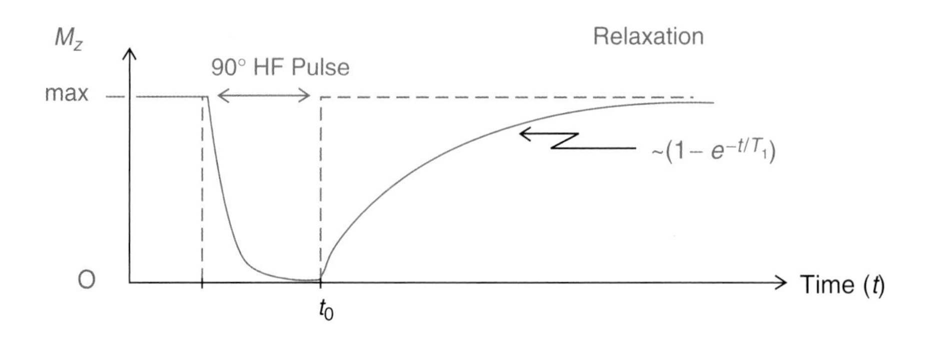


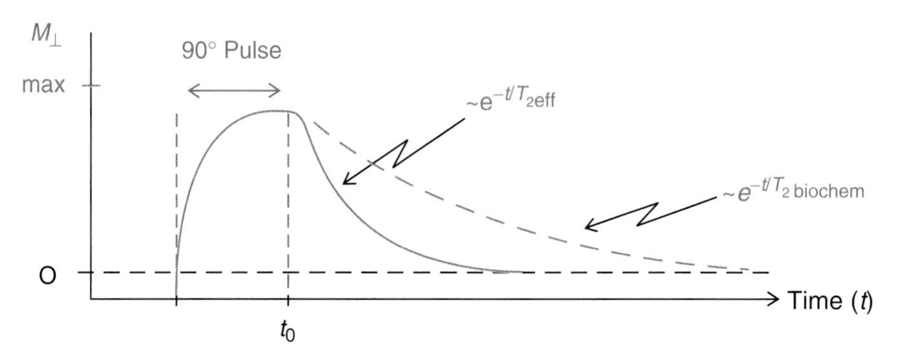


**Figure 17.20.** The time dependence of the sample magnetization along and transverse to the $z$ axis as a function of time during the application of a "90° pulse," which tips the magnetization to the $x$–$y$ plane, and during the time after this pulse is switched off and the spin relaxation processes start. $T_1$ and $T_2$ are, respectively, the characteristic timescales of the spin–lattice relaxation process, by which the magnetization along the $z$ axis is restored, and the spin–spin relaxation process, due to which the net magnetization in the $x$–$y$ plane vanishes exponentially. For $T_2$ one distinguishes the time constant based on the biochemical properties of the sample itself ($T_{2\mathrm{biochem}}$) and $T_{2\mathrm{eff}}$, which is the effective time constant for the spin–spin relaxation process, which is shorter than $T_{2\mathrm{biochem}}$ due to local inhomogeneities of the magnetic field resulting in faster development of the decoherence of the spins.

spin density in the sample slice by slice in the $z$ direction. This gradient magnetic field is applied during the RF pulse. In the other two cases, magnetic fields with gradients in the $x$ or $y$ direction ($G_x$ and $G_y$) are applied shortly after the RF pulse when the transverse magnetization $M_\perp$ is still sizable. In the laboratory frame, the vector of

**Table 17.5.** *The timescales for the spin–lattice and spin–spin relaxation processes in different types of biological tissue*

| Kind of tissue | $T_1$ (ms) | $T_2$ (ms) |
|---|---|---|
| Fat | 250 | 60 |
| Muscle | 900 | 50 |
| Blood | 1400 | 100–200 |
| Brain | | |
| Gray matter | 950 | 100 |
| White matter | 600 | 80 |
| Cerebrospinal fluid | 2000 | 250 |

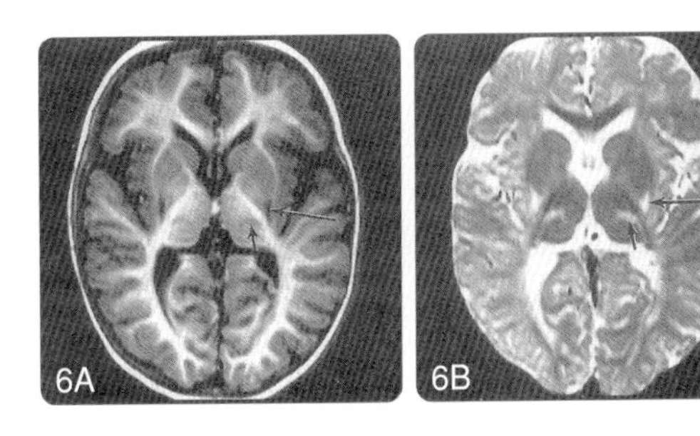


**Figure 17.21.** NMR images of the brain using the $T_1$ (6A) and $T_2$ (6B) weighting methods, showing how different parts of the brain are revealed. Abnormalities (shown by arrows) are more recognizable with $T_2$ weighting. (Courtesy of M. Rutherford.)

$M_\perp$ is rotating with the Larmor frequency of the spins in the magnetic field $B_0$. This induces an oscillating signal in the receiver coil, which is the NMR response from the spin density in the sample. The signal ($s(t)$) is characterized by its amplitude ($A$), the frequency $\omega$, and possibly by a time-independent phase constant $\phi$:

$$s(t) = A\sin(\omega t + \phi). \tag{17.8}$$

The sine function involves the phase $\Phi(t) = \omega t + \phi$. In the absence of any disturbance and interaction with the environment, all spins in a subvolume and hence the magnetization vector $M_\perp$ are moving "in phase" with the same $\Phi(t)$. This is illustrated in Figure 17.22 by the hands of the "stopwatches," which symbolize precessing spins and are all in the same angular position. Since the Larmor frequency, and hence the frequency of the NMR signal, depends linearly on the effective local field, it may be varied in a controlled and position-dependent way for a short time $\Delta t$. After that, the spins in different subvolumes are again rotating with the same frequency but are no longer "in phase." These phase shifts depend on the magnetic-field gradient $G_x$, $G_y$ and the duration $(t_x, t_y)$ of $G_x$, $G_y$: $\Phi_x = g\mu_K G_x x t_x$ (where the gradient is the variation of the extra magnetic field in the $x$ direction, $G_x = dB/dx$) (similarly for $\Phi_y$). Usually the gradient is linear and, for a more transparent mathematical treatment, the variables $k_x = \Phi_x/x$ and $k_y = \Phi_y/y$ are introduced. Within a given $z$ slice of the sample the NMR signal $s(t_x, t_y) = s(k_x, k_y)$ is recorded for many values of $\Phi_x$ and $\Phi_y$, i.e. many points in so-called $k$-space. The quantity of interest for NMRI is the spin-density function $\rho(x, y)$. This is related to the recorded NMR signal $s(k_x, k_y)$ by a Fourier transformation:

$$\rho(x, y) \overset{\text{Fourier}}{\longrightarrow} s(k_x, k_y), \tag{17.9}$$

$$\rho(x, y) \overset{\text{inverse Fourier}}{\longleftarrow} s(k_x, k_y). \tag{17.10}$$

In this way the two-dimensional spin density can easily be reconstructed from the NMR signal sampled in $k$-space. The larger the range of the $k$ values, the better the image quality. By slicing in the $z$ direction, a three-dimensional NMR image of the spin density $\rho(x, y, z)$ is reconstructed and provides structure information for diagnostics. The phase- and gradient-encoding procedures require some time. On the one hand, the magnetization $M_\perp$ in the $x$–$y$ plane causes the NMR signal to decrease rapidly because of spin–spin relaxation (timescale $T_2$) and dephasing due to the inhomogeneous local magnetic field (timescale $T_2'$). This is called the "free-induced-decay" (FID) signal. On the other hand, it is desirable to use the response of one RF pulse several times, because the next RF pulse could create a sizable signal only if the equilibrium magnetization were largely restored. But this takes much longer, due to the timescale for spin–lattice relaxation $T_1$. One way of achieving this is the spin-echo technique. At a certain time $\tau$ after the RF pulse ("$90^\circ$" pulse) another RF signal, this time a $180^\circ$ pulse, is applied. The spin vectors are flipped in the $x$–$y$ plane (i.e. mirrored at the $y'$ axis if the RF pulse is along the $x'$ axis in Figure 17.23.

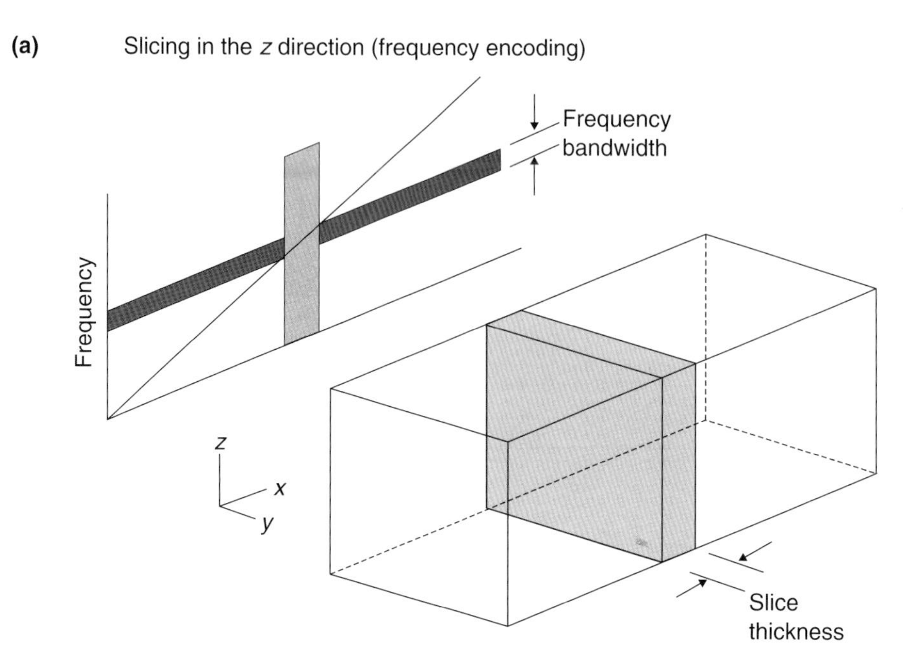


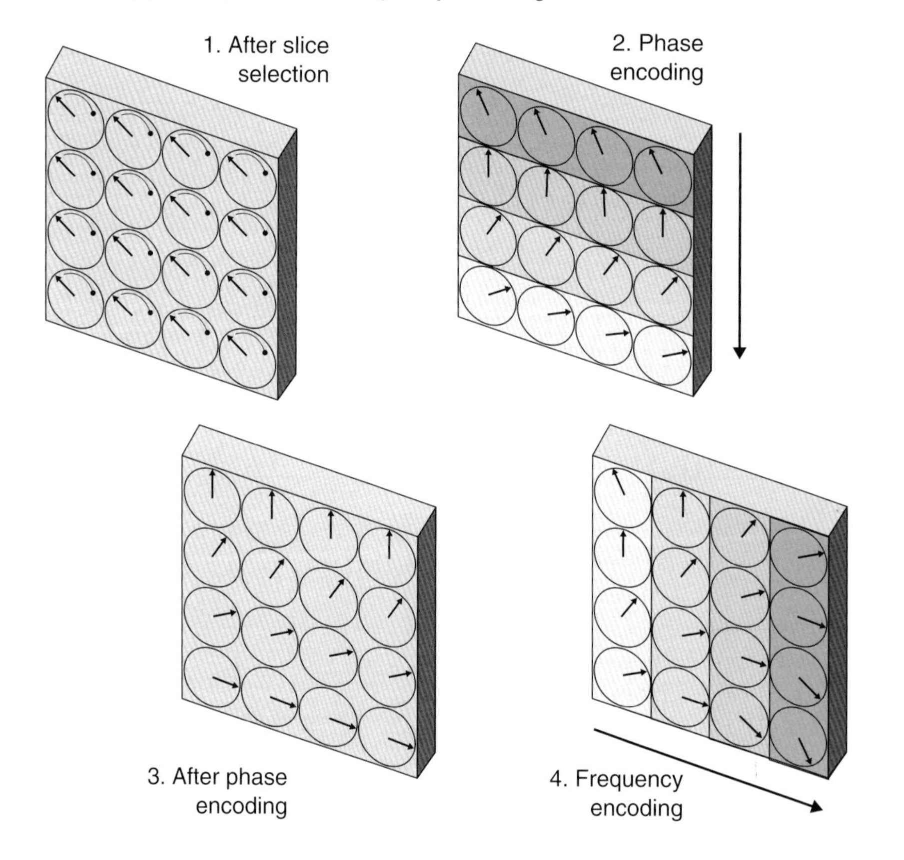


**Figure 17.22.** The three basic techniques used to extend NMR to NMR imaging and improve spatial information on the spin density. (1) Slicing or frequency encoding: superimposing a gradient of the magnetic field on the static field $B_0$ at the time of the radio-frequency (RF) pulse and choosing an appropriate RF frequency $\omega_{\rm rf}(z)$ to induce resonant excitation of spins in a slice around the $z$ position (see part (a)). (2) Gradient encoding: applying a gradient of the magnetic field perpendicular to the $z$ axis for a short time $t_y$, e.g. in the $y$ direction ($G_y = dB_{\rm encoding}/dy$), shortly after the RF pulse. This modifies the phase of the spin vectors rotating in the $x$–$y$ plane in a $y$-dependent manner. (3) Phase encoding: superimposing a time-dependent magnetic field perpendicular to the $z$ axis and the other ($y$) axis, i.e. in the $x$ direction, for fixed time $t_x$ shortly after the RF pulse. This modifies the phase of the spin vectors rotating in the $x$–$y$ plane in an $x$-dependent manner. In cases (2) and (3) the phase shift of the spin rotation, symbolically indicated by the position of the hands of the stopwatches in part (b), also gives a phase shift in the NMR signal and thus allows the signals from different positions to be differentiated.

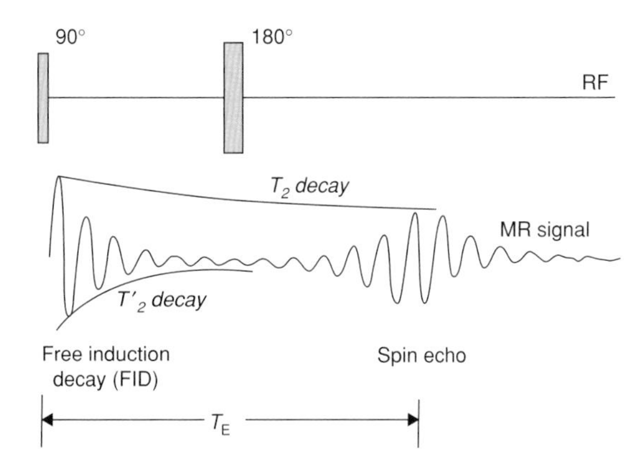


**Figure 17.23.** The principle of the spin-echo technique. Just after the radio-frequency (RF) pulse, the spin vectors rotate in phase, producing a net magnetization in the $x'$-$y'$ plane of the reference frame, which rotates with the Larmor frequency in the laboratory frame. This results in a high-frequency signal in the receiver coil. After some time $\tau$ the spin vectors are dephased and the NMR signal has essentially vanished. An additional RF pulse flips the spins by 180° around the $y'$ axis and the dephasing of spins is reverted. After time $\tau$ the spins are again in phase, yielding a NMR signal – the spin-echo signal.

The effect is that the spin-dephasing process, which has been going on for time $\tau$ after the 90° pulse, is restored and $2\tau$ after the RF pulse the spins are in phase for a short while, producing a NMR signal in the receiver coil, the spin-echo signal. The amplitude of the spin echo is only slightly smaller than the original as long as $T_1 \gg T_2$. As well as allowing fast imaging, this technique is also needed in order to measure the relaxation time $T_2$. The relaxation time $T_1$ can be measured by the "inverted-response" method, so this information, which is needed for the time structures of $T_1$- and $T_2$-weighting, can be accurately determined for tissue. The time structures of the various magnetic-field pulses needed to perform three-dimensional NMRI with a single spin echo are shown and explained in Figure 17.24. This figure demonstrates the extraordinary demands on high-frequency technology for precision and reproducibility. The raw data sampled in $k$-space have to be Fourier transformed, and a great deal of noise filtering and image processing is needed in order to minimize effects due to incomplete coverage of $k$-space and from edge effects.

Although standard NMRI provides structural information with an astonishing resolution in space and time and a high specificity, it cannot provide functional information like SPECT and PET. Recently, however, promising results have suggested ways of inducing spin resonance that depend on metabolic functions or other dynamic processes. This is called functional NMRI (fNMRI). One simple technique is the use of conventional tracer kinetics, i.e. ligands or chelators, with a paramagnetic atom, such as gadolinium (Gd), which produce a local magnetic-field gradient and thus a strong dephasing of spin and substantial loss of the NMR signal. Suitable $T_1$- and $T_2$-weighting gives a contrast enhancement between different parts of an organ. Another technique exploits the different magnetic properties of hemoglobin with and without oxygen (oxyhemoglobin and desoxyhemoglobin). The former has no unpaired electrons and is diamagnetic, whereas the latter molecule has a "hole" in its structure and therefore an imperfect shielding of its central iron atom. Therefore desoxyhemoglobin is paramagnetic and reduces the NMR signal substantially, like Gd tracers, but with the advantage that the blood acts as a contrast agent. This Blood-Oxygen-Level-Dependent (BOLD) imaging is extensively used in brain imaging to reveal oxygen metabolism correlated with brain activity. A striking example is shown in

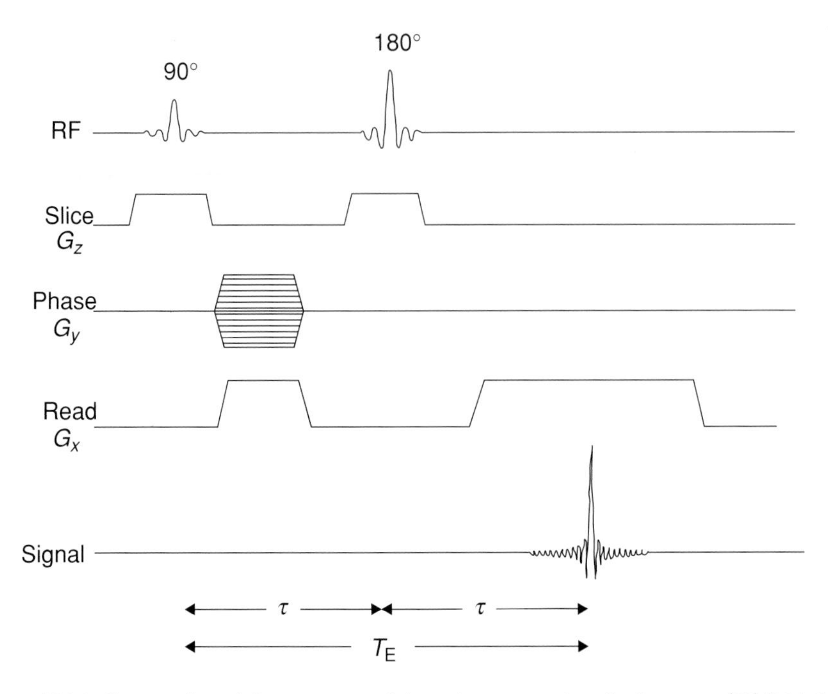


**Figure 17.24.** The complicated time structure of the various magnetic radio-frequency (RF) fields for NMRI with single spin echo: first line, RF signal (90° pulse along the *x* axis to induce a free-induced-decay (FID) signal, and a 180° pulse along the *y* axis to induce a spin echo); second line, the gradient field for selection of a section (slice) in the *z* direction; third line, the gradient field in the *y* direction (multiple lines indicate that the field in the *y* direction is varied in a stepwise fashion within a suitable time for phase encoding in the *y* direction); fourth line, the gradient field for position encoding in the *x* direction; fifth line, NMR signal: first the FID signal and then after $2\tau$ the spin echo. The initial NMR signal and spin echo are used for differentiating along the *y* and *x* directions, respectively. $\tau$ is the time after which the refocusing 180° pulse is applied. The spin echo appears after the time $T_E = 2\tau$.

Figure 17.25, where a clear localization of the brain's response to a visual stimulus can be seen and its time dependence studied. BOLD imaging has become a powerful and very attractive technique since it is completely non-invasive.

Magnetic-resonance angiography (MRA) is another MRI derivative, specializing in the measurement of blood-vessel volume and blood flux (velocity). The increasing variety of techniques for more specific fNMRI and fast MRI result from and require the close interaction among biology, medicine, and nuclear, atomic, and molecular physics.

## 17.6 Measurement of bioelectricity and biomagnetism – tracing motion and emotion

In the human organism, most information transfer through the nervous system and data processing occurs via electromagnetic processes due to ion currents in nerve cells. In the eighteenth century, Volta found that external pulses of electric current applied to a frog's leg stimulate muscle contraction. The more ambitious task of measuring such stimulated currents was first accomplished by P. Berger in 1929. Nowadays muscle activity in the heart is monitored by electro-cardiography (ECG) and currents

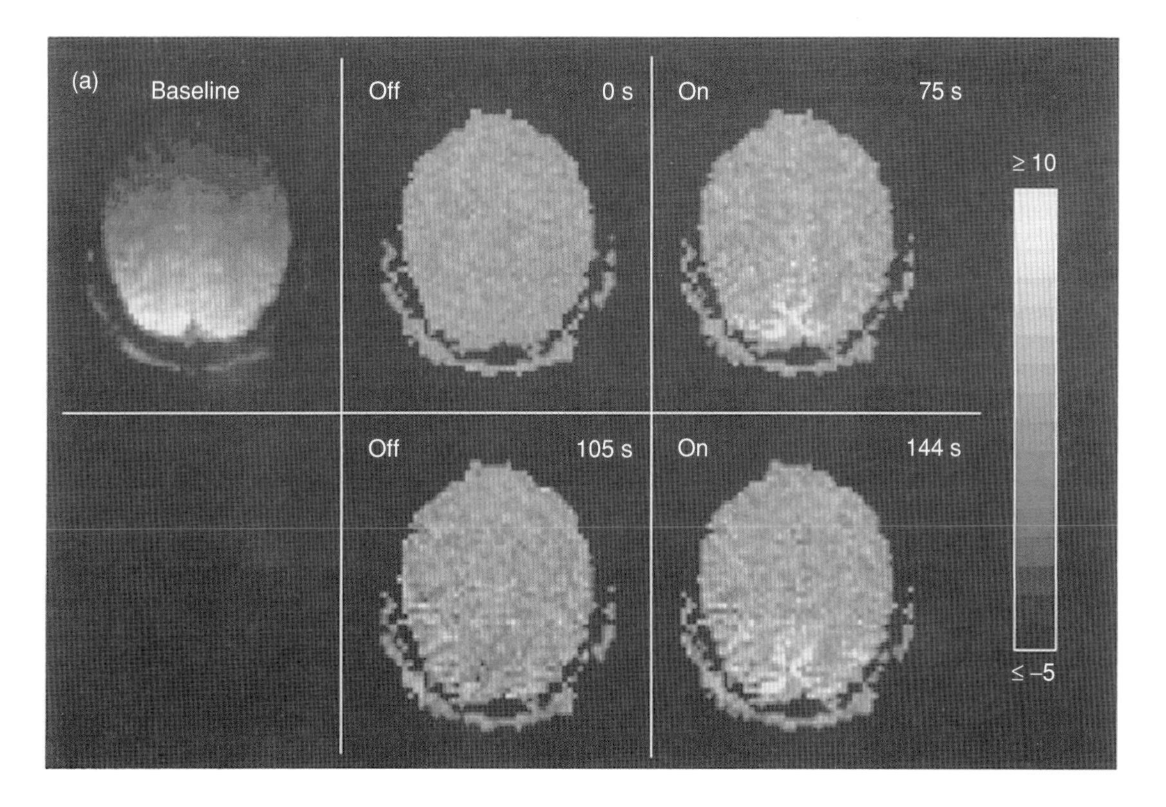


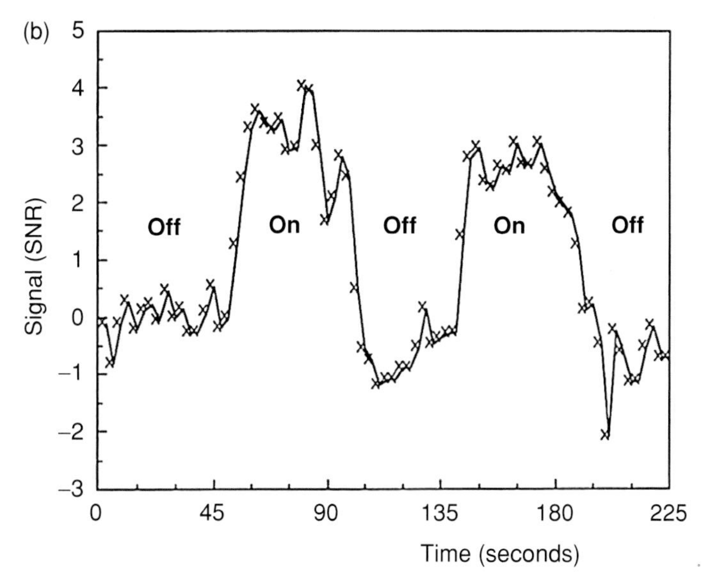


**Figure 17.25.** Blood-oxygen-level-dependent (BOLD) imaging. Using blood oxygen level as a contrast agent, the brain's response to a visual stimulus (flashing light/no flashing light) can be localized and its time dependence studied. The diagram below shows the effect in terms of the signal-to-noise ratio (SNR) as a function of time, where On/Off refers to time with/without flashing light. (Courtesy of M. S. Cohen.)

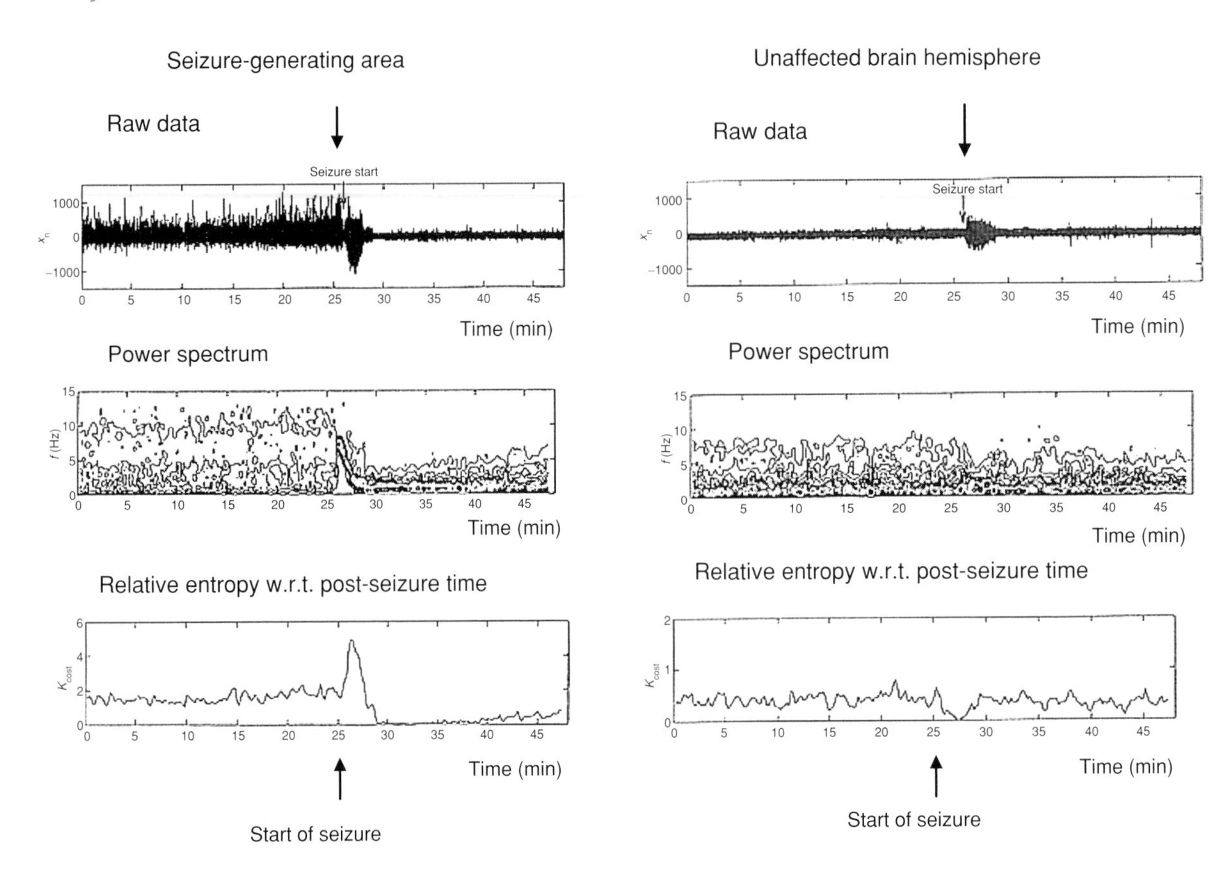


**Figure 17.26.** EEG recordings with electrodes partially implanted in the skull prior to, during, and after an epileptic seizure. The data were taken in the center of the generating area (left-hand side) and from an unaffected area for comparison (right-hand side). The upper plot shows the raw data, the middle one the power spectrum (a measure of amplitude range versus frequency), and the lower plot a measure of entropy in the data normalized with respect to a time window shortly after the seizure. (Courtesy of R. Quian Quiroga, J. Arnold, K. Lehnertz, and P. Grassberger.)

due to brain activity (perception, internally/externally stimulated motion control) are monitored by electro-encephalography (EEG). The activity of other muscles is the subject of electro-myography (EMG). Compared with ECG and EMG, the interpretation of EEG signals requires more sophisticated data analysis. One active EEG research area is epileptology, where a primary goal is to anticipate seizures and allow preventive measures to be taken. Figure 17.26 shows EEG recordings prior to, during, and after an epileptic seizure, together with quantitative analysis. For reliable seizure prognosis, mathematical methods from statistical physics are being investigated, some of them shown in the figure.

There are limitations to EEG and ECG due to systematic errors arising from the contact of electrodes on the skin, which limit precision (implanted electrodes are used only in serious cases). The measurement of biomagnetism by magneto-encephalography (MEG), magneto-cardiography (MCG), and magneto-myography (MMG) offers an elegant solution to this problem, since it requires no contact with the skin. It also offers fascinating new possibilities. The biomagnetic fields induced by currents in nervous cells are generally very small. The strongest is produced by the heart but is as weak as the magnetic field of an electric screwdriver at 5 m ($10^{-10}$ Tesla). The brain's magnetic field is weaker by a factor of 1000, many million times weaker than the geomagnetic field ($10^{-13}$ Tesla compared with $10^{-4}$ Tesla). Therefore measurements require a well-shielded room and an extremely high-sensitivity magnetometer. Here the properties of superconductivity can be exploited very effectively, particularly using superconducting quantum-interference detectors (SQUIDs, see Chapter 8). A SQUID is basically a superconducting ring with two Josephson contacts (Figure 17.27).

In practice SQUID measurements are limited due to the finite value of the self-induction of the ring and to electronic noise. MEG and MCG use a magneto-gradometer

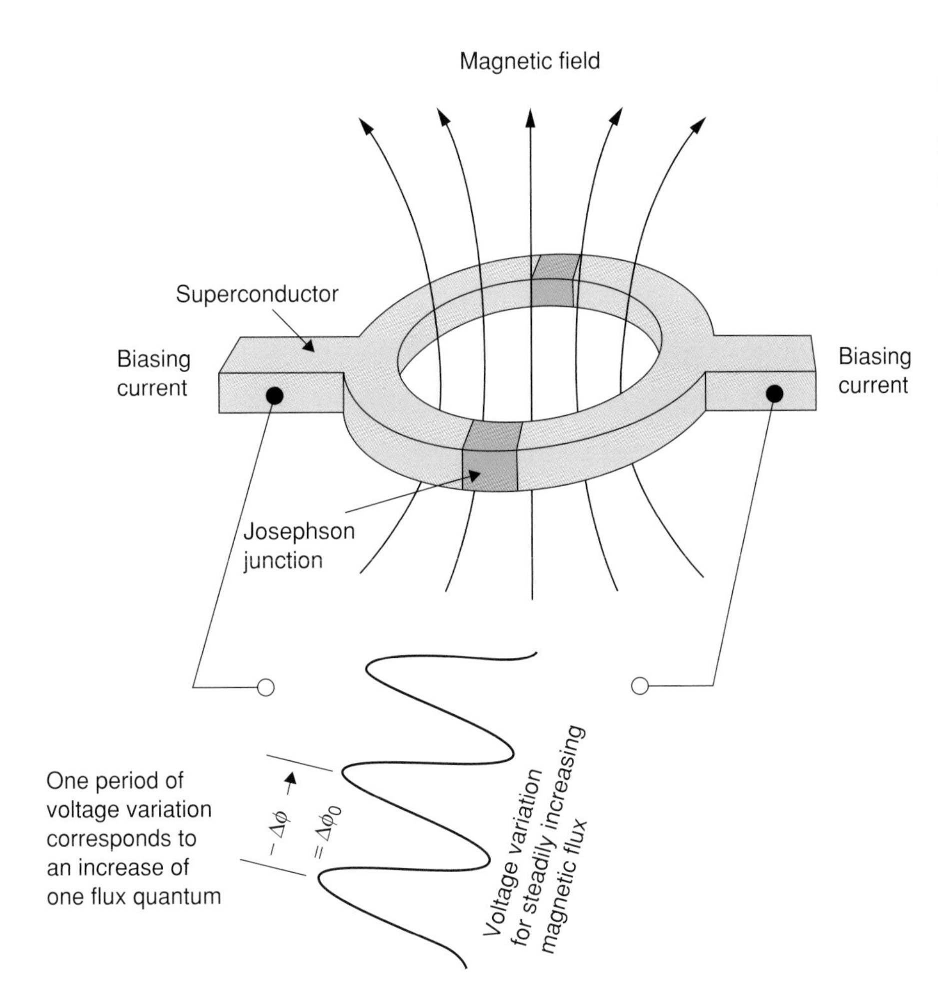


**Figure 17.27.** A superconducting quantum interference detector – SQUID – is a ring of superconducting material with two Josephson contacts, thin layers of insulating material that separate the ring into two pieces with a fixed but independent phase.

sensitive to local changes of magnetic field. For this the SQUID is coupled to superconducting transformer loops. The measurements on a fetal heart described below used a detector with 31 channels, each consisting of a tiny SQUID and two transformer coils of 18–20-mm diameter 70 mm apart. The apparatus is magnetically shielded and embedded in a cryostat (Figure 17.28).

The upper part of Figure 17.29 shows the response from 31 channels averaged over many heart signals, giving a rough spatial distribution of the source of the signals. The raw data from all channels had to be superimposed and a trigger time giving maximum correlation between the signals chosen. The signals also had to be filtered to suppress lower frequencies from the mother's heart.

One astonishing example from medical physics reveals differences in brain activity between pianists and non-pianists! Figure 17.30 compares the reactions of a trained pianist and a non-pianist when both listen to a piece of well-known piano music. The difference in the MEG can be easily interpreted as involuntary motor activity evoked by audible perception. This can be proved by independent tests of the MEG signal associated with controlled finger motion, which are part of brain mapping (see Section 17.8). The next objective is a similar study for violin players!

Although these results employed low-temperature SQUIDs, high-temperature superconducting materials are well suited for MEG. Systems using only high-temperature superconductors for SQUID and transformer coils have successfully been tested. The reduced cooling is a significant advantage and will make MEG even more desirable for

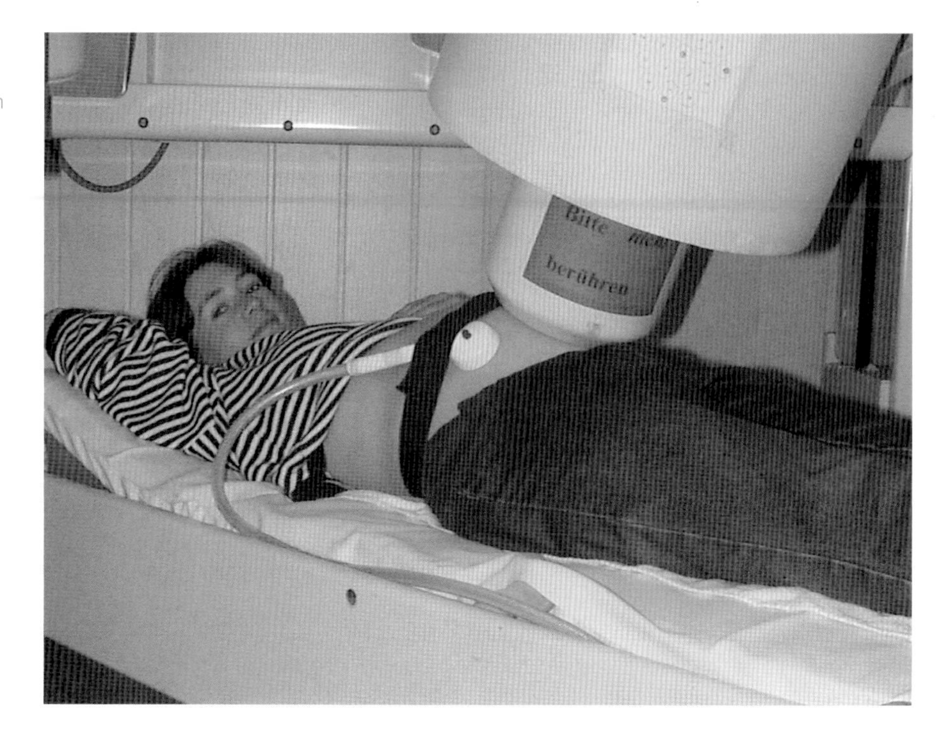


**Figure 17.28.** A 31-channel SQUID apparatus for fetal cardiography. This investigation is performed in a magnetically shielded room in a hospital. (Courtesy of Dr. Ekkehard Schelußner, Friedrich Schiller University, Jena.)

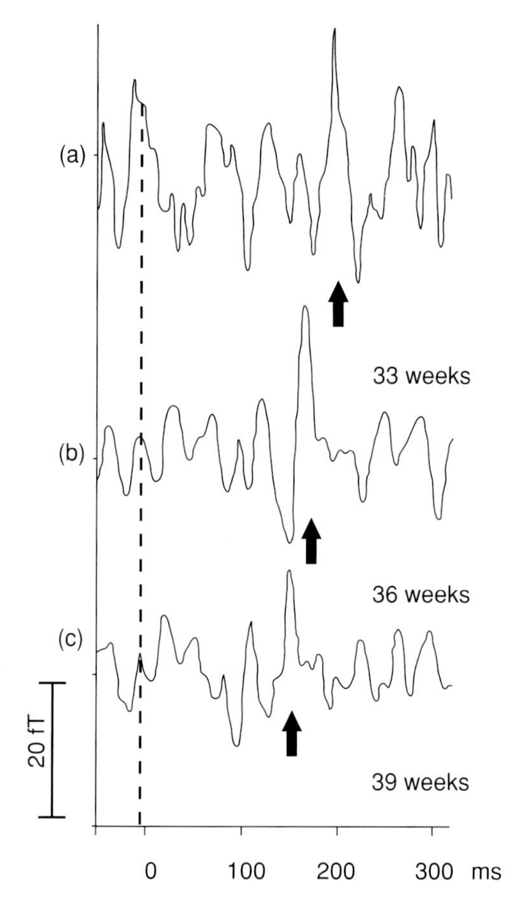


**Figure 17.29.** The fetal magneto-cardiographic (MCG) data are analyzed as a time-averaged signal from each of the 31 SQUID channels, showing also the approximate location of the fetal heart. (Courtesy of E. Schleußner *et al.*)

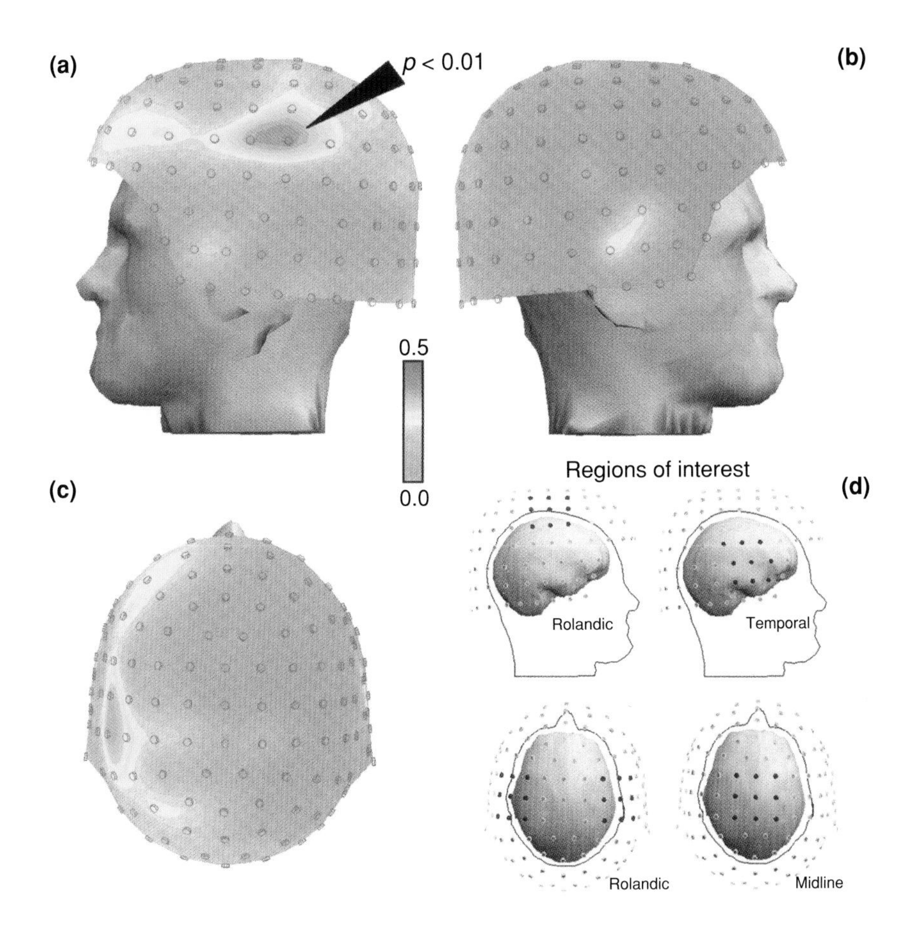


**Figure 17.30.** Magneto-encephalographic (MEG) signals from a trained pianist (a) and (c), and a non-pianist (b). Color indicates the relative magnitude of the surface gradient (variation with location) of the MEG signal. In part (d) the positions of the detectors are shown, with the channel in the region of interest in blue. The sensors have no contact with the head. (Courtesty of J. Haueisen and Th. Knoesche.)

diagnostics. Most of the effort goes into determining the influence of the surrounding tissue on the anisotropy of the signal. This is essential for improving spatial resolution and correct interpretation.

## 17.7 Medical applications of ultrasonics and lasers

Both ultrasonics and lasers are widely used in medicine today. Ultrasound is appreciated because of its strictly non-invasive character, and the laser because of its outstanding qualities of intensity, brilliance, monochromaticity, and coherence. Research mainly focuses on technological improvements and better exploitation of physical effects.

Ultrasound (US) waves – acoustic waves in the frequency range 1–25 MHz – are used both for diagnostics and for therapy. Their penetration depth varies from a few centimeters for 20 MHz to about 50 cm for 1 MHz. The energy of incident US waves is partially absorbed and transformed into heat and mechanical pressure, which are useful for therapy. One example is lithotripsy, whereby suitably shaped US shockwaves are superimposed to create extreme pressure peaks in small volumes so that gallstones and kidney stones can be destroyed.

Ultrasound imaging is also completely non-invasive and does not cause pain or any adverse side effects. A single piezoelectric element may serve as emitter and receiver. Several methods are used to display ultrasonic information, one popular method being the "brightness mode" whereby the amplitude of the echo is converted into a modulation of brightness on a screen. The echo signals are ordered according to their delay

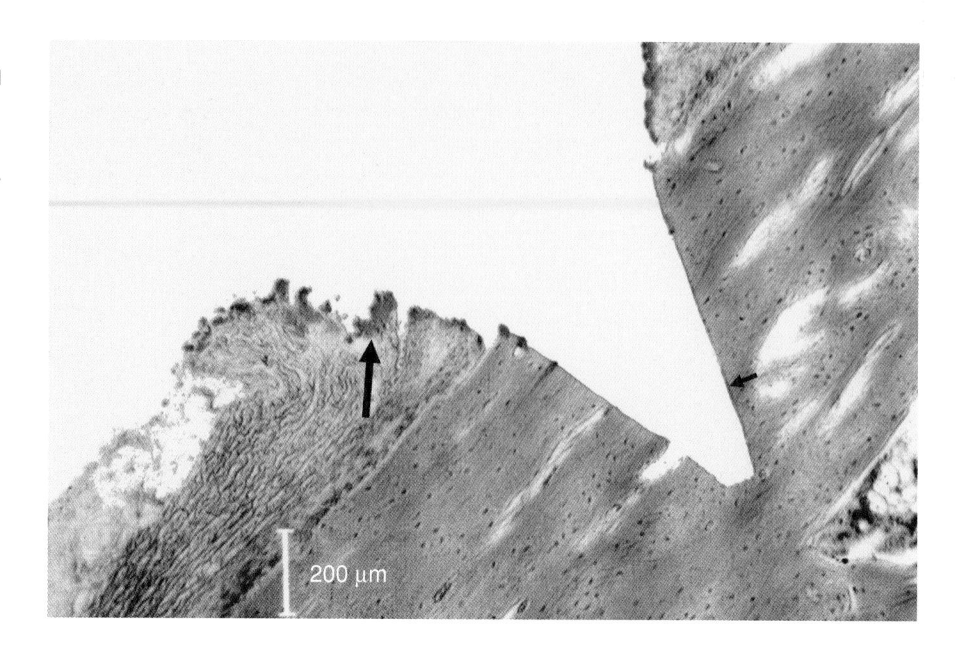


**Figure 17.31.** The profile of a cut of bone (pig rib) performed with a $CO_2$ laser operated in pulsed mode (pulse frequency 5 kHz and pulse width 300 ns). The large arrow indicates the region of coagulated tissue in the periosteum. (Courtesy of M. Ivanenkc and P. Hering.)

relative to the time of emission ($t_0$), which depends on spatial localization inside the body. The lateral spatial resolution is of the order of a few millimeters, and axial resolution along the direction of the wave can be in the sub-millimeter range. Moving objects such as ventricular heart valves and the blood flux can be measured, since the frequency of US waves reflected from moving objects is altered due to Doppler shifts. The frequency shift can easily be analyzed by standard signal-processing techniques, e.g. the demodulation techniques used in radio engineering. The ultrasonic emitter and receiver can be very compact and robust. Therefore all signal generation and signal processing is done with AC signals in the frequency range 1–50 MHz, for which the technology is well developed. Ultrasonic diagnostics are economical and easily used by medical doctors without specialist assistance.

The characteristic properties of laser light can be used for various medical purposes. The most common application is for the laser to replace the scalpel. Here the extremely low divergence of a laser beam, its sharpness and small diameter (down to about 10 μm), high power density, and excellent timing control are exploited. Laser surgery for soft tissue, e.g. in ophthalmology, has become common practice. The next goal is to use laser light for ablation of hard tissue. The problem in this case is not the power density of a laser – with laser light much thicker material can easily be cut – but the minimization of side effects due to necrosis. The laser light rapidly heats up material in a very small subvolume and causes sudden explosive evaporation. During a laser pulse of duration typically 300 ns, the temperature reaches up to 400 °C, and the resulting micro-explosions tear off pieces of hard tissue (e.g. bone).

It is thus important to minimize heat production. Therefore pulsed laser light is used and current experiments employ additional external cooling. Figure 17.31 shows a microscopic view of a cut profile in bone performed with a pulsed $CO_2$ laser. There is a zone of thermal necrosis of the bone surface (periosteum), over about 1000 μm. One of the primary research goals in this field is to decrease this necrosis by varying parameters such as the pulse frequency, pulse length, and focal diameter, sharpening the beam profile, improving external cooling, etc.

Another recent innovative laser application is holographic imaging of a patient's face. These three-dimensional images are used to prepare for maxillo-facial surgery in pathological cases when plastic surgery is requested because deformations cause severe

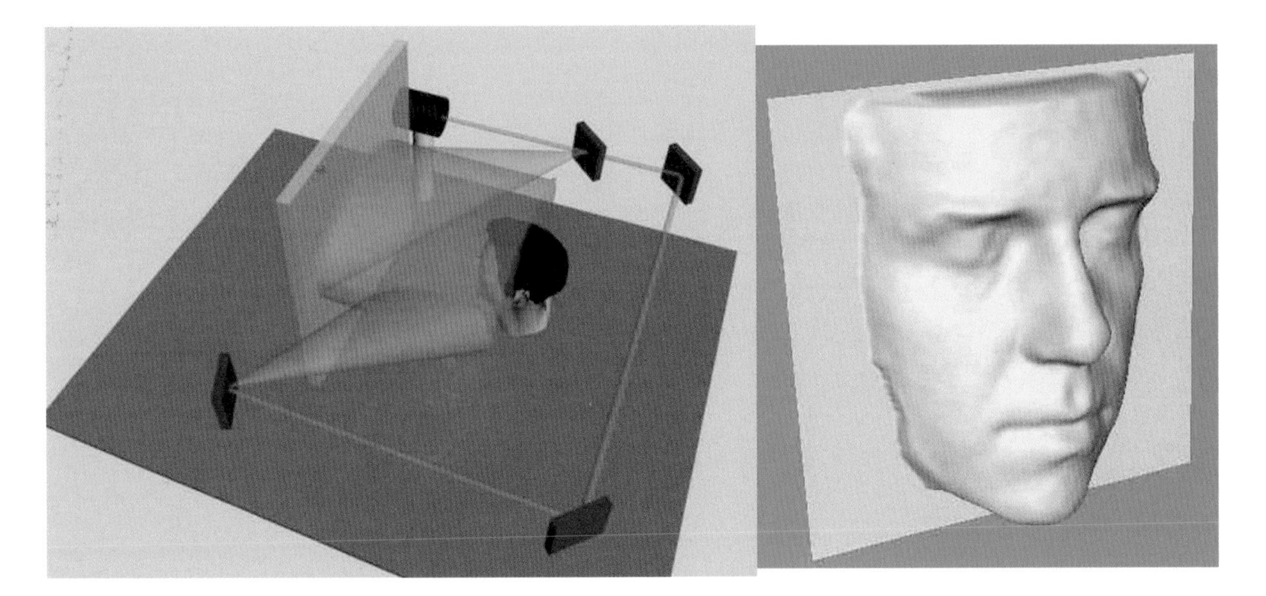

Figure 17.32. Left – the set up for taking a hologram of a patient's portrait prior to maxillo-facial surgery. The actual data taking is done within 25 ns using pulsed laser light, thus avoiding systematic uncertainties due to motion of the patient. The reconstruction of the hologram (right) is more time-consuming but can be done later. (Courtesy of P. Hering and A. Thelen.)

health problems. Traditional methods use laser triangulation or phase measurement but take a relatively long time (seconds), during which time motion of the patient cannot be avoided. With new methods, data taking can be accomplished within 25 ns, using a pulsed laser for the holographic recording so that effects due to motion are completely avoided. Figure 17.32 shows laser holography of a face, with the experimental set up and a reconstructed three-dimensional image. The actual reconstruction of the hologram can be done later with a precision better than 0.5 mm.

## 17.8 Brain research – a complex application

This survey concludes with a look at how medical physics contributes to one of the most fascinating and challenging branches of medical research, brain research. Several complementary techniques are needed. The brain, the most complex organ, controls all vital biomedical processes, both internally and externally stimulated; enables us to communicate with our environment, processing information from the body's sensors and reacting accordingly; and, last but not least, allows us to process, associate, and store information and create emotions and thoughts. The brain's "hardware" consists of some $10^{10}$ neurons (nerve cells) with inter-neuronal connections (synapses), which outnumber neurons by a factor of about 10–100 and may be partly controlled by brain activity. Most of these cells are in the cerebrum, which controls all the characteristics which distinguish humans from animals, in particular in such complex processes as consciousness, self-reflection, and creative and abstract thinking. All information processing involves interrelated electromagnetic processes and biochemical processes.

Centuries ago several distinct parts of the brain were identified anatomically. Later, systematic evaluations of the effect of brain injuries or changes of brain function allowed certain functions to be associated with identifiable parts of the brain, but these methods gave results that were incomplete and imprecise. Moreover, such information could not be obtained from a patient before or during treatment; instead, one had to rely on information from other studies – mostly of dead brains – with the inferences subject to many systematic uncertainties. Medical physics now offers some 30 years' experience of an increasing number of different methods for investigating brain structure in vivo

**Table 17.6.** *Classes of diagnostic tools provided by medical physics for brain research and medical treatment, together with some important benchmarks and the type of information which is provided by the given method (all these methods have been explained in the preceding sections)*

| | | Some general benchmarks | | |
|---|---|---|---|---|
| Method | Kind of information | Time resolution | Localization | Functional specificity |
| NMR imaging | Precise three-dimensional structural information | Down to minutes | mm | None |
| $^{18}$FDG, PET, SPECT, fNMRI BOLD | Glucose metabolism Blood oxygen concentration | Down to ≈10 min | cm | Limited |
| PET, SPECT with neurotransmitter as tracer | Biochemical activity | 10–60 min | 2–5 cm | High |
| MEG, EEG | Direct fast brain activity | ms | ≈cm | Limited |

and studying its function in detail, either as part of anamnesis and diagnosis, or under controlled laboratory conditions. Table 17.6 gives a rough classification of the methods used for brain studies and their specificity, goals, and advantages.

The power of these techniques lies in the combination of methods for treatment and the complementarity of the extracted information. Computer-based fusion of data from different detector systems now enables doctors to combine information in a very quantitative and accurate way. For example, the MEG signal of the brain's surface current density due to finger motion is combined with NMRI of the outer cerebrum (Figure 17.33), allowing precise association of function with brain region. This results in part from the comparison of pianists and non-pianists described in Section 17.6 and is only one example of the very active field of brain mapping.

As well as being essential for research, functional brain imaging is needed as input for surgery or radiation therapy. Since brain maps are very much individual, in particular regarding the cerebrum, where the formation of function centers depends on

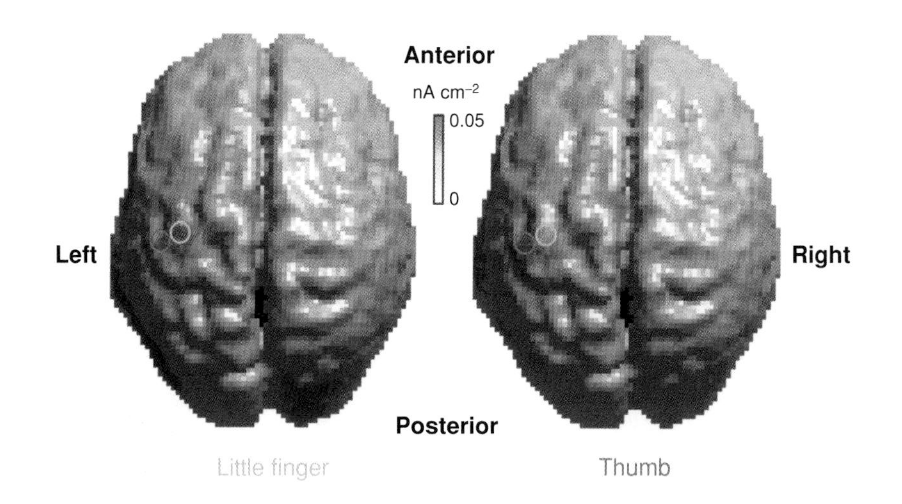


**Figure 17.33.** Brain-surface current density recorded with MEG for controlled movement of little finger and thumb, showing differences between trained pianists and non-pianists. The cerebrum structure from a NMR image allows a precise assignment of activity to brain region. (Courtesty of J. Haueisen and Th. Knoesche.)

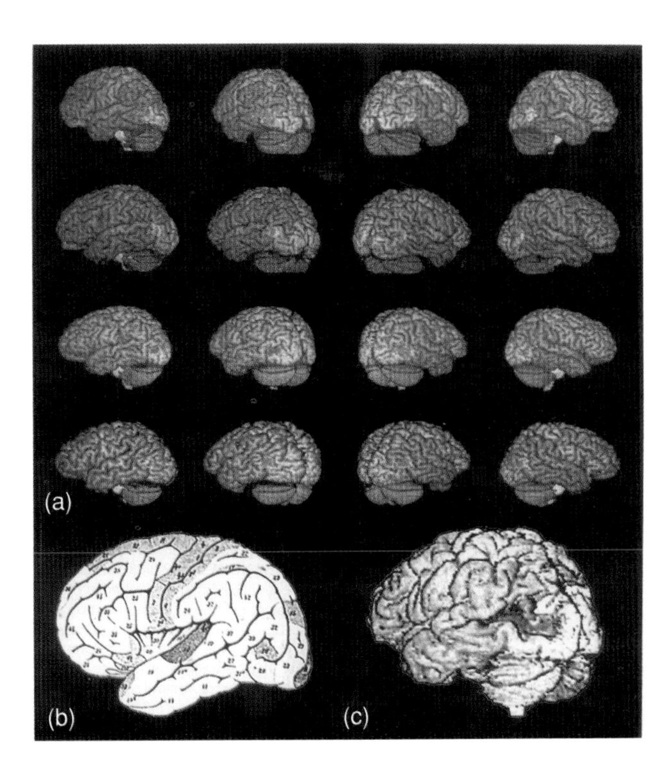


**Figure 17.34.** Functional diagnostic imaging to differentiate high-grade from low-grade tumors before surgery. The tumor is visible in a $T_2$-weighted NMR image (a), but only faintly visible after administration of contrast agent for NMRI (paramagnetic atoms) (b). Cerebral-blood-flux- sensitive measurement (c) suggests the presence of a high-grade tumor, whereas a measurement sensitive to glucose metabolism indicates only a low-grade tumor. The result could be confirmed only by excising the tumor. (Courtesy of J. Mazziotta *et al.*)

subjective impressions, evolution, and possible diseases, the process cannot be performed once and for all, but has to be cross-checked and fine-tuned each time. Another goal is improved tumor classification prior to surgery or biopsy investigation. In an investigation, normal NMRI showed a tumor candidate in the right temporal lobe. To establish whether this is a high-grade tumor, with rapid growth and high malignance, various techniques sensitive to different biomedical characteristics were used. The measurement of the relative blood flux (by NMRI) suggested a high-grade tumor, whereas PET with $^{18}$FDG as tracer measured a low glucose metabolism, indicating a low-grade tumor (Figure 17.34). The excised tumor proved to be of high-grade type, illustrating that systematic study is needed before safe and correct decisions can be taken. It also shows the huge potential for improving the medical treatment of cancer. Active research also tries to find correlations between the occurrence of afflictions such as Creutzfeldt–Jacob disease (CJD), Alzheimer's, and Parkinson's diseases and any structural or functional physiological abnormalities in the brain cells, which would allow early diagnosis and make treatment possible, as well as hinting on the origin of such diseases. Figure 17.35 shows NMR images of a patient in the early stages of CJD, where progressive dementia has already been diagnosed. $T_1$- and $T_2$-weighted images do not show any abnormalities; however, another contrast-enhancing technique, diffusion-weighted NMR imaging, shows abnormally high signal intensities in the right cerebral cortex. It is not clear how these high intensities can be explained and with which physiological effect they are associated. This is the subject of ongoing research.

Other methods also measure brain phenomena such as the formation of plaques (NMRI), the change of cerebral blood flux (PET and fNMRI), and neurotransmitter concentration (PET and SPECT). However, the major problem is to establish a significant and safe correlation between such observations and clinical and pathological findings. This would provide new and powerful tools to fight diseases that today cannot be treated at all. These examples illustrate how improvement of medical treatment is intimately linked to innovation and improvement over a wide field of physics. Industrial and

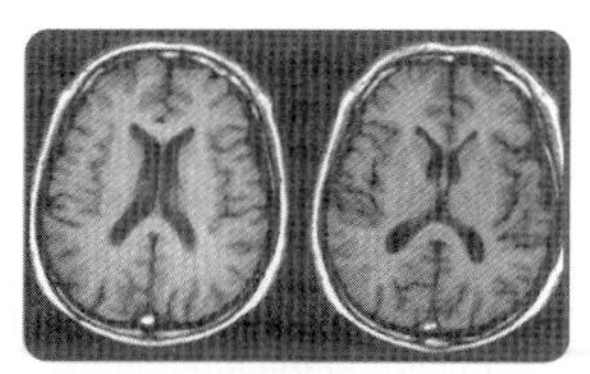
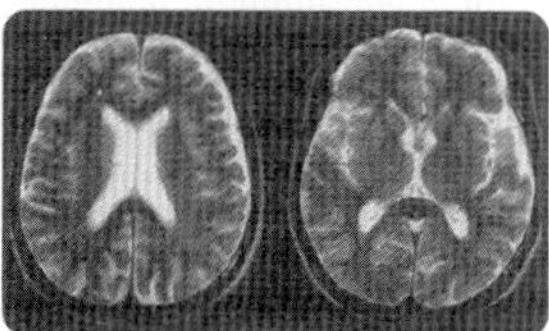
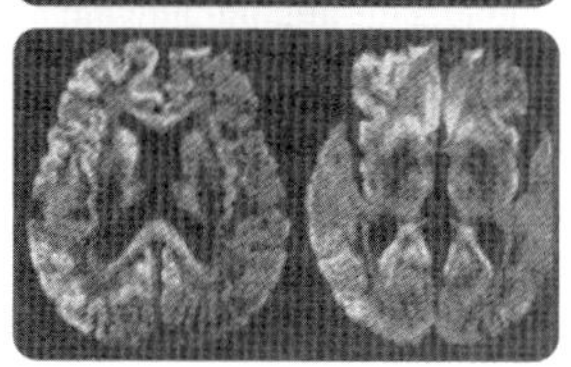

**Figure 17.35.** Searching for early indications of Creutzfeldt–Jacob disease (CJD) by non-invasive diagnostics. $T_1$- and $T_2$-weighted NMR images (upper and middle parts) do not reveal any abnormalities in the brain of a patient who already had progressive dementia and was found later to have CJD. However, diffusion-weighted NMR imaging shows abnormally high signals in the right cerebral cortex. Diffusion-weighted NMRI is based on the effect that the signal is decreased by the Brownian motion of molecules, and thus of their associated spins. The spins are dephased by the spin–spin relaxation process, leading to an exponential decrease of the initial NMR signal after the radio-frequency (RF) pulse. This dephasing can be manipulated by applying repeated RF pulses, which can flip the spin in a similar way to a spin-echo signal. This enables regions with different degrees of diffusion to be distinguished. (Courtesy of P. Gould.)

academic research invest much effort in this field. Progress in medical physics depends on innovation in other fields, such as detector physics, nuclear physics, and laser technology. It profits from a better understanding of basic physical processes, and also of processes that at first glance do not even appear to have any relevance for medical applications. Such is the nature of applied physics research.

## FURTHER READING


1. W. R. Hendee and E. R. Ritenour, *Medical Imaging Physics*, 4th edn., New York, Wiley, 2002.
   Covers a broad range of techniques – NMRI, PET, SPECT, ultrasound.
2. G. B. Saha, *Physics and Radiobiology of Nuclear Medicine*, 2nd edn., Berlin, Springer-Verlag, 2001.
   Covers the physics behind radiation therapy, radiology, nuclear medicine, radionuclides for tracers, and therapy.
3. A. W. Toga and J. C. Mazziotta, *Brain Mapping – The Methods* and *Brain Mapping – The System*, New York, Academic Press, 1996.
   The first book gives a nice historical introduction, with a comprehensive overview on diagnostics like NMRI, PET, SPECT, fluorescence, and optical methods; the second book covers applications with interesting examples for medical doctors.
4. C. Schiepers (ed.), *Diagnostic Nuclear Medicine*, Berlin, Springer-Verlag, 2000.
5. E. M. Haacke (ed.), *Magnetic Resonance Imaging: Physics Principles and Sequence Design*, New York, Wiley, 1999.
6. D. R. Bernier (ed.), *Nuclear Medicine: Technology and Techniques*, St. Louis, Mosby DuPont Pharma, 1993.
7. I. Khalkhali (ed.), *Nuclear Oncology: Diagnosis and Therapy*, Philadelphia, Lippincott, Williams and Wilkins, 2001. For medical doctors.
8. C. A. Puliafito (ed.), *Laser Surgery and Medicine: Principles and Practice*, New York, Wiley, 1996.
9. G. P. Lask, (ed.), *Lasers in Cutaneous and Cosmetic Surgery*, London, Churchill Livingstone, 2000. For practical work, explains basic principles.
10. D. R. Vij, *Medical Applications of Lasers*, Dordrecht, Kluwer, 2002. Modern and comprehensive.
11. G. M. Baxter (ed.), *Clinical Diagnostics – Ultrasound*, London, Blackwell Science, 1999.
12. M. G. Hennerici (ed.), *Cerebrovascular Ultrasound: Theory, Practice, Future Development*, Cambridge, Cambridge University Press, 2001.
13. S. H. Heywang-Koebrunner and R. Beck, *Contrast Enhanced MRI of the Breast*, Berlin, Springer-Verlag, 1995.
14. C. T. W. Moonen (ed.), *Functional MRI*, Berlin, Springer-Verlag, 1999.
15. A. R. Smith (ed.), *Radio Therapy Physics*, Berlin, Springer-Verlag, 1995.


# Nikolaj Pavel

After doctoral work on experimental elementary particle physics at CERN and Wuppertal, Nikolaj Pavel became assistant professor at Hamburg, working in experimental high-energy physics and computational physics.

In 1996–1997 he was a research fellow and guest lecturer at Imperial College, London. From 1998 until 2003 he was a temporary professor at Siegen, working in applied detector physics, with emphasis on applications in medicine and biology, and on detector development for experiments on astroparticle physics. In 2003 he moved to take up the Chair of Experimental Elementary Particle Physics and Applications in Other Fields at the Humboldt University in Berlin.

# 18 Physics of materials

Robert Cahn

## 18.1 Introduction

It is a universally acknowledged truth that any useful artifact is made of materials. A *material* is a substance, or combination of substances, which has achieved utility. To convert a mere substance into a material requires the imposition of appropriate *structure*, which is one of the subtlest of scientific concepts; indeed, the control of structure is a central skill of the materials scientist, and that skill makes new artifacts feasible.

Any structure involves two constituents: building blocks and laws of assembly. *Crystal structure* will serve as the key example, with wide ramifications. Any solid element, pure chemical compound, or solid solution ideally consists of a crystal, or an assembly of crystals that are internally identical. (The qualification "ideally" is necessary because some liquids are kinetically unable to crystallize and turn instead into glasses, which are simply congealed liquids.) Every crystal has a structure, which is defined when (1) the size and shape of the repeating unit, or unit cell, and (2) the position and species of each atom located in the unit cell have all been specified. At one level, the unit cells are the building blocks; at another level, the individual atoms are. The laws of assembly are implicit in the laws of interatomic force: once we know how the strength of attraction or repulsion between two atoms depends on their separation and what angles between covalent bonds issuing from the same atom are stable then, in principle, we can predict the entire arrangement of the atoms in the unit cell. Once the crystal structure has been established, many physical properties, such as the cleavage plane, melting temperature, and thermal expansion coefficient, are thereby rendered determinate. In the fullest sense, *structure determines behavior*.

Structure in a different sense is another central concept in the science of materials. This is *microstructure*. Alloys, ceramics, semiconductors, and minerals all consist of small grains, either of a single crystalline species or else of two or more distinct species, each constituting a *phase*. The compositions, sizes, shapes, and mutual geometric disposition in space of these phases are the variables of a microstructure, and in their totality determine the properties of the material. To take a simple example, a pure metal consists of crystal grains, typically about 0.1 mm across, abutting on each other at plane or slightly curved grain boundaries. Such a microstructure results from solidification of the molten metal. Unlike most crystal structures, however, this kind of *polycrystal* is not stable, because the three boundaries meeting at an edge, or the

*The New Physics for the Twenty-First Century*, ed. Gordon Fraser.

four boundaries meeting at a point, cannot be in configurational equilibrium at all such edges or points (i.e. nodes). At best, equilibrium will obtain at a small minority of such nodes. An example of a stable node would be three identical grain boundaries meeting along a straight edge at mutual inclinations of 120°; if the specific interfacial tensions are equal, they will be in balance for this special configuration only. Gradual evolution of unstable microstructures is a frequent concern for materials scientists; such evolution can be exploited to create desired microstructures.

When two phases are present together, the range of possible microstructures increases very sharply, and the study and control of such microstructures is a central concern of the science of materials. Control depends primarily on precise heating schedules and on modification of minor constituents (impurities) in the material. We shall encounter a number of such forms of control in what follows.

One way of classifying materials, as we have just seen, is in terms of crystal structure and microstructure. Another is to classify them in terms of the nature of their applications. The big divide is between *structural materials* and *functional materials.* A structural material is used for its resistance to mechanical loads of various kinds (steel is an example), whereas a functional material is valued for its response to electrical, magnetic, optical, or chemical stimuli (silicon for transistors is an example). In this concise overview, I shall focus on functional materials, because these link more closely to physics than do structural materials.

In the foregoing, I have repeatedly referred to "materials scientists." Materials science is now a well-established profession that incorporates aspects of metallurgy, ceramic science, polymer science, physics, and chemistry, in a variable mix. In this chapter, however, I am focusing firmly on physics as exemplified in the study and production of materials.

## 18.2 Impurities and dopants

John Chipman Gray, a nineteenth-century American lawyer, remarked that "dirt is only matter out of place"; if he were writing today, while knowing some science, he might well add "and having matter in the right place is crucial for getting materials to behave as we wish." That became very clear in December 1947 when the transistor – a solid-state amplifying device – was created by John Bardeen, Walter Brattain, and William Shockley at the Bell Telephone Laboratories in New Jersey. Transistors depend absolutely for their operation on the presence of two kinds of regions, n-type and p-type, in which, respectively, electrical charge is carried predominantly by freely moving (negatively charged) electrons and by positively charged "holes," which are the embodiment of missing electrons. Silicon atoms have four valence electrons each; alloying silicon with minute amounts of phosphorus (with five valence electrons) means that extra electrons are injected into the silicon crystal and an n-type region results, whereas on alloying with boron (with three valence electrons) holes are injected and a p-type region is formed. These regions were discovered at Bell Telephone Laboratories in 1939 by a closely collaborating team of metallurgists, physicists, and chemists (Russell Ohl, George Southworth, Jack Scaff, and Henry Theuerer) – this was before the concept of an overarching science of materials had taken root. Their crucial finding had to be put aside because of wartime pressures and only came to fruition almost a decade later when the first transistor was invented. A transistor, whatever its exact geometry (and there are numerous different types), has three electrodes, like the triode vacuum tube which was one of the devices used before the solid-state revolution which began in 1947; at least one of these electrodes must be located in an n-type region, at least one in a p-type region. The two types of region are formed in the right places by dissolving the appropriate additive in the appropriate regions of a semiconductor crystal. Intentional additives like these are no longer impurities, matter out of place; instead, they are called *dopants*, minority matter in the right place.

In the early days of the transistor, more attention was paid to germanium (an element similar in several respects to silicon, also of valence four); silicon took over almost completely in the 1950s. In 1950, Bell Labs was faced by a crisis: their best germanium was not pure enough. They recognized that adding tiny amounts of boron or phosphorus to a base element that was contaminated by various elements, including those which were to be added locally, would generate irregular, unintended p and n regions, and the transistor would not then function properly. William Pfann, another Bell Labs stalwart, saved the situation in mid 1950. He had invented a process called *zone-refining*: a rod of the material to be purified was fitted with narrow heating zones, which were made to move slowly along the rod: because solid and liquid in equilibrium with each other have different compositions, the passage of many narrow molten zones causes unwanted impurities to be progressively swept to the end of the rod, which is eventually cut off. The remaining germanium is of an unprecedented degree of purity, ready to be "doped" effectively with boron or phosphorus. Zone-refining was used for a decade or so, and transistor technology would have been impossible without its use. Then, methods of purifying germanium and silicon by purely chemical means were improved so dramatically that zone-refining was no longer needed. Nowadays, elements at 99.999% purity are regarded as nothing special (less than ten parts per million of impurities), and semiconductors are made very much purer than that. The crucial lesson from the early days of the transistor was that impurities (dopants) cannot usefully be added to materials that have not first been purified to a very high degree.

The experience of purifying semiconductors brought about a sea-change in attitudes to "dirt and doping." In the 1930s, most solid-state physicists had turned up their noses at the thought of investigating semiconductors, which were regarded as dirty, entirely irreproducible materials to be avoided by respectable scientists. As late as the 1940s, one eminent physicist declared that his university department would never concern itself "with the physics of dirt." As he spoke, the Bell Laboratories were preparing to announce the discovery of the (doped) transistor.

The result of what happened at that time is that, in materials science, the derogatory concept of "dirt" or "impurity" has given way to the benevolent concept of "additive" or "dopant," and the study and control of dopants has become one of the central skills of materials scientists.

Two other examples of the crucial importance of dopants, among many that could be adduced, are *phosphors* and *varistors*. Phosphors are materials that absorb light of one wavelength – or alternatively X-rays or electrons – and emit light of another wavelength. This process is known as *fluorescence*. Phosphors are crucial, for instance, for coating the faceplates of oscilloscopes and television tubes, and for making medical X-ray radiographs visible to the examining physician. In the first case, a traveling electron beam generates the visible image on the cathode-ray tube; in the second case, X-rays generate visible light. Another familiar use of phosphors is in the coating of fluorescent light tubes; here light of several wavelengths is emitted so that the tubes emit an approximation to white light. The development of highly efficient phosphors, as well as of improved emulsions for photographic films, can be regarded as a byproduct of a long series of fundamental researches carried out in the laboratory of Robert Pohl in the University of Göttingen in the first half of the twentieth century on "color centers" in insulating ionic crystals such as common salt, locations where incident radiation brought about changes in color. It took a long time to understand the roles of the various impurities, linked to crystal defects such as sites where atoms are missing, that generated the color centers. The American Frederick Seitz (regarded by many as a

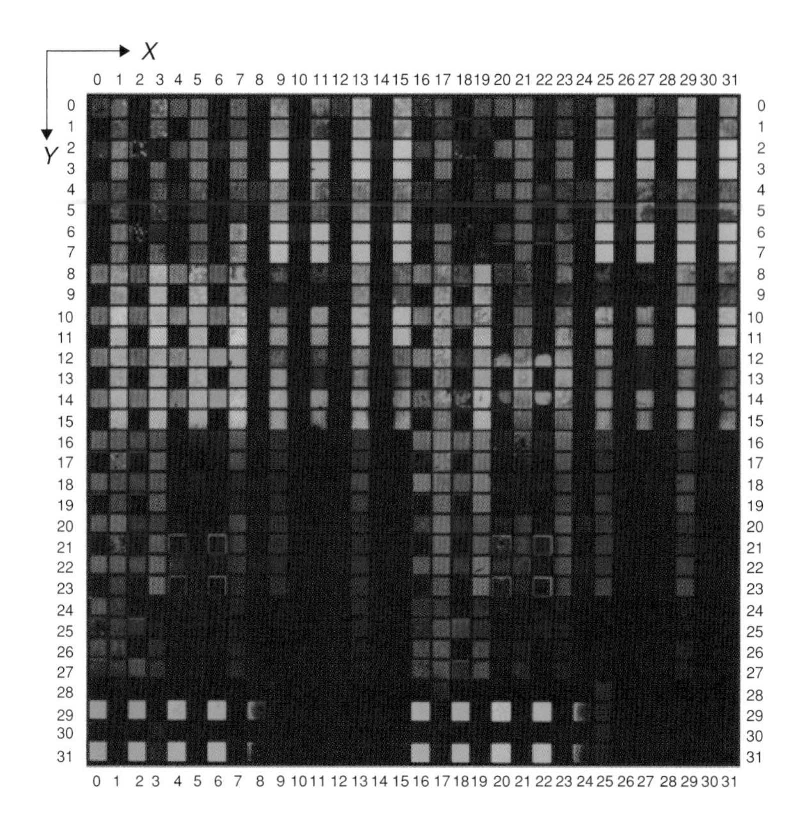


**Figure 18.1.** A "library" of phosphor samples made by systematically varying the proportions of four metal oxides, excited by ultraviolet radiation. The individual samples were deposited from the vapor phase under computer control. (Courtesy X.-D. Xiang.)

father of materials science) in 1946 and 1954 assembled all that was known about color centers in two major review papers; such reviews are a central vehicle for progress in all branches of physics, and notably in materials science. The short period between Seitz's two reviews has been called "the golden age of crystal defects," in particular the study of crystal vacancies, sites where constituent atoms or ions are missing from a crystal.

Even earlier, when American industry was trying to pluck up its collective courage to design, manufacture, and distribute the first television sets, in the 1930s, one of the decisive considerations was the search for a phosphor for the cathode-ray screens. Zinc silicate, made from a scarce natural mineral, was found to perform well, but it was not until the mid 1920s, when Hobart Kraner at Westinghouse Research Laboratories found that this compound could be synthesized and doped with a small, crucial addition of manganese ions, that a plentiful *and reproducible* supply of phosphor became available and television became practicable. At this time, phosphors were still being developed empirically and it was to be some years later that their design became based on clear scientific understanding. Recently, underlying theory has been complemented by a new kind of empiricism – *combinatorial screening*. Here, a large number of compounds varying in composition by small steps is deposited from the vapor, each deposit being only a millimeter or so in size; the sequence of compositions, and the deposition process, is automatically controlled by computer. The "library" of deposits is then irradiated with ultraviolet light and the type and intensity of luminescence (Figure 18.1) can easily be measured (again automatically) for each deposit. In this way, a process of synthesis and screening that used to take weeks or months can be completed in a matter of hours.

"Varistor" is the nickname of a voltage-dependent resistor, widely used as a safety device in electrical circuits. The most common kind is based on zinc oxide, ZnO. Zinc

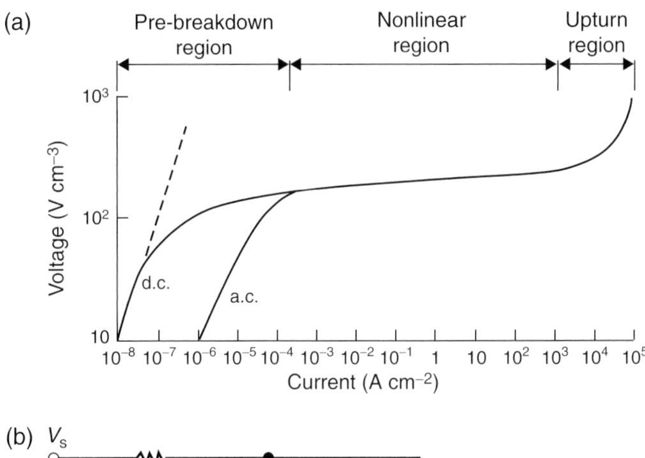


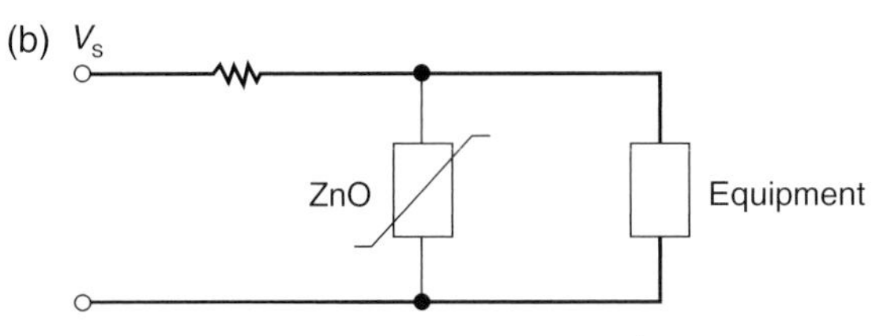


**Figure 18.2.** (a) The typical current–voltage ($I$–$V$) characteristic of a ZnO varistor, for direct current (d.c.) and alternating current (a.c.). In the "nonlinear region" a small increase in voltage generates an enormous increase in current. Both direct-current (d.c.) and alternating-current (a.c.) characteristics are shown. (b) The circuit arrangement used to protect equipment with a varistor.

oxide varistors have to consist of small crystal grains, i.e. they must be polycrystalline (unlike transistors, which until recently have all been made from single crystals, usually of silicon). The essential characteristic of such a varistor is shown in Figure 18.2(a). This "non-ohmic" characteristic in the central (nonlinear) region is very different from that of a normal metal, which would resemble the dashed line on the left of the diagram; in fact, in the pre-breakdown region, the device behaves like a highly resistive metal. In the central region, a small surge in voltage leads to a huge enhancement of the current, and consequently a varistor can be used to prevent overloading of sensitive electronic equipment (Figure 18.2(b)).

Varistors have to consist of small grains; single crystals of ZnO do not have a voltage-dependent resistance, and neither for that matter do high-purity polycrystals. The grain boundaries are populated by dopant molecules, which sharply raise the resistance in their immediate vicinity: the local resistivity of the grain boundaries is larger than that of the grain interiors by a factor of about $10^{12}$ – a huge disparity. The high-resistance region near a grain boundary extends to a distance of 50–100 nm either side of the boundary; because of this symmetry, the device works well with alternating currents as well as with direct currents.

The varistor effect was discovered in the ZnO–$B_2O_3$ system in the mid 1960s by M. Matsuoka in a Japanese industrial laboratory. The company, Matsushita, had long manufactured resistors for electronic circuitry: the resistors were fired in hydrogen, and the company wished to save money by firing in air, so long as this did not ruin the behavior of the resistors. ZnO was one of many resistor materials that were examined. Electrodes were put on the end surfaces of the experimental resistors in the form of firable silver-containing paints. One day, the temperature controller failed and the ZnO now no longer obeyed Ohm's law. Further research showed that the silver paint contained bismuth as an impurity, and this had diffused into the body of the resistor. Matsushita recognized that this chance discovery was interesting and "threw the periodic table at it," with around 100 staff members becoming involved, and many patents taken out. Parts per million of dopants made a crucial difference, just as had been found earlier in the studies on germanium and silicon for transistors. Many different dopants are used now, but bismuth, in the form of its oxide, still has pride of place. The impure grain boundaries receive a positive charge, which makes them effective obstacles to the passage of electrons unless the electron energy passes a critical value. Physical understanding of the precise mode of action of varistors came after the empirical discoveries, as often happens. It is, incidentally, very common for dopants in polycrystals – ceramics, semiconductors, and metals – to segregate to grain boundaries; indeed, the study of the magnitude of such segregation in various materials is a major theme in modern materials science.

Accidental discoveries of major importance arising from the breakdown of a temperature controller have been made in various kinds of materials; for instance, in the late 1950s, Donald Stookey at the Corning Glass Company's research laboratory in New York State discovered a glass that crystallized into a high-strength glass ceramic when accidentally overheated. The key in such episodes is that someone with good background knowledge and, crucially, with the gift of observing the unexpected and pouncing on it should be present. As Louis Pasteur remarked in quite another connection, accident favors the prepared mind.

## 18.3 Isotopes

A hundred years ago, Ernest Rutherford and Frederick Soddy discovered, to their considerable surprise, that the end products of various radioactive cascades, such as radium G, actinium E, and thorium E, were all just variants of lead with slightly different relative atomic masses. Soddy called these "isotopes," meaning that the variants had the same "topos," or place, in the periodic table, with the same atomic number. They are not, of course, restricted to radioactive species.

More recently, the arrival of nuclear energy led to the detailed study of the uranium isotopes, notably those present in nature, $^{235}$U and $^{238}$U. These break down radioactively, with (different) long half lives, to yield lead isotopes, $^{207}$Pb and $^{206}$Pb, respectively. Figure 18.3 shows these breakdowns schematically. One brilliant scientist, Clair Patterson (1922–1995), at the California Institute of Technology succeeded in using these "clocks" to determine the age of the Earth, an issue that had led to decades of intense dispute among geologists and physicists, going right back to Lord Kelvin more than a century ago; several others had struggled with these radioactive clocks, but it was not until Patterson put his hand to the matter in the early 1950s that a reliable result (4.55 billion years) could be derived from the "lead–lead" approach based on the quantity ratios of three lead isotopes, including the invariant stable isotope $^{204}$Pb. He compared the "radioactive" ages of various geological specimens, including deep-sea sediments, and the ages of certain meteorites, which were taken to date from the original formation of the Earth. The difficulty arose from the contamination of background $^{204}$Pb by lead from industrial and other human injection of lead into the biosphere. Patterson found that this contamination began with the smelting of lead ores 5000 years ago, and became much worse with the addition of tetraethyl lead as an antiknock agent in petrol for internal combustion engines, which led to atmospheric contamination. He compared $^{204}$Pb concentrations in the biosphere with those of its biochemical analog, calcium, in pristine and contaminated tissues. He showed that humans contained lead concentrations 500–1000 times pre-industrial levels, and he changed his focus to proving and combating this dangerous buildup of a seriously toxic element. He was mainly responsible for the near-abolition of tetraethyl lead from petrol in recent decades.

Patterson pioneered the use of "clean" techniques to eliminate the contaminant lead from his samples. Clean rooms, using ultrafilters and sealed work spaces, are now used for such purposes as analyzing ice cores from the Antarctic to chart the increase in

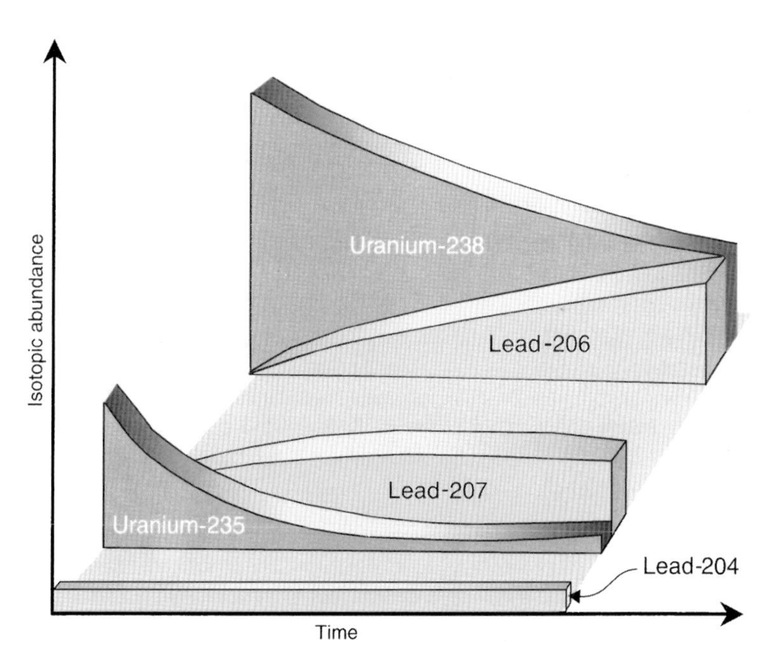


**Three geological clocks.**
Clock 1, uranium-238 → lead-206. Clock 2, uranium-235 → lead-207.
Clock 3, the growth of lead-206 and -207, relative to -204.

**Figure 18.3.** Three geological "clocks" based on the radioactivity of uranium isotopes. Two clocks are based on uranium-238 decaying into lead-206, and on uranium-235 decaying into lead-207. The time axis covers more than a billion years. The third, linked, clock is the growth in concentration of lead-206 and lead-207 relative to lead-204, which is stable and does not change with time. (From *The Dating Game*, by Cherry Lewis, Cambridge University Press, 2000.)

concentrations of pollutants in the air over time, as well as carbon dioxide levels many centuries ago; ice cores are built up of snow falling over the millennia. Clean rooms have also become essential devices for making integrated (micro)circuits in which one dust particle can easily spell malfunction.

Modern techniques of isotope separation have allowed elements that in their natural states contain two different isotopes to be isotopically purified. The techniques all rely on slight differences in mass or mobility of the isotopes; chemical methods are clearly not applicable since isotopes of the same element are chemically identical. Separation was first achieved with uranium during the Manhattan Project of World War II, simultaneously with heavy hydrogen (deuterium) to make heavy water (a crucial neutron moderator for early nuclear reactors), and extended later to boron, of which a pure isotope was needed for electronic counters to detect neutrons.

The most recent element to attract the attention of the isotope separators was carbon. Natural carbon contains 98.93% of $^{12}C$ and 1.07% of $^{13}C$. The small proportion of the heavier isotope suffices to impair the thermal conductivity of diamond, the best conductor of heat in Nature. This is because thermal conduction is limited by the scattering of phonons, vibrational quasiparticles, as they move through a lattice made irregular by containing a few randomly distributed heavy carbon atoms. This is a matter of industrial concern because, since 1984, thin strips of diamond have been used as heat sinks for such devices as power transistors. In 1990, W. F. Banholzer at the General Electric Company's research center grew synthetic diamonds from isotopically purified carbon, with the $^{13}C$ removed; the removal of the mere 1% of the heavy isotope led to a 50% increase in the conduction of heat.

A more recent proposal is the use of deuterium instead of ordinary natural hydrogen to neutralize "dangling" chemical bonds in silicon and silicon dioxide in electronic devices to greatly reduce the likelihood of circuit failure, because deuterium is held more firmly than hydrogen.

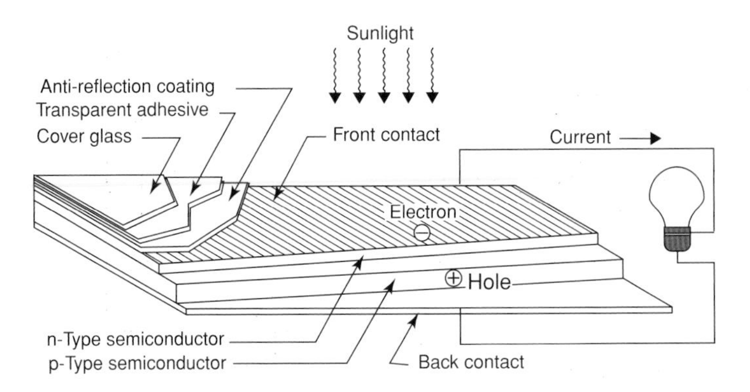


**Figure 18.4.** The structure of a semiconductor solar cell. The total thickness of the active (semiconducting) layers is typically a few tenths of a millimeter.

## 18.4 Solar (photovoltaic) cells

Ultimately, all our energy supply on Earth comes from nuclear fusion of hydrogen isotopes, the source that powers the Sun. Even the radioactivity that keeps the interior of the Earth hotter than it would otherwise be is a pale residue of earlier solar processes. Coal and petroleum represent fossilized solar energy, and therefore doubly fossilized fusion energy. Industrial alcohol made from crops and used as a fuel supplement for cars in some parts of the world again represents stored solar energy. Not surprisingly, physicists have long sought a practical way of harnessing solar radiation *directly*; the search was for direct, as distinct from fossil, energy derived from the Sun.

When I took up photography in the 1930s, I was given a photoelectric exposure meter to tell me what aperture and exposure speeds to use on a bright or a cloudy day. This operated with a semiconductor, elementary selenium, assembled in a very simple circuit that turned light into direct current, which passed through an ammeter. This strategy went back to a study of selenium performed in 1877, not so long after Michael Faraday had studied silver sulfide, one of the very first known semiconductors. Nowadays, exposure meters are built into cameras and are more sophisticated devices that can automatically control camera settings.

Then, in 1954, three scientists working at the Bell Laboratories in New Jersey constructed the first *solar cell* based on p–n junctions in doped silicon, with supplying power for satellites and other space vehicles in mind. Such a cell differs from a transistor in that it has only two electrodes, either side of the junction, unlike the three junctions required for a transistor. A schematic view of such a solar cell is shown in Figure 18.4. The back electric contact is a simple metal sheet, while the front contact is a grid of fine vapor-deposited metal that occupies only about 10% of the surface, so as not to absorb too much of the incident sunlight. The key characteristic of the semiconductor is its *band gap*, the range of kinetic energies of electrons that are excluded because of interference from the crystal structure of the semiconductor. A simplified way of looking at this is that the moving electrons behave like waves (because of the principle of wave/particle duality) and at certain energies – and therefore wavelengths – these waves are totally reflected at sheets of atoms in the crystal and then reflected again . . . in effect, they cannot move at these speeds. For crystalline silicon, this band gap has a magnitude of 1.14 eV. Sunlight at the Earth's surface has a range of wavelengths approximately as shown in Figure 18.5 (the exact plot depends on the atmospheric pressure, humidity, and cloudiness). Such a plot can be expressed in terms of wavelength, frequency, or quantum energy (which is proportional to frequency); in Figure 18.5, the quantum energy of the photons, or "particles" of light, is the measure employed.

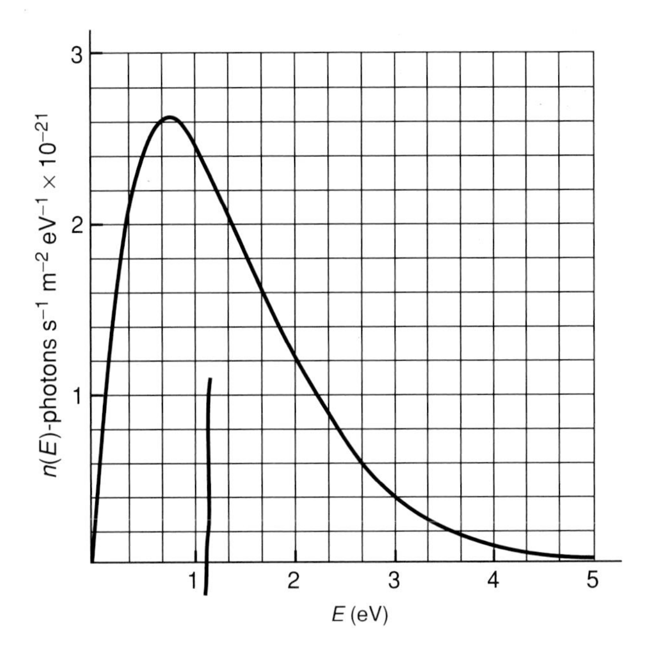


**Figure 18.5.** The solar spectrum at the Earth's surface for a standard air mass, expressed as the number of photons per second per square meter, $n(E)$, falling within a unit band of photon energy, against the photon energy, $E$. The thick vertical line marks the band gap of crystalline silicon.

The vertical line in Figure 18.5 marks the upper-band-gap energy of silicon; any light color of quantum energy less than this (i.e. of wavelength greater than the critical wavelength) cannot generate a current in the solar cell; in effect, only the blue end of the visible spectrum, and ultraviolet beyond that, is useful. Again, photons with quantum energy exceeding 1.14 eV are usefully absorbed in the silicon, but the excess energy above 1.14 eV turns into waste heat. So, allowing both for light that is too red and for the excess energy of useful photons, the maximum theoretical efficiency of a silicon solar cell is ≈44%. An absorbed photon generates an electron–hole pair (where a hole is the name given to a missing electron of a specific kinetic energy; this entity behaves as a well-defined current-carrying medium). If one now allows for the finite electrical resistance of the cell, the imperfect separation of electron–hole pairs at the p–n junction, and certain other electrical factors, the limiting theoretical efficiency of a silicon solar cell finishes up being around 15%. This value was calculated in a UK Open University teaching module in 1979. A few years later, in 1990, much higher theoretical values were estimated, as shown in Figure 18.6, both for silicon and for some

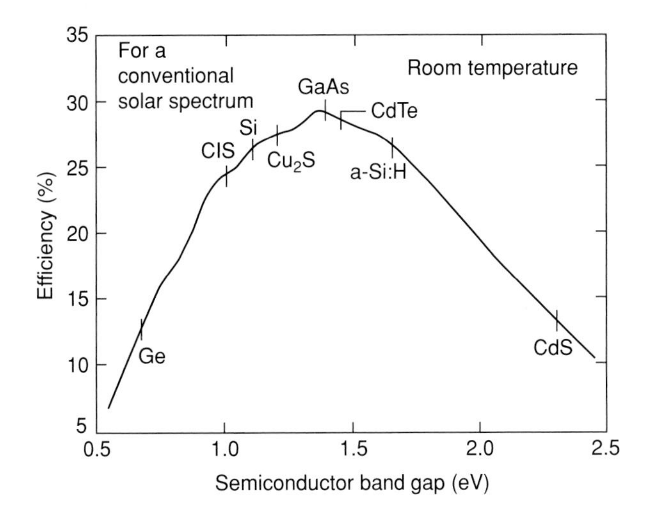


**Figure 18.6.** The theoretical efficiencies of solar-radiation-absorbing materials with band-gap energies ranging from 0.5 to 2.4 eV. The peak efficiency corresponds to a band gap of 1.4 eV, but there are many candidate materials yielding values above 20%. Data from Strategies Unlimited. (After Ken Zweibel, *Harnessing Solar Power: The Photovoltaic Challenge*, Plenum Press, New York, 1990.)

other rival semiconductors; this must be attributed to a reduction of the electrical-loss mechanisms, so that the fundamental limit of 44% is more nearly approached. After several decades of technical improvement, practical efficiencies of up to 25% are now achieved in production silicon solar cells. A typical cell has an area of about 100 $cm^2$ and generates about 3 A at 0.5 V; cells are linked in series to produce useful voltages.

In 2002, Martin Green at the University of New South Wales, Sydney, proposed incorporating into a solar cell another semiconducting device called a "down-converter," which can split a high-energy photon into two photons, each of lower energy. If this were done with, say, ultraviolet photons in the solar spectrum, each of the lower-energy photons resulting from the split could be utilized by the solar cell with much less waste of available energy.

Ken Zweibel, in his book *Harnessing Solar Power* (see Further reading), considers how this level of efficiency should be regarded. Conventional power-plant efficiencies, using superheated steam in turbines, can nowadays be made as high as 55%: at first sight, solar cells look poor value in comparison. However, as Zweibel points out, the original conversion efficiency of the process, millions of years ago, that converted sunlight into coal or petroleum, via photosynthesis followed by decay, was perhaps 1%. Zweibel believes that photovoltaics offer "the most efficient means of transforming the primeval fuel – sunlight – into electricity," being much more efficient than rivals such as biomass growth and conversion into alcohol.

Some of the semiconductors shown in Figure 18.6 are also used in manufacturing solar cells. In practice, the optimum candidate, gallium arsenide, is simply too expensive to make, but "a-Si–H," or hydrogenated amorphous silicon, has in recent years acquired many supporters. Thin-film solar cells made of this material are made by vapor deposition, under circumstances that incorporate about 10% of hydrogen chemically. After years during which it was argued by some that an amorphous – i.e. noncrystalline – material cannot have a well-defined energy gap, and therefore would not respond properly to doping, W. E. Spear and P. G. Le Comber at the University of Dundee, Scotland, showed in 1975 that such hydrogenated a-Si can be doped successfully by the usual dopants (B and P), leading to huge changes in conductivity, and the following year the first a-Si solar cell was demonstrated. Soon afterwards, the precise mechanism of the enabling role played by hydrogen was unraveled: a-Si solar cells are somewhat less efficient than those using crystalline silicon, but are much cheaper to make, and – together with other kinds of cell made of cheap polycrystalline silicon – are taking over a progressively larger market share.

A number of other types of thin-film cell, a current favorite being $Cu(Ga, In)Se_2$, are candidates for future exploitation; the latest approach is based on semiconducting organic materials, which are very cheap but with very low efficiencies. In the long run, cost/efficiency tradeoffs will establish distinct ecological niches for many of these alternative forms . . . but for the present, high-efficiency single-crystal silicon cells still hold the lion's share. The capital cost per watt of output capacity has gone down in the past 20 years from about $12 to about $5, and further economies of scale in future seem assured; accordingly, the spread of solar cells in the economy has now become rapid, especially in the domestic market. Figure 18.7 shows how the price per unit of power has diminished as sales have leapt.

Several large companies, such as Kyocera, Sharp, BP, and Shell, have set up large solar subsidiaries. Some of these are also focusing their energies on hydrogen production, with the forthcoming large-scale introduction of automotive fuel cells very much in their minds. Large banks of solar cells will, in the view of some forecasters, be used to electrolyze water, so that in effect future cars will run *almost* directly on the Sun's rays. A senior manager of one of these companies has claimed that, by 2060, 30%–40%

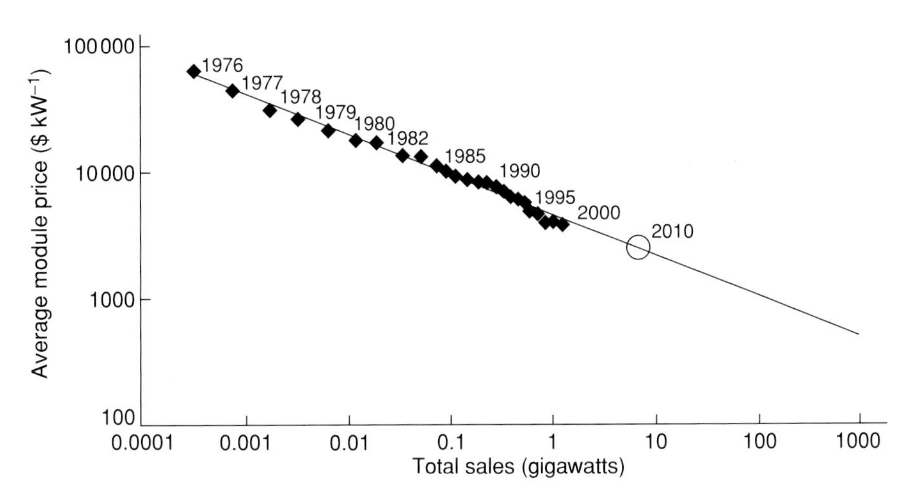


**Figure 18.7.** The average price of a photovoltaic module divided by its peak power versus the total power generated by all modules sold to date. The trend line shows the cost of modules falling about 20% each time that sales double. (From "Solar power to the people," *Phys. World*, July 2002, p. 35, by Terry Peterson and Brian Fies.)

of the world's energy usage will come from renewable sources (solar energy clearly leading this). This extraordinary claim is not entirely beyond possibility.

## 18.5 Liquid crystals and flat-panel displays

For half a century, electronic computers have used cathode-ray tubes to display text and images; originally these were black-and-white images, but more recently color versions became standard. Although their quality has reached very high levels, they do have some drawbacks, especially the need for substantial power and their large "footprint," that is, the space required to accommodate one on a desk. Accordingly, flat displays that need very little power and occupy only a small space are now taking a rapidly growing share of the computer market; for laptop computers, such displays have been the only option from the beginning.

Although alternatives are beginning to appear, flat-panel displays have hitherto depended wholly upon the use of liquid crystals. Liquid crystals are a very curious state of matter, first discovered by Friedrich Reinitzer at the University of Prague in 1888; the first such material had organic molecules in the form of long chains, and its peculiar feature was that it had two melting points (145 and 179 °C). At ambient temperature, the compound, cholesteryl benzoate, is a normal crystalline solid, and above 179 °C it is a normal transparent liquid; in between, the material is a turbid liquid that reacts to polarized light. At the time, no one knew the structure of this turbid liquid, but we now know that the molecules are roughly aligned with each other and parallel to a master line called the *director*, but do not form a periodic structure like a crystal. The first physicist to study this compound, Otto Lehmann, at Aachen in the 1890s, was so intrigued by its properties that he invented the self-contradictory name *liquid crystal*; this descriptor has stuck ever since.

During the first few decades after this discovery, other liquid-crystal compounds were discovered but the existence of liquid crystals met with resolute skepticism from most scientists of the time, some of whom were convinced that they were suspensions of minute particles. This lasted until about 1925. During the next 35 years, liquid crystals were accepted as "real," but interest was limited to a few academic enthusiasts. From 1960 onwards, developments were rapid and constantly accelerating because of the invention of technological applications. In 1974, about 5000 different liquid-crystalline compounds were known, by 1997 this had increased to about 70 000 compounds,

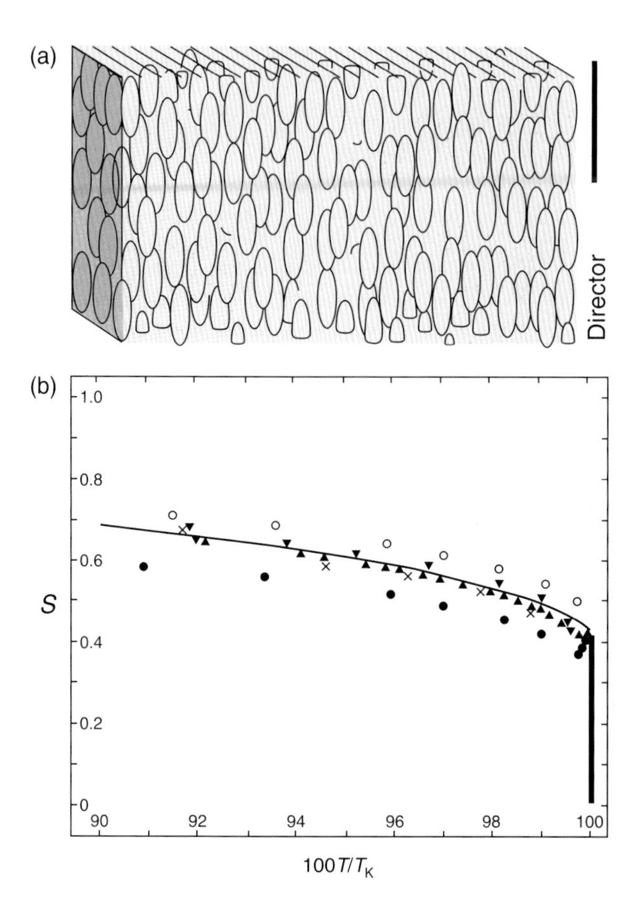


**Figure 18.8.** (a) An idealized view of the molecular organization within a nematic liquid crystal. (b) The nematic liquid-crystal orientational order parameter, $S$, plotted as a function of the reduced temperature, $T/T_K$, for various materials. $S$ is defined as the average value of $(3\cos^2\theta - 1)/2$, where $\theta$ is the angle between the axis of each long-chain molecule and the director. (From *Bull. Inst. Phys.*, p. 279, 1972.)

most but by no means all made up of long molecular chains. Originally, the running was made by chemists who diligently sought new compounds, but more recently, as applications have developed, physicists have become increasingly involved.

Figure 18.8(a) shows the idealized structure of a nematic liquid crystal, where the elongated ellipses represent the long-chain molecules. The position of the center of gravity of each molecule is random (there are other types of liquid crystal that are more highly ordered; for instance, with molecules arranged in well-defined layers). In this sketch, the ellipses are all perfectly aligned, which would be the ideal situation at absolute zero temperature if the compound did not freeze into a crystal first. In reality, each chain is inclined at a variable *small* angle $\theta$ to the director, because of thermal motion, and this leads to a variation of the quality of alignment with temperature, as shown in Figure 18.8(b). It can be seen that the plot falls on or close to the same master curve for a number of different compounds; each compound loses its alignment totally at a sharply defined temperature, $T_K$. (Near the left-hand edge of Figure 18.8(b), the liquid crystal freezes into a proper crystal.) It can be seen that the alignment is never perfect. In this, it resembles the behavior of magnetic spins in ferromagnetic metals, or of order in the distribution of atoms in an intermetallic compound such as $Cu_3Au$. Any molecule that is not spherical should in principle turn into a liquid crystal (an ordered liquid) at a sufficiently low temperature *if* it can be discouraged from freezing first. The freezing points of candidate materials have steadily been lowered, and today some nematic phases are known that are liquid crystals at room temperature. This is crucial for display applications.

A liquid-crystal display (LCD) depends on three characteristics of a nematic liquid crystal: first, that the alignment of the director, and therefore of the undisturbed

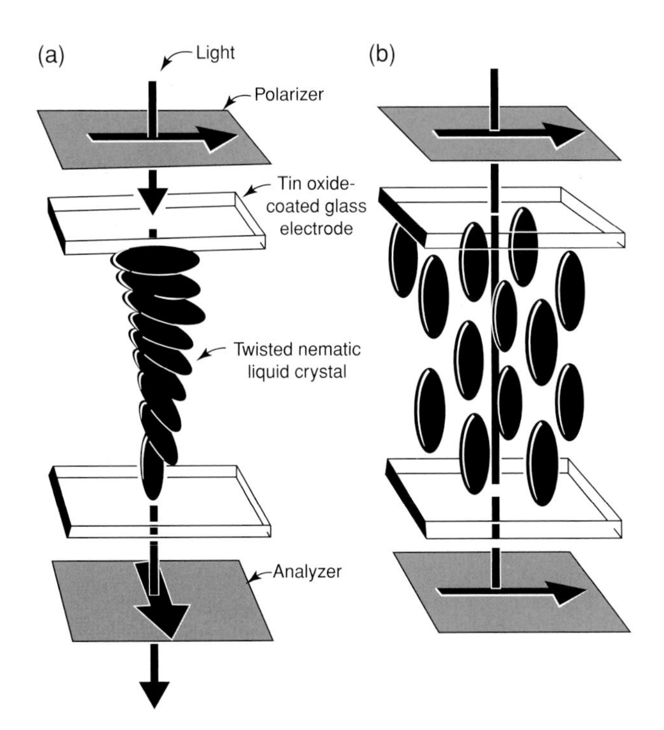


**Figure 18.9.** Operation of the twisted nematic mode in a liquid-crystal display. The thick arrows show the plane of polarization of the light before and after passing through the liquid-crystal layer; the planes of polarization of the polarizer and analyzer are shown by the striations. (a) With no voltage applied the molecules naturally twist between the top and bottom plates, guiding the light through 90° and allowing it to be transmitted. (b) A voltage applied across the two plates causes the molecules to align in the electrical field, thus preventing rotation of the light and transmission through the analyzer.

molecules, can be physically controlled; second, that an applied electric field will deflect the molecules from their undisturbed alignment; and third, that the transmission of polarized light through such a liquid crystal depends sensitively on the alignment of the molecules.

A LCD consists of two glass sheets held 5–10 μm apart, with liquid crystal between. The inside of the sheets is coated with a special chemical layer, which is then brushed with a soft cloth. This quasi-magical process creates minute grooves in the surface layer and the director, and the long molecules align themselves with these grooves. One of several ways – in fact, the most common way – of arranging things is to have the imposed directors on the two glass sheets perpendicular to each other, as shown in Figure 18.9; of necessity, in the undisturbed state the molecules rotate through 90° on going from one glass sheet to the other (the left-hand of part Figure 18.9). The incident, plane-polarized light coming through the first glass sheet (from a uniform white-light source behind the screen) is in effect rotated as it passes along the array of molecules and, when it emerges, it can pass the crossed analyzing screen beyond the second glass sheet, so the image is bright. If now a potential difference of only 1–2 V is applied normal to the sheets, the molecules become aligned with the electric field, as shown on the right-hand side of Figure 18.9, and now the plane of polarization of the transmitted light is unrotated and thus the light is extinguished by the analyzing screen.

All liquid-crystal displays are controlled by arrays of mutually perpendicular conductors, so that each element of the display – a *pixel* – corresponds to one intersection of the array. Small displays, such as for calculators and cheap watches, can be controlled passively: this means that a pixel is turned on (light passes through at that pixel) when *both* of the perpendicular conductors which define that pixel are passing enough current: when that happens, the electric field resulting from the currents suffices to create a situation as in Figure 18.9, right-hand side. The drawback to such a passive mode of operation is that the currents must be "on" for a substantial fraction of a second, and since real-time control of the total image requires each pixel to receive

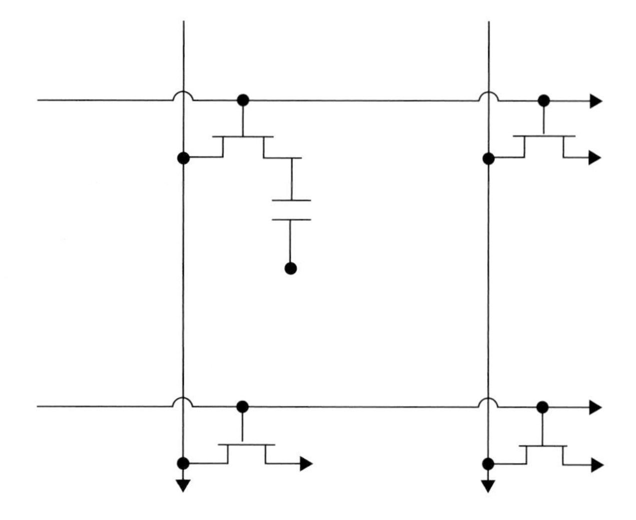

**Figure 18.10.** Control circuitry of a twisted nematic liquid-crystal display with thin-film amorphous-silicon transistors connected to each pixel, acting as switches. The tiny controlling transistor occupies an insignificant fraction of the area of its pixel.

attention several times a second, only a few rows and columns can be controlled in this manner.

For computer displays, *active* control is universally used: this requires a separate transistor to act as a switch for each individual pixel, as shown schematically in Figure 18.10. For this purpose, minute thin-film transistors made of amorphous silicon are used; each transistor is only on the order of a micrometer across, whereas individual pixels are on the order of a tenth of a millimeter wide. Accordingly, the controlling transistor occupies an insignificant fraction of the area of its pixel. Each pulse of current to control a pixel need only take on the order of microseconds and the entire addressing sequence can be repeated several times a second. The state of the light passing, or not passing, each pixel – on or off – stays put between repeat electrical signals. The capacitor shown attached to each transistor in Figure 18.10 is intended to extend the time constant of each pixel so that it does indeed “stay put” between repeats of the addressing cycle; the time constant is further enhanced by using a liquid crystal with a particularly low electrical conductivity. Even with the much faster operation of an active display, rapidly changing images leave blurred trails on the screen. In practice, displays with several hundred pixels along each axis can be made, implying hundreds of thousands of pixels in all; liquid-crystal displays up to about 50 cm along the diagonal are currently being manufactured.

To make a color display possible, each pixel is divided into three rectangles, each with its own transistor, fitted with blue, red, and green filters. Since the percentage transmission of light at a pixel depends sensitively on the voltage difference across the pixel, a wide range of colors can be matched by controlling the relative transmission through the three adjacent colored filters.

The manufacture of a LCD requires techniques of the type employed in the “foundries” used to make integrated microcircuits: in particular, manufacture of the micro-transistors requires very precise control. Enormous efforts have gone into perfecting these transistors and they have an undisputed lead in display technology. Huge numbers of transistors are currently manufactured each year (see Box 18.1).

In the past few years, there has been rapid progress in developing a completely different type of flat-screen display based on light-emitting polymer (plastic) sheets controlled by electric fields. These are beginning to be used for such things as mobile-telephone mini-screens, but are not yet in use for computer displays. There can be little doubt, however, that polymer displays are fated to become big business.

BOX 18.1 SEMICONDUCTORS AND RICE

Transistors are made as parts of integrated circuits (microprocessors), of dynamic random-access memories (DRAMs) for computers, and of liquid-crystal displays. In a study by Randy Goodall and colleagues presented at International SEMATECH, Texas, in 2002 of "long-term productivity mechanisms of the semiconductor industry" it was memorably calculated that there are now far more transistors produced per year in the world than grains of rice.

Five hundred million tons of rice per year, or 450 billion kg per year, is equivalent to some $27 \times 10^{15}$ grains of rice. The number of DRAM "bits" alone, each with a transistor, manufactured in 2002 was around $1000 \times 10^{15}$, or about 40 per rice grain. Furthermore, Goodall asserted that the price of each rice grain today can buy several hundred transistors!

## 18.6 Optical fibers

When the telephone was introduced into public service in 1876, only a few privileged citizens could afford to be connected to the service, and it took many years before automatic dialing was introduced. At first, twisted pairs of wires did all that was required. Transcontinental calls in the USA were not feasible at first; that had to await the invention of the thermionic valve by means of which speech could be amplified at intervals on its way along the wires. Initially, long-distance calls were regarded as an extreme luxury to be made only in severe emergencies. Then, when in the middle of the twentieth century transistors replaced thermionic valves, the telephone spread widely in the population, first with coaxial cables and then with short radio waves beamed between towers or to and from satellites, enabling many more channels to be carried in one cable or beam, encouraging users to speak between continents. Finally, the Internet brought about an explosion in the amount of information carried between distant points.

The need was for very large numbers of individual circuits to be carried on one beam, so that thousands of people could speak to each other, send email messages, or transmit television or Internet signals; many distinct channels had to be squeezed into a "master beam," each differing only very slightly in frequency from its neighbors. That implies the use of visible light, with much higher frequencies (shorter wavelengths) than radio waves. The burgeoning growth of international telecommunications traffic led inexorably to the use of light to carry messages. To do this, "analog" messages, in which light intensity is modulated in accordance with incident sound waves, were replaced by "digital" messages consisting of two kinds of "bits," long and short, which in effect carried the sound. There was just one problem . . . how was this light to be transported across the world? Searchlights or even laser beams sent through the atmosphere were not a practical answer. To solve that problem required many years of painstaking innovation: both microstructural and purity considerations played an essential part. The solution was "optical" fibers made of glass.

The fact that light can be transported along glass fibers over distances of a few meters, by repeated internal reflection at the fiber surface, was known in the nineteenth century and even exploited for simple medical instruments (e.g. to examine the inside of the stomach visually by using bundles of fibers). But the notion of using glass fibers for long-distance telecommunications was conceived and developed by a resolute Chinese immigrant to Britain, Charles Kao (born 1933), against a great deal of

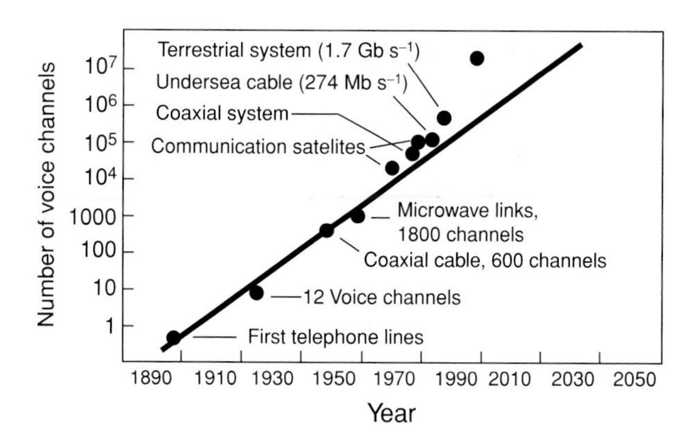


**Figure 18.11.** Chronology of message capacity, showing an exponential growth with time. The three right-hand points, referring to messages transmitted by optical fibers, show a new trend.

initial skepticism. Kao worked in the Standard Telephone Laboratories; the company decided in the early 1960s that none of the communications media used up to then, including millimetric microwaves (ultra-short radio waves), was capable of offering sufficient message capacity, and Kao's manager reluctantly concluded that "the only thing left is optical fiber." The reluctance stemmed from the alarming manufacturing problems involved. One problem was to build in a core and a coating with distinct optical properties, to ensure that no light escapes from the fiber surface; another, the inadequate transparency of glass fibers then available; in 1964 this meant that, in the best fibers, after 20 m, only 1% of the original light intensity remained. This was wholly inadequate. Kao took on the challenge; the result, in terms of message capacity up to the end of the twentieth century, can be seen in Figure 18.11.

Figure 18.12 shows three distinct ways in which optical fibers can be designed. In each variant, there is a central core with low refraction surrounded by a coating with a stronger refraction. The transition between them can be stepped or gradual. After years of uncertainty and research, the variant shown at the bottom of Figure 18.12 is now the preferred form: the core is so thin (on the order of a light wavelength, around a micrometer) that only a single "mode" – that is, only one form of total internal reflection – survives, and this means that an incoming pulse is scarcely distorted at all even after kilometers of propagation. However, such single-mode fibers are by no means easy to make or to handle.

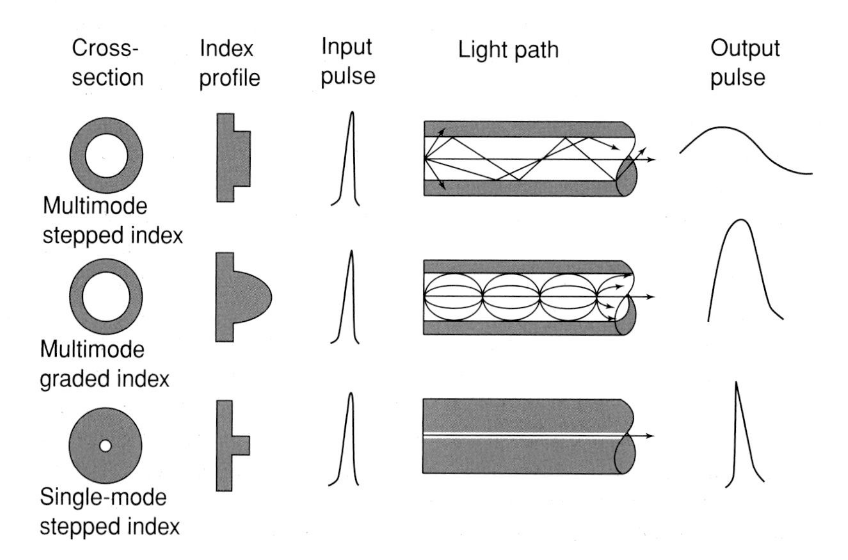


**Figure 18.12.** Forms of optical-communications fiber, showing light trajectories for various refractive-index profiles.

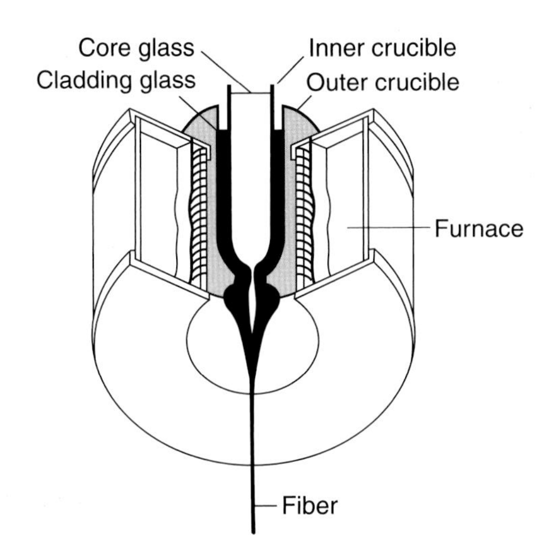


**Figure 18.13.** The early double-crucible process for making multicomponent glass fiber.

Kao began to study the theory of light absorption and concluded that the only promising way to improve transmission drastically was to aim for extreme purity, using silica – silicon dioxide – as the transmitting medium. This material has excellent intrinsic transparency in the infrared range. Kao, together with his assistant Hockham, also looked at the theory of multi-message communication in considerable depth and in 1966 they published a paper (based on the mere assumption that the practicalities of fiber design and manufacture could be solved). This paper excited very widespread interest and brought others into the hunt for a practical system. Many of these quickly became discouraged by the technical difficulties but Kao persisted – he was the quintessential "product champion" – and so the first demonstration of optical communication was achieved by his company and by the British Post Office, which was a close collaborator.

Attempts by Kao, and others, to enhance transparency by chemically removing impurities from glass were not successful; the level of purity needed was extreme, comparable to that needed in silicon destined for semiconductor devices. It was thus logical that the required purification was eventually achieved in the same way as that in which semiconductor-grade silicon is now made, namely by going through the gas phase (silicon tetrachloride, which can be physically separated from the halides of impurity atoms, of transition metals in particular, because of differences in vapor pressure). Silicon tetrachloride can be oxidized to generate silica particles by heating to a very high temperature, and if necessary can be doped by admixing gaseous compounds of, for instance, germanium and phosphorus.

Figure 18.13 shows one early way of making a single-mode fiber from melted precursors, which was originally a popular approach, but when it was realized that starting from the vapor phase enabled much greater purity to be achieved, high temperatures began to be used to generate a "soot" of silica, pure or doped, from a gaseous precursor; the soot melts and congeals into a cored fiber. Variants of this approach were developed during the 1970s in the USA by the Corning Glass Company and by a Japanese consortium. Corning, the major American producer of glass, completely transformed its product mix and is now the leading producer of communications fiber. In the Corning process, a $SiO_2$–$GeO_2$ core is deposited first, followed by a pure silica cladding. The fiber is eventually drawn down to a small outside diameter with a micrometer-sized core (a small overall diameter is requisite so that the fiber can be bent round quite small radii). Today, using the technical measure preferred in the industry, production fiber has a loss of about a decibel per kilometer, which is equivalent to a loss of about 20% of the incident intensity at the end of the kilometer of fiber; this degree of attenuation is still several times greater than the ultimate theoretical limit. The level of transparency now achieved allows transmission runs of as much as a hundred kilometers before a built-in amplifier has to be used to restore signal strength; accordingly, it is now feasible to build a transatlantic cable with an optical-fiber core. (The design of the amplifiers, which are purely optical and involve no electronics, is in itself a triumph of applied physics.)

In parallel with these triumphs of materials development, a whole range of other problems had to be solved, by physicists and by electrical engineers. The physicists developed the semiconductor lasers used to create the input light beam, while the engineers developed the circuits allowing the laser output to be modulated in order to carry each message at a different carrier frequency (color), and to unscramble the separate messages at the end of the journey and reconvert them into electronic signals.

The difficulties do not stop with these issues of re-amplification and unscrambling. When two fibers are to be connected end to end, the two minute cores have to be in register to within a micrometer. One recent way of achieving this is to use a technique developed to shape micrometer-sized mechanisms by chemically machining silicon single crystals. In this way, a tiny V-shaped groove can be machined into silicon and used as an ultra-precise "kinematic" template along which two fibers, their ends accurately polished, can be moved to register exactly.

Of all the major categories of materials, excepting only semiconductors, glass has for some years been the most research-intensive: the laboratories, experts, and attitudes were in place in the 1960s and 1970s to take on the challenge of using light to carry multiple messages.

## 18.7 Magnetic memories

Magnetic materials rival semiconductors for sheer practical indispensability. They cover a huge gamut of uses, from fixing labels to doors of domestic refrigerators, via components of electric generators in cars, magnetic "lenses" in electron microscopes where electron beams take the place of light, and giant magnets to guide electrons into orbits in particle accelerators used to generate X-ray beams of unprecedented brightness, to components of computers. To give some impression of the role of magnetic materials, I will outline the use of magnetic memories for computers.

In practical terms this involves just two of the several families of magnetic materials – ferr*o*magnetic and ferr*i*magnetic materials. The single letter that distinguishes these two families hides major differences. They have in common, however, the features that most of them are crystalline, that individual atoms act as minute magnets, and that these atomic micromagnets line up their tiny fields parallel to a specific direction in the crystal. A crystal has several equivalent directions because of the built-in symmetry: in iron, which has a cubic crystal structure, the atomic magnets are constrained to line their fields up parallel to any one of the three cube edges. Different regions in the same crystal – they are called *domains* – are magnetized along one or other of those three cube edges, in either forward or backward directions, and the result of that is that the overall magnetization is zero: the different domains quench each other completely. When now such a material is put in an external magnetic field, then some domains shrink and others expand in response, and the material overall becomes substantially magnetized. Some materials need only a small field to achieve this – they are called soft magnets – others need a large exciting field before they become magnetized, and these are called hard, or permanent, magnets. Hard magnets cling obstinately to their magnetization when they are removed from the field; the domains remain unbalanced, and we have the kind of magnet that sticks to refrigerator doors.

Now we come to the differences implied by *o* and *i*. A ferr*o*magnet contains atomic magnets that all point the same way inside a domain, whereas a ferr*i*magnet has two populations of chemically distinct atomic magnets that point *opposite* ways; because they are different kinds of atoms, their magnetic strengths are not the same, so a residual magnetization remains in each domain. However, ferr*i*magnets are obviously weaker magnets overall than are ferr*o*magnets. There is another difference: ferromagnets are all metallic and conduct electricity, while most ferrimagnets are insulating compounds. There are cases in which a crystal has only a single kind of magnetic atom – manganese is the best known example, but the interaction between neighboring atoms is such that half of them point one way, half the opposite: these are the antiferromagnets, and their macroscopic magnetization is nil.

Ferromagnets are based on one of the three important ferromagnetic metallic elements – iron, nickel, and cobalt. These metals have (a) the right electronic structure and (b) the right physical separation between adjacent atoms for the atomic magnets to line up parallel. Individually they are all soft magnets, but their magnetic hardness or softness can be controlled by alloying: especially if a fine dispersion of nonmagnetic inclusions can be created, the walls separating adjacent domains find it hard to cross such inclusions, and the material becomes an effective permanent magnet. The metallurgy and physics of ferromagnets, including the subtle factors that determine whether a particular metal is ferromagnetic at all, were well developed by the middle of the twentieth century.

Ferrimagnets were discovered and studied more recently. The key figure was a remarkable, intuitive French physicist, Louis Néel (1904–2000), who in 1936 predicted the existence of antiferromagnetism – his prediction was confirmed two years later in MnO – and in 1948, he explained how it was that certain compounds called *ferrites* exhibited fairly strong magnetism, either soft or hard. These compounds are all variants on lodestone, alternatively magnetite, $FeO{\cdot}Fe_2O_3$ (usually written $Fe_3O_4$), which occurs in Nature and is believed to have been used by early voyagers in primitive compasses. Two kinds of iron ion with different chemical valencies go to make up the structure of lodestone. The first scientist to attempt to develop different variants of lodestone – that is to say, mixed magnetic oxides, incorporating metal ions with different chemical valencies – was the German physicist Hilpert, in 1909: he investigated barium, calcium, and lead ferrites, but his materials did not have useful properties. Things changed in 1932, when two Japanese, Kato and Takei, made a solid solution of cobalt ferrite in lodestone: $Fe_3O_4{\cdot}3(CoO{\cdot}Fe_2O_3)$. This material was a hard ferrimagnet and was duly marketed. Soon afterwards, in 1936, a Dutch physicist, Snoek, and a chemist, Verwey, who worked in the Philips Company's industrial laboratory in the Netherlands, investigated ferrites, at first those containing oxides of Cu, Zn, Co, and Ni. They examined the structure of these crystals, using diffraction of X-rays, and found that some of them had the "spinel" structure, which is shown in Figure 18.14. Others had distinct crystal structures containing different proportions of two oxides. Some were very hard magnets; others (especially those with the structure shown in Figure 18.14) were soft. Soon after World War II, many of these ferrites began to be manufactured on a large scale both in the Netherlands and in Japan. At this stage, French and American physicists also began to study this family of materials and the magnetic interactions between neighboring atoms. But the basic work done in Japan and the Netherlands depended on a close cooperation between physicists and chemists, and indeed this is generally essential in the development of novel materials. Investigators who have been trained in a mix of physics, chemistry and metallurgy join in too . . . they are the materials scientists.

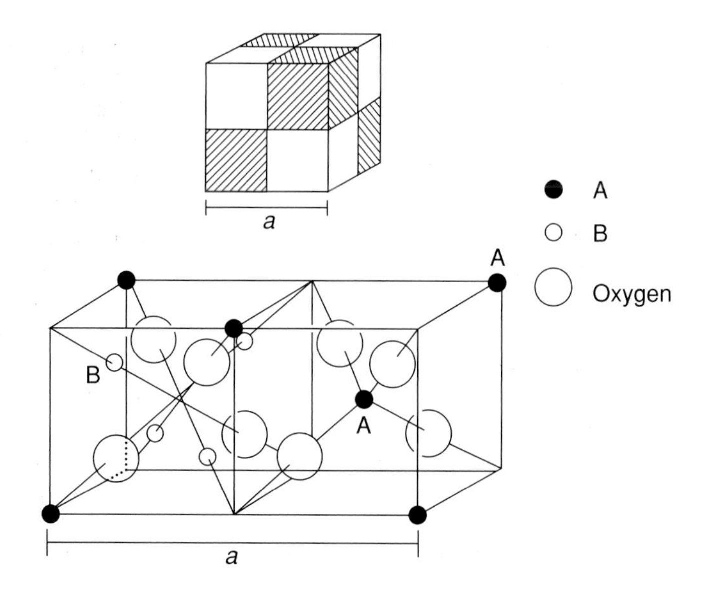


**Figure 18.14.** The spinel structure. (The compound is $MgO{\cdot}Al_2O_3$.) The repeating unit (top) can be divided into octants, half of them containing A metals, the other half B metals, all containing oxygen atoms (shown as large spheres). The separation of A and B atoms is just right for them to line up their magnetic strengths in mutual opposition. *a* is the crystal structure's unit cube length.

Enter the computer. Ever since the first electronic computers were developed soon after World War II, there has been a need for two kinds of built-in memory. The first is the working memory which contains the program (software, in modern parlance) in use, together with auxiliary programs such as the driver for a printer and any documents that are currently being written. The second kind is for long-term memory: many programs, including some that are used only rarely, all the documents that have been written, and forms of entertainment such as games (which are also programs). The working memory (nowadays called a random-access memory, or RAM) needs to operate very rapidly but is closed out at the end of the working day, whereas the long-term memory can be slower but needs huge capacity and its contents have to last for years.

From an early stage, magnetism came into the design of RAMs. An early version, which was current in the 1960s and 1970s, is illustrated in Figure 18.15. The little rings, or "cores," a millimeter or so in diameter, were made of a particular kind of ferrite, containing Mn and Mg oxides. This kind of ferrite has a "square hysteresis loop," which simply means that, as an external magnetic field is gradually increased, the magnetization in the ferrite scarcely changes until a critical field is attained, and then the magnetization leaps in a giant step. A mutually perpendicular array of insulated

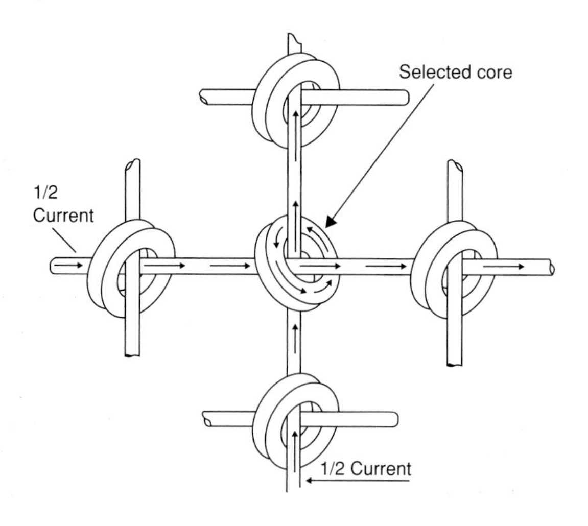


**Figure 18.15.** A small portion of a magnetic core memory, *circa* 1967.

wires is threaded through the cores; the resemblance to the technique used to control electric fields in liquid-crystal displays (Figure 18.10) is apparent. The cores are magnetized circularly, as indicated in Figure 18.15. A single "bit of information" is changed when one core is induced to change its magnetization from clockwise to anticlockwise, or vice versa. *Both* the wires threaded through that core must have currents flowing to bring that about; the current is so chosen that a current in only one wire is not sufficient to reach the critical magnetic field mentioned above. (Each wire is surrounded by a circular magnetic field when the current flows.) To read out the state of the memory the magnetic state of each core has to be destroyed, and after readout the state of the array has to be rewritten at once; so the operation of such a memory is complicated and only moderately fast (with an access time for information of about a microsecond). Threading such an array of cores was a slow and laborious manual process, mostly done by women with small hands and unusual patience. Because of the way they were assembled, core memories were very expensive; they cost around $10 000 per megabyte. It was in these early days that the convention that years should be denoted by just their last two digits was established – thus, 1968 was recorded as "68" – just to save two precious cores in the memory for other purposes. Forty years later, when the century was about to change and "19" was to become "20" hysterical alarm was generated about the disaster which would strike computers when memories became confused about dates. Of course, nothing happened . . . but today, memory has become so cheap that this kind of penny-pinching has become a distant . . . memory.

Today, dynamic random-access memories (DRAMs), made by the techniques of integrated-circuit manufacture, are semiconductor devices based on minute transistor/capacitor units, again controlled by a network of perpendicular microconductors. The plants in which they are made are hugely expensive but such large numbers of units, each corresponding to a unit of information, are made in one operation that such memories are massively cheaper than core memories were. Nowadays, a memory with 48 million bytes, say, can be bought for well under $100; such a memory is almost 10 000 times cheaper, per byte, than a core memory was . . . and a dollar is now worth a good deal less than it was in the 1960s. Also the new memories take up very much less space.

The long-term memory was based on magnetism ever since the earliest computers which used punched paper tape became obsolete, and it remains so based today. The most common form of memory consists of a rapidly rotating disk of some nonmagnetic material – a "hard disk" – coated with a very thin layer of ferromagnetic or ferrimagnetic material. A tiny electromagnet is mounted on the tip of a long arm that is freely tilted, and the tip is shaped to float a matter of micrometers above the magnetic layer because of aerodynamic lift: in effect, the tip is a minute aircraft. The electromagnet is used both to write information, when a current is passed through its coil, and to read it from sites on the magnetic layer: when a site is magnetized, its passing under the electromagnet induces a momentary current. Elaborate procedures are needed to direct the tip to just the right spot on the disk, and the density of information that can be written and read depends greatly on how close the tip can be induced to float above the recording medium without crashing.

Over the past few years, the spatial resolution and speed of writing and reading of hard disks have steadily been improved, partly by improving the mechanism but partly by using improved magnetic layers and electromagnetic tips. The first hard disk was made by IBM in the USA in 1956, incorporated 60 aluminum disks each 60 cm in diameter, weighed a ton and occupied more than a square meter of floor space. It could store 5 million "characters," or bytes; that amount of information would today occupy less than a square centimeter of a single hard disk. The magnetic coating used for that first IBM hard disk was derived from the primer used to paint the Golden Gate Bridge in San Francisco; today, magnetic layers are rather more sophisticated, and the density of packing of information has increased. Once again, the factor of improvement over the decades comes to roughly 10 000.

There are various other kinds of hard disk, each depending on some feature of magnetic behavior; the most recent depends on the phenomenon of magnetoresistance . . . a change in electrical resistance of a metal when a magnetic field is applied to it. Louis Néel, with his characteristic uncanny insight, predicted in 1954 that, in very thin magnetic films (only a few atoms thick), magnetization may unusually be perpendicular to the film, this being attributed, in a way that is still not well understood, to the reduced symmetry of atoms lying on the surface. This was confirmed experimentally, and the next stage was to examine multilayer thin films, with magnetic and nonmagnetic metals alternating. In 1988, M. N. Baibich found that Co/Cu or Fe/Cr multilayers, if the individual layers are thin enough, undergo substantial changes in electrical resistance when a magnetic field is applied normal to the film. A small field of only one oersted suffices to produce a change of several percent in resistance, so these ultra-thin films, coated onto a hard disk, can be used to store information at very high spatial densities. The writing/sensing tip now contains a tiny electrical resistor instead of an electromagnet.

## 18.8 Diamonds, fullerenes, and nanotubes

"Carbon is really peculiar" is a remark by Harold Kroto in his 1997 Nobel Lecture; he shared the Nobel Prize for Chemistry in 1996 for his discovery in 1985, with Robert Smalley and Robert Curl, of a totally new form of carbon, which they whimsically named buckminsterfullerene, or fullerene, or buckyball, for short. (Buckminster Fuller, an architect, designed domes composed of five-fold and six-fold polygons, just like the new molecules shown in Figure 18.16(a).) Fullerenes come in various sizes, of which the two most common are shown. A few years later, S. Iijima in Japan discovered a tubular variant, nanotubes as shown in Figure 18.16(b); these in turn can exist in single-walled form, as shown in the figure, or with multiple walls like a Russian doll. These are chemical achievements, but some of the properties of these new forms of carbon are very much in the physicist's domain. In the study of materials, one really cannot erect firewalls between the physicist and the chemist!

In spite of an unprecedented burst of research by a wide range of scientists since 1985, fullerenes have yet to offer widespread applications; so fullerenes belong to the category of new substances that have not yet become materials. However, nanotubes have a variety of potential uses. The semiconducting variant is being investigated as a constituent of a new kind of tiny transistor. All types of nanotubes are immensely strong and compete with spider silk (natural or synthetic) for huge strength in relation to density; the only problem is that of how to attach a nanotube to a load-transmitting device, because the tubes are so much stronger (and stiffer) than competing materials. A third kind of use is very timely: as fuel cells that produce electricity directly from the oxidation of hydrogen, without combustion, an approach that could be of widespread use to drive automobiles without emission of any $CO_2$, for which the efficient storage of hydrogen has become the central problem; one possibility is to store hydrogen molecules under pressure inside a nanotube. There is intense dispute among researchers about the efficacy of this process; the optimists claim that two atoms (one molecule) of hydrogen can be stored per carbon atom, which would be distinctly interesting.

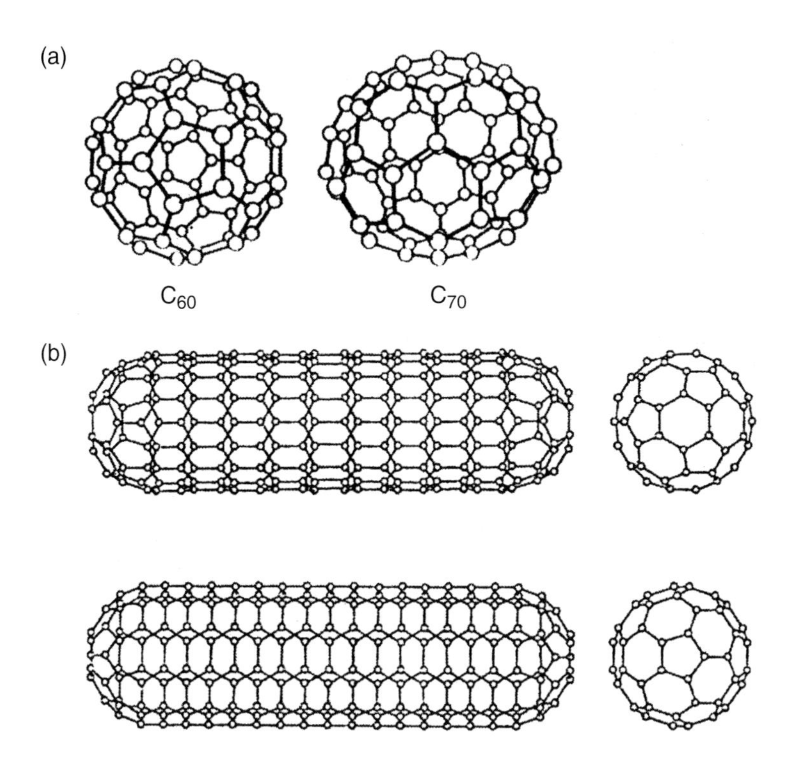


**Figure 18.16.** (a) Exotic carbon: two fullerene molecules, $C_{60}$ and $C_{70}$. (b) Two types of single-walled carbon nanotubes, one with metal-like electrical conductivity, the other, semiconducting. Right: end-on views of the folded graphite sheets.

The recent history of fullerenes and nanotubes shows that even the most familiar of materials can spring major surprises, but these variants do not exhaust the recent developments in carbon. Since 1952, diamond is no longer just a girl's best friend but has also become a synthetic material with a range of scientific uses. In 1952, William Eversole, working in American industry, made the first thin films of diamond by reaction in, and deposition from, a mixture of vapors. By slow stages of trial and error, centered in the USA and Japan, this approach has become steadily faster and more controllable, until today a range of sheets, of area 100 $cm^2$ or even more, and up to a millimeter in thickness, can now be made reproducibly. (One Japanese use for such large sheets is to make exceptionally efficient diaphragms for loudspeakers.) The diamonds enriched with $^{12}C$ for electronic heat sinks, mentioned earlier, can be synthesized by this approach. The current excitement arises from the prospect of using such synthetic diamond as a semiconductor for devices, but for the present this is still a dream. While p-type material can be readily prepared, no way of doping the material to make it n-type has yet been discovered.

Soon after Eversole had made the first diamond films, researchers in the laboratories of the General Electric Company in the USA (GE), exploiting the classical researches of the Harvard physicist Percy Bridgman studying materials under extremely high pressures, found a way to grow single crystals of diamond from non-diamond carbon precursors with a metal catalyst and patented their approach in 1955. To do this required, simultaneously, both high pressure and very high temperatures, and it took years to develop industrially viable methods of doing this on a large scale. One of the leaders of the GE team, Robert Wentorf, in 1956 showed that his favorite snack, peanut butter, could be turned into diamond, and 30 years later, Y. Hirose in Japan showed that diamond could be made by the vapor route by using as a precursor sake, that is, rice wine. Clearly, each nation uses its own favorite snack or tipple to make that most extreme of treasures, diamond.

## FURTHER READING


1. M. Riordan and L. Hoddeson, *Crystal Fire: The Birth of the Information Age*, New York, W. W. Norton & Co., 1997. A splendid account of the discovery of the transistor.
2. J. Hecht, *City of Light: The Story of Fibre Optics*, Oxford, Oxford University Press, 1999.
3. C. E. Yeack-Scranton, Computer memories, *Encyclopedia of Applied Physics*, Vol. 10, New York, VCH, 1994, p. 61.
4. H. B. G. Casimir, *Haphazard Reality: Half a Century of Science*, New York, Harper and Row, 1983.
   This racy autobiography includes an account of the beginnings of the development of ferrites.
5. P. J. F. Harris, *Carbon Nanotubes and Related Structures*, Cambridge, Cambridge University Press, 1999.
6. R. M. Hazen, *The Diamond Makers: A Compelling Drama of Scientific Discovery*, Cambridge, Cambridge University Press, 1999.
7. K. Zweibel, *Harnessing Solar Power: The Photovoltaic Challenge*, New York, Plenum Press, 1990.
8. A. S. Nowick, The golden age of crystal defects, *Annu. Rev. Mater. Sci.* **26** (1996) 1.


# Robert Cahn

Robert Cahn took his first degree in natural sciences at Cambridge, specializing in metallurgy, in 1945. His Ph.D. research, on the physics of metals, was undertaken at the Cavendish Laboratory. Subsequently he joined the newly established Harwell (atomic-energy) Laboratory where he finished his doctoral research and turned to the crystallography of twinning in metallic uranium. In 1951 he moved to the Department of Physical Metallurgy at the University of Birmingham and spent the rest of his career at various universities. His research interests focused on order–disorder transformations, the formation of crystal twins in metals, and mechanical properties of alloys. After a brief sojourn at the University of North Wales he joined the new University of Sussex as Britain's first Professor of Materials Science, and designed novel undergraduate courses. At Sussex, he began research on metastable forms of alloys, including metallic glasses – subjects at the margin of metallurgy and physics. In 1981 he moved for two years to the University of Paris, and returned in nominal retirement to Cambridge.

In 1959 Robert Cahn began an extensive editorial career: the journals he founded include the *Journal of Nuclear Materials* (1959), the *Journal of Materials Science* (1966), and *Intermetallics* (1993). From 1967, he wrote around 100 short articles on many aspects of materials science for the "News and views" columns of *Nature*. He edited a series of solid-state monographs, first for Cambridge University Press (1972–1992) and subsequently for Elsevier Science, and also edited several encyclopedias concerned with materials science. In 1991–2001, he was one of the editors in chief, for the German publisher VCH, of a series of 25 books under the series title *Materials Science and Technology: A Comprehensive Treatment*, which has now been reissued as a softback series. He is a Fellow of the Royal Society and of four other academies around the world.

# 19 Physics and Society

Ugo Amaldi

## 19.1 Introduction

All scientists reading this would agree that scientific research in general, and Physics in particular, deserves to be pursued if only for the sake of understanding and enjoying the world in which we live. They share Albert Einstein's opinion: "Why do we devise theories at all? The answer is simply: because we enjoy 'comprehending' . . . There exists a passion for comprehending, just as there exists a passion for music" [1]. At the same time, all scientists also believe that new basic knowledge will continue to bring concrete benefits to Society.

However, this intimate conviction, well-founded on past experience, is no longer sufficient. For about twenty years, politicians and the public have increasingly been asking scientists, and in particular physicists, to

(i) better describe what they do and what they learn;
(ii) explain what advantages Society has gained and can expect to receive in the future from the funds allocated to fundamental research; and
(iii) organize the production and dissemination of fundamental research in such a way as to maximize its benefits to Society.

It is my opinion that in such an enterprise other scientists perform better than physicists, possibly (but not only) because the subject of their research helps them: advances in medicine and, more recently, in molecular biology are naturally close to human life and are thus easily perceived as "useful."

Among physicists those who study condensed matter – by working at ordinary energy levels and dimensions – have the best arguments and for a long time have known how to use them. The physicists working at energies very far from ambient thermal motion – either so cold as to produce quantum condensates or so large as to create weak bosons – have to struggle. In spite of the difficulties, most physicists have understood the message and have improved on the three points listed above, as I illustrate in this chapter – a (biased) assessment of the consequences for Society of the vision and the results of Physics.

Physicists have also acquired a new awareness of the many influences that Society has on the development of Physics. The next section briefly covers some of these issues.

*The New Physics for the Twenty-First Century*, ed. Gordon Fraser.

## 19.2 Influences of Society on Physics

The first, almost trivial, influence is through the *funding levels* of the various public agencies, foundations, universities, and companies that finance research. In sharing resources among the different fields of science, politicians, directors, presidents, and CEOs have agendas that change with time according to the past records of each field and to their perception of the expected future developments and "benefits." Thus the large funds allocated after World War II to nuclear physics leveled off in the 1990s (witness the demise in the USA of the SSC), whereas biomedical sciences are receiving more and more resources, as best shown by the doubling of the US National Health Institute's budget in the years 1998–2003 to the record level of about 27 million dollars [2] while its number of staff increased by less than 10% [3].

In the 1990s a new player became increasingly important: philanthropy. In the USA the total spending of non-profit organizations on research increased from 8.7 to 19.5 billion dollars between 1990 and 1998, and in the UK at the end of the decade the Wellcome Trust invested more in biomedical research than did the UK Research Councils [4]. Most foundation money is granted to biomedical research, so the shift of public money from physics and engineering toward biology and medicine has been accentuated by the increasing weight of private donations.

However, there are exceptions. At the turn of the century Canadian businessman Mike Lazaridis created the Perimeter Institute for Theoretical Physics in Ontario, which is devoted to quantum gravity and quantum computing [5]. In 2003 physicist Fred Kavli, who founded and led one of the largest suppliers of sensors for aeronautics and industrial applications, created the Kavli Institute for Particle Astrophysics and Cosmology that will collaborate with the SLAC on fundamental research [6]. In 2004 software pioneer Paul Allen – following an initial investment for R&D – announced his commitment to the construction of 206 radio telescopes out of the 320 that will eventually form the Allen Telescope Array (ATA), a general-purpose facility that will also have the capabilities to search for possible signals from technologically advanced civilizations elsewhere in our Galaxy. In spite of such examples, philanthropists still favor by far biomedical research [7].

Within Physics research there has been a shift in funding from nuclear and particle physics toward condensed-matter physics and material sciences, which has been accompanied, particularly in the USA, by acrimonious public debates among physicists. Noteworthy at the end of the 1980s was that on the construction of the Superconducting Super Collider (SSC), which was eventually stopped by a congressional decision [8]. Champions of the two camps – "small science" versus "big science" – were Philip Anderson and Steven Weinberg [9] and the debate was the concrete expression of two different visions of Physics to which I will return in the next section: emergentism and reductionism.

Are these discussions useful to the development of Physics? Superficially not, but, looking deeper, the answer has to be yes. Certainly the public is not impressed by the sight of prestigious scientists battling for a bigger piece of government cake. However, the debate obliges the contenders to sharpen their arguments and to present them in a simpler and more coherent form to the public and to politicians. Moreover, external pressures and challenges oblige the scholars in each field to better identify and promote their results and the techniques that can be used by other sciences and by industry. In a changing environment, with fierce debate among scientists, the best institutions adapt to survive. The international particle-physics laboratory CERN is a good example. Its initial charter did not envisage patenting results and, until the end of the 1980s, very few CERN scientists or engineers considered patents and applications as part of their mission. By 1999 the laboratory had established a division to promote technological transfer.

Then the *intellectual atmosphere* is less recognizable but not less important to the development of Physics. The underlying "philosophical" ideas that dominate the intellectual debate are the background against which new visions of the natural world are conceived and gain approval among scientists and the public.

This is slippery ground in view of the harsh debate that has raged in the past twenty years on the well-known dilemma: is scientific endeavor a progressing quest for an external truth (as defended by the "realists"); or merely a social construct (as advocated by the "relativists" or "constructivists")? To attack the defenders of the latter view, Alan Sokal chose for his now famous spoof paper of 1996 a physics subject: "Towards a hermeneutics of quantum gravity" [10].

In this debate, extreme attitudes have been taken by scholars on both sides of the fence, thus overlooking the balanced perspective put forward by Lewis S. Feuer at the beginning of the 1980s, when the controversy had not yet exploded. One chapter of his *Einstein and The Generations of Science* [11] bears the title "The social roots of Einstein's relativity theory." Suffice it to say that detailed historical reconstructions, such as that of Feuer, indicate that, even if the theories and the models of Physics cover more and more ground with ever fewer ad-hoc hypotheses – so that one can speak of a progressing knowledge of the inner workings of Nature – they are usually structured and presented in a form that is consonant with the contemporary intellectual atmosphere.

Of course, Physics contributes greatly to the intellectual atmosphere. It is enough to recall the debates about determinism and realism and the reverberations that relativity and the uncertainty principle had, and still have. Bell inequalities are today known to philosophers as much as Gödel's incompleteness theorem and Turing's universal machine, at least by name. There are many issues in today's astrophysics that constantly appear in debates on philosophy and, particularly, on religion. The origins of the Universe and of time are the most quoted ones, but more subtle arguments appear in the discussions on free will and on the validity and meaning of the so-called "anthropic principle," the subject of a well-known book by John Barrow and Frank Tipler [12]. Fifteen years later Martin Rees summarized the puzzles put to scientists by the remarkable bio-friendliness of our Universe in *Our Cosmic Habitat* [13]. To discuss them, he even organized in his Cambridge home a meeting of leading cosmologists on "Anthropic arguments in cosmology and fundamental physics" (Figure 19.1) [14, 15]. The conference debate was introduced by Brandon Carter, who coined the term "anthropic principle" in 1974. The subject is of particular interest to religious people, and the conference was supported in part by the Templeton Foundation. This foundation, set up in 1972 by financier John Templeton, has awarded to physicists four prestigious Templeton Prizes "to encourage and honour those who advance spiritual matters" [16]: the recipients were Paul Davies (1995), Ian Barbour (1999), Freeman Dyson (2000), and John Polkinghorne (2002).

A third important theme in which Society has influenced, and is influencing, Physics is in the trend toward a wider *participation of women* in science [17]. I doubt that, without the pressure of the feminist movement and its pervasive effects in Society, male physicists would have moved in such a direction. However, the trend is too slow: by 1998 in the USA women constituted 13% of Physics Ph.D. recipients and only 8% of Physics faculty members [18]. This was in spite of the fact that, as shown in a 1992 study by the American Institute of Physics [19], women graduates who chose not to remain in Physics had performed as well as their male colleagues who did stay in the field.

Figure 19.1. Participants in the conference on "Anthropic Arguments in Cosmology and Fundamental Physics" held in Cambridge from August 30 to September 1, 2001 at the home of Cambridge cosmologist Martin Rees, seen here standing just to the left of Stephen Hawking. (Courtesy Anna N. Zytkon, Institute of Astronomy, University of Cambridge.)

To remedy this state of affairs, in 1972 the American Institute of Physics created a Committee on The Status of Women in Physics that has recommended changes to make departments more comfortable for women faculty members and students. The European Physical Society has a Women in Physics Group, as do most countries, even if the situation varies greatly from country to country. In 2002, the first IUPAP conference "Women in Physics" was held in Paris. Interesting enough for those who think that in Mediterranean countries women devote themselves mainly to the family, in the LHC Atlas Collaboration – formed of about 2000 physicists – the fractions of women scientists are 23% for the groups from Italy and Spain, 11% for Germany and France and 8% for the UK and the USA [20].

## 19.3 Spinoffs from Physics to Society

Figure 19.2 depicts four output streams – "knowledge," "technologies," "methods," and "people" – of any scientific activity. In this context "knowledge" is a term that includes both "basic knowledge" – experimental laws, theories, and models constructed to interpret experimental observations – and "usable knowledge," the scientific and technical information that, once acquired, can bring direct consequences for Society. In any specific case it is usually quite easy to agree whether some knowledge is basic by considering whether the main motivation behind it is curiosity. Sometimes, however, there occurs a curiosity-driven discovery that right from the beginning is either "directly" applicable to other fields of science or usable for new technologies. A striking example

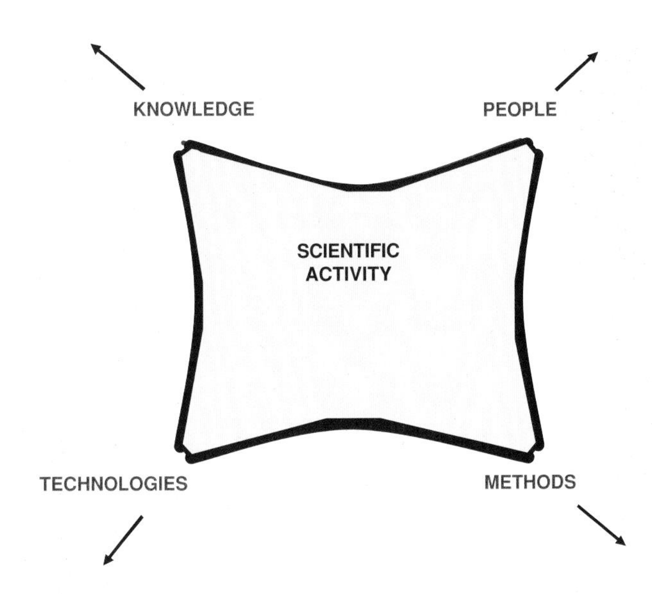


**Figure 19.2.** Any scientific activity has four types of spinoff to Society. Their impact and importance varies with the type of science and with time.

is the discovery of X-rays that, from the very first day, was clearly both basic and usable knowledge.

The development of new "technologies" is instead driven by specific interests – be they humanitarian, commercial, power-seeking, or a mixture of these – and usually combines some basic knowledge with already available techniques. At other times a technology arises from the needs of fundamental research, to attack a purely scientific problem. A good example is the cyclotron, invented as a nuclear-physics instrument at the end of the 1920s and by 1936 used for neutron irradiation – Ernest Lawrence's mother Gunda had an abdominal tumor and this made him and his brother John, a medical doctor, very keen to investigate the medical use of radiation [21].

In the following, the term "technologies" will refer to techniques that become available in the literature or on the market either (i) as a not immediately obvious consequence of a scientific discovery; or (ii) as instruments developed to help gather new basic knowledge.

## 19.4 Knowledge

Physics knowledge has consequences for all the other natural sciences. To understand why it is so pervasive it is enough to consider the following simple question: what is the most important scientific discovery that has to be passed on to future creatures? I believe that the large majority of scientists would agree with Feynman's pithy reply: "All things are made of atoms" [22].

Because matter is made of atoms, without the understanding brought by modern Physics then chemistry, molecular biology, geology, paleontology, neurology . . . could not flourish. Think of quantum calculations of structures of molecules, of protein folding, of sample dating with radioactive isotopes, and of functional magnetic resonance of the brain. During the past thirty years also classical Physics and its modern developments – the Physics of chaos and of the emergent properties of complex systems (see Chapters 13 and 14) – have become more and more relevant to other sciences.

**Table 19.1.** *Examples of direct spinoffs of scientific knowledge and time elapsed from discovery to practical application*

| Discovery | Uses | Time to application |
|---|---|---|
| X-rays | Medical imaging | 0.1 years |
| Nuclear fission | Energy production | 3 years |
| The transistor effect | Information technology | 7 years |
| Muon-induced fission | Energy production | Infinite (?) |
| High-temperature superconductivity | Current transmission and energy storage | 10 years |

Even meteorology needs quantum mechanics, for instance in the understanding of the exact causes of global warming. Indeed, the climatic effects of water vapor are little known, even if in the lower 10 km of the atmosphere water vapor contributes more to the greenhouse effect than does carbon dioxide. The modeling of this phenomenon is difficult because the spectroscopy of water and its dimers is not well known experimentally and difficult to measure [23]. *Ab initio* quantum calculations that use high-performance computers [24] are crucial to the solution of this meteorological problem.

Physics is at the root of all sciences mainly because any piece of matter is made of atoms and, indirectly, of its smallest components. This does not imply that Physics encompasses all sciences, because complex systems have their own behaviors that are not derivable from the behaviors of their components. The epistemological debate will continue to rage for ever and physicists have conflicting views, as epitomized by the book *Facing-up: Science and its Cultural Adversaries* published in 2001 by Weinberg [25] and its critical review written for *Nature* by Philip Anderson [26]. Anderson's dictum "More is different" reverberates with the "Less is more" of architect Mies Van der Rohe and synthesizes the tenet that emergent complex phenomena do not violate microscopic laws but nonetheless do not appear as their logical consequences [27].

Passing to what I called "*usable knowledge*," some examples are listed in Table 19.1. The nuclear-fission case is well known. In the discovery of the point-contact transistor by Walter Brattain and John Bardeen, the far-reaching consequences for the development of electronics had been obvious from the beginning, but the invention of the integrated circuit was needed in order to bring transistors to the masses [28].

In 1956 the fusion of a proton with a deuteron to form $^3$He was discovered [29], and Luis Alvarez wrote that "We had a short but exhilarating experience when we thought we had solved all of the fuel problems of mankind for the rest of time" [30]. Unfortunately, it was later proved that, for the more energetically favorable reaction $d + {}^3H \rightarrow {}^4He + n$, the muon has a high probability of sticking to the helium nucleus produced [31]. Even if the muon mass were smaller than 100 MeV, its much longer lifetime would help somewhat, but it would not really solve the energy problem of humankind, due to this sticking probability.

The immediate promises of cheap transmission and storage of energy are such that the discovery of high-temperature superconductivity is the most clear recent example of immediately "usable" fundamental knowledge, and in fact high-temperature superconductivity had long been sought. However, the first application to small systems took 10 years and the transmission of energy over reasonable distances had to wait about 15 years. It is not yet clear whether this technology will have a large-scale impact.

## 19.5 Technologies

The tremendous development in the engineering sector has been an asset for Physics (and for chemistry) but, at the same time, has meant a loss of public support since engineering – which occupies many people and generates a huge economic output – is not seen by the layman as applied Physics. The situation is very different in the biological sciences because everything is "biology." In particular, students who want to switch later to molecular biology "applied" to medicine take the same university courses as those who will became "pure" scientists.

A Physics spinoff of the past is frequently the engineering of today, so an engineering school is the standard route for those who are preparing to apply basic knowledge mostly gathered by physicists and chemists. Because of this I sense that Physics has lost part of its visibility as an important source of spinoffs, but this state of affairs is at the same time a measure of the considerable successes in creating new products. Extrapolating this trend, I am waiting for the moment when polytechnic schools will offer degrees in "quantum engineering" to prepare those who will develop nanodevices for industry and medicine. (Wolfgang Pauli described Enrico Fermi as a "quantum engineer" (E. Amaldi, private communication); let us hope that the quantum engineers of the future will be as inventive and knowledgeable as Fermi!)

The list of technologies born and developed during the past 400 years in Physics and, later, engineering laboratories is long: optics, mechanics, electromechanics, radiofrequency waves and microwaves, superconductivity, cryogenics, ultra-high vacuum, radiation detection, electronics, information technologies, robotics . . . It is impossible to cover even a small fraction of them, but some examples can be given by distinguishing – and it is not always easy – those spinoffs that are consequences of *a particular discovery* (or *invention* or *research need*) from those for which the product is *a complete integrated system*. (To avoid repetition, in this section I deliberately use examples not found elsewhere in this book.)

As far as *nuclear and subnuclear physics* is concerned, nobody doubts that the needs of constructing huge accelerators and particle detectors endowed with large-volume magnetic fields have been instrumental in the development of large and sophisticated systems based on low-temperature superconductivity. Among these, one can quote the routine use of superconducting solenoids for hospital-based magnetic-resonance imaging (MRI) and demonstration *maglevs*, magnetically levitated trains built during the 1990s in Germany and Japan [32].

The sciences which have most benefited from the developments of radiation detectors are biology and medicine. For the contemporary application of this technique, it is enough to recall positron-emission tomography (PET) based on scintillating crystals, such as BGO, which became practical gamma-ray detectors as a consequence of their use in subatomic-physics experiments. The beginning came in 1948, when Robert Hofstadter discovered that sodium iodide, activated by thallium, made an excellent photon detector. Today, radiation detectors – coupled with a source of high-energy X-rays – are mounted on trucks and, as advertised on the Web [33], are the core of "highly advanced, self contained mobile inspection systems for screening trucks (30 per hour), cargo containers and passenger vehicles for contraband and explosives": an extremely useful application in insecure airports and harbors.

Subnuclear physics, together with meteorology, has also given an enormous boost to the development of parallel computers as a result of the computing power necessary for predictions and for quantum chromodynamics calculations, as explained in Chapter 15.

In *condensed-matter physics*, the number of notable spinoffs of a particular scientific advance is very large. Before giving some examples, it is worth remarking that condensed matter has always been important for human civilization, as is emphasized by the fact that historical epochs are defined by the most valuable type of condensed matter used by humankind: the Stone Age, Bronze Age, and Iron Age. For some decades we have been in the Silicon Age and the progress in condensed-matter physics is such that new materials can be designed from scratch, creating products that fit specific needs. As stressed more than twenty years ago by Cyril Stanley Smith in his *A Search for Structure* [34] this approach goes back to the philosophy of medieval alchemists who tried to transform common metals into gold. Physicists and chemists (the modern alchemists) use computers and complex processing to "transmute" the multi-leveled microstructure of materials, guiding them even to "self-assemble," and achieve the essential property of gold, namely economic value. An age of empirical exploration of new materials and of advances by trial and error is perhaps being superseded by an age of design in which transmutations beyond the alchemists' dreams are achieved: the creation of materials from thought [35] (see Chapter 18).

However, this is not always the case. The high-temperature cuprate superconductors were not designed but discovered, and even magnesium diboride had to wait on the shelf until the January 2001 announcement of its high critical temperature (39 K) [36]. Then fifty preprints had been posted on the Web before the original paper was even published [37]. Recent advances in superconductivity target organic superconductors, with a great broadening of the range of materials that can be made superconductors by doping [38]. High-temperature superconductors are on the market for short-distance energy transportation and for current leads, as in the superconducting magnets of the Large Hadron Collider at CERN. The possibilities being made available by novel organic superconductors are certainly very numerous but difficult to predict.

Nanotechnology started in 1959 when Richard Feynman – delivering his famous visionary lecture "There is plenty of room at the bottom" (www.zyvex.com/nanotech/feynman/htlm)[1] – issued a public challenge by offering $1000 to the first person to create an electrical motor "smaller than 1/64th of an inch." But he was not happy when he had to pay William McLellan, who had built the device by conventional means, i.e. with tweezers and a microscope [39]. The new frontier is the construction of nanoelectromechanical systems (NEMSs) that are sensitive to the presence of a few atoms and in some circumstances feel quantum-mechanical effects (see Chapter 12).

The progress from microelectronics toward nanoelectronics is a triumph of solid-state physics and opens the door to numerous spinoffs. One can say that the first step was taken in 1991 when, in the NEC Fundamental Research Laboratory in Tsukuba, Sumio Iijima saw with an electron microscope the first thin and long macromolecule now called a "nanotube" [40]. When a nanotube, nanocrystal, quantum dot, or individual molecule [41] is attached to metallic electrodes via tunnel barriers, electron transportation is dominated by single-electron charging and its quantum properties [42]. The physics is well advanced, but electrodes have to be placed with accuracies of the order of one nanometer and this cannot be achieved in the very large systems needed to replace standard silicon transistors [43]. Self-assembly is the magic word, but molecular electronics is not yet close to the market [44].

There are some novelties even in the old field of optics. In 1873, Ernst Abbe – who had been commissioned by Carl Zeiss in the early 1870s to find ways of improving the performance of optical microscopes – concluded that the smallest distance that can be resolved by an optical instrument has a physical limit determined by the wavelength of the light [45]. To build very accurate microscopes, the new path around this limit

[1] In the same lecture Feynman offered $1000 to "the first guy who can take the information on the page of a book and put it on an area 1/25,000 smaller in linear scale in such manner that it can be read by an electron microscope." The challenge is still there.

is to confine light in space by sending it through a sub-wavelength aperture placed very close to the object [46] and to combine this trick with the nonlinear phenomena induced by two converging femtosecond laser pulses [47].

For the *spinoffs of integrated systems*, it has to be emphasized that, when "systems" are considered, many "technologies" and many "methods" are automatically involved.

Lasers of petawatt power are finding new applications ranging from the production of extreme-ultraviolet photons through nonlinear phenomena [48] to the creation of intense magnetic fields, reaching 100 000 tesla [49], the testing of inertial-fusion systems (at the National Ignition Facility in the USA and the Laser Mégajoule in Bordeaux), and the acceleration of electrons [50] and ions to hundreds of MeV. To move toward application of the last-mentioned technology, in 2003 high-energy protons produced at the VULCAN facility of the Rutherford Appleton Laboratory were used to produce fluorine-18, which was later used in the PET of a patient [51]. In the longer term these proton beams could become intense enough to be used in the precise proton therapy of deep-seated tumors [52], which now needs cyclotrons or synchrotrons, as discussed below.

My other examples come from nuclear and particle physics, since the systems concerned are most often particle accelerators used in other fields of science, medicine, and industry. Other applications can be read off in Figure 19.3. The list is long: synchrotron-radiation sources, free-electron lasers (FELs), neutron-spallation sources, inertial-fusion plants based on the bombardment of pellets by ion beams, accelerators for waste incineration, production of medical isotopes, and hadron therapy.

To put the numbers in perspective, there are about 20 000 accelerators running in the world and half of them are linacs in which electrons are accelerated to 5–20 MeV and produce – by hitting a heavy target – photon beams used in the treatment of cancer [53]. Synchrotron-light sources are much less numerous, but around each of them tens of experiments are running in fields as diverse as solid-state physics, molecular biology, medicine, environmental sciences, and material sciences. The new game in town is the construction of tunable X-ray FELs based on the "self-amplification of spontaneous emission" (SASE) concept [54]. The high-energy electron bunches wiggling in a long ondulator radiate in unison, so much so that Claudio Pellegrini (who, with Rodolfo Bonifacio and Lorenzo Narducci, introduced self-amplified spontaneous emission FELs) likens the interplay between X-rays and electrons to the effects of a musical conductor: "Instead of a big room with everybody talking, you have a choir" [55]. The photon beams have brilliances eight orders of magnitude larger than those of synchrotron-light sources and, with electron beams of very high energies, wavelengths as small as 0.1 nm. The FELs at DESY [56] and SLAC [57] will produce coherent pulses of hard X-rays lasting only a hundred femtoseconds. This will allow scientists to make movies of chemical reactions, watch surfaces melt in real time, and take photographs of individual molecules [58].

Two other examples are the integrated systems dedicated to the incineration of radioactive wastes and the acceleration of protons and light nuclei to treat some solid tumors.

Accelerator-driven systems (ADSs) – pioneered by Charles Bowman and colleagues at the Los Alamos National Laboratory in the early 1990s and championed by Carlo Rubbia – are planned to use milliamperes of GeV protons to transmute long-lived nuclides produced in the reprocessing of nuclear-reactor fuels. The transmutation is induced by the high neutron flux produced in a target, which has to dissipate megawatts. The most advanced project is in Japan, where the Japan Proton Accelerator Research Complex (J-PARC) is being built for basic research in nuclear and subnuclear physics, for neutron scattering in fundamental and applied sciences, and, last but not least, for the study of transmutation. By the beginning of the 2010s,

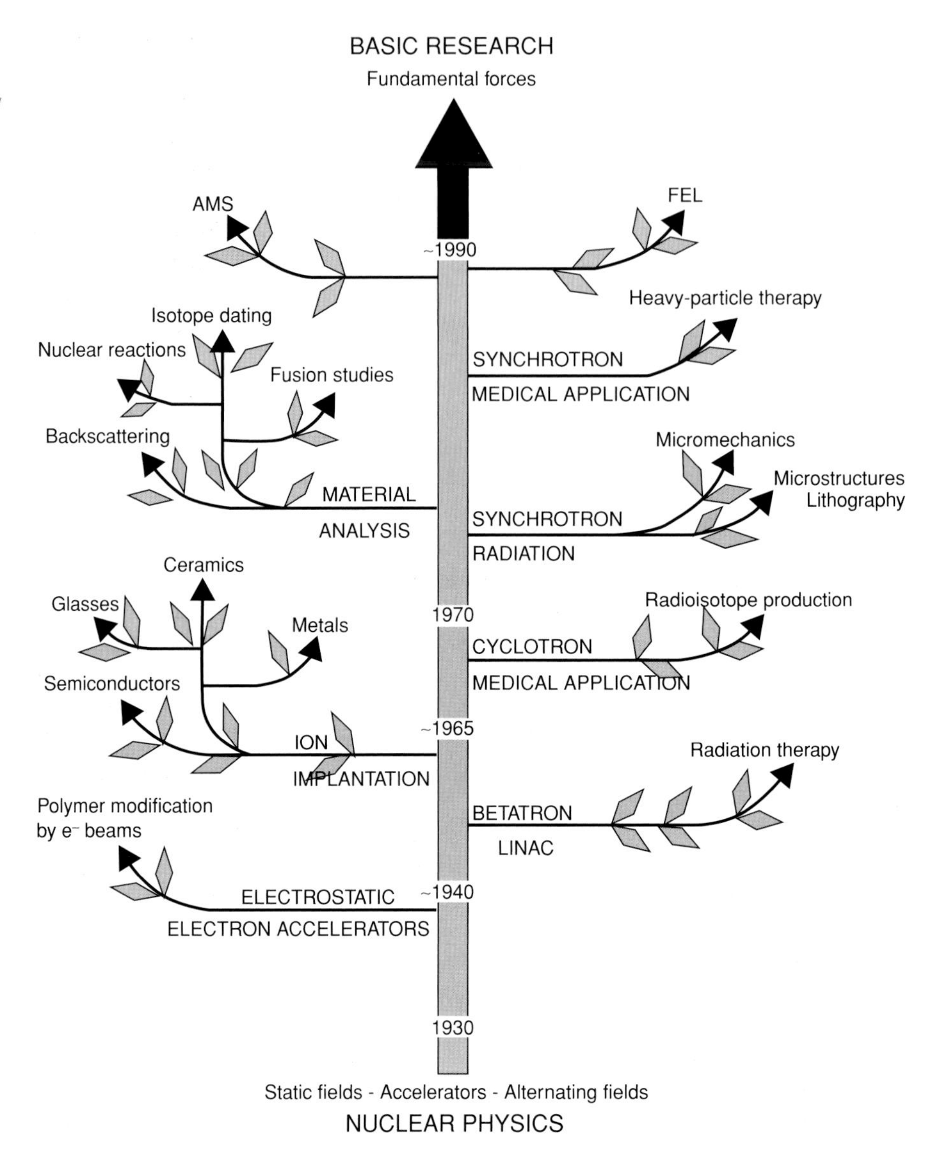


**Figure 19.3.** In this figure, prepared by the author for *Europhysics News* [53] the many applications of particle accelerators are represented as leaves of a flourishing tree.

at its Accelerator-Driven Transmutation Experimental Facility, the reduction of the nuclear-waste inventory will be experimentally studied for the first time. The American Spallation Neutron Source (SNS) will be completed earlier at Oak Ridge (Tennessee) and will serve a large community of neutron-scattering experts, in both fundamental and applied research, but – in its first phase – will not feature a transmutation station. Europe has much less ambitious programs [59].

In 1946 Bob Wilson – then a young researcher in Berkeley and later founder and first director of Fermilab – proposed the radiotherapeutic use of proton and carbon-ion beams because of their favorable deposition of energy in matter with respect to beams of high-energy photons (which are often still called X-rays by radiotherapists) [60]. Fifty years later patients started to be treated in hospital-based facilities, rather than in physics laboratories, and by now there are about twenty such facilities either running or being built (see http://pteog.mgh.harvard.edu). If accelerators producing 200-MeV proton beams were as expensive as linacs accelerating electrons to 10 MeV, protons would be used for 80%–90% of the 20 000 patients now being treated with X-rays for a typical population of 10 million. With the present costs (a factor of 2–3 per patient treated) this fraction is expected to rise to about 15%.

Carbon ions at 400 MeV per nucleon have the same range as 200-MeV protons (27 cm in water) but the ionization produced is so dense that, when an ion traverses a cell nucleus, it produces irreparable nearby double-strand breaks, causing radio-biological effects that are very different from those due to X-rays and protons. The first clinical results on radioresistant tumors – obtained at the Heavy Ion Medical Accelerator Facility (HIMAC) in Japan and at the GSI (Darmstadt) – are very encouraging. A second Japanese carbon-ion (and proton) facility (HIBMC, located close to the synchrotron-radiation source SPRING 8, at Hyogo) began treating patients in 2001. In Europe in 2005 two centers based on projects developed by a GSI group and the TERA foundation are under construction; respectively, HIT in Heidelberg [61] and CNAO in Pavia [62].

## 19.6 Methods

The history of the application of physics methods to other sciences is as long as the history of Western science, since the scientific method was initially itself a method for constructing physics models and theories. Galileo, Bacon, and Newton developed it to investigate the phenomena of Physics, while others applied it to wider scientific endeavors. Concentrating on today's applications, it is interesting to start by recalling that in anthropology Luca Cavalli Sforza and collaborators successfully applied the matrix method, as employed in quantum mechanics, to determine the various waves of migration of humans during the Neolithic Age using as input the frequencies of about a hundred genes present in the Eurasian populations of today [63].

Another example is the application to cancer diagnosis of algorithms developed in astrophysics to quantify the irregular and highly structured distribution of galaxies in the Universe. A group at the Max Planck Institute for Extraterrestrial Physics in Garching has introduced a nonlinear analysis based on fractal dimensions in an appropriate "state space." This scaling-index method (SIM) has been applied to the diagnosis of skin cancer, achieving a detection efficiency of early skin cancer of 90%, whereas experts can detect "only" about 80% of them [64]. The patented method has been transferred to a small company.

Methods of Physics have also found applications in traffic modeling. Microscopic and macroscopic models of car traffic have been developed over the past 50 years. The main characteristic of a traffic-flow model is the dependence of the average vehicle speed on the density of vehicles on the road. To model the dependence of the flow rate on the vehicle density, two qualitatively different scenarios are used [65]. In the first, developed at General Motors in the USA, ideas from statistical physics, such as instabilities and critical points, are used to explain why traffic jams form. In the second, the concept of metastable states has been introduced by physicists working at Daimler Chrysler in Germany. These states cover a broad range of low vehicle densities for which the traffic flow is "free." However, if in a metastable state a local fluctuation exceeds some critical amplitude, it will grow and lead to the formation of a jam.

There is a stream of young physicists moving from the laboratories to finance and applying Physics methods there. Some work on long-term models of stock-market behavior by finding, for instance, an analogy between the spreading of pessimism that drives the stock market down and the way in which tiny cracks spread in a solid [66]. A pessimistic trader, who goes against the trend, is equivalent to an isolated microcrack in a solid; when these microcracks reach a critical density they combine to cause catastrophic failures.

There are also applications of Physics methods in politics and economy. Take, for instance, the application of the Ising model to policy with the aim of understanding the effects of peer pressure during elections. The gist of the matter is that in an election the distribution of votes has much to do with people's tendency to be influenced more by their neighbors than by the details of a candidate's policies [67]. The simulated system shows the behavior that in Physics is called "frustration" and is typical of protein folding, when the energy minimum is difficult to find because of the existence of many states having almost equal energies. Of course, such an approach cannot predict an election result, but it can illuminate the types of processes behind it.

Through these sophisticated descriptions of social behavior, Physics is, in a sense, just returning what it received about 150 years ago. To develop the kinetic theory of gases, James Clerk Maxwell applied the techniques of contemporary social statisticians who were displaying regularities by averaging over large numbers of people. He wrote that "Those uniformities which we observe in our experiments with quantities of matter containing millions and millions of molecules are uniformities of the same kind as those explained by Laplace and wondered by Buckle [in his *History of Civilization in England*, first published in 1857] arising from the clumping together of multitudes of causes each of which is by no means uniform with the others" [68]. Even the term "statistics" originated from this movement to quantify social phenomena.

Physics-based simulation techniques are being introduced in socio-economy to help predict how economic, political, and scientific decisions will influence climate change. Climate scientists now have to interact with economists, sociologists, and political scientists in order to assess the full implications of climate change for sustainable development. This is the next step in the development of high-resolution, coupled ocean–atmosphere carbon-cycle models that run on today's supercomputers. The challenge is the modeling of the nonlinear relationships among three subsystems: the climate system, the socioeconomic system, and governments. Hans Schellnhuber and other scientists of the European Climate Forum are working toward a community-integrated assessment system (CIAS), a versatile open system to investigate the impact of various policy measures under different socioeconomic assumptions [69].

The World Wide Web is an appropriate example to conclude this section. The method of combining the Internet (which started as an American military project) with hypertext by means of user-friendly software was invented at CERN by Tim Berners-Lee to help particle physicists communicate effectively when working in collaborations of hundreds of scientists [70]. The roots of this development are very basic and are found in the fact that the collision energies provided by particle accelerators must be increased with time in order to satisfy physicists' curiosity. However, the interesting cross-sections decrease as the square of the center-of-mass energy because of Heisenberg's uncertainty principle. Greater and greater investments are thus needed to construct higher-energy colliders with ever-increasing luminosities and their correspondingly larger detectors composed of many different sub-detectors, and this requires many groups bringing together their expertise and funds. However, huge scientific collaborations have problems of communication. With some imagination one could say that the Web, invented in the 1990s to solve them, is a consequence of the uncertainty principle. However, CERN has influenced Society more through the three small 'w's' of website addresses than through the W and Z Nobel Prize-winning bosons. The next step will be the "Semantic Web," which – with the contribution of Berners-Lee and many others – will bring structure to the meaningful content of Web pages, creating an environment where software agents roaming from page to page can readily carry out sophisticated tasks for the users [71].

Since hadrons are composite particles, from a similar chain of arguments it follows that data-acquisition rates at the Large Hadron Collider have to increase by an enormous factor, while there are fewer dollars/euros to spend per flop. The solution to this second conundrum is the Grid [72], which is also urgently needed in many other fields (see Chapter 15).

## 19.7 People

The three output streams discussed above (knowledge, technologies, and methods) are homogeneous in the sense that they can be written down either on paper or as computer files. They are eventually condensed in articles published in peer-reviewed journals, conference proceedings, seminars, and patents. The fourth stream is completely different, but no less important, because it has to do with human beings.

We are here considering *a very large community.* At the beginning of the millennium, a few hundred thousand scientists could be classified under the name of "physicists" in so much as they had published at least once in a peer-reviewed physics journal, but the differences appear when looking at a citation index. Among the 100 most-cited physicists of the last 35 years, 75 are from the USA, 6 from Switzerland, 4 from Germany, 3 from Canada and 2 each from France, Italy, Japan, the UK, and the Netherlands. In 2003, Bell Laboratory innovations accounted for five names on this list.[2] As Arnold Wolfendale, former president of the European Physical Society, remarked in 2001, this imbalance could be partially due to the brain drain but also to the reluctance in the USA to refer to papers produced by physicists outside America: "The lack of reference to European [and rest-of-the-world] publications by US scientists shows no sign of improvement. The brain drain is, perhaps, the only example of appreciation that the US community shows for European physicists and other scientists" [73].

However, European physicists are not helping themselves. Of the 90 000 articles published in Physics journals in 2000, about 39% came from Europe, whereas only 29% originated from the USA. However, European journals have only a 27% share because more and more European physicists publish in American journals, possibly because of the larger impact factors [74]. This produces a snowball effect that is difficult to invert.

This is the present state of Physics. Unfortunately, the perspectives of even keeping it at the same level are not good. It is well known that in the past fifteen years the *number of physics degrees awarded has been decreasing.* This decrease is part of a much more general trend: since 1996 the number of science and engineering Ph.D.s delivered by US universities has been decreasing, primarily because the numbers of doctoral degrees granted to US citizens and to non-US citizens have leveled off and decreased, respectively [75]. Since 1995, the number of European Ph.D.s awarded in science and engineering has become larger than that in the USA, but the number of Physics degrees is decreasing. In Germany between 1991 and 1999 the number of first-year Physics students fell from 10 000 to just over 5000 [76]. Between 1995 and 2001, in France the number of students studying physics at first-degree level dropped by 45% [77]. Recently there have been indications that this tendency is reversing, but something should be done if the flux of well-prepared physicists toward other fields of activity has also to be increased (article by Y. Battacharjee in *Science* **305** (2004), 173).

To attract passionate youngsters to universities one can influence the public at large and school pupils in particular. Certainly, popular books are extremely useful both to convince high-school students to choose Physics and to involve laymen in its fascination and relevance to Society. Fortunately the number and quality of these books is increasing, since many eminent scientists endeavor to produce best-sellers whose titles become icons of our times, even to those who have never read them. Steven

[2] Updated information on highly cited scientists can be obtained from http://isihighlycited.com.

Weinberg's *The First Three Minutes* [78] and Stephen Hawking's *A Brief History of Time* [79] are the best-known examples. These and other books have sold millions of copies but, as far as I know, no information is available on the influence of this literature – and of television series devoted to Physics and science – either on the choice of the university curriculum by high-school students or on the image that the public has of science, and of Physics in particular.

As far as school pupils are concerned, a success story comes from France. In the 1990s Georges Charpak – with the help of the Science Academy – pioneered "La main à la pâte," an early approach to science in which girls and boys at infant and primary schools work with their hands on simple experiments. The effect has been the introduction in 2002 of new teaching programs in science as a part of a "plan de renovation" launched by the French Ministry of Education [80].

At the high-school level, in the USA Leon Lederman has for many years been battling against the traditional American (and worldwide) system by emphasizing that "A National Committee of Ten in 1893 chaired by a Harvard President recommended that high-school children be instructed in science in the sequence Biology, then Chemistry, and then Physics. The logic was not wholly alphabetic since Physics was thought to require a more thorough grounding in mathematics" [81]. But now science has changed and has changed the world, while scientific illiteracy prevails. What is needed, according to Lederman, is a revolution in science education – put Physics first [82]. However, in 2000 a survey conducted by the American Institute of Physics found that 61% of school principals and Physics teachers reject the "Physics-first" approach [83]. As an author of textbooks – through which more than a third of Italian high-school students learn Physics – I am well placed to add that in Europe this is a battle that cannot be won. A more realistic solution is to improve the curricula, oblige the schools to offer hands-on courses, and educate a better generation of teachers, helping them to achieve higher salaries and better training. For instance, in the UK, to support Physics teachers the Institute of Physics has set up a network of regional coordinators [84].

Also special programs can be rewarding. In 2004 the winner of the Altran Foundation Innovation Award was HISPARC, the "high-school project on astrophysics research with cosmics" (*sic*) initiated by physicists at the University of Nijmegen [85]. A cluster of schools was chosen to host a network of plastic detectors combined with accurate time measurements based on the Global Positioning System. The students were so enthusiastic about being able to study extended showers with easy-to-understand instrumentation that the promoters intend to spread the system outside the Netherlands.

Physicists are useful outside Physics, in particular in other sciences and in industrial research. The most striking examples of the effect of physicists in other fields of science mainly concern biology. It is enough to name Francis Crick, Max Delbrück, and Manfred Eigen to recall some of the major influences that physicists have had on the development of modern biology. Moreover, molecular biology initiated its tumultuous development with the discovery of the structure of DNA, which was made possible by the application of X-ray scattering – a Physics technique.

Industrial companies are interested in hiring physicists (and engineers) who have worked on Physics projects. Various enquiries have been made to ascertain the reasons why. They have been found to be [86] that companies value

analytical thought and a systematic approach to new problems;
experience in designing and carrying out complex projects;
the habit of documenting and presenting the work done;
experience in working in international teams at the edge of knowledge; and
specific knowledge of fundamental sciences and technologies.

(Note that knowledge of Physics is not the main reason.)

This is so not only for students, but also for junior staff, mostly scientists and engineers, since a relatively large number of them leave (or have to leave) Physics after 5–6 years of employment. They have had time to absorb the many technological and behavioral aspects of research in Physics and are ready to disseminate novel approaches in the laboratories and companies they may join.

Last, but not least, one has to consider well-known physicists, who have left the subject to take up influential roles in industry and government, becoming ambassadors of Physics.

In Europe Hendrik Casimir is the first name that comes to mind [87]. He studied and worked with Ehrenfest, Bohr, and Pauli in the burgeoning field of quantum mechanics. During World War II he moved to the Philips Research Laboratories in Eindhoven, where he was responsible for research and development [88]. He remained an active scientist and in 1948 published a celebrated paper with Dik Polder; on the "Casimir effect," the attraction between two neutral conducting plates due to vacuum fluctuations, which was observed experimentally much later. The Casimir effect has practical consequences – because it influences the performance of cantilever nanodevices – and theoretical ones – because it is relevant to the bag model of hadrons and in surface-critical phenomena [89]. Casimir managed to bring the Philips Laboratory to the frontier of science application. He theorized his approach with the "science–technology" spiral and was always a staunch defender of fundamental Physics: "I have heard statements that the role of academic research in innovation is slight. It is about the most blatant piece of nonsense it has been my fortune to stumble upon" [90].

In the USA Richard Garwin's career had a somewhat similar profile. After a period at the University of Chicago, during which he was one of the co-discoverers of parity nonconservation in muon decay, he left the academic world and in 1952 joined IBM's Watson Laboratory in New York. Here he continued to do fundamental research and, in parallel, contributed to many patents in magnetic-resonance imaging, high-speed laser printers, and superconducting computing. During his career with IBM, Garwin divided his time between corporate responsibilities and work as an expert on nuclear, chemical, and biological weapons for the government. He was one of the most active American promoters of nuclear nonproliferation and of arms control [91]. With such a long and diversified experience, he stated that "over the years I have been impressed by the many instances in which tools and techniques developed for pure physics research have been applicable to important problems in industry and in government programs."

These are the good ambassadors of Physics. Positive, but sometimes negative, other ambassadors are the characters in plays centered on Physics. The most famous is the *Life of Galileo* written in German by Bertold Brecht before World War II and first performed in 1947 in Los Angeles, staged by Charles Laughton, who had translated the text with Brecht. This was about Science and Society, a very sensitive subject following the atomic bombing of Japan. Such plays focus on the benefit or harm of Science for Society and on Society's attitude toward Science. They can also convey scientific ideas, as Brecht does through discussions between Galileo and Sagredo. Another such play is *The Physicists* written during the cold war by the Swiss author Friedrich Dürrenmatt. The protagonist is a physicist who, having discovered unexplained 'physics principles' that could destroy humanity, hides in a mental institution. Two spies following him claim to be Einstein and Newton. Here the subject is the moral responsibility of the scientists rather than mysterious physics. For plays focusing on scientists as people, a good example is *QED*, about Richard Feynman, by Peter Parnell, which opened in New York City in 2002 with excellent reviews. For Science as the very fabric of the play, the most notable example is *Copenhagen* by Michael Frayn, the well-known novelist, translator of Chekhov, and author of many successful comedies. Here, Physics subtleties are mixed with the drama as, for instance, when Bohr's wife Margrethe says to Heisenberg "Your talent is for skiing too fast for anyone to see where you are; for always being in more than one position at the same time, like one of your particles." The play suggests that no-one knew for sure whether Heisenberg sabotaged Hitler's atomic bomb or was simply unable to make it, and that nobody will know what happened that evening in Copenhagen. For Frayn, intentions are like the movements of a particle under Heisenberg's Uncertainty Principle. However, as a result of the popularity of the play, in 2002 the Bohr family released documents that shed new light on the episode (see www.nba.nbi.dk).

## 19.8 The pathways from fundamental research to technology transfer

The four output streams of spinoffs percolate to Society through different paths. *Knowledge, methods*, and *technologies* pass naturally through scientific and technical publications, conference proceedings, seminars, and, very often, private communications. *People* carry with them specific "basic" and "usable" knowledge, methods, and technologies together with models of actions and general know-how.

Most of these transfers to Society are not predictable and cannot be programmed and, for this, it may be more appropriate to speak of "*spill-overs*," thus emphasizing the somewhat casual and random nature of these processes. Can something be done to promote this? At least three such pathways can be better programmed and this section is devoted to them: patenting, procurements, and joint development projects.

*Patenting* (and the consequent creation of spinoff companies) is the natural means for biomedical researchers to transfer their scientific and technological advances. In the material sciences, patenting is very frequent, particularly in the USA, whereas in subatomic physics, and particularly in Europe, this has not been the custom, and many senior European physicists would add "fortunately." Quantitatively, about 1% of the American-authored papers published between 1993 and 1995 and included in the Science Citation Index were cited in 1997 US-invented patents, while fewer than 0.01% of the papers in subatomic, fluid, and plasma physics were [92].

In Europe the subject is very delicate since the rule "patent first and then talk about it and publish it" (which is not applied in the USA) threatens the atmosphere of openness which characterizes Physics collaborations. However, it is difficult to resist the pressure of policy makers and administrators, who want to see "practical" results from the money invested in basic research and consider the number of patents as the only valid indicator. Particle physicists and institute directors are now moving in an unfamiliar world in which condensed-matter physicists and biologists feel much more at home.

For the second pathway, *procurement*, industries that receive an order for a high-tech product from a research laboratory interact with people at the frontier of research and, while providing and delivering the goods ordered, acquire new knowledge and new technologies. Passing knowledge through procurement is well known to astrophysicists, fusion physicists, and particle physicists, who with the help of industry build large and complex apparatus, which is not normally the case in condensed-matter physics.

Four studies have been published on the subject, two (in 1975 and 1984) concerning CERN [93, 94] and two (in 1980 and 1988) concerning the European Space Agency (ESA) [95, 96]. By interviewing a large number of companies it was shown that the "economic utility" of the procurement process ranges between 2.7 (for the ESA) and 3.7 (for CERN). The four analyses taken together indicate that, for every euro spent in a high-tech contract, the company will receive and/or save in total 3–4 euros in the form of increased turnover and/or cost savings. A further detailed study of the effects of CERN procurement activity was published in 2003 [97]. The enquiry involved 612 companies that had to produce for CERN high-tech components involving a noticeable technological development. The contracts summed to 700 million euros during the period 1997–2001, corresponding to 50% of the overall procurement budget. Some of the main conclusions were that (i) 38% of the companies developed new products, the total number of new products being more than 500; (ii) 60% of the firms acquired new customers, the total number of new customers being about 4400; (iii) 44% of the companies indicated that they had undergone technological learning, and 52% would have had poorer sales performances without CERN; and (iv) 40% would have had worse technological performance.

*Common projects* involving a research institution and one or more companies constitute the third pathway. The purpose is the development of a product that does not yet exist on the market. Through such collaboration, methods and technologies are transferred to the industrial partners. Again using CERN experience, one can quote the work by M. Haehnle who studied 21 common projects originated by CERN [98] and concluded that learning and gaining know-how is the main motivation for "large" companies (those with more than 500 employees) taking part in R&D collaborations. As could be expected, the inquiry showed that "small" companies are almost exclusively interested in products as the end result of the collaboration.

With the above numbers, one could argue that practically all the money that tax payers invest in a Big-Science laboratory goes to improve productivity and sales of high-tech industry. Indeed, almost one-quarter of the budget is invested, on average, in high-tech industries. But this investment has a utility factor of more than three, so that the turnover generated roughly equals the overall budget. Distributing the same amount of money to high-tech industries would not have such an impact.

In an attempt to rationalize the complex environment of spinoffs of basic science, four streams (knowledge, people, methods, and technologies) and many pathways (publications, conferences, seminars, private communications, people themselves, patents, procurement activities, and joint projects) have been identified. To improve the spillovers, I venture to put forward three suggestions to those responsible for Physics departments, institutes, and laboratories.

(i) They should organize, update, and distribute (also through the Internet) a "*technology portfolio*" describing the application potentialities of their basic research.

(ii) They should be *very selective in the patenting process*, without promising large revenues to funding organizations, in order to avoid painful disillusionment.

(iii) They should *encourage head-hunters* to follow the work of younger people working in their laboratories, in order to facilitate turnover and optimize the use of human resources for Society. Some potentially good scientists may be lost, but good careers will be found for many – who anyway cannot remain in science – and strong arguments will be found to defend the view that Physics has, through well-prepared people, a positive impact on Society.

## 19.9 A daring look into the crystal ball

Before tackling the impossible task of predicting some of the future spinoffs of Physics, recall that physicists, even great ones, do not have a very good record here. The most-quoted failed predictions are the following [99]:

Lord Kelvin (1885): *Heavier-than-air flying machines are impossible.*
Lord Rayleigh (1889): *I have not the smallest molecule of faith in aerial navigation other than ballooning.*
Lord Rutherford (1933): *Anyone who expects a source of power from the transformation of [the nuclei of] atoms is talking moonshine.*
John von Neumann (1956): *A few decades hence, energy may be free, just like unmetered air.*
Sir George Thomson (1956): *The possibility of travel in space seems at present to appeal to schoolboys more than to scientists.*

Well aware of such failures, I move onto the slippery slope of predicting some possible future spinoffs from Physics (and applied Physics, i.e. engineering) in fields of interest to Society at large.

One can already see that *electromagnetism* has some surprises in store for the not-too-distant future. In the electromagnetic spectrum used in applications there is a gap – the terahertz gap – where semiconductor devices do not work: transistors work up to about 300 GHz (wavelengths of the order of 1 mm) while lasers cannot go below about 30 THz (wavelengths of 10 μm). New types of synthetic crystals, made from many superimposed semiconducting layers, promise to become intense sources of terahertz radiations [100]. T-rays will be used to reveal the molecular compositions of surfaces by measuring the reflected signal, detect cancer, see through walls and floors, and pick out metal objects beneath clothing [101, 102]. Privacy issues will have to be resolved before T-cameras arrive on the market!

The invention of the scanning tunneling microscope (STM) by Gerd Binnig and Heinrich Rohrer in the early 1980s initiated a technological revolution that has already had far-reaching consequences in imaging on the nanometer scale with many types of scanning-probe microscopes (collectively called STMs) and in bringing about the possibility of moving single atoms around [103]. For the development of *nanotechnology* condensed-matter physicists, engineers, chemists, and molecular biologists are working and will work more and more together. Ten years after the first step along the new road, K. Eric Drexler in his *Engines of Creation* depicts a future world in which nanomachines will replicate themselves, produce any material good we may need, cure cancer, eliminate aging, and repair cells in the human body [104]. It was a fascinating vision, even if less attractive than Raquel Welsh, who – shrunk to minute size with her companions and their submarine – destroys a brain clot with a laser beam in the 1966 film *Fantastic Voyage*, which was based on an idea by Isaac Asimov.

The really novel applications of nanotechnologies will be, I believe, artifacts at *the frontier between biology and Physics*. After all, the building blocks of cells are nanoscale physical structures (proteins, nucleic acids, lipids, etc.) and magnetotactic bacteria are already mixing Physics with biology since they move by sensing the direction of the Earth's magnetic field through twenty magnetic crystals, each 50 nm long [105]. Inspiration for predicting future developments can be found in the September 2001 special issue of *Scientific American* on nanotechnology [106]. My own list is as follows: the curing of cancer by nanorobots attached to monoclonal antibodies, new diagnostic tools based on active biomechanical complexes that target the organs at risk [107], and biological and physical analysis of minute quantities of liquids and gases in self-controlling nanofluidic devices [108].

In *particle physics* theorists are grooming their Standard Model, victims of their own success, but it is difficult today to see what "directly usable" knowledge it contains. Still, there is a facet of the Standard Model that may lead in the far future to practical applications: the neutrino sector. In 1983 A. de Rújula, S. L. Glashow, R. R. Wilson, and G. Charpak proposed four startling ideas with imaginative acronyms [109]. The

GEOTRON is a 10-TeV proton synchrotron, possibly floating in deep water, which could direct a TeV neutrino beam through the Earth towards a specific site. GENIUS stands for "Geological Explorations by Neutrino-Induced Underground Sounds," while GEMINI means "Geological Exploration by Muons Induced by Neutrino Interactions," a scheme that employs large muon detectors mounted on trucks to survey the upper layers of the Earth and could be used to find underground oil reserves. Finally, such a source of neutrinos would also have direct scientific applications with GEOSCAN, a neutrino scanner for measuring the density profile of the Earth.

Could some "usable knowledge" also come from new discoveries that would show that there is new *Physics beyond the Standard Model?* Some not entirely unfounded possibilities, mainly related to energy production, can be listed. For instance, if magnetic monopoles exist, they could catalyze proton decay [110]. In general, many types of long-lived charged particles could catalyze fusion, as first proposed by Frank and Sakharov in the 1940s. Moreover, non-topological stable solutions of quantum chromodynamics would have a global baryon number and could produce energy when fed with nucleons [111].

While obtaining energy from non-standard particles depends on the generosity of Nature, the final prediction of this section follows from chaos theory in meteorology. I share the opinion of Ross Hoffman in his article "Controlling hurricanes" in the October 2004 issue of *Scientific American* – "Modest trials could be instituted in perhaps 10 to 20 years. With success there, larger-scale weather control using space-based heating may become a reasonable goal". See also Chapter 13.

## 19.10 The image of Physics in Society

From all the above, one could conclude that – apart from the difficulties in attracting young people to university courses and in acquiring increases in funding to match those of the biomedical sciences – Physics and physicists are doing fine. However, such a conclusion would hide the fact that the image that the public has of Physics is not as good as it merits, and this is so despite the very positive attitude of laymen toward the sciences and scientists.

A 2002 survey by the Wellcome Trust and the UK Office for Science and Technology showed that in the UK more than 70% of the public claim to be interested in science and 72% believe that the government should support advances in science even if there are no immediate benefits [112]. Of course, the main interest is in biomedicine, but between 50% and 60% of the respondents were interested in climate change, telecommunications, computing, and new modes of transport. Physics is not at the same level.

The roots of the problem go back to the development of the atomic bomb and to the Chernobyl disaster. The opposition to the storage of nuclear waste compounds public fears, especially in times of possible terrorist attacks. Radiation and radioactivity, which physicists let out of their Pandora's box, are perceived as negative and uncontrollable, in spite of the fact that – as every scientist knows – most chemical contaminations are much more difficult to check and control.

What could be done? On the subject of generalized fear of radiation, it has to be recognized that many concerns are related to the obscure technical language used, and to the fact that radiation exposures are expressed in units that non-specialists find difficult to comprehend. On this point, an interesting proposal [113] has been put forward by Georges Charpak and Richard Garwin, who suggested a new unit of equivalent dose – the DARI – the annual dose provided to a human being by the naturally occurring radioactivity of human tissues, mainly $^{40}$K. Numerically 1 DARI (in French: *dose annuelle due aux radiations internes*) would correspond to 0.2 millisievert.

Expressed in this very "natural" unit, for a Briton the dose due to the native soil is on average 10 DARI while the dose received by a Frenchman from his nuclear-power industry is only 0.1 DARI. A computer-tomography scan gives 40 DARI and the lethal dose for a whole-body irradiation is 25 000 DARI. Finally, for whole-body irradiation and a supposed linear response to low doses, 1 DARI causes eight lethal cancers in a population of a million. If the unit were adopted, it would help people to gain an intuitive feeling for the effects and dangers of ionizing radiations.

To boost the image of Physics, its links with developments that have enhanced the quality of life have to be recognized by the public. Today even the patients who profit from medical examinations and cancer cures do not realize that Physics and its technological spinoffs have put at our disposal – unfortunately, only in the more developed nations – sophisticated apparatus for diagnostics (computer tomography with X-rays, magnetic-resonance imaging, single-photon-emission tomography, positron-emission tomography, etc.) and therapy (with photons produced by electron linacs, with "low-dose" and "high-dose" brachytherapy, with radiation immunotherapy, etc.). Over the 15 years I have devoted to the introduction, in Italy and in Europe, of the treatment of cancer based on energetic light ions, I have learned that even a cursory explanation of these medical techniques positively affects the image that laymen have of Physics.

Unfortunately, even a wider knowledge of the medical spinoffs – and of those in energy production, transportation, and information technology – would not be enough to convince every citizen of the benefits that Physics brings to Society. The poor image of Physics is indeed part of a generalized criticism of all basic sciences, which are accused of living in an ivory tower, avoiding their responsibilities by abusing the very convenient distinction between pure science (pursued without control and mainly with public money) and applied science (which has other control mechanisms and is funded mainly by industry). The 2000-year-old Hippocratic oath has percolated through Society so that people, without even knowing who Hippocrates was, "know" that a doctor has the care of the patient as his foremost duty and (usually) have confidence in him and in the system. The scientists are perceived in a different way, i.e. as individuals ready to put humanity and the environment in danger in order to reach their scientific and technological goals. Maybe a step out of this impasse could be made by following a suggestion of Joseph Rotblat (Figure 19.4), a physicist who has devoted his life to promoting the sense of responsibility of fellow scientists by being the prime mover behind the Pugwash Conference movement. He proposed that young scientists when graduating should take the pledge introduced by the Student Pugwash Group in the USA [114]:

> I promise to work for a better world, where science and technology are used in socially responsible ways. I will not use my education for any purpose intended to harm human beings or the environment. Throughout my career, I will consider the ethical implications of my work before I take action. While the demands placed upon me may be great, I sign this declaration because I recognize that individual responsibility is the first step on the path to peace.

Ethical considerations enter also into the production and dissemination of scientific results. In Physics, frauds had been quite rare, but, unfortunately, in a single year (2002) it was discovered that physicists in two famous US laboratories had falsified data that at the time of publication had been considered very important. At the Lawrence Berkeley Laboratory the discovery of element 118 had to be retracted by its authors, and at the Bell Laboratories it was found that many papers on nanotechnology and the superconductivity of organic semiconductors were based on (poorly) fabricated data. The two physicists responsible for the frauds were dismissed [115], but the many

**Figure 19.4.** Joseph Rotblat (right) and Francesco Calogero, Secretary General of the Pugwash movement, after receiving the 1995 Nobel Peace Prize, awarded jointly to Rotblat and the Pugwash movement.

well-known co-authors were only mildly reprimanded [116]. These episodes have been very detrimental to the image of Physics. Avoiding similar such episodes has been seen as a duty of the learned societies, as is emphasized by the fact that guidelines for professional conduct have been issued by the American Physical Society (APS) [117].

## 19.11 A general outlook

As shown in Figure 19.5, the four streams of spinoffs are related to four activities typical of humankind in the twenty-first century: the building up of a "scientific culture," the increase of human "wetware" ability (i.e. the ability of brains to produce new ideas), the development of new "software," and the introduction of novel "hardware" technologies. The four streams of spinoffs come from any field of science to Society, but their relative fluxes are very different. For instance, "usable knowledge" is at present abundant for basic research in biology, while it languishes for particle physics. This is best illustrated by the competition that went on in the decoding of the human genome. Money was raised to create a company that would complete the decoding before the publicly funded consortium, with motivations that were only partially scientific. On the other hand, a private company that would invest money in an accelerator laboratory dedicated to the discovery of the Higgs boson has yet to be created!

Condensed matter is doing much better than particle physics as far as knowledge spill-overs are concerned, but the technologies developed for subatomic physics have large effects on society, as witnessed by the use of accelerators in medicine.

The famous *Fountain of the Four Rivers* (1650; Figure 19.6) by Gian Lorenzo Bernini in Rome's Piazza Navona can be used as a metaphor. The four streams (representing the Nile, the Danube, the Ganges, and the Plate) are far from being equal at all times

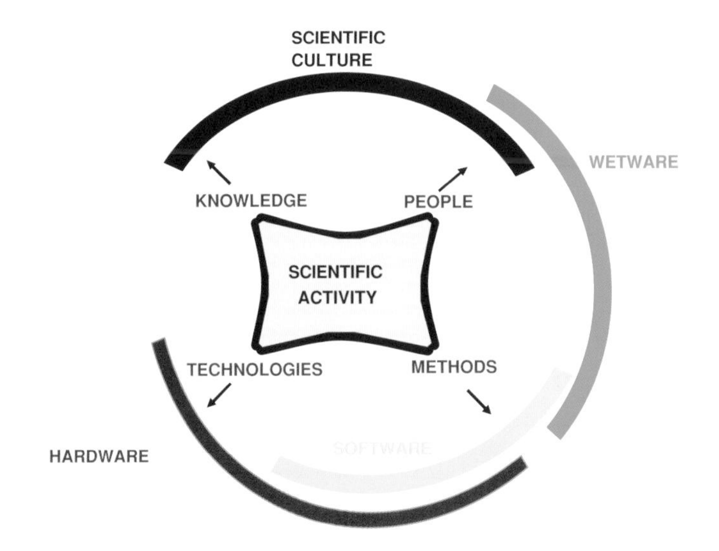


**Figure 19.5.** The four output streams of any scientific endeavor are connected with four major activities of humankind: the construction of a unique scientific culture, and the growth of the world patrimony of wetware (the ability of brains to produce new ideas), software, and hardware technologies.

**Figure 19.6.** Gian Lorenzo Bernini's *Fountain of the Four Rivers* can be seen as a metaphor of the spinoffs of the various fields of Physics to Society.

but still, both in winter and in summer, the fountain is full of fresh water and beautiful. Similarly, different fields of Physics produce four streams of unequal fluxes. However, they can all contribute to the beauty of "Rome" (our Society) if physicists of all fields are concerned not only with the production of basic knowledge but also with spinoffs from the other three streams: technologies, methods, and people.

In 2002 the General Assembly of the International Union of Pure and Applied Physics accepted the European Physical Society's proposal to make 2005 the "World Year of Physics," a hundred years after Einstein's *annus mirabilis* [118]. The major goal is the illustration of the physical sciences, their achievements, and their spinoffs; the expected result is an improved image of Physics.

## Acknowledgement

I am grateful to Gordon Fraser for very useful comments and suggestions.

## FURTHER READING


1. U. Amaldi and B. Larsson (eds.), *Hadrontherapy in Oncology*, Amsterdam, Elsevier, 1994.
2. E. Autio, M. Bianchi-Streit, and A. P. Hameri, *Technology Transfer and Technological Learning through CERN's Procurement Activity*, CERN Yellow Report, Geneva, CERN, 2003.
3. J. D. Barrow and F. J. Tipler, *The Anthropic Cosmological Principle*, Oxford, Oxford University Press, 1986.
4. H. Casimir, *Haphazard Reality – Half a Century of Science*, New York, Harper and Row, 1983.
5. L. L. Cavalli Sforza, *Genes, Peoples, and Languages*, New York, North Point Press, 2000.
6. L. S. Feuer, *Einstein and the Generations of Science*, New Brunswick, Transaction Inc., 1982.
7. R. Feynman, *Six Easy Pieces*, New York, Basic Books, 1994.
8. R. L. Garwin and G. Charpak, *Megawatts and Megatons: Turning Point for the Nuclear Age?*, Chicago, University of Chicago Press, 2002.
9. W. Heisenberg, *Physics and Philosophy: The Revolution in Modern Science*, New York, Harper and Row, 1958.
10. R. B. Laughlin, *A Different Universe: Reinventing Physics from the Bottom Down*, New York, Basic Books, 2005.
11. N-P. Ong and R. Bhatt (eds.), *More is Different: Fifty Years of Condensed Matter Physics*, Princeton, Princeton University Press, 2001.
12. M. Rees, *Our Cosmic Habitat*, London, Weidenfeld and Nicholson, 2001.
13. C. S. Smith, *A Search for Structure*, Cambridge, MA, MIT Press, 1981.
14. J. Stewart and D. LaVaque-Manty, *Women in Science: Career Processes and Outcomes*, Cambridge, MA, Harvard University Press, 2003.
15. C. Tobias and I. Tobias, *People and Particles*, San Francisco, San Francisco Press, 1997.
16. S. Weinberg, *Facing-up: Science and Its Cultural Adversaries*, Cambridge, MA, Harvard University Press, 2001.


## Ugo Amaldi

Ugo Amaldi studied physics at Rome University where he obtained the Laurea in 1957, and completed his postgraduate studies in 1960. At the physics laboratory of the Rome Istituto Superiore di Sanità (ISS) he then worked in atomic and nuclear physics, introducing a new method, namely coincidence electron scattering on atoms and nuclei, which is now called electron-momentum spectrosopy and (e, e′p) reactions, respectively. His subsequent research activities have concentrated on the physics of elementary particles. In 1973 he moved from the ISS to CERN, where many successful experiments followed. In 1975 he co-founded the CHARM collaboration, which performed many fundamental experiments on neutrino scattering. For 13 years he served as spokesman for the DELPHI collaboration, which he formed directly from the preparations for experiments at the LEP electron–positron storage ring at CERN in 1980. In 1991, using the first LEP data, he made a widely recognized contribution to the understanding of the unification of the electroweak and strong forces. Since the 1980s, first with his father Edoardo Amaldi and later on his own, he has written physics textbooks: at present more than a third of Italian high-school pupils use his books. The hadron cancer-therapy project of the TERA Foundation of Novara, founded in 1992, is Amaldi's most recent initiative. In 2001 the Italian government financed the construction of the Centro Nazionale di Adroterapia Oncologica (CNAO) in Pavia. The CNAO was proposed and designed by TERA in collaboration with CERN, the GSI, INFN, and other European institutes. At present he is Professor of Medical Physics at the Università Milano Bicocca.

## References

[1] J. Jackson, The question, *Fermi News* 25 (2002) 7.
[2] J. Kaiserin, House bill signals the end of NIH's double-digit growth, *Science* 300 (2003) 2019.
[3] A call from downtown to shed staff, authority . . ., *Science* 300 (2003) 877.
[4] Philanthropy rising tide lifts science, *Science* 286 (1999) 214.
[5] C. Seife, At Canada Perimeter Institute, "Waterloo" means "Shangri-La," *Science* 302 (2003) 1650.
[6] Stanford to host the institute for particle astrophysics and cosmology, *CERN Courier*, March 2003, 33.
[7] Press release of March 18, 2004, www.seti.org.
[8] M. Riordan, A tale of two cultures: building the Superconducting Super Collider, *Hist. Stud. Phys. Biol. Sci.* 32 (2001) 125.
[9] S. Weinberg, Newtonianism, reductionism and the art of Congressional testimony, *Nature* 330 (1987) 433.
[10] A. Sokal, Towards a hermeneutics of quantum gravity, *Social Text* 46/7 (1996) 217.
[11] L. S. Feuer, *Einstein and The Generations of Science*, New Brunswick, Transaction Inc., 1982.
[12] J. D. Barrow and F. J. Tipler, *The Anthropic Cosmological Principle*, Oxford, Oxford University Press, 1986.
[13] M. Rees, *Our Cosmic Habitat*, London, Weidenfeld and Nicholson, 2001.
[14] B. Carr, Life, the cosmos and everything, *Phys. World*, October 2001, 23. This article can be found at http:/physicsweb.org/article/world/14/10/3.
[15] H. Oberhummer, Report of the Conference 'Anthropic Arguments in Cosmology and Fundamental Physics,' *Nucl. Phys. News*, 12 (2) (2002) 35.
[16] M. Durrani, Physicist scoops religion prize, *Phys. World*, March 14, 2002.
[17] A. J. Stewart and D. LaVaque-Manty, *Women in Science: Career Processes and Outcomes*, Cambridge, MA, Harvard University Press, 2003.
[18] B. Whitten, S. Foster, and M. Duncombe, What works for women in undergraduate physics?, *Phys. Today*, September 2003.
[19] M. Fehrs and R. Czujko, Gender disparity in physics, *Phys. Today*, August 1992, p. 33.
[20] G. Pancheri, Women and their contribution to modern and contemporary physics: barriers and perspectives, *Analysis*, January–February 2002, 4.
[21] C. Tobias and I. Tobias, *People and Particles*, San Francisco, San Francisco Press, 1997.
[22] R. Feynman, *Six Easy Pieces*, New York, Basic Books, 1994.
[23] A. Maurellis and J. Tennyson, The climate effects of water vapour, *Phys. World*, May 2003, 29.
[24] O. Polyansky, A. Csaszar, S. Shirin *et al.*, High-accuracy *ab initio* rotation-vibration transitions for water, *Science* 299 (2003) 539.
[25] S. Weinberg, *Facing-up: Science and its Cultural Adversaries*, Cambridge, MA, Harvard University Press, 2001.
[26] P. W. Anderson, *Phys. Today*, July 2002.
[27] P. W. Anderson, in *More is Different: Fifty Years of Condensed Matter Physics*, N-P. Ong and R. Bhatt, eds., Princeton University Press, Princeton, 2001, p. 1.
[28] T. R. Reid, *The Chip: How Two Americans Invented the Microchip and Launched a Revolution*, New York, Random House, 2001. (The two Americans who in 1958–59 did the work were Jack Kilby and Robert Noyce.)
[29] L. Alvarez, Catalysis of nuclear reactions by muons, *Phys. Rev.* 105 (1957) 1127.
[30] L. Alvarez, *Evolution of Particle Physics*, New York, Academic Press, 1970, pp. 1–49.
[31] M. Leon ed., *Proceedings of the Workshop on Low Energy Muon Science – LEMS 93*, Los Alamos, Los Alamos National Laboratory, 1994.
[32] Reports of the Railway Technical Research Institute: www.rtri.or.jp.
[33] See for instance www.airport-technology.com/contractors/security/waterbridge.
[34] C. Stanley Smith, *A Search for Structure*, Cambridge, MA, MIT Press, 1981.
[35] G. B. Olson, Designing a new material world, *Science* 288 (2000) 993.
[36] J. Nagamatsu, N. Nakagawa *et al.*, *Nature*, 410 (2001) 63.

[37] A. M. Campbell, How could we miss it?, *Science* **292** (2001) 65.
[38] P. Phillips, From insulator to superconductor, *Nature* **406** (2000) 687.
[39] M. Roukes, Nanoelectromechanical systems face the future, *Phys. World*, February 2001, 25.
[40] S. Iijima, *Nature* **354** (1991) 56.
[41] M. Reed and J. M. Tour, Computing with molecules, *Scient. Am.*, June 2000.
[42] W. Llang, M. Shores *et al.*, Kondo resonance in a single-molecule transistor, *Nature* **417** (2002) 725.
[43] S. De Franceschi and L. Kouwenhoven, Electronics and the single atom, *Nature* **417** (2002) 701.
[44] R. F. Service, Can chemists assemble a future for molecular electronics?, *Science* **295** (2002) 2398.
[45] E. H. K. Stelzer, Beyond the diffraction limit?, *Nature* **417** (2002) 806.
[46] A. Lewis, A marriage made for the nanoworld, *Phys. World*, December 2001, 25.
[47] J. R. Guest, T. H. Stiehvater, Gang Chen *et al.*, Near-field coherent spectroscopy and microscopy of a quantum dot system, *Science* **293** (2001) 2224.
[48] J. Hecht, Extreme laser shed light on nanoworld, *New Scientist*, **27** July 2002, 17.
[49] P. Noyes, Physics with petawatt lasers, *Phys. World*, September 2002, 39.
[50] V. Malka, A new and exciting optically induced electron source: extreme acceleration gradients beyond 1 TV/m, *Europhys. News*, March/April 2004, 43.
[51] J. Hogan, How to create and destroy elements in a flash of light, *New Scientist*, August 23, 2003, 10.
[52] K. W. D. Ledingham, P. McKenna, and R. P. Singhal, Applications for nuclear phenomena generated by ultra-intense lasers, *Science* **300** (2003) 1107.
[53] U. Amaldi, The importance of particle accelerators, *Europhys. News*, **31** (6) (2000).
[54] P. G. O'Shea and H. Freund, Free-electron lasers: status and applications, *Science* **292** (2001) 1853.
[55] A. Cho, The ultimate bright idea, *Science* **296** (2002) 1008.
[56] E. Plönjes, J. Feldhaus, and T. Möller, Taking free-electron lasers into the X-ray regime, *Phys. World*, July 2003, 33.
[57] R. F. Service, Battle to become the next-generation X-ray sources, *Science* **298** (2002) 1356.
[58] N. Patel, Shorter, brighter, better, *Nature* **415** (2002) 110.
[59] J. E. Cartlidge, Nuclear alchemy, *Phys. World*, June 2003, 8.
[60] U. Amaldi and B. Larsson, eds., *Hadrontherapy in Oncology*, Amsterdam, Elsevier, 1994.
[61] K. D. Gross and M. Pavlovic, eds., *Proposal for a Dedicated Ion Beam Facility for Cancer Therapy*, Darmstadt, GSI, 1998.
[62] U. Amaldi, ed., The path to the national center for ion therapy, Vercelli, Mercurio, 2005.
[63] L. L. Cavalli Sforza, *Genes, Peoples, and Languages*, New York, North Point Press, 2000.
[64] G. Morfill and W. Bunk, New design on complex pattern, *Europhys. News*, May/June 2001, 77.
[65] B. S. Kerner, The physics of traffic, *Phys. World*, August 1999, 25.
[66] J. Hogan, Will physics crack the market? *New Scientist*, 7 December 2002, 16. (The work quoted is due to D. Sornette and Wei-Xing Zhou.)
[67] B. Schechter, Push me pull me, *New Scientist*, 24 August 2002, 41. (The work quoted was done by J. Sznajd, K. Sznajd-Weron, D. Stauffer, and collaborators.)
[68] S. Strogatz, The physics of crowds, *Nature* **428** (2004) 367.
[69] K. Hasselmann, H. J. Schellnhuber, and O. Edenhofer, Climate change: complexity in action, *Phys. World*, June 2004, 31.
[70] A. Gaines, *Tim Berners-Lee and The Development of The World Wide Web*, Hockessin, Mitchell Lane Publisher, 2002.
[71] T. Berners-Lee, J. Hendler, and O. Lassila, The Semantic Web, *Scient. Am.*, May 2001, 29.
[72] I. Foster, The GRID: computing without bounds, *Scient. Am.*, April 2003, 61.
[73] P. Rodgers, US physicists dominate world top 100, *Phys. World*, June 2001, 7.

[74] D. Jérome and J.-M. Raimond, US threats to European journals, *Phys. World*, May 2004, 20.
[75] R. Doyle, Filling the pipeline – The 2002 edition of "Science and Engineering indicators," *Scient. Am.*, July 2002, 15.
[76] H. McCabe, Physicists unite to combat a crisis of falling numbers, *Nature* **401** (1999), 102.
[77] Pure science courses fail to entice French students, *Nature* **417** (2002) 9.
[78] S. Weinberg, *The First Three Minutes*, New York, Basic Books, 1977.
[79] S. Hawking, *A Brief History of Time*, London, Bantam Press, 1988.
[80] *Enseigner les sciences à l'école, un outil pour la mise en oeuvre des programmes 2002*, Paris, Centre National de Documentation Pédagogique, 2002.
[81] L. M. Lederman, What is today's most important unreported story?, to be found in www.edge.org/3rd_culture/story/story_69.html.
[82] L. M. Lederman, Revolution in science: put physics first, *Phys. Today*, November 2002, 55.
[83] Flocking to physics, *Science* **303** (2004) 166.
[84] C. Wilson, Expanding the teacher network, *Phys. World*, November 2003, 56.
[85] Schools at $10^{20}$ eV and beyond, *Nature* **429** (2004) 685.
[86] *Basic Science and Technology Transfer – Means and Methods in the CERN Environment*, F. Bourgeois, ed., Geneva, CERN report CERN/BLIT/98–101, February 1998.
[87] Hendrick Brugt, Gerhard Casimir, *Phys. Today*, September 2000, 80.
[88] Hendrik Casimir, *Haphazard Reality – Half A Century of Science*, New York, Harper and Row, 1983.
[89] T.T. Nguyen, Casimir effect and vacuum fluctuations, Caltech lecture notes, http://www.hep.caltech.edu/~phys199/lectures/lect5_6_cas.pdf.
[90] Quoted by D. L. Tyrrell, H. G. B. Casimir, Contribution to Symposium on Technology and World Trade, US Department of Commerce, November 16, 1966.
[91] Richard L. Garwin receives the National Medal of Science, *IBM Research News*, 2003.
[92] R. N. Kostoff and J. A. del Rio, The impact of physics research, *Phys. World*, June 2001, 47.
[93] H. Schmied, *A Study of Economic Utility Resulting from CERN Contracts*, CERN Yellow Report, Geneva, CERN, 1975.
[94] M. Bianchi-Streit, N. Blackburn, H. Schmied *et al.*, *Economic Utility Resulting from CERN Contracts*, CERN Yellow Report, Geneva, CERN, 1984.
[95] P. Brendle *et al.*, *Les effets "économiques" induits de l'ESA*, ESA Contracts Report Vol. 3, 1980.
[96] L. Bach *et al.*, *Study of the economic effects of European space expenditure*, ESA Contract No. 7062/87/F/RD.
[97] R. E. Autio, M. Bianchi-Streit, and A. P. Hameri, *Technology Transfer and Technological Learning through CERN's Procurement Activity*, CERN Yellow Report, Geneva, CERN, 2003.
[98] M. Haehnle, *R&D Collaboration between CERN and Industrial Companies. Organisational and Spatial Aspects.* Dissertation, September 1997, Vienna University of Economy and Business Administration.
[99] Sir J. M. Thomas, *Predictions, Notes Rec. R. Soc. London* 55–1 (2001) 105.
[100] C. Sirtori, Bridge for the terahertz gap, *Nature* **417** (2002) 132.
[101] J. Mullins, Forbidden zone, *New Scientist*, September 2002, 34.
[102] I. Osborne, Revealing the invisible, *Science* **297** (2002) 1097.
[103] A. Yazdani and C. M. Lieber, Up close and personal atoms, *Nature* **401** (1999) 227.
[104] K. E. Drexler, *Engines of Creation: The Coming Era of Nanotechnology*, London, Fourth Estate, 1990.
[105] A. P. Alivisatos, Less is more in medicine, *Scient. Am.*, September 2001, 59.
[106] *Nanotech, The Science of The Small Gets Down to Business, Scient. Am.* special issue, September 2001.
[107] Small particles of gold and iron, attached to antibodies, have already been used by chemist C. Mirkin and collaborators to detect prostate-specific antigen: J-M. Nam, C. S. Thaxton, and C. A. Mirkin, Nanoparticle-based bio-bar codes for the ultrasensitive detection of proteins, *Science* **301** (2003) 1884.

[108] Bacteria and viruses have already been detected with cantilevers a few micrometers long, see for instance B. Ilic, H. G. Craighead, S. Krylov *et al.*, Attogram detection using nanoelectromechanical oscillators, *J. Appl. Phys.* **95** (2004) 3694.
[109] A. de Rújula, S. L. Glashow, R. R. Wilson, and G. Charpak, *Phys. Reports* **99**(6) (1983) 341.
[110] C. Callan, *Catalysis of Baryon Decay*, Princeton, Princeton University Press, 1983.
[111] A. Kusenko and M. Shaposhnikov, Baryogenesis and the baryon to dark matter ratio, *Phys. Lett.* B **418** (1998) 46.
[112] J. King, Exciting times and challenges ahead, *Phys. World*, November 2002, 53.
[113] G. Charpak and R. L. Garwin, The DARI, a unit of measure suitable to the practical appreciation of the effect of low doses of ionizing radiation, *Europhys. News*, January/February 2002, 14.
[114] J. Rotblat, The social conscience of scientists, *Phys. World*, December 1999, 65.
[115] D. Goodstein, In the matter of J. Hendrick Schoen, *Phys. World*, November 2002, 17.
[116] G. Trilling, Co-authors are responsible too, *Phys. World*, June 2003, 16.
[117] Quoted in G. Trilling, Co-authors are responsible too, *Phys. World*, June 2003, 16.
[118] M. Ducloy, The World Year of Physics, *Europhys. News*, January/February 2003, 26.

# Index

Page numbers in *italic* refer to figures